physica status solidi (b)

0370-1972 • SUBJECT AND AUTHOR INDEX (b) VOLUMES 79 TO 84 • 1977

0031-8965 • SUBJECT AND AUTHOR INDEX (a) VOLUMES 39 TO 44 • 1977

physica status solidi (a)

Classification Scheme

1. Structure of Crystalline Solids
   1.1 Perfectly Periodic Structure
   1.2 Solid-State Phase Transformations
   1.3 Alloys. Metallurgy
   1.4 Microstructure (Magnetic Domains See 18; Ferroelectric Domains See 14.4.1)
   1.5 Films
   1.6 Surfaces
2. Non-Crystalline State
3. Crystal Growth
4. Bonding Properties
5. Mössbauer Spectroscopy
6. Lattice Dynamics. Phonons
7. Acoustic Properties
8. Thermal Properties
9. Diffusion
10. Defect Properties (Irradiation Defects See 11)
    10.1 Metals
    10.2 Non-Metals
11. Irradiation Effects (X-Ray Diffraction Investigations See 1 and 10)
12. Mechanical Properties (Plastic Deformations See 10)
    12.1 Metals
    12.2 Non-Metals
13. Electron States
    13.1 Band Structure
    13.2 Fermi Surfaces
    13.3 Surface and Interface States
    13.4 Impurity and Defect States
    13.5 Elementary Excitations (Phonons See 6)
        13.5.1 Excitons
        13.5.2 Plasmons
        13.5.3 Polarons
        13.5.4 Magnons
14. Electrical Properties. Transport Phenomena
    14.1 Metals. Semi-Metals
    14.2 Superconductivity. Superconducting Materials and Devices
    14.3 Semiconductors
        14.3.1 Films
        14.3.2 Surfaces and Interfaces
        14.3.3 Devices. Junctions (Contact Problems See 14.3.4)
        14.3.4 High-Field Phenomena, Space-Charge Effects, Inhomogeneities, Injected Carriers (Electroluminescence See 20. 3; Junctions See 14.3.3)
    14.4 Dielectrics
        14.4.1 Ferroelectrics
15. Thermoelectric and Thermomagnetic Properties
16. Photoconductivity. Photovoltaic Effects
17. Emission of Electrons and Ions
    17.1 Field Emission Microscope Investigations
18. Magnetic Properties
    18.1 Paramagnetic Properties
    18.2 Ferromagnetic Properties
        18.2.1 Ferromagnetic Films
    18.3 Ferrimagnetic Properties
    18.4 Antiferromagnetic Properties

(Continued on cover three)

physica status solidi (b)
basic research

physica status solidi (a)
applied research

SUBJECT AND AUTHOR INDEX

Volumes 79 to 84, 1977 | Volumes 39 to 44, 1977

ISSN 0370-1972/0031-8965

Edited by
E. BORCHARDT

AKADEMIE-VERLAG · BERLIN
and
ACADEMIC PRESS · LONDON-NEW YORK

Subscriptions and orders for this Subject and Author Index
or single copies of physica status solidi (b) and physica status solidi (a)
for the British Commonwealth and USA should be directed to:

ACADEMIC PRESS INC. (London) Limited
24 – 28 Oval Road
London N. W. 1

or

ACADEMIC PRESS INC., 111, Fifth Avenue, New York 10003, USA
whichever address is the most convenient.

For all other countries should be directed to:
AKADEMIE-VERLAG, DDR-108 Berlin, Leipziger Straße 3 – 4.

Distributed in Great Britain by Academic Press, London.

Application to mail at second-class postage rates is pending at Jamaica, New York 11431, USA.

Herausgeber und verantwortlich für den Inhalt des Registers:
Dipl.-Phys. E. Borchardt, DDR-102 Berlin, Neue Schönhauser Straße 20.
Verlag: Akademie-Verlag, DDR-108 Berlin, Leipziger Straße 3 – 4; Fernruf 22 36 221 und 22 36 229; Telex-Nr. 114 420;
Bank: Staatsbank der DDR, Berlin, Kto.-Nr.: 6836-26-20712.
Veröffentlicht unter der Lizenznummer 1310 und 1620
des Presseamtes beim Vorsitzenden des Ministerrates der Deutschen Demokratischen Republik.
Gesamtherstellung: VEB Druckerei „Thomas Müntzer", DDR-582 Bad Langensalza.
Bestellnummer des Registers 1068/85 R 10.

AN(EDV) 20831/20735

Perface

The tenth index volume contains volumes 79 to 84 of "physica status solidi (b) — basic research" as well as volumes 39 to 44 of "physica status solidi (a) — applied research", i.e. all volumes having appeared in 1977 containing 1481 papers.

The general scheme of the last index volume was maintained. The basis for quoting is the subject classification given for every paper in the respective issue. Nevertheless, before taking over these subject classifications into the index volume they were once more carefully revised and in some cases improved and completed in order to get a most uniform and clear classification for all papers. The average number of quotations per paper in this index volume is nearly the same as in the last one, namely about 3.0.

Again I have to express my thanks to the readers for their interest, and I will always readily take up any critical remarks and proposals for improvements. I am especially obliged to Prof. Dr. P. Görlich, Prof. Dr. E. Gutsche, Dr. H.-J. Hänsch, Dr. H. Lange, Dr. S. Oberländer, and Dipl. Phys. K. Wannewitz for giving me advice and unselfish help. Furthermore I have to thank Mrs. E. Schwan for her exact and reliable secretary's work at the manuscript.

Berlin, January 1978

Eva Borchardt

Classification Scheme

1. Structure of Crystalline Solids
   1.1 Perfectly Periodic Structure
   1.2 Solid-State Phase Transformations
   1.3 Alloys. Metallurgy
   1.4 Microstructure (Magnetic Domains See 18; Ferroelectric Domains See 14.4.1)
   1.5 Films
   1.6 Surfaces

2. Non-Crystalline State

3. Crystal Growth

4. Bonding Properties

5. Mössbauer Spectroscopy

6. Lattice Dynamics. Phonons

7. Acoustic Properties

8. Thermal Properties

9. Diffusion

10. Defect Properties (Irradiation Defects See 11)
    10.1 Metals
    10.2 Non-Metals

11. Irradiation Effects (X-Ray Diffraction Investigations See 1 and 10)

12. Mechanical Properties (Plastic Deformations See 10)
    12.1 Metals
    12.2 Non-Metals

13. Electron States
    13.1 Band Structure
    13.2 Fermi Surfaces
    13.3 Surface and Interface States
    13.4 Impurity and Defect States
    13.5 Elementary Excitations (Phonons See 6)
        13.5.1 Excitons
        13.5.2 Plasmons
        13.5.3 Polarons
        13.5.4 Magnons

14. Electrical Properties. Transport Phenomena
    14.1 Metals. Semi-Metals
    14.2 Superconductivity. Superconducting Materials and Devices
    14.3 Semiconductors
        14.3.1 Films
        14.3.2 Surfaces and Interfaces
        14.3.3 Devices. Junctions (Contact Problems See 14.3.4)
        14.3.4 High-Field Phenomena, Space-Charge Effects, Inhomogeneities, Injected Carriers (Electroluminescence See 20.3; Junctions See 14.3.3)
    14.4 Dielectrics
        14.4.1 Ferroelectrics

15. Thermoelectric and Thermomagnetic Properties

16. Photoconductivity. Photovoltaic Effects

17. Emission of Electrons and Ions
    17.1 Field Emission Microscope Investigations

18. Magnetic Properties
    18.1 Paramagnetic Properties
    18.2 Ferromagnetic Properties
        18.2.1 Ferromagnetic Films
    18.3 Ferrimagnetic Properties
    18.4 Antiferromagnetic Properties

19. Magnetic Resonance

20. Optical Properties
    20.1 Spectra. Optical Constants (X-Ray Spectra See 20)
    20.2 Lasers
    20.3 Luminescence (X-Ray Spectra See 20)

21. Metals. Alloys
    21.1 Transition Metals (Group Sc to Zn) and Their Alloys
        21.1.1 Fe and Fe Alloys
    21.2 Alkali Metals
    21.3 Alkali-Earth Metals
    21.4 Rare-Earth Metals
    21.5 Actinides
    21.6 Noble Metals
    21.7 Semi-Metals (Semiconducting Alloys See 22.7)

22. Semiconductors and Ionic Crystals
    22.1 Elements
        22.1.1 Germanium
        22.1.2 Silicon
        22.1.3 Group VI Elements
    22.2 III–V Compounds
        22.2.1 Arsenides
        22.2.2 Phosphides
        22.2.3 Antimonides
        22.2.4 Mixed Crystals
    22.3 IV–IV, II–V, and III–VI Compounds
        22.3.1 Mixed Crystals
    22.4 II–VI Compounds
        22.4.1 Sulphides
        22.4.2 Selenides
        22.4.3 Tellurides
        22.4.4 Mixed Crystals
    22.5 Halides
        22.5.1 Silver Halides
        22.5.2 Alkali Halides
        22.5.3 Alkali-Earth Halides
        22.5.4 Mixed Crystals
    22.6 Simple Oxides (CdO, ZnO See 22.4)
        22.6.1 Mixed Crystals
    22.7 Semiconducting Intermetallic Compounds
    22.8 Tertiary and Higher Compounds
        22.8.1 Salts of Oxy-Acids (Carbonates, Phosphates, Silicates Including Mica etc.)
        22.8.2 Spinels, Garnets, Ferrites
    22.9 Organic Semiconductors

23. Solidified Gases

For comments on the use of the classification scheme see p. 6.

Some Comments on the Use of the Subject and Author Index

The classification system[1]) comprises 20 broad sections each devoted to a major field of solid state physics and 3 broad sections devoted to the main groups of substances. A section may contain a number of subsections dealing with aspects of this field or special substance, respectively, which tend to receive particular attention. A paper dealing with this field will either be classified under one or more of the subsections or, if none of these are relevant, it will just appear under the general section heading (i.e. sections and subsections are equivalent in this sense). For example, a paper dealing with problems of paramagnetism would appear under 18.1 but not 18, whereas a paper on diamagnetism would be quoted under 18; a paper dealing with GaP would appear under 22.2.2, not under 22 or 22.2, whereas a paper on AlN would be quoted under 22.2, and a paper on a II–IV or a IV–VI compound would be quoted under 22.

All papers will be quoted under each relevant section. There is made a distinction between one or two principal entries (refering to the main subject matter of the paper) and secondary entries (refering to subsidiary subject matter and substance, respectively). In the prinicipal entries the title of the paper is given in its original language and in English. In addition, every principal entry is marked by * in front of the title of the paper. In the secondary entries the title is only given in English.

Each entry gives the volume, the number of the issue, the initial and final page numbers, and the year in which the paper appeared. A paper of physica status solidi (b) or physica status solidi (a) is marked by (b) or (a) in front of the line, respectively. The letter "K" before the page numbers indicates a short note published in the offset part of the journal. Each entry is followed by references to the other entries. These references give the subject classifications and the pages in the index where the other entries appear. The references to the principal entries are given in italics.

Review articles appear at the head of the list of entries under a given subject and are given the subtitle "Review Articles". These articles are listed in the chronological order of their publication, irrespective of being a paper of physica status solidi (b) or physica status solidi (a). The subject classification for a review article consisting of several parts is based on the article as a whole rather than on the individual parts.

The entries for original papers, short notes, and errata are made together in a single list for each subject and this follows the entries for review articles. These are again ordered according to data of publication, irrespective of being a paper of physica status solidi (b) or physica status solidi (a). Original papers are sometimes published in several parts under the same principal title (in general with special subtitles). The various parts of papers of this type are treated separately and each part is classified according to its particular contents.

In quoting errata the following rule is obeyed. If the erratum refers to a paper quoted in the preceding index, it is listed in the chronological order of its publication, otherwise the erratum is quoted together with the paper to which it refers.

---

[1]) See cover two and three, and pages 4 and 5. For the case of losing the cover by bookbinding we repeat the classification scheme on pages 4 and 5.

In using the subject index attention should be given to the following details:

subject	quoted in	remarks
representations of space groups	1 and 13.1	
Alfvén waves, helicon waves	13 and 20	
hot electrons	13 and 14.3.4	
transport phenomena	14 or subsection of 14	if not clearly to be assigned to 8 or 15
galvanomagnetic effects	14 or subsection of 14	not in 18
electro-optical effects	14 and 20 (or subsection of 14 and 20)	
piezoelectricity, electrostriction	14 or subsection of 14	not in 12 or subsection of 12; different electromechanical effects are quoted in 12 or subsection of 12
ionic conduction	14.4	not in 14.3
magneto-optical effects	18 and 20 (or subsection of 18 and 20)	possibly in additional sections, e. g. Faraday effect in 13
ferromagnetic films	18.2.1	not additionally in 1.5
X-ray spectra	20	but not X-ray studies of structure
MIS and MOS structures	the section assigned to the semiconductor	not in the section assigned to the oxide or the metal
MIM structures	the section assigned to the insulator	generally not in the section assigned to the metal
epitaxial layers	the section assigned to the layer	not in the section assigned to the substrate

The author index gives the names of the authors in alphabetical order. Each name is followed by the bibliographic details of the paper and the numbers of the pages on which the paper appears in the subject index. A paper of physica status solidi (b) or physica status solidi (a) is again marked by (b) or (a) in front of the number of the volume, respectively. The principal entry is again given in italics. When several papers have been published by the same author they are arranged according to the dates of publication, irrespective of being a paper of physica status solidi (b) or physica status solidi (a).

Subject Index

1. Structure of Crystalline Solids

Original Papers and Short Notes

Y. KHAN, K. NAHM, M. ROSENBERG, and H. WILLNER
Superconductivity and Semiconductor–Metal Phase Transition in the System $BaPb_{1-x}Bi_xO_3$
(a) Vol. **39**, No. 1, p. 79–88 (1977).
See also 13.1, p. 114; *14.2*, p. *161*; 20.1, p. 248; 22.6.1, p. 393.

J. C. PHILLIPS
*A Simple Explanation for the "$4k_F$ Anomaly" in TTF-TCNQ
(b) Vol. **80**, No. 1, p. 395–398 (1977).
See also *14*, p. *155*; 22.9, p. 415.

F. EICHHORN, K. HENNIG, B. LIPPOLD, and W. MATZ
*Diffraction of a Pulsed Thermal Neutron Beam on an Elastically Bent Quartz Crystal Plate
(a) Vol. **40**, No. 1, p. 205–210 (1977).
See also *11*, p. *95*; 22.6, p. 387.

S. G. ABDULLAEVA and H. S. MAMEDOV
*X-Ray Investigation of Single Crystals of $GaSe_xS_{1-x}$ $(0 \leqq x \leqq 1)$
(a) Vol. **40**, No. 1, p. K7–K9 (1977).
See also 22.3.1, p. 360.

P. A. BEZIRGANYAN
*A Three-Wave Method for the Measurement of the Length of a Wave-Train of X-Ray Radiation
(a) Vol. **40**, No. 1, p. K77–K80 (1977).

M. V. EREMIN and YU. V. RAKITIN
Channel Model in Isotropic Exchange Theory
(b) Vol. **80**, No. 2, p. 579–587 (1977).
See also *13*, p. *108*.

G. L. KRASKO and A. B. MAKHNOVETSKII
*Pseudopotential Theory and the Problem of Covalent Bonding (I): General Theory of Covalent Charge
(b) Vol. **80**, No. 2, p. 713–725 (1977).
See also *4*, p. *44*.

R. BONNET and E. E. LAUFER
*Precise Determination of the Relative Orientation of Two Crystals from the Analysis of Spot Diffraction Patterns
(a) Vol. **40**, No. 2, p. 599–603 (1977).
See also *1.4*, p. *25*; 21.1, p. 292.

B. B. SHARMA, R. AYYAR, and H. SINGH
*Stability of the Tetrahedral Phase in the $A_2^IB^{IV}C_3^{IV}$ Group of Compounds
(a) Vol. **40**, No. 2, p. 691–696 (1977).
See also 1.4, p. 25; 22.8, p. 397.

A. B. MAKHNOVETSKII and G. L. KRASKO
*Pseudopotential Theory and the Problem of Covalent Bonding (II): Crystals of Si, Ge, and α–Sn as–Ordered Systems
(b) Vol. **81**, No. 1, p. 263–273 (1977).
See also *4*, p. *45*; 21.7, p. 319; 22.1.1, p. 334; 22.1.2, p. 339.

L. BAURIEGEL and G. OTTO
*The Influence of Magnetization of a Monocrystalline Target on the Channeling and Blocking Effect
(a) Vol. **41**, No. 2, p. K193–K195 (1977).
See also 18.2, p. 220; 18.3, p. 229; 21.1, p. 294; 22.8.2, p. 411.

M. V. Eremin and Yu. V. Rakitin
Channel Model in Isotropic Exchange Theory (II)
(b) Vol. **82**, No. 1, p. 221–228 (1977).
See also *13*, p. *110*; 22.5.4, p. 385; 22.6, p. 389.

Y. Muraoka, M. Shiga, and Y. Nakamura
Magnetic Properties and Mössbauer Effect of $A(Fe_{1-x}B_x)_2$ (A = Y or Zr, B = Al or Ni) Laves Phase Intermetallic Compounds
(a) Vol. **42**, No. 1, p. 369–374 (1977).
See also *5*, p. *48*; *18*, p. *211*; 21.1.1, p. 304.

A. M. Afanasev, M. V. Kovalchuk, E. K. Kovev, and V. G. Kohn
*X-Ray Diffraction in a Perfect Crystal with Disturbed Surface Layer
(a) Vol. **42**, No. 1, p. 415–422 (1977).
See also *1.6*, p. *31*; 22.1.2, p. 341.

M. W. Muller
*Site Ordering and Magnetriction in Garnet
(b) Vol. **83**, No. 1, p. 177–185 (1977).
See also *18.3*, p. *230*; 22.8.2, p. 412.

M. Pernet, C. Berthet-Colominas, M. A. Alario Franco et A. N. Christensen
*Etude par diffraction neutronique de l'oxyhydroxyde de chrome β-CrOOH
(Study of Chromium Oxyhydroxide (β-CrOOH) by Neutron Diffraction)
(a) Vol. **43**, No. 1, p. 81–88 (1977).
See also *18.4*, p. *233*; 22.9, p. 418.

Yu. M. Tairov, V. F. Tsvetkov, M. A. Chernov, and V. A. Taranets
*Investigation of Phase Transformations and Polytype Stability of β-SiC
(a) Vol. **43**, No. 1, p. 363–369 (1977).
See also 1.2, p. 18; 22.3, p. 358.

G. M. Aladzhadzhyan, P. A. Bezirganyan, O. S. Semerdzhyan, and D. M. Vardanyan
*X-Ray Spectrometers with Diffractional Focussing
(a) Vol. **43**, No. 2, p. 399–406 (1977).

P. A. Bezirganyan, R. Z. Gabrielyan und F. O. Eiramshyan
*Ein Dreistrahl-Röntgeninterferometer
(A Three-Beam X-Ray Interferometer)
(a) Vol. **43**, No. 2, p. K183–K186 (1977).

N. K. Radhakrishnan and A. B. Biswas
*A Neutron Diffraction Study of the Spinel Oxide $CuMn_2O_4$
(a) Vol. **44**, No. 1, p. 45–49 (1977).
See also 22.8.2, p. 412.

V. F. Kaminskii, M. L. Khidekel, R. B. Lyubovskii, I. F. Shchegolev, R. P. Shibaeva, E. B. Yagubskii, A. V. Zvarykina, and G. L. Zvereva
*Metal–Insulator Phase Transition in TTT_2I_3 Affected by Iodine Concentration
(a) Vol. **44**, No. 1, p. 77–82 (1977).
See also *14*, p. *156*; 18.1, p. 215; 22.9, p. 419.

E. A. van den Bergen, R. Delhez, and E. J. Mittemeijer
An X-Ray Diffraction Study of Diffusional Homogenization in RbCl–KCl Powder Blends
(a) Vol. **44**, No. 2, p. 517–524 (1977).
See also *9*, p. *71*; 22.5.2, p. 382.

J. J. Bara
*Cation Distributions in Nickel Ferrite-Aluminates
(a) Vol. **44**, No. 2, p. 737–740 (1977).
See also 22.8.2, p. 413.

U. Bonse and W. Graeff
*On the Possibility of Measuring the Length of X-Ray or Neutron Wave Trains with the Three-Beam Case Interferometer
(a) Vol. **44**, No. 2, p. K131–K132 (1977).

1.1 Perfectly Periodic Structure

Original Papers and Short Notes

M. David, R. Gevers, and R. Serneels
*The Critical Voltage Effect in Transmission Electron Microscopy (VII): The Influence of Absorption
(b) Vol. **79**, No. 1, p. 215–230 (1977).

H.-G. Brühl, L. Hildisch, H. Morwinski, W. Schmidt, and E. Schubert
*Diffuse X-Ray Scattering and Compositional Disorder in $Ga_xIn_{1-x}P$
(a) Vol. **39**, No. 1, p. 133–139 (1977).
See also 1.2, p. 14; 22.2.4, p. 357.

D. Bloor and F. H. Preston
Optical Properties of Bis (p-Toluene Sulphonate) Diacetylene (II): Low-Temperature Spectra
(a) Vol. **39**, No. 2, p. 607–614 (1977).
See also *20.1*, p. *250*; 22.9, p. 414.

Meera Rao and O. N. Srivastava
*Interpolytypic Transformations and the Electrical Behaviour of PbI_2 Crystals
(a) Vol. **39**, No. 2, p. K81–K84 (1977).
See also *1.2*, p. *15*; 14.3, p. 166; 22, p. 322.

V. Geist, R. Flagmeyer, D. Stephan, and H.-J. Ullrich
*Kossel Interferences of Proton-Induced AlKα Radiation
(a) Vol. **40**, No. 1, p. 113–117 (1977).
See also 21, p. 279; 21.1, p. 290.

M. David and R. Gevers
*The Critical Voltage Effect in Transmission Electron Microscopy (VIII): The Qualitative Influence of Non-Systematic Reflections on the Critical Voltage
(b) Vol. **80**, No. 2, p. 477–490 (1977).

H. Föll and K. Ulmer
*A New Approach for an Absolute Determination of 2*d* Values of Spectrometer Crystals
(a) Vol. **41**, No. 1, p. 113–116 (1977).
See also 22.8, p. 398.

T. Matsushita and J. Hayashi
*X-Ray Structure Factors and the Debye-Waller Factor of Gallium Arsenide Crystals Determined from Full Widths at Half Maximum of Bragg Case Diffraction Curves
(a) Vol. **41**, No. 1, p. 139–145 (1977).
See also 6, p. 56; 22.2.1, p. 351.

I. I. Tupizin and I. V. Abarenkov
*The Crystal Potential and Summation of Diverging Series
(b) Vol. **82**, No. 1, p. 99–105 (1977).
See also 22, p. 325.

V. K. Garg and M. P. Verma
*Evaluation of Some Important Lattice Sums on NaCl and CsCl Structures
(b) Vol. **82**, No. 1, p. 139–143 (1977).
See also 12.2, p. 105; 22.5.2, p. 379.

M. A. Fortes
*Coincidence Site Lattices in Non-Cubic Lattices
(b) Vol. **82**, No. 1, p. 377–382 (1977).
See also *1.4*, p. *26*.

V. G. Baryshevskii and I. Ya. Dubovskaya
*Coherent Radiation of the Chanelling Positron (Electron)
(b) Vol. **82**, No. 1, p. 403–412 (1977).
See also *11*, p. *96*.

M. Suzuki and H. Okazaki
*The Structure of α-AgI
(a) Vol. **42**, No. 1, p. 133–140 (1977).
See also 22.5.1, p. 373.

M. Galtier, B. Pardo et D. Renard
*Etude de la validité des termes réels et imaginaires du potentiel crystallin par une méthode de microscopie électronique
(Study of the Validity of the Real and Imaginary Terms of the Lattice Potential by an Electron Microscopic Method)
(a) Vol. **42**, No. 1, p. 141–148 (1977).
See also 21.6, p. 316.

E. W. Kreutz and N. Sotnik
*Kikuchi Patterns of Spherical Ag Crystals in the Bragg Case
(a) Vol. **42**, No. 1, p. K11–K14 (1977).
See also *1.6*, p. *31*; 21.6, p. 316.

L. B. Coleman, A. M. Hermann, R. Williams, and R. B. Somoano
*On the Crystal Phases of (DEPE) $(TCNQ)_4$
(b) Vol. **82**, No. 2, p. K117–K121 (1977).
See also 22.9, p. 417.

J. Härtwig
*X-Ray Diffraction in the Extremely Asymmetric Laue Case with a Small Angle between the Crystal Surface and the Incident Beam
(a) Vol. **42**, No. 2, p. 495–500 (1977).
See also 22.1.2, p. 341.

J. van Landuyt, G. van Tendeloo, and S. Amelinckx
*Electron Diffraction Study of Inter- and Intrapolytypic Phase Transitions in Transition Metal Dichalcogenides (IV): Cinematographic Observations of the $1T_\beta \rightarrow 1T_\gamma$-Phase Transition in TaS_2
(a) Vol. **42**, No. 2, p. 565–570 (1977).
See also *1.2*, p. *18*; 22, p. 327.

P. J. Walsh, G. C. Vezzoli, and L. W. Doremus
*Thermodynamic Basis for Generalization of Group VI-A Melting Curves
(a) Vol. **43**, No. 1, p. 305–310 (1977).
See also *1.2*, p. *18*; 8, p. 67; 22.1.3, p. 346.

A. G. Moldovan, S. K. Malik, and W. E. Wallace
Structural and Magnetization Studies on $ThCo_{5-x}Mn_x$ and $ThNi_{5-x}Mn_x$ Ternaries
(a) Vol. **43**, No. 1, p. 317–320 (1977).
See also *18.2*, p. *222*; 21.1, p. 297; 21.5, p. 314.

B. K. Jap
*On the Dependence of the Diffracted Beam Intensity Features with Electron Accelerating Voltage
(a) Vol. **43**, No. 1, p. K37–K40 (1977).
See also 21, p. 285.

T. Yamamoto
*A Note on the Contrast Mechanism of Electron Channeling Patterns in Scanning Electron Microscopy
(a) Vol. **43**, No. 1, p. K53–K56 (1977).
See also *1.4*, p. *27*.

M. Sauvage, J. F. Petroff, and P. Skalicky
*Synchroton Radiation Topographic Evidence for the Interaction of Plane-Polarized X-Ray Photons with Perfect Crystals
(a) Vol. **43**, No. 2, p. 473—477 (1977).
See also 22.1.2, p. 342.

U. Bonse, W. Graeff, R. Teworte, and H. Rauch
*Oscillatory Structure of Laue Case Rocking Curves
(a) Vol. **43**, No. 2, p. 487—492 (1977).
See also 22.1.2, p. 342.

J. A. Kozubowski
*Determination of the Crystal Orientation from Intersections of Kikuchi Lines
(a) Vol. **43**, No. 2, p. 535—539 (1977).

G. van Tendeloo, J. van Landuyt, and S. Amelinckx
*Electron Diffraction Evidence for a One-Dimensional Peierls Distortion in a Transition Metal Trichalcogenide: TaS_3
(a) Vol. **43**, No. 2, p. K137—K140 (1977).
See also 1.4, p. 28; 22, p. 329.

M. David and R. Gevers
*The Critical Voltage Effect in Transmission Electron Microscopy (IX): The Calculation of Critical Voltages and Experimental Extinction Distances in Complicated Many-Beam Systems
(b) Vol. **84**, No. 1, p. 133—147 (1977).

G. Albertini, A. Boeuf, B. Klar, S. Lagomarsino, S. Mazkedian, S. Melone, P. Puliti, and F. Rustichelli
*Dynamical Neutron Diffraction by Curved Crystals in the Laue Geometry
(a) Vol. **44**, No. 1, p. 127—136 (1977).
See also 22.1.2, p. 343.

J. J. Bara, A. T. Pędziwiatr, Z. M. Stadnik, A. Szytuła, J. Todorović, Z. Tomkowicz, and W. Zarek
*Investigations of Crystal and Magnetic Properties of Nickel Ferrite-Aluminates
(a) Vol. **44**, No. 1, p. 325—331 (1977).
See also 5, p. 49; *18.3*, p. *230*; 22.8.2, p. 413.

J. M. Corbett and S. S. Sheinin
*A Determination of the Reflections Required to Describe Electron Diffraction Phenomena in Gold
(a) Vol. **44**, No. 1, p. 353—361 (1977).
See also 21.6, p. 317.

J. Beauvillain et R. Ayroles
*Etude des diagrammes de pseudo-lignes de Kikuchi obtenues par une méthode d'observation en condenseur-objectif
(Study of the Kikuchi Pseudo-Line Patterns Obtained by an Observation Method in Condenser Objective)
(a) Vol. **44**, No. 2, p. 485—497 (1977).

1.2 Solid-State Phase Transformations

Review Article

V. A. Buravikhin, V. V. Litvintsev, and A. I. Ushakov
*Polymorphic Transformations and Physical Properties of Thin Films
(a) Vol. **40**, No. 1, p. 11—27 (1977).
See also 1.4, p. 24; *1.5*, p. *29*; 18.2, p. 217; 21.1, p. 288; 21.1.1, p. 302.

Original Papers and Short Notes

R. A. Tahir-Kheli
Green's Function Treatment of Ordered Ternary Alloy
(b) Vol. **79**, No. 1, p. 93—103 (1977).
See also *1.3*, p. *21*.

L. Gouskov, A. Gouskov, V. Lemos, W. May, and H. Sampaio
Electrical Properties of $GaTe_xSe_{1-x}$ Crystals
(a) Vol. **39**, No. 1, p. 65—71 (1977).
See also *13.4*, p. *129*; *14.3*, p. *165*; 22.3.1, p. 360.

H.-G. Brühl, L. Hildisch, H. Morwinski, W. Schmidt, and E. Schubert
Diffuse X-Ray Scattering and Compositional Disorder in $Ga_xIn_{1-x}P$
(a) Vol. **39**, No. 1, p. 133—139 (1977).
See also *1.1*, p. *11*; 22.2.4, p. 357.

V. T. Bublik and A. A. Zaitsev
Determination of the Energy of Mixing of HgTe–CdTe Solid Solutions by Diffuse X-Ray Scattering
(a) Vol. **39**, No. 1, p. 345—349 (1977).
See also *1.3*, p. *21*; 22.4.4, p. 368.

J. Kaczér and P. Novotný
High-Field Magnetization of U_3As_4
(a) Vol. **39**, No. 1, p. 351—355 (1977).
See also *18.2*, p. *217*; 22, p. 321.

S. I. Subbotin, V. V. Panfilov, and R. T. Molchanova
*Pressure-Induced Polytypic Transformations and Edge Absorption of GaSe
(a) Vol. **39**, No. 1, p. 357—362 (1977).
See also 12.2, p. 104; 13.5.1, p. 143; *20.1*, p. *248*; 22.3, p. 358.

W. Klement, Jr.
*Phase Transitions in Ag_2O to 39 kbar
(a) Vol. **39**, No. 1, p. K45—K48 (1977).
See also 22.6, p. 386.

J. Albers and H. Küppers
*Dielectric and Thermal Properties of Ammonium Hydrogen Oxalate Hemihydrate
(a) Vol. **39**, No. 1, p. K49—K51 (1977).
See also 8, p. 64; 14.4, p. 189; 22.9, p. 414.

B. N. Ganguly and M. Nicol
Effect of Hydrostatic Pressure on the Vibrational Properties and the Structure of $SrWO_4$ and $PbMoO_4$
(b) Vol. **79**, No. 2, p. 617—622 (1977).
See also *6*, p. *52*; 12.2, p. 104; 20.1, p. 249; 22.8.1, p. 402.

I. Ya. Korenblit
Phase Transition and Instability of the Homogeneous State in Ferromagnetic Semiconductors
(b) Vol. **79**, No. 2, p. K95—K98 (1977).
See also *13*, p. *107*; *18.2*, p. *218*; 22, p. 321.

V. I. Litvinov and K. D. Tovstyuk
*On the Existence of the Off-Diagonal Long-Range Order in the Hubbard Model
(b) Vol. **79**, No. 2, p. K131—K135 (1977).
See also *13*, p. *108*.

R. de Ridder, D. van Dyck, G. van Tendeloo, J. van Landuyt, and S. Amelinckx
*Microstructural Study of Copper-Intercalated Niobium Disulphide and Tantalum Disulphide Cu_xNbS_2 and Cu_xTaS_2 (III): Study of the Diffuse Scattering: the Transition State
(a) Vol. **39**, No. 2, p. 383—399 (1977).
See also 1.4, p. 24; 22, p. 322.

P. P. Seregin, V. P. Sivkov, F. S. Nasredinov, L. N. Vasilev, Yu. V. Krylnikov, and Yu. P. Kostikov
*The Influence of the Crystal-to-Glass Transition on the Local Structure of Semiconductors
(a) Vol. **39**, No. 2, p. 437—444 (1977).
See also 2, p. 34; 5, p. 47; 22, p. 322.

R. Wang and M. D. Merz
Polymorphic Bonding and Thermal Stability of Elemental Noncrystalline Solids
(a) Vol. **39**, No. 2, p. 697—703 (1977).
See also *2*, p. *34*; *4*, p. *44*; 21, p. 278; 22.1, p. 332.

Meera Rao and O. N. Srivastava
*Interpolytypic Transformation and the Electrical Behaviour of PbI_2 Crystals
(a) Vol. **39**, No. 2, p. K81—K84 (1977).
See also *1.1*, p. *11*; 14.3, p. 166; 22, p. 322.

K. R. Allakhverdiev and M. M. Tagyev
Infrared Investigation of the GaS_xSe_{1-x} System
(a) Vol. **39**, No. 2, p. K111—K113 (1977).
See also *6*, p. *53*; *20.1*, p. *250*; 22.3.1, p. 360.

F. G. Brady Moreira, I. P. Fittipaldi, S. M. Rezende, R. A. Tahir-Kheli, and B. Žekš
Field-Induced Phase Transitions in a Randomly Diluted Antiferromagnet. Mean-Field Description of Classical Spin Model
(b) Vol. **80**, No. 1, p. 385—394 (1977).
See also *18.1*, p. *213*; *18.4*, p. *231*.

A. L. Roitburd
*Predominant Effect of the Internal Stress Relaxation on Microstructural and Kinetic Features of Martensitic Transformation
(a) Vol. **40**, No. 1, p. 333—342 (1977).
See also 1.4, p. 25; 21.1.1, p. 303.

R. de Ridder, G. van Tendeloo, J. van Landuyt, D. van Dyck, and S. Amelinckx
*Erratum to: Microstructural Study of Copper-Intercalated Niobium Disulphide and Tantalum Disulphide Cu_xNbS_2 and Cu_xTaS_2 (I)
(a) Vol. **40**, No. 1, p. 343 (1977).
See also 1.4, p. 25; 22, p. 323.

A. K. Agyei and J. L. Birman
*Transition-Active $\boldsymbol{k} = \boldsymbol{0}$ Representations and the Microscopic Description of Phase Transitions
(b) Vol. **80**, No. 2, p. 509—520 (1977).

V. Winterfeldt and G. Schaack
Phonon Contributions to the Anomalous Specific Heat of Triglycine Sulfate near T_c
(b) Vol. **80**, No. 2, p. 559—567 (1977).
See also *6*, p. *54*; *8*, p. *65*; 14.4.1, p. 194; 22.9, p. 415.

M. CRIŞAN and P. PÎRVU
Intersection Lines in the Magnetic–Superconducting Phase Transition
(b) Vol. **80**, No. 2, p. K121—K125 (1977).
See also *14.2*, p. *162*; *18.2*, p. *219*.

A. KIRFEL, W. SCHÄFER, G. WILL, and K. H. J. BUSCHOW
*The High-Temperature Polymorph β-$FeSO_4$ Stability, Magnetic Structure, and Magnetic Properties
(a) Vol. **40**, No. 2, p. 447—453 (1977).
See also *18.4*, p. *232*; 22.8.1, p. 403.

U. GAHN
*Short- and Long-Range Order of B.C.C. Solid Solutions
(a) Vol. **40**, No. 2, p. 463—470 (1977).
See also *1.3*, p. *21*; 21, p. 279.

W. BRÜCKNER, U. GERLACH, and B. THUSS
*Phase Diagram of $V_{1-x}Al_xO_2$
(a) Vol. **40**, No. 2, p. K131—K134 (1977).
See also *18.1*, p. *214*; 22.6.1, p. 393.

R. KROGGEL, H. LÖFFLER, and O. KABISCH
Experimental Investigations on the Position of the Coherency Spinodal Curve in Al–Zn (15, 16, and 18 at%) Alloys
(a) Vol. **40**, No. 2, p. K163—K164 (1977).
See also *10.1*, p. *77*; 21, p. 280; 21.1, p. 292.

E. I. TERUKOV, W. REICHELT, M. P. VOLOTZKOI, F. A. CHUDNOVSKII, and H. OPPERMANN
*Hydrostatic Pressure Dependence of the Phase Transition of V_3O_5
(a) Vol. **41**, No. 1, p. 125—128 (1977).
See also 14.3, p. 168; 22.6, p. 388.

L. VAN GOETHEM, J. VAN LANDUYT, and S. AMELINCKX
*The α → β Transition in Amethyst Quartz as Studied by Electron Microscopy and Diffraction. The Interaction of Dauphiné with Brazil Twins
(a) Vol. **41**, No. 1, p. 129—137 (1977).
See also *1.4*, p. *25*; 22.6, p. 388.

A. NAUDON and A. M. FLANK
*Spinodal Curve and Metastable Miscibility Gap in Low-Concentrated Al–Ag Alloys
(a) Vol. **41**, No. 1, p. 207—211 (1977).
See also *1.3*, p. *21*; 21, p. 280; 21.6, p. 316.

S. POPOVIĆ, B. ČELUSTKA, Ž. RUŽIĆ-TOROŠ, and D. BROZ
*X-Ray Diffraction Study and Semiconducting Properties of the System Ga_2Se_3–In_2Se_3
(a) Vol. **41**, No. 1, p. 255—262 (1977).
See also 14.3, p. 168; 22.3, p. 358.

M. COUZI, A. DAOUD, and R. PERRET
*Low-Frequency Raman Spectra and Phase Transitions in Perovskite-Type Layer Compounds. $(CH_3NH_3)_2MnCl_4$ and $(CH_3NH_3)_2CdCl_4$
(a) Vol. **41**, No. 1, p. 271—282 (1977).
See also 6, p. 56; 20.1, p. 254; 22.9, p. 416.

T. KIICHI, M. NOMOTO, N. NAKAMURA, and H. CHIHARA
*Phase Transitions in Tetramethylammonium Chlorate
(a) Vol. **41**, No. 1, p. K17—K19 (1977).
See also *19*, p. *239*; 22.9, p. 416.

E. V. KOZLOV, V. N. EMELYANOV, and V. A. ANTONOV
*Quantum Theory of the Order–Disorder Transition Energy in $AuCu_3$
(b) Vol. **81**, No. 2, p. 719—725 (1977).
See also *1.3*, p. *21*; 21.1, p. 293; 21.6, p. 316.

J. J. Couderc, J. Bras et M. Fagot
*Précipitation dans le système fer–gallium au voisinage de 25 at % Ga
(Precipitation in the Iron–Gallium System in the Vicinity of 25 at % Ga)
(a) Vol. **41**, No. 2, p. 595—605 (1977).
See also *1.3*, p. *22*; 21, p. 281; 21.1.1, p. 304.

G. Dolino and J. P. Bachheimer
*Fundamental and Second-Harmonic Light Scattering by the α–β Coexistence State of Quartz
(a) Vol. **41**, No. 2, p. 673—677 (1977).
See also 22.6, p. 388.

N. B. Nešković, B. Babić, and J. Konstantinović
*High Temperature Anomalous Behaviour of the Crystal Lattice of Hematite
(a) Vol. **41**, No. 2, p. K133—K136 (1977).
See also 18.4, p. 233; 22.6, p. 388.

J. Pelzl, P. Engels, and R. Florian
*Raman Spectroscopic Study of the Structural Phase Transitions in K_2SnCl_6
(b) Vol. **82**, No. 1, p. 145—148 (1977).
See also 6, p. 57; 20.1, p. 256; 22.8, p. 399.

V. V. Vorobev, V. S. Belovol, and V. A. Finkel
*X-Ray Diffraction Study of Magnetostriction and Phase Transitions in a Magnetic Field in Rare Earth Metals and Some Alloys
(b) Vol. **82**, No. 1, p. 213—219 (1977).
See also *18.2*, p. *221*; 21.4, p. 312.

H. Spiering, H. Vogel, and W. Irler
*The Phase Transformation in Ferreous Fluosilicate Hexahydrate. Mössbauer Study
(b) Vol. **82**, No. 1, p. 243—250 (1977).
See also 5, p. 48; 18.1, p. 214; 22.8, p. 399.

L. Billard, P. Villemain, and A. Chamberod
*Interaction of Atomic Order with Localized Ferromagnetism
(b) Vol. **82**, No. 1, p. 269—278 (1977).
See also *18.2*, p. *221*; 21.1, p. 295; 21.1.1, p. 304.

W. Brückner, U. Gerlach, H.-P. Brückner, W. Moldenhauer, and H. Oppermann
*Influence of Nonstoichiometry on the Phase Transitions in Ga-, Al-, and Fe-Doped VO_2
(a) Vol. **42**, No. 1, p. 295—303 (1977).
See also *3*, p. *42*; 14.3, p. 170; 18.1, p. 214; 22.6.1, p. 393.

S. K. Chakrabarti
*On the Polymorphic Phase Transition in Alkali Halides
(b) Vol. **82**, No. 2, p. 557—564 (1977).
See also *12.2*, p. *105*; 22.5.2, p. 379.

A. K. Agyei and J. L. Birman
*A New Group-Theoretical Approach to Cell-Preserving Phase Transitions in Crystals
(b) Vol. **82**, No. 2, p. 565—575 (1977).
See also 22, p. 327.

S. Trimper
*Phase Transition in Ferroelectric Liquid Crystals
(b) Vol. **82**, No. 2, p. K75—K79 (1977).
See also 2, p. 38; 14.4.1, p. 196.

S. K. Hoffmann, L. S. Szczepaniak, and J. Stankowski
*Microscopic Mechanism of the Temperature Dependence of the Spontaneous Polarization in TGS Crystals
(b) Vol. **82**, No. 2, p. K85—K87 (1977).
See also *14.4.1*, p. *196*; 19, p. 241; 22.9, p. 417.

M. Rohr and P. Müller
*Superconducting Properties of Cold-Worked Core-Wires on the Basis of A15-Phases
(a) Vol. **42**, No. 2, p. 477—482 (1977).
See also *14.2*, p. *162*; 21, p. 283.

J. van Landuyt, G. van Tendeloo, and S. Amelinckx
*Electron Diffraction Study of Inter- and Intrapolytypic Phase Transitions in Transition Metal Dichalcogenides (IV): Cinematographic Observations of the $1T_\beta \rightarrow 1T_\gamma$-Phase Transition in TaS_2
(a) Vol. **42**, No. 2, p. 565—570 (1977).
See also *1.1*, p. *12*; 22, p. 327.

E. E. Matyas
*Phase Diagram of the InAs–AlAs Pseudobinary System
(a) Vol. **42**, No. 2, p. K129—K131 (1977).
See also 22.2.4, p. 357.

D. Bäuerle, W. B. Holzapfel, A. Pinczuk, and Y. Yacoby
*Temperature and Hydrostatic Pressure Dependence of Vibrational Modes in $PbTi_{1-x}Zr_xO$
(b) Vol. **83**, No. 1, p. 99—107 (1977).
See also 6, p. 58; *14.4.1*, p. *196*; 20.1, p. 259; 22.8.1, p. 406.

J. C. Phillips
*The "Extra" Phase Transition in TTF-TCNQ
(b) Vol. **83**, No. 1, p. 123—127 (1977).
See also 1.4, p. 27; 12.2, p. 105; 22.9, p. 417.

S. Hodorowicz, M. Ciechanowicz-Rutkowska, J. M. Janik, and J. A. Janik
*The X-Ray Powder Investigation of Phase Transitions in $[Ni(NH_3)_6](ClO_4)_2$
(a) Vol. **43**, No. 1, p. 53—57 (1977).
See also 22.8, p. 399.

G. Chapuis
*X-Ray Study of the First-Order Phase Transition Pcab–Bmab in $(CH_3CH_2NH_3)_2CdCl_4$
(a) Vol. **43**, No. 1, p. 203—212 (1977).
See also 22.9, p. 418.

F. Reynaud
A Mean Field Theory of Ordering of Alloys of AB Stoichiometry
(a) Vol. **43**, No. 1, p. 239—250 (1977).
See also *1.3*, p. *22*.

H. P. Aubauer
*Morphological Stability Study in Binary Alloys (I): Discrete Lattice Theory and Application to Interfacial Zone
(a) Vol. **43**, No. 1, p. 261—268 (1977).
See also 4, p. 45.

P. J. Walsh, G. C. Vezzoli, and L. W. Doremus
*Thermodynamic Basic for Generalization of Group VI-A Melting Curves
(a) Vol. **43**, No. 1, p. 305—310 (1977).
See also *1.1*, p. *12*; 8, p. 67; 22.1.3, p. 346.

Yu. M. Tairov, V. F. Tsvetkov, M. A. Chernov, and V. A. Taranets
Investigation of Phase Transformations and Polytype Stability of β-SiC
(a) Vol. **43**, No. 1, p. 363—369 (1977).
See also *1*, p. *10*; 22.3, p. 358.

P. Müller and M. Rohr
*Lattice Instability of Superconducting Cold-Worked Core-Wires on the Base of $PbMo_5S_6$
(a) Vol. **43**, No. 1, p. K19—K22 (1977).
See also *14.2*, p. *163*; 22.8, p. 399.

S. N. VAIDYA, D. K. JOSHI, M. D. KARKHANAVALA, and I. K. GOPALAKRISHNAN
*High Pressure Phase Transformation in La_2CuO_4
(a) Vol. **43**, No. 1, p. K31—K35 (1977).
See also *14.3*, p. *170*; 22.8.1, p. 406.

I. O. BASHKIN, E. G. PONYATOVSKII, and M. E. KOST
*Phase Transformations in Hydrides of Rare-Earth Metals under Pressure (I): Structural Transition in $NdH_{2.82}$
(b) Vol. **83**, No. 2, p. 517—520 (1977).
See also 14.1, p. 159; 21.4, p. 313.

B. N. GANGULY, M. NICOL, and W. D. ELLENSON
*Raman Spectroscopic Studies of the High Pressure Phase Transformation of $NaClO_3$
(b) Vol. **83**, No. 2, p. K115—K120 (1977).
See also *6*, p. *59*; 12.2, p. 106; 20.1, p. 260; 22.8.1, p. 407.

C. JOURDAN and J. GASTALDI
X-Ray Topographic Investigation of Dislocations in Titanium Single Crystals Grown by Recrystallization
(a) Vol. **43**, No. 2, p. 425—435 (1977).
See also *1.4*, p. *27*; *10.1*, p. *80*; 21.1, p. 297.

T. SCHOBER and A. CARL
*A Differential Thermal Analysis Study of the Vanadium–Hydrogen System
(a) Vol. **43**, No. 2, p. 443—449 (1977).
See also 1.3, p. 23; 8, p. 67; 21.1, p. 297.

H. P. AUBAUER
*Morphological Stability Study in Binary Alloys (II): Stability of Coherent Concentration Modulations
(a) Vol. **43**, No. 2, p. 601—610 (1977).
See also 4, p. 46; 21.1, p. 298.

A. T. MACRANDER and R. K. CRAWFORD
*Density of Solid Argon at Melting
(a) Vol. **43**, No. 2, p. 611—617 (1977).
See also 23, p. 420.

Ö. SÄVBORG and M. NYGREN
*Magnetic, Electrical, and Thermal Studies of the $V_{1-x}Re_xO_2$ System with $0 \leqq x \leqq 0.15$
(a) Vol. **43**, No. 2, p. 645—652 (1977).
See also 8, p. 67; 14.3, p. 171; 18.1, p. 215; 22.6.1, p. 393.

N. ŌTSUKA, Y. HANAWA, and S. NAGAKURA
*Crystal Structure and Phase Transition of Mn_6N_5 Studied by Electron Diffraction
(a) Vol. **43**, No. 2, p. K127—K129 (1977).
See also 21.1, p. 298.

Y. KIMURA
*On the Relation between the Density of States in the Cubic State and Phase Transition Temperatures (T_c, T_m) of V–Zr–Nb Ternary Laves Compounds
(a) Vol. **43**, No. 2, p. K141—K146 (1977).
See also 8, p. 67; 13, p. 112; *14.2*, p. *163*; 21.1, p. 298.

M. SCHWAB and H. G. KAHLE
Specific Heat of $TbPO_4$
(b) Vol. **84**, No. 1, p. 167—170 (1977).
See also *8*, p. *67*; *18.4*, p. *234*; 22.8.1, p. 407.

S. LIMMER, S. GRANDE, and A. LÖSCHE
*NMR Investigations of the Melting Behaviour of Mesogen Compounds
(b) Vol. **84**, No. 1, p. 283—292 (1977).
See also 19, p. 242; 22.9, p. 419.

I. SAVATINOVA and E. ANACHKOVA
Raman Spectrum and Dielectric Relaxation in $K_4Fe(CN)_6 \cdot 3\,H_2O$
(b) Vol. **84**, No. 1, p. 401—406 (1977).
See also 6, p. 60; *14.4.1*, p. *197*; *20.1*, p. *262*; 22.8, p. 400.

M. STUBIČAR, E. BABIĆ, D. SUBAŠIĆ, D. PAVUNA, and Ž. MAROHNIĆ
*Phase Transformations during Isochronal Annealing of $F_{40}Ni_{40}B_{20}$ Glass
(a) Vol. **44**, No. 1, p. 339—344 (1977).
See also *2*, p. *40*; 21.1, p. 299; 21.1.1, p. 306.

U. ROY, K. PETROV, I. TSOLOVSKI, and P. PESHEV
*On the Structural Phase Transition and the Thermal Expansion of Spinel Lithium Titanate
(a) Vol. **44**, No. 1, p. K25—K27 (1977).
See also 8, p. 67; 22.8.1, p. 407.

W. BRÜCKNER, E. WIESER, W. MOLDENHAUER, and W. REICHELT
*On the Conductivity Transition in V_3O_5
(a) Vol. **44**, No. 1, p. K95—K98 (1977).
See also 5, p. 49; 14.3, p. 172; 22.6, p. 391.

E. SCHUBERTH
*First Order Raman Scattering from Ar Crystals: Evidence for the H.C.P. Phase
(b) Vol. **84**, No. 2, p. K91—K94 (1977).
See also 6, p. 60; 20.1, p. 263; 23, p. 420.

M. SAKOH and M. SHIMIZU
*Thermodynamics of Phase Transitions in Iron
(a) Vol. **44**, No. 2, p. 511—516 (1977).
See also *8*, p. *68*; 21.1.1, p. 307.

R. KIND
*On the Soft Modes in $(CH_3NH_3)_2CdCl_4$
(a) Vol. **44**, No. 2, p. 661—667 (1977).
See also 6, p. 61; 22.9, p. 419.

T. I. CHEKMASOVA, I. S. KABANOV, and V. I. YUZVAK
*The Pressure–Temperature Phase Diagram of $LiNH_4SO_4$
(a) Vol. **44**, No. 2, p. K155—K157 (1977).
See also *14.4.1*, p. *197*; 19, p. 243; 22.8.1, p. 408.

E. I. TERUKOV, H. WAGNER, W. REICHELT und H. OPPERMANN
*IR-spektroskopische Untersuchungen zum Halbleiter–Metall-Phasenübergang des V_3O_5
(IR Spectroscopic Investigations on the Semiconductor–Metal-Phase Transition of V_3O_5)
(a) Vol. **44**, No. 2, p. K187—K190 (1977).
See also 6, p. 61; *20.1*, p. *264*; 22.6, p. 392.

1.3 Alloys. Metallurgy

Original Papers and Short Notes

R. A. Tahir-Kheli
*Green's Function Treatment of Ordered Ternary Alloy
(b) Vol. **79**, No. 1, p. 93—103 (1977).
See also 1.2, p. 14.

P. Rennert and A. M. Radwan
*Structural Investigation of the Laves Phase $MgZn_2$ with Model Potential Calculations
(b) Vol. **79**, No. 1, p. 167—173 (1977).
See also 10.1, p. 75; 13, p. 107; 21.1, p. 288; 21.3, p. 310.

V. T. Bublik and A. A. Zaitsev
*Determination of the Energy of Mixing of HgTe–CdTe Solid Solutions by Diffuse X-Ray Scattering
(a) Vol. **39**, No. 1, p. 345—349 (1977).
See also 1.2, p. 14; 22.4.4, p. 368.

A. B. Makhnovetskii and G. L. Krasko
The Pseudopotential Method and the Problem of Ordering in Alloys (III): Intermetallic Compounds LiAl, LiTl, and NaTl
(b) Vol. **80**, No. 1, p. 341—351 (1977).
See also *4*, p. *44*; 21, p. 278.

G. Athanassiadis, G. Le Caer, J. Foct, and L. Rimlinger
*Study of Ternary Ordered Solid Solutions Derived from Fe_3Al by Substitution
(a) Vol. **40**, No. 2, p. 425—435 (1977).
See also 5, p. 48; 21, p. 279; 21.1.1, p. 303.

U. Gahn
*Short- and Long-Range Order of B.C.C. Solid Solutions
(a) Vol. **40**, No. 2, p. 463—470 (1977).
See also *1.2*, p. *16*; 21, p. 279.

G. van Tendeloo, R. Wolf, and S. Amelinckx
*The Microstructure of the Alloy Au_5Mn_2: A Domain Structure with 84 Variants
(a) Vol. **40**, No. 2, p. 531—550 (1977).
See also *1.4*, p. *25*; 21.1, p. 292; 21.6, p. 315.

A. Naudon and A. M. Flank
*Spinodal Curve and Metastable Miscibility Gap in Low-Concentrated Al–Ag Alloys
(a) Vol. **41**, No. 1, p. 207—211 (1977).
See also *1.2*, p. *16*; 21, p. 280; 21.6, p. 316.

J. Sołtys
*Order–Disorder Phase Transitions in the Heusler Alloy Ni_2MnIn
(a) Vol. **41**, No. 1, p. K51—K54 (1977).
See also 21, p. 280; 21.1, p. 293.

E. V. Kozlov, V. N. Emelyanov, and V. A. Antonov
*Quantum Theory of the Order–Disorder Transition Energy in $AuCu_3$
(b) Vol. **81**, No. 2, p. 719—725 (1977).
See also *1.2*, p. *16*; 21.1, p. 293; 21.6, p. 316.

H. D. Wiesinger
*Metallkundliche Möglichkeiten zur Erhöhung der Sprungtemperatur zur Supraleitung von A15-Phasen
(Possibilities for the Increase of the Transition Temperature to Superconductivity from the Metal-Scientific Point of View)
(a) Vol. **41**, No. 2, p. 465—479 (1977).
See also *14.2*, p. *162*; 21, p. 281.

J. J. Couderc, J. Bras et M. Fagot
*Précipitation dans le système fer–gallium au voisinage de 25 at % Ga
(Precipitation in the Iron–Gallium System in the Vicinity of 25 at % Ga)
(a) Vol. **41**, No. 2, p. 595—605 (1977).
See also *1.2*, p. *17*; 21, p. 281; 21.1.1, p. 304.

M. Kh. Ashurov, Yu. K. Voronko, V. V. Osiko, A. A. Sobol, and M. I. Timoshechkin
*Spectroscopic Study of Stoichiometry Deviation in Crystals with Garnet Structure
(a) Vol. **42**, No. 1, p. 101—110 (1977).
See also 13.4, p. 136; 20.1, p. 257; 20.3, p. 271; 22.8.2, p. 411.

G. van Tendeloo, R. de Ridder, L. van Goethem, D. van Dyck, J. van Landuyt, and S. Amelinckx
Microstructural Study of Octahedral Intercalation Compounds of Tantalum and Niobium Sulphide
(a) Vol. **42**, No. 1, p. 319—335 (1977).
Erratum
(a) Vol. **44**, No. 1, p. 411 (1977).
See also *1.4*, p. *26*; 22, p. 326.

F. A. Khvaya, A. A. Katsnelson, V. M. Silonov, and M. M. Khrushchov
*A Pseudopotential Approach to the Electronic Theory of Short-Range Order
(b) Vol. **82**, No. 2, p. 701—704 (1977).
See also 21.1, p. 296.

Y. Aoki and Y. Gotoh
*Electron-Microscopic Study on the Low-Temperature Phase of the Co-Rich Co–V Alloy System
(a) Vol. **42**, No. 2, p. 469—475 (1977).
See also *1.4*, p. *27*; 21.1, p. 296.

M. Fumeron, J. P. Guillot et A. Dauger
*Modèle de diffusion des rayons X par des zones de Guinier et Preston ellipsoïdales dans le cas de l'Al–Zn
(Model X-Ray Scattering by Ellipsoidal Giunier-Preston-Zones in the Case of Al–Zn)
(a) Vol. **42**, No. 2, p. 687—697 (1977).
See also 12.1, p. 102; 21, p. 283; 21.1, p. 296.

R. de Ridder, G. van Tendeloo, and S. Amelinckx
*The Ordering of Vacancies in $Ni_{1-x}Al$. Study of the Transition State by Electron Diffraction
(a) Vol **43**, No. 1, p. 133—139 (1977).
See also 10.1, p. 80; 21, p. 284; 21.1, p. 297.

F. Reynaud
*A Mean Field Theory of Ordering of Alloys of AB Stoichiometry
(a) Vol. **43**, No. 1, p. 239—250 (1977).
See also 1.2, p. 18.

B. K. Jain, A. K. Singh, and K. Chandra
*Effect of Alloying Elements in Plain Carbon Steels
(a) Vol. **43**, No. 1, p. 357—361 (1977).
See also 5, p. 49; 21.1.1, p. 305.

K. Mukasa and M. Maeda
*Effects of Annealing on Magnetostrictive Vibrations of Co–Fe Alloys
(a) Vol. **43**, No. 2, p. 393—398 (1977).
See also *18.2*, p. *223*; 21.1, p. 297; 21.1.1, p. 306.

T. SCHOBER and A. CARL
A Differential Thermal Analysis Study of the Vanadium–Hydrogen System
(a) Vol. **43**, No. 2, p. 443—449 (1977).
See also *1.2*, p. *19*; 8, p. 67; 21.1, p. 297.

G. VAN TENDELOO and S. AMELINCKX
*Long-Period Antiphase Boundary Structures in the Ordered Alloy Au_4Zn (I): One-Dimensional Long-Period Antiphase Boundary Structure
(a) Vol. **43**, No. 2, p. 553—564 (1977).
See also *1.4*, p. *28*; 21.1, p. 298; 21.6, p. 317.

C. GAULT, A. DAUGER, and P. BOCH
Variations of the Elastic Constants of Aluminium–Magnesium Single Crystals with Guinier-Preston Zones
(a) Vol. **43**, No. 2, p. 625—632 (1977).
See also *12.1*, p. *102*; 21, p. 286; 21.3, p. 310.

A. J. MORTON
The Displacement Vector Associated with Inversion Boundaries in γ-Brass Type Phases — A Reassessment
(a) Vol. **43**, No. 2, p. K123—K125 (1977).
See also *1.4*, p. *28*; 21.1, p. 298.

L. BILLARD, P. VILLEMAIN, and A. CHAMBEROD
*Interaction of Atomic Order with Itinerant Ferromagnetism
(b) Vol. **84**, No. 1, p. 335—343 (1977).
See also *18.2*, p. *223*; 21.1.1, p. 306.

A. J. MORTON
*Inversion Domains in γ-Brass Type Phases. Stabilisation Mechanism–The Role of Electron Concentration
(a) Vol. **44**, No. 1, p. 205—214 (1977).
See also *1.4*, p. *28*; 21.1, p. 299.

M. JURISCH, K.-H. BERTHEL, and H.-J. ULLRICH
Influence of Chemical Composition within the Range of Homogeneity on Phase Transition and Transition Temperature of V_3Si Single Crystals (I): Preparation and Characterization of Single Crystals
(a) Vol. **44**, No. 1, p. 277—284 (1977).
See also *3*, p. *43*; 14.2, p. 164; 21.1, p. 299.

C. KORN and D. TEITEL
*Influence of Vanadium on Hydrogen Diffusion and Phase Constitution in Ti–H
(a) Vol. **44**, No. 2, p. 755—766 (1977).
See also 19, p. 243; 21.1, p. 300.

1.4 Microstructure
(Magnetic Domains See 18; Ferroelectric Domains See 14.4.1)

Review Article

V. A. Buravikhin, V. V. Litvintsev, and A. I. Ushakov
Polymorphic Transformations and Physical Properties of Thin Films
(a) Vol. **40**, No. 1, p. 11–27 (1977).
See also *1.2*, p. *14*; *1.5*, p. *29*; 18.2, p. 217; 21.1, p. 288; 21.1.1, p. 302.

Original Papers and Short Notes

J. D. Stephenson
*Some Applications of Characteristic Square Etch Pits Formed on {100} Surfaces of Fe-3wt% Si Crystals
(a) Vol. **39**, No. 1, p. 89–101 (1977).
See also 10.1, p. 75; 21.1.1, p. 302.

K. Borggreen and A. R. Thölén
*Diffraction Contrast from Misfitting Spherical Particles Precipitated Close Together on a Dislocation (II): The Cylinder as a Converged Model
(a) Vol. **39**, No. 1, p. 223–227 (1977).
See also 10, p. 72.

H. B. Haanstra and H. Ihrig
*Voltage Contrast Imaging of PTC-Type $BaTiO_3$ Ceramics Having Low and High Titanium Excess
(a) Vol. **39**, No. 1, p. K7–K10 (1977).
See also *14.4.1*, p. *194*; 22.8.1, p. 402.

R. de Ridder, D. van Dyck, G. van Tendeloo, J. van Landuyt, and S. Amelinckx
Microstructural Study of Copper-Intercalated Niobium Disulphide and Tantalum Disulphide Cu_xNbS_2 and Cu_xTaS_2 (III): Study of the Diffuse Scattering: the Transition State
(a) Vol. **39**, No. 2, p. 383–399 (1977).
See also *1.2*, p. *15*; 22, p. 322.

J. Bohm, H.-D. Kürsten, P. Reiche und R. Schalge
*Morphologie von Versetzungen und Ausscheidungen in Gadoliniumgalliumgranat (GGG) (Morphology of Dislocations and Segregations in Gadolinium Garnet (GGG))
(a) Vol. **39**, No. 2, p. 517–523 (1977).
See also *10.2*, p. *84*; 22.8.2, p. 409.

J. N. Boland, B. E. Hobbs, and A. C. McLaren
*The Defect Structure in Natural and Experimentally Deformed Kyanite
(a) Vol. **39**, No. 2, p. 631–641 (1977).
See also *10.2*, p. *84*; 22.8, p. 396.

M. Yacamán, L. W. Hobbs, and M. J. Goringe
*The Direct Observation of Suzuki-Phase Precipitates in $NaCl:MnCl_2$ by Transmission Electron Microscopy
(a) Vol. **39**, No. 2, p. K85–K90 (1977).
See also *10.2*, p. *85*; 22.5.2, p. 376.

H. P. Geserich, W. Möller, G. Schreiber, and L. Pintschovius
Optical Investigation of the Electrical Anisotropy of $(SN)_x$ Single Crystals
(b) Vol. **80**, No. 1, p. 119–123 (1977).
See also *14.1*, p. *157*; *20.1*, p. *250*; 22, p. 322.

W. A. Szilva and L. E. Murr
*Variation of Laser Mirror Metal Microstructure and Its Effect on Reflectivity at 10.6 μm
(a) Vol. **40**, No. 1, p. 211–224 (1977).
See also *20.2*, p. *265*; 21.1, p. 291.

A. L. Roitburd
Predominant Effect of the Internal Stress Relaxation on Microstructural and Kinetic Features of Martensitic Transformation
(a) Vol. **40**, No. 1, p. 333—342 (1977).
See also *1.2*, p. *15*; 21.1.1, p. 303.

R. de Ridder, G. van Tendeloo, J. van Lannduyt, D. van Dyck, and S. Amelinckx
Erratum to: Microstructural Study of Copper-Intercalated Niobium Disulphide and Tantalum Disulphide Cu_xNbS_2 and Cu_xTaS_2 (I)
(a) Vol. **40**, No. 1, p. 343 (1977).
See also *1.2*, p. *15*; 22, p. 323.

G. van Tendeloo, R. Wolf, and S. Amelinckx
*The Microstructure of the Alloy Au_5Mn_2: A Domain Structure with 84 Variants
(a) Vol. **40**, No. 2, p. 531—550 (1977).
See also *1.3*, p. *21*; 21.1, p. 292; 21.6, p. 315.

R. Urao
*On {211} Twins in B.C.C. Cr Films Electrodeposited onto {100} Cu Substrates
(a) Vol. **40**, No. 2, p. 567—574 (1977).
See also *1.5*, p. *29*; 21.1, p. 292.

R. Bonnet and E. E. Laufer
*Precise Determination of the Relative Orientation of Two Crystals from the Analysis of Spot Diffraction Patterns
(a) Vol. **40**, No. 2, p. 599—603 (1977).
See also *1*, p. *9*; 21.1, p. 292.

R. de Ridder, D. van Dyck, G. van Tendeloo, and S. Amelinckx
*A Cluster Model for the Transition State and Its Study by Means of Electron Diffraction (II): Application to Some Particular Systems
(a) Vol. **40**, No. 2, p. 669—683 (1977).
See also 10.2, p. 87; 22, p. 324.

B. B. Sharma, R. Ayyar, and H. Singh
Stability of the Tetrahedral Phase in the $A_2^IB^{IV}C_3^{VI}$ Group of Compounds
(a) Vol. **40**, No. 2, p. 691—696 (1977).
See also *1*, p. *9*; 22.8, p. 397.

J. J. Legendre, M. Huber, and M. Sauvage
*Observation of Linear and Planar Defects in $TiS_{1.7}$ Polytypic Crystals
(a) Vol. **40**, No. 2, p. K101—K103 (1977).
See also 22, p. 324.

C. Lacour and M. Paulus
*The Effect of Equilibrium Segregation at the Boundaries in the Discontinuous Growth Region
(a) Vol. **40**, No. 2, p. K179—K181 (1977).
See also *3*, p. *42*; 22.8.2, p. 410.

M. Snykers, R. Serneels, P. Delavignette, R. Gevers, J. van Landuyt, and S. Amelinckx
*Diffraction Contrast between Inversion Domains and at Inversion Domain Boundaries
(a) Vol. **41**, No. 1, p. 51—63 (1977).
See also 22.7, p. 394.

L. van Goethem, J. van Landuyt, and S. Amelinckx
*The $\alpha \rightarrow \beta$ Transition in Amethyst Quartz as Studied by Electron Microscopy and Diffraction. The Interaction of Dauphiné with Brazil Twins
(a) Vol. **41**, No. 1, p. 129—137 (1977).
See also *1.2*, p. *16*; 22.6, p. 388.

M. J. Marcinkowski
*The Interrelationship between Internal Boundaries of Large and Small Misfit
(a) Vol. **41**, No. 1, p. 213—224 (1977).
See also *10*, p. *72*.

D. B. Marshall and A. C. McLaren
*Elastic Twinning in Experimentally Deformed Plagioclase Feldspars
(a) Vol. **41**, No. 1, p. 231—240 (1977).
See also 10.2, p. 87; 22.8, p. 398.

M. Condat and M. Fayard
*On the Formation of Prismatic Loops in Quenched F.C.C. Metals and Alloys
(a) Vol. **41**, No. 1, p. K5—K8 (1977).
See also *10.1*, p. *78*; 21, p. 280.

A. Authier and Y. Epelboin
*Variation of Stacking Fault Contrast with the Value of the Phase Shift in X-Ray Topography
(a) Vol. **41**, No. 1, p. K9—K12 (1977)
See also *10*, p. *72*.

R. de Ridder, G. van Tendeloo, D. van Dyck, and S. Amelinckx
*Application of the Cluster Model to the Study of Ordering of Atom Strings in Ternary W–Nb Oxides
(a) Vol. **41**, No. 2, p. 555—560 (1977).
See also *10.2*, p. *88*; 22.6, p. 388.

G. Laslaz, G. Kostorz, M. Roth, P. Guyot, and R. J. Stewart
*Interparticlc Interference in Neutron Small-Angle Scattering of Al–6.8 at % Zn during Ageing at 133 °C
(a) Vol. **41**, No. 2, p. 577—583 (1977).
See also 10.1, p. 78; 21, p. 281; 21.1, p. 294.

M. A. Fortes
*Coincidence Site Lattices in Non-Cubic Lattices
(b) Vol. **82**, No. 1, p. 377—382 (1977).
See also *1.1*, p. *11*.

K. Scheerschmidt and J. Heydenreich
*On the Computer Simulation of the Electron Microscopic Diffraction Contrast of Dislocations in Jog and Dipole Configurations
(a) Vol. **42**, No. 1, p. 47—59 (1977).
See also 10, p. 73.

F. Appel, H. Bethge, and U. Messerschmidt
*Dislocation Motion and Multiplication at the Deformation of MgO Single Crystals in the High Voltage Electron Microscope
(a) Vol. **42**, No. 1, p. 61—71 (1977).
See also 10.2, p. 89; 22.6, p. 389.

M. J. Marcinkowski
*The Differential Geometry of Twisting
(a) Vol. **42**, Xo. 1, p. 209—220 (1977).
See also 10, p. 73.

G. van Tendeloo, R. de Ridder, L. van Goethem, D. van Dyck, J. van Landuyt, and S. Amelinckx
*Microstructural Study of Octahedral Intercalation Compounds of Tantalum and Niobium Sulphide
(a) Vol. **42**, No. 1, p. 319—335 (1977).
Erratum
(a) Vol. **44**, No. 1, p. 411 (1977).
See also 1.3, p. 22; 22, p. 326.

Y. Aoki and Y. Gotoh
*Electron-Microscopic Study on the Low-Temperature Phase of the Co-Rich Co–V Alloy System
(a) Vol. **42**, No. 2, p. 469–475 (1977).
See also *1.3*, p. *22*; 21.1, p. 296.

M. J. Witcomb
*Microstructure and Superconducting Properties of Niobium and Tantalum Films Sputtered onto Copper Substrates
(a) Vol. **42**, No. 2, p. 595–607 (1977).
See also *14.2*, p. *162*; 21, p. 283.

H. H. Mende and W. Hagen
*X-Ray Topographic Investigations of Deformed Iron Whiskers
(a) Vol. **42**, No. 2, p. K95–K98 (1977).
See also *10.1*, p. *80*; 21.1.1, p. 305.

J. C. Phillips
The "Extra" Phase Transition in TTF-TCNQ
(b) Vol. **83**, No. 1, p. 123–127 (1977).
See also *1.2*, p. *18*; 12.2, p. 105; 22.9, p. 417.

Yu. I. Kharkats
*The Thermodynamic Theory of Domain Formation in Superionic Crystals
(b) Vol. **83**, No. 1, p. 313–322 (1977).
See also *10*, p. *73*.

U. Essmann, W. Wiethaup, and H. U. Habermeier
The Intermediate State of Superconducting Pure Tantalum
(a) Vol. **43**, No. 1, p. 151–158 (1977).
See also *14.2*, p. *163*; 21, p. 284.

H. Raidt, G. Kästner, H. Menninger, and G. Voigt
*On Real Structure Investigations of $Ga_xIn_{1-x}P$ by Electron Microscopic Methods
(a) Vol. **43**, No. 1, p. 167–173 (1977).
See also *3*, p. *43*; 20.3, p. 273; 22.2.4, p. 357.

H. M. Eissa
*Hexagonal Networks in the Basal Plane of Zn Single Crystals
(a) Vol. **43**, No. 1, p. 343–349 (1977).
See also *10.1*, p. *80*; 21.1, p. 297.

T. Yamamoto
A Note on the Contrast Mechanism of Electron Channeling Patterns in Scanning Electron Microscopy
(a) Vol. **43**, No. 1, p. K53–K56 (1977).
See also *1.1*, p. *12*.

C. Jourdan and J. Gastaldi
*X-Ray Topographic Investigation of Dislocations in Titanium Single Crystals Grown by Recrystallization
(a) Vol. **43**, No. 2, p. 425–435 (1977).
See also 1.2, p. 19; *10.1*, p. *80*; 21.1, p. 297.

G. Vanderschaeve and T. Sarrazin
*On the Nature of Deformation Twins in Ordered Ni_3V
(a) Vol. **43**, No. 2, p. 459–464 (1977).
See also 10.1, p. 80; 21.1, p. 298.

D. van Dyck, R. de Ridder, G. van Tendeloo, and S. Amelinckx
A Cluster Model for the Transition State and Its Study by Means of Electron Diffraction (III): Generalisations of the Theory and Relation to the SRO Parameters
(a) Vol. **43**, No. 2, p. 541–552 (1977).
See also 10.2, p. 91; 22, p. 329.

G. VAN TENDELOO and S. AMELINCKX
*Long-Period Anitphase Boundary Structures in the Ordered Alloy Au_4Zn (I): One-Dimensional Long-Period Antiphase Boundary Structure
(a) Vol. **43**, No. 2, p. 553–564 (1977).
See also *1.3*, p. *23*; 21.1, p. 298; 21.6, p. 317.

A. J. MORTON
*The Displacement Vector Associated with Inversion Boundaries in γ-Brass Type Phases — A Reassessment
(a) Vol. **43**, No. 2, p. K123–K125 (1977).
See also 1.3, p. 23; 21.1, p. 298.

G. VAN TENDELOO, J. VAN LANDUYT, and S. AMELINCKX
Electron Diffraction Evidence for a One-Dimensional Peierls Distortion in a Transition Metal Trichalcogenide: TaS_3
(a) Vol. **43**, No. 2, p. K137–K140 (1977).
See also *1.1*, p. *13*; 22, p. 329.

E. ZIELIŃSKA-ROHOZIŃSKA
*X-Ray Diffraction Investigations of Annealing Characteristics in P^+-Implanted Silicon Crystals
(a) Vol. **44**, No. 1, p. 59–64 (1977).
See also 11, p. 99; 22.1.2, p. 343.

T. YAMAMOTO
*Experimental Aspects of Electron Channeling Patterns in Scanning Electron Microscopy (I): Measurement of Contrast
(a) Vol. **44**, No. 1, p. 137–146 (1977).
See also *10.1*, p. *81*; 21, p. 287; 21.1, p. 299; 21.6, p. 317; 22.1.2, p. 343.

M. WILKENS and P. RAPPS
*Electron Diffraction in Crystalline Specimens Containing a High Density of Lattice Defects (I): Development of a Statistical Theory
(a) Vol. **44**, No. 1, p. 173–182 (1977).
See also *10*, p. *73*.

A. J. MORTON
*Inversion Domains in γ-Brass Type Phases. Stabilisation Mechanism–The Role of Electron Concentration
(a) Vol. **44**, No. 1, p. 205–214 (1977).
See also *1.3*, p. *23*; 21.1, p. 299.

J. D. LIVINGSTON
*Grain Size in A-15 Reaction Layers
(a) Vol. **44**, No. 1, p. 295–301 (1977).
See also 14.2, p. 164; 21, p. 287.

T. YAMAMOTO
*Experimental Aspects of Electron Channeling Patterns in Scanning Electron Microscopy (II): Estimation of Contrast Depth
(a) Vol. **44**, No. 2, p. 467–476 (1977).
See also *10.1*, p. *82*; 21, p. 287; 21.1, p. 300; 21.6, p. 318; 22.1.2, p. 344.

R. Z. VALIEV and O. A. KAIBYSHEV
Mechanism of Superplastic Deformation in a Magnesium Alloy (II): The Role of Grain Boundaries
(a) Vol. **44**, No. 2, p. 477–484 (1977).
See also *10.1*, p. *82*; 21, p. 287; 21.3, p. 310.

1.5 Films

Review Articles

V. A. Buravikhin, V. V. Litvintsev, and A. I. Ushakov
*Polymorphic Transformations and Physical Properties of Thin Films
(a) Vol. **40**, No. 1, p. 11—27 (1977).
See also *1.2*, p. *14*; 1.4, p. 24; 18.2, p. 217; 21.1, p. 288; 21.1.1, p. 302.

L. N. Aleksandrov
*Transition Regions in Epitaxial Films
(a) Vol. **44**, No. 1, p. 11—43 (1977).
See also *3*, p. *42*; 14.3.1, p. 173; 22.1.1, p. 333; 22.1.2, p. 337; 22.2.1, p. 350.

Original Papers and Short Notes

J. P. Strømsheim, T. Olsen, and D. J. Ruzicka
*Analyses of Surface Preparation, Anodic Oxidation, and Epitaxial Layers of GaAs Using Nuclear Backscattering
(a) Vol. **39**, No .1, p. 167—172 (1977).
See also *1.6*, p. *31*; 22.2.1, p. 350.

J. Bauer, L. Biste, and D. Bolze
*Optical Properties of Aluminium Nitride Prepared by Chemical and Plasmachemical Vapour Deposition
(a) Vol. **39**, No. 1, p. 173—181 (1977).
See also 13.1, p. 114; 20.1, p. 248; 22.2, p. 348.

N. C. Halder, J. M. Montante, and W. E. Swartz, Jr.
*Evidence of a Polycrystalline Phase in Amorphous Ge Films Evaporated at Room Temperature
(a) Vol. **39**, No. 1, p. 213—222 (1977).
See also *2*, p. *33*; 13.1, p. 115; 14.3.1, p. 173; 22.1.1, p. 333.

G. M. Mladenov and T. N. Yakimov
*Electron Beam Scattering during Thin-Film Processing
(a) Vol. **39**, No. 1, p. 321—326 (1977).

L. N. Aleksandrov and I. A. Entin
*On the Kinetics of Thin-Film Growth under High Adatom Concentrations
(a) Vol. **39**, No. 1, p. 327—330 (1977).
See also *3*, p. *42*; 22.1.1, p. 333.

A. Hadni, R. Thomas, and C. Erhard
*An Unusual Type of Epitaxial Growth
(a) Vol. **39**, No. 2, p. 419—424 (1977).
See also *3*, p. *42*; 22.9, p. 414.

R. Urao
*On {211} Twins in B.C.C. Cr Films Electrodeposited onto {100} Cu Substrates
(a) Vol. **40**, No. 2, p. 567—574 (1977).
See also *1.4*, p. *25*; 21.1, p. 292.

Á. Barna, P. B. Barna, G. Radnóczi, L. Tóth, and P. Thomas
*A Comparative Study of the Structure of Evaporated and Glow Discharge Silicon
(a) Vol. **41**, No. 1, p. 81—84 (1977).
See also 2, p. 36; 22.1.2, p. 339.

E. A. Matson and S. A. Polyakov
*On the "Evolutionary Selection" Principle in Relation to the Growth of Polycrystalline Silicon Films
(a) Vol. **41**, No. 1, p. K93—K95 (1977).
See also *3*, p. *42*; 22.1.2, p. 339.

H. Bialas and Li-Shing Li
Hypersound Excitation at 8.9 GHz by Exploiting Magnetic Anisotropy in Oriented Nickel Films
(a) Vol. **42**, No. 1, p. 125—132 (1977).
See also *7*, p. *62*; *19*, p. *240*; **21.1**, p. **295**.

M. J. Yacamán and T. Ocaña Z.
*High-Resolution Dark-Field Electron Microscopy of Small Metal Particles
(a) Vol. **42**, No. 2, p. 571—577 (1977).
See also **21.6**, p. **316**.

G. Oertel and E.-H. Weber
Growth of Thin SiO_2 Films on Clean Si (111) Surfaces by Low-Pressure Oxidation and Their Evaporation
(a) Vol. **43**, No. 1, p. 141—150 (1977).
See also *17*, p. *207*; *20.1*, p. *260*; **22.6**, p. **390**.

O. P. Pchelyakov, E. B. Gorokhov, S. V. Pokrovskaya, L. V. Sokolov, and S. I. Stenin
*Structure of SiO_2 Pyrolitic Films on Ge Substrates
(a) Vol. **43**, No. 2, p. 659—666 (1977).
See also **2**, p. **39**; **22.1.1**, p. **336**; **22.6**, p. **391**.

M. Németh-Sallay, Á. Barna, A. Lörinczy, and I. C. Szép
Induced Crystallization of Amorphous Silicon Oxide
(a) Vol. **43**, No. 2, p. K135—K136 (1977).
See also *1.6*, p. *32*; **2**, p. **39**; **22.6**, p. **391**.

S. Al-Dallal
*Epitaxial Growth of CdSe on ZnTe (111)
(a) Vol. **44**, No. 1, p. 183—189 (1977).
See also *3*, p. *43*; **22.4.2**, p. **366**.

U. Gradmann and P. Tillmanns
*Supersaturation and Mode of Growth for Fe Films on Cu(111) — An Experimental Study Using LEED and AES
(a) Vol. **44**, No. 2, p. 539—547 (1977).
See also *3*, p. *43*; 21.1.1, p. 307.

N. L. P. Sinha, A. K. Chaudhuri, and H. N. Bose
*Structural Analysis of Lead Telluride Films and Their Sensitization
(a) Vol. **44**, No. 2, p. K127—K129 (1977).
See also **14.3.1**, p. **175**; **16**, p. **205**; **22**, p. **331**.

F. K. Naehring, A. Schmidt, and J. Schöneich
Ion Induced Carbon Contamination and Recoil Implantation
(a) Vol. **44**, No. 2, p. K141—K145 (1977).
See also *11*, p. *99*; **22.1.2**, p. **344**.

1.6 Surfaces

Original Papers and Short Notes

J. P. Strømsheim, T. Olsen, and D. J. Ruzicka
*Analyses of Surface Preparation, Anodic Oxidation, and Epitaxial Layers of GaAs Using Nuclear Backscattering
(a) Vol. **39**, No. 1, p. 167—172 (1977).
See also *1.5*, p. *29*; 22.2.1, p. 350.

J.-M. Derochette and J. Marien
*Field Emission Study of Hydrogen Adsorption on Smooth and Perfectly Ordered Iridium Surfaces (111) and (100) — Influence of Steps and Iridium Clusters on the Surface Potential of the Hydrogen Film
(a) Vol. **39**, No. 1, p. 281—289 (1977).
See also *17.1*, p. *208*; 21, p. 277.

V. I. Savenko, L. A. Kachanova, and E. D. Shchukin
*The Generation of Dislocations near the Surfaces of Crystals under Microindentation
(a) Vol. **39**, No. 1, p. K11—K15 (1977).
See also *10.2*, p. *83*; 12.2, p. 104; 22.5.2, p. 375.

K. Kojima and M. Meshii
*Surface Film Softening of Iron Single Crystals Oriented for Single Slip
(a) Vol. **39**, No. 2, p. 491—498 (1977).
See also *10.1*, p. *75*; 21.1.1, p. 302.

R. Erlandsson, U. Birkholz, and S.-E. Karlsson
*Surface Investigation of Si_2Te_3 with Auger Spectroscopy
(a) Vol. **41**, No. 2, p. K163—K165 (1977).
See also 22, p. 325.

R. Lenk, H. Solbrig, H. Wolf, and K.-H. Sommer
*Multiple Scattering of Electrons by an Infinite Lattice Plane (I): Derivation of the Basic Equations Using a Scattering Amplitude Differential Operator
(b) Vol. **82**, No. 1, p. 303—310 (1977).

Diep-The-Hung
Effect of the Next-Nearest Neighbour Coupling in a Semi-Infinite Ferrimagnet
(b) Vol. **82**, No. 1, p. 383—389 (1977).
See also *13.5.4*, p. *153*; *18.3*, p. *229*.

A. M. Afanasev, M. V. Kovalchuk, E. K. Kovev, and V. G. Kohn
*X-Ray Diffraction in a Perfect Crystal with Disturbed Surface Layer
(a) Vol. **42**, No. 1, p. 415—422 (1977).
See also *1*, p. *10*; 22.1.2, p. 341.

E. W. Kreutz and N. Sotnik
*Kikuchi Patterns of Spherical Ag Crystals in the Bragg Case
(a) Vol. **42**, No. 1, p. K11—K14 (1977).
See also *1.1*, p. *12*; 21.6, p. 316.

H. Solbrig
*Multiple Scattering of Electrons by an Infinite Lattice Plane (II): On the Properties of the Planar Scattering Matrix
(b) Vol. **82**, No. 2, p. 429—437 (1977).

D. I. Blokhintsev and N. M. Plakida
*On the Heating of Ultracold Neutrons
(b) Vol. **82**, No. 2, p. 627—632 (1977).
See also 11, p. 96.

H. PFEIFFER, W. HAUBENREISSER, and TH. KLUPSCH
*Analytic Treatment of Stochastic Models of Crystal Growth. Comparison of Two Different Interface Dynamics
(b) Vol. **83**, No. 1, p. 129—138 (1977).
See also *3*, p. *42*.

A. MOSSER, S. SHINDO, and Y. ADACHI
*Angular and Temperature Dependence of the LEED Kikuchi Pattern Intensity
(a) Vol. **43**, No. 1, p. 295—304 (1977).
See also 22.6, p. 390.

V. SIEGEL and W. RASP
The TSEE of Lapped and Unlapped BeO Ceramics
(a) Vol. **43**, No. 2, p. K117—K121 (1977).
See also 13.3, p. 128; *17*, p. *207*; 22.6, p. 391.

M. NÉMETH-SALLAY, Á. BARNA, A. LÖRINCZY, and I. C. SZÉP
*Induced Crystallization of Amorphous Silicon Oxide
(a) Vol. **43**, No. 2, p. K135—K136 (1977).
See also 1.5, p. 30; 2, p. 39; 22.6, p. 391.

DIEP-THE-HUNG
Effect of Induced Surface Magnetic Moment on Spin-Wave Spectrum
(b) Vol. **84**, No. 2, p. K159—K161 (1977).
See also *13.5.4*, p. *154*; 18.4, p. 234.

2. Non-Crystalline State

Original Papers and Short Notes

D. Engemann and R. Fischer
*Photoluminescence in Amorphous Silicon
(b) Vol. **79**, No. 1, p. 195—202 (1977).
See also *20.3*, p. *267*; 22.1.2, p. 337.

E. Kolley and W. Kolley
*On the Theory of Phonon-Modulated Electronic Transport in Off-Diagonal Random Alloys
(b) Vol. **79**, No. 1, p. 325—337 (1977).
See also 6, p. 51; *14*, p. *155*; 21, p. 276.

E. Majerníková
On the Theory of Exciton Absorption in Amorphous Semiconductors
(b) Vol. **79**, No. 1, p. 339—345 (1977).
See also *13.5.1*, p. *143*; *20.1*, p. *248*; 22, p. 320.

R. Kishore
*Exactly Solvable Off-Diagonal Disorder
(b) Vol. **79**, No. 1, p. K9—K11 (1977).
See also *13*, p. *107*; 22, p. 320.

J. Szczyrbowski
*The Shape of Absorption Edge of Amorphous Ge, Si, InSb, GaSb, GaAs, and GaP
(b) Vol. **79**, No. 1, p. K43—K48 (1977).
See also *20.1*, p. *248*; 22.1, p. 332; 22.2, p. 348.

J. Livage, P. Pineau, M. C. Leroy, and M. Michaud
*Semiconducting Vanadium Phosphate Glasses
(a) Vol. **39**, No. 1, p. 73—78 (1977).
See also 19, p. 235; 22.6.1, p. 393.

N. C. Halder, J. M. Montante, and W. E. Swartz, Jr.
*Evidence of a Polycrystalline Phase in Amorphous Ge Films Evaporated at Room Temperature
(a) Vol. **39**, No. 1, p. 213—222 (1977).
See also *1.5*, p. *29*; 13.1, p. 115; 14.3.1, p. 173; 22.1.1, p. 333.

R. M. Catchings
*Magnetic Susceptibility Studies of Iron-Dysprosium Sodium Borate Glasses
(a) Vol. **39**, No. 1, p. K71—K74 (1977).
See also *18*, p. *209*; 22.8, p. 396.

E. Kolley and W. Kolley
*Calculation of the Temperature-Dependent DC Conductivity of Disordered Alloys in Off-Diagonal CPA
(b) Vol. **79**, No. 2, p. 461—472 (1977).
See also 6, p. 52; 13, p. 107; *14.1*, p. *157*; 21, p. 277.

W. Rehm, R. Fischer, J. Stuke, and H. Wagner
*Photo and Dark Conductivity of Doped Amorphous Silicon
(b) Vol. **79**, No. 2, p. 539—547 (1977).
See also 14.3.1, p. 173; *16*, p. *201*; 22.1.2, p. 337.

G. Cubiotti, E. Donato, E. S. Giuliano, and R. Ruggeri
*On the Influence of Disorder on the Electron Density of States in a Simplified Model for a Quasi-Onedimensional Conductor
(b) Vol. **79**, No. 2, p. 655—661 (1977).
See also *13*, p. *107*; 14, p. 155 ; 22, p. 321.

J. Richter
*Influence of Off-Diagnonal Disorder on Density of States and Itinerant Ferromagnetism
(b) Vol. **79**, No. 2, p. K139—K142 (1977).
See also 13, p. 108; *18.2*, p. *218*.

J. Bauer
*Optical Properties, Band Gap, and Surface Roughness of Si_3N_4
(a) Vol. **39**, No. 2, p. 411—418 (1977).
See also 13.1, p. 115; 20.1, p. 249; 22, p. 322.

P. P. Seregin, V. P. Sivkov, F. S. Nasredinov, L. N. Vasilev, Yu. V. Krylnikov, and Yu. P. Kostikov
*The Influence of the Crystal-to-Glass Transition on the Local Structure of Semiconductors
(a) Vol. **39**, No. 2, p. 437—444 (1977).
See also *1.2*, p. *15*; 5, p. 47; 22, p. 322.

R. M. Hill
The Frequency Dependence of Variable Range Hopping
(a) Vol. **39**, No. 2, p. 615—620 (1977).
See also *13.4*, p. *131*; *14.3.1*, p. *173*.

R. Wang and M. D. Merz
*Polymorphic Bonding and Thermal Stability of Elemental Noncrystalline Solids
(a) Vol. **39**, No. 2, p. 697—703 (1977).
See also 1.2, p. 15; *4*, p. *44*; 21, p. 278; 22.1, p. 332.

R. M. Catchings
*EPR Study of the Ferric Oxide–Nickel Oxide–Sodium–Lithium–Borosilicate Glass System
(a) Vol. **39**, No. 2, p. K101—K103 (1977).
See also *19*, p. *236*; 22.8, p. 396.

L. Kraus and J. Schneider
*Magnetostriction of Amorphous $(Fe_{1-x}Ni_x)_{80}P_{10}B_{10}$ Alloys
(a) Vol. **39**, No. 2, p. K161—K163 (1977).
See also *18.2*, p. *218*; 19, p. 236; 22.8, p. 397.

N. C. Halder
*Calculation of Nuclear Quadrupolar Relaxation in Liquid Metals
(b) Vol. **80**, No. 1, p. 273—278 (1977).
See also *19*, p. *236*; 21, p. 278.

R. Kishore
*Semiclassical Theory of Spin Waves in Amorphous Antiferromagnets
(b) Vol. **80**, No. 1, p. K33—K35 (1977).
See also 13.5, p. 141; *18.4*, p. *232*.

J. A. Sawicki and B. D. Sawicka
Comparability of ^{57}Fe Mössbauer Data in Ion-Implanted and Amorphous FeGe
(b) Vol. **80**, No. 1, p. K41—K44 (1977).
See also *5*, p. *47*; 21.1.1, p. 302.

E. Jäger
*Structure Fluctuations in the Molecular Field Theory of Amorphous Ferrimagnets
(b) Vol. **80**, No. 1, p. K81—K83 (1977).
See also *18.3*, p. *228*; 21.1, p. 290; 21.4, p. 311.

H.-R. Müller, E. Jäger, and M. Bitterlich
*Influence of Impurities on the Magnetic Properties of Amorphous Gd–Co Thin Films
(b) Vol. **80**, No. 1, p. K85—K87 (1977).
See also *18.3*, p. *228*; 21.1, p. 290; 21.4, p. 311.

W. I. Kinney and G. S. Cargill III
*Angle-of-Deplosition Effects in Evaporated Amorphous Germanium
(a) Vol. **40**, No. 1, p. 37—42 (1977).
See also 22.1.1, p. 333.

B. Fromm and W. Teubner
*Screening Effects in Amorphous Semiconductors Influencing the Poole-Frenkel Conduction
(a) Vol. **40**, No. 1, p. 63—68 (1977).
See also *14.3.1*, p. *173*; 22, p. 323.

K. Hübner
Chemical Bond and Related Properties of SiO_2 (I): Character of the Chemical Bond
(a) Vol. **40**, No. 1, p. 133—140 (1977).
See also *4*, p. *44*; *13.1*, p. *116*; 14.4, p 189; 20.1, p. 251; 22.6, p. 387.

B. Boucher
*Bulk Magnetic Properties of Amorphous $Tb_{52}Ag_{48}$. Evidence of a Strong Local Anisotropy
(a) Vol. **40**, No. 1, p. 197—203 (1977).
See also *18.2*, p. *219*; 21.4, p. 311; 21.6, p. 315.

K. L. Chopra, A. P. Thakoor, S. K. Barthwal, and P. Nath
*Electron Transport Properties of Amorphous Cu–Ag Films
(a) Vol. **40**, No. 1, p. 247—255 (1977).
See also *14.3.1*, p. *173*; 15, p. 198; 21.1, p. 291; 21.6, p. 315.

O. Uemura and T. Satow
*The Amorphous Structure of $Cd(Ge_{1-x}Sn_x)As_2$ with $x = 0.0$, 0.1, and 0.2
(a) Vol. **40**, No. 1, p. 303—308 (1977).
See also 18.1, p. 213; 22.8, p. 397.

H. Häusler and T. Pusch
*Photoconductivity Measurements on Amorphous GeSe Films
(a) Vol. **40**, No. 1, p. K15—K17 (1977).
See also *16*, p. *202*; 22.7, p. 394.

B. Movaghar and L. Schweitzer
ESR and Conductivity in Amorphous Germanium and Silicon
(b) Vol. **80**, No. 2, p. 491—498 (1977).
See also *14.3*, p. *167*; *19*, p. *237*; 22.1.1, p. 334; 22.1.2, p. 338.

J. P. Badiali, A. Defrain, M. Dupont, and C. Regnaut
*Effective Pair Potential and Thermodynamical Properties of Liquid Gallium
(b) Vol. **80**, No. 2, p. 533—541 (1977).
See also 8, p. 65; 21, p. 279.

H. Böttger and V. V. Bryksin
A Theory of the Hall Effect in the Hopping Region in Disordered Systems (I): Rate Equation in Presence of Electrical and Magnetic Fields
(b) Vol. **80**, No. 2, p. 569—578 (1977).
See also 6, p. 54; 13, p. 108; 13.5.3, p. 151; *14.3*, p. *167*; 22, p. 323.

F. P. Koffyberg
Optical and Modulated Absorption Spectra in the Band Gap Region of Amorphous Vanadium Zinc Borate Semiconductors
(b) Vol. **80**, No. 2, p. 669—673 (1977).
See also 13.1, p. 116; *20.1*, p. *252*; 22.8, p. 397.

N. Kishimoto, K. Morigaki, K. Murakami, A. Shimizu, and A. Hiraki
*Electron Spin Resonance in the Amorphous Si–Au System
(b) Vol. **80**, No. 2, p. K113—K115 (1977).
Erratum
(b) Vol. **82**, No. 2, p. K123 (1977).
See also *19*, p. *237*; 21.6, p. 315; 22.1.2, p. 339.

R. L. Havill, J. Marsden, and J. M. Titman
*Nuclear Quadrupole Relaxation in Liquid Mercury
(b) Vol. **80**, No. 2, p. K141—K143 (1977).
See also *19*, p. *237*; 21, p. 279.

K. Hübner
Chemical Bond and Related Properties of SiO_2 (II): Structural Trends
(a) Vol. **40**, No. 2, p. 487—495 (1977).
See also *4*, p. *45*; *13.1*, p. *117*; 14.4, p. 190; 20.1, p. 252; 22.6, p. 387.

R. Perthel, W. Keilig, R. Kosicik, U. Röpke, and C.-G. d'Ambly
*Determination of the Magnetic Anisotropy in Amorphous Gd–Co Films by Torque Measurement
(a) Vol. **40**, No. 2, p. K135—K138 (1977).
See also *18.3*, p. *229*; 21, p. 280; 21.1, p. 292.

A. Buroff, E. Nebauer, P. Süptitz, and I. Willert
*Migration of Silver and Gold in Amorphous As_2S_3
(a) Vol. **40**, No. 2, p. K195—K198 (1977).
See also *9*, p. *70*; 22, p. 324.

H. Böttger and V. V. Bryksin
A Theory of the Hall Effect in the Hopping Region in Disordered Systems (II): Random Resistor Network and Hall Effect
(b) Vol. **81**, No. 1, p. 97—106 (1977).
See also 6, p. 55; 13, p. 109; 13.5.3, p. 151; *14.3*, p. *167*; 22, p. 324.

F. Kammler and K. Elk
Numerical Results of the Hubbard III Approximation for Arbitrary Values of the Correlation Energy and the Electron Density (with Application to Disordered Alloys)
(b) Vol. **81**, No. 1, p. 179—184 (1977).
See also 13, p. 109; *18.2*, p. *219*.

Y. Waseda and J. G. Wright
*The Resistivity of Amorphous Transition Elements
(b) Vol. **81**, No. 1, p. K37—K40 (1977).
See also *14.1*, p. *158*; 21.1, p. 293.

M. Kumeda, Y. Jinno, and T. Shimizu
*Electron Spin Resonance of Amorphous Si–Ge Alloys
(b) Vol. **81**, No. 1, p. K71—K74 (1977).
See also *19*, p. *238*; 22.1.1, p. 334; 22.1.2, p. 339.

Á. Barna, P. B. Barna, G. Radnóczi, L. Tóth, and P. Thomas
A Comparative Study of the Structure of Evaporated and Glow Discharge Silicon
(a) Vol. **41**, No. 1, p. 81—84 (1977).
See also *1.5*, p. *29*; 22.1.2, p. 339.

J. Sawicki, B. Sawicka, and O. Gzowski
Conversion Electron Mössbauer Analysis of Glass Surfaces
(a) Vol. **41**, No. 1, p. 173—176 (1977).
See also *5*, p. *48*; 22.8.1, p. 404.

K. K. Srivastava, D. R. Goyal, A. Kumar, K. N. Lakshminarayan, O. S. Panwar, and I. Krishan
Dielectric Properties of Glassy $As_{10}Te_{75}Ge_{15}$
(a) Vol. **41**, No. 1, p. 323—329 (1977).
See also *14.4*, p. *190*; 22.8, p. 398.

H. Kresse, D. Demus, and S. König
*Dielectric Relaxation of Liquid Crystals with Nematic and Smectic Modifications
(a) Vol. **41**, No. 1, p. K67—K69 (1977).
See also *14.4*, p. *191*; 22.9, p. 416.

H. Böttger and V. V. Bryksin
A Theory of the Hall Effect in the Hopping Region in Disordered Systems (III): DC Hall Effect
(b) Vol. **81**, No. 2, p. 433–441 (1977).
See also 6, p. 56; 13, p. 109; 13.5.3, p. 151; *14.3*, p. *168*; 22, p. 325.

K. S. Dy
*Generalized Friedel Sum Rule
(b) Vol. **81**, No. 2, p. K111–K113 (1977).
See also *13*, p. *110*.

M. Matyáš
*Magnetic Properties of As_xSe_{100-x} Glasses in the Temperature Range 77 to 300 K
(b) Vol. **81**, No. 2, p. K115–K117 (1977).
See also *19*, p. *239*; 22, p. 325.

O. Uemura, Y. Suzuki, and T. Satow
*The Amorphous Structure of the Fe–Ge System
(a) Vol. **41**, No. 2, p. 417–421 (1977).
See also 21.1.1, p. 303; 22.1.1, p. 334.

N. A. Blum, C. Feldman, and F. G. Satkiewicz
Infrared Absorption of Amorphous Boron Films Containing Carbon and Hydrogen
(a) Vol. **41**, No. 2, p. 481–486 (1977).
See also *20.1*, p. *256*; 22.1, p. 332.

K. Yasuda, A. Yoshida, and T. Arizumi
*The Effects of Annealing on Mott's Parameters for Hopping Conduction in Amorphous Ge
(a) Vol. **41**, No. 2, p. K181–K184 (1977).
See also *14.3.1*, p. *174*; 22.1.1, p. 334.

R. M. Mehra, P. C. Mathur, A. K. Kathuria, and Radhey Shyam
*Dependence of Conductivity and Activation Energy on Tellurium Concentration in Amorphous Selenium
(a) Vol. **41**, No. 2, p. K189–K192 (1977).
See also *14.3*, p. *169*; 22.1.3, p. 345.

I. V. Chepeleva, E. A. Zhilinskaya, V. N. Lazukin, A. P. Chernov, and V. I. Olkhovskii
Mn^{2+} EPR Study of Cu–As–Se, As–Se–I and Cu–As–Se–I Glass Systems
(b) Vol. **82**, No. 1, p. 189–192 (1977).
See also *19*, p. *240*; 22.8, p. 399.

J. E. Lowther
*Proton Transfer by Dynamic Interactions and the Quantum Theory of Mutations in DNA
(b) Vol. **82**, No. 1, p. K41–K44 (1977).
See also *6*, p. *57*; 22.9, p. 416.

M. Zieliński and M. Kryszewski
*Thermal Sampling Technique for the Thermally Stimulated Discharge in Polymers. Model Calculations
(a) Vol. **42**, No. 1, p. 305–314 (1977).
See also 14.4, p. 191; 22.9, p. 417.

T. Suzuki, M. Hirose, S. Ogose, and Y. Osaka
Electrical Transport and Photoconductivity in Amorphous Silicon
(a) Vol. **42**, No. 1, p. 337–344 (1977).
See also 13.1, p. 120; *14.3.1*, p. *174*; 15, p. 199; *16*, p. *203*; 22.1.2, p. 340.

E. Kittinger
*Temperature Dependence of Sound Velocity in Vitreous Selenium
(a) Vol. **42**, No. 1, p. K33–K35 (1977).
See also *7*, p. *62*; 22.1.3, p. 346.

K. Hattori
*Magnetic Properties of a Random Binary System
(b) Vol. **82**, No. 2, p. 489—494 (1977).
See also 18.1, p. 215; 18.2, p. 221; 18.3, p. 229.

S. Trimper
Phase Transition in Ferroelectric Liquid Crystals
(b) Vol. **82**, No. 2, p. K75—K79 (1977).
See also *1.2*, p. *17*; 14.4.1, p. 196.

W. John
*Comment on the Paper "Exactly Solvable Off-Diagonal Disorder"
(b) Vol. **82**, No. 2, p. K111—K112 (1977).
See also *13*, p. *110*; 22, p. 327.

K. Hübner
Chemical Bond and Related Properties of SiO_2-(III): Core Level Shifts in SiO_x
(a) Vol. **42**, No. 2, p. 501—509 (1977).
See also *4*, p. *45*; *13.1*, p. *120*; 14.4, p. 191; 20.1, p. 258; 22.6, p. 390.

M. Janai and P. S. Rudman
The Kinetics of the Light-Enhanced Vaporization of Amorphous As_2S_3 Thin Films
(a) Vol. **42**, No. 2, p. 729—737 (1977).
See also 14.3.1, p. 174; *16*, p. *203*; 22, p. 327.

B. Boucher
*Propriétés magnétiques de l'alliage TbAu amorphe et massif
(Magnetic Properties of Amorphous and Bulk TbAu Alloy)
(a) Vol. **42**, No. 2, p. K165—K169 (1977).
See also *18.2*, p. *222*; 21.4, p. 313; 21.6, p. 316.

I. P. Zvyagin
Correlation Effects in Hopping Conduction
(b) Vol. **83**, No. 1, p. 63—70 (1977).
See also 13, p. 110; *14.3*, p. *170*; 15, p. 199; 22, p. 327.

R. S. Tripathi, L. L. Moseley, and T. Lukes
Optical Absorption in Non-Crystalline Semiconductors
(b) Vol. **83**, No. 1, p. 197—207 (1977).
See also *13*, p. *110*; *20.1*, p. *259*; 22, p. 327.

F. Bechstedt and R. Enderlein
*Theory of Resonance Raman Scattering in Disordered Solids. Smooth Random Field Model
(b) Vol. **83**, No. 1, p. 239—247 (1977).
See also *6*, p. *58*; 13.1, p. 121; *20.1*, p. *259*; 22, p. 328.

W. Löser
Thermodynamic Properties of a Finite Disordered Hubbard Cluster
(b) Vol. **83**, No. 1, p. 299—304 (1977).
See also *8*, p. *66*; *18.4*, p. *233*.

B. I. Shklovskii
*Thickness Dependence of the Hopping Conduction in Amorphous Films. Experiment versus Theory
(b) Vol. **83**, No. 1, p. K11—K15 (1977).
See also *14.3.1*, p. *175*; 22.1, p. 332.

I. I. Fishchuk
Theory of the Phononless High Frequency Conductivity in Disordered Systems
(b) Vol. **83**, No. 1, p. K29—K33 (1977).
See also *14.3*, p. *170*.

N. SALEH and M. GAWISH
Effect of Heat Treatment on the Electronic Properties of Calcium Borate Glasses Containing Iron Oxide
(a) Vol. **43**, No. 1, p. 99—102 (1977).
See also *14.4*, p. *192*; 22.8.1, p. 406.

J. C. MALAURENT and J. DIXMIER
*Random Chain Assembly Model for Amorphous Sulfur, Selenium, and Tellurium
(a) Vol. **43**, No. 1, p. K61—K64 (1977).
See also 22.1.3, p. 346.

M. ZÁVĚTOVÁ and I. GREGORA
*Refractive Index of Amorphous Arsenic
(a) Vol. **43**, No. 1, p. K109—K111 (1977).
See also *20.1*, p. *260*; 22.1, p. 332.

N. N. MEDVEDEV and YU. I. NABERUKHIN
*Asymptotic Oscillations of the Radial Distribution Function in a Quasi-Crystalline Model of Liquids
(b) Vol. **83**, No. 2, p. 405—411 (1977).

A. K. JONSCHER
A Many-Body Model of Dielectric Polarisation in Solids (I): Experimental Data and Existing Interpretations
(b) Vol. **83**, No. 2, p. 585—597 (1977).
See also *14.4*, p. *192*; 14.4.1, p. 196; 22, p. 329; 22.9, p. 418.

M. P. DAS
An Effective Periodic Potential for Disordered Binary Alloys
(b) Vol. **83**, No. 2, p. K165—K168 (1977).
See also *13*, p. *111*; 21, p. 285.

A. BOBÁK
*Influence of an Inhomogeneity of the Effective Field on the Properties of a Disordered Heisenberg Ferromagnet
(b) Vol. **83**, No. 2, p. K169—K171 (1977).
See also *18.2*, p. *223*; 21, p. 285.

E. NEBAUER, P. SÜPTITZ, and I. WILLERT
*Determination of the Charge of Migrating Impurities from Electrodiffusion Profiles: the System a-As_2Se_3:Cu
(a) Vol. **43**, No. 2, p. 451—458 (1977).
See also *9*, p. *71*; 14.3.4, p. 187; 22, p. 329.

O. P. PCHELYAKOV, E. B. GOROKHOV, S. V. POKROVSKAYA, L. V. SOKOLOV, and S. I. STENIN
Structure of SiO_2 Pyrolitic Films on Ge Substrates
(a) Vol. **43**, No. 2, p. 659—666 (1977).
See also *1.5*, p. 30; 22.1.1, p. 336; 22.6, p. 391.

M. NÉMETH-SALLAY, Á. BARNA, A. LÖRINCZY, and I. C. SZÉP
Induced Crystallization of Amorphous Silicon Oxide
(a) Vol. **43**, No. 2, p. K135—K136 (1977).
See also 1.5, p. *30*; *1.6*, p. *32*; 22.6, p. 391.

A. K. JONSCHER
A Many-Body Model of Dielectric Polarisation in Solids (II): The Universal Model
(b) Vol. **84**, No. 1, p. 159—166 (1977).
See also *14.4*, p. *192*; 14.4.1, p. 196; 22, p. 329; 22.9, p. 419.

O. UEMURA and T. SATOW
*The Magnetic Susceptibility of Antimony and Bismuth Chalcogenides in the Liquid State
(b) Vol. **84**, No. 1, p. 353—360 (1977).
See also *18*, p. *212*; 22, p. 330.

H. S. Reehal and C. B. Thomas
Poole-Frenkel Conduction in Amorphous Solids for Centres with a Screened Coulomb Potential
(b) Vol. 84, No. 1, p. K13—K16 (1977).
See also *13.4*, p. *138*; *14.3.4*, p. *187*; 22.8, p. 400.

E. J. Unterhorst, V. Müller, and G. Schanz
Nuclear Acoustic Resonance in Liquid Gallium
(b) Vol. 84, No. 1, p. K53—K56 (1977).
See also *19*, p. *243*; 21, p. 286.

C. Viger, C. Vautier, and D. Carles
Recombination and Trapping Processes in Amorphous Materials in Presence of an Exponential Distribution of Traps
(a) Vol. 44, No. 1, p. 97—106 (1977).
See also 13.4, p. 139; 14.3.1, p. 175; *16*, p. *204*; 22.1.3, p. 347.

W. Nowak
*Scattered Light Investigations on Amorphous Systems
(a) Vol. 44, No. 1, p. 265—275 (1977).
See also 10.2, p. 92; 22.6, p. 391.

M. Stubičar, E. Babić, D. Subašić, D. Pavuna, and Ž. Marohnić
*Phase Transformations during Isochronal Annealing of $Fe_{40}Ni_{40}B_{20}$ Glass
(a) Vol. 44, No. 1, p. 339—344 (1977).
See also *1.2*, p. *20*; 21.1, p. 299; 21.1.1, p. 306.

S. R. Herd
*TEM Observation on the Source of Perpendicular Anisotropy in Amorphous $GdCo(O_2)$
(a) Vol. 44, No. 1, p. 363—380 (1977).
See also *18.2.1*, p. *227*; 21.4, p. 313.

E. Kittinger
*Time Dependent Thermal Expansion of Vitreous Selenium
(a) Vol. 44, No. 1, p. K35—K37 (1977).
See also *8*, p. *68*; 22.1.3, p. 347.

A. I. Popov, I. Kh. Geller, and V. K. Shemetova
*Memory and Threshold Switching Effects in Amorphous Selenium
(a) Vol. 44, No. 1, p. K71—K73 (1977).
See also *14.3.4*, p. *187*; 22.1.3, p. 347.

H. Ratajczak and I. Łabędzka
*Anomalous Hall Hysteresis Loops in the Vicinity of the Compensation Point in HoCo Amorphous Films
(a) Vol. 44, No. 1, p. K103—K105 (1977).
See also *14.1*, p. *160*; *18.2*, p. *224*; 21.1, p. 299; 21.4, p. 313.

R. Leihkauf and H. Böttger
*Dispersion Relation of Phonon-Like Excitations in a Simple Model of a Noncrystalline Solid
(b) Vol. 84, No. 2, p. 575—580 (1977).
See also *6*, p. *60*.

N. R. Mitra, J. Thakur, and P. L. Srivastava
Resistivity and Thermoelectric Power of Liquid Lithium and Sodium from Harrison's First Principle
(b) Vol. 84, No. 2, p. 797—801 (1977).
See also *14.1*, p. *160*; *15*, p. *200*; 21, p. 287.

E. Majerníková
*Quantum Shift of the Electron Density of States in a Gaussian Random Potential
(b) Vol. 84, No. 2, p. K109—K111 (1977).
See also *13*, p. *113*; 22, p. 330.

G. Buchheim, B. Junghans, and H. Lippmann
Application of the Kelvin Method for Oxide Charge Evaluation in Si–SiO_2 Structures
(a) Vol. **44**, No. 2, p. 585–591 (1977).
See also *14.3.1*, p. *175*; *14.3.2*, p. *177*; 22.1.2, p. 344.

T. Satow, O. Uemura, and S. Watanabe
*The Amorphous Structure of $CdGe_{1-x}M_xAs_2$ (M = Si, Sn, and Pb) with $x = 0.0, 0.1$, and 0.2
(a) Vol. **44**, No. 2, p. 731–736 (1977).
See also 18.1, p. 216; 22.8, p. 401.

S. Ohnuma, K. Watanabe, and T. Masumoto
*Low Coercivity and Zero Magnetostriction of Amorphous Fe–Co–Ni System Alloys
(a) Vol. **44**, No. 2, p. K151–K154 (1977).
See also *18.2*, p. *225*; 21.1, p. 301; 21.1.1, p. 307.

V. A. Gritsenko, E. E. Meerson, K. P. Mogilnikov, and S. P. Sinitsa
*On the Conductivity of Amorphous Silicon Nitride in High Electric Fields
(a) Vol. **44**, No. 2, p. K167–K170 (1977).
See also *14.3.4*, p. *188*; 22, p. 331.

R. Durný and D. Barančok
*EPR Investigation of Ag-Doped Ge–S Glasses
(a) Vol. **44**, No. 2, p. K177–K179 (1977).
See also *19*, p. *243*; 22, p. 331.

K. Hoselitz
*Magnetization of Fe–Si–B Metallic Glass
(a) Vol. **44**, No. 2, p. K191–K193 (1977).
See also *18.2*, p. *225*; 21.1.1, p. 307.

3. Crystal Growth

Review Article

L. N. ALEKSANDROV
*Transition Regions in Epitaxial Films
(a) Vol. **44**, No. 1, p. 11—43 (1977).
See also *1.5*, p. *29*; 14.3.1, p. 173; 22.1.1, p. 333; 22.1.2, p. 337; 22.2.1, p. 350.

Original Papers and Short Notes

L. N. ALEKSANDROV and I. A. ENTIN
*On the Kinetics of Thin-Film Growth under High Adatom Concentrations
(a) Vol. **39**, No. 1, p. 327—330 (1977).
See also *1.5*, p. *29*; 22.1.1, p. 333.

A. HADNI, R. THOMAS, and C. ERHARD
*An Unusual Type of Epitaxial Growth
(a) Vol. **39**, No. 2, p. 419—424 (1977).
See also *1.5*, p. *29*; 22.9, p. 414.

C. LACOUR and M. PAULUS
*The Effect of Equilibrium Segregation at the Boundaries in the Discontinuous Growth Region
(a) Vol. **40**, No. 2, p. K179—K181 (1977).
See also *1.4*, p. *25*; 22.8.2, p. 410.

G. WENDE and P. GÖRNERT
*Study of ACRT Influence on Crystal Growth in High-Temperature Solutions by the "High-Resolution Induced Striation Method"
(a) Vol. **41**, No. 1, p. 263—270 (1977).
See also 22.8.2, p. 411.

E. A. MATSON and S. A. POLYAKOV
*On the "Evolutionary Selection" Principle in Relation to the Growth of Polycrystalline Silicon Films
(a) Vol. **41**, No. 1, p. K93—K95 (1977).
See also *1.5*, p. *29*; 22.1.2, p. 339.

P. GÖRNERT, S. BORNMANN, F. VOIGT, and M. WENDT
*Study of the Liquid Phase Epitaxy Process of Garnet Layers by Induced Striations
(a) Vol. **41**, No. 2, p. 505—511 (1977).
See also 22.8.2, p. 411.

W. BRÜCKNER, U. GERLACH, H.-P. BRÜCKNER, W. MOLDENHAUER, and H. OPPERMANN
Influence of Nonstoichiometry on the Phase Transitions in Ga-, Al- and Fe-Doped VO_2
(a) Vol. **42**, No. 1, p. 295—303 (1977).
See also *1.2*, p. *17*; 14.3, p. 170; 18.1, p. 214; 22.6.1, p. 393.

H. PFEIFFER, W. HAUBENREISSER, and TH. KLUPSCH
*Analytic Treatment of Stochastic Models of Crystal Growth. Comparison of Two Different Interface Dynamics
(b) Vol. **83**, No. 1, p. 129—138 (1977).
See also *1.6*, p. *32*.

G. H. SCHWUTTKE
Low Cost Silicon for Solar Energy Application
(a) Vol. **43**, No. 1, p. 43—51 (1977).
See also 14.3.3, p. 181; *16*, p. *204*; 20.1, p. 260; 22.1.2, p. 341.

A. A. KAMINSKII, S. E. SARKISOV, J. BOHM, P. REICHE, D. SCHULTZE, and R. UECKER
Growth, Spectroscopic and Laser Properties of Crystals in the $K_5Bi_{1-x}Nd_x(MoO_4)_4$ System
(a) Vol. **43**, No. 1, p. 71—79 (1977).
See also 13.4, p. 137; 20.1, p. 260; *20.2*, p. *266*; *20.3*, p. *272*; 22.8, p. 399.

H. Raidt, G. Kästner, H. Menninger, and G. Voigt
*On Real Structure Investigations of $Ga_xIn_{1-x}P$ by Electron Microscopic Methods
(a) Vol. **43**, No. 1, p. 167—173 (1977).
See also *1.4*, p. *27*; 20.3, p. 273; 22.2.4, p. 357.

D. Linzen, S. Bornmann, P. Görnert, R. Hergt, H. Pfeiffer, W. Schüppel, F. Voigt, and M. Wendt
Growth-Induced Magnetic Anisotropy of (Sm, Ga)-Substituted YIG Films Grown on {111} Substrates
(a) Vol. **43**, No. 1, p. 277—281 (1977).
See also *18.3*, p. *230*; 22.8.2, p. 412.

J. Eisner
The Physical Properties of TGS Single Crystals, Grown from Aqueous TGS Solutions Containing Aniline
(a) Vol. **43**, No. 1, p. K1—K4 (1977).
See also *14.4*, p. *192*; 22.9, p. 418.

S. Al-Dallal
*Epitaxial Growth of CdSe on ZnTe (111)
(a) Vol. **44**, No. 1, p. 183—189 (1977).
See also *1.5*, p. *30*; 22.4.2, p. 366.

H. Pfeiffer
*Growth Rate Dependence of the Growth-Induced Anisotropy in Garnets. A Simple Theoretical Model
(a) Vol. **44**, No. 1, p. 221—227 (1977).
See also 22.8.2, p. 413.

M. Jurisch, K.-H. Berthel, and H.-J. Ullrich
*Influence of Chemical Composition within the Range of Homogeneity on Phase Transition and Transition Temperature of V_3Si Single Crystals (I): Preparation and Characterization of Single Crystals
(a) Vol. **44**, No. 1, p. 277—284 (1977).
See also 1.3, p. 23; 14.2, p. 164; 21.1, p. 299.

M. K. Anis and A. R. Piercy
*Growth of Single Crystals of GaSe with Natural Facets at Large Angles to the Layers
(a) Vol. **44**, No. 1, p. K5—K6 (1977).
See also 22.3, p. 359.

B. Etlinger, R. Trojko, and U. V. Desnica
*Thermal Decomposition of $Al_2In_{12}S_{21}$
(a) Vol. **44**, No. 1, p. K29—K30 (1977).
See also 22.3.1, p. 360.

E. V. Shvedov, V. S. Postnikov, and V. M. Ievlev
*Nucleus Saturation Density and Epitaxy in the Condensation of Metals on Alkali-Halide Crystals (I): Saturation Density
(a) Vol. **44**, No. 2, p. 423—427 (1977).
See also *10.2*, p. *92*; 22.5.2, p. 381.

U. Gradmann and P. Tillmanns
*Supersaturation and Mode of Growth for Fe Films on Cu(111) — An Experimental Study Using LEED and AES
(a) Vol. **44**, No. 2, p. 539—547 (1977).
See also *1.5*, p. *30*; 21.1.1, p. 307.

H. Neumann, K. Jacobs, Nguyen Van Nam, W. Koj, and C. Krause
*Properties of Liquid Phase Epitaxial Ge-Doped GaAs
(a) Vol. **44**, No. 2, p. 675—678 (1977).
See also *10.2*, p. *93*; 22.2.1, p. 353.

4. Bonding Properties

Original Papers and Short Notes

M. N. Sharma, A. Kant, and K. D. Misra
*A New Approach to Overlap Repulsion
(b) Vol. **79**, No. 1, p. 359–364 (1977).
See also 12.2, p. 104; 22.5.2, p. 375.

T. Soma
*Electronic Screening Effect on the Structure Stability of Face-Centered Tetragonal In
(b) Vol. **79**, No. 1, p. K53–K56 (1977).
See also *13*, p. *107*; 21, p. 276.

A. Milchev and R. Pickenhain
*The Quadratic Response of a Fermi Gas. Building-Up of the Covalent Bonding in Zincblende Semiconductors. A Nondiagonal Density Matrix Treatment
(b) Vol. **79**, No. 2, p. 549–558 (1977).
See also 13, p. 107; 22.1, p. 332; 22.1.1, p. 333; 22.1.2, p. 338; 22.2.1, p. 350.

R. A. Evarestov and V. A. Lovchikov
*Large Unit Cell Calculations of Solids in the CNDO Approximation
(b) Vol. **79**, No. 2, p. 743–751 (1977).
See also *13*, p. *107*; 22, p. 321.

M. V. Eremin and A. A. Kornienko
The Superposition Model in Crystal Field Theory
(b) Vol. **79**, No. 2, p. 775–785 (1977).
See also *13.4*, p. *131*; 20.1, p. 249; 22.5.3, p. 383.

R. Wang and M. D. Merz
*Polymorphic Bonding and Thermal Stability of Elemental Noncrystalline Solids
(a) Vol. **39**, No. 2, p. 697–703 (1977).
See also 1.2, p. 15; *2*, p. *34*; 21, p. 278; 22.1, p. 332.

A. A. Haav, E. Peljo, and P. Suortti
Thermal Vibration Amplitudes of TlCl between 296 and 534 K
(b) Vol. **80**, No. 1, p. 255–264 (1977).
See also *6*, p. *53*; 22.5, p. 370.

A. B. Markhnovetskii and G. L. Krasko
*The Pseudopotential Method and the Problem of Ordering in Alloys (III): Intermetallic Compounds LiAl, LiTl, and NaTl
(b) Vol. **80**, No. 1, p. 341–351 (1977).
See also 1.3, p. 21; 21, p. 278.

K. Hübner
*Chemical Bond and Related Properties of SiO_2 (I): Character of the Chemical Bond
(a) Vol. **40**, No. 1, p. 133–140 (1977).
See also 2, p. 35; *13.1*, p. *116*; 14.4, p. 189; 20.1, p. 251; 22.6, p. 387.

P. N. Dyachkov and A. A. Levin
*Bond Orbital Model for Tetrahedral Semiconductors on the Base of Molecular Parameters. Covalent Crystals and Perturbative Treatment of Partially Ionic Semiconductors
(b) Vol. **80**, No. 2, p. 681–690 (1977).
See also *13.1*, p. *116*; 22.1.2, p. 339; 22.1.3, p. 345; 22.2, p. 348; 22.4, p. 361.

G. L. Krasko and A. B. Maknhovetskii
*Pseudopotential Theory and the Problem of Covalent Bonding (I): General Theory of Covalent Charge
(b) Vol. **80**, No. 2, p. 713–725 (1977).
See also *1*, p. *9*.

R. YAMAMOTO, N. OWADA, O. TAKAI, and M. DOYAMA
*Calculations of the Heat of Solution of the Impurities in Aluminium and the Cohesive Energies of Simple Metals
(b) Vol. **80**, No. 2, p. K131—K135 (1977).
See also *10.1*, p. *77*; 21, p. 279.

K. HÜBNER
*Chemical Bond and Related Properties of SiO_2 (II): Structural Trends
(a) Vol. **40**, No. 2, p. 487—495 (1977).
See also 2, p. 36; *13.1*, p. *117*; 14.4, p. 190; 20.1, p. 252; 22.6, p. 387.

A. B. MAKHNOVETSKII and G. L. KRASKO
*Pseudopotential Theory and the Problem of Covalent Bonding (II): Crystals of Si, Ge, and α-Sn as-Ordered Systems
(b) Vol. **81**, No. 1, p. 263—273 (1977).
See also *1*, p. *9*; 21.7, p. 319; 22.1.1, p. 334; 22.1.2, p. 339.

F. BROSENS, J. CORNELIS, D. C. WALLACE, and J. T. DEVREESE
*A Method to Derive Interatomic Potentials in Metals from Experimental Phonon Spectra
(b) Vol. **81**, No. 2, p. 557—564 (1977).
See also 6, p. 56; 21, p. 280; 21.2, p. 308.

V. K. GARG, D. S. PURI, and M. P. VERMA
*Many-Body Effects on the Fourth-Order Elastic Constants of Ionic Solids (I): Theory
(b) Vol. **82**, No. 1, p. 325—334 (1977).
See also *12*, p. *100*.

O. P. SHARMA, A. P. GUPTA, and JAI SHANKER
*Effect of the Three Body Interactions on the Cohesive Energies of Alkali Halides
(b) Vol. **82**, No. 1, p. K13—K16 (1977).
See also 12.2, p. 105; 22.5.2, p. 379.

V. K. GARG, D. S. PURI, and M. P. VERMA
*Many-Body Effects on the Fourth-Order Elastic Constants of Ionic Solids (II): Applications
(b) Vol. **82**, No. 2, p. 481—488 (1977).
See also *12.2*, p. *105*; 22.5.2, p. 379.

I. GYÉMÁNT and G. SOLT
*Formation Energy of Spherical Cavities in Metals
(b) Vol. **82**, No. 2, p. 651—655 (1977).
See also *10.1*, p. *79*; 21.2, p. 309.

K. HÜBNER
*Chemical Bond and Related Properties of SiO_2 (III): Core-Level Shifts in SiO_x
(a) Vol. **42**, No. 2, p. 501—509 (1977).
See also 2, p. 38; *13.1*, p. *120*; 14.4, p. 191; 20.1, p. 258; 22.6, p. 390.

S. K. SARKAR and S. SENGUPTA
On Born-Huang Invariance Conditions
(b) Vol. **83**, No. 1, p. 263—271 (1977).
See also *6*, p. *58*.

G. R. UMAROV
*Isoelectronic Series and Trends in Band Structures
(b) Vol. **83**, No. 1, p. K93—K95 (1977).
See also *13.1*, p. *121*; 22, p. 328.

H. P. AUBAUER
Morphological Stability Study in Binary Alloys (I): Discrete Lattice Theory and Application to Interfacial Zone
(a) Vol. **43**, No. 1, p. 261—268 (1977).
See also *1.2*, p. *18*.

S. C. GOYAL and R. PRAKASH
*Many-Body Effects on the Second-Order Elastic Constants of Solids with ZnS Structure
(b) Vol. **83**, No. 2, p. 509–515 (1977).
See also *12.2*, p. *106*; 22, p. 328; 22.1, p. 332; 22.1.1, p. 335; 22.1.2, p. 342; 22.2, p. 348; 22.3, p. 358; 22.4, p. 361.

R. N. SINGH, P. L. SRIVASTAVA, and N. R. MITRA
Phonon Frequency, Electrical Resistivity, and Inter-Ionic Pair Potential of Li
(b) Vol. **83**, No. 2, p. 651–657 (1977).
See also *6*, p. *59*; 14.1, p. 159; 21.2, p. 309.

H. P. AUBAUER
Morphological Stability Study in Binary Alloys (II): Stability of Coherent Concentration Modulations
(a) Vol. **43**, No. 2, p. 601–610 (1977).
See also *1.2*, p. *19*; 21.1, p. 298.

L. DAGENS
*Calculation of a Resonant Model Potential for Copper, Silver, and Gold
(b) Vol. **84**, No. 1, p. 311–324 (1977).
See also *13*, p. *112*; 21.1, p. 299; 21.6, p. 317.

B. MESTNIK and Z. OGORELEC
*Madelung Constants of Stoichiometric and Non-Stoichiometric Average Crystals
(b) Vol. **84**, No. 1, p. K37–K41 (1977).
See also 22, p. 330.

M. N. SHARMA and A. KANT
*A New Approach to Lattice Properties of Alkali Halide Crystals
(b) Vol. **84**, No. 2, p. 427–434 (1977).
See also *8*, p. *68*; 20.1, p. 262; 22.5.2, p. 381.

L. P. MOKHRACHEVA, V. A. TSKHAI, and P. V. GELD
Chemical Binding and Energy Band Structure of Vanadium Monocarbide
(b) Vol. **84**, No. 2, p. 465–471 (1977).
See also *13.1*, p. *122*; 22.3, p. 359.

K. TAKANAKA and R. YAMAMOTO
*Temperature-Dependent Multi-Ion Interactions in Simple Metals
(b) Vol. **84**, No. 2, p. 813–816 (1977).
See also 21, p. 287.

5. Mössbauer Spectroscopy

Original Papers and Short Notes

H. ORTBAUER, W. STEINER, and R. HAFERL
*Mößbaueruntersuchungen an $Y(Fe_xCo_{1-x})_2$ im Bereich kleiner Fe-Konzentrationen
(Mössbauer Studies of $Y(Fe_xCo_{1-x})_2$ in the Region of Small Fe Concentrations
(a) Vol. **39**, No. 1, p. 157–165 (1977).
See also 18.1, p. 213; 18.2, p. 217; 21, p. 276; 21.1, p. 288; 21.1.1, p. 302,

J. BARTON, E. WIESER, and M. MÜLLER
*Investigation of Magnetic Textures in an Fe–Mn Base Alloy by Means of Mössbauer Spectroscopy
(a) Vol. **39**, No. 1, p. 259–264 (1977).
See also *18.2*, p. *217*; 21.1, p. 288; 21.1.1, p. 302.

E. ANTONČÍK
*Calibration of the Isomer Shift for ^{119}Sn
(b) Vol. **79**, No. 2, p. 605–611 (1977).
See also 13.4, p. 130; 21.7, p. 319.

G. N. BELOZERSKII, YU. P. KHIMICH, and V. N. GITSOVICH
*Determination of the Parameters of Hyperfine Interaction from Mössbauer Spectra of Polycrystals in an Externally Applied Magnetic Field
(b) Vol. **79**, No. 2, p. K125–K130 (1977).
See also 18, p. 209; 22.8.2, p. 409.

R. KMIEĆ, K. ŁĄTKA, and K. RUEBENBAUER
*Resonant Self-Absorption Correction for the $Ba^{119m}SnO_3$ Mössbauer Sources
(b) Vol. **79**, No. 2, p. K137–K138 (1977).
See also 22.8.1, p. 402.

M. WAUTELET, A. GÉRARD, F. GRANDJEAN, K. DE STROOPER, and G. ROBBRECHT
Magnetic Properties of Fe_2P
(a) Vol. **39**, No. 2, p. 425–430 (1977).
See also *18.2*, p. *218*; 22, p. 322.

P. P. SEREGIN, V. P. SIVKOV, F. S. NASREDINOV, L. N. VASILEV, YU. V. KRYLNIKOV, and YU. P. KOSTIKOV
The Influence of the Crystal-to-Glass Transition on the Local Structure of Semiconductors
(a) Vol. **39**, No. 2, p. 437–444 (1977).
See also *1.2*, p. *15*; 2, p. 34; 22, p. 322.

J. PIEKOSZEWSKI, L. DĄBROWSKI, J. SUWALSKI, and S. MAKOLĄGWA
*Mössbauer Study of Local Spin Structure in Singly Substituted YIG Ferrites
(a) Vol. **39**, No. 2, p. 643–650 (1977).
See also 22.8.2, p. 409.

F. SCHNEIDER, K. MELZER, H. MEHNER, and G. DEHE
^{119}Sn Hyperfine Fields in α-Fe_2O_3
(a) Vol. **39**, No. 2, p. K115–K117 (1977).
See also *18.3*, p. *228*; 22.6, p. 386.

J. A. SAWICKI and B. D. SAWICKA
*Comparability of ^{57}Fe Mössbauer Data in Ion-Implanted and Amorphous FeGe
(b) Vol. **80**, No. 1, p. K41–K44 1977).
See also 2, p. 34; 21.1.1, p. 302.

M. G. TOWNSEND, J. R. GOSSELIN, J. L. HORWOOD, L. G. RIPLEY, and R. J. TREMBLAY
*Violarite, a Metallic Natural Spinel
(a) Vol. **40**, No. 1, p. K25–K29 (1977).
See also 15, p. 198; *18.1*, p. *213*; 22.8.2, p. 410.

G. ATHANASSIADIS, G. LE CAER, J. FOCT, and L. RIMLINGER
Study of Ternary Ordered Solid Solutions Derived from Fe_3Al by Substitution
(a) Vol. **40**, No. 2, p. 425—435 (1977).
See also *1.3*, p. *21*; 21, p. 279; 21.1.1, p. 303.

G. P. CHUGUNOVA and A. V. MITIN
*Dynamic Theory of Conversion Electron Mössbauer Spectroscopy
(b) Vol. **81**, No. 1, p. 69—78 (1977).

R. NAGARAJAN and J. K. SRIVASTAVA
*Anomalous Mössbauer Lineshapes in the Magnetically Ordered Ga_2O_3–Fe_2O_3 System
(b) Vol. **81**, No. 1, p. 107—117 (1977).
See also 22.6.1, p. 393.

K. HAMILTON
*Theory of the Nuclear Electric Field Gradient in Trigonal Tellurium
(b) Vol. **81**, No. 1, p. K23—K26 (1977).
See also 22.1.3, p. 345.

J. SAWICKI, B. SAWICKA, and O. GOZWSKI
*Conversion Electron Mössbauer Analysis of Glass Surfaces
(a) Vol. **41**, No. 1, p. 173—176 (1977).
See also 2, p. 36; 22.8.1, p. 404.

J. SUWALSKI, L. DĄBROWSKI, J. PIEKOSZEWSKI, and Z. TUCHOLSKI
*Study of Order–Disorder Transformation in Fe–Al–Si Alloys by Mössbauer Spectroscopy
(a) Vol. **41**, No. 1, p. 191—195 (1977).
See also 18.2, p. 219; 21.1.1, p. 303.

S. S. BHANDARI and J. VARMA
*Anharmonicity Effect in the Oxides of Pb, Sn, and Bi
(a) Vol. **41**, No. 1, p. K89—K92 (1977).
See also 21, p. 280; 21.7, p. 319.

R. TAHARA, Y. NAKAMURA, M. INAGAKI, and Y. IWAMA
*Mössbauer Study of Spinodal Decomposition in Fe–Cr–Co Alloy
(a) Vol. **41**, No. 2, p 451—458 (1977).
See also 18.2, p. 220; 21.1, p. 294; 21.1.1, p. 303.

S. K. ARIF, I. SIGALAS, and D. ST. P. BUNBURY
*Mössbauer Effect of Fe^{57} Situated in (Rare-Earth) Ni_2 Laves Phases
(a) Vol. **41**, No. 2, p. 585—589 (1977).
See also 18.2, p. 220; 21.1, p. 294; 21.4, p. 312.

H. SPIERING, H. VOGEL, and W. IRLER
The Phase Transformation in Ferreous Fluosilicate Hexahydrate. Mössbauer Study
(b) Vol. **82**, No. 1, p. 243—250 (1977).
See also *1.2*, p. *17*; 18.1, p. 214; 22.8, p. 399.

Y. MURAOKA, M. SHIGA, and Y. NAKAMURA
*Magnetic Properties and Mössbauer Effect of $A(Fe_{1-x}B_x)_2$ (A = Y or Zr, B = Al or Ni) Laves Phase Intermetallic Compounds
(a) Vol. **42**, No. 1, p. 369—374 (1977).
See also 1, p. 10; *18*, p. *211*; 21.1.1, p. 304.

A. NARAYANASAMY, T. ERICSSON, T. NAGARAJAN, and P. MUTHUKUMARASAMY
*Mössbauer Studies on the $Fe_{1-x}Ti_{1+x}$ System
(a) Vol. **42**, No. 1, p. K65—K69 (1977).
See also 21.1, p. 296; 21.1.1, p. 304.

Y. SATO
*Mössbauer Effect of γ-Irradiated CaF_2 Doped with ^{57}Co
(b) Vol. **82**, No. 2, p. 611—616 (1977).
See also 11, p. 96; 22.5.3, p. 384.

W. STEINER and R. HAFERL
*Hyperfine Interactions in Y_2Fe_{17}
(a) Vol. **42**, No. 2, p. 739–750 (1977).
See also *18.2*, p. *222*; 21.1.1, p. 305; 21.4, p. 312.

R. CHANDRA and S. LOKANATHAN
*Electric Field Gradient in Biotite Mica
(b) Vol. **83**, No. 1, p. 273–280 (1977).
See also 22.8, p. 399.

R. K. PURI and L. R. GUPTA
*Computation of Force Constant Change of Dilute Mössbauer Impurities in Metallic Hosts from Mössbauer Parameters
(b) Vol. **83**, No. 1, p. 293–298 (1977).
See also 6, p. 58; 21, p. 284.

W. STEINER, R. HAFERL, and B. K. CHOI
*Hyperfine Interactions in $RE_2(Fe, Co)_{17}$ (RE = Y, Gd, Dy)
(a) Vol. **43**, No. 1, p. 111–117 (1977).
See also *18.2*, p. *222*; 21.1.1, p. 305; 21.4, p. 313.

B. K. JAIN, A. K. SINGH, and K. CHANDRA
Effect of Alloying Elements in Plain Carbon Steels
(a) Vol. **43**, No. 1, p. 357–361 (1977).
See also *1.3*, p. *22*; 21.1.1, p. 305.

J. J. BARA, B. BRANDT, A. GUTSZE, A. T. PĘDZIWIATR, Z. M. STADNIK, and A. WRONKOWSKI
*The Mössbauer Effect and Magnetometric Investigations of Iron-Doped Sodium A-Type Zeolite
(a) Vol. **43**, No. 1, p. K41–K44 (1977).
See also *18.1*, p. *215*; 22.8.1, p. 407.

W. STEINER, R. HAFERL, and H. WELSER
*Hyperfine Interactions in $RE_2(Fe, Ni)_{17}$ (RE = Y, Gd, Dy)
(a) Vol. **43**, No. 2, p. 437–442 (1977).
See also *18.2*, p. *223*; 21.1.1, p. 306; 21.4, p. 313.

J. DE SITTER, A. GOVAERT, E. DE GRAVE, D. CHAMBAERE, and G. ROBBRECHT
*A Mössbauer Study of Ca^{2+}-Containing Magnetites
(a) Vol. **43**, No. 2, p. 619–624 (1977).
See also 22.8.2, p. 412.

H. R. ANAND
*Mössbauer Study of Dispersed Iron Ions in Alkali Halides
(b) Vol. **84**, No. 1, p. 227–233 (1977).
See also 22.5.2, p. 381.

V. V. GAGULIN, N. V. FADEEVA, A. G. BELOUS, L. A. SEVASTIANOVA, A. V. TITOV, M. V. PLOTNIKOVA, K. P. MITROFANOV, E. V. ZUBOVA, S. P. SOLOVEV, and YU. N. VENEVTSEV
New Rhenium Containing Seignettomagnets and Ferroelectrics
(a) Vol. **44**, No. 1, p. 247–257 (1977).
See also *14.4.1*, p. *197*; 18.2, p. 223; *18.4*, p. *234*; 22.8, p. 400.

J. J. BARA, A. T. PĘDZIWIATR, Z. M. STADNIK, A. SZYTUŁA, J. TODOROVIČ, Z. TOMKOWICZ, and W. ZAREK
Investigations of Crystal and Magnetic Properties of Nickel Ferrite-Aluminates
(a) Vol. **44**, No. 1, p. 325–331 (1977).
See also *1.1*, p. *13*; *18.3*, p. *230*; 22.8.2, p. 413.

W. BRÜCKNER, E. WIESER, W. MOLDENHAUER, and W. REICHELT
On the Conductivity Transition in V_3O_5
(a) Vol. **44**, No. 1, p. K95–K98 (1977).
See also *1.2*, p. *20*; 14.3, p. 172; 22.6, p. 391.

J. J. Bara and B. F. Bogacz
*Investigations of Thickness Dependences of Mössbauer Absorption and Scattering Lines
(a) Vol. **44**, No. 1, p. K107—K109 (1977).
See also 22.8.1, p. 408.

K. Ruebenbauer, J. Fink, H. Schmidt, G. Czjzek, and K. Tomala
*An Investigation of Magnetic Ordering in $Gd_xCe_{1-x}Ru_2$ by ^{155}Gd Mössbauer Spectroscopy
(b) Vol. **84**, No. 2, p. 611—618 (1977).
See also 14.2, p. 164; *18.2*, p. *224*; 21.4, p. 313.

A. K. Singh, B. K. Jain, and K. Chandra
*Mössbauer Studies of Natural Goethite and Bog Iron Ore
(a) Vol. **44**, No. 2, p. 443—447 (1977).
See also 22.8.2, p. 413.

M. Henry and F. Varret
*Mössbauer Polarimetry Using Fluosilicates. Double Motion Drive and Effective Thickness Measurements
(a) Vol. **44**, No. 2, p. 601—607 (1977).
See also 22.8.1, p. 408.

P. J. Ouseph and H. E. Groskreutz
*Mössbauer Study of Iron in Mica
(a) Vol. **44**, No. 2, p. K181—K185 (1977).
See also 22.8.1, p. 408.

6. Lattice Dynamics. Phonons

Review Articles

G. Borstel and H. J. Falge
*Surface Phonon-Polaritons at Semi-Infinite Crystals
(b) Vol. **83**, No. 1, p. 11—45 (1977).
See also *13.5*, p. *141*; 20.1, p. 247; 22, p. 320; 22.6, p. 386.

V. T. Agekyan
Spectroscopic Properties of Semiconductor Crystals with Direct Forbidden Energy Gap
(a) Vol. **43**, No. 1, p. 11—42 (1977).
See also *13.5.1*, p. *143*; 20.1, p. 247; 20.3, p. 267; 22.6, p. 386.

Original Papers and Short Notes

G. F. Syrykh, A. P. Zhernov, M. G. Zemlyanov, N. A. Chernoplekov, and G. I. Solovev
*Deformation of the Vanadium Phonon Spectrum Due to Additions of Uranium Atoms
(b) Vol. **79**, No. 1, p. 105—109 (1977).
See also 21.1, p. 288.

K. A. Long
*A Theoretical Investigation of the Specific Heats and Transport Properties of the Actinides
(b) Vol. **79**, No. 1, p. 155—165 (1977).
See also 8, p. 64; *14.1*, p. *157*; 15, p. 198; 21.5, p. 314.

S. Onari, M. Cardona, E. Schönherr, and W. Stetter
*Effect of Stress on the Raman Spectra of Mg_2Si and Mg_2Sn
(b) Vol. **79**, No. 1, p. 269—276 (1977).
See also 12.2, p. 104; *20.1*, p. *247*; 22, p. 320.

H. Eschrig, K. Feldmann, K. Hennig, W. Matz, and P. Paufler
*Phonon Spectra of the Laves Phase Intermetallic Compound $CaMg_2$
(b) Vol. **79**, No. 1, p. 283—288 (1977).
See also 21.3, p. 310.

R. C. Rai and M. P. Hemkar
*A Simple Phenomenological Model for the Lattice Dynamics of B.C.C. Metals
(b) Vol. **79**, No. 1, p. 289—298 (1977).
See also 12.1, p. 101; 21.2, p. 308.

P. Parikh and D. R. Hay
*Crystallographic Aspects of Elastic Dispersion in Body-Centered Cubic Transition Metals
(b) Vol. **79**, No. 1, p. 299—305 (1977).
See also 21.1, p. 288.

E. Kolley and W. Kolley
On the Theory of Phonon-Modulated Electronic Transport in Off-Diagonal Random Alloys
(b) Vol. **79**, No. 1, p. 325—337 (1977).
See also *2*, p. *33*; *14*, p. *155*; 21, p. 276.

E. V. Chisler, V. Yu. Davydov, I. N. Goncharuk, and E. A. Ivanova
*The "Destructive" Fermi Resonance
(b) Vol. **79**, No. 1, p. 347—357 (1977).
See also *20.1*, p. *248*; 22, p. 320.

L. Bányai and A. Aldea
Semi-Classical and Quantum-Mechanical Theory of Hopping Conduction
(b) Vol. **79**, No. 1, p. 365—377 (1977).
Erratum
(b) Vol. **81**, No. 1, p. 409 (1977).
See also 13.4, p. 129; *14.3*, p. *165*; 22, p. 320.

T. IIDA and R. MONNIER
Erratum to: Static Properties of the Self-Trapped Hole in Simple Cubic Cesium Halides
(b) Vol. **79**, No. 1, p. 391 (1977).
See also *13.4*, p. *129*; 20.1, p. 248; 22.5.2, p. 375.

M. A. BELOGOLOVSKII, YU. M. IVANCHENKO, E. V. KOCHERGIN, A. A. LISYANSKII, and YU. V. MEDVEDEV
Interaction of Electrons with Phonons and Magnetic Impurities in Tunneling Junctions
(b) Vol. **79**, No. 1, p. K17—K21 (1977).
See also *14.1*, p. *157*; *14.2*, p. *161*; 21, p. 276.

P. R. R. PEREIRA and L. C. M. MIRANDA
*Magnon Contribution to the Phonon Damping in Ferromagnetic Semiconductors
(b) Vol. **79**, No. 1, p. K49—K52 (1977).
See also *13.5.4*, p. *153*; 18.2, p. 217; 22.8, p. 396.

S. D. STOJANOVIĆ, J. P. ŠETRAJČIĆ, M. I. ŠKRINJAR, and B. S. TOŠIĆ
*On Some Possible Consequences of Exciton-Phonon Interaction
(b) Vol. **79**, No. 2, p. 433—439 (1977).
See also *13.5.1*, p. *143*.

E. KOLLEY and W. KOLLEY
Calculation of the Temperature-Dependent DC Conductivity of Disordered Alloys in Off-Diagonal CPA
(b) Vol. **79**, No. 2, p. 461—472 (1977).
See also *2*, p. *33*; 13, p. 107; *14.1*, p. *157*; 21, p. 277.

H. HAPP and G. ROTHER
Investigation on the Far-Infrared Response of DKDP by Asymmetric Fourier Spectroscopy
(b) Vol. **79**, No. 2, p. 473—477 (1977).
See also *14.4.1*, p. *194*; 20.1, p. 248; 22.8.1, p. 402.

J. VON BORZESZKOWSKI and J. SCHOLZ
On the Electric Field Dependence of the Conductivity of Holes in Si at Low Temperatures (II): The Influence of the Variation of Steady-State Hole Density with Electric Field on the Conductivity
(b) Vol. **79**, No. 2, p. 525—531 (1977).
See also 13.1, p. 115; *14.3.4*, p. *183*; 22.1.2, p. 337.

B. N. GANGULY and M. NICOL
*Effect of Hydrostatic Pressure on the Vibrational Properties and the Structure of $SrWO_4$ and $PbMoO_4$
(b) Vol. **79**, No. 2, p. 617—622 (1977).
See also 1.2, p. 14; 12.2, p. 104; 20.1, p. 249; 22.8.1, p. 402.

A. SCALABRIN, A. S. CHAVES, D. S. SHIM, and S. P. S. PORTO
*Temperature Dependence of the A_1 and E Optical Phonons in $BaTiO_3$
(b) Vol. **79**, No. 2, p. 731—742 (1977).
See also 14.4, p. 189; 20.1, p. 249; 22.8.1, p. 402.

G. A. GLEDHILL and J. F. ANGRESS
*Temperature Dependence of the Infrared Reflection Spectra of $RbCl_yBr_{1-y}$ Alloys
(b) Vol. **79**, No. 2, p. K107—K110 (1977).
See also *20.1*, p. *249*; 22.5.4., p. 385.

R. P. GOYAL and S. C. GOYAL
*Dispersion of Phonons in Transition Metal Oxides (CoO and NiO)
(b) Vol. **79**, No. 2, p. K115—K117 (1977).
See also 22.6, p. 386.

K. S. Dubey
*Contribution to the Lattice Thermal Conductivity Due to the Correction Term in the Frame of the Generalized Callaway Integral: Application to Ge
(b) Vol. **79**, No. 2, p. K119—K124 (1977).
See also *8*, p. *64*; 22.1.1, p. 333.

Th. F. Nonnenmacher and R. Wunderle
*Thermal Conductivity for Defect and Boundary Scattering of Phonons at Low Temperatures
(b) Vol. **79**, No. 2, p. K153—K155 (1977).
See also *8*, p. *64*.

K. R. Allakhverdiev and M. M. Tagyev
*Infrared Investigation of the GaS_xSe_{1-x} System
(a) Vol. **39**, No. 2, p. K111—K113 (1977).
See also 1.2, p. 15; *20.1*, p. *250*; 22.3.1, p. 360.

W. Müller-Lierheim, T. Suski, and H. H. Otto
*Factor Group Analysis of the Raman Spectrum of $Pb_5Ge_3O_{11}$
(b) Vol. **80**, No. 1, p. 31—41 (1977).
See also 13.5.3, p. 151; 14.4.1, p. 194; 20.1, p. 250; 22.8.1, p. 403.

A. A. Adbumalikov and A. A. Klochikhin
*The Double Resonance in Two-Phonon Raman Scattering
(b) Vol. **80**, No. 1, p. 43—50 (1977).
See also 13.1, p. 115; *20.1*, p. *250*.

K. H. Weilacher, H. Roth-Seefrid, and H. Bross
*Lattice Dynamics of Lithium. A Theoretical Microscopic Calculation
(b) Vol. **80**, No. 1, p. 137—142 (1977).
See also 21.2, p. 308.

A. Pierrefeu, E. F. Steigmeier, and B. Dorner
*Inelastic Neutron Scattering in SbSI near the Ferroelectric Phase Transformation
(b) Vol. **80**, No. 1, p. 167—171 (1977).
See also 14.4.1, p. 194; 22.8, p. 397.

R. Enderlein and F. Bechstedt
*Theory of Allowed Resonance Raman Scattering of First and Second Order in a Magnetic Field
(b) Vol. **80**, No. 1, p. 225—234 (1977).
See also 13.1, p. 115; 13.5.1, p. 144; *20.1*, p. *251*.

A. A. Haav, E. Peljo, and P. Suortti
*Thermal Vibration Amplitudes of TlCl between 296 and 534 K
(b) Vol. **80**, No. 1, p. 255—264 (1977).
See also 4, p. 44; 22.5, p. 370.

W. Gębicki and W. Nazarewicz
*Long-Wavelength Optical Phonons in Mn_xHg_{1-x} Te Mixed Crystals
(b) Vol. **80**, No. 1, p. 307—311 (1977).
See also *20.1*, p. *251*; 22.8, p. 397.

H. J. L. van der Valk and C. Haas
*Anisotropy of Lattice Vibrations of Layered Compounds
(b) Vol. **80**, No. 1, p. 321—332 (1977).
See also *20.1*, p. *251*; 22, p. 322; 22.5, p. 370.

I. J. Gupta and S. K. Trikha
*Analysis of Lattice Thermal Conductivity of Solid Argon at Low Temperatures
(b) Vol. **80**, No. 1, p. 353—359 (1977).
See also *8*, p. *64*; 23, p. 420.

B. PRASAD and R. S. SRIVASTAVA
*Lattice Dynamics and Thermal Properties of Simple Metals Based on Model Pseudopotential Calculations
(b) Vol. **80**, No. 1, p. 379–383 (1977).
See also *8*, p. *64*; 21.2, p. 308.

J. BERGMANN, K. KNEIPP, and H. E. PONATH
*Raman-Induced Kerr Effect in $LiIO_3$ Single Crystals
(b) Vol. **80**, No. 1, p. K55–K58 (1977).
See also *20*, p. *244*; 22.8.1, p. 403.

K. TAKASAKI and S. TANAKA
Tunnelling Study of $Pb_{1-x}Sn_xTe$
(a) Vol. **40**, No. 1, p. 173–180 (1977).
See also 13.1, p. 116; *14.3.2*, p. *176*; 22.8, p. 397.

W. LOCHMANN
*Phonon-Assisted Auger Recombination in Semiconductors
(a) Vol. **40**, No. 1, p. 285–292 (1977).
See also *13.1*, p. *116*; 20, p. 244; 22.2.1, p. 351.

M. V. BELOUSOV, D. E. POGAREV, and A. A. SHULTIN
*Vibrations of Isotopic Impurities in Crystals Built up by Complex Ions
(b) Vol. **80**, No. 2, p. 417–424 (1977).
See also 20.1, p. 252; 22.8.1, p. 403.

G. L. MAILYAN and N. M. PLAKIDA
*Fluctuations of Order Parameter and Anharmonic Interaction in the Vibronic Model of Ferroelectrics
(b) Vol. **80**, No. 2, p. 543–547 (1977).
See also *14.4.1*, p. *194*.

V. WINTERFELDT and G. SCHAACK
*Phonon Contributions to the Anomalous Specific Heat of Triglycine Sulfate near T_c
(b) Vol. **80**, No. 2, p. 559–567 (1977).
See also 1.2, p. 15; *8*, p. *65*; 14.4.1, p. 194; 22.9, p. 415.

H. BÖTTGER and V. V. BRYKSIN
A Theory of the Hall Effect in the Hopping Region in Discordered Systems (I): Rate Equation in Presence of Electrical and Magnetic Fields
(b) Vol. **80**, No. 2, p. 569–578 (1977).
See also 2, p. 35; 13, p. 108; 13.5.3, p. 151; *14.3*, p. *167*; 22, p. 323.

S. S. KUBAKADDI and B. S. KRISHNAMURTHY
*The Electron–Two Short-Wavelength Phonon Scattering in Non-Polar Semiconductors
(b) Vol. **80**, No. 2, p. 603–609 (1977).
See also *14.3.4*, p. *184*; 22.1.2, p. 338.

D. IHLE
Electrical Conductivity in the Hubbard Model Including Electron–Phonon Interaction
(b) Vol. **80**, No. 2, p. 619–628 (1977).
See also *14*, p. *155*.

V. MÜLLER, G. SCHANZ, E. FISCHER, and E. J. UNTERHORST
Nuclear Acoustic Resonance Absorption and Dispersion in Vanadium via Coupling to the Magnetic Dipole Moment
(b) Vol. **80**, No. 2, p. 629–639 (1977).
See also *19*, p. *237*; 21.1, p. 291.

G. P. SRIVASTAVA
*Hypercircle Approach and Complementary Variational Principles for Lattice Thermal Conductivity
(b) Vol. **80**, No. 2, p. 657–660 (1977).
See also *8*, p. *65*.

S. Stamenković, S. D. Stojanović, and R. B. Žakula
*The Influence of Lattice Vibrations on the Parametric Resonance in Hydrogen-Bonded Ferroelectrics by Means of Parallel Electric Microwave Pumping
(b) Vol. **80**, No. 2, p. 661–668 (1977).
See also *14.4.1*, p. *195*.

G. M. Shmelev, I. A. Chaikovskii, V. V. Pavlovich, and E. M. Epshtein
*Electron–Phonon Interaction in a Superlattice
(b) Vol. **80**, No. 2, p. 697–701 (1977).
See also 22, p. 323.

Y. E. Perlin, A. A. Kaminskii, S. I. Klokishner, V. N. Enakii, Kh. S. Bagdasarov, G. A. Bogomolova, and D. N. Vylegzhanin
Thermal Broadening and Shift of the Energy Levels of TR^{3+} Ions in Crystals
(a) Vol. **40**, No. 2, p. 643–653 (1977).
See also *13.4*, p. *133*; 20.3, p. 269; 22.8.1, p. 403.

H. Böttger and V. V. Bryksin
A Theory of the Hall Effect in the Hopping Region in Disordered Systems (II): Random Resistor Network and Hall Effect
(b) Vol. **81**, No. 1, p. 97–106 (1977).
See also 2, p. 36; 13, p. 109; 13.5.3, p. 151; *14.3*, p. *167*; 22, p. 324.

J. Laugsch and G. A. Gehring
Optical Side Bands in a Cooperative Jahn-Teller Material: $TmAsO_4$
(b) Vol. **81**, No. 1, p. 167–178 (1977).
See also 13.5.1, p. 145; *20.1*, p. *253*; 22.8.1, p. 404.

T. Arai and T. P. Martin
Absorption Spectra of S_2 Molecules Embedded in van der Waals Solids
(b) Vol. **81**, No. 1, p. 185–189 (1977).
See also *20.1*, p. *253*; 23, p. 420.

O. P. Gupta, H. L. Kharoo, and M. P. Hemkar
*Ideal Thermal Resistivity of Lithium, Potassium, and Rubidium
(b) Vol. **81**, No. 1, p. 275–280 (1977).
See also *8*, p. *65*; 21.2, p. 308.

J. Azoulay, D. Gerlich, E. Wiener-Avnear, and I. Pelah
Piezoelectric Coupled Soft Mode in CsH_2AsO_4 and CsD_2AsO_4 Studied by Brillouin Scattering
(b) Vol. **81**, No. 1, p. 295–306 (1977).
See also *14.4.1*, p. *195*; 22.8, p. 398.

A. Freiberg and L. A. Rebane
*Peculiarities of the Electron–Phonon Interaction in the Spectra of the Molecular Ions O_2^-, S_2^-, and NO_2^- in Alkali Halide Crystals
(b) Vol. **81**, No. 1, p. 359–369 (1977).
See also *20.3*, p. *270*; 22.5.2, p. 377.

C. M. Goel and B. Dayal
*Lattice Dynamics of Rubidium Halide Crystals
(b) Vol. **81**, No. 1, p. 403–407 (1977).
See also 22.5.2, p. 377.

R. P. Goyal, K. K. Sarkar, and S. C. Goyal
*Erratum to: Dispersion of Phonons in Transition Metal Oxides (MnO)
(b) Vol. **81**, No. 1, p. 409 (1977).
See also 22.6, p. 387.

T. P. Martin and H. Schaber
Infrared Absorption of Aggregated and Isolated Microcrystals
(b) Vol. **81**, No. 1, p. K41–K45 (1977).
See also *20.1*, p. *254*; 22.5.2, p. 378.

A. K. Goyal and S. C. Goyal
*Dispersion of Phonons in NbC
(b) Vol. **81**, No. 1, p. K63—K65 (1977).
See also 12.2, p. 104; 22, p. 324.

T. Matsushita and J. Hayashi
X-Ray Structure Factors and the Debye-Waller Factor of Gallium Arsenide Crystals Determined from Full Widths at Half Maximum of Bragg Case Diffraction Curves
(a) Vol. **41**, No. 1, p. 139—145 (1977).
See also *1.1*, p. *11*; 22.2.1, p. 351.

M. Couzi, A. Daoud, and R. Perret
Lower-Frequency Raman Spectra and Phase Transitions in Perovskite-Type Layer Compounds. $(CH_3NH_3)_2MnCl_4$ and $(CH_3NH_3)_2CdCl_4$
(a) Vol. **41**, No. 1, p. 271—282 (1977).
See also *1.2*, p. *16*; 20.1, p. 254; 22.9, p. 416.

H. Böttger and V. V. Bryksin
A Theory of the Hall Effect in the Hooping Region in Disordered Systems (III): DC Hall Effect
(b) Vol. **81**, No. 2, p. 433—441 (1977).
See also 2, p. 37; 13, p. 109; 13.5.3, p. 151; *14.3*, p. *168*; 22, p. 325.

D. Durand and F. Luty
*Raman Scattering from Tunnelling and Hindered Rotational States of CN^- Molecules in KCl
(b) Vol. **81**, No. 2, p. 443—450 (1977).
See also *20.1*, p. *254*; 22.5.2, p. 378.

M. Rueff, E. Sigmund, and M. Wagner
*Resonance Scattering of Phonons at Trigonal Jahn-Teller Centres
(b) Vol. **81**, No. 2, p. 511—520 (1977).
See also *13.4*, p. *134*.

F. Brosens, J. Cornelis, D. C. Wallace, and J. T. Devreese
A Method to Derive Interatomic Potentials in Metals from Experimental Phonon Spectra
(b) Vol. **81**, No. 2, p. 557—564 (1977).
See also *4*, p. *45*; 21, p. 280; 21.2, p. 308.

Vijay A. Singh, C. Weigel, J. W. Corbett, and L. M. Roth
Vibrational and Electronic Structure of Hydrogen-Related Defects in Silicon Calculated by the Extended Hückel Theory
(b) Vol. **81**, No. 2, p. 637—646 (1977).
See also *13.4*, p. *134*; 22.1.2, p. 340.

K. S. Dubey
*Contribution to the Lattice Thermal Conductivity Due to the Correction Term in the Callaway Integral
(b) Vol. **81**, No. 2, p. K83—K86 (1977).
See also *8*, p. *65*; 22, p. 325.

A. Holas
*One-Phonon Scattering of Ultra Cold Neutrons in Copper
(b) Vol. **81**, No. 2, p. K107—K110 (1977).
See also 21.1, p. 294.

P. L. Vuillermoz, J. Jouglar, A. Laugier, and H. R. Winteler
Phonon-Defect Interaction in Dislocation-Free GaAs
(a) Vol. **41**, No. 2, p. 561—567 (1977).
See also 8, p. 66; *10.2*, p. *88*; 22.2.1, p. 352.

Yu. I. Balkarei and V. A. Nikitov
*On Magnon–Acoustic Phonon Interaction in Magnetic Semiconductors
(b) Vol. 82, No. 1, p. 127—131 (1977).
See also *13.5.4*, p. *153*; 18.2, p. 221; 22, p. 326.

J. Pelzl, P. Engels, and R. Florian
Raman Spectroscopic Study of the Structural Phase Transitions in K_2SnCl_6
(b) Vol. 82, No. 1, p. 145—148 (1977).
See also *1.2*, p. *17*; 20.1, p. 256; 22.8, p. 399.

M. Mijatović and S. Milošević
*The Linear Ising Chain in the Three-Dimensional Phonon Field
(b) Vol. 82, No. 1, p. 193—203 (1977).
See also *18*, p. *211*.

Th. F. Nonnenmacher and R. Wunderle
*Size Effect on the Thermal Conductivity of Small Dielectric Crystals
(b) Vol. 82, No. 1, p. 257—262 (1977).
See also *8*, p. *66*.

T. Soma
*Temperature Dependence of the Grüneisen Constant of Si and Ge
(b) Vol. 82, No. 1, p. 319—324 (1977).
See also 8, p. 66; 22.1.1, p. 335; 22.1.2, p. 340.

J. E. Lowther
*Proton Transfer by Dynamic Interactions and the Quantum Theory of Mutations in DNA
(b) Vol. 82, No. 1, p. K41—K44 (1977).
See also *2*, p. *37*; 22.9, p. 416.

K. S. Dubey
Role of Point-Defect Scattering in the Lattice Thermal Conductivity of an Insulator at Low Temperatures
(b) Vol. 82, No. 1, p. K63—K67 (1977).
See also *8*, p. *66*; 22, p. 326.

W. Lochmann
*Scattering Mechanisms in Phonon-Assisted Auger Recombination
(a) Vol. 42, No. 1, p. 181—185 (1977).
See also *20*, p. *245*; 22.2.1, p. 352.

R. Ramji Rao and A. Ramanand
Third-Order Elastic Constants, Thermal Expansion, and Bulk Modulus of Holmium
(a) Vol. 42, No. 1, p. 247—255 (1977).
See also *8*, p. *66*; *12.1*, p. *102*; 21.4, p. 312.

W. F. W. M. van Heugten
An Interatomic Potential for Molybdenum from Phonon Spectra
(b) Vol. 82, No. 2, p. 501—506 (1977).
See also *10.1*, p. *79*; 21, p. 282.

H. Herrmann
*Infrared Studies of the Layered Compound $ZnIn_2S_4$
(b) Vol. 82, No. 2, p. 513—521 (1977).
See also *20.1*, p. *258*; 22.8, p. 399.

H. Paerschke, K.-E. Süsse, D.-G. Welsch, and B. Wilhelmi
*Theory of Absorption and Dispersion Profiles of Molecular Transitions in Dense Media
(b) Vol. 82, No. 2, p. 633—641 (1977).
See also 20.1, p. 258.

I. Savatinova and E. Anachkova
*Raman Study of Dynamic Disorder and the Ferroelectric Transition in $K_4Fe(CN)_6 \cdot 3\,H_2O$
(b) Vol. 82, No. 2, p. 677—683 (1977).
See also 14.4.1, p. 196; *20.1*, p. *258*; 22.8, p. 399.

G. L. Belenkii, L. N. Alieva, R. Kh. Nani, and E. Yu. Salaev
*Vibrational Spectrum of InSe
(b) Vol. **82**, No. 2, p. 705—709 (1977).
See also *20.1*, p. *258*; 22.3, p. 358.

A. V. Lelyakov
Luminescence Band Shape of Free Excitons in AgBr Crystals at High Exciton Density
(b) Vol. **82**, No. 2, p. 711—716 (1977).
See also *13.5.1*, p. *146*; *20.3*, p. *271*; 22.5.1, p. 373.

V. Hizhnyakov and I. Tehver
*On the Theory of Hot Luminescence and Resonant Raman Effect of Impurity Centres
(b) Vol. **82**, No. 2, p. K89—K93 (1977).
See also 10.2, p. 89; *20.3*, p. *271*.

P. Čapková and J. Šedivý
*X-Ray Debye Characteristic Temperatures of Magnesium and α-Titanium
(a) Vol. **42**, No. 2, p. 681—686 (1977).
See also *8*, p. *66*; 21, p. 283; 21.3, p. 310.

D. Bäuerle, W. B. Holzapfel, A. Pinczuk, and Y. Yacoby
Temperature and Hydrostatic Pressure Dependence of Vibrational Modes in $PbTi_{1-x}Zr_xO_3$
(b) Vol. **83**, No. 1, p. 99—107 (1977).
See also *1.2*, p. *18*; *14.4.1*, p. *196*; 20.1, p. 259; 22.8.1, p. 406.

G. L. Chen, Shi-yu Wu, and M. Schwartz
*A Classical Analysis of the Resonance Modes in a Randomly Disordered Chain
(b) Vol. **83**, No. 1, p. 115—121 (1977).
See also *10*, p. *73*.

B. Khelifa, A. Delahaigue, and P. Jouve
Soft Mode and Debye Relaxations in Thiourea
(b) Vol. **83**, No. 1, p. 139—145 (1977).
See also *14.4.1*, p. *196*; 20.1, p. 259; 22.9, p. 417.

M. B. Haxton
*The Phonon Spectra β-CuI (I): The Model
(b) Vol. **83**, No. 1, p. 147—154 (1977).
See also 22.5, p. 371.

M. Kumar, P. S. Mahesh, and T. P. Sharma
*Second-Order Raman Spectra and Phonon Spectrum of CsF
(b) Vol. **83**, No. 1, p. 155—159 (1977).
See also *20.1*, p. *259*; 22.5.2, p. 380.

F. Bechstedt and R. Enderlein
*Theory of Resonance Raman Scattering in Disordered Solids. Smooth Random Field Model
(b) Vol. **83**, No. 1, p. 239—247 (1977).
See also 2, p. 38; 13.1, p. 121; *20.1*, p. *259*; 22, p. 328.

S. K. Sarkar and S. Sengupta
*On Born-Huang Invariance Conditions
(b) Vol. **83**, No. 1, p. 263—271 (1977).
See also 4, p. 45.

R. K. Puri and L. R. Gupta
Computation of Force Constant Change of Dilute Mössbauer Impurities in Metallic Hosts from Mössbauer Parameters
(b) Vol. **83**, No. 1, p. 293—298 (1977).
See also *5*, p. *49*; 21, p. 284.

J. Kuriata, N. Guskos, and T. Rewaj
*Temperature Dependence of the Second-Order Spin–Lattice Coefficients for Eu^{2+} in CaF_2 Single Crystals
(b) Vol. **83**, No. 1, p. K79—K82 (1977).
See also *19*, p. *241*; 22.5.3, p. 384.

S. Guha and P. K. Sen
Two-Stream Instability in n-Type Gallium Arsenide
(a) Vol. **43**, No. 1, p. 89—97 (1977).
See also 13, p. 111; *14.3.4*, p. *186*; 22.2.1, p. 353.

G. Massé, J. P. Aicardi, and J. P. Leyris
Green Emission of Mercury Sulphide (α-HgS)
(a) Vol. **43**, No. 1, p. 191—201 (1977).
See also 13.4, p. 137; *20.3*, p. *273*; 22.4.1, p. 363.

Cao-Xuân An
Determination of the Elastic Constants of $SrCl_2$ by Brillouin Scattering and Study of Their Variations with Temperature
(a) Vol. **43**, No. 1, p. K69—K72 (1977).
See also *12.2*, p. *106*; 22.5.3, p. 384.

Shi-yu Wu, J. A. Cocks, and V. Chiu
*Frequency Spectra of Disordered Chains with Short-Range Order
(b) Vol. **83**, No. 2, p. 365—375 (1977).
See also *10*, p. *73*.

Vo Hong Anh and Nguyen Van Trong
*On the Theory of Parametric Excitation in Quantum Electron–Phonon Systems
(b) Vol. **83**, No. 2, p. 395—404 (1977).
See also *13.5.2*, p. *150*; 20, p. 246; 22, p. 328.

M. B. Haxton
*The Phonon Spectra of β-CuI (II): The Numerical Evaluation
(b) Vol. **83**, No. 2, p. 529—536 (1977).
See also 22.5, p. 372.

S. K. Sarkar, S. K. Das, D. Roy, and S. Sengupta
*Three-Body Interaction and Lattice Dynamics of Metals
(b) Vol. **83**, No. 2, p. 615—624 (1977).
See also 21, p. 285; 21.1, p. 297; 21.6, p. 317.

R. N. Singh, P. L. Srivastava, and N. R. Mitra
*Phonon Frequency, Electrical Resistivity, and Inter-Ionic Pair Potential of Li
(b) Vol. **83**, No. 2, p. 651—657 (1977).
See also 4, p. 46; 14.1, p. 159; 21.2, p. 309.

B. N. Ganguly, M. Nicol, and W. D. Ellenson
*Raman Spectroscopic Studies of the High Pressure Phase Transformation of $NaClO_3$
(b) Vol. **83**, No. 2, p. K115—K120 (1977).
See also *1.2*, p. *19*; 12.2, p. 106; 20.1, p. 260; 22.8.1, p. 407.

T. O. Gegechkori and Z. S. Kachlishvili
Hot Electron Current–Voltage Curves in a Transverse Magnetic Field with Account of Phonon Heating
(a) Vol. **43**, No. 2, p. 513—518 (1977).
See also *13*, p. *111*; *14.3.4*, p. *187*.

K. S. Dubey and R. H. Misho
*Three-Phonon Scattering Relaxation Rate and Phonon Conductivity. Application to Mg_2Ge
(b) Vol. **84**, No. 1, p. 69—81 (1977).
See also 8, p. 67; 22.7, p. 395.

Vo Hong Anh and Nguyen Nhu Dat
*Parametric Action of a Strong Radiation Field on Anisotropic Semiconductors
(b) Vol. 84, No. 1, p. 115–123 (1977).
See also *13.5*, p. *142*; 20, p. 246; 22.4.1, p. 364.

Nguyen Van Trong and Ha Vinh Tan
*On the Quantum Theory of Anomalous Electromagnetic Fluctuations in Electron–Phonon Systems
(b) Vol. 84, No. 1, p. 345–352 (1977).
See also *13*, p. *112*.

M. Dhwaj Singh
*The Effect of Internal Strain and Electrical Field of an Impurity on Electron–Phonon Interaction in n-Type Germanium and Silicon
(b) Vol. 84, No. 1, p. 375–384 (1977).
See also 8, p. 67; 22.1.1, p. 336; 22.1.2, p. 343.

I. Savatinova and E. Anachkova
Raman Spectrum and Dielectric Relaxation in $K_4Fe(CN)_6 \cdot 3\,H_2O$
(b) Vol. 84, No. 1, p. 401–406 (1977).
See also 1.2, p. 20; *14.4.1*, p. *197*; *20.1*, p. *262*; 22.8, p. 400.

R. Schwabe, F. Thuselt, and T. Hänsel
*Exciton–Phonon Energy Transfer in Semiconductors
(b) Vol. 84, No. 1, p. K43–K47 (1977).
See also *13.5.1*, p. *148*; 20.3, p. 274; 22.3, p. 359.

K. S. Dubey
*Phonon Conductivity of Sapphire
(b) Vol. 84, No. 1, p. K57–K60 (1977).
See also 8, p. 67; 22.8.1, p. 407.

I. V. Bodnar, A. G. Karoza, and G. F. Smirnova
*Lattice Reflectivity Spectra of $CuGaS_2$–$CuGaSe_2$ Solid Solutions
(b) Vol. 84, No. 1, p. K65–K69 (1977).
See also *20.1*, p. *262*; 22.8, p. 400.

M. M. Babaev and T. M. Gasymov
Thermopower and the Transverse Nernst-Ettingshausen Effect in Semiconductors in a Strong Electric Field
(b) Vol. 84, No. 2, p. 473–483 (1977).
See also 13.4, p. 139; *14.3.4*, p. *188*; *15*, p. *199*; 22, p. 330.

R. Leihkauf and H. Böttger
*Dispersion Relation of Phonon-Like Excitations in a Simple Model of a Noncrystalline Solid
(b) Vol. 84, No. 2, p. 575–580 (1977).
See also *2*, p. *40*.

W. Richter, H. Köhler, and C. R. Becker
*A Raman and Far-Infrared Investigation of Phonons in the Rhombohedral V_2–VI_3 Compounds: Bi_2Te_3, Bi_2Se_3, Sb_2Te_3, and $Bi_2(Te_{1-x}Se_x)_3$ $(0 < x < 1)$, $(Bi_{1-y}Sb_y)_2Te_3$ $(0 < y < 1)$
(b) Vol. 84, No. 2, p. 619–628 (1977).
See also *20.1*, p. *262*; 22.7, p. 395.

P. Kocevar
*Drag Effects in a Heated Electron–Phonon System. n-Germanium and n-Silicon
(b) Vol. 84, No. 2, p. 681–690 (1977).
See also 22.1.1, p. 336; 22.1.2, p. 344.

E. Schuberth
First Order Raman Scattering from Ar Crystals: Evidence for the H.C.P. Phase
(b) Vol. 84, No. 2, p. K91–K94 (1977).
See also *1.2*, p. *20*; 20.1, p. 263; 23, p. 420.

I. J. Gupta and S. K. Trikha
*On the Low Temperature Lattice Thermal Conductivity of Solid Argon
(b) Vol. **84**, No. 2, p. K95—K99 (1977).
See also *8*, p. *68*; 23, p. 420.

A. M. Lejus and D. Michel
*Raman Spectrum of Er_2O_3 Sesquioxide
(b) Vol. **84**, No. 2, p. K105—K108 (1977).
See also *20.1*, p. *263*; 22.6, p. 391.

V. Valvoda and J. Ječný
*X-Ray Investigation of the Anisotropic Thermal Motion of Atoms in Tellurium
(a) Vol. **44**, No. 2, p. 557—565 (1977).
See also 22.1.3, p. 347.

R. Kind
On the Soft Modes in $(CH_3NH_3)_2CdCl_4$
(a) Vol. **44**, No. 2, p. 661—667 (1977).
See also *1.2*, p. *20*; 22.9, p. 419.

E. I. Terukov, H. Wagner, W. Reichelt und H. Oppermann
IR Spectroscopic Investigations on the Semiconductor–Metal-Phase Transition of V_3O_5
(a) Vol. **44**, No. 2, p. K187—K190 (1977).
See also *1.2*, p. *20*; *20.1*, p. *264*; 22.6, p. 392.

7. Acoustic Properties

Original Papers and Short Notes

B.-P. Kietis, R. Reksnys, and A. Sakalas
*Ohmic and Acoustoelectric Currents in the Hall Effect for the Investigation of Bipolar Conductivity
(a) Vol. **39**, No. 2, p. 601—605 (1977).
See also *14.3.4*, p. *183*; 22.4.1, p. 362.

A. Vincent, S. M. Seyed Reihani, and J. Perez
The Nature of Dislocation-Pinning Defects in Cold-Worked Aluminium Studied by Ultrasonic Waves
(a) Vol. **39**, No. 2, p. 651—660 (1977).
See also *10.1*, p. *76*; 21, p. 278.

C. A. S. Lima and L. C. M. Miranda
Plasmon-Enhanced Sound Amplification in Piezoelectric Semiconductors under a Strong Magnetic Field
(b) Vol. **80**, No. 1, p. 57—62 (1977).
See also *13.5.2*, p. *149*; 18, p. 210; 22.2.1, p. 351.

H. Kuzmanny and Z. Szalmassy
*Systematic Study of the Acoustoelectric Interaction of Surface Waves with Inhomogeneously Distributed Free Carriers in a Photoconductor
(a) Vol. **40**, No. 1, p. 75—82 (1977).
See also *14.3.2*, p. *176*; 14.3.3, p. 178; 16, p. 201; 22.4.1, p. 361; 22.8.1, p. 403.

J. A. Jordão, A. Sakamoto, and M. de Souza
The Ultrasonic Attenuation due to Cu^+ Substitutional Ions in KCl
(a) Vol. **40**, No. 1, p. 109—112 (1977).
See also *10.2*, p. *85*; 22.5.2, p. 376.

J. M. Peransin, E. Groubert, and B. Goutiers
*Time Investigation of Parametric Interactions in Acoustoelectric Oscillators
(a) Vol. **40**, No. 2, p. 583—588 (1977).
See also *14.3.4*, p. *184*; 22.4.1, p. 362.

O. Singh
*Ultrasonic Absorption in the Intermediate State of Gallium and Indium
(b) Vol. **82**, No. 1, p. 179—188 (1977).
See also 14.2, p. 162; 21, p. 282.

H. Bialas and Li-Shing Li
*Hypersound Excitation at 8.9 GHz by Exploiting Magnetic Anisotropy in Oriented Nickel Films
(a) Vol. **42**, No. 1, p. 125—132 (1977).
See also 1.5, p. 30; *19*, p. *240*; 21.1, p. 295.

B. V. Petukhov and Yu. Z. Estrin
Thermal Instability of the Amplitude-Dependent Internal Friction
(a) Vol. **42**, No. 1, p. 391—396 (1977).
See also *10*, p. *73*; *12*, p. *100*.

V. S. Sardaryan, A. V. Shekoyan, and E. M. Kazaryan
*Self-Focusing of an Intense Ultra- and Hypersonic Beam in an Inhomogeneous and Absorbing Solid
(a) Vol. **42**, No. 1, p. 409—414 (1977).
See also *10*, p. *73*.

E. Kittinger
*Temperature Dependence of Sound Velocity in Vitreous Selenium
(a) Vol. **42**, No. 1, p. K33—K35 (1977).
See also *2*, p. *37*; 22.1.3, p. 346.

I. Pelant and P. Sladký
The Ratio of Photoelastic Constants p_{11}/p_{12} of AgCl
(a) Vol. **42**, No. 1, p. K57–K59 (1977).
See also *12.2*, p. *105*; *20.1*, p. *257*; 22.5.1, p. 373.

T. Hata and M. Suzuki
*Mechanism of Optical Modulation Caused by Acoustic Domains in CdS
(a) Vol. **42**, No. 1, p. K71–K74 (1977).
See also *20.1*, p. *257*; 22.4.1, p. 363.

U. Kh. Kopvillem and S. V. Prants
Octupole and Hexadecapole Dynamics in Solid State Physics
(b) Vol. **83**, No. 1, p. 109–114 (1977).
See also *14*, p. *156*.

Y. Li and D. Y. Chung
*A Method for the Determination of the Elastic Constants for Crystals with Low Trigonal Symmetry
(a) Vol. **43**, No. 1, p. 329–332 (1977).
See also *12.2*, p. *106*; 22.6.1, p. 393.

V. Kunigėlis, A. Jucys, E. Garška, and V. Vitkauskas
*Influence of the Surface Potential on the Acoustoelectric Interaction in Semiconductor–Piezoelectric Structures
(a) Vol. **43**, No. 2, p. 699–703 (1977).
See also *14.3.2*, p. *177*; 22.4.1, p. 364.

B.-P. Kietis, R. Reksnys, A. Sakalas, and P. Höschl
*Investigation of the Bipolar Photoconductivity in CdTe
(a) Vol. **43**, No. 2, p. 705–709 (1977).
See also *14.3*, p. *171*; 16, p. 204; 22.4.3, p. 367.

8. Thermal Properties

Original Papers and Short Notes

K. A. Long
A Theoretical Investigation of the Specific Heats and Transport Properties of the Actinides
(b) Vol. **79**, No. 1, p. 155–165 (1977).
See also *6*, p. *51*; *14.1*, p. *157*; 15, p. 198; 21.5, p. 314.

F. K. Schulte
Electronic Density of States, Heat Capacity, and Spin Susceptibility of Thin Metal Films
(b) Vol. **79**, No. 1, p. K5–K8 (1977).
See also *13.3*, p. *126*; 18, p. 209; 21, p. 276.

M. Yahagi and K. Iwamura
*Temperature Dependences of Electrical Resistivity and Thermal Expansion in the Intermetallic Compound LiIn
(a) Vol. **39**, No. 1, p. 189–192 (1977).
See also 10.1, p. 75; *14.1*, p. *157*; 21.2, p. 308.

J. Albers and H. Küppers
Dielectric and Thermal Properties of Ammonium Hydrogen Oxalate Hemihydrate
(a) Vol. **39**, No. 1, p. K49–K51 (1977).
See also *1.2*, p. *14*; 14.4, p. 189; 22.9, p. 414.

S. C. Goel, Y. R. Midha, and V. S. Nanda
*Applications of a Simple Equation of State for Metals
(b) Vol. **79**, No. 2, p. K103–K105 (1977).
See also 21, p. 277; 21.1, p. 289; 21.6, p. 315.

K. S. Dubey
*Contribution to the Lattice Thermal Conductivity Due to the Correction Term in the Frame of the Generalized Callaway Integral: Application to Ge
(b) Vol. **79**, No. 2, p. K119–K124 (1977).
See also *6*, p. *53*; 22.1.1, p. 333.

Th. F. Nonnenmacher and R. Wunderle
*Thermal Conductivity for Defect and Boundary Scattering of Phonons at Low Temperatures
(b) Vol. **79**, No. 2, p. K153–K155 (1977).
See also *6*, p. *53*.

P. Delhaes, G. Keryer, S. Flandrois et J. P. Manceau
*Etude de la chaleur spécifique de sels doubles de TCNQ à cations non-radicalaires
(Study of the Specific Heat of Double Salts of TCNQ with Non-Radical Cations)
(b) Vol. **80**, No. 1, p. 125–136 (1977).
See also 18.1, p. 213; 22.9, p. 414.

D. I. Pushkarov
*On the Theory of Thermal Conductivity in Quantum Crystals
(b) Vol. **80**, No. 1, p. 295–300 (1977).
See also 10, p. 72; 13.5, p. 141.

I. J. Gupta and S. K. Trikha
*Analysis of Lattice Thermal Conductivity of Solid Argon at Low Temperatures
(b) Vol. **80**, No. 1, p. 353–359 (1977).
See also *6*, p. *53*; 23, p. 420.

B. Prasad and R. S. Srivastava
*Lattice Dynamics and Thermal Properties of Simple Metals Based on Model Pseudopotential Calculations
(b) Vol. **80**, No. 1, p. 379–383 (1977).
See also *6*, p. *54*; 21.2, p. 308.

D. SEGERS, C. DAUWE, M. DORIKENS, and L. DORIKENS-VANPRAET
*On the Effect of Thermal Expansion on Positron Annihilation Lineshape Factors
(a) Vol. **40**, No. 1, p. 153–157 (1977).
See also 21, p. 279.

J. P. BADIALI, A. DEFRAIN, M. DUPONT, and C. REGNAUT
Effective Pair Potential and Thermodynamical Properties of Liquid Gallium
(b) Vol. **80**, No. 2, p. 533–541 (1977).
See also *2*, p. *35*; 21, p. 279.

V. WINTERFELDT and G. SCHAACK
*Phonon Contributions to the Anomalous Specific Heat of Triglycine Sulfate near T_c
(b) Vol. **80**, No. 2, p. 559–567 (1977).
See also 1.2, p. 15; *6*, p. *54*; 14.4.1, p. 194; 22.9, p. 415.

S. KRZEMIŃSKI
Random One-Dimensional Classical Magnet
(b) Vol. **80**, No. 2, p. 651–656 (1977).
See also *13*, p. *108*; *18.2*, p. *219*; 18.4, p. 232.

G. P. SRIVASTAVA
*Hypercircle Approach and Complementary Variational Principles for Lattice Thermal Conductivity
(b) Vol. **80**, No. 2, p. 657–660 (1977).
See also *6*, p. *54*.

T. F. M. KORTEKAAS and J. J. M. FRANSE
*Thermal Expansion of Some Ni_3Al and Ni–Pt Alloys
(a) Vol. **40**, No. 2, p. 479–485 (1977).
See also 10.1, p. 77; 21.1, p. 292.

V. G. KVACHADZE, M. F. ZHVANIYA, and T. SH. DAVITASHVILI
On the Nature of the Radiative Centres of the Halogen Impurity
(a) Vol. **40**, No. 2, p. K113–K115 (1977).
See also *10.2*, p. *87*; *20.1*, p. *253*; 22.5.2, p. 377.

T. KONDOW, U. MIZUTANI, and T. B. MASSALSKI
*Low Temperature Specific Heats of TCNQ Complexes
(b) Vol. **81**, No. 1, p. 157–165 (1977).
See also 22.9, p. 415.

O. P. GUPTA, H. L. KHAROO, and M. P. HEMKAR
*Ideal Thermal Resistivity of Lithium, Potassium, and Rubidium
(b) Vol. **81**, No. 1, p. 275–280 (1977).
See also *6*, p. *55*; 21.2, p. 308.

G. HAUSCH
*Invar-Like Thermal Expansion Anomaly in Antiferromagnetic Mn–Ni Alloys
(a) Vol. **41**, No. 1, p. K35–K38 (1977).
See also *18.4*, p. *232*; 21.1, p. 293.

M. MALINOWSKI, A. PIETRASZKO, and M. POŁOMSKA
*Thermal Expansion and Dielectric Permittivity in the Vicinity of Transition Points of Lead Germanate
(a) Vol. **41**, No. 1, p. K55–K58 (1977).
See also *14.4*, p. *191*; 22.8.1, p. 404.

K. S. DUBEY
*Contribution to the Lattice Thermal Conductivity Due to the Correction Term in the Callaway Integral
(b) Vol. **81**, No. 2, p. K83–K86 (1977).
See also *6*, p. *56*; 22, p. 325.

P. L. Vuillermoz, J. Jouglar, A. Laugier, and H. R. Winteler
Phonon–Defect Interaction in Dislocation-Free GaAs
(a) Vol. **41**, No. 2, p. 561—567 (1977).
See also 6, p. 56; *10.2*, p. *88*; 22.2.1, p. 352.

Yu. V. Rakitin, V. T. Kalinnikov, and W. E. Hatfield
*Magnetic and Thermodynamic Properties of Heisenberg Open Finite Chains (II): Specific Heat
(b) Vol. **82**, No. 1, p. 69—73 (1977).
See also *18*, p. *211*.

Th. F. Nonnenmacher and R. Wunderle
*Size Effect on the Thermal Conductivity of Small Dielectric Crystals
(b) Vol. **82**, No. 1, p. 257—262 (1977).
See also *6*, p. *57*.

T. Soma
Temperature Dependence of the Grüneisen Constant of Si and Ge
(b) Vol. **82**, No. 1, p. 319—324 (1977).
See also *6*, p. *57*; 22.1.1, p. 335; 22.1.2, p. 340.

M. N. Sharma
*A Simple Expression for the Anderson-Grüneisen Parameter
(b) Vol. **82**, No. 1, p. K53—K58 (1977).
See also 22.5.2, p. 379.

K. S. Dubey
*Role of Point-Defect Scattering in the Lattice Thermal Conductivity of an Insulator at Low Temperatures
(b) Vol. **82**, No. 1, p. K63—K67 (1977).
See also 6, p. 57; 22, p. 326.

R. Ramji Rao and A. Ramanand
*Third-Order Elastic Constants, Thermal Expansion, and Bulk Modulus of Holmium
(a) Vol. **42**, No. 1, p. 247—255 (1977).
See also 6, p. 57; *12.1*, p. *102*; 21.4, p. 312.

K. Bartkowski, D. Włosewicz, and J. Rafałowicz
*Anisotropy of Thermal Conductivity of Tin Monocrystals Depending on Chemical Admixtures in the Temperature Range 5 to 70 K
(a) Vol. **42**, No. 1, p. 351—357 (1977).
See also 21.7, p. 319.

F. Söffge, E. Steichele, and K. Stierstadt
*Thermal Expansion Anomaly of Nickel near the Curie Point
(a) Vol. **42**, No. 2, p. 621—627 (1977).
See also 12.1, p. 102; 18.2, p. 221; 21.1, p. 296.

P. Čapková and J. Šedivý
*X-Ray Debye Characteristic Temperatures of Magnesium and α-Titanium
(a) Vol. **42**, No. 2, p. 681—686 (1977).
See also *6*, p. *58*; 21, p. 283; 21.3, p. 310.

C. L. Reynolds, Jr.
*Influence of Pressure on the Melting Temperature of Alkali Halides
(a) Vol. **42**, No. 2, p. K151—K153 (1977).
See also 22.5.2, p. 379.

W. Löser
*Thermodynamic Properties of a Finite Disordered Hubbard Cluster
(b) Vol. **83**, No. 1, p. 299—304 (1977).
See also 2, p. 38; *18.4*, p. *233*.

P. J. WALSH, G. C. VEZZOLI, and L. W. DOREMUS
Thermodynamic Basis for Generalization of Group VI-A Melting Curves
(a) Vol. **43**, No. 1, p. 305—310 (1977).
See also *1.1*, p. *12*; *1.2*, p. *18*; 22.1.3, p. 346.

T. SCHOBER and A. CARL
A Differential Thermal Analysis Study of the Vanadium–Hydrogen System
(a) Vol. **43**, No. 2, p. 443—449 (1977).
See also *1.2*, p. *19*; 1.3, p. 23; 21.1, p. 297.

Ö. SÄVBORG and M. NYGREN
Magnetic, Electrical, and Thermal Studies of the $V_{1-x}Re_xO_2$ System with $0 \leq x \leq 0.15$
(a) Vol. **43**, No. 2, p. 645—652 (1977).
See also *1.2*, p. *19*; 14.3, p. 171; 18.1, p. 215; 22.6.1, p. 393.

Y. KIMURA
On the Relation between the Density of States in the Cubic State and Phase Transition Temperatures (T_c, T_m) of V–Zr–Nb Ternary Laves Compounds
(a) Vol. **43**, No. 2, p. K141—K146 (1977).
See also *1.2*, p. *19*; 13, p. 112; *14.2*, p. *163*; 21.1, p. 298.

C. B. SATTERTHWAITE
On Isotope Effects in $PdH(D)_x$
(a) Vol. **43**, No. 2, p. K147—K149 (1977).
See also 13, p. 112; *14.2*, p. *163*; 21, p. 286.

K. S. DUBEY and R. H. MISHO
Three-Phonon Scattering Relaxation Rate and Phonon Conductivity. Application to Mg_2Ge
(b) Vol. **84**, No. 1, p. 69—81 (1977).
See also *6*, p. *59*; 22.7, p. 395.

M. SCHWAB and H. G. KAHLE
*Specific Heat of $TbPO_4$
(b) Vol. **84**, No. 1, p. 167—170 (1977).
See also 1.2, p. 19; *18.4*, p. *234*; 22.8.1, p. 407.

W. WASSERBÄCH
*Low-Temperature Thermal Conductivity of Pure and Impure Niobium and Tantalum Single Crystals
(b) Vol. **84**, No. 1, p. 205—214 (1977).
See also 10.1, p. 81; 21, p. 286.

M. DHWAJ SINGH
The Effect of Internal Strain and Electrical Field of an Impurity on Electron–Phonon Interaction in n-Type Germanium and Silicon
(b) Vol. **84**, No. 1, p. 375—384 (1977).
See also *6*, p. *60*; 22.1.1, p. 336; 22.1.2, p. 343.

K. S. DUBEY
Phonon Conductivity of Sapphire
(b) Vol. **84**, No. 1, p. K57—K60 (1977).
See also *6*, p. *60*; 22.8.1, p. 407.

Z. BIEGAŃSKI and M. DRULIS
*Low-Temperature Phase Transitions in Lanthanum and Neodymium Trihydrides. Specific Heat Measurements
(a) Vol. **44**, No. 1, p. 91—96 (1977).
See also 18, p. 212; 21.4, p. 313.

U. ROY, K. PETROV, I. TSOLOVSKI, and P. PESHEV
On the Structural Phase Transition and the Thermal Expansion of Spinel Lithium Titanate
(a) Vol. **44**, No. 1, p. K25—K27 (1977).
See also *1.2*, p. *20*; 22.8.1, p. 407.

E. Kittinger
*Time Dependent Thermal Expansion of Vitreous Selenium
(a) Vol. **44**, No. 1, p. K35—K37 (1977).
See also *2*, p. *40*; 22.1.3, p. 347.

N. E. Alekseevskii, G. Wolf, N. M. Dobrovolskii, and C. Hohlfeld
*Thermal and Magnetic Properties of the Superconducting Compound $YbMo_6S_8$ at Low Temperatures
(a) Vol. **44**, No. 1, p. K79—K82 (1977).
See also *14.2*, p. *164*; 18, p. 212; 22.8, p. 400.

S. C. Goel, Y. R. Midha, and V. S. Nanda
*A Low Temperature Equation of State for Monovalent Metals
(b) Vol. **84**, No. 2, p. 423—426 (1977).
See also 21.1, p. 300.

M. N. Sharma and A. Kant
*A New Approach to Lattice Properties of Alkali Halide Crystals
(b) Vol. **84**, No. 2, p. 427—434 (1977).
See also *4*, p. *46*; 20.1, p. 262; 22.5.2, p. 381.

J. B. Hartmann and J. J. Martin
*Thermal Conductivity of Irradiated $RbMgF_3$
(b) Vol. **84**, No. 2, p. 721—724 (1977).
See also 11, p. 99; 22.5.4, p. 385.

I. G. Gochev and K. I. Grozdev
*On the Heat Capacity of $Fe(HCOO)_2 \cdot 2\,H_2O$
(b) Vol. **84**, No. 2, p. 785—789 (1977).
See also *18.4*, p. *234*; 22.9, p. 419.

I. J. Gupta and S. K. Trikha
*On the Low Temperature Lattice Thermal Conductivity of Solid Argon
(b) Vol. **84**, No. 2, p. K95—K99 (1977).
See also *6*, p. *61*; 23, p. 420.

M. Sakoh and M. Shimizu
*Thermodynamics of Phase Transitions in Iron
(a) Vol. **44**, No. 2, p. 511—516 (1977).
See also *1.2*, p. *20*; 21.1.1, p. 307.

9. Diffusion

Review Articles

K. Shahi
Transport Studies on Superionic Conductors
(a) Vol. **41**, No. 1, p. 11–44 (1977).
See also 10.2, p. 83; *14.4*, p. *189*; 15, p. 198; 22.5.1, p. 373; 22.8, p. 396; 22.9, p. 414.

T. D. Dzhafarov
*Atomic Diffusion in Semiconductor Epitaxial Structures
(a) Vol. **42**, No. 1, p. 11–45 (1977).
See also 10.2, p. 83; 14.3.1, p. 173; 14.3.3, p. 178; 20.2, p. 265; 22.1, p. 332; 22.2, p. 348; 22.4, p. 361.

Original Papers and Short Notes

H. Kronmüller, N. König, and F. Walz
*Isotope Effects of Diatomic C–H and C–D Complexes in Ni Single Crystals
(b) Vol. **79**, No. 1, p. 237–247 (1977).
See also 10.1, p. 75; 18.2, p. 217; 21.1, p. 288.

T. Ōmura, M. Kitagawa, and Y. H. Ohtsuki
*Theory of Axial Dechanneling
(b) Vol. **79**, No. 1, p. 321–324 (1977).
See also 11, p. 94; 22.1.2, p. 337.

N. A. Stolwijk and H. Bakker
*On Computer Simulation for the Calculation of Correlation Factors for Diffusion
(b) Vol. **79**, No. 1, p. K1–K4 (1977).

Udayan De
Erratum to: The Effect of Thermal Expansion on the Vacancy Mechanism of Diffusion of Impurities in Metals
(b) Vol. **79**, No. 1, p. K87 (1977).

V. Lantto
*A Theoretical Study of the Absorption of Photoelectrons Created by X-Rays in Semi-Infinite Targets
(b) Vol. **79**, No. 2, p. 699–711 (1977).
See also *17*, p. *206*; 21, p. 277; 21.1, p. 289; 21.6, p. 315.

B. Tuck and M. D. Zahari
*"Up-Hill" Diffusion of Zinc in the III–V Semiconductors
(a) Vol. **39**, No. 2, p. 509–515 (1977).
See also 22.2, p. 348.

F. S. Shishiyanu, V. Gh. Gheorghiu, and S. K. Palazov
*Diffusion, Solubility, and Electrical Activity of Co and Fe in InP
(a) Vol. **40**, No. 1, p. 29–35 (1977).
See also 10.2, p. 85; 14.3, p. 167; 22.2.2, p. 354.

S. E. Ormrod and D. L. Kirk
*The Anomalous Behaviour of Na^+ Ions in Polycrystalline β″-Alumina
(a) Vol. **40**, No. 1, p. 265–270 (1977).
See also 22.6, p. 387.

Yu. V. Konobeev, V. I. Tscherbak, V. N. Bykov, and A. V. Subbotin
*Determination of the Vacancy Diffusivity in Molybdenum and Nickel from Data on the Width of the Grain Boundary Void Denuded Zone
(a) Vol. **40**, No. 2, p. K89–K92 (1977).
See also *11*, p. *95*; 21, p. 280; 21.1, p. 292.

A. Buroff, E. Nebauer, P. Süptitz, and I. Willert
*Migration of Silver and Gold in Amorphous As_2S_3
(a) Vol. **40**, No. 2, p. K195—K198 (1977).
See also *2*, p. *36*; 22, p. 324.

F. N. Tavadze, G. M. Surmava, and M. M. Gvichiya
*Investigation of the Diffusion Process of Zinc in Nickel Whiskers
(a) Vol. **41**, No. 1, p. K43—K45 (1977).
See also 21.1, p. 293.

D. Powers and M. Wuttig
*Low Order Correction for Subsurface Contributions in Diffusion Experiments
(a) Vol. **41**, No. 1, p. K79—K81 (1977).

H.-D. Geiler, G. Götz, K.-D. Klinge, and N. Triem
Investigation of Laser Induced Diffusion and Annealing Processes of Arsenic-Implanted Silicon Crystals
(a) Vol. **41**, No. 2, p. K171—K173 (1977).
See also *11*, p. *96*; 22.1.2, p. 340.

R. Sh. Malckovich and V. A. Pokoeva
*Impurity Diffusion with Complex Formation
(b) Vol. **82**, No. 2, p. 421—428 (1977).

H. J. de Bruin, H. Bakker, and L. P. van der Mey
*Diffusion Correlation in Random Primitive Cubic Alloys
(b) Vol. **82**, No. 2, p. 581—588 (1977).
See also 10.1, p. 79; 21, p. 283.

N. A. Stolwijk, T. Spruijt, M. A. Hoetjes-Eijkil, and H. Bakker
*Measurements of Diffusion Coefficients of ^{60}Co in the Intermetallic Compound CoGa
(a) Vol. **42**, No. 2, p. 537—544 (1977).
See also 21, p. 283; 21.1, p. 296.

G. Müller, M. Haubold, R. Schimko, C.-E. Richter, and G. Schwarz
*The Diffusion of Ion-Implanted Zinc in $GaAs_{1-x}P_x$
(a) Vol. **42**, No. 2, p. 579—583 (1977).
See also *11*, p. *97*; 22.2.4, p. 357.

M. Halbwachs
*The Mobilities of Self-Interstitials and Vacancies in a Concentrated Ag–Zn Alloy
(a) Vol. **42**, No. 2, p. 707—712 (1977).
See also *10.1*, p. *80*; 21.1, p. 296; 21.6, p. 316.

Hideo Nakajima and Ken-ichi Hirano
*Influence of the Addition of Nickel and Chromium on the Electromigration of Carbon in γ-Iron
(a) Vol. **43**, No. 1, p. K57—K60 (1977).
See also 21.1.1, p. 305.

E. Ildstad, I. Svare, and T. A. Fjeldly
*Diffusion in Lanthanum Fluoride Studied with ^{19}F Relaxation
(a) Vol. **43**, No. 1, p. K65—K68 (1977).
See also *19*, p. *241*; 22.5, p. 371.

I. V. Aleksandrov and R. G. Gabrielyan
On the Theory of Spin–Lattice Relaxation Due to Diffusion to Impurities in Solids and Liquids
(b) Vol. **83**, No. 2, p. K173—K177 (1977).
See also *13*, p. *111*; *18*, p. *212*.

E. NEBAUER, P. SÜPTITZ, and I. WILLERT
*Determination of the Charge of Migrating Impurities from Electrodiffusion Profiles: the System a-As_2Se_3:Cu
(a) Vol. **43**, No. 2, p. 451—458 (1977).
See also *2*, p. *39*; 14.3.4, p. 187; 22, p. 329.

S. BHARATI, S. BADRINARAYANAN, and A. P. B. SINHA
*On the Diffusion of Iron in Silver
(a) Vol. **43**, No. 2, p. 653—657 (1977).
See also 21.6, p. 317.

V. PAIDAR
*Diffusion in the Force Field in Solid Solutions
(a) Vol. **43**, No. 2, p. K165—K168 (1977).
See also 21.1, p. 306.

S. BHARATI and A. P. B. SINHA
*Isotope Effect and Correlation Factor for the Diffusion of Cadmium in Silver
(a) Vol. **44**, No. 1, p. 391—399 (1977).
See also 21.6, p. 317.

S. I. BEN-ABRAHAM, A. RABINOVITCH, and J. PELLEG
Relations between Vacancy Migration and Formation Energies, Debye Temperature and Melting Point
(b) Vol. **84**, No. 2, p. 435—441 (1977).
See also *10*, p. *74*.

YU. H. KALNIN and F. V. PIROGOV
*On Rate Equations in the Theory of Diffusion-Controlled Reactions
(b) Vol. **84**, No. 2, p. 521—527 (1977).
See also *10*, p. *74*.

E. A. VAN DEN BERGEN, R. DELHEZ, and E. J. MITTEMEIJER
*An X-Ray Diffraction Study of Diffusional Homogenization in RbCl–KCl Powder Blends
(a) Vol. **44**, No. 2, p. 517—524 (1977).
See also 1, p. 10; 22.5.2, p. 382.

K. MAIER
*Self-Diffusion in Copper at "Low" Temperatures
(a) Vol. **44**, No. 2, p. 567—576 (1977).
See also 10.1, p. 82; 21.1, p. 300.

D. L. KIRK and S. E. ORMROD
*The Mechanism of Charge Compensation Occurring in Polycrystalline β''-Alumina
(a) Vol. **44**, No. 2, p. 643—650 (1977).
See also 10.2, p. 93; *14.4*, p. *193*; 22.6, p. 392.

10. Defect Properties (Irradiation Defects See 11)

Original Papers and Short Notes

K. Borggreen and A. R. Thölén
Diffraction Contrast from Misfitting Spherical Particles Precipitated Close Together on a Dislocation (II): The Cylinder as a Converged Model
(a) Vol. **39**, No. 1, p. 223–227 (1977).
See also *1.4*, p. *24*.

W. R. Thorpe and I. O. Smith
*The Double Exponential Rate Equation for Thermally Activated Dislocation Motion
(a) Vol. **39**, No. 2, p. 583–587 (1977).

D. I. Pushkarov
On the Theory of Thermal Conductivity in Quantum Crystals
(b) Vol. **80**, No. 1, p. 295–300 (1977).
See also *8*, p. *64*; 13.5, p. 141.

J. L. Farvacque and P. Lenglart
*Inhomogeneous Strain Field Scattering Deformation Potential (I): Theory
(b) Vol. **80**, No. 1, p. 361–367 (1977).
See also 13.1, p. 115; *14*, p. *155*.

A. Korner, P. Svoboda, and H. O. K. Kirchner
*Finite Dislocation Elements in Finite Media (I): Basic Theory
(b) Vol. **80**, No. 2, p. 441–450 (1977).

N. K. Gilra
*Behaviour of Like Dislocations near an Interface in the Presence of Lattice Frictional Forces
(a) Vol. **40**, No. 2, p. 685–689 (1977).

M. Condat
*Dilatation Due to a Dislocation Made of Straight Segments
(a) Vol. **40**, No. 2, p. K183–K184 (1977).

A. Korner, H. P. Karnthaler, and H. O. K. Kirchner
*Finite Dislocation Elements in Finite Media (II): Trapezoidal Splitting of Partial Dislocations in Thin Foils
(b) Vol. **81**, No. 1, p. 191–198 (1977).

M. J. Marcinkowski
*The Interrelationship between Internal Boundaries of Large and Small Misfit
(a) Vol. **41**, No. 1, p. 213–224 (1977).
See also *1.4*, p. *26*.

A. Authier and Y. Epelboin
*Variation of Stacking Fault Contrast with the Value of the Phase Shift in X-Ray Topography
(a) Vol. **41**, No. 1, p. K9–K12 (1977).
See also *1.4*, p. *26*.

M. H. Gradowczyk and F. A. Rojas
*A Theory of Dislocation Dynamics with Internal Stresses
(b) Vol. **81**, No. 2, p. 647–655 (1977).

B. Michel
*Influence of Solid Surfaces on the Interaction Energy of Point Defects
(b) Vol. **81**, No. 2, p. K87–K90 (1977).

K. Scheerschmidt and J. Heydenreich
On the Computer Simulation of the Electron Microscopic Diffraction Contrast of Dislocations in Jog and Dipole Configurations
(a) Vol. **42**, No. 1, p. 47—59 (1977).
See also *1.4*, p. *26*.

M. J. Marcinkowski
The Differential Geometry of Twisting
(a) Vol. **42**, No. 1, p. 209—220 (1977).
See also *1.4*, p. *26*.

B. V. Petukhov and Yu. Z. Estrin
*Thermal Instability of the Amplitude-Dependent Internal Friction
(a) Vol. **42**, No. 1, p. 391—396 (1977).
See also 7, p. 62; *12*, p. *100*.

V. S. Sardaryan, A. V. Shekoyan, and E. M. Kazaryan
*Self-Focusing of an Intense Ultra- and Hypersonic Beam in an Inhomogeneous and Absorbing Solid
(a) Vol. **42**, No. 1, p. 409—414 (1977).
See also *7*, p. *62*.

W. Scheu, W. Frank, and H. Kronmüller
*Elastic Interaction between Crowdions and Extended Atomic Defects in Isotropic Media
(b) Vol. **82**, No. 2, p. 523—530 (1977).
See also *12*, p. *100*.

K. Jagannadham and M. J. Marcinkowski
*Dislocation Model of the Plastic Zone at a Tensile Crack
(a) Vol. **42**, No. 2, p. 439—447 (1977).

M. Grosbras, E. Dedieu, and M. Cahoreau
*Some Considerations about Stress Relaxation Experiments
(a) Vol. **42**, No. 2, p. 449—457 (1977).
See also *12*, p. *100*.

G. L. Chen, Shi-yu Wu, and M. Scmwartz
*A Classical Analysis of the Resonance Modes in a Randomly Disordered Chain
(b) Vol. **83**, No. 1, p. 115—121 (1977).
See also *6*, p. *58*.

Yu. I. Kharkats
*The Thermodynamic Theory of Domain Formation in Superionic Crystals
(b) Vol. **83**, No. 1, p. 313—322 (1977).
See also *1.4*, p. *27*.

A. Melander
*The Influence of the Finite Size of Inpenetrable Obstacles on the Critical Resolved Shear Stress
(a) Vol. **43**, No. 1, p. 223—230 (1977).

Shi-yu Wu, J. A. Cocks, and V. Chiu
*Frequency Spectra of Disordered Chains with Short-Range Order
(b) Vol. **83**, No. 2, p. 365—375 (1977).
See also *6*, p. *59*.

M. Wilkens and P. Rapps
*Electron Diffraction in Crystalline Specimens Containing a High Density of Lattice Defects (I): Development of a Statistical Theory
(a) Vol. **44**, No. 1, p. 173—182 (1977).
See also *1.4*, p. *28*.

S. I. Ben-Abraham, A. Rabinovitch, and J. Pelleg
*Relations between Vacancy Migration and Formation Energies, Debye Temperature and Melting Point
(b) Vol. **84**, No. 2, p. 435—441 (1977).
See also **9**, p. 71.

Yu. H. Kalnin and F. V. Pirogov
*On Rate Equations in the Theory of Diffusion-Controlled Reactions
(b) Vol. **84**, No. 2, p. 521—527 (1977).
See also *9*, p. *71*.

D. M. Barnett
*On the Validity of Using the Classical Elastic Line Force Solution in Lattice Dislocation Calculations
(b) Vol. **84**, No. 2, p. 741—746 (1977).

K. Sadananda, K. Jagannadham, and M. J. Marcinkowski
*Discrete Dislocation Analysis of a Plastic Tensile Crack
(a) Vol. **44**, No. 2, p. 633—642 (1977).

M. J. Marcinkowski
*Dislocations in Two-Dimensional Spaces
(a) Vol. **44**, No. 2, p. 703—712 (1977).

V. V. Tokii
*Dislocation Multiplication (I): On the Theory of Dislocation Multiplication
(a) Vol. **44**, No. 2, p. 767—770 (1977).

10.1 Metals

Original Papers and Short Notes

P. Rennert and A. M. Radwan
Structural Investigation of the Laves Phase $MgZn_2$ with Model Potential Calculations
(b) Vol. **79**, No. 1, p. 167—173 (1977).
See also *1.3*, p. *21*; 13, p. 107; 21.1, p. 288; 21.3, p. 310.

H. Kronmüller, N. König, and F. Walz
Isotope Effects of Diatomic C–H and C–D Complexes in Ni Single Crystals
(b) Vol. **79**, No. 1, p. 237—247 (1977).
See also *9*, p. *69*; 18.2, p. 217; 21.1, p. 288.

J. D. Stephenson
Some Applications of Characteristic Square Etch Pits Formed on {100} Surfaces of Fe-3wt% Si Crystals
(a) Vol. **39**, No. 1, p. 89—101 (1977).
See also *1.4*, p. *24*; 21.1.1, p. 302.

M. Wilkens, M. L. Jenkins, and K.-H. Katerbau
*TEM Diffraction Contrast of Lattice Defects Causing Strain Contrast and Structure Factor Contrast Simultaneously
(a) Vol. **39**, No. 1, p. 103—107 (1977).
See also 21.1, p. 288.

M. Yahagi and K. Iwamura
Temperature Dependences of Electrical Resistivity and Thermal Expansion in the Intermetallic Compound LiIn
(a) Vol. **39**, No. 1, p. 189—192 (1977).
See also *8*, p. *64*; *14.1*, p. *157*; 21.2, p. 308.

H. U. Habermeier
The Penetration Field and the Magnetization Curve of Type-I Superconductors Containing Lattice Defects
(a) Vol. **39**, No. 1, p. 203—212 (1977).
See also *14.2*, p. *161*; 18, p. 209; 21, p. 276.

E. Ronander and S. Kritzinger
*The Structure of Multiple-Faulted Loop Configurations in Quenched Aluminium and Aluminium Alloys
(a) Vol. **39**, No. 1, p. 239—248 (1977).
See also 21, p. 276.

G. Dlubek, O. Brümmer, and P. Sickert
*Study of the Nature of the Processes Occurring in the Recovery Stages III and IV of Plastically Deformed Ni of Different Purity by Means of the Positron Annihilation Method
(a) Vol. **39**, No. 2, p. 401—410 (1977).
See also 21.1, p. 289.

K. Kojima and M. Meshii
*Surface Film Softening of Iron Single Crystals Oriented for Single Slip
(a) Vol. **39**, No. 2, p. 491—498 (1977).
See also *1.6*, p. *31*; 21.1.1, p. 302 .

C. P. Chang and C. W. Chen
*Observations on Interactions of Glissile Dislocations in Niobium by Transmission Electron Microscopy
(a) Vol. **39**, No. 2, p. 549—560 (1977).
See also 21, p. 277.

H. Föll and M. Wilkens
*Transmission Electron Microscope Studies of Dislocation Loops in Heavy-Ion Irradiated H.C.P. Cobalt
(a) Vol. **39**, No. 2, p. 561—571 (1977).
See also **11**, p. **94**; **21.1**, p. **289**.

A. Vincent, S. M. Seyed Reihani, and J. Perez
*The Nature of Dislocation-Pinning Defects in Cold-Worked Aluminium Studied by Ultrasonic Waves
(a) Vol. **39**, No. 2, p. 651—660 (1977).
See also **7**, p. **62**; **21**, p. **278**.

E. Bonetti, E. Evangelista, P. Gondi, and R. Tognato
*Deformation and Annealing Effects on High-Temperature Internal Friction Peaks of Aluminium Single and Polycrystals
(a) Vol. **39**, No. 2, p. 661—667 (1977).
See also ***12.1***, p. ***101***; **21**, p. **278**.

G. Dlubek, O. Brümmer, and N. Meyendorf
*A Study of the Influence of Mono- and Multivacancies on the Peak Height of the Angular Correlation Curve of the Positron Annihilation Radiation in Al, Cu, and Ni
(a) Vol. **39**, No. 2, p. K95—K99 (1977).
See also **21**, p. **278**; **21.1**, p. **289**.

E. V. Kornelsen, A. van Veen, L. M. Caspers, and H. Bakker
*Thermal Release of Injected Helium from $Co_{56}Ga_{41}$
(a) Vol. **39**, No. 2, p. K143—K146 (1977).
See also ***11***, p. ***94***; **21.1**, p. **290**.

R. Schmucker
The Influence of Plastic Deformation of the Flux-Line Lattice on Flux Transport in Hard Superconductors
(b) Vol. **80**, No. 1, p. 89—97 (1977).
See also ***14.2***, p. ***161***.

O. Takai, R. Yamamoto, and M. Doyama
*The Formation Energies of a Vacancy and a Divacancy and the Binding Energy of a Divacancy in Simple Metals
(b) Vol. **80**, No. 1, p. 193—199 (1977).
See also **21**, p. **278**.

G. A. Hassan
*Very High Ductility of Technically Pure Copper by Bidirectional Twisting at Room Temperature
(a) Vol. **40**, No. 1, p. 43—47 (1977).
See also ***12.1***, p. ***101***; **21.1**, p. **290**.

J. Victoria, J. Piqueras, R. González, and L. Brú
*Temperature Dependence of Flow Stress in Fatigued High-Purity Copper Polycrystals
(a) Vol. **40**, No. 1, p. 167—171 (1977).
See also **21.1**, p. **290**.

A. van Veen, L. M. Cacspers, E. V. Kornelsen, R. Fastenau, A. van Gorkum, and A. Warnaar
*Vacancy Creation by Helium Trapping at Substitutional Krypton in Tungsten
(a) Vol. **40**, No. 1, p. 235—246 (1977).
See also **21**, p. **279**.

B. Wielke, W. Tikvic, and G. Schoeck
*Cottrell-Stokes Law in Cadmium and Zinc
(a) Vol. **40**, No. 1, p. 271—278 (1977).
See also **21**, p. **279**; **21.1**, p. **291** .

J. Th. M. de Hosson
*On the Formation of Argon–Vacancy Clusters in Copper Irradiated with 4 to 6 kV Argon Ions
(a) Vol. **40**, No. 1, p. 293—301 (1977).
See also 21.1, p. 291.

J. Victoria, E. González, and J. Piqueras
*Influence of Purity on the Temperature Dependence of Flow Stress in Fatigued Copper Polycrystals
(a) Vol. **40**, No. 1, p. K19—K123 (1977).
See also 21.1, p. 291.

R. Yamamoto, N. Owada, O. Takai, and M. Doyama
*Calculations of the Heat of Solution of the Impurities in Aluminium and the Cohesive Energies of Simple Metals
(b) Vol. **80**, No. 2, p. K131—K135 (1977).
See also *4*, p. *45*; 21, p. 279.

Th. Müller and P. Paufler
*Yield Strength of the Monocrystalline Intermetallic Compound $MgZn_2$
(a) Vol. **40**, No. 2, p. 471—477 (1977).
See also 21.1, p. 291; 21.3, p. 310.

T. F. M. Kortekaas and J. J. M. Franse
Thermal Expansion of Some Ni_3Al and Ni–Pt Alloys
(a) Vol. **40**, No. 2, p. 479—485 (1977).
See also *8*, p. *65*; 21.1, p. 292.

J. Merlin, M. Kohen et P. F. Gobin
*Conditions d'apparition de hachures dans les courbes effort–déformation des alliages substitutionnels
(Conditions for the Appeareance of Striations in the Stress–Strain Curves of Substitutional Alloys)
(a) Vol. **40**, No. 2, p. 519—530 (1977).
See also 21.1, p. 292.

S. Hanada and O. Izumi
*Orientation Dependence of Yield Stress and Operative Slip Systems of β-CuZn Single Crystals at Low Temperatures
(a) Vol. **40**, No. 2, p. 589—598 (1977).
See also *12.1*, p. *101*; 21.1, p. 292.

R. Kroggel, H. Löffler, and O. Kabisch
*Experimental Investigations on the Position of the Coherency Spinodal Curve in Al–Zn (15, 16, and 18 at%) Alloys
(a) Vol. **40**, No. 2, p. K163—K164 (1977).
See also 1.2, p. 16; 21, p. 280; 21.1, p. 292.

P. Kratochvíl
*The Stress Dependence of the Activation Volume in Copper Base Alloy Single Crystals
(a) Vol. **40**, No. 2, p. K165—K167 (1977).
See also 12.1, p. 101; 21.1, p. 293; 21.6, p. 315.

F. Perrot
*A Model for Vacancies in Simple Metals Including an Ionic Potential in the Self-Consistent Density Calculation
(b) Vol. **81**, No. 1, p. 205—210 (1977).
See also 21.2, p. 308.

B. Hertel, K.-H. Katerbau, and R. Bair
*A Possible Source of Error in the Determination of the Type of Small Latice Defects from Electron Micrographs
(a) Vol. **41**, No. 1, p. K1—K4 (1977).
See also 11, p. 95; 21.1, p. 293.

M. Condat and M. Fayard
*On the Formation of Prismatic Loops in Quenched F.C.C. Metals and Alloys
(a) Vol. **41**, No. 1, p. K5—K8 (1977).
See also *1.4*, p. *26*; 21, p. 280.

S. H. van den Brink, A. van den Beukel, and P. G. McCormick
*The Influence of Specimen Dimensions and Machine Stiffness on the Portevin-Le Chatelier Effect in an Al–Cu Alloy
(a) Vol. **41**, No. 2, p. 513—521 (1977).
See also 21, p. 281; 21.1, p. 294.

G. A. Hassan
*Superplastic Behaviour of an Aluminium–Magnesium Alloy Effected by Bidirectional Twisting
(a) Vol. **41**, No. 2, p. 541—546 (1977).
See also 21, p. 281; 21.3, p. 310.

G. Laslaz, G. Kostorz, M. Roth, P. Guyot, and R. J. Stewart
Interparticle Interference in Neutron Small-Angle Scattering of Al–6.8 at% Zn during Ageing at 133 °C
(a) Vol. 41, No. 2, p. 577—583 (1977).
See also *1.4*, p. *26*; 21, p. 281; 21.1, p. 294.

J. P. Simon
*Etude par trempe des interactions lacunes–impuretés dans les alliages dilués AlZr et AlCr (Annealing Study of the Vacancy–Impurity Interaction in the Diluted Alloys Al$\overline{\text{Zr}}$ and Al$\overline{\text{Cr}}$)
(a) Vol. **41**, No. 2, p. K107—K112 (1977).
See also 21, p. 281; 21.1, p. 294.

R. Kamel, S. A. Mahmoud, and F. M. Mansy
*A Dislocation–Carbon Atom Relaxation in 4% Carbon Iron
(a) Vol. **41**, No. 2, p. K113—K117 (1977).
See also *12.1*, p. *101*; 21.1.1, p. 304.

D. Wagner and F. Stangler
*Activation Energy and Primary Characteristics of the Bordoni Relaxation Peak of Highest Purity Polycrystalline Lead
(a) Vol. **41**, No. 2, p. K137—K140 (1977).
See also *12.1*, p. *101*; 21, p. 281.

S. Takács and A. Pevala
The Angular Dependence of Critical Currents in Superconductors with Long Narrow Defects
(a) Vol. **41**, No. 2, p. K175—K179 (1977).
See also *14.2*, p. *162*; 21, p. 281.

I. Stulíková, P. Lukáč, and G. I. Shklarevskaya
*The Critical Resolved Shear Stress of Cd–Ag Single Crystals at Low Temperatures
(a) Vol. **41**, No. 2, p. K197—K198 (1977).
See also 21, p. 282; 21.6, p. 316.

A. Lodder and J. C. C. de Ruiter
*The Anisotropy of the Lattice Distortion Structure Factor in the Electron–Impurity Scattering Rate
(b) Vol. **82**, No. 1, p. K23—K28 (1977).
See also 21, p. 282.

V. B. Vykhodets, A. Ya. Fishman, and V. B. Demin
*An Analysis of Deviations from Sievert's Law for Hydrogen in Metals
(b) Vol. **82**, No. 1, p. K59—K61 (1977).
See also 21, p. 282.

S. TAKAHASHI
*Compression-Induced Magnetic Anisotropy and Superlattice Dislocations in Ni_3Fe (I): Calculated Results
(a) Vol. **42**, No. 1, p. 201—207 (1977).
See also *18.2*, p. *221*; 21.1, p. 295; 21.1.1, p. 304.

L. T. MIYADA, S. WATANABE, K. TANAKA, and A. ISORĒ
*Internal Friction Peaks in Cold-Worked and Irradiated Titanium
(a) Vol. **42**, No. 1, p. 235—239 (1977).
See also 11, p. 96; *12.1*, p. *101*; 21.1, p. 295.

O. BUCK and A. SEEGER
*Self-Interstitial Migration and Crowdion Conversion in Nickel Studied by Dislocation Pinning Experiments
(a) Vol. **42**, No. 1, p. 285—293 (1977).
See also 11, p. 96; 12.1, p. 102; 21.1, p. 295.

G. DLUBEK, O. BRÜMMER, N. MEYENDORF, I. YA. DEKHTYAR und R. G. FEDCHENKO
*Positronenannihilationsuntersuchungen zur Ausheilung von Kristallbaufehlern in plastisch verformten PdFe-Legierungen
(Positron Annihilation Studies of Annealing of Crystal Defects in Plastically Deformed PdFe Alloys)
(a) Vol. **42**, No. 1, p. K15—K18 (1977).
See also 21, p. 282.

A. A. ADAMENKO and A. M. SHALAEV
*On the Application of the Positron Annihilation Method to the Dislocation Density Evaluation in Metals
(a) Vol. **42**, No. 1, p. K29—K31 (1977).
See also 21.1, p. 295.

YA. M. SOIFER and V. G. SHTEINBERG
*Low Temperature Internal Friction and Microplastic Strain in Copper
(a) Vol. **42**, No. 1, p. K87—K90 (1977).
See also *12.1*, p. *102*; 21.1, p. 296.

W. F. M. VAN HEUGTEN
*An Interatomic Potential for Molybdenum from Phonon Spectra
(b) Vol. **82**, No. 2, p. 501—506 (1977).
See also 6, p. 57; 21, p. 282.

H. J. DE BRUIN, H. BAKKER, and L. P. VAN DER MEY
Diffusion Correlation in Random Primitive Cubic Alloys
(b) Vol. **82**, No. 2, p. 581—588 (1977).
See also *9*, p. *70*; 21, p. 283.

I. GYÉMÁNT and G. SOLT
*Formation Energy of Spherical Cavities in Metals
(b) Vol. **82**, No. 2, p. 651—655 (1977).
See also *4*, p. *45*; 21.2, p. 309.

V. HIVERT, P. GROH, P. MOSER, and W. FRANK
*Internal Friction Peaks Due to Dislocation Relaxations in Plastically Deformed and/or Irradiated High-Purity α-Iron (I): Experiments
(a) Vol. **42**, No. 2, p. 511—518 (1977).
See also 11, p. 97; *12.1*, p. *102*; 21.1.1, p. 304.

S. TAKAHASHI
*Compression-Induced Magnetic Anisotropy and Superlattice Dislocations in Ni_3Fe (II): Experimental Results
(a) Vol. **42**, No. 2, p. 529—536 (1977).
See also *18.2*, p. *221*; 21.1, p. 296; 21.1.1, p. 304.

M. Halbwachs
*The Mobilities of Self-Interstitials and Vacancies in a Concentrated Ag–Zn Alloy
(a) Vol. **42**, No. 2, p. 707–712 (1977).
See also *9*, p. *70*; 21.1, p. 296; 21.6, p. 316.

H. H. Mende and W. Hagen
*X-Ray Topographic Investigations of Deformed Iron Whiskers
(a) Vol. **42**, No. 2, p. K95–K98 (1977).
See also *1.4*, p. *27*; 21.1.1, p. 305.

M. L. Johnson, S. Panchanadeeswaran, S. Saterlie, and J. G. Byrne
*Positron Measurements of Aging in Complex Aluminium Alloys
(a) Vol. **42**, No. 2, p. K175–K177 (1977).
See also 21, p. 284.

R. de Ridder, G. van Tendeloo, and S. Amelinckx
The Ordering of Vacancies in $Ni_{1-x}Al$. Study of the Transition State by Electron Diffraction
(a) Vol. **43**, No. 1, p. 133–139 (1977).
See also *1.3*, p. *22*; 21, p. 284; 21.1, p. 297.

S. Takeuchi, K. Maeda, and T. Suzuki
*Temperature and Strain-Rate Dependence of Flow Stress of Tantalum in the Superconducting and Normal States
(a) Vol. **43**, No. 1, p. 289–294 (1977).
See also *14.2*, p. *163*; 21, p. 285.

H. M. Eissa
*Hexagonal Networks in the Basal Plane of Zn Single Crystals
(a) Vol. **43**, No. 1, p. 343–349 (1977).
See also *1.4*, p. *27*; 21.1, p. 297.

V. A. Pavlov and N. L. Pecherkina
*Cold-Rolling Cavity Formation in Armco-Fe
(a) Vol. **43**, No. 1, p. K5–K8 (1977).
See also 21.1.1, p. 305.

E. Ya. Mikhlin and V. V. Nelaev
*The Effect of Uniaxial Compression upon Dimensions and Shape of the Frenkel Defect Recombination Zone in α-Iron
(a) Vol. **43**, No. 1, p. K23–K27 (1977).
See also 21.1.1, p. 305.

F. Walz, H. Schreyer, and H. Kronmüller
The Magnetic After-Effect Spectrum of α-Iron after Low-Temperature Electron Irradiation
(a) Vol. **43**, No. 1, p. K81–K85 (1977).
See also *11*, p. *97*; *18.2*, p. *222*; 21.1.1, p. 306.

L. H. Dunleavy and H. J. Blythe
Investigation of the Magnetic After-Effect in α-Iron after Neutron Irradiation at 77 K
(a) Vol. **43**, No. 1, p. K87–K90 (1977).
See also *11*, p. *97*; *18.2*, p. *222*; 21.1.1, p. 306.

C. Jourdan and J. Gastaldi
*X-Ray Topographic Investigation of Dislocations in Titanium Single Crystals Grown by Recrystallization
(a) Vol. **43**, No. 2, p. 425–435 (1977).
See also 1.2, p. 19; *1.4*, p. *27*; 21.1, p. 297.

G. Vanderschaeve and T. Sarrazin
*On the Nature of Deformation Twins in Ordered Ni_3V
(a) Vol. **43**, No. 2, p. 459–464 (1977).
See also *1.4*, p. *27*; 21.1, p. 298.

H. NEUHÄUSER and G. SCHULZ
*Etch Pit Study of Dislocation Breakaway in Pre-Yield Deformation of Copper Single Crystals
(a) Vol. **43**, No. 2, p. 493—503 (1977).
See also 21.1, p. 298.

S. MYHRA
*Defect Recovery in Zn after Low Temperature Electron Irradiation
(a) Vol. **43**, No. 2, p. 579—587 (1977).
See also *11*, p. *98*; 21.1, p. 298.

M. YAMAGUCHI and Y. UMAKOSHI
*Slip Geometry and Anomalous Strengthening in β-Cu (ZnMn) Single Crystals
(a) Vol. **43**, No. 2, p. 667—674 (1977).
See also 12.1, p. 102; 21.1, p. 298.

S. D. MAMCHICH
*Phenomenological Dependence of the Deformation Activation Volume upon the Deformation Value
(a) Vol. **43**, No. 2, p. K155—K158 (1977).
See also 21.1, p. 298.

W. MENSCH and J. DIEHL
The Magnetic After-Effect Spectrum of α-Iron after Low Temperature Neutron Irradiation
(a) Vol. **43**, No. 2, p. K175—K178 (1977).
See also 11, p. 98; *18.2*, p. *223*; 21.1.1, p. 306.

J. DIEHL, W. MENSCH, and M. WELLER
Correlations between Magnetic After-Effect and Internal Friction in α-Iron after Low Temperature Neutron Irradiation
(a) Vol. **43**, No. 2, p. K179—K182 (1977).
See also *11*, p. *98*; 12.1, p. 102; *18.2*, p. *223*; 21.1.1, p. 306.

J. VICTORIA, R. GONZÁLES, J. PIQUERAS, and L. BRÚ
*Relationship between Point Defect Clusters and the Temperature Dependence of Yield Stress in Fatigued Copper Polycrystals
(a) Vol. **43**, No. 2, p. K187—K191 (1977).
See also 21.1, p. 298.

W. WASSERBÄCH
Low-Temperature Thermal Conductivity of Pure and Impure Niobium and Tantalum Single Crystals
(b) Vol. **84**, No. 1, p. 205—214 (1977).
See also *8*, p. *67*; 21, p. 286.

R. Z. VALIEV and O. A. KAIBYSHEV
*Mechanism of Superplastic Deformation in a Magnesium Alloy (I): Structural Changes and Operative Deformation Mechanisms
(a) Vol. **44**, No. 1, p. 65—76 (1977).
See also 21, p. 286; 21.3, p. 310.

J. BUCHAR and Z. BÍLEK
*Amplitude and Frequency Response of Copper Single Crystals to One-Dimensional Transient Stress Waves
(a) Vol. **44**, No. 1, p. 117—125 (1977).
See also 12.1, p. 103; 21.1, p. 299.

T. YAMAMOTO
*Experimental Aspects of Electron Channeling Patterns in Scanning Electron Microscopy (I): Measurement of Contrast
(a) Vol. **44**, No. 1, p. 137—146 (1977).
See also *1.4*, p. *28*; 21, p. 287; 21.1, p. 299; 21.6, p. 317; 22.1.2, p. 343.

E. B. Leiko, A. Luft, and E. M. Nadgornii
*Slip Rosettes and Dislocation Structure Formed by Microindentations in Molybdenum Single Crystals
(a) Vol. **44**, No. 1, p. 285—293 (1977).
See also 21, p. 287.

P. Gondi, C. Patuelli, and R. Tognato
*Dislocation Source Activation by Load Impulses
(a) Vol. **44**, No. 1, p. 333—338 (1977).
See also 12.1, p. 103; 21, p. 287.

E. Bonetti, E. Evangelista, and P. Gondi
*Kê Peak Analysis by Means of Internal Friction Measurements during Deformation
(a) Vol. **44**, No. 1, p. K31—K33 (1977).
See also *12.1*, p. *103*; 21, p. 287.

R. Kamel, E. Aziz, and K. H. Georgy
*Strain Hardening of Pre-Fatigued Ag–5 at% Cu Solid Solution at High Temperatures
(a) Vol. **44**, No. 1, p. K53—K57 (1977).
See also 21.6, p. 317.

J. Degauque, B. Astie, and C. Levade
*A Study on the Interaction between Magnetic Domain Walls and Dislocations in Pure Iron by Measurements at Low Amplitudes of Magnetoelastic Damping
(a) Vol. **44**, No. 1, p. K59—K62 (1977).
See also *18.2*, p. *224*; 21.1.1, p. 306.

F. Kroupa
*Vibration Modes of Split Screw Dislocations in B.C.C. Metals
(b) Vol. **84**, No. 2, p. 725—730 (1977).
See also 21.1.1, p. 307.

T. Yamamoto
*Experimental Aspects of Electron Channeling Patterns in Scanning Electron Microscopy (II): Estimation of Contrast Depth
(a) Vol. **44**, No. 2, p. 467—476 (1977).
See also *1.4*, p. *28*; 21, p. 287; 21.1, p. 300; 21.6, p. 318; 22.1.2, p. 344.

R. Z. Valiev and O. A. Kaibyshev
*Mechanism of Superplastic Deformation in a Magnesium Alloy (II): The Role of Grain Boundaries
(a) Vol. **44**, No. 2, p. 477—484 (1977).
See also 1.4, p. 28; 21, p. 287; 21.3, p. 310.

P. Paufler, E. Zedler, H.-J. Ullrich, K.-H. Berthel, U. Krämer, M. Jurisch, K. Richter, and K. Eichler
*Influence of Chemical Composition within the Range of Homogeneity on Phase Transition and Transition Temperature of V_3Si Single Crystals (II): On the Defect Structure of V_3Si Single Crystals
(a) Vol. **44**, No. 2, p. 499—504 (1977).
See also 14.2, p. 164; 21.1, p. 300.

J. Parisot and G. Ferron
*A Calculation of Dislocation Densities in Quenched α-Iron from Internal Friction Experiments
(a) Vol. **44**, No. 2, p. 549—555 (1977).
See also 12.1, p. 103; 21.1.1, p. 307.

K. Maier
Self-Diffusion in Copper at "Low" Temperatures
(a) Vol. **44**, No. 2, p. 567—576 (1977).
See also *9*, p. *71*; 21.1, p. 300.

10.2 Non-Metals

Review Articles

K. Shahi
Transport Studies on Superionic Conductors
(a) Vol. **41**, No. 1, p. 11—44 (1977).
See also 9, p. 69; *14.4*, p. *189*; 15, p. 198; 22.5.1, p. 373; 22.8, p. 396; 22.9, p. 414.

T. D. Dzhafarov
Atomic Diffusion in Semiconductor Epitaxial Structures
(a) Vol. **42**, No. 1, p. 11—45 (1977).
See also *9*, p. *69*; 14.3.1, p. 173; 14.3.3, p. 178; 20.2, p. 265; 22.1, p. 332; 22.2, p. 348; 22.4, p. 361.

Original Papers and Short Notes

P. Gosar
*Quantum Mechanics of Dislocation Kinks in Covalent Semiconductors
(b) Vol. **79**, No. 1, p. 175—183 (1975).
See also 22, p. 320.

J. M. de Siebenthal and H. Bill
*A Perturbed O_2^- Molecule Ion in $SrCl_2$ Single Crystals. EPR Results
(b) Vol. **79**, No. 1, p. 259—267 (1977).
See also 19, p. 235; 22.5.3, p. 383.

J. Giber, A. Hübler, and D. Marton
*On the Local Energy Levels of a Silicon Vacancy. A Model of the Defect-Molecule in the Wannier Approximation
(b) Vol. **79**, No. 1, p. K71—K74 (1977).
See also *13.4*, p. *129*; 22.1.2, p. 337.

A. Vanderhaeghe and G. Jacobs
The Lifetime of the F_H Centre in CsBr:Cl and CsBr:F Crystals
(b) Vol. **79**, No. 1, p. K75—K79 (1977).
See also *20.3*, p. *267*; 22.5.3, p. 383.

V. Belzner and F. Granzer
*Atomistic Calculations on the Dissociation of an Edge Dislocation in NaCl
(a) Vol. **39**, No. 1, p. 183—187 (1977).
See also 22.5.2, p. 375.

M. E. van Hulle and W. Maenhout-van der Vorst
*Ionic Conductivity and Space Charge Region in Silver Bromide (111) Microcrystals
(a) Vol. **39**, No. 1, p. 253—258 (1977).
See also *14.4*, p. *189*; 22.5.1, p. 373.

G. Berg, M. Dubiel, and F. Fröhlich
*T_1-Measurements in Ca-Doped NaCl Crystals
(a) Vol. **39**, No. 1, p. 265—269 (1977).
See also 19, p. 235; 22.5.2, p. 375.

B. J. Pletka, T. E. Mitchell, and A. H. Heuer
*Solid Solution Hardening of Sapphire (α-Al_2O_3)
(a) Vol. **39**, No. 1, p. 301—311 (1977).
See also 22.6, p. 386.

V. I. Savenko, L. A. Kachanova, and E. D. Shchukin
*The Generation of Dislocations near the Surfaces of Crystals under Microindentation
(a) Vol. **39**, No. 1, p. K11—K15 (1977).
See also *1.6*, p. *31*; 12.2, p. 104; 22.5.2, p. 375.

J. A. Baldwin, Jr.
*Force of Interaction between a 180° Domain Wall and a Dislocation in YIG
(a) Vol. **39**, No. 1, p. K17—K19 (1977).
See also *18.3*, p. *228*; 22.8.2, p. 409.

M. Ikeya and J. H. Crawford, Jr.
Recombination Luminescence in Alkali Halides Doped with Monovalent Cations. (V_K + e)A-Type Emission
(b) Vol. **79**, No. 2, p. 559—566 (1977).
See also *20.3*, p. *267*; 22.5.2, p. 375.

A. Y. S. Kung and J. M. Vail
A Molecular Model for the Emission Process of F_A(II)-Centres in Alkali Halides
(b) Vol. **79**, No. 2, p. 663—672 (1977).
See also 13.4, p. 130; 20.3, p. 268; 22.5.2, p. 375.

J. Giber, P. Deák, and D. Marton
*The Mechanical Stress Deriving from Molecular Fitting on the Si–SiO_2 Interface and Its Effect upon the Development and Distribution of Vacancies
(b) Vol. **79**, No. 2, p. K89—K93 (1977).
See also 22.1.2, p. 338.

J. F. Baumard and E. Tani
Thermoelectric Power in Reduced Pure and Nb-Doped TiO_2 Rutile at High Temperature
(a) Vol. **39**, No. 2, p. 373—382 (1977).
See also *15*, p. *198*; 22.6, p. 386.

S. K. Maksimov and D. I. Piskunov
*Correct Calculation of the Stress Relaxation on the Foil Surface in Constructing Theoretical Electron Micrographs
(a) Vol. **39**, No. 2, p. 459—466 (1977).
See also 22.1.2, p. 338.

W. Bollmann and H.-J. Stöhr
*Incorporation and Mobility of OH^- Ions in $LiNbO_3$ Crystals
(a) Vol. **39**, No. 2, p. 477—484 (1977).
See also 14.3, p. 166; 20.1, p. 250; 22.8.1, p. 402.

J. Prakash and F. Fischer
*Ionic Thermocurrent and Optical Absorption Studies of S^{--}–Vacancy Dipoles in KI
(a) Vol. **39**, No. 2, p. 499—507 (1977).
See also *14.4*, p. *189*; 20.1, p. 250; 22.5.2, p. 375.

J. Bohm, H.-D. Kürsten, P. Reiche und R. Schalge
*Morphologie von Versetzungen und Ausscheidungen in Gadoliniumgalliumgranat (GGG) (Morphology of Dislocations and Segregations in Gadolinium Garnet (GGG))
(a) Vol. **39**, No. 2, p. 517—523 (1977).
See also *1.4*, p. *24*; 22.8.2, p. 409.

R. Albrecht, V. Schmidt, and V. I. Betekhtin
*The Damage Process Preceding Semi-Brittle Fracture in Dependence on Deformation (I): Scanning Microscopic In-Situ Studies of Microcracks in NaCl Crystals
(a) Vol. **39**, No. 2, p. 621—630 (1977).
See also *12.2*, p. *104*; 22.5.2, p. 376.

J. N. Boland, B. E. Hobbs, and A. C. McLaren
*The Defect Structure in Natural and Experimentally Deformed Kyanite
(a) Vol. **39**, No. 2, p. 631—641 (1977).
See also *1.4*, p. *24*; 22.8, p. 396.

K. H. Lee, G. E. Holmberg, and H. J. Crawford, Jr.
*Optical and ESR Studies of Holes Centers in γ-Irradiated Al_2O_3
(a) Vol. **39**, No. 2, p. 669—674 (1977).
See also 19, p. 236; *20.3*, p. *268*; 22.6, p. 386.

T. Bedyńska, R. Bubáková, and Z. Šourek
*Erratum to: Comparison between Experimental and Theoretical Dislocation Image for the Bragg Case
(a) Vol. **39**, No. 2, p. 725 (1977).
See also 22.1.2, p. 338.

M. Yacamán, L. W. Hobbs, and M. J. Goringe
*The Direct Observation of Suzuki-Phase Precipitates in NaCl:$MnCl_2$ by Transmission Electron Microscopy
(a) Vol. **39**, No. 2, p. K85—K90 (1977).
See also *1.4*, p. *24*; 22.5.2, p. 376.

V. K. Garg, D. S. Puri, and M. P. Verma
*On the Third-Order Elastic Constants of Ionic Solids with NaCl Structure
(b) Vol. **80**, No. 1, p. 63—72 (1977).
See also *12.2*, p. *104*; 22.5.2, p. 376.

H. J. Hackelöer, H. Selbach, O. Kanert, A. W. Sleeswyk, and G. Hut
*Determination of the Velocity of Mobile Dislocations by Nuclear Spin Relaxation Measurements
(b) Vol. **80**, No. 1, p. 235—243 (1977).
See also 19, p. 236; 22.5.2, p. 376.

G. S. Zavt
*Dynamical Properties of NO_2^- Impurity Molecules in Alkali Halides (I): Theory
(b) Vol. **80**, No. 1, p. 399—408 (1977).
See also 20.1, p. 251; 22.5.2, p. 376.

F. S. Shishiyanu, V. Gh. Gheorghiu, and S. K. Palazov
Diffusion, Solubility, and Electrical Activity of Co and Fe in InP
(a) Vol. **40**, No. 1, p. 29—35 (1977).
See also *9*, p. *69*; 14.3, p. 167; 22.2.2, p. 354.

W. Bollmann
*Evaluation of the Association Parameters of Quadripoles (Dimers of Dipoles) in Ionic Crystals from Conductivity Investigations (I): General
(a) Vol. **40**, No. 1, p. 49—53 (1977).
See also 22.5.1, p. 373; 22.5.2, p. 376; 22.5.3, p. 383.

W. Bollmann
*The Origin of Photoelectrons and the Concentration of Point Defects in $LiNbO_3$ Crystals
(a) Vol. **40**, No. 1, p. 83—91 (1977).
See also *16*, p. *201*; 22.8.1, p. 403.

J. A. Jordão, A. Sakamoto, and M. de Souza
*The Ultrasonic Attenuation due to Cu^+ Substitutional Ions in KCl
(a) Vol. **40**, No. 1, p. 109—112 (1977).
See also 7, p. 62; 22.5.2, p. 376.

R. Albrecht, V. Schmidt, and V. I. Betekhtin
*The Damage Process Preceding Semi-Brittle Fracture in Dependence on Deformation (II): A Discussion on Results of Microcrack Development in NaCl Crystals
(a) Vol. **40**, No. 1, p. 147—152 (1977).
See also *12.2*, p. *104*; 22.5.2, p. 376.

D. Bloor, D. N. Batchelder, and F. H. Preston
Optical Properties of Bis (p-Toluene Sulphonate) Diacetylene (III): Defect-Induced Spectra
(a) Vol. **40**, No. 1, p. 279—284 (1977).
See also *20.1*, p. *252*; 20.3, p. 269; 22.9, p. 415.

A. A. Urusovskaya, N. L. Sizova, and I. A. Rachkov
*Nature of the Yield Point in CsI Crystals
(a) Vol. **40**, No. 1, p. K1—K6 (1977).
See also 22.5.2, p. 377.

Amitabh Jain and B. J. Smith
*A New Method for Demonstrating the Bourgoin Mechanism in Semiconductors
(a) Vol. **40**, No. 1, p. K31—K33 (1977).
See also 14.3.3, p. 179; 22.1.2, p. 338.

M. E. van Hulle
*Dielectric Loss Measurements of AgBr (100) Emulsions
(a) Vol. **40**, No. 1, p. K57—K60 (1977).
See also *14.4*, p. *190*; 22.5.1, p. 373.

J. L. Farvacque and P. Lenglart
*Inhomogeneous Strain Field Scattering Deformation Potential (II): Free-Carrier Mobility Associated with Dislocation Strain Fields in Semiconductors
(b) Vol. **80**, No. 2, p. 433—439 (1977).
See also *14.3*, p. *167*; 22, p. 323; 22.1.3, p. 345.

C. J. Delbecq, J. P. Stokes, and P. H. Yuster
*Thermal Reorientation Kinetics of I_2^- in RbI
(b) Vol. **80**, No. 2, p. 471—475 (1977).
See also 20.1, p. 252; 22.5.2, p. 377.

S. C. Zílio and M. de Souza
*The Dielectric Relaxation of Impurity–Vacancy Complexes in KI Doped with Cl^- Anions
(b) Vol. **80**, No. 2, p. 597—601 (1977).
See also 14.4, p. 190; 22.5.2, p. 377.

A. I. Anselm and I. G. Lang
*Theory of the Excitonic Mechanism of Crystal Defect Formation
(b) Vol. **80**, No. 2, p. 691—695 (1977).
See also *13.5.1*, p. *144*; 22.5.2, p. 377.

A. Watterich and K. Raksányi
*Optical and ESR Investigations of Mg Centres in X-Irradiated and F-Bleached KCl Crystals at RT
(b) Vol. **80**, No. 2, p. K105—K108 (1977).
See also 19, p. 237; *20.1*, p. *252*; 22.5.2, p. 377.

L. Kemter and H. Strunk
*Dislocation Densities in Deformed NaCl Single Crystals Determined by Transmission Electron Microscopy
(a) Vol. **40**, No. 2, p. 385—391 (1977).
See also 22.5.2, p. 377.

J. di Persio and B. Escaig
*Initiation of Dislocation Slip in the B.C.C. Lattice. The Case of the van der Waals B.C.C. Hexamine
(a) Vol. **40**, No. 2, p. 393—401 (1977).
See also 22.9, p. 415.

W. Bollmann
*Evaluation of the Association Parameters of Quadripoles (Dimers of Dipoles) in Ionic Crystals from Conductivity Investigations (II): Association Parameters of $(O^{--}–F_V^-)$ Dipoles and Their Dimers in CaF_2 Crystals
(a) Vol. **40**, No. 2, p. 409—414 (1977).
See also 14.4, p. 190; 22.5.3, p. 383.

S. Tsunekawa, M. Suezawa, and H. Takei
*New Type Ferroelastic Behaviour of $NdNbO_4$ Crystals
(a) Vol. **40**, No. 2, p. 437—446 (1977).
See also *14.4.1*, p. *195*; 22.8.1, p. 403.

K. A. Shoaib, F. H. Hashmi, M. Ali, S. J. H. Bukhari, and C. A. Majid
Thermoluminescence from X-Ray-Irradiated Stabilized ZrO_2 Single Crystals
(a) Vol. **40**, No. 2, p. 605—612 (1977).
See also 13.4, p. 133; *20.3*, p. *269*; 22.6, p. 387.

G. Chassagne, D. Durand, J. Serughetti, and L. W. Hobbs
*The Behaviour of Alkali Metal Colloids in Additively Coloured Potassium Chloride Crystals (I): Study by Optical Absorption and Surface Replicas — Colloid Size Distribution
(a) Vol. **40**, No. 2, p. 629—642 (1977).
See also 20.1, p. 253; 22.5.2, p. 377.

R. de Ridder, D. van Dyck, G. van Tendeloo, and S. Amelinckx
A Cluster Model for the Transition State and Its Study by Means of Electron Diffraction (II): Application to Some Particular Systems
(a) Vol. **40**, No. 2, p. 669—683 (1977).
See also *1.4*, p. *25*; 22, p. 324.

V. G. Kvachadze, M. F. Zhvaniya, and T. Sh. Davitashvili
*On the Nature of the Radiative Centres of the Halogen Impurity
(a) Vol. **40**, No. 2, p. K113—K115 (1977).
See also 8, p. 65; *20.1*, p. *253*; 22.5.2, p. 377.

H. Röppischer and J. Jacobs
*ESR Method for Investigating Associated Point Defects in ZnSe:Mn Single Crystals
(a) Vol. **40**, No. 2, p. K117—K120 (1977).
See also *19*, p. *238*; 22.4.2, p. 365.

B. Jarosz
*Dislocation Charge in X-Irradiated NaCl
(a) Vol. **40**, No. 2, p. K149—K151 (1977).
See also 22.5.2, p. 377.

M. E. van Hulle and W. Maenhout-van der Vorst
*Ionic Conductivity of AgBr and AgBr:Pb^{2+} Single Crystals
(a) Vol. **40**, No. 2, p. K173—K177 (1977).
See also *14.4*, p. *190*; 22.5.1, p. 373.

L. A. Rebane, G. S. Zavt, and K. E. Haller
*Dynamical Properties of NO_2^- Impurity Molecules in Alkali Halides (II): Vibrational Spectra
(b) Vol. **81**, No. 1, p. 57—68 (1977).
See also 20.1, p. 253; 22.5.2, p. 377.

H. Overeijnder, M. Szymoński, A. Haring, and A. E. de Vries
*The Mechanism of the Electron Sputtering Process of Alkali Halides
(b) Vol. **81**, No. 1, p. K11—K14 (1977).
See also *17*, p. *206*; 22.5.2, p. 378.

V. A. Antonov, P. A. Arsenev, Sh. A. Vakhidov, E. M. Ibragimova, and D. S. Petrova
Spectroscopic Properties of Neodymium Ions in $LiYO_2$ Monocrystals
(a) Vol. **41**, No. 1, p. 45—50 (1977).
See also 13.4, p. 133; *20.1*, p. *254*; *20.3*, p. *270*; 22.8.1, p. 404.

D. I. M. Knottnerus, A. Lanjouw, and H. W. den Hartog
*Electron-Excess Centres in Calciumchlorapatite
(a) Vol. **41**, No. 1, p. 95—101 (1977).
See also 20.1, p. 254; 22.8, p. 398.

G. Chassagne, D. Durand, J. Serughetti, and L. W. Hobbs
*The Behaviour of Alkali Metal Colloids in Additively Coloured Potassium Chloride Crystals (II): Study by Transmission Electron Microscopy — Colloid Nucleation and Growth
(a) Vol. **41**, No. 1, p. 183—190 (1977).
See also 22.5.2, p. 378.

D. B. Marshall and A. C. McLaren
Elastic Twinning in Experimentally Deformed Plagioclase Feldspars
(a) Vol. **41**, No. 1, p. 231—240 (1977).
See also *1.4*, p. *26*; 22.8, p. 398.

K. Ozawa and A. Kikuchi
*Induced Flow Stress and Optical Absorption in NaCl Crystals Irradiated with Neutrons at Low Temperatures
(a) Vol. **41**, No. 1, p. 289—294 (1977).
See also 11, p. 95; *20.1*, p. *254*; 22.5.2, p. 378.

J. W. Mattwews and T. S. Plaskett
*Erratum to: Optical Birefringence Images of Edge Dislocations
(a) Vol. **41**, No. 1, p. 347 (1977).
See also 22.8.2, p. 411.

G. Kühn, M. Hein and S. Wartewig
*Manganese-Oxygen Impurities in GaAs
(a) Vol. **41**, No. 1, p. K13—K15 (1977).
See also *19*, p. *239*; 22.2.1, p. 351.

L. Bartelsen
The EPR Fine Structure Spectrum of Dislocations in Silicon
(b) Vol. **81**, No. 2, p. 471—478 (1977).
See also *19*, p. *239*; 22.1.2, p. 340.

T. Goworek and C. Rybka
*Impurities and Positronium Formation in Molecular Crystals
(b) Vol. **81**, No. 2, p. 565—569 (1977).
See also 22.9, p. 416.

A. A. Urusovskaya, N. L. Sizova, I. A. Rachkov, Ya. A. Zakharin, and V. M. Dobryak
*Influence of Impurities on the Mechanical Properties of CsI Crystals
(a) Vol. **41**, No. 2, p. 443—450 (1977).
See also 22.5.2, p. 378.

R. Visocekas and A. Geoffroy
*Tunnelling Afterglows and Retrapping in Calcite
(a) Vol. **41**, No. 2, p. 499—503 (1977).
See also *20.3*, p. *270*; 22.8.1, p. 405.

R. de Ridder, G. van Tendeloo, D. van Dyck, and S. Amelinckx
*Application of the Cluster Model to the Study of Ordering of Atom Strings in Ternary W–Nb Oxides
(a) Vol. **41**, No. 2, p. 555—560 (1977).
See also *1.4*, p. *26*; 22.6, p. 388.

P. L. Vuillermoz, J. Jouglar, A. Laugier, and H. R. Winteler
*Phonon–Defect Interaction in Dislocation-Free GaAs
(a) Vol. **41**, No. 2, p. 561—567 (1977).
See also 6, p. 56; 8, p. 66; 22.2.1, p. 352.

R. D. Newton and W. A. Sibley
*Deformation Induced Damage in MgO
(a) Vol. **41**, No. 2, p. 569—575 (1977).
See also 20.1, p. 256; 20.3, p. 270; 22.6, p. 388.

Y.-H. Lee, J. W. Corbett, and K. L. Brower
*EPR of a Carbon-Oxygen–Divacancy Complex in Irradiated Silicon
(a) Vol. **41**, No. 2, p. 637—647 (1977).
See also 11, p. 95; *19*, p. *240*; 22.1.2, p. 340.

S. Radhakrishna and S. Haridoss
*Methods of Analysis of Piezo-Stimulated Current Spectra
(a) Vol. **41**, No. 2, p. 649—652 (1977).
See also *14.4*, p. *191*; 22.5.2, p. 378.

G. A. Andreev, M. Hartmanová, and V. A. Klimov
*Precipitation in Nickel-Doped Sodium Chloride Crystals
(a) Vol. **41**, No. 2, p. 679—685 (1977).
See also 22.5.2, p. 378.

R. Rauch
*Stark Effect on Sodium-Disturbed F_2 Centres in CaF_2
(a) Vol. **41**, No. 2, p. K97—K99 (1977).
See also *20.1*, p. *256*; 20.3, p. 271; 22.5.3, p. 384.

J. Peña and J. M. Cabrera
*About Additive Colouration of KCl:Pb
(a) Vol. **41**, No. 2, p. K119—K121 (1977).
See also 20.1, p. 256; 22.5.2, p. 378.

H.-J. Kos and R. Nink
*Z_1 Centres in Mg-Doped Lithium Fluoride
(a) Vol. **41**, No. 2, p. K157—K161 (1977).
See also 13.4, p. 135; *20.1*, p. *256*; 22.5.2, p. 378.

B. Frick and D. Siebert
*ESR in Semi-Insulating GaAs:Cr and Colorimetric Determination of the Cr Content
(a) Vol. **41**, No. 2, p. K185—K187 (1977).
See also *19*, p. *240*; 22.2.1, p. 352.

T. Nagarajan, P. Ramasamy, and S. Ramasamy
*Positron Annihilation in Colour Centres
(b) Vol. **82**, No. 1, p. 75—81 (1977).
See also 22.5.2, p. 379.

L. Bosi and M. Nimis
*On the Decay of the M_A Centre in KCl:Li and KCl:Na
(b) Vol. **82**, No. 1, p. K49—K51 (1977).
See also *13.4*, p. *136*; 20.3, p. 271; 22.5.2, p. 379.

F. Appel, H. Bethge, and U. Messerschmidt
Dislocation Motion and Multiplication at the Deformation of MgO Single Crystals in the High Voltage Electron Microscope
(a) Vol. **42**, No. 1, p. 61—71 (1977).
See also *1.4*, p. *26*; 22.6, p. 389.

G. Berg, F. Fröhlich, and D. Schneider
*On the Existence of Charged Aggregates of Divalent Impurities in NaCl-Type Crystals (Point Charge Approximation)
(a) Vol. **42**, No. 1, p. 73—77 (1977).
See also 22.5.2, p. 379.

K. Mori
*Transient Colour Centres by UV Light Irradiation in Yttrium Aluminium Garnet Crystals
(a) Vol. **42**, No. 1, p. 375—384 (1977).
See also 20.1, p. 257; 20.2, p. 265; 20.3, p. 271; 22.8.2, p. 412.

T. Ohgaku and N. Takeuch
Dielectric Loss of KCl-KBr Mixed Crystals Doped with Sr^{2+}
(a) Vol. **42**, No. 1, p. K83—K85 (1977).
See also *14.4*, p. *191*; 22.5.4, p. 385.

V. Hizhnyakov and I. Tehver
On the Theory of Hot Luminescence and Resonant Raman Effect of Impurity Centres
(b) Vol. **82**, No. 2, p. K89—K93 (1977).
See also *6*, p. *58*; *20.3*, p. *271*.

M. T. Carlos, M. Siu Li, and M. de Souza
*The F Centre Half-Width in KCN–KCl Mixed Crystals
(b) Vol. **82**, No. 2, p. K103–K105 (1977).
See also *20.1*, p. *258*; 22.5.4, p. 385.

V. A. Lisovenko, L. A. Khutornaya, M. T. Shpak, and E. N. Velikaya
*Basal Dislocations in Anthracene Crystals
(a) Vol. **42**, No. 2, p. 433–437 (1977).
See also *20.3*, p. *272*; 22.9, p. 417.

H. Oppolzer
*Reflection X-Ray Topography of Oxidation-Induced Stacking Faults in Silicon
(a) Vol. **42**, No. 2, p. K91–K93 (1977).
See also 22.1.2, p. 341.

R. K. Gartia
*Optical Bleaching – a Technique to Clean Thermoluminescence Peaks
(a) Vol. **42**, No. 2, p. K155–K158 (1977).
See also *20.3*, p. *272*; 22.5.2, p. 380.

G. R. Augst
*Distortion of Ionic Crystals by Vacancies
(b) Vol. **83**, No. 1, p. 55–61 (1977).
See also 22.5.2, p. 380.

T. Wosiński and T. Figielski
*Spin-Dependent Recombination at Exchange-Coupled Dislocation Centres in Silicon
(b) Vol. **83**, No. 1, p. 93–98 (1977).
See also 13.4, p. 137; *16*, p. *204*; 22.1.2, p. 341.

J. M. Herreros and F. Jaque
*F-Thermal Bleaching of X-Irradiated NaCl
(b) Vol. **83**, No. 1, p. K21–K23 (1977).
See also 22.5.2, p. 380.

U. V. Desnica and N. B. Urli
*An Approach to the Problem of the Displacement Energy Threshold in Semiconductors
(b) Vol. **83**, No. 1, p. K41–K44 (1977).
See also 11, p. 97; 22, p. 328.

D. Brinkmann, G. Finger, and H. Arend
*On the Motion of Br^- Ions in $K_2Pt(CN)_4Br_{0.3} \cdot 3.2\ H_2O$ (KCP)
(b) Vol. **83**, No. 1, p. K45–K48 (1977).
See also *19*, p. *241*; 22.8, p. 399.

T. Goworek and C. Rybka
*Positronium Formation in p-Terphenyl at Elevated Temperatures
(b) Vol. **83**, No. 1, p. K91–K92 (1977).
See also 22.9, p. 418.

W. Bollmann
*Extrema of the Temperature-Dependent Degree of Association of Point Defects in Ionic Crystals
(a) Vol. **43**, No. 1, p. 175–184 (1977).
See also 22.5.2, p. 380; 22.5.3, p. 384.

N. N. Voinova and V. G. Govorkov
*Temperature–Time Dependence of the Yield Point of Ruby by Compression and Extension
(a) Vol. **43**, No. 1, p. K49–K52 (1977).
See also 22.6, p. 390.

A. Vanderhaeghe and G. Jacobs
*On F- and F_H-Centres in Cesium Halides
(b) Vol. **83**, No. 2, p. 471–477 (1977).
See also 20.3, p. 273; 22.5.2, p. 380.

S. Paul and S. Sengupta
*A Simple Model for the Calculation of Heat of Formation for High-Concentration Substitutional Defects in Alkali Halide Crystals
(b) Vol. **83**, No. 2, p. 645—650 (1977).
See also 22.5.2, p. 380.

K. O. Kechechyan and A. A. Kirakossyan
Wannier-Mott Exciton in the Field of Screw Dislocations
(b) Vol. **83**, No. 2, p. K105—K109 (1977).
See also *13.5.1*, p. *147*; 22, p. 329.

N. A. Drozdov, A. A. Patrin, and V. D. Tkachev
On the Nature of the Dislocation Luminescence in Silicon
(b) Vol. **83**, No. 2, p. K137—K139 (1977).
See also *20.3*, p. *273*; 22.1.2, p. 342.

V. A. Grazhulis, V. V. Kveder, and V. Yu. Mukhina
Investigation of the Energy Spectrum and Kinetic Phenomena in Dislocated Si Crystals (I)
(a) Vol. **43**, No. 2, p. 407—415 (1977).
See also *13.4*, p. *138*; *14.3*, p. *171*; 22.1.2, p. 342.

D. van Dyck, R. de Ridder, G. van Tendeloo, and S. Amelinckx
A Cluster Model for the Transition State and Its Study by Means of Electron Diffraction (III): Generalisations of the Theory and Relation to the SRO Parameters
(a) Vol. **43**, No. 2, p. 541—552 (1977).
See also *1.4*, p. *27*; 22, p. 329.

A. S. Vishnevskii, I. Ya. Dekhtyar, E. G. Madatova, V. S. Mikhalenkov, and A. Čížek
*Positron Annihilation in Boron Nitride Crystals
(a) Vol. **43**, No. 2, p. 589—592 (1977).
See also 22, p. 329.

H. W. den Hartog and H. Benning
*Paramagnetic Color Centers in Calciumchlorapatite Doped with F^- Ions
(a) Vol. **43**, No. 2, p. 633—638 (1977).
See also 19, p. 242; 22.8, p. 400.

J. Härtwig, B. Dietrich, R. Böttger, and G. Lübke
*Connection between Lattice Deformation and Coloration Pattern in Synthetic Quartz Monocrystals
(a) Vol. **43**, No. 2, p. K131—K133 (1977).
See also *20.1*, p. *261*; 22.6, p. 391.

I. Chernov, Yu. Kryuchkov, A. Mamontov, N. Slavin, and Yu. Timoshnikov
*Lattice Location of ^{16}O Implanted in Si Crystals by Backscattering
(a) Vol. **43**, No. 2, p. K201—K203 (1977).
See also *11*, p. *98*; 22.1.2, p. 342.

A. Cavallini, P. Gondi, and A. Castaldini
Dislocation Energy Levels in Ge
(a) Vol. **43**, No. 2, p. K205—K208 (1977).
See also *13.4*, p. *138*; 14.3, p. 171; 22.1.1, p. 336.

T. Goworek, C. Rybka, and J. Wawryszczuk
*Positronium Lifetimes in Molecular Crystal with Impurities
(b) Vol. **84**, No. 1, p. K49—K52 (1977).
See also 22.9, p. 419.

V. A. Grazhulis, V. V. Kveder, and V. Yu. Mukhina
Investigation of the Energy Spectrum and Kinetic Phenomena in Dislocated Si Crystals (II): Microwave Conductivity
(a) Vol. **44**, No. 1, p. 107—115 (1977).
See also *13.4*, p. *139*; *14.3*, p. *171*; 22.1.2, p. 343.

M. E. VAN HULLE and W. MAENHOUT-VAN DER VORST
*Space Charge Characteristics of Silver Halide Microcrystals
(a) Vol. 44, No. 1, p. 229–236 (1977).
See also *14.4*, p. *193*; 22.5.1, p. 374.

H.-J. KOS and D. NEUBERT
Two-Step Photoconductivity by Dislocations in Silicon
(a) Vol. 44, No. 1, p. 259–264 (1977).
See also *16*, p. *205*; 22.1.2, p. 343.

W. NOWAK
Scattered Light Investigations on Amorphous Systems
(a) Vol. 44, No. 1, p. 265–275 (1977).
See also *2*, p. *40*; 22.6, p. 391.

S. RADHAKRISHNA and V. S. SIVASANKAR
*F_{Z_1} Centres in Cesium Halides
(a) Vol. 44, No. 1, p. 345–351 (1977).
See also 20.1, p. 262; 22.5.2, p. 381.

D. KROPMAN, A. SÜGIS, and M. VINNAL
EPR Investigation of Si–SiO_2 Bonding Layer
(a) Vol. 44, No. 1, p. K1–K3 (1977).
See also *19*, p. *243*; 22.1.2, p. 343.

H. ROSIN, G. FREYDANK, R. KLEIN, M. BRUCHHOLZ, and V. GOTTSCHALCH
*Reveal of Dislocation Etch Pits on {001} GaP with Hot Phosphoric Acid
(a) Vol. 44, No. 1, p. K13–K14 (1977).
See also 22.2.2, p. 355.

R. K. GARTIA
*Thermoluminescence of Z_3 Centres in X-Irradiated Mg-Doped LiF Crystals
(a) Vol. 44, No. 1, p. K21–K24 (1977).
See also *20.3*, p. *274*; 22.5.2, p. 381.

V. I. ZAITSEV, V. I. BARBASHOV, and YU. B. TKACHENKO
*Effect of High Hydrostatic Pressure on the Dislocation Mobility in NaCl Crystals
(a) Vol. 44, No. 1, p. K39–K41 (1977).
See also *12.2*, p. *106*; 22.5.2, p. 381.

M. AGUILAR, F. JAQUE, and F. AGULLÓ-LÓPEZ
X-Ray Induced Luminescence in NaCl:Ca^{++}
(b) Vol. 84, No. 2, p. 595–601 (1977).
See also *20.3*, p. *275*; 22.5.2, p. 381.

A. TRUŢIA, F. IOVA, and M. VODĂ
*Concentration of "Impurity" Ion Clusters in a Two-Dimensional Host Lattice
(b) Vol. 84, No. 2, p. 817–824 (1977).
See also *20.1*, p. *263*; 22.5, p. 372.

P. HENNL
*Low Temperature Photochemistry on Chalcogen–Vacancy Centres in Alkali Halide Crystals
(b) Vol. 84, No. 2, p. K151–K154 (1977).
See also *20.1*, p. *263*; 22.5.2, p. 381.

V. SHVEDOV, V. S. POSTNIKOV, and V. M. IEVLEV
*Nucleus Saturation Density and Epitaxy in the Condensation of Metals on Alkali-Halide Crystals (I): Saturation Density
(a) Vol. 44, No. 2, p. 423–427 (1977).
See also *3*, p. *43*; 22.5.2, p. 381.

V. Z. Bengus and T. P. Kovalenko
*Mean Mobile Dislocation Velocity in LiF Crystals under Strain
(a) Vol. **44**, No. 2, p. 457—465 (1977).
See also 22.5.2, p. 381.

H.-J. Kos and R. Nink
*Mechanism of Restored Thermoluminescence in Lithium Fluoride
(a) Vol. **44**, No. 2, p. 505—510 (1977).
See also 20.1, p. 263; *20.3*, p. *275*; 22.5.2, p. 381.

D. L. Kirk and S. E. Ormrod
The Mechanism of Charge Compensation Occurring in Polycrystalline β''-Alumina
(a) Vol. **44**, No. 2, p. 643—650 (1977).
See also *9*, p. *71*; *14.4*, p. *193*; 22.6, p. 392.

Y. Epelboin, M. Gandais, and C. Willaime
*Influence of the Elastic Anisotropy on the Contrast of Dislocation Images in Electron Microscopy and X-Ray Topography. Application to Feldspars
(a) Vol. **44**, No. 2, p. 651—659 (1977).
See also *12.2*, p. *106*; 22.8, p. 400.

H. Neumann, K. Jacobs, Nguyen van Nam, W. Koj, and C. Krause
*Properties of Liquid Phase Epitaxial Ge-Doped GaAs
(a) Vol. **44**, No. 2, p. 675—678 (1977).
See also *3*, p. *43*; 22.2.1, p. 353.

S. K. Mehta, J. F. Merklin, and H. J. Donnert
*Thermoluminescence Related Z Centers in LiF:Mg, Ti
(a) Vol. **44**, No. 2, p. 679—685 (1977).
See also *20.1*, p. *264*; 22.5.2, p. 382.

V. M. Orera and R. Alcalá
*Optical Properties of Cation Colloidal Particles in CaF_2 and SrF_2
(a) Vol. **44**, No. 2, p. 717—723 (1977).
See also *20.1*, p. *264*; 22.5.3, p. 384.

Ah Mee Hor and P. W. M. Jacobs
*Thermal Depolarization in Crystals of Rubidium Bromide Doped with Ba^{2+} and Sn^{2+}
(a) Vol. **44**, No. 2, p. 725—729 (1977).
See also 22.5.2, p. 382.

V. S. Sivasankar and S. Radhakrishna
*Electric Field Gradient at the F_{Z_1} Electron Site in Cesium Halides
(a) Vol. **44**, No. 2, p. 741—745 (1977).
See also 13.4, p. 140; 22.5.2, p. 382.

V. R. Dumke
*Image Effect in Previously Electrolytically Coloured KCl–OH^-
(a) Vol. **44**, No. 2, p. K133—K134 (1977).
See also 20.1, p. 264; 22.5.2, p. 382.

M. M. Kaila
*Effect of Magnetic Field on Electrical Effects of Dislocations in InSb
(a) Vol. **44**, No. 2, p. K135—K139 (1977).
See also *14.3*, p. *172*; 22.2.3, p. 356.

11. Irradiation Effects
(X-Ray Diffraction Investigations See 1 and 10)

Original Papers and Short Notes

T. Ōmura, M. Kitagawa, and Y. H. Ohtsuki
Theory of Axial Dechanneling
(b) Vol. **79**, No. 1, p. 321—324 (1977).
See also *9*, p. *69*; 22.1.2, p. 337.

H. Runge
*Device Geometry Dependent Effects of Deep Levels on $C(U)$ Measurements of Ion-Implanted MOS Diodes
(a) Vol. **39**, No. 1, p. 141—146 (1977).
See also *14.3.3*, p. *178*; 22.1.2, p. 337.

U. V. Desnica
*Indirect Doping as a Possibility for Obtaining p-Type Conductivity CdS
(a) Vol. **39**, No. 1, p. K33—K36 (1977).
See also *14.3*, p. *165*; 22.4.1, p. 362.

E. Klugmann, Yu. F. Bydin, and V. A. Volpyas
*Effect of Ion Implantation on the Magnetic Properties of Ferromagnetic Metals
(a) Vol. **39**, No. 1, p. K53—K57 (1977).
See also *18.2*, p. *217*; 21.1, p. 289.

F. Burkhardt and C. Wagner
*Depth Distribution of Phosphorus Implanted into Silicon
(a) Vol. **39**, No. 1, p. K63—K65 (1977).
See also 22.1.2, p. 337.

A. F. Burenkov and F. F. Komarov
*The Effect of Atomic Chemical Bonding under Fast Ion Stopping in Solids
(b) Vol. **79**, No. 2, p. K161—K164 (1977).
See also 22.5.2, p. 375.

H. Föll and M. Wilkens
Transmission Electron Microscope Studies of Dislocation Loops in Heavy-Ion Irradiated H.C.P. Cobalt
(a) Vol. **39**, No. 2, p. 561—571 (1977).
See also *10.1*, p. *76*; 21.1, p. 289.

H. Runge
*Distribution of Implanted Ions under Arbitrarily Shaped Mask Edges
(a) Vol. **39**, No. 2, p. 595—599 (1977).
See also 22.1.2, p. 338.

P. N. Murty, C. R. K. Murty, and S. Subramanian
Electron Paramagnetic Resonance Study of X-Irradiated Arsenate-Doped Phosphate: KH_2AsO_4/KH_2PO_4
(a) Vol. **39**, No. 2, p. 675—682 (1977).
See also *19*, p. *236*; 22.8.1, p. 402.

E. V. Kornelsen, A. van Veen, L. M. Caspers, and H. Bakker
*Thermal Release of Injected Helium from $Co_{56}Ga_{44}$
(a) Vol. **39**, No. 2, p. K143—K146 (1977).
See also *10.1*, p. *76*; 21.1, p. 290.

W. Dander and H. E. Schaefer
*Electrical Resistivity Recovery of Cobalt after Low-Temperature Electron Irradiation
(b) Vol. **80**, No. 1, p. 173—179 (1977).
See also *14.1*, p. *157*; 21.1, p. 290.

J. N. Daou, J. E. Bonnet, P. Vajda, M. Biget, A. Lucasson, and P. Lucasson
*The Effect of Low-Temperature Electron Irradiation on Lutetium and Its Hydrogen Solid Solutions
(a) Vol. **40**, No. 1, p. 101—108 (1977).
See also 21.4, p. 311.

F. Eichhorn, K. Hennig, B. Lippold, and W. Matz
*Diffraction of a Pulsed Thermal Neutron Beam on an Elastically Bent Quartz Crystal Plate
(a) Vol. **40**, No. 1, p. 205—210 (1977).
See also *1*, p. *9*; 22.6, p. 387.

M. Binder, N. E. Geacintov, M. Pope, and C. E. Swenberg
Magnetic Field Modulated Scintillation from Tetracene-Doped Anthracene Crystals under α-Particle Irradiation
(a) Vol. **40**, No. 2, p. 621—628 (1977).
See also *13.5.1*, p. *144*; *20.3*, p. *269*; 22.9, p. 415.

Yu. V. Konobeev, V. I. Tscherbak, V. N. Bykov, and A. V. Subbotin
*Determination of the Vacancy Diffusivity in Molybdenum and Nickel from Data on the Width of the Grain Boundary Void Denuded Zone
(a) Vol. **40**, No. 2, p. K89—K92 (1977).
See also *9*, p. *69*; 21, p. 280; 21.1, p. 292.

V. V. Gann, A. I. Skvortsov, and A. van den Bosch
*The Superconducting Transition Temperature of $U_6Fe_{0.98}$ Reduced by Neutron Irradiation
(a) Vol. **41**, No. 1, p. 225—229 (1977).
See also *14.2*, p. *162*; 18.1, p. 214; 21.1.1, p. 303; 21.5, p. 314.

D. Goldschmidt, T. Bernstein, and P. S. Rudman
*The Kinetics of Photodissolution of Silver in Amorphous As_2S_3 Films
(a) Vol. **41**, No. 1, p. 283—287 (1977).
See also 14.3.2, p. 176; 22, p. 324.

K. Ozawa and A. Kikuchi
Induced Flow Stress and Optical Absorption in NaCl Crystals Irradiated with Neutrons at Low Temperatures
(a) Vol. **41**, No. 1, p. 289—294 (1977).
See also *10.2*, p. *88*; *20.1*, p. *254*; 22.5.2, p. 378.

Yu. I. Gorkun, V. L. Korobanov, V. S. Lysenko, V. A. Novominskii, and V. I. Turchanikov
Effect of Bombardment with Low-Energy Ar^+ Ions on Electron Scattering in Accumulation Layers of n-Ge
(a) Vol. **41**, No. 1, p. 339—346 (1977).
See also *13.4*, p. *134*; *14.3.3*, p. *179*; 22.1.1, p. 334.

B. Hertel, K.-H. Katerbau, and R. Bair
A Possible Source of Error in the Determination of the Type of Small Lattice Defects from Electron Micrographs
(a) Vol. **41**, No. 1, p. K1—K4 (1977).
See also *10.1*, p. *77*; 21.1, p. 293.

I. K. Amanis and J. G. Kliava
Models of Paramagnetic Centres Formed by Copper and Silver Atoms in Quartz
(a) Vol. **41**, No. 2, p. 385—392 (1977).
See also *19*, p. *239*; 22.6, p. 388.

Y.-H. Lee, J. W. Corbett, and K. L. Brower
EPR of a Carbon-Oxygen–Divacancy Complex in Irradiated Silicon
(a) Vol. **41**, No. 2, p. 637—647 (1977).
See also *10.2*, p. *88*; *19*, p. *240*; 22.1.2, p. 340.

Yu. V. Bogatyrev and F. P. Korshunov
*The Effect of High-Speed Electrons of Different Energy on the Charge Storage in SiO_2 and Si_3N_4 Dielectric Films
(a) Vol. **41**, No. 2, p. K151—K154 (1977).
See also *14.4*, p. *191*; 22, p. 325; 22.6, p. 389.

H.-D. Geiler, G. Götz, K.-D. Klinge, and N. Triem
*Investigation of Laser Induced Diffusion and Annealing Processes of Arsenic-Implanted Silicon Crystals
(a) Vol. **41**, No. 2, p. K171—K173 (1977).
See also 9, p. 70; 22.1.2, p. 340.

S. Tanigawa, K. Hinode, and M. Doyama
*Phenomenological Interpretation of Positron Lifetimes in Metals
(b) Vol. **82**, No. 1, p. 369—376 (1977).
See also 21, p. 282.

V. G. Baryshevskii and I. Ya. Dubovskaya
*Coherent Radiation of the Channelling Positron (Electron)
(b) Vol. **82**, No. 1, p. 403—412 (1977).
See also *1.1*, p. *12*.

K.-D. Ufert
*Doping of VO_2 Thin Films by Ion Implantation
(a) Vol. **42**, No. 1, p. 187—190 (1977).
See also *14.3.1*, p. *174*; 22.6, p. 389.

L. T. Miyada, S. Watanabe, K. Tanaka, and A. Isoré
Internal Friction Peaks in Cold-Worked and Irradiated Titanium
(a) Vol. **42**, No. 1, p. 235—239 (1977).
See also *10.1*, p. *79*; *12.1*, p. *101*; 21.1, p. 295.

H. Mutka and O. J. A. Tiainen
*Annealing Experiments on Fast Neutron Irradiated Selenium Single Crystals
(a) Vol. **42**, No. 1, p. 241—245 (1977).
See also 14.3, p. 169; 20.1, p. 257; 22.1.3, p. 346.

O. Buck and A. Seeger
Self-Interstitial Migration and Crowdion Conversion in Nickel Studied by Dislocation Pinning Experiments
(a) Vol. **42**, No. 1, p. 285—293 (1977).
See also *10.1*, p. *79*; 12.1, p. 102; 21.1, p. 295.

R. Barrio, D. Ríos, and G. Torres
*Radiation Damage in Cu_2O
(a) Vol. **42**, No. 1, p. K1—K5 (1977).
See also 22.6, p. 389.

Y. Sato
Mössbauer Effect of γ-Irradiated CaF_2 Doped with ^{57}Co
(b) Vol. **82**, No. 2, p. 611—616 (1977).
See also *5*, p. *48*; 22.5.3, p. 384.

D. I. Blokhintsev and N. M. Plakida
On the Heating of Ultracold Neutrons
(b) Vol. **82**, No. 2, p. 627—632 (1977).
See also *1.6*, p. *31*.

H.-D. Hattendorff
*Transition Radiation from Gold and Silver Films Bombarded with 80 keV Electrons
(a) Vol. **42**, No. 2, p. 489—494 (1977).
See also *20*, p. *245*; 21.6, p. 316.

V. Hivert, P. Groh, P. Moser, and W. Frank
Internal Friction Peaks Due to Dislocation Relaxations in Plastically Deformed and/or Irradiated High-Purity α-Iron (I): Experiments
(a) Vol. **42**, No. 2, p. 511—518 (1977).
See also *10.1*, p. *79*; *12.1*, p. *102*; 21.1.1, p. 304.

G. Müller, M. Haubold, R. Schimko, C.-E. Richter, and G. Schwarz
*The Diffusion of Ion-Implanted Zinc in $GaAs_{1-x}P_x$
(a) Vol. **42**, No. 2, p. 579—583 (1977).
See also *9*, p. *70*; 22.2.4, p. 357.

H. Kato, T. Kanda, K. Yasuda, A. Yoshida, and T. Arizumi
*Electron Beam Irradiation Effects on SnO_2–Si Photodiodes
(a) Vol. **42**, No. 2, p. 769—774 (1977).
See also *14.3.2*, p. *177*; 16, p. 203; 22.1.2, p. 341.

G. S. White, K. H. Lee, and J. H. Crawford, Jr.
Effects of γ-Irradiation upon the Optical Behavior of Spinel
(a) Vol. **42**, No. 2, p. K137—K141 (1977).
See also *20.1*, p. *259*; *20.3*, p. *272*; 22.8.2, p. 412.

U. V. Desnica and N. B. Urli
An Approach to the Problem of the Displacement Energy Threshold in Semiconductors
(b) Vol. **83**, No. 1, p. K41—K44 (1977).
See also *10.2*, p. *90*; 22, p. 328.

V. I. Roslyakov, A. S. Rudnev, E. I. Sirotinin, A. F. Tulinov, and V. A. Khodyrev
*Analysis of Backscattering Energy Spectra and Determination of Energy Loss of Channelling Particles
(a) Vol. **43**, No. 1, p. 59—70 (1977).
See also 21, p. 284.

G. Sixt, H. Kappert, and G. H. Schwuttke
*Minority Carrier Lifetime and Defect Structure in Silicon after Cesium Implantation
(a) Vol. **43**, No. 1, p. 119—131 (1977).
See also *14.3*, p. *170*; 22.1.2, p. 341.

H.-J. Fitting, H. Glaefeke, and W. Wild
*Electron Penetration and Energy Transfer in Solid Targets
(a) Vol. **43**, No. 1, p. 185—190 (1977).
See also 21, p. 285; 21.1, p. 297; 21.3, p. 310; 21.6, p. 317.

Y. Uchida and F. Fujimoto
*Electron Channeling in ⟨111⟩ Direction of Tungsten Crystals (I): Energy Range of 200 to 1000 keV
(a) Vol. **43**, No. 1, p. 269—275 (1977).
See also 13, p. 111; 21, p. 285.

F. Walz, H. Schreyer, and H. Kronmüller
*The Magnetic After-Effect Spectrum of α-Iron after Low-Temperature Electron Irradiation
(a) Vol. **43**, No. 1, p. K81—K85 (1977).
See also 10.1, p. 80; *18.2*, p. *222*; 21.1.1, p. 306.

L. H. Dunleavy and H.J. Blythe
*Investigation of the Magnetic After-Effect in α-Iron after Neutron Irradiation at 77 K
(a) Vol. **43**, No. 1, p. K87—K90 (1977).
See also 10.1, p. 80; *18.2*, p. *222*; 21.1.1, p. 306.

L. Meyer, M. Klein, and R. Wedell
*The Energy–Angle Distribution of Heavy Particles Penetrating Solids
(b) Vol. **83**, No. 2, p. 451—463 (1977).

K. KOYAMA and T. KANEKO
Radiative Electron Capture by Backscattered Ions from Solid Surfaces
(b) Vol. **83**, No. 2, p. 485– 492 (1977).
See also 13.3, p. 127; 21.6, p. 317.

K. ELLMER and R. WEDELL
*Total Energy Loss Distributions of Low Energy Heavy Particles after Penetration through Matter
(b) Vol. **83**, No. 2, p. K129–K131 (1977).

A. F. AKKERMAN
*Correlations between the Elastic Energy Transferred in Solids by Implanted Hydrogen Ions and Hydrogen Trapping
(b) Vol. **83**, No. 2, p. K145–K147 (1977).
See also 21, p. 285; 22.1.2, p. 342.

H. NAKATA and H. MORISHIMA
An ESR Study of Annealing Characteristics for Heavily Irradiated Silicon
(a) Vol. **43**, No. 2, p. 417–423 (1977).
See also *19*, p. *241*; 22.1.2, p. 342.

J. HÜLLER and M. T. PHAM
*Characterization of Ion Implanted Silicon by the Electrolytic Reverse Current
(a) Vol. **43**, No. 2, p. 505–512 (1977).
See also *14.3.2*, p. *177*; 22.1.2, p. 342.

S. MYHRA
*Defect Recovery in Zn after Low Temperature Electron Irradiation
(a) Vol. **43**, No. 2, p. 579–587 (1977).
See also *10.1*, p. *81*; 21.1, p. 298.

W. MENSCH and J. DIEHL
*The Magnetic After-Effect Spectrum of α-Iron after Low Temperature Neutron Irradiation
(a) Vol. **43**, No. 2, p. K175–K178 (1977).
See also 10.1, p. 81; *18.2*, p. *223*; 21.1.1, p. 306.

J. DIEHL, W. MENSCH, and M. WELLER
*Correlations between Magnetic After-Effect and Internal Friction in α-Iron after Low Temperature Neutron Irradiation
(a) Vol. **43**, No. 2, p. K179–K182 (1977).
See also 10.1, p. 81; 12.1, p. 102; *18.2*, p. *223*; 21.1.1, p. 306.

I. CHERNOV, YU. KRYUCHKOV, A. MAMONTOV, N. SLAVIN, and YU. TIMOSHNIKOV
*Lattice Location of ^{16}O Implanted in Si Crystals by Backscattering
(a) Vol. **43**, No. 2, p. K201–K203 (1977).
See also *10.2*, p. *91*; 22.1.2, p. 342.

M. A. KUMAKHOV
*Theory of Radiation of Charged Particles Channeled in a Crystal
(b) Vol. **84**, No. 1, p. 41–54 (1977).
See also 13, p. 112; 22.1.2, p. 342.

M. FÄHNLE
*The Influence of Fast Neutron Irradiation on the Critical Current Densities in Nb_3Sn (I): Low Neutron Doses ($\Phi < 2 \times 10^{18}$ neutrons/cm^2)
(b) Vol. **84**, No. 1, p. 245–251 (1977).
See also *14.2*, p. *163*; 21, p. 286.

A. A. BEDNYAKOV, F. BERNHARD, YU. V. BULGAKOV, V. L. CHERNOV, and U. MÜLLER-JAHREIS
*Energy Loss Straggling of Protons and Helium Ions in Carbon
(b) Vol. **84**, No. 1, p. K79–K81 (1977).
See also 22.1, p. 332.

E. Zielińska-Rohozińska
X-Ray Diffraction Investigations of Annealing Characteristics in P^+-Implanted Silicon Crystals
(a) Vol. **44**, No. 1, p. 59–64 (1977).
See also *1.4*, p. *28*; 22.1.2, p. 343.

M. Fähnle
*The Influence of Fast-Neutron Irradiation on the Critical Current Densities in Nb_3Sn (II): High Neutron Doses ($\Phi > 2 \times 10^{18}$ neutrons/cm^2)
(b) Vol. **84**, No. 2, p. 541–549 (1977).
See also *14.2*, p. *164*; 21, p. 287.

M. A. Kumakhov and R. Wedell
*Theory of Radiation of Relativistic Channelled Particles
(b) Vol. **84**, No. 2, p. 581–593 (1977).
See also 13, p. 112; 22.1.2, p. 344.

J. B. Hartmann and J. J. Martin
Thermal Conductivity of Irradiated $RbMgF_3$
(b) Vol. **84**, No. 2, p. 721–724 (1977).
See also *8*, p. *68*; 22.5.4, p. 385.

V. P. Dobrego, O. P. Ermolaev, and V. D. Tkachev
*Investigation of Shallow Radiation Defect Levels Produced in Ge by Fast-Neutron Irradiation
(a) Vol. **44**, No. 2, p. 435–442 (1977).
See also *13.4*, p. *140*; 22.1.1, p. 336.

M. Wurlitzer and U. Reinhold
*X-Ray Induced Change of Initial Permeability in Polycrystalline YIG (Me^{4+})
(a) Vol. **44**, No. 2, p. 531–538 (1977).
See also 18.3, p. 230; 22.8.2, p. 413.

F. K. Naehring, A. Schmidt, and J. Schöneich
*Ion Induced Carbon Contamination and Recoil Implantation
(a) Vol. **44**, No. 2, p. K141–K145 (1977).
See also 1.5, p. 30; 22.1.2, p. 344.

12. Mechanical Properties (Plastic Deformations See 10)

Original Papers and Short Notes

V. K. Garg, D. S. Puri, and M. P. Verma
*Many-Body Effects on the Fourth-Order Elastic Constants of Ionic Solids (I): Theory
(b) Vol. **82**, No. 1, p. 325—334 (1977).
See also *4*, p. *45*.

F. Fröhlich, P. Grau, and W. Grellmann
*Performance and Analysis of Recording Microhardness Tests
(a) Vol. **42**, No. 1, p. 79—89 (1977).

B. V. Petukhov and Yu. Z. Estrin
*Thermal Instability of the Amplitude-Dependent Internal Friction
(a) Vol. **42**, No. 1, p. 391—396 (1977).
See also 7, p. 62; *10*, p. *73*.

W. Scheu, W. Frank, and H. Kronmüller
*Elastic Interaction between Crowdions and Extended Atomic Defects in Isotropic Media
(b) Vol. **82**, No. 2, p. 523—530 (1977).
See also *10*, p. *73*.

M. Grosbras, E. Dedieu, and M. Cahoreau
*Some Considerations about Stress Relaxation Experiments
(a) Vol. **42**, No. 2, p. 449—457 (1977).
See also *10*, p. *73*.

12.1 Metals

Original Papers and Short Notes

R. C. RAI and M. P. HEMKAR
A Simple Phenomenological Model for the Lattice Dynamics of B.C.C. Metals
(b) Vol. **79**, No. 1, p. 289–298 (1977).
See also *6*, p. *51*; 21.2, p. 308.

E. BONETTI, E. EVANGELISTA, P. GONDI, and R. TOGNATO
*Deformation and Annealing Effects on High-Temperature Internal Friction Peaks of Aluminium Single and Polycrystals
(a) Vol. **39**, No. 2, p. 661–667 (1977).
See also *10.1*, p. *76*; 21, p. 278.

G. A. HASSAN
*Very High Ductility of Technically Pure Copper by Bidirectional Twisting at Room Temperature
(a) Vol. **40**, No. 1, p. 43–47 (1977).
See also *10.1*, p. *76*; 21.1, p. 290.

G. HAUSCH and E. TÖRÖK
*Thermal Expansivity and Elastic Constants of CrFe Alloys
(a) Vol. **40**, No. 1, p. 55–62 (1977).
See also 18.4, p. 232; 21.1, p. 290; 21.1.1, p. 303.

S. HANADA and O. IZUMI
*Orientation Dependence of Yield Stress and Operative Slip Systems of β-CuZn Single Crystals at Low Temperatures
(a) Vol. **40**, No. 2, p. 589–598 (1977).
See also *10.1*, p. *77*; 21.1, p. 292.

P. KRATOCHVÍL
The Stress Dependence of the Activation Volume in Copper Base Alloy Single Crystals
(a) Vol. **40**, No. 2, p. K165–K167 (1977).
See also *10.1*, p. *77*; 21.1, p. 293; 21.6, p. 315.

K. SALAMA and G. A. ALERS
*The Composition Dependence of the Third-Order Elastic Constants of the Cu–Ni System
(a) Vol. **41**, No. 1, p. 241–247 (1977).
See also 21.1, p. 293.

S. SATERLIE, M. L. JOHNSON, P. ALEXOPOULOS, F. SEPPI, R. URE, and J. G. BYRNE
*Positron Annihilation Observations of Shot Peened Aluminium Alloys
(a) Vol. **41**, No. 1, p. K75–K78 (1977).
See also 21, p. 280.

R. KAMEL, S. A. MAHMOUD, and F. M. MANSY
*A Dislocation–Carbon Atom Relaxation in 4% Carbon Iron
(a) Vol. **41**, No. 2, p. K113–K117 (1977).
See also *10.1*, p. *78*; 21.1.1, p. 304.

D. WAGNER and F. STANGLER
*Activation Energy and Primary Characteristics of the Bordoni Relaxation Peak of Highest Purity Polycrystalline Lead
(a) Vol. **41**, No. 2, p. K137–K140 (1977).
See also *10.1*, p. *78*; 21, p. 281.

L. T. MIYADA, S. WATANABE, K. TANAKA, and A. ISORĒ
*Internal Friction Peaks in Cold-Worked and Irradiated Titanium
(a) Vol. **42**, No. 1, p. 235–239 (1977).
See also *10.1*, p. *79*; 11, p. 96; 21.1, p. 295.

R. Ramji Rao and A. Ramanand
*Third-Order Elastic Constants, Thermal Expansion, and Bulk Modulus of Holmium
(a) Vol. **42**, No. 1, p. 247—255 (1977).
See also 6, p. 57; *8*, p. *66*; 21.4, p. 312.

O. Buck and A. Seeger
Self-Interstitial Migration and Crowdion Conversion in Nickel Studied by Dislocation Pinning Experiments
(a) Vol. **42**, No. 1, p. 285—293 (1977).
See also *10.1*, p. *79*; 11, p. 96; 21.1, p. 295.

C. L. Reynolds, Jr. and P. R. Couchman
*Tait's Equation of State at High Pressures and the Isothermal Bulk Modulus
(a) Vol. **42**, No. 1, p. K47—K52 (1977).
See also *12.2*, p. *105*; 21, p. 282; 22.5.2, p. 379; 23, p. 420.

Ya. M. Soifer and V. G. Shteinberg
*Low Temperature Internal Friction and Microplastic Strain in Copper
(a) Vol. **42**, No. 1, p. K87—K90 (1977).
See also *10.1*, p. *79*; 21.1, p. 296.

V. Hivert, P. Groh, P. Moser, and W. Frank
*Internal Friction Peaks Due to Dislocation Relaxations in Plastically Deformed and/or Irradiated High-Purity α-Iron (I): Experiments
(a) Vol. **42**, No. 2, p. 511—518 (1977).
See also *10.1*, p. *79*; 11, p. 97; 21.1.1, p. 304.

F. Söffge, E. Steichele, and K. Stierstadt
Thermal Expansion Anomaly of Nickel near the Curie Point
(a) Vol. **42**, No. 2, p. 621—627 (1977).
See also *8*, p. *66*; 18.2, p. 221; 21.1, p. 296.

M. Fumeron, J. P. Guillot and A. Dauger
Model X-Ray Scattering by Ellipsoidal Giunier-Preston Zones in the Case of Al-Zn
(a) Vol. **42**, No. 2, p. 687—697 (1977).
See also *1.3*, p. *22*; 21, p. 283; 21.1, p. 296.

A. Machová, V. Paidar, and F. Kroupa
*Theory of Elastic Constants of Fe–Si Alloys at 0 K
(a) Vol. **42**, No. 2, p. 713—720 (1977).
See also 21.1.1, p. 305.

G. A. Hassan
Effect of Pre-Torsional Deformation at Different Temperatures on the Creep Characteristics of Copper
(a) Vol. **42**, No. 2, p. K115—K117 (1977).
See also 21.1, p. 297.

C. Gault, A. Dauger, and P. Boch
*Variations of the Elastic Constants of Aluminium–Magnesium Single Crystals with Guinier-Preston Zones
(a) Vol. **43**, No. 2, p. 625—632 (1977).
See also 1.3, p. 23; 21, p. 286; 21.3, p. 310.

M. Yamaguchi and Y. Umakoshi
Slip Geometry and Anomalous Strengthening in β-Cu (ZnMn) Single Crystals
(a) Vol. **43**, No. 2, p. 667—674 (1977).
See also *10.1*, p. *81*; 21.1, p. 298.

J. Diehl, W. Mensch, and M. Weller
Correlations between Magnetic After-Effect and Internal Friction in α-Iron after Low Temperature Neutron Irradiation
(a) Vol. **43**, No. 2, p. K179—K182 (1977).
See also 10.1, p. 81; *11*, p. *98*; *18.2*, p. *223*; 21.1.1, p. 306.

J. Buchar and Z. Bílek
Amplitude and Frequency Response of Copper Single Crystals to One-Dimensional Transient Stress Waves
(a) Vol. **44**, No. 1, p. 117—125 (1977).
See also *10.1*, p. *81*; 21.1, p.299

P. Gondi, C. Patuelli, and R. Tognato
Dislocation Source Activation by Load Impulses
(a) Vol. **44**, No. 1, p. 333—338 (1977).
See also *10.1*, p. *82*; 21, p. 287.

E. Bonetti, E. Evangelista, and P. Gondi
*Kê Peak Analysis by Means of Internal Friction Measurements during Deformation
(a) Vol. **44**, No. 1, p. K31—K33 (1977).
See also *10.1*, p. *82*; 21, p. 287.

J. Parisot and G. Ferron
A Calculation of Dislocation Densities in Quenched α-Ion from Internal Friction Experiments
(a) Vol. **44**, No. 2, p. 549—555 (1977).
See also *10.1*, p. *82*; 21.1.1, p. 307.

N. Rusović and H. Warlimont
*The Elastic Behaviour of β_2-NiAl Alloys
(a) Vol. **44**, No. 2, p. 609—619 (1977).
See also 21.1, p. 300.

A. Wolfenden
*Mechanical Damping at 40 kHz in Mn–Cu Alloys
(a) Vol. **44**, No. 2, p. K171—K175 (1977).
See also 21.1, p. 301.

12.2 Non-Metals

Original Papers and Short Notes

S. ONARI, M. CARDONA, E. SCHÖNHERR, and W. STETTER
Effect of Stress on the Raman Spectra of Mg_2Si and Mg_2Sn
(b) Vol. **79**, No. 1, p. 269–276 (1977).
See also *6*, p. *51*; *20.1*, p. *247*; 22, p. 320.

M. N. SHARMA, A. KANT, and K. D. MISRA
A New Approach to Overlap Repulsion
(b) Vol. **79**, No. 1, p. 359–364 (1977).
See also *4*, p. *44*; 22.5.2, p. 375.

S. I. SUBBOTIN, V. V. PANFILOV, and R. T. MOLCHANOVA
Pressure-Induced Polytypic Transformations and Edge Absorption of GaSe
(a) Vol. **39**, No. 1, p. 357–362 (1977).
See also *1.2*, p. *14*; 13.5.1, p. 143; *20.1*, p. *248*; 22.3, p. 358.

V. I. SAVENKO, L. A. KACHANOVA, and E. D. SHCHUKIN
The Generation of Dislocations near the Surfaces of Crystals under Microindentation
(a) Vol. **39**, No. 1, p. K11–K15 (1977).
See also *1.6*, p. *31*; *10.2*, p. *83*; 22.5.2, p. 375.

B. N. GANGULY and M. NICOL
Effect of Hydrostatic Pressure on the Vibrational Properties and the Structure of $SrWO_4$ and $PbMoO_4$
(b) Vol. **79**, No. 2, p. 617–622 (1977).
See also 1.2, p. 14; *6*, p. *52*; 20.1, p. 249; 22.8.1, p. 402.

R. ALBRECHT, V. SCHMIDT, and V. I. BETEKHTIN
*The Damage Process Preceding Semi-Brittle Fracture in Dependence on Deformation (I): Scanning Microscopic In-Situ Studies of Microcracks in NaCl Crystals
(a) Vol. **39**, No. 2, p. 621–630 (1977).
See also *10.2*, p. *84*; 22.5.2, p. 376.

V. K. GARG, D. S. PURI, and M. P. VERMA
*On the Third-Order Elastic Constants of Ionic Solids with NaCl Structure
(b) Vol. **80**, No. 1, p. 63–72 (1977).
See also *10.2*, p. *85*; 22.5.2, p. 376.

A. D. BROTHERS and D. M. WIELICZKA
Temperature and Hydrostatic Pressure Measurements of the Exciton Peaks in BiI_3 and SbI_3
(b) Vol. **80**, No. 1, p. 201–206 (1977).
See also 13.1, p. 115; *13.5.1*, p. *144*; 22.5, p. 370.

R. ALBRECHT, V. SCHMIDT, and V. I. BETEKHTIN
*The Damage Process Preceding Semi-Brittle Fracture in Dependence on Deformation (II): A Discussion on Results of Microcrack Development in NaCl Crystals
(a) Vol. **40**, No. 1, p. 147–152 (1977).
See also *10.2*, p. *85*; 22.5.2, p. 376.

B. P. CHANDRA
*Stress Dependence of Triboluminescence in Crystals
(a) Vol. **40**, No. 2, p. 403–408 (1977).
See also *20.3*, p. *269*; 22.9, p. 415.

A. K. GOYAL and S. C. GOYAL
Dispersion of Phonons in NbC
(b) Vol. **81**, No. 1, p. K63–K65 (1977).
See also *6*, p. *56*; 22, p. 324.

J. Cisowski and W. Żdanowicz
Pressure and Temperature Dependence of the Hall Coefficient in $Cd_{3-x}Zn_xAs_2$ Solid Solutions
(a) Vol. **41**, No. 1, p. K59—K62 (1977).
See also *14.3*, p. *168*; 22.3.1, p. 360.

V. K. Garg and M. P. Verma
Evaluation of Some Important Lattice Sums on NaCl and CsCl Structures
(b) Vol. **82**, No. 1, p. 139—143 (1977).
See also *1.1*, p. *11*; 22.5.2, p. 379.

O. P. Sharma, A. P. Gupta, and Jai Shanker
Effect of the Three Body Interactions on the Cohesive Energies of Alkali Halides
(b) Vol. **82**, No. 1, p. K13—K16 (1977).
See also *4*, p. *45*; 22.5.2, p. 379.

V. B. Anzin, M. I. Eremets, Yu. V. Kosichkin, A. I. Nadezhdinskii, and A. M. Shirokov
Measurement of the Energy Gap in Tellurium under Pressure
(a) Vol. **42**, No. 1, p. 385—390 (1977).
See also 13.1, p. 120; *16*, p. *203*; *20.1*, p. *257*; 22.1.3, p. 346.

C. L. Reynorlds, Jr. and P. R. Couchman
*Tait's Equation of State at High Pressures and the Isothermal Bulk Modulus
(a) Vol. **42**, No. 1, p. K47—K52 (1977).
See also *12.1*, p. *102*; 21, p. 282; 22.5.2, p. 379; 23, p. 420.

I. Pelant and P. Sladký
*The Ratio of Photoelastic Constants p_{11}/p_{12} of AgCl
(a) Vol. **42**, No. 1, p. K57—K59 (1977).
See also 7, p. 63; *20.1*, p. *257*; 22.5.1, p. 373.

V. K. Garg, D. S. Puri, and M. P. Verma
*Many-Body Effects on the Fourth-Order Elastic Constants of Ionic Solids (II): Applications
(b) Vol. **82**, No. 2, p. 481—488 (1977).
See also *4*, p. *45*; 22.5.2, p. 379.

S. K. Chakrabarti
*On the Polymorphic Phase Transition in Alkali Halides
(b) Vol. **82**, No. 2, p. 557—564 (1977).
See also *1.2*, p. *17*; 22.5.2, p. 379.

V. A. Voloshin, and M. G. Krimnus
*Effect of Pressure on the Spectrum of Praseodymium-Doped Single Crystals
(b) Vol. **82**, No. 2, p. 695—700 (1977).
See also *20.1*, p. *258*; 22.5, p. 371; 22.8.1, p. 406.

W. Rehwald, J. R. Sandercock, and M. Rossinelli
*Elastic Properties of KCN and $K(CN)_{1-x}Cl_x$
(a) Vol. **42**, No. 2, p. 699—705 (1977).
See also 22, p. 327.

J. C. Phillips
The "Extra" Phase Transition in TTF-TCNQ
(b) Vol. **83**, No. 1, p. 123—127 (1977).
See also *1.2*, p. *18*; 1.4, p. 27; 22.9, p. 417.

G. N. Neilo, A. D. Prokhorov, and G. A. Tsintsadze
Electron Spin-Lattice Relaxation of Rare-Earth Ions in Lanthanium Magnesium Double Nitrate under Pressure
(b) Vol. **83**, No. 1, p. K59—K61 (1977).
See also *19*, p. *241*; 22.8.1, p. 406.

Y. Li and D. Y. Chung
*A Method for the Determination of the Elastic Constants for Crystals with Low Trigonal Symmetry
(a) Vol. **43**, No. 1, p. 329—332 (1977).
See also *7*, p. *63*; 22.6.1, p. 393.

Cao-Xuân An
Determination of the Elastic Constants of $SrCl_2$ by Brillouin Scattering and Study of Their Variations with Temperature
(a) Vol. **43**, No. 1, p. K69—K72 (1977).
See also 6, p. 59; 22.5.3, p. 384.

G. Busse and A. Pfaller
Stress Spectroscopy of a Resonant Bandmode with a Far Infrared Laser
(a) Vol. **43**, No. 1, p. K103—K107 (1977).
See also *20.1*, p. *260*; 22.5.2, p. 380.

S. C. Goyal and R. Prakash
*Many-Body Effects on the Second-Order Elastic Constants of Solids with ZnS Structure
(b) Vol. **83**, No. 2, p. 509—515 (1977).
See also *4*, p. *46*; 22, p. 328; 22.1, p. 332; 22.1.1, p. 335; 22.1.2, p. 342; 22.2, p. 348; 22.3, p. 358; 22.4, p. 361.

B. N. Ganguly, M. Nicol, and W. D. Ellenson
Raman Spectroscopic Studies of the High Pressure Phase Transformation of $NaClO_3$
(b) Vol **83**, No. 2, p. K115—K120 (1977).
See also *1.2*, p. *19*; *6*, p. *59*; 20.1, p. 260; 22.8.1, p. 407.

V. I. Zaitsev, V. I. Barbashov, and Yu. B. Tkachenko
*Effect of High Hydrostatic Pressure on the Dislocation Mobility in NaCl Crystals
(a) Vol. **44**, No. 1, p. K39—K41 (1977).
See also *10.2*, p. *92*; 22.5.2, p. 381.

S. Lara, R. Moreira Xavier, and C. A. Taft
*Remarks on the Effects of Pressure on the Energy Gaps in EuO and EuS
(b) Vol. **84**, No. 2, p. K143—K145 (1977).
See also *13.1*, p. *123*; 22, p. 330; 22.6, p. 391.

Y. Epelboin, M. Gandais, and C. Willaime
*Influence of the Elastic Anisotropy on the Contrast of Dislocation Images in Electron Microscopy and X-Ray Topography. Application to Feldspars
(a) Vol. **44**, No. 2, p. 651—659 (1977).
See also *10.2*, p. *93*; 22.8, p. 400.

13. Electron States

Original Papers and Short Notes

P. RENNERT and A. M. RADWAN
Structural Investigation of the Laves Phase $MgZn_2$ with Model Potential Calculations
(b) Vol. **79**, No. 1, p. 167—173 (1977).
See also *1.3*, p. *21*; 10.1, p. 75; 21.1, p. 288; 21.3, p. 310.

R. KISHORE
*Exactly Solvable Off-Diagonal Disorder
(b) Vol. **79**, No. 1, p. K9—K11 (1977).
See also *2*, p. *33*; 22, p. 320.

H.-G. HOLZHÜTTER
*On the Theory of Disorder Scattering in Mixed Crystal Systems
(b) Vol. **79**, No. 1, p. K13—K15 (1977).
See also 22.3.1, p. 360; 22.4.4, p. 368.

T. SOMA
*Electronic Screening Effect on the Structure Stability of Face-Centered Tetragonal In
(b) Vol. **79**, No. 1, p. K53—K56 (1977).
See also *4*, p. *44*; 21, p. 276.

D. P. BHATTACHARYA and Z. S. KACHLISHVILI
*On the Theory of the Impurity Breakdown in a Magnetic Field
(a) Vol. **39**, No. 1, p. 49—56 (1977).
See also *14.3.4*, p. *183*; 22, p. 321.

V. I. GREBENNIKOV, YU. A. BABANOV, and O. B. SOKOLOV
*Extra-Atomic Relaxation and X-Ray Spectra of Narrow-Band Metals (I): Formalism
(b) Vol. **79**, No. 2, p. 423—432 (1977).
See also 20, p. 244; 21, p. 277.

E. KOLLEY and W. KOLLEY
Calculation of the Temperature-Dependent DC Conductivity of Disordered Alloys in Off-Diagonal CPA
(b) Vol. **79**, No. 2, p. 461—472 (1977).
See also *2*, p. *33*; 6, p. 52; *14.1*, p. *157*; 21, p. 277.

A. MILCHEV and R. PICKENHAIN
The Quadratic Response of a Fermi Gas. Building Up of the Covalent Bonding in Zinc-blende Semiconductors. A Nondiagonal Density Matrix Treatment
(b) Vol. **79**, No. 2, p. 549—558 (1977).
See also *4*, p. *44*; 22.1, p. 332; 22.1.1, p. 333; 22.1,2, p. 338; 22.2.1, p. 350.

G. CUBIOTTI, E. DONATO, E. S. GIULIANO, and R. RUGGERI
*On the Influence of Disorder on the Electron Density of States in a Simplified Model for a Quasi-Onedimensional Conductor
(b) Vol. **79**, No. 2, p. 655—661 (1977).
See also *2*, p. *33*; 14, p. 155; 22, p. 321.

R. A. EVARESTOV and V. A. LOVCHIKOV
*Large Unit Cell Calculations of Solids in the CNDO Approximation
(b) Vol. **79**, No. 2, p. 743—751 (1977).
See also *4*, p. *44*; 22, p. 321.

I. YA. KORENBLIT
*Phase Transition and Instability of the Homogeneous State in Ferromagnetic Semiconductors
(b) Vol. **79**, No. 2, p. K95—K98 (1977).
See also 1.2, p. 14; *18.2*, p. *218*; 22, p. 321.

V. I. Litvinov and K. D. Tovstyuk
*On the Existence of the Off-Diagonal Long-Range Order in the Hubbard Model
(b) Vol. **79**, No. 2, p. K131—K135 (1977).
See also *1.2*, p. *15*.

J. Richter
Influence of Off-Diagonal Disorder on Density of States and Itinerant Ferromagnetism
(b) Vol. **79**, No. 2, p. K139—K142 (1977).
See also *2*, p. *34*; *18.2*, p. *218*.

I. M. Chapnik
*Compounds of Transition Metals with Possible Closed Electron Shells outside the Inert Gas Shells
(a) Vol. **39**, No. 2, p. K135—K138 (1977).
See also 21.1, p. 289; 21.1.1, p. 302; 22.7, p. 394.

V. I. Grebennikov, Yu. A. Babanov, and O. B. Sokolov
*Extra-Atomic Relaxation and X-Ray Spectra of Narrow-Band Metals (II): Results
(b) Vol. **80**, No. 1, p. 73—82 (1977).
See also 20, p. 244; 21, p. 278.

M. Guillot and M. Le Gall
Superexchange Field and Intense Magnetic Field Dependences of the Faraday Rotation Induced by the Electric Dipole Transitions in YIG
(b) Vol. **80**, No. 1, p. 279—287 (1977).
See also *18.3*, p. *228*; *20*, p. *244*; 22.8.2, p. 410.

G. V. Ionova, S. P. Ionov, O. M. Pachev, and E. F. Makarov
*Orbital-Charge Ordering in Crystals
(b) Vol. **80**, No. 1, p. K9—K12 (1977).
See also 22, p. 323.

S. P. Ionov, V. S. Lubimov, L. A. Manakova, and E. F. Makarov
*The Charge-Ordered State and Magnetic Collective Excitations
(b) Vol. **80**, No. 1, p. K13—K15 (1977).
See also 13.5, p. 141; 22, p. 323.

H. Böttger and V. V. Bryksin
A Theory of the Hall Effect in the Hopping Region in Disordered Systems (I): Rate Equation in Presence of Electrical and Magnetic Fields
(b) Vol. **80**, No. 2, p. 569—578 (1977).
See also 2, p. 35; 6, p. 54; 13.5.3, p. 151; *14.3*, p. *167*; 22, p. 323.

M. V. Eremin and Yu. V. Rakitin
*Channel Model in Isotropic Exchange Theory
(b) Vol. **80**, No. 2, p. 579—587 (1977).
See also 1, p. 9.

H. Neddermeyer
*Charge Transfer in Aluminium–Magnesium Alloys
(b) Vol. **80**, No. 2, p. 611—614 (1977).
See also 20, p. 244; 21, p. 279; 21.3, p. 310.

S. Krzemiński
*Random One-Dimensional Classical Magnet
(b) Vol. **80**, No. 2, p. 651—656 (1977).
See also 8, p. 65; *18.2*, p. *219*; 18.4, p. 232.

A. Murasik, J. Leciejewicz, and Z. Żołnierek
*An Observation of the Crystal-Field Transition in UCl_4 by Neutron Spectroscopy
(b) Vol. **80**, No. 2, p. K137—K140 (1977).
See also *20.1*, p. *252*; 22.5, p. 370.

H. Böttger and V. V. Bryksin
A Theory of the Hall Effect in the Hopping Region in Disordered Systems (II): Random Resistor Network and Hall Effect
(b) Vol. **81**, No. 1, p. 97–106 (1977).
See also 2, p. 36; 6, p. 55; 13.5.3, p. 151; *14.3*, p. *167*; 22, p. 324.

H.-G. Holzhütter and M. Mocker
*Generalized Boltzmann Equation with Field-Dependent Collison Terms Inclusive of Interband Transitions
(b) Vol. **81**, No. 1, p. 139–146 (1977).
See also *14.3.4*, p. *184*.

M. Brauer
*Dynamical Response of Electrons in GaAs at 300 K
(b) Vol. **81**, No. 1, p. 147–152 (1977).
See also *14.3.4*, p. *185*; 22.2.1, p. 351.

F. Kammler and K. Elk
Numerical Results of the Hubbard III Approximation for Arbitrary Values of the Correlation Energy and the Electron Density (with Application to Disordered Alloys)
(b) Vol. **81**, No. 1, p. 179–184 (1977).
See also 2, p. 36; *18.2*, p. *219*.

B. S. Tošić, M. M. Marinković, and S. Berar
*The Boson Field Analysis of the Critical Temperature Region in Ferromagnets
(b) Vol. **81**, No. 1, p. 245–252 (1977).
See also 18.2, p. 219.

E. F. Kustov
*Reduction Symmetry Method in Crystal Field Theory (III)
(b) Vol. **81**, No. 2, p. 421–432 (1977).

H. Böttger and V. V. Bryksin
A Theory of the Hall Effect in the Hopping Region in Disordered Systems (III): DC Hall Effect
(b) Vol. **81**, No. 2, p. 433–441 (1977).
See also 2, p. 37; 6, p. 56; 13.5.3, p. 151; *14.3*, p. *168*; 22, p. 325.

F. M. Galperin
*Hyperfine Field on the Nucleus of an Atom of a Transition d Element (Metal), Magnetic Moment and Electron Structure of a Ferromagnetic Metal Atom
(b) Vol. **81**, No. 2, p. 545–550 (1977).
See also 18.2, p. 220; 21.1, p. 293.

T. Chiba, G. B. Dürr, and W. Brandt
*Enhancement Effect on Positron Annihilation with Tightly Bound Electrons
(b) Vol. **81**, No. 2, p. 609–614 (1977).
See also 22.5.2, p. 378; 22.6, p. 388.

G. V. Ionova, E. F. Makarov, and S. P. Ionov
*Statical Charge Waves in Crystals
(b) Vol. **81**, No. 2, p. 671–679 (1977).

S. P. Ionov, V. S. Lubimov, L. A. Manakova, and E. F. Makarov
*The Electronic Spectra of Charge-Ordered Crystals
(b) Vol. **81**, No. 2, p. 691–701 (1977).
See also 13.5, p. 141; 22, p. 325.

E. Kolley and W. Kolley
*Coherent Ladder Approximation for the Random Hubbard Model
(b) Vol. **81**, No. 2, p. 735–745 (1977).
See also 21, p. 281.

A. Balogh and I. Dézsi
*Further Positron Lifetimes of Some Elements
(b) Vol. **81**, No. 2, p. K81—K82 (1977).
See also 21, p. 281; 22.1.3, p. 345.

V. S. Ostrovskii and V. N. Kharkyanen
*On a Possibility of Direct Orbital Ordering in Crystals Containing Degenerate Centres
(b) Vol. **81**, No. 2, p. K103—K106 (1977).

K. S. Dy
*Generalized Friedel Sum Rule
(b) Vol. **81**, No. 2, p. K111—K113 (1977).
See also *2*, p. *37*.

M. V. Eremin and Yu. V. Rakitin
*Channel Model in Isotropic Exchange Theory (II)
(b) Vol. **82**, No. 1, p. 221—228 (1977).
See also 1, p. 10; 22.5.4, p. 385; 22.6, p. 389.

N. Nhu Dat and N. van Hieu
*On the High-Order Harmonic Generation by Free Charge Carriers in Solids
(b) Vol. **82**, No. 1, p. K39—K40 (1977).
See also *20*, p. *245*.

B. Szpunar and P.-A. Lindgård
Theory of Temperature Dependence of the Magnetization in Rare-Earth–Transition-Metal Alloys
(b) Vol. **82**, No. 2, p. 449—456 (1977).
See also *18.1*, p. *214*; *18.4*, p. *233*; 21.1, p. 296; 21.4, p. 312.

E. El-Mamoun and G. Röpke
*Exchange Energy of a Strongly Inhomogeneous Electron Gas
(b) Vol. **82**, No. 2, p. 617—625 (1977).

W. John
*Comment on the Paper "Exactly Solvable Off-Diagonal Disorder"
(b) Vol. **82**, No. 2, p. K111—K112 (1977).
See also *2*, p. *38*; 22, p. 327.

I. P. Zvyagin
Correlation Effects in Hopping Conduction
(b) Vol. **83**, No. 1, p. 63—70 (1977).
See also 2, p. 38; *14.3*, p. *170*; 15, p. 199; 22, p. 327.

R. S. Tripathi, L. L. Moseley, and T. Lukes
*Optical Absorption in Non-Crystalline Semiconductors
(b) Vol. **83**, No. 1, p. 197—207 (1977).
See also 2, p. 38; *20.1*, p. *259*; 22, p. 327.

L. G. Ferreira, S. R. Salinas, and M. J. Oliveira
*On a Variational Procedure for Obtaining the Thermodynamic Properties of Statistical Models
(b) Vol. **83**, No. 1, p. 229—237 (1977).
See also *18*, p. *212*.

O. B. Sokolov, V. I. Grebennikov, and E. A. Turov
*A New Approach to the Theory of X-Ray Spectra of Metals with Allowance for the Finite Lifetime of the Deep Hole
(b) Vol. **83**, No.1, p. 281—291 (1977).
See also 20, p. 246; 21, p. 284.

I. Ya. Dekhtyar and V. I. Shevchenko
*Positron Annihilation in Hydrides of Transition Metals
(b) Vol. **83**, No. 1, p. 323—330 (1977).
See also 21, p. 284 .

S. A. Vorontsov, D. G. Dolgopolov, V. I. Poltoratskii, and I. V. Svechkarev
*Magnetic Susceptibility Anisotropy of Holes in Tellurium
(b) Vol. **83**, No. 1, p. 341–346 (1977).
See also *18.1*, p. *215*; 22.1.3, p. 346.

S. Guha and P. K. Sen
Two-Stream Instability in n-Type Gallium Arsenide
(a) Vol. **43**, No. 1, p. 89–97 (1977).
See also 6, p. 59; *14.3.4*, p. *186*; 22.2.1, p. 353.

Y. Uchida and F. Fujimoto
Electron Channeling in ⟨111⟩ Direction of Tungsten Crystals (I): Energy Range of 200 to 1000 keV
(a) Vol. **43**, No. 1, p. 269–275 (1977).
See also *11*, p. *97*; 21, p. 285.

G. U. Reinsperger and W. Ludwig
*The Contribution of Once Scattered Electrons to the Transmission Coefficient of Hot Electrons in Tunnel Triodes
(a) Vol. **43**, No. 1, p. K77–K80 (1977).
See also *14.3.4*, p. *186*; 22.6, p. 391.

O. B. Sokolov, V. I. Grebennikov, and E. A. Turov
*Resonance Scattering of X-Rays in Metals
(b) Vol. **83**, No. 2, p. 383–393 (1977).
See also 20, p. 246; 21, p. 285.

G. Paasch and P. Rennert
*Quadratic Electronic Polarizability of the Interacting Electron Gas
(b) Vol. **83**, No. 2, p. 501–508 (1977).
See also *14.4*, p. *192*.

V. G. Aleshin, T. I. Serebryakova and A. I. Kharlamov
*Ladungszustand von Bor in Boriden der Übergangsmetalle
(Charge State of Boron in Transition Metal Borides)
(b) Vol. **83**, No. 2, p. 537–542 (1977).
See also 17, p. 207; 22, p. 328.

Yu. G. Borodko, M. G. Kaplunov, T. M. Moravskaya, and K. I. Pokhodnya
X-Ray Photoelectron Spectra of Single Crystals of Tetracyanoquinodimethane and Its Salts
(b) Vol. **83**, No. 2, p. K141–K144 (1977).
See also 17, p. 207; *20*, p. *246*; 22.9, p. 418.

M. P. Das
*An Effective Periodic Potential for Disordered Binary Alloys
(b) Vol. **83**, No. 2, p. K165–K168 (1977).
See also 2, p. 39; 21, p. 285.

I. V. Aleksandrov and R. G. Gabrielyan
*On the Theory of Spin–Lattice Relaxation Due to Diffusion to Impurities in Solids and Liquids
(b) Vol. **83**, No. 2, p. K173–K177 (1977).
See also 9, p. 70; *18*, p. *212*.

T. O. Gegechkori and Z. S. Kachlishvili
*Hot Electron Current-Voltage Curves in a Transverse Magnetic Field with Account of Phonon Heating
(a) Vol. **43**, No. 2, p. 513–518 (1977).
See also 6, p. 59; *14.3.4*, p. *187*.

Y. Kimura
On the Relation between the Density of States in the Cubic State and Phase Transition Temperatures (T_c, T_m) of V–Zr–Nb Ternary Laves Compounds
(a) Vol. **43**, No. 2, p. K141—K146 (1977).
See also *1.2*, p. *19*; 8, p. 67; *14.2*, p. *163*; 21.1, p. 298.

C. B. Satterthwaite
On Isotope Effects in $PdH(D)_x$
(a) Vol. **43**, No. 2, p. K147—K149 (1977).
See also 8, p. 67; *14.2*, p. *163*; 21, p. 286.

M. A. Kumakhov
Theory of Radiation of Charged Particles Channeled in a Crystal
(b) Vol. **84**, No. 1, p. 41—54 (1977).
See also *11*, p. *98*; 22.1.2, p. 342.

L. A. Manakova, E. F. Makarov, and S. P. Ionov
*The Electronic Spectra and Electromagnetic Properties of Charge-Ordered Systems
(b) Vol. **84**, No. 1, p. 55—67 (1977).
See also 13.5, p. 142; 22, p. 329.

V. L. Bonch-Bruevich
*Coulomb Gap in a Disordered Semiconductor
(b) Vol. **84**, No. 1, p. 99—105 (1977).
See also 14.3, p. 171; 22, p. 329.

A. A. Kornyshev, A. I. Rubinshtein, and M. A. Vorotyntsev
*Image Potential near a Dielectric–Plasma-Like Medium Interface
(b) Vol. **84**, No. 1, p. 125—132 (1977).
See also *14.4*, p. *192*.

L. Dagens
*Calculation of a Resonant Model Potential for Copper, Silver, and Gold
(b) Vol. **84**, No. 1, p. 311—324 (1977).
See also *4*, p. *46*; 21.1, p. 299; 21.6, p. 317.

Nguyen van Trong and Ha Vinh Tan
*On the Quantum Theory of Anomalous Electromagnetic Fluctuations in Electron–Phonon Systems
(b) Vol. **84**, No. 2, p. 345—352 (1977).
See also *6*, p. *60*.

L. A. Manakova, E. F. Makarov, V. S. Lubimov, and S. P. Ionov
*Some Peculiarities of Charge-Ordered Systems at Constant Electric Field
(b) Vol. **84**, No. 1, p. K1—K6 (1977).
See also 13.5, p. 142; 22, p. 330.

J. L. Vicent
Conduction Electrons and s–d Exchange Interaction
(b) Vol. **84**, No. 1, p. K71—K74 (1977).
See also *14.1*, p. *160*; *18.2.1*, p. *226*; 21.1, p. 299.

J. Mizia
*CPA Treatment of the Extendet Hubbard Model. Antiferromagnetic Solution
(b) Vol. **84**, No. 2, p. 449—455 (1977).
See also *18.4*, p. *234*.

G. V. Ionova, S. P. Ionov, O. M. Pachev, and E. F. Makarov
*Orbital–Charge Ordering in Crystals
(b) Vol. **84**, No. 2, p. 493—501 (1977).
See also 13.4, p. 139.

M. A. Kumakhov and R. Wedell
Theory of Radiation of Relativistic Channelled Particles
(b) Vol. **84**, No. 2, p. 581—593 (1977).
See also *11*, p. *99*; 22.1.2, p. 344.

H. J. Fischbeck
*The Diagonal Vertex Approximation for Gaussian Random Systems as a Further Test of the CPA
(b) Vol. **84**, No. 2, p 603—609 (1977).
See also **14**, p. 156.

K. A. Chao, J. Spałek, and A. M. Oleś
*The Kinetic Exchange Interaction in Doubly Degenerate Narrow Bands
(b) Vol. **84**, No. 2, p. 747—759 (1977).
See also **18.2**, p. 224; **18.4**, p. 234.

V. I. Grebennikov, O. B. Sokolov, and E. A. Turov
*Response of Metal Electrons to Local Fields of Holes Arising in X-Ray Processes
(b) Vol. **84**, No. 2, p. 773—784 (1977).
See also **20**, p. 246; **21**, p. 287.

E. Majerníková
*Quantum Shift of the Electron Density of States in a Gaussian Random Potential
(b) Vol. **84**, No. 2, p. K109—K111 (1977).
See also *2*, p. *40*; **22**, p. 330.

E. Kolley and W. Kolley
*On the Spin Susceptibility for Itinerant Electrons with Energy-Dependent Effective Interaction
(b) Vol. **84**, No. 2, p. K147—K150 (1977).
See also **18.1**, p. 216; **18.2**, p. 224.

13.1 Band Structure

Review Articles

M. I. Klinger and A. A. Samokhvalov
Electron Conduction in Magnetite and Ferrites
(b) Vol. **79**, No. 1, p. 9—48 (1977).
Erratum
(b) Vol. **83**, No. 1, p. 353 (1977).
See also *13.5.3*, p. *151*; *14.3*, p. *165*; 15, p. 198; 18.3, p. 228; 22.8.2, p. 409.

Yu. V. Knyazev and M. M. Noskov
The Optical Properties of Rare Earth Metals
(b) Vol. **80**, No. 1, p. 11—29 (1977).
See also *20.1*, p. *247*; 21.4, p. 311.

A. Goldmann
*Band Structure and Optical Properties of Tetrahedrally Coordinated Cu- and Ag-Halides
(b) Vol. **81**, No. 1, p. 9—47 (1977).
See also 13.5.1, p. 143; *20.1*, p. *247*; 22.5, p. 370; 22.5.1, p. 373.

Original Papers and Short Notes

D. Glötzel and L. Fritsche
*Volume Dependence of the Electronic Structure of Lanthanum and Cerium
(b) Vol. **79**, No. 1, p. 85—92 (1977).
See also 13.2, p. 124; 21.4, p. 311.

D. Theis
*Wavelength-Modulated Reflectivity Spectra of ZnSe and ZnS from 2.5 to 8 eV
(b) Vol. **79**, No. 1, p. 125—130 (1977).
See also 13.5.1, p. 143; *20.1*, p. *247*; 22.4.1, p. 362; 22.4.2, p. 365.

I. Pollini
*Ultraviolet Reflectivity Spectra of $CoCl_2$ and $CoBr_2$ Crystals
(b) Vol. **79**, No. 1, p. 231—236 (1977).
See also *20.1*, p. *247*; 22.5, p. 370.

P. O. Nilsson, I. Curelaru, and T. Jarlborg
*XPS and APS Studies of V_3Si
(b) Vol. **79**, No. 1, p. 277—281 (1977).
See also 20, p. 244; 21.1, p. 288.

A. N. Chakravarti and A. K. Chowdhury
Influence of Band Non-Parabolicity on the Oscillatory Dependence of the Diffusivity–Mobility Ratio in Degenerate Semiconductors on Magnetic Quantization
(b) Vol. **79**, No. 1, p. K31—K34 (1977).
See also *14.3*, p. *165*; 18, p. 209; 22, p. 320.

Y. Khan, K. Nahm, M. Rosenberg, and H. Willner
Superconductivity and Semiconductor–Metal Phase Transition in the System $BaPb_{1-x}Bi_xO_3$
(a) Vol. **39**, No. 1, p. 79—88 (1977).
See also 1, p. 9; *14.2*, p. *161*; 20.1, p. 248; 22.6.1, p. 393.

J. Bauer, L. Biste, and D. Bolze
Optical Properties of Aluminium Nitride Prepared by Chemical and Plasmachemical Vapour Deposition
(a) Vol. **39**, No. 1, p. 173—181 (1977).
See also *1.5*, p. *29*; 20.1, p. 248; 22.2, p. 348.

N. C. HALDER, J. M. MONTANTE, and W. E. SWARTZ, JR.
Evidence of a Polycrystalline Phase in Amorphous Ge Films Evaporated at Room Temperature
(a) Vol. **39**, No. 1, p. 213—222 (1977).
See also *1.5*, p. *29*; *2*, p. *33*; 14.3.1, p. 173; 22.1.1, p. 333.

J. VON BORZESZKOWSKI and J. SCHOLZ
On the Electric Field Dependence of the Conductivity of Holes in Si at Low Temperatures (II): The Influence of the Variation of Steady-State Hole Density with Electric Field on the Conductivity
(b) Vol. **79**, No. 2, p. 525—531 (1977).
See also 6, p. 52; *14.3.4*, p. *183*; 22.1.2, p. 337.

G. BASTARD, C. RIGAUX, and A. MYCIELSKI
*Giant Spin Splitting Induced by Exchange Interactions in $Hg_{1-k}Mn_kTe$ Mixed Crystals
(b) Vol. **79**, No. 2, p. 585—593 (1977).
See also 18, p. 209; 20.1, p. 249; 22.4.4, p. 368.

W. SZUSZKIEWICZ
Optical Absorption inHgTe
(b) Vol. **79**, No. 2, p. 691—698 (1977).
See also *20.1*, p. *249*; 22.4.3, p. 367.

S. H. EL-MAHALAWY and B. L. EVANS
Temperature Dependence of the Electrical Conductivity and Hall Coefficient in 2H-MoS_2, $MoSe_2$, WSe_2, and $MoTe_2$
(b) Vol. **79**, No. 2, p. 713—722 (1977).
See also *14.3*, p. *166*; 22, p. 321.

J. BAUER
Optical Properties, Band Gap, and Surface Roughness of Si_3N_4
(a) Vol. **39**, No. 2, p. 411—418 (1977).
See also *2*, p. *34*; 20.1, p. 249; 22, p. 322.

A. A. ADBUMALIKOV and A. A. KLOCHIKHIN
The Double Resonance in Two-Phonon Raman Scattering
(b) Vol. **80**, No. 1, p. 43—50 (1977).
See also *6*, p. *53*; *20.1*, p. *250*.

A. D. BROTHERS and D. M. WIELICZKA
Temperature and Hydrostatic Pressure Measurements of the Exciton Peaks in BiI_3 and SbI_3
(b) Vol. **80**, No. 1, p. 201—206 (1977).
See also 12.2, p. 104; *13.5.1*, p. *144*; 22.5, p. 370.

R. HERRMANN and P. GOY
*Investigation of Plasma Effects in Bismuth–Antimony Alloys in the Submillimeter Wave Range
(b) Vol. **80**, No. 1, p. 207—213 (1977).
See also *19*, p. *236*; 21.7, p. 319.

R. ENDERLEIN and F. BECHSTEDT
Theory of Allowed Resonance Raman Scattering of First and Second Order in a Magnetic Field
b) Vol. **80**, No. 1, p. 225—234 (1977).
See also *6*, p. *53*; 13.5.1, p. 144; *20.1*, p. *251*.

J. L. FARVACQUE and P. LENGLART
Inhomogeneous Strain Field Scattering Deformation Potential (I): Theory
(b) Vol. **80**, No. 1, p. 361—367 (1977).
See also *10*, p. *72*; *14*, p. *155*.

V. K. Bashenov, Kh. S. Bagdasarov, and V. V. Timofeenko
*Universal Valence Bands for Fluorite-Type Crystals
(b) Vol. **80**, No. 1, p. K17—K20 (1977).
See also 20.1, p. 251; 22.5.3, p. 383.

W. Hörig, H. Neumann, H.-J. Höbler, and G. Kühn
Temperature Dependence of the Absorption Edge in $CuInSe_2$
(b) Vol. **80**, No. 1, p. K21—K24 (1977).
See also *20.1*, p. *251*; 22.8, p. 397.

H. Köhler
Investigation of the Magnetoresistance of n-Type Bi_2Te_3 under Uniaxial Stress
(b) Vol. **80**, No. 1, p. K29—K32 (1977).
See also *14.3*, p. *166*; 20.1, p. 251; 22.7, p. 394.

K. Hübner
*Chemical Bond and Related Properties of SiO_2 (I): Character of the Chemical Bond
(a) Vol. **40**, No. 1, p. 133—140 (1977).
See also 2, p. 35; *4*, p. *44*; 14.4, p. 189; 20.1, p. 251; 22.6, p. 387.

K. Takasaki and S. Tanaka
Tunneling Study of $Pb_{1-x}Sn_xTe$
(a) Vol. **40**, No. 1, p. 173—180 (1977).
See also 6, p. 54; *14.3.2*, p. *176*; 22.8, p. 397.

W. Lochmann
*Phonon-Assisted Auger Recombination in Semiconductors
(a) Vol. **40**, No. 1, p. 285—292 (1977).
See also *6*, p. *54*; 20, p. 244; 22.2.1, p. 351.

F. L. Battye, A. Goldmann, and L. Kasper
*Ultraviolet Photoelectron Valence Band Studies on Phthalocyanine Compounds
(b) Vol. **80**, No. 2, p. 425—432 (1977).
See also 17, p. 206; 22.9, p. 415.

K. Bärner, R. Braunstein, and E. Chock
The Optical Properties of MnAs and $MnAs_{0.85}P_{0.15}$
(b) Vol. **80**, No. 2, p. 451—460 (1977).
See also *14.4*, p. *190*; *20.1*, p. *252*; 22.8, p. 397.

U. Büchner and W. Guse
The Dielectric Tensor Components of Czochralski-Grown 2:1Mullite Single Crystals
(b) Vol. **80**, No. 2, p. 615—618 (1977).
See also *14.4*, p. *190*; 22.6.1, p. 393.

F. P. Koffyberg
Optical and Modulated Absorption Spectra in the Band Gap Region of Amorphous Vanadium Zinc Borate Semiconductors
(b) Vol. **80**, No. 2, p. 669—673 (1977).
See also 2, p. 35; *20.1*, p. *252*; 22.8, p. 397.

P. N. Dyachkov and A. A. Levin
*Bond Orbital Model for Tetrahedral Semiconductors on the Base of Molecular Parameters. Covalent Crystals and Perturbative Treatment of Partially Ionic Semiconductors
(b) Vol. **80**, No. 2, p. 681—690 (1977).
See also *4*, p. *44*; 22.1.2, p. 339; 22.1.3, p. 345; 22.2, p. 348; 22.4, p. 361.

V. K. Bashenov, M. Bardashova, and A. M. Mutal
*Baldereschi Points for Some Noncubic Lattices
(b) Vol. **80**, No. 2, p. K89—K93 (1977).
See also 22, p. 324.

F. WERFEL, G. DRÄGER, O. BRÜMMER, and M. JURISCH
*XS and XPS Investigation of the V_3Si Electron Structure
(b) Vol. **80**, No. 2, p. K95—K99 (1977).
See also 17, p. 206; 20, p. 245; 22.7, p. 394.

K. HÜBNER
*Chemical Bond and Related Properties of SiO_2 (II): Structural Trends
(a) Vol. **40**, No. 2, p. 487—495 (1977).
See also 2, p. 36; *4*, p. *45*; 14.4, p. 190; 20.1, p. 252; 22.6, p. 387.

R. KATZSCHMANN, R. KRANOLD, and A. REHFELD
*Optical Anomaly of Small Particles in Glasses
(a) Vol. **40**, No. 2, p. K161 (1977).
See also *20.1*, p. *253*; 22.4.2, p. 365.

L. RESCA and R. RESTA
*Energy Bands in Cubic Ice. Ab Initio Calculation Using the Method of Linear Combination of Molecular Orbitals
(b) Vol. **81**, No. 1, p. 129—138 (1977).
See also 22.6, p. 387.

H. VENGHAUS
Infrared Reflectance and Dielectric Properties of Pyrolytic Graphite for $\boldsymbol{E} \parallel \boldsymbol{c}$ Polarization
(b) Vol. **81**, No. 1, p. 221—225 (1977).
See also 13.5.2, p. 149; *20.1*, p. *253*; 22.1, p. 332.

U. BÜCHNER
Wave-Vector Dependence of the Electron Energy Losses of Boron Nitride and Graphite
(b) Vol. **81**, No. 1, p. 227—234 (1977).
See also *20.1*, p. *253*; 22.1, p. 332; 22.2, p. 348.

O. GOEDE
Deformation Effects of Biexcitons in Wurtzite-Type Crystals
(b) Vol. **81**, No. 1, p. 235—244 (1977).
See also *13.5.1*, p. *145*; *20.1*, p. *253*; 22.4.1, p. 363.

D. CHATTOPADHYAY
Average Energy of Conduction Electrons in Non-Parabolic Semiconductors
(b) Vol. **81**, No. 1, p. K51—K53 (1977).
See also *14.3*, p. *168*; 22, p. 324.

V. K. BASHENOV, I. BAUMANN, and V. V. TIMOFEENKO
*Upper Valence Bands of Rutile-Type Fluorides
(b) Vol. **81**, No. 1, p. K55—K58 (1977).
See also 22.5, p. 371.

H. WEINERT, H. NEUMANN, H.-J. HÖBLER, G. KÜHN, and NGUYEN VAN NAM
*Infrared Faraday Effect in n-Type $CuInSe_2$
(b) Vol. **81**, No. 1, p. K59—K61 (1977).
See also *20.1*, p. *254*; 22.8, p. 398.

K. SIERAŃSKI, J. SZATKOWSKI, J. M. PAWLIKOWSKI, E. MAJCHROWSKA, and E. PŁACZEK-POPKO
*Fermi Level in $Cd_xHg_{1-x}Te$
(b) Vol. **81**, No. 1, p. K67—K70 (1977).
See also 14.3, p. 168; 22.4.4, p. 368.

M. L. YOUNG and C. P. FENDER
Sub-Threshold Velocity–Field Characteristics of $InAs_{1-x}P_x$ Alloys
(a) Vol. **41**, No. 1, p. 85—94 (1977).
See also *14.3.4*, p. *185*; 22.2.4, p. 357.

V. A. STUKAN, V. I. TRIFONOV, V. YA. SHEVCHENKO, and G. I. GONCHARENKO
Electrical Properties of p-CdSb at Low Temperatures
(a) Vol. **41**, No. 1, p. 307—315 (1977).
See also 13.4, p. 134; *14.3*, p. *168*; 22.3, p. 358.

S. Chatterjee and P. Sinha
*Orthogonalized Augmented Plane-Wave Method for the Solution of the Energy Band Problem in Solids
(b) Vol. **81**, No. 2, p. 493—502 (1977).
See also 22.4.1, p. 363.

R. Nürnberger, F.-J. Himpsel, E. E. Koch, and N. Schwentner
*Band Formation in Xenon–Argon Alloys Studied by Photoelectron Spectroscopy
(b) Vol. **81**, No. 2, p. 503—509 (1977).
See also 17, p. 206; 23, p. 420.

H.-R. Trebin
Polarons in the Degenerate-Band Case (II): Weak to Intermediate Hole-Phonon Coupling
(b) Vol. **81**, No. 2, p. 527—534 (1977).
See also *13.5.3*, p. *151*; 22.2, p. 348; 22.4, p. 361.

W. Blau, S. Mager, and E. Wieser
Determination of Localized Magnetic Moments in Fe–Cr–Al Alloys and the Electron Structure
(b) Vol. **81**, No. 2, p. 535—544 (1977).
See also *18.2*, p. *220*; *18.4*, p. *233*; 21.1.1, p. 303.

Y. Guldner, C. Rigaux, A. Mycielski, and Y. Couder
Magnetooptical Investigation of $Hg_{1-x}Cd_xTe$ Mixed Crystals (I): Semimetallic Configuration
(b) Vol. **81**, No. 2, p. 615—627 (1977).
See also *18*, p. *210*; *20.1*, p. *255*; 22.4.4, p. 368.

W. Thielemann, and F. Bergauer
Test of the Random Fluctuations of the Conduction Band Edge in Heavily Doped n-Type GaAs by Means of the Magnetic Circular Dichroism
(b) Vol. **81**, No. 2, p. K91—K94 (1977).
See also *18*, p. *211*; *20.1*, p. *255*; 22.2.1, p. 352.

W. Szuszkiewicz
*Direct Determination of the Negative Energy Gap E_0 for Semiconductors with Inverted Band Structure
(b) Vol. **81**, No. 2, p. K119—K123 (1977).
See also 20.1, p. 255; 22.4.3, p. 367.

M. Zoaeter, A. Conan, and D. Delaunay
Transport Measurements and Band Structure of Bulk $MoSe_{2-x}$ Compounds
(a) Vol. **41**, No. 2, p. 629—635 (1977).
See also *14.3*, p. *169*; *15*, p. *199*; 22, p. 325.

N. N. Sirota, I. V. Bodnar, and G. F. Smirnova
*Lattice Reflection Spectra of $(GaP)_x(InAs)_{1-x}$ Solid Solutions
(a) Vol. **41**, No. 2, p. 669—672 (1977).
See also *20.1*, p. *256*; 22.2.4, p. 357.

K. M. Demchuk and I. M. Tsidilkovskii
*Scattering of Electrons by the Deformation Potential in Doped InSb
(b) Vol. **82**, No. 1, p. 59—67 (1977).
See also 14.3, p. 169; 15, p. 199; 22.2.3, p. 356.

Y. Guldner, C. Rigaux, A. Mycielski, and Y. Couder
Magnetooptical Investigation of $Hg_{1-x}Cd_xTe$ Mixed Crystals (II): Semiconducting Configuration and Semimetal → Semiconductor Transition
(b) Vol. **82**, No. 1, p. 149 — 158 (1977).
See also *18*, p. *211*; *20.1*, p. *256*; 22.4.4, p. 368.

E. Gratz and C. A. Poldy
The Influence of Antiferromagnetic Order on the Evaluation of Spin Disorder Resistivity
(b) Vol. **82**, No. 1, p. 159—167 (1977).
See also 14.1, p. 158; *18.2*, p. *221*; *18.4*, p. *233*; 21.1, p. 295; 21.4, p. 312.

W. Schülke
*The One-Dimensional Fourier Transform of Compton Profiles. Theorems and Applications
(b) Vol. 82, No. 1, p. 229–235 (1977).
See also *13.2*, p. *124*; 21.2, p. 308; 22.1.2, p. 340.

Jai Singh
*Calculation of the Energy Band Structures and Carrier Mobilities in γ-Platinum Phthalocyanine Single Crystals
(b) Vol. 82, No. 1, p. 263–268 (1977).
See also 14.3, p. 169; 22.9, p. 416.

I. Gorczyca
*Spin-Dependent Wave Functions and Band Parameters at the Γ Point of InSb
(b) Vol. 82, No. 1, p. 279–288 (1977).
See also 22.2.3, p. 356.

J. Mertsching
*Transformations of the Hubbard Model
(b) Vol. 82, No. 1, p. 289–296 (1977).
See also *18*, p. *211*.

J. R. Chelikowsky, M. Schlüter, and M. L. Cohen
*Semi-Empirical Pseudopotential Calculation for the Electronic Structure of Polymeric Sulfur Nitride
(b) Vol. 82, No. 1, p. 357–367 (1977).
See also 14, p. 155; 22, p. 326.

M. v. Ortenberg and L. G. Rubin
*Submillimeter Cyclotron Resonance of Electrons in Stressed Silicon at High Magnetic Fields
(b) Vol. 82, No. 1, p. K1–K4 (1977).
See also *19*, p. *240*; 22.1.2, p. 340.

R. Taranko
*The Effect of the Model Hybridized Energy Bands on the Field Emitted Electrons from Metals
(b) Vol. 82, No. 1, p. K5–K7 (1977).
See also *17*, p. *206*; 21, p. 282.

J. Zieliński
*Delocalization of Electrons in a Double d-Band Model of Transition Metals
(b) Vol. 82, No. 1, p. K17–K19 (1977).
See also *18.2*, p. *221*; 21.1, p. 295.

S. Lara, R. Moreira Xavier, and C. A. Taft
*On the Indirect Exchange in Intrinsic Semiconductors
(b) Vol. 82, No. 1, p. K21–K22 (1977).
See also 22, p. 326.

W. Schnitker and F. R. Kessler
Piezoreflectance at the E_1 and $E_1 + \Delta_1$ Edges of Ge in High Magnetic Fields
(b) Vol. 82, No. 1, p. K35–K37 (1977).
See also *18*, p. *211*; *20.1*, p. *257*; 22.1.1, p. 335.

G. Ambrazevičius, G. Babonas, and A. Šileika
*Thermoreflectance Spectra of $CdSiP_2$
(b) Vol. 82, No. 1, p. K45–K48 (1977).
See also *20.1*, p. *257*; 22.8, p. 399.

D. Mergel and R. Labusch
Optical Excitation of Dislocation States in Germanium (II): Analysis of the Experimental Results
(a) Vol. 42, No. 1, p. 165–171 (1977).
See also *13.4*, p. *136*; *16*, p. *203*; 22.1.1, p. 335.

R. K. Jain
*Calculation of the Fermi Level, Minority Carrier Concentration, Effective Intrinsic Concentration, and Einstein Relation in n- and p-Type Germanium and Silicon
(a) Vol. 42, No. 1, p. 221–226 (1977).
See also *14.3*, p. *169*; 22.1.1, p. 335; 22.1.2, p. 340.

T. Suzuki, M. Hirose, S. Ogose, and Y. Osaka
Electrical Transport and Photoconductivity in Amorphous Silicon
(a) Vol. 42, No. 1, p. 337–344 (1977).
See also 2, p. 37; *14.3.1*, p. *174*; 15, p. 199; *16*, p. *203*; 22.1.2, p. 340.

V. B. Anzin, M. I. Eremets, Yu. V. Kosichkin, A. I. Nadezhdinskii, and A. M. Shirokov
Measurement of the Energy Gap in Tellurium under Pressure
(a) Vol. 42, No. 1, p. 385–390 (1977).
See also 12.2, p. 105; *16*, p. *203*; *20.1*, p. *257*; 22.1.3, p. 346.

H. Stolze, M. Lutz, and P. Grosse
*The Optical Activity of Tellurium
(b) Vol. 82, No. 2, p. 457–466 (1977).
See also *20.1*, p. *257*; 22.1.3, p. 346.

L. C. R. Alfred
*Charge Screening in an Electron Gas of Arbitrary Degeneracy
(b) Vol. 82, No. 2, p. 467–471 (1977).
See also *14*, p. *156*; 21, p. 282; 22, p. 326.

W. Niederlag, W. Blau, and K. Kleinstück
X-Ray Spectroscopical Investigation of the Electronic Structure of Fe–Cr
(b) Vol. 82, No. 2, p. 495–499 (1977).
See also *20*, p. *245*; 21.1, p. 296; 21.1.1, p. 304.

M. Taut and A. M. Radwan
*Soft X-Ray Spectra of Al and the Laves Phase $CaAl_2$
(b) Vol. 82, No. 2, p. 507–511 (1977).
See also *20*, p. *245*; 21, p. 283; 22, p. 326.

A. Jędrzejczak and T. Dietl
*Erratum to: Thermomagnetic Properties of n-Type and p-Type HgTe
(b) Vol. 82, No. 2, p. 717 (1977).
See also 14.3, p. 170; *15*, p. *199*; 22.4.3, p. 367.

V. K. Bashenov, S. A. Garyainov, and D. I. Marvakov
*Band Structure of Layer Gallium Selenide
(b) Vol. 82, No. 2, p. K95–K97 (1977).
See also 22.3, p. 358.

V. Ya. Altshul, V. K. Bashenov, and V. V. Timofeenko
*Upper Valence Bands of Perobskite- and Tysonite-Type Fluorides
(b) Vol. 82, No. 2, p. K99–K101 (1977).
See also 22.5, p. 371.

G. Jungk
*Critical Points in the E_0'-Region of Ge from Ellipsometric Data
(b) Vol. 82, No. 2, p. K113–K116 (1977).
See also 20.1, p. 258; 22.1.1, p. 335.

K. Hübner
*Chemical Bond and Related Properties of SiO_2 (III): Core-Level Shifts in SiO_x
(a) Vol. 42, No. 2, p. 501–509 (1977).
See also 2, p. 38; *4*, p. *45*; 14.4, p. 191; 20.1, p. 258; 22.6, p. 390.

G. Massé, J.-P. Aicardi, and C. Butti
*Study of Band-Gap Shrinkage in Cinnabar (α-HgS) by Cathodoluminescence under Very Strong Excitation
(a) Vol. **42**, No. 2, p. 545—552 (1977).
See also *20.3*, p. *272*; 22.4.1, p. 363.

J. Szatkowski, K. Sierański, J. M. Pawlikowski, E. Płaczek-Popko, P. Becla, and E. Dudziak
Electrical Transport Properties of Epitaxial Graded-Gap $Cd_xHg_{1-x}Te$ Layers
(a) Vol. **42**, No. 2, p. 721—727 (1977).
See also *14.3.1*, p. *174*; 22.4.4, p. 369.

C. M. Vodenicharov
*A Simple Method for the Determination of the Effective Electron Mass from Current–Voltage Characteristics in the Thermionic Region
(a) Vol. **42**, No. 2, p. 785—789 (1977).
See also *14.3.1*, p. *174*; 14.3.4, p. 186; 22, p. 327; 22.6, p. 390; 22.9, p. 417.

S. Lewonczuk, J. G. Gross, M. A. Khan, and J. Ringeissen
*Optical Properties of Cuprous Halides in the 4 to 30 eV Range
(b) Vol. **83**, No. 1, p. 161—168 (1977).
See also *20.1*, p. *259*; 22.5, p. 371.

F. Bechstedt and R. Enderlein
Theory of Resonance Raman Scattering in Disordered Solids. Smooth Random Field Model
(b) Vol. **83**, No. 1, p. 239—247 (1977).
See also 2, p. 38; *6*, p. *58*; *20.1*, p. *259*; 22, p. 328.

F. Werfel, G. Dräger, and O. Brümmer
*Electronic Structure Investigation of Metallic Vanadium by XS and XPS
(b) Vol. **83**, No. 1, p. 257—262 (1977).
See also 17, p. 207; 20, p. 245; 21.1, p. 297.

C. H. Chen
*Electron Energy Loss Studies of Direct Nonvertical Interband Transitions in Silicon
(b) Vol. **83**, No. 1, p. 347—351 (1977).
See also 14.4, p. 192; 22.1.2, p. 341.

E. N. Mgbenu
*The Lithium Band Structure
(b) Vol. **83**, No. 1, p. K75—K77 (1977).
See also 13.2, p. 125; 21.2, p. 309.

G. R. Umarov
*Isoelectronic Series and Trends in Band Structures
(b) Vol. **83**, No. 1, p. K93—K95 (1977).
See also *4*, p. *45*; 22, p. 328.

Y. Khwaja and M. Nauciel-Bloch
Off-Stoichiometry Effects in the Antiferromagnetic FeRh Alloy
(b) Vol. **83**, No. 2, p. 413—424 (1977).
See also *18.2*, p. *223*; *18.4*, p. *233*; 21.1.1, p. 306.

A. Blumen and C. Merkel
*Energy Band Calculations on Helical Systems
(b) Vol. **83**, No. 2, p. 425—431 (1977).
See also 14.4.1, p. 196; 22.5, p. 371.

U. Büchner
*Wave-Vector Dependence of the Electron Energy Losses of PbS, PbSe, and PbTe
(b) Vol. **83**, No. 2, p. 493–500 (1977).
See also 14.4, p. 192; 22, p. 328.

F. G. dos Reis, Hilda A. Cerdeira, and R. Luzzi
Dielectric Response of Saturated Semiconductors
(b) Vol. **83**, No. 2, p. 607–614 (1977).
See also *14.4*, p. *192*; *20*, p. *246*; 22, p. 329.

S. I. Radautsan, E. K. Arushanov, L. N. Lukyanova, and A. N. Nateprov
The Thermoelectric Power of $Cd_3(As_xP_{1-x})_2$
(a) Vol. **43**, No. 2, p. K159–K164 (1977).
See also *15*, p. *199*; 22.3.1, p. 360.

L. Pasemann
*A Non-Diagonally Screened Model Potential for Semiconductors Applied to the Band Structure Calculation of Si
(b) Vol. **84**, No. 1, p. 175–188 (1977).
See also 22.1.2, p. 343.

H. Köhler and A. Freudenberger
*Investigation of the Highest Valence Band in $(Bi_{1-x}Sb_x)_2Te_3$ Crystals
(b) Vol. **84**, No. 1, p. 195–203 (1977).
See also *13.2*, p. *125*; 14.3, p. 171; 22.7, p. 395.

J. Kudrnovský and J. Petrů
Electrical Conductivity of Two-Band Models for Disordered Binary Substitutional Alloys. Theory
(b) Vol. **84**, No. 1, p. 325–333 (1977).
See also *14.1*, p. *159*; 21, p. 286.

K. Bärner
*Energy Bands and Electron Correlation Effects in Manganese Pnictides
(b) Vol. **84**, No. 1, p. 385–392 (1977).
See also 14.1, p. 159; 18.4, p. 234; *20.1*, p. *261*; 21.1, p. 299.

Nguyen van Hieu and Nguyen Ai Viet
Interband Electronic Raman Scattering by a Three-Band Semiconductor in the Presence of a Resonant Electromagnetic Field
(b) Vol. **84**, No. 2, p. 417–422 (1977).
See also *14.3.4*, p. *188*; *20.1*, p. *262*; 22.2, p. 349; 22.4, p. 361.

L. P. Mokhracheva, V. A. Tskhai, and P. V. Geld
*Chemical Binding and Energy Band Structure of Vanadium Monocarbide
(b) Vol. **84**, No. 2, p. 465–471 (1977).
See also 4, p. 46; 22.3, p. 359.

H. Lange and I. A. Iliev
*Evidence of Coulomb Interaction at the E_1 Interband Transition of Hexagonal CdSe
(b) Vol. **84**, No. 2, p. 569–574 (1977).
See also *13.5.1*, p. *148*; *20.1*, p. *262*; 22.4.2, p. 366.

K. Heidrich, W. Schäfer, M. Schreiber, G. Trendel, and J. Treusch
*Optical Properties of Polymorphous Thallous Halide Mixed Crystal Layers
(b) Vol. **84**, No. 2, p. 629–635 (1977).
See also *20.1*, p. *263*; 22.5.4, p. 385.

U. Behn and H. Zinke
*On the Density of States at the Fermi Level for Normal Systems
(b) Vol. **84**, No. 2, p. 649–652 (1977).
See also 14.1, p. 160.

J. Monecke
*One-Particle Spectrum of Correlated Electrons near the Free-Electron Limit
(b) Vol. **84**, No. 2, p. 665—672 (1977).
See also 14, p. 156; 18.4, p. 234.

A. Mauger
*Indirect Exchange in Europium Chalcogenides
(b) Vol. **84**, No. 2, p. 761—771 (1977).
See also 18.1, p. 216; 18.2, p. 224; 22, p. 330.

F. Lukeš
Electroreflectance Spectrum of InAs in the Range of E_0 and $E_0 + \Delta_0$ Transitions
(b) Vol. **84**, No. 2, p. K113—K117 (1977).
See also *14.3.4*, p. *188*; *20.1*, p. *263*; 22.2.1, p. 353.

S. Lara, R. Moreira Xavier, and C. A. Taft
*Remarks on the Effects of Pressure on the Energy Gaps in EuO and EuS
(b) Vol. **84**, No. 2, p. K143—K145 (1977).
See also *12.2*, p. *106*; 22, p. 330; 22.6, p. 391.

13.2 Fermi Surfaces

Original Papers and Short Notes

D. Glötzel and L. Fritsche
Volume Dependence of the Electronic Structure of Lanthanum and Cerium
(b) Vol. **79**, No. 1, p. 85—92 (1977).
See also *13.1*, p. *114*; 21.4, p. 311.

W. Braune, G. Kuka, S. Hess, H.-U. Müller, and T. Jung
*Cyclotron Resonance and Quantum Oscillations in Semimetallic Bismuth–Antimony Alloys with $E_g = 0$
(b) Vol. **79**, No. 2, p. 501—508 (1977).
See also 14.1, p. 157; 21.7, p. 319.

K. Saermark and J. Lebech
*Dispersion of Cyclotron Waves in Metals and Observability of Many-Body Effects
(b) Vol. **79**, No. 2, p. 723—730 (1977).
See also 14.1, p. 157; 21.2, p. 308; 21.6, p. 315.

W. Schülke
*Fermi Surface Mapping from the One-Dimensional Fourier Transform of Compton Profiles
(b) Vol. **80**, No. 1, p. K67—K70 (1977).
See also 21.2, p. 308.

K. Saermark and J. Lebech
A Finite $\omega\tau$-Calculation of the Dispersion Relation for Ordinary Cyclotron Waves for a Spherical Fermi Surface
(b) Vol. **81**, No. 1, p. K31—K36 (1977).
See also *14.1*, p. *158*.

J. Lakner
The Effect of Non-Transition Metals on the Magnetic Susceptibility of Aluminium Solid Solutions
(b) Vol. **81**, No. 2, p. 593—601 (1977).
See also *18*, p. *210*; 21, p. 280.

G. Oelgart and R. Herrmann
*Non-Ellipsoidal Fermi Surface in Semiconducting $Bi_{1-x}Sb_x$
(b) Vol. **81**, No. 2, p. K135—K138 (1977).
See also 22.7, p. 394.

W. Schülke
*The One-Dimensional Fourier Transform of Compton Profiles. Theorems and Applications
(b) Vol. **82**, No. 1, p. 229—235 (1977).
See also *13.1*, p. *119*; 21.2, p. 308; 22.1.2, p. 340.

Mahendra Kumar and M. P. Hemkar
*Density of States and Fermi Energy of Simple Metals
(b) Vol. **82**, No. 1, p. K29—K34 (1977).
See also 21, p. 282; 21.2, p. 309.

G. K. Bristow, T. R. Finlayson, and T. F. Smith
*Hall Coefficient Sign Reversals in 2H Transition Metal Dichalcogenides
(b) Vol. **82**, No. 2, p. K81—K84 (1977).
See also *14.1*, p. *159*; 21, p. 283.

K. Saermark, J. Lebech, and H. Johansen
*On the Concept of Cyclotron Resonance in Metals
(b) Vol. **83**, No. 1, p. 187—195 (1977).
See also *14.1*, p. *159*; 21, p. 284.

E. N. Mgbenu
The Lithium Band Structure
(b) Vol. **83**, No. 1, p. K75—K77 (1977).
See also *13.1*, p. *121*; 21.2, p. 309.

M. Goodarz and R. D. Barnard
*Deviations from Matthiessen's Rule in Silver Containing Low-Density Defects
(b) Vol. **83**, No. 2, p. 555—564 (1977).
See also 13.4, p. 137; 14.1, p. 159; 15, p. 199; 21.6, p. 317.

H. Köhler and A. Freudenberger
*Investigation of the Highest Valence Band in $(Bi_{1-x}Sb_x)_2Te_3$ Crystals
(b) Vol. **84**, No. 1, p. 195—203 (1977).
See also *13.1*, p. *122*; 14.3, p. 171; 22.7, p. 395.

P. Franzosi
*The Conduction Electron *g*-Factor on the Fermi Surface of Copper
(b) Vol. **84**, No. 2, p. K131—K135 (1977).
See also 21.1, p. 300.

P. Franzosi
*The Geometrical Curvature Factor on the Fermi Surface of the Noble Metals
(b) Vol. **84**, No. 2, p. K137—K141 (1977).
See also 21.1, p. 300; 21.6, p. 318.

H. P. Fischer, K.-H. Steinmetz, and E. Nembach
*A New High-Resolution, High-Intensity Apparatus for the Study of Fermi Surfaces of Alloys by Positron Annihilation
(a) Vol. **44**, No. 2, p. 669—673 (1977).
See also 21.1, p. 300.

13.3 Surface and Interface States

Original Papers and Short Notes

F. K. Schulte
*Energies and Fermi Level of Eectrons in Thin Size-Quantized Metal Films
(b) Vol. **79**, No. 1, p. 149—153 (1977).
See also 21, p. 276.

G. S. Agarwal and H. D. Vollmer
*Surface Polariton Effects in Spontaneous Emission
(b) Vol. **79**, No. 1, p. 249—257 (1977).
See also 13.5.2, p. 149; 20.1, p. 247; 21, p. 276; 22.2.2, p. 354.

F. K. Schulte
*Electronic Density of States, Heat Capacity, and Spin Susceptibility of Thin Metal Films
(b) Vol. **79**, No. 1, p. K5—K8 (1977).
See also 8, p. 64; 18, p. 209; 21, p. 276.

I. Munteanu
*The Effect of Effective Lifetime Modulation in Semiconductors
(a) Vol. **39**, No. 1, p. 291—300 (1977).
See also *14.3.3*, p. *178*; 22, p. 321.

U. Heilig
*Determination of the Concentration of Slow States at Real Germanium Surfaces by Means of Field Effect Measurements
(a) Vol. **39**, No. 1, p. K29—K31 (1977).
See also 22.1.1, p. 333.

G. Iadonisi, V. Marigliano Ramaglia, and G. P. Zucchelli
*Atomic Hydrogen Levels near a Metallic Surface
(b) Vol. **80**, No. 1, p. 245—254 (1977).
See also 21, p. 278.

M. Kolář
*Surface Density of States in a Simple Model of Nickel
(b) Vol. **80**, No. 2, p. 521—531 (1977).
See also 21.1, p. 291.

E. W. Kreutz
*Magneto-Surface Experiments on the III–V Compound Semiconductor InSb
(a) Vol. **40**, No. 2, p. 415—423 (1977).
See also *14.3.2*, p. *176*; 22.2.3, p. 356.

J. M. Pawlikowski
Schematic Energy Band Diagram of Metal–$Cd_xHg_{1-x}Te$ Contacts
(a) Vol. **40**, No. 2, p. 613—620 (1977).
See also *14.3.4*, p. *184*; 22.4.4, p. 368.

Th. Englert, K. von Klitzing, R. Silbermann, and G. Landwehr
*Influence of the Surface on the Galvanomagnetic Properties of Tellurium
(b) Vol. **81**, No. 1, p. 119—127 (1977).
See also 14.3.2, p. 176; 22.1.3, p. 345.

F.-W. Breitbarth and H.-J. Tiller
Tunnelling in Metal–Glow-Polymer–Metal Sandwich Structures
(a) Vol. **41**, No. 1, p. 151—156 (1977).
See also 14.3.2, p. 176; *14.3.4*, p. *185*; 22.9, p. 415.

K. Bhattacharya, P. Basu, and H. Saha
Back Wall Schottky Barrier Solar Cell with an Interfacial Layer
(a) Vol. **41**, No. 1, p. 317—321 (1977).
See also *14.3.3*, p. *179*; *16*, p. *202*; 22.1.2, p. 339; 22.2.1, p. 351.

AMAR NATH NIGAM and RATNA RANI
*Surface States of Chemisorbed Hydrogen and Electrical Conduction in Thin Hydrogenated Nickel Foils
(a) Vol. **41**, No. 2, p. 493—498 (1977).
See also *14.1*, p. *158*; 21.1, p. 294.

S. N. KOZLOV
*Slow Relaxation Kinetics on a Heterogeneous Semiconductor Surface
(a) Vol. **42**, No. 1, p. 115—124 (1977).
See also 14.3.2, p. 176; 22, p. 326.

M. HIROSE, S. YOKOYAMA, and Y. OSAKA
*Surface States in GaAs Tunnel MIS Structures
(a) Vol. **42**, No. 2, p. 483—488 (1977).
See also 14.3.3, p. 180; 22.2.1, p. 352.

A. E. SAPEGA
*Determination of Surface States in Selenium Diodes by Conductance and Capacitance Studies
(a) Vol. **42**, No. 2, p. 585—594 (1977).
See also *14.3.3*, p. *180*; 22.1.3, p. 346.

A. T. ÅBERG
*Low-Temperature Annealing of Al-Implanted Si–SiO_2 Structures
(a) Vol. **42**, No. 2, p. 639—645 (1977).
See also *14.3.2*, p. *177*; 22.1.2, p. 341.

L. M. BASKIN, O. I. LVOV, and G. N. FURSEY
On the Theory of Field Emission from p-Type Semiconductors
(a) Vol. **42**, No. 2, p. 757—767 (1977).
See also *14.3.4*, p. *186*; 22, p. 327.

L. I. JOHANSSON, A. L. HAGSTRÖM, A. PLATAU, and S.-E. KARLSSON
*Adsorption of Oxygen at Room Temperature on Polycrystalline Titanium Studied by Ultraviolet and X-Ray Photoelectron Spectroscopy
(b) Vol. **83**, No. 1, p. 77—84 (1977).
See also 17, p. 207; 21.1, p. 297.

G. PAASCH and M. HIETSCHOLD
*A Layer Summation for Electrostatic Surface Problems. Application to the Classical Cleavage Energy
(b) Vol. **83**, No. 1, p. 209—222 (1977).
See also 21, p. 284.

H. PUSZKARSKI
*Surface-State "Swapping" Effect in Thin Magnetic Films
(b) Vol. **83**, No. 1, p. K25—K27 (1977).
See also *18.2.1*, p. *226*; 22.8.2, p. 412.

H. FLIETNER, W. FÜSSEL, and N. D. SINH
*Interface States of the Si/SiO_2-System and Their Separation in Groups
(a) Vol. **43**, No. 1, p. K99—K101 (1977).
See also 22.1.2, p. 342.

V. V. MAKSIMENKO, A. J. SIMONOV, and A. A. LUSHNIKOV
*Far-Infrared Absorption in Small Metallic Particles
(b) Vol. **83**, No. 2, p. 377—382 (1977).
See also 13.5.2, p. 150; *20.1*, p. *260*.

K. KOYAMA and T. KANEKO
Radiative Electron Capture by Backscattered Ions from Solid Surfaces
(b) Vol. **83**, No. 2, p. 485—492 (1977).
See also *11*, p. *98*; 21.6, p. 317.

M. Kolář
*On the Calculation of the Surface Density of States in the Tight-Binding Formalism
(b) Vol. **83**, No. 2, p. 625–631 (1977).

A. M. Brodskii and M. I. Urbakh
*On the Dependence of Light Reflection from Metals on Adatom Characteristics
(b) Vol. **83**, No. 2, p. 633–644 (1977).
See also 20.1, p. 260; 21, p. 285.

V. Zhelev and J. Malinowski
Ionic Field Effect in Epitaxial AgBr Layers
(a) Vol. **43**, No. 2, p. 675–683 (1977).
See also *14.3.4*, p. *187*; 22.5.1, p. 374.

V. Siegel and W. Rasp
The TSEE of Lapped and Unlapped BeO Ceramics
(a) Vol. **43**, No. 2, p. K117–K121 (1977).
See also 1.6, p. 32; *17*, p. *207*; 22.6, p. 391.

G. Kaden and H. Reimer
*Bestimmung der Dichte und des Einfangquerschnittes von Oberflächenzuständen mit dem Oberflächenladungspumpeffekt
(Determination of the Density and Capture Cross Section of Surface States by the Surface Charge Pumping Effect)
(a) Vol. **44**, No. 2, p. K117–K121 (1977).
See also *14.3.3*, p. *182*; 22, p. 330.

13.4 Impurity and Defect States

Review Articles

SH. M. KOGAN and T. M. LIFSHITS
Photoelectric Spectroscopy — A New Method of Analysis of Impurities in Semiconductors
(a) Vol. **39**, No. 1, p. 11—39 (1977).
See also *16*, p. *201*; 22, p. 320; 22.1.1, p. 333; 22.1.2, p. 337; 22.2.1, p. 350.

G. N. RAO and A. K. SINGHVI
*Dilute Impurity Hyperfine Fields. Systematics and Comparison with Models
(b) Vol. **84**, No. 1, p. 9—21 (1977).
See also *18.2*, p. *217*; 21.1, p. 288; 21.1.1, p. 302; 21.4, p. 311.

Original Papers and Short Notes

M. F. TRINKLER and I. S. ZOLOVKINA
*A-Luminescence of Alkali Halides Activated by Monovalent Mercury-Like Ions
(b) Vol. **79**, No. 1, p. 49—59 (1977).
See also *20.3*, p. *267*; 22.5.2, p. 375.

S. I. GOLUBOV and YU. V. KONOBEEV
Energy Transfer by Excitons in a Tetracene-Doped Anthracene Crystal
(b) Vol. **79**, No. 1, p. 79—84 (1977).
See also *13.5.1*, p. *143*; 20.3, p. 267; 22.9, p. 414.

G. B. SCOTT and J. L. PAGE
The Absorption Spectra of $Y_3Fe_5O_{12}$ and $Y_3Ga_5O_{12}:Fe^{3+}$ to 5.5 eV
(b) Vol. **79**, No. 1, p. 203—213 (1977).
See also *20.1*, p. *247*; 22.8.2, p. 409.

L. BÁNYAI and A. ALDEA
Semi-Classical and Quantum-Mechanical Theory of Hopping Conduction
(b) Vol. **79**, No. 1, p. 365—377 (1977).
Erratum
(b) Vol. **81**, No. 1, p. 409 (1977).
See also 6, p. 51; *14.3*, p. *165*; 22, p. 320.

T. IIDA and R. MONNIER
*Erratum to: Static Properties of the Self-Trapped Hole in Simple Cubic Cesium-Halides
(b) Vol. **79**, No. 1, p. 391 (1977).
See also 6, p. 52; 20.1, p. 248; 22.5.2, p. 375.

F. THUSELT and K. UNGER
*Exchange Splitting of Donor–Acceptor Pairs
(b) Vol. **79**, No. 1, p. K39—K42 (1977).
See also 13.5.1, p. 143; 22.2, p. 348.

I. J. R. BAUMVOL, A. VASQUEZ, J. MARTINEZ, and F. C. ZAWISLAK
*Temperature Dependence of the Electric Field Gradient at Ta Nuclei in $CaHfO_3$
(b) Vol. **79**, No. 1, p. K65—K69 (1977).
See also 14.4, p. 189; 22.8.1, p. 402.

J. GIBER, A. HÜBLER, and D. MARTON
*On the Local Energy Levels of a Silicon Vacancy. A Model of the Defect-Molecule in the Wannier Approximation
(b) Vol. **79**, No. 1, p. K71—K74 (1977).
See also *10.2*, p. *83*; 22.1.2, p. 337.

L. GOUSKOV, A. GOUSKOV, V. LEMOS, W. MAY, and H. SAMPAIO
*Electrical Properties of $GaTe_xSe_{1-x}$ Crystals
(a) Vol. **39**, No. 1, p. 65—71 (1977).
See also 1.2, p. 14; *14.3*, p. *165*; 22.3.1, p. 360.

J.-P. Aicardi, J.-P. Leyris, and G. Masse
Cathodoluminescence of Cinnabar. Evidence of a Donor–Acceptor-Type Transition
(a) Vol. **39**, No. 1, p. 125–131 (1977).
See also *20.3*, p. *267*; 22.4.1, p. 362.

J. C. Gâcon, J. C. Souillat, J. Sériot, and B. Di Bartolo
Radiative and Radiationless Transition Rate Measurements Using a Tunable Pulsed Dye Laser. Application to Sm^{2+} Fluorescence in BaClF
(a) Vol. **39**, No. 1, p. 147–156 (1977).
See also *20.3*, p. *267*; 22.5.3, p. 383.

W. Barth, K. Elsaesser, W. Güth, and E. Kamieniecki
*Photoconductivity in Correlation to Absorption in Plastically Deformed Germanium
(a) Vol. **39**, No. 1, p. 249–252 (1977).
See also 16, p. 201; 20.1, p. 248; 22.1.1, p. 333.

K. Irie
*Temperature Dependence of the Shift of the Acceptor Level in Ferroelectric and Photoconductive SbSI
(a) Vol. **39**, No. 1, p. 313–320 (1977).
See also *14.4.1*, p. *194*; 22.5.4, p. 385.

A. Samoć, M. Samoć, and J. Sworakowski
*Interpretation of One-Carrier Thermally Stimulated Currents and Isothermal Decay Currents (III): Influence of Blocking Contacts
(a) Vol. **39**, No. 1, p. 337–344 (1977).
See also 14.3.4, p. 183; 14.4, p. 189; 22.9, p. 414.

C. Tsallis
*Conduction Electron Spin Resonance Due to Exchange and Spin–Orbit Scatterings by Dilute Impurities in Superconductors
(b) Vol. **79**, No. 2, p. 451–460 (1977).
See also *14.2*, p. *161*.

Martin Schmidt
*Acceptor Ground States in Nearly Cubic Semiconductors
(b) Vol. **79**, No. 2, p. 533–538 (1977).
See also 22.2.1, p. 350.

A. Q. Sarker
*Spherical Effective Impurity Electron Mass in Si
(b) Vol. **79**, No. 2, p. 567–571 (1977).
See also 14.3, p. 165; 22.1.2, p. 338.

E. Antončík
Calibration of the Isomer Shift for ^{119}Sn
(b) Vol. **79**, No. 2, p. 605–611 (1977).
See also *5*, p. *47*; 21.7, p. 319.

S. Winter
*Bound Electron States Close to the Conduction Band in Germanium Due to 60° Dislocations
(b) Vol. **79**, No. 2, p. 637–644 (1977).
See also 22.1.1, p. 333.

A. Y. S. Kung and J. M. Vail
A Molecular Model for the Emission Process of F_A(II)-Centres in Alkali Halides
(b) Vol. **79**, No. 2, p. 663–672 (1977).
See also *10.2*, p. *84*; 20.3, p. 268; 22.5.2, p. 375.

S. W. Biernacki and M. Hamera
*LCAO Approach to Mn^{2+} in ZnSe
(b) Vol. **79**, No. 2, p. 683–689 (1977).
See also 19, p. 236; 22.4.2, p. 365.

J. L. Farvacque, J. C. Doukhan, U. von Alpen, and E. Gmelin
*Screw-Dislocation-Induced Scattering Processes and Acceptor States in Te
(b) Vol. **79**, No. 2, p. 763–773 (1977).
See also *14.3*, p. *166*; 20.1, p. 249; 22.1.3, p. 345.

M. V. Eremin and A. A. Kornienko
*The Superposition Model in Crystal Field Theory
(b) Vol. **79**, No. 2, p. 775–785 (1977).
See also 4, p. 44; 20.1, p. 249; 22.5.3, p. 383.

N. Singh, J. Singh, and S. Prakash
Residual Resistivity of Noble Metals
(b) Vol. **79**, No. 2, p. 787–794 (1977).
See also *14.1*, p. *157*; 21.6, p. 315.

E. Loh, G. Samoggia, E. Reguzzoni, and L. Nosenzo
*Thermally Modulated Ultraviolet Absorption Spectra of CaF_2:Ce^{3+} Crystals
(b) Vol. **79**, No. 2, p. 795–799 (1977).
Addendum
(b) Vol. **81**, No. 1, p. 411 (1977).
See also *20.1*, p. *249*; 22.5.3, p. 383.

I. N. Douglas
Magnetic Circular Dichroism of Co^{2+} Ions in CdF_2 and CaF_2
(b) Vol. **79**, No. 2, p. K99–K101 (1977).
See also *18*, p. *209*; *20.1*, p. *249*; 22.5, p. 370; 22.5.3, p. 383.

G. K. Averkieva, V. S. Grigoreva, I. A. Maltseva, V. D. Prochukhan, and Yu. V. Rud
Photoluminescence of p-Type $ZnGeP_2$ Crystals
(a) Vol. **39**, No. 2, p. 453–457 (1977).
See also *20.3*, p. *268*; 22.8, p. 396.

K. Ziegler and U. Birkholz
*Photoelectric Properties of Si_2Te_3 Single Crystals
(a) Vol. **39**, No. 2, p. 467–475 (1977).
See also 14.3, p. 166; 16, p. 201; 20.3, p. 268; 22, p. 322.

A. A. Kaminskii, T. I. Butaeva, V. A. Fedorov, Kh. S. Bagdasarov, and A. G. Petrosyan
*Absorption, Luminescence, and Stimulated Emission Investigations in $Lu_3Al_5O_{12}$–Er^{3+} Crystals
(a) Vol. **39**, No. 2, p. 541–548 (1977).
See also 20.2, p. 265; 20.3, p. 268; 22.8, p. 396.

R. M. Hill
*The Frequency Dependence of Variable Range Hopping
(a) Vol. **39**, No. 2, p. 615–620 (1977).
See also 2, p. 34; *14.3.1*, p. *173*.

L. I. Filina, A. G. Ilyashenko, Yu. V. Vorobev, Yu. I. Karkhanin, and M. K. Sheinkman
*The Thermally Stimulated Conductivity and the Features of the Process of Recombination of Charge Carriers in Semi-Insulating GaAs(Cr)
(a) Vol. **39**, No. 2, p. 717–724 (1977).
See also *14.3*, p. *166*; 22.2.1, p. 350.

M. Rueff and E. Sigmund
*E–e Jahn-Teller System: Optical Response Studies
(b) Vol. **80**, No. 1, p. 215–223 (1977).
See also *20.1*, p. *251*.

L. Mărculescu
Luminescence of Lead-Activated Sodium Chloride Single Crystals
(b) Vol. **80**, No. 1, p. 265–272 (1977).
See also *20.3*, p. *268*; 22.5.2, p. 376.

I. ALSTRUP and S. MARKLUND
*The Electron States Associated with the Core Region of the 60° Dislocation in Silicon
(b) Vol. **80**, No. 1, p. 301–306 (1977).
See also 22.1.2, p. 338.

D. D. NEDEOGLO
*Formation and Properties of the Impurity Band in n-ZnSe
(b) Vol. **80**, No. 1, p. 369–377 (1977).
See also *14.3*, p. *166*; 22.4.2, p. 365.

V. KUMAR, N. DASS, and K. CHANDRA
*On the Investigations of the Crystalline Field of Nd^{3+} and Yb^{3+} Ions at O_h Sites in Calcium Fluoride
(b) Vol. **80**, No. 1, p. K1–K4 (1977).
See also 22.5.3, p. 383.

H. S. REEHAL
*Poole-Frenkel Effect with a Screened Coulomb Potential
(b) Vol. **80**, No. 1, p. K63–K65 (1977).
See also *14.3.4*, p. *184*.

V. CAPOZZI, G. MARIOTTO, M. MONTAGNA, A. CINGOLANI, and A. MINAFRA
*Space-Charge-Limited Currents in GaSe at Different Temperatures
(a) Vol. **40**, No. 1, p. 93–100 (1977).
See also *14.3.4*, p. *184*; 22.3, p. 358.

M. KRYSTEK
*On the Evaluation Method of Balarin and Zetzsche
(a) Vol. **40**, No. 1, p. K65–K68 (1977).
See also *20.3*, p. *269*; 22, p. 323.

P. KRISPIN
*Impurity Detection by the Admittance–Frequency Dispersion
(a) Vol. **40**, No. 1, p. K81–K84 (1977).
See also *14.3.3*, p. *179*; 22.2.2, p. 354.

F. THUSELT
*Spin-Dependent Luminescence of Donor–Acceptor Pairs in Semiconductors (I): General Theory
(b) Vol. **80**, No. 2, p. 499–507 (1977).
See also *20.3*, p. *269*; 22, p. 323.

A. N. GEORGOBIANI, YU. V. OZEROV, E. E. STRUMBAN, and V. S. VINOGRADOV
*Influence of Random Electric Fields on the Polarization of Deep Centre Electroabsorption
(b) Vol. **80**, No. 2, p. 589–596 (1977).
See also 16, p. 202; 20.1, p. 252; 20.3, p. 269; 22.4.1, p. 362.

G. BASTARD
*Resonant Acceptor Level in Zero-Gap Semiconductors. Single and Multiple Scattering Effects
(b) Vol. **80**, No. 2, p. 641–650 (1977).
See also 14.3, p. 167; 22.4.4, p. 368.

YU. E. PERLIN, B. S. TSUKERBLAT, and T. SINGH DOD
Moments of Electroabsorption of Jahn-Teller Impurity Centres
(b) Vol. **80**, No. 2, p. 703–707 (1977).
See also *14.3.4*, p. *184*; *20.1*, p. *252*; 22, p. 323.

H. RÖPPISCHER, J. JACOBS, and B. V. NOVIKOV
Influence of Preparation Conditions on the Optical Band Gap and the Mn^{2+} ESR Spectrum of ZnSe Single Crystals
(a) Vol. **40**, No. 2, p. 575–581 (1977).
See also *13.5.1*, p. *144*; *19*. p. *237*; 22.4.2, p. 365.

K. A. Shoaib, F. H. Hashmi, M. Ali, S. J. H. Bukhari, and C. A. Majid
Thermoluminescence from X-Ray-Irradiated Stabilized ZrO_2 Single Crystals
(a) Vol. **40**, No. 2, p. 605—612 (1977).
See also 10.2, p. 86; *20.3*, p. *269*; 22.6, p. 387.

Y. E. Perlin, A. A. Kaminskii, S. I. Klokishner, V. N. Enakii, Kh. S. Bagdasarov, G. A. Bogomolova, and D. N. Vylegzhanin
*Thermal Broadening and Shift of the Energy Levels of TR^{3+} Ions in Crystals
(a) Vol. **40**, No. 2, p. 643—653 (1977).
See also 6, p. 55; 20.3, p. 269; 22.8.1, p. 403.

F. Thuselt and K. Unger
*Spin-Dependent Luminescence of Donor–Acceptor Pairs in Semiconductors (II): Application to Inner Strains
(b) Vol. **81**, No. 1, p. 89—95 (1977).
See also *20.3*, p. *269*; 22, p. 324.

E. Gerlach and M. Rautenberg
*Scattering of DC Electrons by a Spherical Macroscopic Potential Including Screening
(b) Vol. **81**, No. 1, p. 153—156 (1977).
See also *14*, p. *155*.

A. A. Antipin, A. V. Vinokurov, M. P. Davydova, S. L. Korableva, A. L. Stolov, and A. A. Fedii
*Optical Spectra, EPR, and Spin–Lattice Relaxation of Yb^{3+} Ions in Crystals Having Perovskite-Type Structure
(b) Vol. **81**, No. 1, p. 287—293 (1977).
See also 19, p. 238; 20.3, p. 270; 22.5.4, p. 385.

W. Karthe
*Zero-Field Splitting in Trigonal Symmetry. Cr^{3+} and Fe^{3+} in α-$LiIO_3$
(b) Vol. **81**, No. 1, p. 323—332 (1977).
See also *19*, p. *238*; 22.8.1, p. 404.

P. J. Alonso and R. Alcalá
*On the Optical Absorption Spectrum of Co^{2+} in CaF_2
(b) Vol. **81**, No. 1, p. 333—339 (1977).
See also *20.1*, p. *254*; 22.5.3, p. 384.

V. V. Eremenko, A. I. Maslennikov, and V. M. Naumenko
*Host and Impurity Excitations in Antiferromagnetic and Paramagnetic $FeCO_3$ in the Far Infrared
(b) Vol. **81**, No. 1, p. 351—357 (1977).
See also 18.4, p. 232; 19, p. 238; 22.8.1, p. 404.

V. S. Korolkov and A. G. Makhanek
*The Effects of Coordination Polyhedron Structure on Rare-Earth Ion Spectra
(b) Vol. **81**, No. 1, p. 387—394 (1977).
See also 20.1, p. 254; 22.8.2, p. 411.

S. W. Biernacki, M. Hamera, and S. Wartewig
*Ligand Hyperfine Structure of Mn^{2+} in ZnSe as Calculated by EHT–SCCC, Scattered-Wave, and LCAO Methods
(b) Vol. **81**, No. 1, p. K5—K9 (1977).
See also 22.4.2, p. 365.

V. A. Antonov, P. A. Arsenev, Sh. A. Vakhidov, E. M. Ibragimova, and D. S. Petova
Spectroscopic Properties of Neodymium Ions in $LiYO_2$ Monocrystals
(a) Vol. **41**, No. 1, p. 45—50 (1977).
See also 10.2, p. 87; *20.1*, p. *254*; *20.3*, p. *270*; 22.8.1, p. 404.

J. M. Moret and H. Bill
Experimental Determination of the Spin-Stress Coefficients of Gd^{3+} in the New Host Crystal Na_2S
(a) Vol. **41**, No. 1, p. 163–171 (1977).
See also *19*, p. *239*; 22, p. 324.

L. Eglinski, A. F. Rudolph, T. Banys, and A. Dargys
Microwave Breakdown in Tellurium
(a) Vol. **41**, No. 1, p. 177–181 (1977).
See also *14.3.4*, p. *185*; 22.1.3, p. 345.

V. A. Stukan, V. I. Trifonov, V. Ya. Shevchenko, and G. I. Goncharenko
Electrical Properties of p-CdSb at Low Temperatures
(a) Vol. **41**, No. 1, p. 307–315 (1977).
See also 13.1, p. 117; *14.3*, p. *168*; 22.3, p. 358.

Yu. I. Gorkun, V. L. Korobanov, V. S. Lysenko, V. A. Novominskii, and V. I. Turchanikov
*Effect of Bombardment with Low-Energy Ar^+ Ions on Electron Scattering in Accumulation Layers of n-Ge
(a) Vol. **41**, No. 1, p. 339–346 (1977).
See also 11, p. 95; *14.3.3*, p. *179*; 22.1.1, p. 334.

C. Guttel, E. Antic, and P. E. Caro
*Crystal Field Parameters for Gd^{3+} in $Gd_2(MoO_4)_3$
(b) Vol. **81**, No. 2, p. 463–470 (1977).
See also 20.1, p. 255; 22.8.2, p. 411.

M. Rueff, E. Sigmund, and M. Wagner
*Resonance Scattering of Phonons at Trigonal Jahn-Teller Centres
(b) Vol. **81**, No. 2, p. 511–520 (1977).
See also *6*, p. *56*.

V. Čápek
*Localized States on Imperfections and Red Shift of the Absorption Edge in Some Ferromagnetic Semiconductors
(b) Vol. **81**, No. 2, p. 571–577 (1977).
See also 20.1, p. 255; 22.8.2, p. 411.

S. Benci, M. P. Fontana, and M. Manfredi
*On the Excited States of the Tl^+ Monomer Centre in NaI
(b) Vol. **81**, No. 2, p. 603–608 (1977).
See also 20.1, p. 255; 20.3, p. 270; 22.5.2, p. 378.

K. Kocot and J. M. Baranowski
*The Resonant States of the $Ti(d)^2$ Impurity in CdSe
(b) Vol. **81**, No. 2, p. 629–635 (1977).
See also 16, p. 202; 20.1, p. 255; 22.4.2, p. 366.

Vijay A. Singh, C. Weigel, J. W. Corbett, and L. M. Roth
*Vibrational and Electronic Structure of Hydrogen-Related Defects in Silicon Calculated by the Extended Hückel Theory
(b) Vol. **81**, No. 2, p. 637–646 (1977).
See also 6, p. 56; 22.1.2, p. 340.

M. Lannoo and G. Allan
*Contribution of the d-Electron Screening to the Properties of Impurities in Transition Metals
(b) Vol. **81**, No. 2, p. 681–690 (1977).
See also 14.4, p. 191; 21.1, p. 293.

L. T. Ho
*Zeeman Effect of the Excitation Spectrum of Neutral Magnesium Donors in Silicon
(b) Vol. **81**, No. 2, p. K99–K101 (1977).
See also *20.1*, p. *255*; 22.1.2, p. 340.

K. B. Tolpygo and V. A. Telezhkin
*Role of Interelectronic Correlation in the Explanation of Paramagnetic Properties of the Impurity–Vacancy Complex in Silicon
(b) Vol. **81**, No. 2, p. K139—K140 (1977).
See also 22.1.2, p. 340.

W. Plesiewicz
*Photoelectronic Investigations of Multi-Level Gold Centres in Ge:Sb:Au
(a) Vol. **41**, No. 2, p. 403—409 (1977).
See also 14.3, p. 168; *16*, p. *202*; 22.1.1, p. 334.

D. Mergel and R. Labusch
Optical Excitation of Dislocation States in Germanium (I): Experiments
(a) Vol. **41**, No. 2, p. 431—438 (1977).
See also *16*, p. *202*; 20.1, p. 256; 22.1.1, p. 334.

W. Stręk, C. Szafrański, E. Łukowiak, Z. Mazurak, and B. Jeżowska-Trzebiatowska
Fluorescence Quenching in Neodymium Pentaphosphate
(a) Vol. **41**, No. 2, p. 547—553 (1977).
See also *20.3*, p. *270*; 22.8.1, p. 405.

A. B. van Oosterhout
Tungstate Luminescence in Ordered Perovskites
(a) Vol. **41**, No. 2, p. 607—617 (1977).
See also *20.3*, p. *271*; 22.8.1, p. 405.

S. Nešpurek and J. Sworakowski
A Differential Method of Analysis of Steady-State Space-Charge-Limited Current–Voltage Characteristics
(a) Vol. **41**, No. 2, p. 619—627 (1977).
See also *14.3.4*, p. *185*; 22, p. 325.

K.D. Glinchuk, A. V. Prokhorovich, V. E. Rodionov, and V. I. Vovnenko
*The Scheme of Electronic Transitions via 0.94, 1.0, 1.2, and 1.3 eV Radiative Centres in n-GaAs
(a) Vol. **41**, No. 2, p. 659—668 (1977).
See also *20.3*, p. *271*; 22.2.1, p. 352.

M. Z. Iqbal
*Deep Level Photocapacitance Spectra of Green-Emitting GaP Lamps
(a) Vol. **41**, No. 2, p. K155—K156 (1977).
See also *16*, p. *202*; 22.2.2, p. 354.

H.-J. Kos and R. Nink
Z_1 Centres in Mg-Doped Lithium Fluoride
(a) Vol. **41**, No. 2, p. K157—K161 (1977).
See also *10.2*, p. *89*; *20.1*, p. *256*; 22.5.2, p. 378.

A. F. Ellervee
Luminescence of Pb^{2+} and Bi^{3+} Centres in Alkali-Earth Sulphides and Oxides
(b) Vol. **82**, No. 1, p. 91—98 (1977).
See also *20.3*, p. *271*; 22.4.1, p. 363; 22.6, p. 389.

M. Bacci
*Quadratic Jahn-Teller Effect and Tunneling Splitting in Octahedral Clusters
(b) Vol. **82**, No. 1, p. 169—177 (1977).
See also *20.1*, p. *257*; 22.6, p. 389.

V. D. Korolev, S. A. Peskovatskii, and V. M. Shulga
Spin–Lattice Relaxation in Fe^{3+}-Doped Andalusite Crystals (I): Relaxation Mechanisms
(b) Vol. **82**, No. 1, p. 251—255 (1977).
See also *19*, p. *240*; 22.6, p. 389.

Y. D. Yao, S. Arajs, and E. E. Anderson
Phenomenological "Dilution Model" for Kondo-Type Effects in Concentrated Alloy Systems
(b) Vol. **82**, No. 1, p. 343—347 (1977).
See also *14.1*, p. *158*; 18, p. 211; 21.1, p. 295.

L. Bosi and M. Nimis
*On the Decay of the M_A Centre in KCl:Li and KCl:Na
(b) Vol. **82**, No. 1, p. K49—K51 (1977).
See also *10.2*, p. *89*; 20.3, p. 271; 22.5.2, p. 379.

J. P. Zielinger, C. Noguet, and M. Tapiero
Photomemory Effect in Cu_2O Single Crystals. Kinetics of the Excitation and De-Excitation Processes
(a) Vol. **42**, No. 1, p. 91—100 (1977).
See also *16*, p. *203*; 22.6, p. 389.

M. Kh. Ashurov, Yu. K. Voronko, V. V. Osiko, A. A. Sobol, and M. I. Timoshechkin
Spectroscopic Study of Stoichiometry Deviation in Crystals with Garnet Structure
(a) Vol. **42**, No. 1, p. 101—110 (1977).
See also *1.3*, p. *22*; 20.1, p. 257; 20.3, p. 271; 22.8.2, p. 411.

H. Krause and R. Grünler
*DC Current-Voltage Behaviour of SiO_2 Layers
(a) Vol. **42**, No. 1, p. 149—155 (1977).
See also *14.3.3*, p. *180*; 22.6, p. 389.

D. Mergel and R. Labusch
*Optical Excitation of Dislocation States in Germanium (II): Analysis of the Experimental Results
(a) Vol. **42**, No. 1, p. 165—171 (1977).
See also 13.1, p. 119; *16*, p. *203*; 22.1.1, p. 335.

A. H. Marshak and K. M. van Vliet
Current in Semiconductors with Position-Dependent Band Gap and Electron Affinity
(a) Vol. **42**, No. 1, p. 279—283 (1977).
See also *14.3*, p. *169*; 22, p. 326.

I. M. Kotina, S. R. Novikov, and T. I. Pirozhkova
*Determination of the Parameters of Impurity Centres from the Thermally Stimulated Current of the p-n Junction Space Charge Region
(a) Vol. **42**, No. 1, p. 397—401 (1977).
See also *14.3.3*, p. *180*; 22, p. 326.

E. R. Vance and N. B. Manson
Magnetic Circular Dichroism of Neutron-Irradiated MgO and CaO
(b) Vol. **82**, No. 2, p. 473—479 (1977).
See also *18*, p. *211*; *20*, p. *245*; 22.6, p. 390.

C.-P. Czaya, K. Keck, and H. Rüthlein
*Influence of Surface Carriers upon the Galvanomagnetic Effects in Slightly Doped Tellurium Samples at Low Temperatures
(b) Vol. **82**, No. 2, p. 595—602 (1977).
See also 14.3.2, p. 177; 22.1.3, p. 346.

M. Moreno
*A Study on the Magnetic Properties of Compressed D_{4h} Clusters
(b) Vol. **82**, No. 2, p. 669—676 (1977).
See also *19*, p. *240*; 22.5, p. 371.

I. S. Osadko
*Electron-Vibrational Transitions in the Presence of Several Minima of Adiabatic Potentials
(b) Vol. **82**, No. 2, p. K107—K110 (1977).
See also 20.1, p. 258; 20.3, p. 272.

R. C. Du Varney and A. K. Garrison
*An EPR-ENDOR Study of the [Li]° Center in Crystalline and Ceramic BeO
(a) Vol. **42**, No. 2, p. 609–614 (1977).
See also 19, p. 241; 22.6, p. 390.

A. F. Kravchenko, B. S. Lisenker, Yu. E. Maronchuk, A. P. Sherstyakov, and V. N. Sherstyakova
*Investigation of Recombination of Non-Equilibrium Carriers in GaAs Epitaxial Layers
(a) Vol. **42**, No. 2, p. 647–655 (1977).
See also *14.3.1*, p. *174*; 16, p. 203; 22.2.1, p. 352.

W. Eichler
Thermomagnetic Effects in Pure Imperfect Tellurium Single Crystals
(a) Vol. **42**, No. 2, p. K107–K110 (1977).
See also *15*, p. *199*; 22.1.3, p. 346.

T. Wosiński and T. Figielski
Spin-Dependent Recombination at Exchange-Coupled Dislocation Centres in Silicon
(b) Vol. **83**, No. 1, p. 93–98 (1977).
See also *10.2*, p. *90*; *16*, p. *204*; 22.1.2, p. 341.

R. Pässler
*High-Temperature Limit of the Non-Radiative Multiphonon Carrier-Capture Cross-Sections of Deep Traps in Semiconductors
(b) Vol. **83**, No. 1, p. K55–K58 (1977).
See also 22, p. 328.

A. A. Kaminskii, S. E. Sarkisov, J. Bohm, P. Reiche, D. Schultze, and R. Uecker
Growth, Spectroscopic and Laser Properties of Crystals in the $K_5Bi_{1-x}Nd_x(MoO_4)_4$ System
(a) Vol. **43**, No. 1, p. 71–79 (1977).
See also 3, p. 42; 20.1, p. 260; *20.2*, p. *266*; *20.3*, p. *272*; 22.8, p. 399.

G. Massé, J. P. Aicardi, and J. P. Leyris
Green Emission of Mercury Sulphide (α-HgS)
(a) Vol. **43**, No. 1, p. 191–201 (1977).
See also 6, p. 59; *20.3*, p. *273*; 22.4.1, p. 363.

A. L. Kartuzhanskii, A. G. Kavetskii, L. D. Kekhva, T. E. Kekhva, A. P. Kiselev, and B. T. Plachenov
*Recombination Processes during Excitation of $AgCl:Me^{2+}$ Crystals by Ionizing Radiation
(a) Vol. **43**, No. 1, p. 321–327 (1977).
See also *20.3*, p. *273*; 22.5.1, p. 373.

P. A. Arsenev, A. V. Potemkin, V. V. Fenin, and I. Senff
Investigation of Stimulated Emission of Er^{3+} Ions in Mixed Crystals with Perovskite Structure
(a) Vol. **43**, No. 1, p. K15–K18 (1977).
See also *20.2*, p. *266*; 22.8, p. 399.

G. Duggan, G. J. Morgan, N. Shahtahmasebi, and A. Lettington
*The Effective Screening Function in Insulating Solids
(b) Vol. **83**, No. 2, p. 543–553 (1977).
See also 14.4, p. 192; 22.1.1, p. 335.

M. Goodarz and R. D. Barnard
Deviations from Matthiessen's Rule in Silver Containing Low-Density Defects
(b) Vol. **83**, No. 2, p. 555–564 (1977).
See also *13.2*, p. *125*; 14.1, p. 159; 15, p. 199; 21.6, p. 317.

R. Pässler
*Low-Temperature Dependences of the Nonradiative Multiphonon Carrier-Capture Cross-Sections of Deep Traps in Semiconductors
(b) Vol. **83**, No. 2, p. K111–K114 (1977).
Erratum
(b) Vol. **84**, No. 2, p. K163 (1977).
See also 22, p. 329.

A. ZARĘBA, A. LUBOMIRSKA-WITTLIN, A. SZADKOWSKI, and B. KRUKOWSKA-FULDE
*Photoconversion from the Insulating to the Semiconducting State of $CdF_2:MF_3$
(b) Vol. **83**, No. 2, p. K155–K157 (1977).
See also 16, p. 204; 22.5.3, p. 384.

V. A. GRAZHULIS, V. V. KVEDER, and V. YU. MUKHINA
*Investigation of the Energy Spectrum and Kinetic Phenomena in Dislocated Si Crystals (I)
(a) Vol. **43**, No. 2, p. 407–415 (1977).
See also 10.2, p. 91; *14.3*, p. *171*; 22.1.2, p. 342.

H. EICHELE and M. SCHWOERER
Fluorescence of Localized States in a Partially Polymerized Diacetylene Single Crystal
(a) Vol. **43**, No. 2, p. 465–472 (1977).
See also *20.3*, p. *273*; 22.9, p. 418.

R. BINDEMANN, H. FISCHER, and K. KREHER
*Determination of the Concentration of Substitutional O in GaP by Means of Donor–Acceptor Pair Luminescence
(a) Vol. **43**, No. 2, p. 529–533 (1977).
See also *20.3*, p. *273*; 22.2.2, p. 354.

A. CAVALLINI, P. GONDI, and A. CASTALDINI
*Dislocation Energy Levels in Ge
(a) Vol. **43**, No. 2, p. K205–K208 (1977).
See also 10.2, p. 91; 14.3, p. 171; 22.1.1, p. 336.

M. CASALBONI, U. M. GRASSANO, A. SCACCO, and A. TANGA
*Emission Spectra of $NaBr:Tl^+$ Excited in the A-Band
(b) Vol. **84**, No. 1, p. 23–26 (1977).
See also *20.3*, p. *274*; 22.5.2, p. 381.

A. YA. KORETS, V. P. FEDOROV, A. V. KORSHUNOV, and V. A. NAUMOV
Spectroscopic Investigation of Orientational Disorder in an Acenaphthylene Molecular Crystal
(b) Vol. **84**, No. 1, p. 107–113 (1977).
See also *13.5.1*, p. *148*; *20.1*, p. *261*; 22.9, p. 419.

E. PALANIYANDI and S. BALASUBRAMANIAN
*Polarizabilities of Donors in Elemental Semiconductors
(b) Vol. **84**, No. 1, p. 171–174 (1977).
See also *14.3*, p. *171*; 22.1.1, p. 336; 22.1.2, p. 343.

V. A. KRYLOV and W. ULRICI
*Influence of an Electric Field on the Jahn-Teller Split Absorption Band $^4A_{2g} \rightarrow {}^4T_{1g}(P)$ of Co^{2+} in CdF_2 and CaF_2
(b) Vol. **84**, No. 1, p. 215–225 (1977).
See also *20.1*, p. *261*; 22.5.3, p. 384.

LE MANH HOANG and J. M. BARANOWSKI
Absorption and Luminescence of $V(d^3)$ in II–VI Compound Semiconductors
(b) Vol. **84**, No. 1, p. 361–365 (1977).
See also *20.1*, p. *261*; *20.3*, p. *274*; 22.4, p. 361.

A. RANFAGNI and G. VILIANI
*Spin–Orbit-Allowed Tunneling Splitting in the $T \times \varepsilon_g$ Jahn-Teller Problem. Splitting of the Zero-Phonon Line
(b) Vol. **84**, No. 1, p. 393–399 (1977).
See also 20.1, p. 261; 22.5.4, p. 385; 22.6, p. 391.

H. S. REEHAL and C. B. THOMAS
*Poole-Frenkel Conduction in Amorphous Solids for Centres with a Screened Coulomb Potential
(b) Vol. **84**, No. 1, p. K13–K16 (1977).
See also 2, p. 40; *14.3.4*, p. *187*; 22.8, p. 400.

C. VIGER, C. VAUTIER, and D. CARLES
Recombination and Trapping Processes in Amorphous Materials in Presence of an Expotential Distribution of Traps
(a) Vol. **44**, No. 1, p. 97—106 (1977).
See also 2, p. 40; 14.3.1, p. 175; *16*, p. *204*; 22.1.3, p. 347.

V. A. GRAZHULIS, V. V. KVEDER, and V. YU. MUKHINA
*Investigation of the Energy Spectrum and Kinetic Phenomena in Dislocated Si Crystals (II): Microwave Conductivity
(a) Vol. **44**, No. 1, p. 107—115 (1977).
See also 10.2, p. 91; *14.3*, p. *171*; 22.1.2, p. 343.

H. J. HOFFMANN and W. SOHN
*Analysis of Localized Levels in Semiconducting CdS from Generation–Recombination Noise Spectra
(a) Vol. **44**, No. 1, p. 237—246 (1977).
See also *14.3*, p. *171*; 22.4.1, p. 364.

H. KNAUER, J. RICHTER, and P. SEIDEL
A Direct Calculation of the Resonance Tunneling in Metal–Insulator–Metal Tunnel Junctions
(a) Vol. **44**, No. 1, p. 303—312 (1977).
See also *14.3.3*, p. *181*; 22, p. 330.

V. D. KOROLEV, S. A. PESKOVATSKII, and V. M. SHULGA
Spin–Lattice Relaxation in Fe^{3+}-Doped Andalusite Crystals (II): The Effect of Lattice Oscillator Heating
(b) Vol. **84**, No. 2, p. 443—447 (1977).
See also *19*, p. *243*; 22.6, p. 391.

M. M. BABAEV and T. M. GASYMOV
Thermopower and the Transverse Nernst-Ettingshausen Effect in Semiconductors in a Strong Electric Field
(b) Vol. **84**, No. 2, p. 473—483 (1977).
See also 6, p. 60; *14.3.4*, p. *188*; *15*, p. *199*; 22, p. 330.

G. V. IONOVA, S. P. IONOV, O. M. PACHEV, and E. F. MAKAROV
Orbital–Charge Ordering in Crystals
(b) Vol. **84**, No. 2, p. 493—501 (1977).
See also *13*, p. *112*.

A. P. RADLINSKI
*On the Properties of Cobalt Impurities in Zinc Selenide Crystals (I): Effect of In-Band Position of States on Photoelectric Properties
(b) Vol. **84**, No. 2, p. 503—511 (1977).
See also 16, p. 205; 20.1, p. 262; 22.4.2, p. 366.

A. KASTEN and P. J. BECKER
Spectroscopic Investigation of Optically Excited Doublet States of Ho^{3+} in Magnetically Ordered $(Ho_{0.01}, Gd_{0.99})VO_4$
(b) Vol. **84**, No. 2, p. 551—560 (1977).
See also *18.4*, p. *234*; *20.1*, p. *262*; 22.8.1, p. 408.

J. GARDAVSKÝ, I. BARVÍK, and M. ZVÁRA
*Jahn-Teller Effect on the $^3T_1(^3F)$ State and Resonance of the $^3T_1(^3P)$ State with the Conduction Band in CdTe:Ti^{2+}
(b) Vol. **84**, No. 2, p. 691—698 (1977).
See also *20.1*, p. *263*; 22.4.3, p. 367.

S. W. BIERNACKI
*The Multimode Jahn-Teller Coupling for an Orbital Triplet in Cubic Crystals
(b) Vol. **84**, No. 2, p. 699—708 (1977).
See also 18.1, p. 215; *20.1*, p. *263*.

V. I. MURAVEV and N. B. YUNUSOV
*Covalency and Hyperfine Interaction on the Impurity Ion in Crystals
(b) Vol. 84, No. 2, p. 791—796 (1977).
See also 18.1, p. 216; 19, p. 243; 21.1, p. 300.

M. V. EREMIN and N. I. SILKIN
Theory of the ESR Spectra of Tl^{2+} Ions in K_2SO_4 Single Crystals
(b) Vol. 84, No. 2, p. 803—811 (1977).
See also *19*, p. *243*; 22.8.1, p. 408.

P. BYSZEWSKI and R. R. GAŁĄZKA
*Resonance Acceptor Level in $Hg_{0.98}Mn_{0.02}Te$
(b) Vol. 84, No. 2, p. K83—K86 (1977).
See also *14.3*, p. *172*; 22.4.3, p. 367.

W. ULRÍCI
*CdF_2:Cr^{2+} — an Example for an Orthorhombic Jahn-Teller Distortion
(b) Vol. 84, No. 2, p. K155—K157 (1977).
See also 20.1, p. 263; 22.5.3, p. 384.

V. P. DOBREGO, O. P. ERMOLAEV, and V. D. TKACHEV
*Investigation of Shallow Radiation Defect Levels Produced in Ge by Fast-Neutron Irradiation
(a) Vol. 44, No. 2, p. 435—442 (1977).
See also *11*, p. *99*; 22.1.1, p. 336.

T. L. TANSLEY
Thermally-Assisted Tunnelling in Composite Barrier Heterojunctions
(a) Vol. 44, No. 2, p. 449—455 (1977).
See also *14.3.3*, p. *182*; 22.2.1, p. 353; 22.2.4, p. 357.

A. TONEVA and A. ALAZHAZHIAN
Mobility in p-$Pb_{1-x}Sn_xTe$
(a) Vol. 44, No. 2, p. 621—626 (1977).
See also *14.3*, p. *172*; 22, p. 330.

R. P. KHARE and R. NATH
On the Analysis of Complex TSC Patterns
(a) Vol. 44, No. 2, p. 627—631 (1977).
See also *14.3.1*, p. *175*; 22.6, p. 392.

V. S. SIVASANKAR and S. RADHAKRISHNA
Electric Field Gradient at the F_{Z_1} Electron Site in Cesium Halides
(a) Vol. 44, No. 2, p. 741—745 (1977).
See also *10.2*, p. *93*; 22.5.2, p. 382.

K. D. GLINCHUK and A. V. PROKHOROVICH
*Effect of the 0.94, 1.0, 1.2, and 1.3 eV Radiative Centres on the Intrinsic Luminescence Intensity in n-GaAs
(a) Vol. 44, No. 2, p. 777—785 (1977).
See also *20.3*, p. *275*; 22.2.1, p. 353.

13.5 Elementary Excitations (Phonons See 6)

Review Article

G. Borstel and H. J. Falge
*Surface Phonon-Polaritons at Semi-Infinite Crystals
(b) Vol. **83**, No. 1, p. 11—45 (1977).
See also *6*, p. *51*; 20.1, p. 247; 22, p. 320; 22.6, p. 386.

Original Papers and Short Notes

S. P. Ionov, V. S. Lubimov, L. A. Manakova, V. S. Mironov, and E. F. Makarov
*The Charge-Ordered State and Collective Excitations
(b) Vol. **79**, No. 2, p. 403—411 (1977).

E. Schuller, H. J. Falge, and G. Borstel
*Dispersion of Bulk Phonon Polaritons inside Reststrahlen Bands
(b) Vol. **80**, No. 1, p. 109—117 (1977).
See also *20.1*, p. *250*; 22.6, p. 386.

A. R. Vasconcellos and R. Luzzi
*Light Scattering and Spin Diffusion in Doped Semiconductors
(b) Vol. **80**, No. 1, p. 181—191 (1977).
See also 18.1, p. 213; 20.1, p. 251; 22.2.1, p. 351.

D. I. Pushkarov
On the Theory of Thermal Conductivity in Quantum Crystals
(b) Vol. **80**, No. 1, p. 295—300 (1977).
See also *8*, p. *64*; 10, p. 72.

S. P. Ionov, V. S. Lubimov, L. A. Manakova, and E. F. Makarov
The Charge-Ordered State and Magnetic Collective Excitations
(b) Vol. **80**, No. 1, p. K13—K15 (1977).
See also *13*, p. *108*; 22, p. 323.

R. Kishore
Semiclassical Theory of Spin Waves in Amorphous Antiferromagnets
(b) Vol. **80**, No. 1, p. K33—K35 (1977).
See also *2*, p. *34*; *18.4*, p. *232*.

J. S. Nkoma
*Theory of Two-Photon Absorption by Surface Polaritons
(b) Vol. **81**, No. 1, p. 307—312 (1977).
See also *20.1*, p. *254*; 22.2.2, p. 354.

S. P. Ionov, V. S. Lubimov, L. A. Manakova, and E. F. Makarov
The Electronic Spectra of Charge-Ordered Crystals
(b) Vol. **81**, No. 2, p. 691—701 (1977).
See also *13*, p. *109*; 22, p. 325.

M. Rösler and R. Zimmermann
*Phase Diagram of Electron–Hole Liquids in Polar Semiconductors
(b) Vol. **83**, No. 1, p. 85—91 (1977).
See also 13.5.1, p. 146; 13.5.2, p. 149; 13.5.3, p. 152; 22.2.1, p. 353; 22.4.1, p. 363; 22.4.2, p. 366; 22.4.3, p. 367.

B. Hönerlage, Vu Duy Phach, A. Bivas, and E. Ostertag
*Polariton Effects in Resonant Raman Scattering in CuCl
(b) Vol. **83**, No. 2, p. K101—K103 (1977).
See also *20.3*, p. *273*; 22.5, p. 372.

J. C. Ashley and R. H. Ritchie
*The Influence of Damping on the Mean Free Path of Low Energy Electrons in an Electron Gas
(b) Vol. 83, No. 2, p. K159—K163 (1977).
See also 13.5.2, p. 150; 21, p. 285.

L. A. Manakova, E. F. Makarov, and S. P. Ionov
The Electronic Spectra and Electromagnetic Properties of Charge-Ordered Systems
(b) Vol. 84, No. 1, p. 55—67 (1977).
See also *13*, p. *112*; 22, p. 329.

Vo Hong and Nguyen Nhu Dat
*Parametric Action of a Strong Radiation Field on Anisotropic Semiconductors
(b) Vol. 84, No. 1, p. 115—123 (1977).
See also *6*, p. *60*; 20, p. 246; 22.4.1, p. 364.

L. A. Manakova, E. F. Makarov, V. S. Lubimov, and S. P. Ionov
Some Peculiarities of Charge-Ordered Systems at Constant Electric Field
(b) Vol. 84, No. 1, p. K1—K6 (1977).
See also *13*, p. *112*; 22, p. 330.

M. J. Škrinjar, J. P. Šetrajčić, and D. V. Kapor
*On the Polaritonic Bose Condensation
(b) Vol. 84, No. 1, p. K25—K27 (1977).

V. S. Lubimov, E. F. Makarov, and S. P. Ionov
*The Magnetic Properties of Charge-Ordered Crystals
(b) Vol. 84, No. 2, p. 485—492 (1977).
See also *18*, p. *212*.

13.5.1 Excitons

Review Articles

A. Goldmann
*Band Structure and Optical Properties of Tetrahedrally Coordinated Cu- and Ag-Halides
(b) Vol. **81**, No. 1, p. 9—47 (1977).
See also *13.1*, p. *114*; *20.1*, p. *247*; 22.5, p. 370; 22.5.1, p. 373.

V. T. Agekyan
*Spectroscopic Properties of Semiconductor Crystals with Direct Forbidden Energy Gap
(a) Vol. **43**, No. 1, p. 11—42 (1977).
See also 6, p. 51; 20.1, p. 247; 20.3, p. 267; 22.6, p. 386.

Original Papers and Short Notes

P. Petelenz
*Mixing of Frenkel Excitons and Ionic Excited States of a Linear Molecular Crystal with Two Molecules in the Unit Cell (II): Physical Consequences
(b) Vol. **79**, No. 1, p. 61—70 (1977).
See also 20.1, p. 247; 22, p. 320.

S. I. Golubov and Yu. V. Konobeev
*Energy Transfer by Excitons in a Tetracene-Doped Anthracene Crystal
(b) Vol. **79**, No. 1, p. 79—84 (1977).
See also 13.4, p. 129; 20.3, p. 267; 22.9, p. 414.

D. Theis
Wavelength-Modulated Reflectivity Spectra of ZnSe and ZnS from 2.5 to 8 eV
(b) Vol. **79**, No. 1, p. 125—130 (1977).
See also *13.1*, p. *114*; *20.1*, p. *247*; 22.4.1, p. 362; 22.4.2, p. 365.

E. Majerníková
*On the Theory of Exciton Absorption in Amorphous Semiconductors
(b) Vol. **79**, No. 1, p. 339—345 (1977).
See also 2, p. 33; *20.1*, p. *248*; 22, p. 320.

F. Thuselt and K. Unger
Exchange Splitting of Donor–Acceptor Pairs
(b) Vol. **79**, No. 1, p. K39—K42 (1977).
See also *13.4*, p. *129*; 22.2, p. 348.

F. Henneberger, K. Henneberger, and J. Voigt
*On the Emission Lineshape of Excitonic Molecules in Wurtzite-Type Semiconductors
(b) Vol. **79**, No. 1, p. K81—K85 (1977).
See also *20.1*, p. *248*; 22.4.1, p. 362.

S. I. Subbotin, V. V. Panfilov, and R. T. Molchanova
Pressure-Induced Polytypic Transformations and Edge Absorption of GaSe
(a) Vol. **39**, No. 1, p. 357—362 (1977).
See also *1.2*, p. *14*; 12.2, p. 104; *20.1*, p. *248*; 22.3, p. 358.

S. D. Stojanović, J. P. Šetrajčić, M. J. Škrinjar, and B. S. Tošić
*On Some Possible Consequences of Exciton–Phonon Interaction
(b) Vol. **79**, No. 2, p. 433—439 (1977).
See also *6*, p. *52*.

G. O. Berim and A. R. Kessel
*The Model Hamiltonian and Some Properties of Magnetic Resonance of Triplet Excitons (II): Kubo-Tomita Approximation
(b) Vol. **79**, No. 2, p. 489—494 (1977).
See also *19*, p. *235*.

B. Gerlach
*On the Binding Energy of Wannier Excitons in Strong Anisotropic Crystals
(b) Vol. **79**, No. 2, p. 613—616 (1977).
See also 22, p. 321.

A. P. Dzhotyan
*Bose-Einstein Condensation of Excitons in Thin Semiconductors
(b) Vol. **79**, No. 2, p. K143—K146 (1977).
See also 22, p. 322.

W. Gorzkowski
*Excitons Bound to Ionized Donors in Polar Semiconductors
(b) Vol. **79**, No. 2, p. K151—K152 (1977).
See also 22, p. 322; 22.4.1, p. 362; 22.4.2, p. 365.

W. Ekardt and M. I. Sheboul
*LO-Phonon-Assisted Two-Photon Absorption of the First Excited State of the Biexciton in CuCl
(b) Vol. **80**, No. 1, p. 51—56 (1977).
See also 20.1, p. 250; 20.3, p. 268; 22.5, p. 370.

A. D. Brothers and D. M. Wieliczka
*Temperature and Hydrostatic Pressure Measurements of the Exciton Peaks in BiI_3 and SbI_3
(b) Vol. **80**, No. 1, p. 201—206 (1977).
See also 12.2, p. 104; 13.1, p. 115; 22.5, p. 370.

R. Enderlein and F. Bechstedt
Theory of Allowed Resonance Raman Scattering of First and Second Order in a Magnetic Field
(b) Vol. **80**, No. 1, p. 225—234 (1977).
See also *6*, p. *53*; 13.1, p. 115; *20.1*, p. *251*.

J. B. Aladekomo, S. Arnold, and M. Pope
*Triplet Exciton Diffusion and Double Photon Absorption in Tetracene
(b) Vol. **80**, No. 1, p. 333—340 (1977).
See also *20.3*, p. *268*; 22.9, p. 414.

J. Collet, J. Barrau, M. Brousseau, and H. Maaref
*Simultaneous Kinetics of Drops and Excitons in Silicon during the Drop Formation
(b) Vol. **80**, No. 2, p. 461—470 (1977).
See also 20.3, p. 269; 22.1.2, p. 338.

A. I. Anselm and I. G. Lang
*Theory of the Excitonic Mechanism of Crystal Defect Formation
(b) Vol. **80**, No. 2, p. 691—695 (1977).
See also *10.2*, p. *86*; 22.5.2, p. 377.

H. Röppischer, J. Jacobs, and B. N. Novikov
*Influence of Preparation Conditions on the Optical Band Gap and the Mn^{2+} ESR Spectrum of ZnSe Single Crystals
(a) Vol. **40**, No. 2, p. 575—581 (1977).
See also 13.4, p. 132; *19*, p. *237*; 22.4.2, p. 365.

M. Binder, N. E. Geacintov, M. Pope, and C. E. Swenberg
*Magnetic Field Modulated Scintillation from Tetracene-Doped Anthracene Crystals under α-Particle Irradiation
(a) Vol. **40**, No. 2, p. 621—628 (1977).
See also 11, p. 95; *20.3*, p. *269*; 22.9, p. 415.

E. L. Frankevich, M. M. Tribel, I. A. Sokolik, and B. V. Kotov
Photoconductivity of Charge Transfer Complex Crystals. Anthracene–Dimethylpyromellitimide
(a) Vol. **40**, No. 2, p. 655—662 (1977).
See also *16*, p. *202*; 22.9, p. 415.

P. Petelenz and B. Petelenz
*Exciton Dissociation in the Electric Field of the Charged Defect
(b) Vol. **81**, No. 1, p. 49—56 (1977).
See also 22.9, p. 415.

J. Laugsch and G. A. Gehring
Optical Side Bands in a Cooperative Jahn-Teller Material: $TmAsO_4$
(b) Vol. **81**, No. 1, p. 167—178 (1977).
See also 6, p. 55; *20.1*, p. *253*; 22.8.1, p. 404.

J. Donecker and J. Kluge
Magneto-Optical Investigation of the Ground State of the Indirect Free Exciton in Si and GaP
(b) Vol. **81**, No. 1, p. 199—203 (1977).
See also *18*, p. *210*; *20.1*, p. *253*; 22.1.2, p. 339; 22.2.2, p. 354.

F. Bechstedt and F. Henneberger
*Two-Photon Resonance Raman Scattering by Excitons via an Intermediate Excitonic Molecule (I): Bare-Exciton Approach
(b) Vol. **81**, No. 1, p. 211—220 (1977).
See also *20.3*, p. *270*; 22.4.1, p. 363; 22.5, p. 370.

O. Goede
*Deformation Effects of Biexcitons in Wurtzite-Type Crystals
(b) Vol. **81**, No. 1, p. 235—244 (1977).
See also 13.1, p. 117; *20.1*, p. *253*; 22.4.1, p. 363.

M. Sondergeld
*Two-Photon Absorption by Envelope–Hole Coupled Exciton States in Cubic ZnSe (I): Energy Spectrum of the H_d Term and Valence-Band Parameters
(b) Vol. **81**, No. 1, p. 253—262 (1977).
See also *20.1*, p. *254*; 22.4.2, p. 365.

B. A. Lukiyanets and K. D. Tovstyuk
*The Exciton Contribution to the Dielectric Susceptibility of a Crystal with Extended Defects
(b) Vol. **81**, No. 1, p. K1—K3 (1977).
See also *14.4*, p. *190*.

G. D. Guseinov, S. G. Abdullaeva, E. M. Godzhaev, A. Sh. Mekhtiev, L. A. Alieva, and M. Z. Ismailov
Electroabsorption of $TlInS_2$ Single Crystals
(b) Vol. **81**, No. 1, p. K47—K50 (1977).
See also *14.3.4*, p. *185*; *20.1*, p. *254*; 22.8, p. 398.

M. Sondergeld
*Two-Photon Absorption by Envelope–Hole Coupled Exciton States in Cubic ZnSe (II): Polarization Dependence in the Case of Degenerate Valence Bands
(b) Vol. **81**, No. 2, p. 451—461 (1977).
See also *20.1*, p. *255*; 22.4.2, p. 366.

J. Hoshen and R. Kopelman
*Molecular Exciton Cluster States and Spectra with Application to Benzene
(b) Vol. **81**, No. 2, p. 479—491 (1977).
See also *20.1*, p. *255*; 22.9, p. 416.

P. Schlotter, J. Kalinowski, and H. Bässler
*Photoionization of Triplet Excitons in Tetracene Crystals
(b) Vol. **81**, No. 2, p. 521—526 (1977).
See also 16, p. 202; 22.9, p. 416.

G. Antonioli, D. Bianchi, V. Canevari, U. Emiliani, and P. Podini
Optical Constants of GaSe at the Fundamental Absorption Edge
(b) Vol. **81**, No. 2, p. 665—670 (1977).
See also *20.1*, p. *255*; 22.3, p. 358.

Yu. E. Kotelnikov and B. I. Kochelaev
*On the Bose-Einstein Condensation of Excitons
(b) Vol. **81**, No. 2, p. 747—751 (1977).

S. I. Pekar
*On the Theory of Additional Light Waves and Boundary Conditions in Anisotropic Crystals
(b) Vol. **82**, No. 1, p. 83—89 (1977).
See also *20.1*, p. *256*; 22, p. 325.

E. Ostertag and J. B. Grun
*Subnanosecond Time-Resolved Spectroscopy of the Biexciton Luminescence in CuCl
(b) Vol. **82**, No. 1, p. 335—342 (1977).
See also *20.3*, p. *271*; 22.5, p. 371.

H. Haug and S. Koch
*On the Theory of Laser Action in Dense Exciton Systems
(b) Vol. **82**, No. 2, p. 531—543 (1977).
See also *20.2*, p. *266*; 22.4.1, p. 363.

K. K. Bajaj and C. Aldrich
*Effective Electron–Hole Interaction for Intermediate and Strong Electron–Phonon Coupling — A Comment
(b) Vol. **82**, No. 2, p. 663—668 (1977).
See also 13.5.3, p. 152; 22.5, p. 371; 22.6, p. 390.

A. V. Lelyakov
*Luminescence Band Shape of Free Excitons in AgBr Crystals at High Exciton Density
(b) Vol. **82**, No. 2, p. 711—716 (1977).
See also 6, p. 58; *20.3*, p. *271*; 22.5.1, p. 373.

H. Rüfer and W. Ruppel
Luminescence of Graded $Zn_xCd_{1-x}S$ Single Crystals
(a) Vol. **42**, No. 2, p. 519—524 (1977).
See also *20.2*, p. *266*; *20.3*, p. *272*; 22.4.4, p. 368.

A. Selkin
*Energy Transfer by the Normal Modes in the Exciton Resonance Region
(b) Vol. **83**, No. 1, p. 47—53 (1977).
See also 20.1, p. 259; 22.4.1, p. 363.

M. Rösler and R. Zimmermann
Phase Diagram of Electron–Hole Liquids in Polar Semiconductors
(b) Vol. **83**, No. 1, p. 85—91 (1977).
See also *13.5*, p. *141*; 13.5.2, p. 149; 13.5.3, p. 152; 22.2.1, p. 353; 22.4.1, p. 363; 22.4.2, p. 366; 22.4.3, p. 367.

P. Petelenz
*Effect of the Mixing between Frenkel Excitons and Charge Transfer States on Charge Carrier Transfer in Molecular Crystals
(b) Vol. **83**, No. 1, p. 169—175 (1977).
See also 22.9, p. 417.

J. Burgos, M. Pope, Ch. E. Swenberg, and R. R. Alfano
*Heterofission in Pentacene-Doped Tetracene Single Crystals
(b) Vol. **83**, No. 1, p. 249—256 (1977).
See also *20.3*, p. *272*; 22.9, p. 418.

J. P. WOLFE, R. S. MARKIEWICZ, J. E. FURNEAUX, S. M. KELSO, and C. D. JEFFRIES
*Magnetostriction of an Electron–Hole Drop in Ge
(b) Vol. **83**, No. 1, p. 305–312 (1977).
See also 18.1, p. 215; 22.1.1, p. 335.

F. R. KESSLER and E. BREDNER
Cathodoluminescence of Selenium
(a) Vol. **43**, No. 1, p. 159–166 (1977).
See also *20.3*, p. *272*; 22.1.3, p. 346.

F. HENNEBERGER, K. HENNEBERGER, and J. VOIGT
*Two-Photon Resonance Raman Scattering by Excitons via an Intermediate Excitonic Molecule (II): Polarization Properties and Polariton Effects
(b) Vol. **83**, No. 2, p. 439–450 (1977).
See also *20.3*, p. *273*; 22.4.1, p. 364; 22.5, p. 371.

D. HADŽIAHMETOVIĆ, M.PIRIĆ, and B. S. TOŠIĆ
*Kinematical Exciton Levels and Absorption of Light
(b) Vol. **83**, No. 2, p. 479–483 (1977).
See also 20.1, p. 260; 20.3, p. 273.

A. FORT and A. CORET
*Triplet Absorption of Crystalline Anthracene by High Resolution Photoconductivity Spectroscopy
(b) Vol. **83**, No. 2, p. 599–606 (1977).
See also *16*, p. *204*; 22.9, p. 418.

R. S. MARKIEWICZ
*Acoustic Phonon Scattering by Electron–Hole Droplets in Ge: Deformation Potentials
(b) Vol. **83**, No. 2, p. 659–668 (1977).
See also 14.3, p. 171; 22.1.1, p. 335.

K. O. KECHECHYAN and A. A. KIRAKOSSYAN
*Wannier-Mott Exciton in the Field of Screw Dislocations
(b) Vol. **83**, No. 2, p. K105–K109 (1977).
See also 10.2, p. 91; 22, p. 329.

Ž. M. ŠKRBIČ, M. J. ŠKRINJAR, and D. V. KARPO
*On the Existence of Biexcitonic Level Splitting
(b) Vol. **83**, No. 2, p. K125–K127 (1977).

NGUYEN THE KHOI and J. A. GAJ
Fundamental Absorption Edge of $Cd_{1-x}Mn_xTe$ Mixed Crystals
(b) Vol. **83**, No. 2, p. K133–K135 (1977).
See also *20.1*, p. *261*; 22.4.4, p. 369.

V. P. MUSHINSKII, L. I. PALAKI, and V. V. CHEBOTARU
Optical Absorption in α-Ga_2S_3 Single Crystals
(b) Vol. **83**, No. 2, p. K149–K154 (1977).
See also *20.1*, p. *261*; 22.3, p. 358.

T. MATSUMURA, M. SUDO, C. TATSUYAMA, and S. ICHIMURA
Photoluminescence in GaSe
(a) Vol. **43**, No. 2, p. 685–693 (1977).
See also *20.3*, p. *273*; 22.3, p. 359.

W. BIZZARO, J. ROSENTHAL, N. F. BERK, and L. YARMUS
*The EPR Linewidth of Triplet Excitons in Single-Crystal Anthracene: a Re-Examination
(b) Vol. **84**, No. 1, p. 27–32 (1977).
See also *19*, p. *242*; 22.9, p. 419.

V. A. POPOV and A. A. LOGINOV
*Generalization of the Heisenberg Hamiltonian for the Case of an Atomic System with a Variable Atomic Spin Value
(b) Vol. **84**, No. 1, p. 83–91 (1977).
See also *13.5.4*, p. *154*.

A. Ya. Korets, V. P. Fedorov, A. V. Korshunov, and V. A. Naumov
*Spectroscopic Investigation of Orientational Disorder in an Acenaphthylene Molecular Crystal
(b) Vol. 84, No. 1, p. 107—113 (1977).
See also 13.4, p. 138; *20.1*, p. *261*; 22.9, p. 419.

A. Bivas, Vu Duy Phach, B. Hönerlage, and J. B. Grun
*Induced Absorption Due to Exciton–Biexciton Transitions in CuCl
(b) Vol. 84, No. 1, p. 235—243 (1977).
See also *20.3*, p. *274*; 22.5, p. 372.

E. Schwarzer and H. Haken
*Theory of the Influence of Exciton Motion on Proton Relaxation in Organic Solids
(b) Vol. 84, No. 1, p. 253—267 (1977).
See also 19, p. 242; 22.9, p. 419.

W. Ekardt
*Theory of Excitons in Zincblende-Structure Semiconductors in Intermediate Magnetic Fields
(b) Vol. 84, No. 1, p. 293—304 (1977).
See also 20.1, p. 261; 22.2.1, p. 353; 22.2.2, p. 355.

R. Schwabe, F. Thuselt, and T. Hänsel
*Exciton–Phonon Energy Transfer in Semiconductors
(b) Vol. 84, No. 1, p. K43—K47 (1977).
See also *6*, p. *60*; 20.3, p. 274; 22.3, p. 359.

V. I. Lesin, V. P. Sakun, A. I. Pristupa, and E. L. Frankevich
*Reaction Yield Detected Magnetic Resonance Spectra of Intermediate Pairs of Triplet Excitons. Evolution of Spectra during Degeneracy of Energy Levels
(b) Vol. 84, No. 2, p. 513—520 (1977).
See also *19*, p. *243*; 22.9, p. 419.

W. Schmid
*Auger Lifetimes for Excitons Bound to Neutral Donors and Acceptors in Si
(b) Vol. 84, No. 2, p. 529—540 (1977).
See also *20.3*, p. *275*; 22.1.2, p. 344.

H. Lange and I. A. Iliev
*Evidence of Coulomb Interaction at the E_1 Interband Transition of Hexagonal CdSe
(b) Vol. 84, No. 2, p. 569—574 (1977).
See also 13.1, p. 122; *20.1*, p. *262*; 22.4.2, p. 366.

E. Ostertag, A. Bivas and J. B. Grun
*Polariton Effects on the Recombination of Cold Biexcitons
(b) Vol. 84, No. 2, p. 673—680 (1977).
See also *20.3*, p. *275*; 22.5, p. 372.

Vu Duy Phach, A. Bivas, B. Hönerlage, and J. B. Grun
*Biexciton Resonant Two-Photon Absorption in CuCl
(b) Vol. 84, No. 2, p. 731—740 (1977).
See also *20.1*, p. *263*; 22.5, p. 372.

A. Anedda and E. Fortin
*Excitons in HgI_2
(b) Vol. 84, No. 2, p. K87—K90 (1977).
See also 20.1, p. 263; 22.5, p. 372.

S. D. Stojanović and M. J. Škrinjař
*The Influence of the Repulsion in the Exciton System on the Formation of Exciton Drops
(b) Vol. 84, No. 2, p. K101—K103 (1977).

13.5.2 Plasmons

Original Papers and Short Notes

G. S. Agarwal and H. D. Vollmer
Surface Polariton Effects in Spontaneous Emission
(b) Vol. **79**, No. 1, p. 249—257 (1977).
See also *13.3*, p. *126*; 20.1, p. 247; 21, p. 276; 22.2.2, p. 354.

C. A. S. Lima and L. C. M. Miranda
*Plasmon-Enhanced Sound Amplification in Piezoelectric Semiconductors under a Strong Magnetic Field
(b) Vol. **80**, No. 1, p. 57—62 (1977).
See also 7, p. 62; 18, p. 210; 22.2.1, p. 351.

H. Hora
*Corrections of the Long Beating Length in Quantum Modulated Electron Beams
(b) Vol. **80**, No. 1, p. 143—147 (1977).
Erratum
(b) Vol. **83**, No. 1, p. 355 (1977).
See also *20.2*, p. *265*.

H. J. Stolz, E. Petri, and A. Otto
*Optical Properties and Plasmon Dispersion in Epitaxial $(SN)_x$ by Electron Energy Loss Spectroscopy
(b) Vol. **80**, No. 1, p. 289—294 (1977).
See also 20.1, p. 251; 21, p. 278.

H. Venghaus
Infrared Reflectance and Dielectric Properties of Pyrolytic Graphite for $\boldsymbol{E} \parallel \boldsymbol{c}$ Polarization
(b) Vol. **81**, No. 1, p. 221—225 (1977).
See also 13.1, p. 117; *20.1*, p. *253*; 22.1, p. 332.

B. S. Abramovich and A. A. Ignatov
*Plasma Wave Damping in a Layered Electron Gas with Random Inhomogeneities
(b) Vol. **81**, No. 1, p. 341—349 (1977).

V. B. Campos, O. Hipólito, and R. Lobo
*Correlations and Plasmons in Quasi-One-Dimensional Metals
(b) Vol. **81**, No. 2, p. 657—663 (1977).
See also 14.4, p. 191; 21, p. 281.

G. M. Shmelev, I. A. Chaikovskii, V. V. Pavlovich, and E. M. Epshtein
*Plasma Oscillations in a Superlattice
(b) Vol. **82**, No. 1, p. 391—395 (1977).
See also 22, p. 326.

K. Suzuki, M. Kitagawa, and Y. H. Ohtsuki
*Theory of Surf-Riding Electrons in the Wake Potential of Ion Beams in a Metal
(b) Vol. **82**, No. 2, p. 643—650 (1977).
See also 21, p. 283.

V. V. Maksimenko, A. J. Simonov, and A. A. Lushnikov
*Surface Plasmons in Thin Metal Films, Circular Cylinders, and Spherical Voids
(b) Vol. **82**, No. 2, p. 685—693 (1977).
See also 21, p. 283.

M. Rösler and R. Zimmermann
Phase Diagram of Electron–Hole Liquids in Polar Semiconductors
(b) Vol. **83**, No. 1, p. 85—91 (1977).
See also *13.5*, p. *141*; 13.5.1, p. 146; 13.5.3, p. 152; 22.2.1, p. 353; 22.4.1, p. 363; 22.4.2, p. 366; 22.4.3, p. 367.

B. RHEINLÄNDER, S. HAASE, and H. WEINERT
On the Determination of the Free Carrier Relaxation Time by Transverse Magneto-Plasma Reflectivity Measurements on Sn-Doped n-Type GaAs
(b) Vol. 83, No. 1, p. K83—K86 (1977).
See also *18*, p. *212*; *20.1*, p. *260*; 22.2.1, p. 353.

V. V. MAKSIMENKO, A. J. SIMONOV, and A. A. LUSHNIKOV
Far-Infrared Absorption in Small Metallic Particles
(b) Vol. 83, No. 2, p. 377—382 (1977).
See also *13.3*, p. *127*; *20.1*, p. *260*.

VO HONG ANH and NGUYEN VAN TRONG
*On the Theory of Parametric Excitation in Quantum Electron–Phonon Systems
(b) Vol. 83, No. 2, p. 395—404 (1977).
See also *6*, p. *59*; 20, p. 246; 22, p. 328.

J. C. ASHLEY and R. H. RITCHIE
The Influence of Damping on the Mean Free Path of Low Energy Electrons in an Electron Gas
(b) Vol. 83, No. 2, p. K159—K163 (1977).
See also *13.5*, p. *142*; 21, p. 285.

R. H. RITCHIE, R. N. HAMM, M. W. WILLIAMS, and E. T. ARAKAWA
*Dispersion Relations of Elementary Excitations when Damping Is Present
(b) Vol. 84, No. 1, p. 367—373 (1977).

13.5.3 Polarons

Review Article

M. I. Klinger and A. A. Samokhvalov
*Electron Conduction in Magnetite and Ferrites
(b) Vol. **79**, No. 1, p. 9—48 (1977).
Erratum
(b) Vol. **83**, No. 1, p. 353 (1977).
See also 13.1, p. 114; *14.3*, p. *165*; 15, p. 198; 18.3, p. 228; 22.8.2, p. 409.

Original Papers and Short Notes

R. von Baltz and B. Urban
*Interaction of Holes with Local Modes and One and Two Optical Phonons as Revealed in M–I–Silicon Tunnelling Spectra
(b) Vol. **79**, No. 1, p. 185—194 (1977).
See also 22.1.2, p. 337.

S. P. Reifman
*The Influence of a Strong Magnetic Field on the Self-Trapping of Electron in Homopolar Crystals
(b) Vol. **79**, No. 1, p. K57—K59 (1977).
See also 22, p. 321.

W. Müller-Lierheim, T. Suski, and H. H. Otto
Factor Group Analysis of the Raman Spectrum of $Pb_5Ge_3O_{11}$
(b) Vol. **80**, No. 1, p. 31—41 (1977).
See also *6*, p. *53*; 14.4.1, p. 194; 20.1, p. 250; 22.8.1, p. 403.

H. Böttger and V. V. Bryksin
A Theory of the Hall Effect in the Hopping Region in Disordered Systems (I): Rate Equation in Presence of Electrical and Magnetic Fields
(b) Vol. **80**, No. 2, p. 569—578 (1977).
See also 2, p. 35; 6, p. 54; 13, p. 108; *14.3*, p. *167*; 22, p. 323.

H. Böttger and V. V. Bryksin
A Theory of the Hall Effect in the Hopping Region in Disordered Systems (II): Random Resistor Network and Hall Effect
(b) Vol. **81**, No. 1, p. 97—106 (1977).
See also 2, p. 36; 6, p. 55; 13, p. 109; *14.3*, p. *167*; 22, p. 324.

H. Böttger and V. V. Bryksin
A Theory of the Hall Effect in the Hopping Region in Disordered Systems (III): DC Hall Effect
(b) Vol. **81**, No. 2, p. 433—441 (1977).
See also 2, p. 37; 6, p. 56; 13, p. 109; *14.3*, p. *168*; 22, p. 325.

H.-R. Trebin
*Polarons in the Degenerate-Band Case (II): Weak to Intermediate Hole–Phonon Coupling
(b) Vol. **81**, No. 2, p. 527—534 (1977).
See also 13.1, p. 118; 22.2, p. 348; 22.4, p. 361.

W. J. Huybrechts
*Infrared Absorption by an Optical Polaron in a Magnetic Field
(b) Vol. **81**, No. 2, p. 585—592 (1977).
See also *20.1*, p. *255*; 22, p. 325.

L. F. Lemmens, J. T. Devreese, and F. Brosens
*On the Ground State Energy of a Gas of Interacting Polarons
(b) Vol. **82**, No. 2, p. 439—447 (1977).

W. J. HUYBRECHTS
*Influence of a DC Field on the Free Polaron Infrared Absorption
(b) Vol. **82**, No. 2, p. 589—594 (1977).
See also *20.1*, p. *258*; 22, p. 327.

K. K. BAJAJ and C. ALDRICH
Effective Electron–Hole Interaction for Intermediate and Strong Electron–Phonon Coupling — A Comment
(b) Vol. **82**, No. 2, p. 663—668 (1977).
See also *13.5.1*, p. *146*; 22.5, p. 371; 22.6, p. 390.

B. R. SETHI, P. L. TALWAR, O. P. SHARMA, and P. C. MATHUR
The Electrical Transport Properties of a Zinc Selenide Crystal Highly Annealed in Molten Zinc
(a) Vol. **42**, No. 2, p. 791—795 (1977).
See also *14.3*, p. *170*; 22.4.2, p. 366.

M. RÖSLER and R. ZIMMERMANN
Phase Diagram of Electron–Hole Liquids in Polar Semiconductors
(b) Vol. **83**, No. 1, p. 85—91 (1977).
See also *13.5*, p. *141*; 13.5.1, p. 146; 13.5.2, p. 149; 22.2.1, p. 353; 22.4.1, p. 363; 22.4.2, p. 366; 22.4.3, p. 367.

R. MAŃKA
*The Thermoelectric Power for a Small Polaron in Ferroelectrics
(b) Vol. **83**, No. 1, p. K69—K73 (1977).
See also 14.4.1, p. 196; *15*, p. *199*; 22.8.1, p. 406.

K. PESZ and T. LUTY
*On the Charge Carrier Mobilities in Molecular Crystals
(b) Vol. **83**, No. 2, p. 565—571 (1977).
See also 14.3, p. 171; 22, p. 328.

13.5.4 Magnons

Review Article

P. Röschmann and H. Dötsch
Properties of Magnetostatic Modes in Ferrimagnetic Spheroids
(b) Vol. **82**, No. 1, p. 11—57 (1977).
See also *18.3*, p. *228*; 19, p. 235; 22.8.2, p. 409.

Original Papers and Short Notes

P. R. R. Pereira and L. C. Miranda
*Magnon Contribution to the Phonon Damping in Ferromagnetic Semiconductors
(b) Vol. **79**, No. 1, p. K49—K52 (1977).
See also *6*, p. *52*; 18.2, p. 217; 22.8, p. 396.

A. M. Oleś and J. Spałek
*Spin-Wave Excitations in Narrow s-Band with Inclusion of Intersite Exchange Interactions
(b) Vol. **79**, No. 1, p. K61—K64 (1977).

W. Nolting
*Influence of Electron–Ion Spin Coupling on the Electronic Properties of a Ferromagnetic Semiconductor
(b) Vol. **79**, No. 2, p. 573—584 (1977).
See also 18.2, p. 218; 22, p. 321.

T. Nasu
The Temperature Dependence of the Absorption Spectrum of NiO
(b) Vol. **80**, No. 2, p. 675—679 (1977).
See also *20.1*, p. *252*; 22.6, p. 387.

G. M. Carneiro and P. R. Rodrigues Pereira
*Fermi Liquid Effects in Ferromagnetic Metals
(b) Vol. **81**, No. 1, p. 79—88 (1977).
See also 18.2, p. 219; 21, p. 280.

P. R. R. Pereira, M. A. Amato, and L. C. M. Miranda
*Carrier-Assisted Radiation Pumping of Spin Waves in Ferromagnetic Semiconductors under a Strong Magnetic Field
(b) Vol. **81**, No. 1, p. 395—401 (1977).
See also 14.3, p. 167; 18.2, p. 219; 22, p. 324.

D. I. Pushkarov and Kh. I. Pushkarov
*Solitons in One-Dimensional Ferromagnetic Systems
(b) Vol. **81**, No. 2, p. 703—707 (1977).
See also *18.2*, p. *220*.

Yu. I. Balkarei and V. A. Nikitov
*On Magnon–Acoustic Phonon Interaction in Magnetic Semiconductors
(b) Vol. **82**, No. 1, p. 127—131 (1977).
See also *6*, p. *57*; 18.2, p. 221; 22, p. 326.

Diep-The-Hung
*Effect of the Next-Nearest Neighbour Coupling in a Semi-Infinite Ferrimagnet
(b) Vol. **82**, No. 1, p. 383—389 (1977).
See also 1.6, p. 31; *18.3*, p. *229*.

V. V. Zolotarev
*Erratum to: Magnon Damping Caused by Spin–Spin Exchange Interaction
(b) Vol. **83**, No. 2, p. K179 (1977).
See also 18.2, p. 223.

V. A. Popov and A. A. Loginov
*Generalization of the Heisenberg Hamiltonian for the Case of an Atomic System with a Variable Atomic Spin Value
(b) Vol. 84, No. 1, p. 83—91 (1977).
See also *13.5.1*, p. *147*.

Diep-The-Hung
*Effect of Induced Surface Magnetic Moment on Spin-Wave Spectrum
(b) Vol. 84, No. 2, p. K159—K161 (1977).
See also **1.6**, p. **32**; **18.4**, p. **234**.

14. Electrical Properties. Transport Phenomena

Original Papers and Short Notes

E. Kolley and W. Kolley
*On the Theory of Phonon-Modulated Electronic Transport in Off-Diagonal Random Alloys
(b) Vol. **79**, No. 1, p. 325—337 (1977).
See also *2*, p. *33*; 6, p. 51; 21, p. 276.

G. Cubiotti, E. Donato, E. S. Giuliano, and R. Ruggeri
On the Influence of Disorder on the Electron Density of States in a Simplified Model for a Quasi-Onedimensional Conductor
(b) Vol. **79**, No. 2, p. 655—661 (1977).
See also *2*, p. *33*; *13*, p. *107*; 22, p. 321.

F. Brosens, J. T. Devreese, and L. F. Lemmens
*Frequency-Dependent Exchange Correction to the Dielectric Function of the Electron Gas (II)
(b) Vol. **80**, No. 1, p. 99—107 (1977).

J. L. Farvaque and P. Lenglart
*Inhomogeneous Strain Field Scattering Deformation Potential (I): Theory
(b) Vol. **80**, No. 1, p. 361—367 (1977).
See also *10*, p. *72*; 13.1, p. 115.

J. C. Phillips
*A Simple Explanation for the "$4k_F$ Anomaly" in TTF-TCNQ
(b) Vol. **80**, No. 1, p. 395—398 (1977).
See also *1*, p. *9*; 22.9, p. 415.

D. Ihle
*Electrical Conductivity in the Hubbard Model Including Electron–Phonon Interaction
(b) Vol. **80**, No. 2, p. 619—628 (1977).
See also *6*, p. *54*.

E. Gerlach and M. Rautenberg
*Scattering of DC Electrons by a Spherical Macroscopic Potential Including Screening
(b) Vol. **81**, No. 1, p. 153—156 (1977).
See also *13.4*, p. *133*.

R. B. Somoano, V. Hadek, S. P. S. Yen, A. Rembaum, C. H. Hsu, R. J. Deck, T. Datta, and A. M. Hermann
*Electrical Properties of (DEPE) $(TCNQ)_4$
(b) Vol. **81**, No. 1, p. 281—286 (1977).
See also 15, p. 198; 22.9, p. 415.

F. Brosens, L. F. Lemmens, and J. T. Devreese
*The Pair Correlation Function and the Dielectric Response in the Homogeneous Electron Gas
(b) Vol. **82**, No. 1, p. 117—125 (1977).

W. Schiller, G. Röpke, G. Buhr, and V. Christoph
*The Derivation of a Direct Formula for the Hall Coefficient
(b) Vol. **82**, No. 1, p. 297—302 (1977).

J. R. Chelikowsky, M. Schlüter, and M. L. Cohen
Semi-Empirical Pseudopotential Calculation for the Electronic Structure of Polymeric Sulfur Nitride
(b) Vol. **82**, No. 1, p. 357—367 (1977).
See also *13.1*, p. *119*; 22, p. 326.

L. C. R. Alfred
*Charge Screening in an Electron Gas of Arbitrary Degeneracy
(b) Vol. **82**, No. 2, p. 467—471 (1977).
See also *13.1*, p. *120*; 21, p. 282; 22, p. 326.

U. Kh. Kopvillem and S. V. Prants
*Octupole and Hexadecapole Dynamics in Solid State Physics
(b) Vol. **83**, No. 1, p. 109—114 (1977).
See also *7*, p. *63*.

V. F. Kaminskii, M. L. Khidekel, R. B. Lyubovskii, I. F. Shchegolev, R. P. Shibaeva, E. B. Yagubskii, A. V. Zvarykina, and G. L. Zvereva
*Metal–Insulator Phase Transition in TTT_2I_3 Affected by Iodine Concentration
(a) Vol. **44**, No. 1, p. 77—82 (1977).
See also *1*, p. *10*; 18.1, p. 215; 22.9, p. 419.

H. J. Fischbeck
The Diagonal Vertex Approximation for Gaussian Random Systems as a Further Test of the CPA
(b) Vol. **84**, No. 2, p. 603—609 (1977).
See also *13*, p. *113*.

J. Monecke
One-Particle Spectrum of Correlated Electrons near the Free-Electron Limit
(b) Vol. **84**, No. 2, p. 665—672 (1977).
See also *13.1*, p. *123*; 18.4, p. 234.

14.1 Metals. Semi-Metals

Original Papers and Short Notes

K. A. Long
*A Theoretical Investigation of the Specific Heats and Transport Properties of the Actinides
(b) Vol. **79**, No. 1, p. 155—165 (1977).
See also *6*, p. *51*; 8, p. 64; 15, p. 198; 21.5, p. 314.

M. A. Belogolovskii, Yu. M. Ivanchenko, E. V. Kochergin, A. A. Lisyanskii, and Yu. V. Medvedev
*Interaction of Electrons with Phonons and Magnetic Impurities in Tunneling Junctions
(b) Vol. **79**, No. 1, p. K17—K21 (1977).
See also 6, p. 52; *14.2*, p. *161*; 21, p. 276.

M. Yahagi and K. Iwamura
*Temperature Dependences of Electrical Resistivity and Thermal Expansion in the Intermetallic Compound LiIn
(a) Vol. **39**, No. 1, p. 189—192 (1977).
See also *8*, p. *64*; 10.1, p. 75; 21.2, p. 308.

A. Feher, M. Reiffers, P. Petrovič, and Š. Jánoš
*The Electrical Resistivity of Nb–Ti and Nb–Ti–Zr Alloys at Low Temperatures
(a) Vol. **39**, No. 1, p. K67—K69 (1977).
See also 14.2, p. 161; 21, p. 277.

E. Kolley and W. Kolley
*Calculation of the Temperature-Dependent DC Conductivity of Disordered Alloys in Off-Diagonal CPA
(b) Vol. **79**, No. 2, p. 461—472 (1977).
See also *2*, p. *33*; 6, p. 52; 13, p. 107; 21, p. 277.

W. Braune, G. Kuka, S. Hess, H.-U. Müller, and T. Jung
Cyclotron Resonance and Quantum Oscillations in Semimetallic Bismuth–Antimony Alloys with $E_g = 0$
(b) Vol. **79**, No. 2, p. 501—508 (1977).
See also *13.2*, p. *124*; 21.7, p. 319.

G. J. Ashwell, D. D. Eley, M. R. Willis, and J. Woodward
*Electrical and Magnetic Properties of Two Crystallographic Forms of $(DEPE)^{2+}(TCNQ)_4^{2-}$
(b) Vol. **79**, No. 2, p. 629—636 (1977).
See also *14.3*, p. *165*; 15, p. 198; 18.1, p. 213; 22.9, p. 414.

K. Saermark and J. Lebech
Dispersion of Cyclotron Waves in Metals and Observability of Many-Body Effects
(b) Vol. **79**, No. 2, p. 723—730 (1977).
See also *13.2*, p. *124*; 21.2, p. 308; 21.6, p. 315.

N. Singh, J. Singh, and S. Prakash
*Residual Resistivity of Noble Metals
(b) Vol. **79**, No. 2, p. 787—794 (1977).
See also 13.4, p. 131; 21.6, p. 315.

H. P. Geserich, W. Möller, G. Scheiber, and L. Pintschovius
*Optical Investigation of the Electrical Anisotropy of $(SN)_x$ Single Crystals
(b) Vol. **80**, No. 1, p. 119—123 (1977).
See also 1.4, p. 24; *20.1*, p. *250*; 22, p. 322.

W. Dander and H. E. Schaefer
*Electrical Resistivity Recovery of Cobalt after Low-Temperature Electron Irradiation
(b) Vol. **80**, No. 1, p. 173—179 (1977).
See also *11*, p. *94*; 21.1, p. 290.

V. A. Kovarskii, N. F. Perelman, I. A. Chaikovskii, and V. N. Chebotar
*Conductivity in Thin (Quantizing) Wires
(b) Vol. **80**, No. 1, p. K5—K8 (1977).
See also 21.7, p. 319.

E. Mrosan
*Residual Resistivity for Electrons Scattered by Spin-Dependent Potentials
(b) Vol. **80**, No. 1, p. K49—K53 (1977).
See also 21.1, p. 290.

E. M. Gray and J. H. Smith
Effect of Ageing on the Magnetic Moment of MnCu Micromagnet
(a) Vol. **40**, No. 1, p. K73—K76 (1977).
See also *18.2*, p. *219*; 21.1, p. 291.

V. Christoph and G. Röpke
*On the Relation between the Resistivity Formula and the Boltzmann Equation
(b) Vol. **80**, No. 2, p. K117—K120 (1977).
See also *14.3*, p. *167*.

K. Ozawa and S. Anzai
*Effect of Pressure on a New Magnetic Transition Temperature in the Ferrimagnetic State of $Mn_{1.77}Sn$
(a) Vol. **40**, No. 2, p. K199—K202 (1977).
See also *18.3*, p. *229*; 21.7, p. 319.

Š. Jánoš, A. Feher, and K. Flachbart
*Magnetoresistance of Pr, Nd, Sm, and Eu at Low Temperatures
(b) Vol. **81**, No. 1, p. K19—K22 (1977).
See also 21.4, p. 312.

K. Saermark and J. Lebech
*A Finite $\omega\tau$-Calculation of the Dispersion Relation for Ordinary Cyclotron Waves for a Spherical Fermi Surface
(b) Vol. **81**, No. 1, p. K31—K36 (1977).
See also 13.2, p. 124.

Y. Waseda and J. G. Wright
*The Resistivity of Amorphous Transition Elements
(b) Vol. **81**, No. 1, p. K37—K40 (1977).
See also *2*, p. *36*; 21.1, p. 293.

V. Christoph and K. Elk
*Electrical Conductivity of Disordered Substitutional Alloys with Strong Electron Correlation
(b) Vol. **81**, No. 2, p. K125—K130 (1977).
See also 21, p. 281.

Amar Nath Nigam and Ratna Rani
*Surface States of Chemisorbed Hydrogen and Electrical Conduction in Thin Hydrogenated Nickel Foils
(a) Vol. **41**, No. 2, p. 493—498 (1977).
See also *13.3*, p. *127*; 21.1, p. 294.

E. Gratz and C. A. Poldy
The Influence of Antiferromagnetic Order on the Evaluation of Spin Disorder Resistivity
(b) Vol. **82**, No. 1, p. 159—167 (1977).
See also 13.1, p. 118; *18.2*, p. *221*; *18.4*, p. *233*; 21.1, p. 295; 21.4, p. 312.

Y. D. Yao, S. Arajs, and E. E. Anderson
*Phenomenological "Dilution Model" for Kondo-Type Effects in Concentrated Alloy Systems
(b) Vol. **82**, No. 1, p. 343—347 (1977).
See also *13.4*, p. *136*; 18, p. 211; 21.1, p. 295.

F. ANTONANGELI, A. BALZAROTTI, A. BIANCONI, E. BURATTINI, P. PERFETTI, and A. KISIEL
*Influence of the Hydrogenation on the Electrical Resistance of Palladium Thin Films
(a) Vol. **42**, No. 1, p. K41—K45 (1977).
See also 21, p. 282.

G. K. BRISTOW, T. R. FINLAYSON, and T. F. SMITH
*Hall Coefficient Sign Reversals in 2H Transition Metal Dichalcogenides
(b) Vol. **82**, No. 2, p. K81—K84 (1977).
See also *13.2*, p. *124*; 21, p. 283.

K. SAERMARK, J. LEBECH, and H. JOHANSEN
*On the Concept of Cyclotron Resonance in Metals
(b) Vol. **83**, No. 1, p. 187—195 (1977).
See also *13.2*, p. *124*; 21, p. 284.

H. KRIZEK, K. N. R. TAYLOR, and W. D. CORNER
*Resistivity of Some Light–Heavy Rare-Earth Alloys (I): Resistivity Variation in the Temperature Range 4.2 to 300 K
(a) Vol. **43**, No. 1, p. 251—260 (1977).
See also 21.4, p. 313.

S. CHAUDHURI and A. K. PAL
Thickness Dependence of Curie Temperature of Cu–Ni Films
(a) Vol. **43**, No. 1, p. K9—K13 (1977).
See also *18.1*, p. *215*; *18.2*, p. *222*; 21.1, p. 297.

I. O. BASHKIN, E. G. PONYATOVSKII, and M. E. KOST
Phase Transformations in Hydrides of Rare-Earth Metals under Pressure (I): Structural Transition in $NdH_{2.82}$
(b) Vol. **83**, No. 2, p. 517—520 (1977).
See also *1.2*, p. *19*; 21.4, p. 313.

M. GOODARZ and R. D. BARNARD
Deviations from Matthiessen's Rule in Silver Containing Low-Density Defects
(b) Vol. **83**, No. 2, p. 555—564 (1977).
See also *13.2*, p. *125*; 13.4, p. 137; 15, p. 199; 21.6, p. 317.

R. N. SINGH, P. L. SRIVASTAVA, and N. R. MITRA
Phonon Frequency, Electrical Resistivity, and Inter-Ionic Pair Potential of Li
(b) Vol. **83**, No. 2, p. 651—657 (1977).
See also 4, p. 46; *6*, p. *59*; 21.2, p. 309.

S. M. BORODACHEV, S. I. MASHAROV, and V. A. VOLKOV
*Electrical Resistivity of Antiferromagnetic Alloys. Coherent Potential Approximation
(b) Vol. **84**, No. 1, p. 33—40 (1977).
See also 18.4, p. 233; 21, p. 286.

J. KUDRNOVSKÝ and J. PETRŮ
*Electrical Conductivity of Two-Band Models for Disordered Binary Substitutional Alloys. Theory
(b) Vol. **84**, No. 1, p. 325—333 (1977).
See also 13.1, p. 122; 21, p. 286.

K. BÄRNER
Energy Bands and Electron Correlation Effects in Manganese Pnictides
(b) Vol. **84**, No. 1, p. 385—392 (1977).
See also *13.1*, p. *122*; 18.4, p. 234; *20,1*, p. *261*; 21.1, p. 299.

J. PETRŮ and J. KUDRNOVSKÝ
*Electrical Conductivity of Two-Band Metals for Disordered Binary Substitutional Alloys: Numerical Investigation
(b) Vol. **84**, No. 1, p. K21—K23 (1977).
See also 21, p. 286.

J. L. Vicent
*Conduction Electrons and s–d Exchange Interaction
(b) Vol. 84, No. 1, p. K71–K74 (1977).
See also 13, p. 112; *18.2.1*, p. *226*; 21.1, p. 299.

W. Schnelle and U. Dillner
*Thickness Dependence of Resistivity and Its Temperature Coefficient of Polycrystalline Bismuth Films
(a) Vol. 44, No. 1, p. 197–204 (1977).
See also *14.3.1*, p. *175*; 21.7, p. 319.

H. Ratajczak and I. Łabędzka
*Anomalous Hall Hysteresis Loops in the Vicinity of the Compensation Point in HoCo Amorphous Films
(a) Vol. 44, No. 2, p. K103–K105 (1977).
See also 2, p. 40; *18.2*, p. *224*; 21.1, p. 299; 21.4, p. 313.

U. Behn and H. Zinke
On the Density of States at the Fermi Level for Normal Systems
(b) Vol. 84, No. 2, p. 649–652 (1977).
See also *13.1*, p. *122*.

N. R. Mitra, J. Thakur, and P. L. Srivastava
*Resistivity and Thermoelectric Power of Liquid Lithium and Sodium from Harrison's First Principle
(b) Vol. 84, No. 2, p. 797–801 (1977).
See also 2, p. 40; *15*, p. *200*; 21, p. 287.

14.2 Superconductivity. Superconducting Materials and Devices

Original Papers and Short Notes

YU. P. YUROVSKII
*Tunnelling Effect between Superconductors in an Alternating Field
(b) Vol. **79**, No. 1, p. 71—77 (1977).

M. FÄHNLE and H. KRONMÜLLER
*On the Domain Structure near the Penetration Field of the Intermediate State in Type-I Superconductors
(b) Vol. **79**, No. 1, p. 139—147 (1977).
See also 21, p. 276.

M. A. BELOGOLOVSKII, YU. M. IVANCHENKO, E. V. KOCHERGIN, A. A. LISYANSKII, and YU. V. MEDVEDEV
*Interaction of Electrons with Phonons and Magnetic Impurities in Tunneling Junctions
(b) Vol. **79**, No. 1, p. K17—K21 (1977).
See also 6, p. 52; *14.1*, p. *157*; 21, p. 276.

Y. KHAN, K. NAHM, M. ROSENBERG, and H. WILLNER
*Superconductivity and Semiconductor–Metal Phase Transition in the System $BaPb_{1-x}Bi_xO_3$
(a) Vol. **39**, No. 1, p. 79—88 (1977).
See also 1, p. 9; 13.1, p. 114; 20.1, p. 248; 22.6.1, p. 393.

H. U. HABERMEIER
*The Penetration Field and the Magnetization Curve of Type-I Superconductors Containing Lattice Defects
(a) Vol. **39**, No. 1, p. 203—212 (1977).
See also 10.1, p. 75; 18, p. 209; 21, p. 276.

V. S. POSTNIKOV, V. V. POSTNIKOV, and V. S. ZHELEZNYI
*Superconductivity in Mo–Re System Alloy Films Produced by Electron Beam Evaporation in High Vacuum
(a) Vol. **39**, No. 1, p. K21—K23 (1977).
See also 21, p. 277.

A. FEHER, M. REIFFERS, P. PETROVIČ, and Š. JÁNOŠ
The Electrical Resistivity of Nb–Ti and Nb–Ti–Zr Alloys at Low Temperatures
(a) Vol. **39**, No. 1, p. K67—K69 (1977).
See also *14.1*, p. *157*; 21, p. 277.

C. TSALLIS
*Conduction Electron Spin Resonance Due to Exchange and Spin–Orbit Scatterings by Dilute Impurities in Superconductors
(b) Vol. **79**, No. 2, p. 451—460 (1977).
See also *13.4*, p. *130*.

R. SCHMUCKER and E. H. BRANDT
*On Flux Pinning by Point Pinning Centres in Type-II Superconductors
(b) Vol. **79**, No. 2, p. 479—487 (1977).

M. HOROBIOWSKI, T. SKOŚKIEWICZ, and E. TROJNAR
*Magnetization of the Superconducting Palladium Hydrides
(b) Vol. **79**, No. 2, p. K147—K150 (1977).
See also 18, p. 209; 21, p. 277.

R. SCHMUCKER
*The Influence of Plastic Deformation of the Flux-Line Lattice on Flux Transport in Hard Superconductors
(b) Vol. **80**, No. 1, p. 89—97 (1977).
See also 10.1, p. 76.

M. CRIŞAN and P. PÎRVU
*Intersection Lines in the Magnetic–Superconducting Phase Transition
(b) Vol. **80**, No. 2, p. K121—K125 (1977).
See also 1.2, p. 16; *18.2*, p. *219*.

V. V. GANN, A. I. SKVORTSOV, and A. VAN DEN BOSCH
*The Superconducting Transition Temperature of $U_6Fe_{0.98}$ Reduced by Neutron Irradiation
(a) Vol. **41**, No. 1, p. 225—229 (1977).
See also *11*, p. *95*; 18.1, p. 214; 21.1.1, p. 303; 21.5, p. 314.

G. BURRAFATO, G. GIAQUINTA, N. A. MANCINI, A. PENNISI, F. SIMONE, and S. O. TROIA
*Fluctuations of the Superconducting Order Parameter below T_c by Tunnel Effect
(b) Vol. **81**, No. 2, p. 579—583 (1977).

V. A. KASHCHEI
*Second Critical Current of a Long Superconducting Film in a Perpendicular Magnetic Field
(b) Vol. **81**, No. 2, p. K131—K134 (1977).

A. BARONE, G. PATERNÒ, M. RUSSO, and R. VAGLIO
*Diffraction and Interference Phenomena in Single Josephson Junctions
(a) Vol. **41**, No. 2, p. 393—401 (1977).

H. D. WIESINGER
*Metallkundliche Möglichkeiten zur Erhöhung der Sprungtemperatur zur Supraleitung von A15-Phasen
Possibilities for the Increase of the Transition Temperature to Superconductivity from the Metal-Scientific Point of View
(a) Vol. **41**, No. 2, p. 465—479 (1977).
See also *1.3*, p. *21*; 21, p. 281.

S. TAKÁCS and A. PEVALA
*The Angular Dependence of Critical Currents in Superconductors with Long Narrow Defects
(a) Vol. **41**, No. 2, p. K175—K179 (1977).
See also 10.1, p. 78; 21, p. 281.

O. SINGH
Ultrasonic Absorption in the Intermediate State of Gallium and Indium
(b) Vol. **82**, No. 1, p. 179—188 (1977).
See also *7*, p. *62*; 21, p. 282.

C. SUŁKOWSKI
*Electronic Structure and Superconductivity in Solid Solutions Based on In, Pb, and Tl
(b) Vol. **82**, No. 2, p. 657—662 (1977).
See also 21, p. 283.

M. ROHR and P. MÜLLER
*Superconducting Properties of Cold-Worked Core-Wires on the Basis of A15-Phases
(a) Vol. **42**, No. 2, p. 477—482 (1977).
See also *1.2*, p. *18*; 21, p. 283.

M. J. WITCOMB
*Microstructure and Superconducting Properties of Niobium and Tantalum Films Sputtered onto Copper Substrates
(a) Vol. **42**, No. 2, p. 595—607 (1977).
See also *1.4*, p. *27*; 21, p. 283.

A. M. CUCOLO, S. PACE, and R. D. PARMENTIER
*Surface Effects on the Tunneling Characteristics of Nb–Pb Josephson Junctions
(a) Vol. **42**, No. 2, p. K111—K114 (1977).

W. D. Schmidt and G. Albrecht
*Enhancement of the Instability Current of Superconducting Thin Film Bridges in Perpendicular Magnetic Fields
(a) Vol. **42**, No. 2, p. K143—K145 (1977).
See also 21, p. 283.

J. Sosnowski, M. Dryś, and J. Szymaszek
*Properties of Superconducting Nb–Ga Alloys with A15 Type Structure
(a) Vol. **42**, No. 2, p. K159—K164 (1977).
See also 21, p. 284.

U. Essmann, W. Wiethaup, and H. U. Habermeier
*The Intermediate State of Superconducting Pure Tantalum
(a) Vol. **43**, No. 1, p. 151—158 (1977).
See also 1.4, p. 27; 21, p. 284.

F. Dettmann and B. Klotz
*Critical Josephson Currents of Evaporated Nb–NbO_x–Pb Tunnel Junctions
(a) Vol. **43**, No. 1, p. 283—287 (1977).

S. Takeuchi, K. Maeda, and T. Suzuki
*Temperature and Strain-Rate Dependence of Flow Stress of Tantalum in the Superconducting and Normal States
(a) Vol. **43**, No. 1, p. 289—294 (1977).
See also *10.1*, p. *80*; 21, p. 285.

P. Müller and M. Rohr
*Lattice Instability of Superconducting Cold-Worked Core-Wires on the Base of $PbMo_5S_6$
(a) Vol. **43**, No. 1, p. K19—K22 (1977).
See also *1.2*, p. *18*; 22.8, p. 399.

M. Fähnle
*The Elementary Interaction Force between the Flux-Line Lattice and Precipitates
(b) Vol. **83**, No. 2, p. 433—438 (1977).

G. Ilonca
*Superconductivity in Nb_3Ge Films above 21.9 K
(a) Vol. **43**, No. 2, p. 387—392 (1977).

Y. Kimura
*On the Relation between the Density of States in the Cubic State and Phase Transition Temperatures (T_c, T_m) of V–Zr–Nb Ternary Laves Compounds
(a) Vol. **43**, No. 2, p. K141—K146 (1977).
See also *1.2*, p. *19*; 8, p. 67; 13, p. 112; 21.1, p. 298.

C. B. Satterthwaite
*On Isotope Effects in $PdH(D)_x$
(a) Vol. **43**, No. 2, p. K147—K149 (1977).
See also 8, p. 67; 13, p. 112; 21, p. 286.

M. Fähnle
*The Influence of Fast Neutron Irradiation on the Critical Current Densities in Nb_3Sn (I): Low Neutron Doses ($\Phi < 2 \times 10^{18}$ neutrons/cm^2)
(b) Vol. **84**, No. 1, p. 245—251 (1977).
See also *11*, p. *98*; 21, p. 286.

E. H. Brandt
*Elastic Response of the Fluxoid Lattice with Planar Pinning Forces
(b) Vol. **84**, No. 1, p. 269—281 (1977).

H. R. Khan, K. Lüders, Ch. J. Raub, and Z. Szücs
*Knight Shift of Pt^{195} in A15- and σ-Phases of Nb–Pt
(b) Vol. **84**, No. 1, p. K33—K35 (1977).
See also *19*, p. *242*; 21, p. 286; 21.6, p. 317.

M. JURISCH, K.-H. BERTHEL, and H.-J. ULLRICH
Influence of Chemical Composition within the Range of Homogeneity on Phase Transition and Transition Temperature of V_3Si Single Crystals (I): Preparation and Characterization of Single Crystals
(a) Vol. **44**, No. 1, p. 277–284 (1977).
See also 1.3, p. 23; *3*, p. *43*; 21.1, p. 299.

J. D. LIVINGSTON
Grain Size in A-15 Reaction Layers
(a) Vol. **44**, No. 1, p. 295–301 (1977).
See also *1.4*, p. *28*; 21, p. 287.

N. E. ALEKSEEVSKII, G. WOLF, N. M. DOBROVOLSKII, and C. HOHLFELD
*Thermal and Magnetic Properties of the Superconducting Compound $YbMo_6S_8$ at Low Temperatures
(a) Vol. **44**, No. 1, p. K79–K82 (1977).
See also *8*, p. *68*; 18, p. 212; 22.8, p. 400.

W. RICHTER, K. BLÜTHNER, H.-J. KÖHLER, and G. ALBRECHT
*Thin Film DC SQUID with Transformer
(a) Vol. **44**, No. 1, p. K99–K101 (1977).
See also 21, p. 287.

M. FÄHNLE
*The Influence of Fast Neutron Irradiation on the Critical Current Densities in Nb_3Sn (II): High Neutron Doses ($\Phi < 2 \times 10^{18}$ neutrons/cm^2)
(b) Vol. **84**, No. 2, p. 541–549 (1977).
See also *11*, p. *99*; 21, p. 287.

K. RUEBENBAUER, J. FINK, H. SCHMIDT, G. CZJZEK, and K. TOMALA
An Investigation of Magnetic Ordering in $Gd_xCe_{1-x}Ru_2$ by ^{155}Gd Mössbauer Spectroscopy
(b) Vol. **84**, No. 2, p. 611–618 (1977).
See also *5*, p. *50*; *18.2*, p. *224*; 21.4, p. 313.

E. H. BRANDT
*Instability of the Flux-Line Lattice in a Type-II Superconductor Containing Planar Imperfections
(b) Vol. **84**, No. 2, p. 637–647 (1977).

P. PAUFLER, E. TEDLER, H.-J. ULLRICH, K.-H. BERTHEL, U. KRÄMER, M. JURISCH, K. RICHTER, and K. EICHLER
Influence of Chemical Composition within the Range of Homogeneity on Phase Transition and Transition Temperature of V_3Si Single Crystals (II): On the Defect Structure of V_3Si Single Crystals
(a) Vol. **44**, No. 2, p. 499–504 (1977).
See also *10.1*, p. *82*; 21.1, p. 300.

F. DETTMANN, K. BLÜTHNER, P. PERTSCH, P. WEBER, and G. ALBRECHT
*A Study of V–VO_x–Pb Josephson Tunnel Junctions
(a) Vol. **44**, No. 2, p. 577–584 (1977).
See also 22.6, p. 391.

14.3 Semiconductors

Review Article

M. I. Klinger and A. A. Samokhvalov
*Electron Conduction in Magnetite and Ferrites
(b) Vol. **79**, No. 1, p. 9—48 (1977).
Erratum
(b) Vol. **83**, No. 1, p. 353 (1977).
See also 13.1, p. 114; *13.5.3*, p. *151*; 15, p. 198; 18.3, p. 228; 22.8.2, p. 409.

Original Papers and Short Notes

A. Philipp, F. Kuchar, and K. Seeger
*Submillimeter Conductivity of Thermal and Warm Electrons in GaAs and InP
(b) Vol. **79**, No. 1, p. 115—124 (1977).
See also 22.2.1, p. 350; 22.2.2, p. 354.

V. N. Lugovoi
*Electromagnetic Waves in an Electron Plasma in an Alternating External Magnetic Field
(b) Vol. **79**, No. 1, p. 131—138 (1977).
See also *20*, p. *244*.

L. Bányai and A. Aldea
*Semi-Classical and Quantum-Mechanical Theory of Hopping Conduction
(b) Vol. **79**, No. 1, p. 365—377 (1977).
Erratum
(b) Vol. **81**, No. 1, p. 409 (1977).
See also 6, p. 51; 13.4, p. 129; 22, p. 320.

A. N. Chakravarti and A. K. Chowdhury
*Influence of Band Non-Parabolicity on the Oscillatory Dependence of the Diffusivity-Mobility Ratio in Degenerate Semiconductors on Magnetic Quantization
(b) Vol. **79**, No. 1, p. K31—K34 (1977).
See also 13.1, p. 114; 18, p. 209; 22, p. 320.

L. Gouskov, A. Gouskov, V. Lemos, W. May, and H. Sampaio
*Electrical Properties of $GaTe_xSe_{1-x}$ Crystals
(a) Vol. **39**, No. 1, p. 65—71 (1977).
See also 1.2, p. 14; *13.4*, p. *129*; 22.3.1, p. 360.

S. Cristoloveanu and P. Viktorovitch
*Influence of Surfaces on the Ettingshausen Effect. Application to the Characterization of Narrow-Gap Semiconductors
(a) Vol. **39**, No. 1, p. 109—116 (1977).
See also *15*, p. *198*; 22.2.3, p. 356.

U. V. Desnica
*Indirect Doping as a Possibility for Obtaining p-Type Conductivity CdS
(a) Vol. **39**, No. 1, p. K33—K36 (1977).
See also *11*, p. *94*; 22.4.1, p. 362.

A. Q. Sarker
Spherical Effective Impurity Electron Mass in Si
(b) Vol. **79**, No. 2, p. 567—571 (1977).
See also *13.4*, p. *130*; 22.1.2, p. 338.

G. J. Ashwell, D. D. Eley, M. R. Willis, and J. Woodward
*Electrical and Magnetic Properties of Two Crystallographic Forms of $(DEPE)^{2+}(TCNQ)_4^{2-}$
(b) Vol. **79**, No. 2, p. 629—636 (1977).
See also *14.1*, p. *157*; 15, p. 198; 18.1, p. 213; 22.9, p. 414.

S. H. El Mahalawy and B. L. Evans
*Temperature Dependence of the Electrical Conductivity and Hall Coefficient in 2H-MoS_2, $MoSe_2$, WSe_2, and $MoTe_2$
(b) Vol. **79**, No. 2, p. 713—722 (1977).
See also 13.1, p. 115; 22, p. 321.

B. A. Aronzon and E. Z. Meilikhov
*Helicon Damping in Semiconductors in the Quantum Limit
(b) Vol. **79**, No. 2, p. 753—762 (1977).
See also 18, p. 209; *20*, p. *244*; 22.2.3, p. 356.

J. L. Farvacque, J. C. Doukhan, U. von Alpen, and E. Gmelin
*Screw-Dislocation-Induced Scattering Processes and Acceptor States in Te
(b) Vol. **79**, No. 2, p. 763—773 (1977).
See also *13.4*, p. *131*; 20.1, p. 249; 22.1.3, p. 345.

C. K. Sarkar and P. K. Basu
*Inelastic Acoustic Phonon Scattering in Nonparabolic Semiconductors under Longitudinal Quantizing Magnetic Field
(b) Vol. **79**, No. 2, p. K157—K160 (1977).
See also 18, p. 210; 22, p. 322.

K. Ziegler and U. Birkholz
Photoelectric Properties of Si_2Te_3 Single Crystals
(a) Vol. **39**, No. 2, p. 467—475 (1977).
See also *13.4*, p. *131*; 16, p. 201; 20.3, p. 268; 22, p. 322.

W. Bollmann and H.-J. Stöhr
Incorporation and Mobility of OH^- Ions in $LiNbO_3$ Crystals
(a) Vol. **39**, No. 2, p. 477—484 (1977).
See also *10.2*, p. *84*; 20.1, p. 250; 22.8.1, p. 402.

L. I. Filina, A. G. Ilyashenko, Yu. V. Vorobev, Yu. I. Karkhanin, and M. K. Sheinkman
*The Thermally Stimulated Conductivity and the Features of the Process of Recombination of Charge Carriers in Semi-Insulating GaAs(Cr)
(a) Vol. **39**, No. 2, p. 717—724 (1977).
See also *13.4*, p. *131*; 22.2.1, p. 350.

Meera Rao and O. N. Srivastava
Interpolytypic Transformations and the Electrical Behaviour of PbI_2 Crystals
(a) Vol. **39**, No. 2, p. K81—K84 (1977).
See also *1.1*, p. *11*; *1.2*, p. *15*; 22, p. 322.

K. Takita, Truong Dinh Thanh, and S. Tanaka
*Galvanomagnetic Effects and Alloy Scattering in n-Type $Pb_{1-x}Sn_xTe$ at Low Temperatures
(b) Vol. **80**, No. 1, p. 149—158 (1977).
See also 22.8, p. 397.

D. D. Nedeoglo
*Formation and Properties of the Impurity Band in n-ZnSe
(b) Vol. **80**, No. 1, p. 369—377 (1977).
See also *13.4*, p. *132*; 22.4.2, p. 365.

H. Köhler
*Investigation of the Magnetoresistance of n-Type Bi_2Te_3 under Uniaxial Stress
(b) Vol. **80**, No. 1, p. K29—K32 (1977).
See also 13.1, p. 116; 20.1, p. 251; 22.7, p. 394.

H. van Cong
*Activity Coefficients of Electrons with Small Effective Mass in Heavily Doped Semiconductors
(b) Vol. **80**, No. 1, p. K45—K47 (1977).
See also 22.2.3, p. 356.

F. S. Shishiyanu, V. Gh. Gheorghiu, and S. K. Palazov
Diffusion, Solubility, and Electrical Activity of Co and Fe in InP
(a) Vol. **40**, No. 1, p. 29–35 (1977).
See also *9*, p. *69*; 10.2, p. 85; 22.2.2, p. 354.

S. Mochizuki
*Microwave Conductivity of NiO at High Temperatures
(a) Vol. **40**, No. 1, p. 119–123 (1977).
See also 22.6, p. 387.

D. Kostopoulos and J. Grammatikakis
*Changes in the State of Magnetite due to Magnetic Fields Applied at 78 K Observed in Magnetoresistance Measurements
(a) Vol. **40**, No. 1, p. 141–145 (1977).
See also *18*, p. *210*; 22.8.2, p. 410.

H. R. Bilger and J. P. Nougier
*Perturbation of a Homogeneous Current Path by a Stub
(a) Vol. **40**, No. 1, p. K85–K88 (1977).
See also 22, p. 323.

J. L. Farvacque and P. Lenglart
*Inhomogeneous Strain Field Scattering Deformation Potential (II): Free-Carrier Mobility Associated with Dislocation Strain Fields in Semiconductors
(b) Vol. **80**, No. 2, p. 433–439 (1977).
See also *10.2*, p. *86*; 22, p. 323; 22.1.3, p. 345.

B. Movaghar and L. Schweitzer
*ESR and Conductivity in Amorphous Germanium and Silicon
(b) Vol. **80**, No. 2, p. 491–498 (1977).
See also 2, p. 35; *19*, p. *237*; 22.1.1, p. 334; 22.1.2, p. 338.

H. Böttger and V. V. Bryksin
*A Theory of the Hall Effect in the Hopping Region in Disordered Systems (I): Rate Equation in Presence of Electrical and Magnetic Fields
(b) Vol. **80**, No. 2, p. 569–578 (1977).
See also 2, p. 35; 6, p. 54; 13, p. 108; 13.5.3, p. 151; 22, p. 323.

G. Bastard
Resonant Acceptor Level in Zero-Gap Semiconductors. Single and Multiple Scattering Effects
(b) Vol. **80**, No. 2, p. 641–650 (1977).
See also *13.4*, p. *132*; 22.4.4, p. 368.

V. Christoph and G. Röpke
*On the Relation between the Resistivity Formula and the Boltzmann Equation
(b) Vol. **80**, No. 2, p. K117–K120 (1977).
See also *14.1*, p. *158*.

H. Böttger and V. V. Bryksin
*A Theory of the Hall Effect in the Hopping Region in Disordered Systems (II): Random Resistor Network and Hall Effect
(b) Vol. **81**, No. 1, p. 97–106 (1977).
See also 2, p. 36; 6, p. 55; 13, p. 109; 13.5.3, p. 151; 22, p. 324.

P. R. R. Pereira, M. A. Amato, and L. C. M. Miranda
Carrier-Assisted Radiation Pumping of Spin Waves in Ferromagnetic Semiconductors under a Strong Magnetic Field
(b) Vol. **81**, No. 1, p. 395–401 (1977).
See also *13.5.4*, p. *153*; 18.2, p. 219; 22, p. 324.

D. Chattopadhyay
*Average Energy of Conduction Electrons in Non-Parabolic Semiconductors
(b) Vol. **81**, No. 1, p. K51—K53 (1977).
See also 13.1, p. 117; 22, p. 324.

K. Sierański, J. Szatkowski, J. M. Pawlikowski, E. Majchrowska, and E. Płaczek-Popko
Fermi Level in $Cd_xHg_{1-x}Te$
(b) Vol. **81**, No. 1, p. K67—K70 (1977).
See also *13.1*, p. *117*; 22.4.4, p. 368.

E. I. Terukov, W. Reichelt, M. P. Volotzkoi, F. A. Chudnovskii, and H. Oppermann
Hydrostatic Pressure Dependence of the Phase Transition of V_3O_5
(a) Vol. **41**, No. 1, p. 125—128 (1977).
See also *1.2*, p. *16*; 22.6, p. 388.

W. Pietrenko
*Einfluß von Temperatur und Störstellenkonzentration auf den Piezowiderstandseffekt in n-Silizium
(Influence of Temperature and Impurity Concentration on the Piezoresistance in n-Silicon)
(a) Vol. **41**, No. 1, p. 197—205 (1977).
See also 22.1.2, p. 339.

S. Popović, B. Čelustka, Ž. Ružić-Toroš, and D. Broz
X-Ray Diffraction Study and Semiconducting Properties of the System Ga_2Se_3–In_2Se_3
(a) Vol. **41**, No. 1, p. 255—262 (1977).
See also *1.2*, p. *16*; 22.3, p. 358.

V. A. Stukan, V. I. Trifonov, V. Ya. Shevchenko, and G. I. Goncharenko
*Electrical Properties of p-CdSb at Low Temperatures
(a) Vol. **41**, No. 1, p. 307—315 (1977).
See also 13.1, p. 117; 13.4, p. 134; 22.3, p. 358.

J. Cisowski and W. Żdanowicz
*Pressure and Temperature Dependence of the Hall Coefficient in $Cd_{3-x}Zn_xAs_2$ Solid Solutions
(a) Vol. **41**, No. 1, p. K59—K62 (1977).
See also 12.2, p. 105; 22.3.1, p. 360.

I. M. Chapnik
*On the Possibility of Semiconducting Properties in Diborides ReB_2, TcB_2, OsB_2, and RuB_2
(a) Vol. **41**, No. 1, p. K71—K73 (1977).
See also 22, p. 325.

H. Böttger and V. V. Bryksin
*A Theory of the Hall Effect in the Hopping Region in Disordered Systems (III): DC Hall Effect
(b) Vol. **81**, No. 2, p. 433—441 (1977).
See also 2, p. 37; 6, p. 56; 13, p. 109; 13.5.3, p. 151; 22, p. 325.

P. I. Baranskii, I. S. Buda, and L. A. Shcherbina
*Structure of the Thermomagnetic Tensor in Semiconductors of n-Ge Type
(b) Vol. **81**, No. 2, p. 709—718 (1977).
See also 15, p. 199; 22.1.1, p. 334.

W. Plesiewicz
Photoelectronic Investigations of Multi-Level Gold Centres in Ge:Sb:Au
(a) Vol. **41**, No. 2, p. 403—409 (1977).
See also *13.4*, p. *135*; *16*, p. *202*; 22.1.1, p. 334.

S. Mochizuki and T. Sakurai
*Electrical Conduction of MgO Doped with Cr at High Temperatures in Oxidizing Atmosphere
(a) Vol. **41**, No. 2, p. 411—415 (1977).
See also 22.6, p. 388.

M. Böhm, O. Erb, and A. Scharmann
*Thermally Stimulated Currents in $Ca(NbO_3)_2$
(a) Vol. **41**, No. 2, p. 535—540 (1977).
See also 22.8.1, p. 405.

S. Mochizuki
*Electrical Conductivity of α-Fe_2O_3
(a) Vol. **41**, No. 2, p. 591—594 (1977).
See also 22.6, p. 388.

M. Zoaeter, A. Conan, and D. Delaunay
*Transport Measurements and Band Structure of Bulk $MoSe_{2-x}$ Compounds
(a) Vol. **41**, No. 2, p. 629—635 (1977).
See also 13.1, p. 118; *15*, p. *199*; 22, p. 325.

A. S. Popov and I. Y. Yanchev
*On the Disorder Scattering in $GaAs_{1-x}P_x$ Solid Solutions
(a) Vol. **41**, No. 2, p. 687—691 (1977).
See also 22.2.4, p. 357.

R. M. Mehra, P. C. Mathur, A. K. Kathuria, and Radhey Shyam
*Dependence of Conductivity and Activation Energy on Tellurium Concentration in Amorphous Selenium
(a) Vol. **41**, No. 2, p. K189—K192 (1977).
See also *2*, p. *37*; 22.1.3, p. 345.

K. M. Demchuk and I. M. Tsidilkovskii
Scattering of Electrons by the Deformation Potential in Doped InSb
(b) Vol. **82**, No. 1, p. 59—67 (1977).
See also *13.1*, p. *118*; 15, p. 199; 22.2.3, p. 356.

J. Spałek, M. Lubecka, and A. Węgrzyn
*Carrier Concentration and Magnetic Susceptibility of Doped Ferromagnetic Semiconductors with Application to EuO:Eu
(b) Vol. **82**, No. 1, p. 107—116 (1977).
See also *18.2*, p. *220*; 22.6, p. 389.

G. N. Galkin, K. H. Herrmann, and E. A. Bobrova
*Recombination in Tellurium at High Optical Excitation
(b) Vol. **82**, No. 1, p. 237—242 (1977).
See also *16*, p. *203*; 22.1.3, p. 345.

Jai Singh
Calculation of the Energy Band Structures and Carrier Mobilities in γ-Platinum Phthalocyanine Single Crystals
(b) Vol. **82**, No. 1, p. 263—268 (1977).
See also *13.1*, p. *119*; 22.9, p. 416.

R. K. Jain
*Calculation of the Fermi Level, Minority Carrier Concentration, Effective Intrinsic Concentration, and Einstein Relation in n- and p-Type Germanium and Silicon
(a) Vol. **42**, No. 1, p. 221—226 (1977).
See also *13.1*, p. *120*; 22.1.1, p. 335; 22.1.2, p. 340.

H. Mutka and O. J. A. Tiainen
Annealing Experiments on Fast Neutron Irradiated Selenium Single Crystals
(a) Vol. **42**, No. 1, p. 241—245 (1977).
See also *11*, p. *96*; 20.1, p. 257; 22.1.3, p. 346.

A. H. Marshak and K. M. van Vliet
*Current in Semiconductors with Position-Dependent Band Gap and Electron Affinity
(a) Vol. **42**, No. 1, p. 279—283 (1977).
See also 13.4, p. 136; 22, p. 326.

W. Brückner, U. Gerlach, H.-P. Brückner, W. Moldenhauer, and H. Oppermann
Influence of Nonstoichiometry on the Phase Transitions in Ga-, Al-, and Fe-Doped VO_2
(a) Vol. **42**, No. 1, p. 295—303 (1977).
See also *1.2*, p. *17*; *3*, p. *42*; 18.1, p. 214; 22.6.1, p. 393.

A. Jędrzejczak and T. Dietl
Erratum to: Thermomagnetic Properties of n-Type and p-Type HgTe
(b) Vol. **82**, No. 2, p. 717 (1977).
See also *13.1*, p. *120*; *15*, p. *199*; 22.4.3, p. 367.

G. H. Schwuttke, K. Yang, and H. Kappert
*Lifetime Control in Silicon through Impact Sound Stressing
(a) Vol. **42**, No. 2, p. 553—564 (1977).
See also 14.3.1, p. 174; 22.1.2, p. 341.

D. D. Nedeoglo, Dang Huy Lam, and A. V. Simashkevich
*The Effect of Thermal Treatment and Doping on the Electrical Properties of Zinc Selenide
(a) Vol. **42**, No. 2, p. 675—679 (1977).
See also 16, p. 203; 22.4.2, p. 366.

B. R. Sethi, P. L. Talwar, O. P. Sharma, and P. C. Mathur
*The Electrical Transport Properties of a Zinc Selenide Crystal Highly Annealed in Molten Zinc
(a) Vol. **42**, No. 2, p. 791—795 (1977).
See also 13.5.3, p. 152; 22.4.2, p. 366.

I. P. Zvyagin
*Correlation Effects in Hopping Conduction
(b) Vol. **83**, No. 1, p. 63—70 (1977).
See also 2, p. 38; 13, p. 110; 15, p. 199; 22, p. 327.

I. I. Fishchuk
*Theory of the Phononless High Frequency Conductivity in Disordered Systems
(b) Vol. **83**, No. 1, p. K29—K33 (1977).
See also *2*, p. *38*.

G. Sixt, H. Kappert, and G. H. Schwuttke
*Minority Carrier Lifetime and Defect Structure in Silicon after Cesium Implantation
(a) Vol. **43**, No. 1, p. 119—131 (1977).
See also *11*, p. *97*; 22.1.2, p. 341.

S. Lakeou, S. Cristoloveanu, and A. Chovet
*Magnetoresistance Effect in Near Intrinsic Semiconductors. Influence of Sample Geometry. A New Method for the Determination of Carrier Densities and Mobilities
(a) Vol. **43**, No. 1, p. 213—222 (1977).
See also 22.1.1, p. 335; 22.1.2, p. 341; 22.2.3, p. 356.

A. N. Zyuganov, S. V. Svechnikov, and P. S. Smertenko
*Effect of Integrally Weak Diffusion on the Parameters of Current–Voltage Characteristics of Quasi-Monopolar Semiconductors
(a) Vol. **43**, No. 1, p. 333—341 (1977).
See also 14.3.4, p. 186; 22, p. 328.

S. A. Hamid
*Piezoelectric Properties of $Li_{1-x}H_xIO_3$
(a) Vol. **43**, No. 1, p. K29—K30 (1977).
See also 22.8.1, p. 406.

S. N. Vaidya, D. K. Joshi, M. D. Karkhanavala, and I. K. Gopalakrishnan
*High Pressure Phase Transformation in La_2CuO_4
(a) Vol. **43**, No. 1, p. K31—K35 (1977).
See also *1.2*, p. *19*; 22.8.1, p. 406.

H. NEUMANN, H.-G. ERNST, E. BUTTER, P. KRAMER, and W. KOHL
*Electrical Properties of Liquid Phase Epitaxial GaP Grown from Tin Solution
(a) Vol. **43**, No. 1, p. K95—K98 (1977).
See also 22.2.2, p. 354.

K. PESZ and T. LUTY
On the Charge Carrier Mobilities in Molecular Crystals
(b) Vol. **83**, No. 2, p. 565—571 (1977).
See also *13.5.3*, p. *152*; 22, p. 328.

R. S. MARKIEWICZ
Acoustic Phonon Scattering by Electron–Hole Droplets in Ge: Deformation Potentials
(b) Vol. **83**, No. 2, p. 659—668 (1977).
See also *13.5.1*, p. *147*; 22.1.1, p. 335.

V. A. GRAZHULIS, V. V. KVEDER, and V. YU. MUKHINA
*Investigation of the Energy Spectrum and Kinetic Phenomena in Dislocated Si Crystals (I)
(a) Vol. **43**, No. 2, p. 407—415 (1977).
See also 10.2, p. 91; *13.4*, p. *138*; 22.1.2, p. 342.

Ö. SÄVBORG and M. NYGREN
Magnetic, Electrical, and Thermal Studies of the $V_{1-x}Re_xO_2$ System with $0 \leqq x \leqq 0.15$
(a) Vol. **43**, No. 2, p. 645—652 (1977).
See also *1.2*, p. *19*; 8, p. 67; 18.1, p. 215; 22.6.1, p. 393.

B.-P. KIETIS, R. REKSNYS, A. SAKALAS, and P. HÖSCHL
*Investigation of the Bipolar Photoconductivity in CdTe
(a) Vol. **43**, No. 2, p. 705—709 (1977).
See also *7*, p. *63*; 16, p. 204; 22.4.3, p. 367.

A. CAVALLINI, P. GONDI, and A. CASTALDINI
*Dislocation Energy Levels in Ge
(a) Vol. **43**, No. 2, p. K205—K208 (1977).
See also 10.2, p. 91; *13.4*, p. *138*; 22.1.1, p. 336.

V. L. BONCH-BRUEVICH
Coulomb Gap in a Disordered Semiconductor
(b) Vol. **84**, No. 1, p. 99—105 (1977).
See also *13*, p. *112*; 22, p. 329.

E. PALANIYANDI and S. BALASUBRAMANIAN
*Polarizabilities of Donors in Elemental Semiconductors
(b) Vol. **84**, No. 1, p. 171—174 (1977).
See also *13.4*, p. *138*; 22.1.1, p. 336; 22.1.2, p. 343.

H. KÖHLER and A. FREUDENBERGER
Investigation of the Highest Valence Band in $(Bi_{1-x}Sb_x)_2Te_3$ Crystals
(b) Vol. **84**, No. 1, p. 195—203 (1977).
See also *13.1*, p. *122*; *13.2*, p. *125*; 22.7, p. 395.

V. A. GRAZHULIS, V. V. KVEDER, and V. YU. MUKHINA
*Investigation of the Energy Spectrum and Kinetic Phenomena in Dislocated Si Crystals (II): Microwave Conductivity
(a) Vol. **44**, No. 1, p. 107—115 (1977).
See also 10.2, p. 91; *13.4*, p. *139*; 22.1.2, p. 343.

H. J. HOFFMANN and W. SOHN
*Analysis of Localized Levels in Semiconducting CdS from Generation–Recombination Noise Spectra
(a) Vol. **44**, No. 1, p. 237—246 (1977).
See also *13.4*, p. *139*; 22.4.1, p. 364.

A. SZADKOWSKI, A. LUBOMIRSKA-WITTLIN, A. ZARĘBA, and B. KRUKOWSKA-FULDE
*On Conversion Processes of $CdF_2:MF_3$
(a) Vol. **44**, No. 1, p. K43—K46 (1977).
See also *20.1*, p. *262*; 22.5.3, p. 384.

I. B. BARKAN, M. V. ENTIN, and S. I. MARENNIKOV
*Conductivity of Fe-Doped $LiNbO_3$ Crystals
(a) Vol. **44**, No. 1, p. K91—K94 (1977).
See also 22.8.1, p. 408.

W. BRÜCKNER, E. WIESER, W. MOLDENHAUER, and W. REICHELT
On the Conductivity Transition V_3O_5
(a) Vol. **44**, No. 1, p. K95—K98 (1977).
See also *1.2*, p. *20*; 5, p. 49; 22.6, p. 391.

P. BYSZEWSKI and R. R. GAŁĄZKA
*Resonance Acceptor Level in $Hg_{0.98}Mn_{0.02}Te$
(b) Vol. **84**, No. 2, p. K83—K86 (1977).
See also *13.4*, p. *140*; 22.4.3, p. 367.

A. TONEVA and A. ALAZHAZHIAN
*Mobility in p-$Pb_{1-x}Sn_xTe$
(a) Vol. **44**, No. 2, p. 621—626 (1977).
See also 13.4, p. 140; 22, p. 330.

M. M. KAILA
*Effect of Magnetic Field on Electrical Effects of Dislocations in InSb
(a) Vol. **44**, No. 2, p. K135—K139 (1977).
See also *10.2*, p. *93*; 22.2.3, p. 356.

14.3.1 Films

Review Articles

T. D. Dzhafarov
Atomic Diffusion in Semiconductor Epitaxial Structures
(a) Vol. **42**, No. 1, p. 11—45 (1977).
See also *9*, p. *69*; 10.2, p. 83; 14.3.3, p. 178; 20.2, p. 265; 22.1, p. 332; 22.2, p. 348; 22.4, p. 361.

L. N. Aleksandrov
Transition Regions Epitaxial Films
(a) Vol. **44**, No. 1, p. 11—43 (1977).
See also *1.5*, p. *29*; *3*, p. *42*; 22.1.1, p. 333; 22.1.2, p. 337; 22.2.1, p. 350.

Original Papers and Short Notes

N. C. Halder, J. M. Montante, and W. E. Swartz, Jr.
Evidence of a Polycrystalline Phase in Amorphous Ge Films Evaporated at Room Temperature
(a) Vol. **39**, No. 1, p. 213—222 (1977).
See also *1.5*, p. *29*; *2*, p. *33*; 13.1, p. 115; 22.1.1, p. 333.

M. Chybicki
Electrical Properties of Metal–Polymer–Metal (MPM) and Metal–Polymer–Silicon (MPS) Structures with Thin Polymer Films
(a) Vol. **39**, No. 1, p. 271—279 (1977).
See also *14.3.3*, p. *178*; 22.9, p. 414.

C. M. Vodenicharov, A. A. Stanchev, and M. Lozev
Thermionic Emission in the System M–GeS–M
(a) Vol. **39**, No. 1, p. K1—K5 (1977).
See also *14.3.4*, p. *183*; 22, p. 321.

W. Rehm, R. Fischer, J. Stuke, and H. Wagner
Photo and Dark Conductivity of Doped Amorphous Silicon
(b) Vol. **79**, No. 2, p. 539—547 (1977).
See also *2*, p. *33*; *16*, p. *201*; 22.1.2, p. 337.

R. M. Hill
*The Frequency Dependence of Variable Range Hopping
(a) Vol. **39**, No. 2, p. 615—620 (1977).
See also 2, p. 34; *13.4*, p. *131*.

B. Fromm and W. Teubner
*Screening Effects in Amorphous Semiconductors Influencing the Poole-Frenkel Conduction
(a) Vol. **40**, No. 1, p. 63—68 (1977).
See also *2*, p. *35*; 22, p. 323.

A. Kubový and M. Janda
*Influence of Tunneling through Barriers on Grain Boundaries on Hole Mobility in Polycrystalline Tellurium Films
(a) Vol. **40**, No. 1, p. 225—233 (1977).
See also 22.1.3, p. 345.

K. L. Chopra, A. P. Thakoor, S. K. Barthwal, and P. Nath
*Electron Transport Properties of Amorphous Cu–Ag Films
(a) Vol. **40**, No. 1, p. 247—255 (1977).
See also *2*, p. *35*; 15, p. 198; 21.1, p. 291; 21.6, p. 315.

K.-D. Ufert and E. I. Terukov
*On the Problem of the Phase Transition in Disordered VO_2 Thin Films
(a) Vol. **40**, No. 2, p. K157—K159 (1977).
See also 22.6, p. 387.

M. K. Rao and S. R. Jawalekar
*Hall Voltage Measurements on CdSe Films
(a) Vol. 41, No. 1, p. K31—K33 (1977).
See also 22.4.2, p. 365.

M. Vodenicharova and C. Vodenicharov
*Electrode- and Bulk-Limited Currents in Thin Films of Aliphatic Unsaturated Poly-(1,3,4)-Oxadiazoles
(a) Vol. 41, No. 2, p. 487—491 (1977).
See also *14.3.3*, p. *180*; 22.9, p. 416.

K. Yasuda, A. Yoshida, and T. Arizumi
*The Effects of Annealing on Mott's Parameters for Hopping Conduction in Amorphous Ge
(a) Vol. 41, No. 2, p. K181—K184 (1977).
See also *2*, p. *37*; 22.1.1, p. 334.

W. Ecke
*The Photomagnetoelectric Effect in Heteroepitaxial Silicon Films
(a) Vol. 42, No. 1, p. 157—163 (1977).
See also *16*, p. *203*; 22.1.2, p. 340.

K.-D. Ufert
*Doping of VO_2 Thin Films by Ion Implantation
(a) Vol. 42, No. 1, p. 187—190 (1977).
See also *11*, p. *96*; 22.6, p. 389.

T. Suzuki, M. Hirose, S. Ogose, and Y. Osaka
*Electrical Transport and Photoconductivity in Amorphous Silicon
(a) Vol. 42, No. 1, p. 337—344 (1977).
See also 2, p. 37; 13.1, p. 120; 15, p. 199; *16*, p. *203*; 22.1.2, p. 340.

V. F. Kiselev, V. V. Kurylev, and N. L. Levshin
*The Effect of Adsorption on the Electroconductivity of PcFe and PcCu Films
(a) Vol. 42, No. 1, p. K61—K64 (1977).
See also *14.3.2*, p. *177*; 22.9, p. 417.

G. H. Schwuttke, K. Yang, and H. Kappert
Lifetime Control in Silicon through Impact Sound Stressing
(a) Vol. 42, No. 2, p. 553—564 (1977).
See also *14.3*, p. *170*; 22.1.2, p. 341.

A. F. Kravchenko, B. S. Lisenker, Yu. E. Maronchuk, A. P. Sherstyakov, and V. N. Sherstyakova
*Investigation of Recombination of Non-Equilibrium Carriers in GaAs Epitaxial Layers
(a) Vol. 42, No. 2, p. 647—655 (1977).
See also *13.4*, p. *137*; 16, p. 203; 22.2.1, p. 352.

J. Szatkowski, K. Sierański, J. M. Pawlikowski, E. Płaczek-Popko, P. Becla, and E. Dudziak
*Electrical Transport Properties of Epitaxial Graded-Gap $Cd_xHg_{1-x}Te$ Layers
(a) Vol. 42, No. 2, p. 721—727 (1977).
See also 13.1, p. 121; 22.4.4, p. 369.

M. Janai and P. S. Rudman
The Kinetics of the Light-Enhanced Vaporization of Amorphous As_2S_3 Thin Films
(a) Vol. 42, No. 2, p. 729—737 (1977).
See also *2*, p. *38*; *16*, p. *203*; 22, p. 327.

C. M. Vodenicharov
*A Simple Method for the Determination of the Effective Electron Mass from Current–Voltage Characteristics in the Thermionic Region
(a) Vol. 42, No. 2, p. 785—789 (1977).
See also *13.1*, p. *121*; 14.3.4, p. 186; 22, p. 327; 22.6, p. 390; 22.9, p. 417.

B. I. Shklovskii
*Thickness Dependence of the Hopping Conduction in Amorphous Films. Experiment versus Theory
(b) Vol. **83**, No. 1, p. K11–K15 (1977).
See also *2*, p. *38*; 22.1, p. 332.

U. Hartmann
*Charge Measurements in Thin Insulating Layers of MIS and MIM Structures by Means of Current–Voltage Characteristics and Application to the MOS System
(a) Vol. **43**, No. 1, p. 103–110 (1977).
See also *14.3.3*, p. *181*; 22.6, p. 390.

P. Lilienkamp and H. Pfleiderer
*An ESFI SOS Magnetodiode
(a) Vol. **43**, No. 2, p. 479–486 (1977).
See also *14.3.4*, p. *187*; 22.1.2, p. 342.

K. O. Lee and T. T. Gan
*Space-Charge-Limited Currents in Evaporated Films of Pentacene
(a) Vol. **43**, No. 2, p. 565–571 (1977).
See also *14.3.4*, p. *187*; 22.9, p. 418.

C. Viger, C. Vautier, and D. Carles
Recombination and Trapping Processes in Amorphous Materials in Presence of an Exponential Distribution of Traps
(a) Vol. **44**, No. 1, p. 97–106 (1977).
See also 2, p. 40; 13.4, p. 139; *16*, p. *204*; 22.1.3, p. 347.

W. Schnelle and U. Dillner
*Thickness Dependence of Resistivity and Its Temperature Coefficient of Polycrystalline Bismuth Films
(a) Vol. **44**, No. 1, p. 197–204 (1977).
See also *14.1*, p. *160*; 21.7, p. 319.

C. Lawther and K. V. Anand
*Physical Mechanism of Current Conduction and Light Emission in High-Resistivity ZnS:Mn Thin Films
(a) Vol. **44**, No. 1, p. 313–323 (1977).
See also *20.3*, p. *274*; 22.4.1, p. 364.

G. Buchheim, B. Junghans, and H. Lippmann
*Application of the Kelvin Method for Oxide Charge Evaluation in Si–SiO_2 Structures
(a) Vol. **44**, No. 2, p. 585–591 (1977).
See also 2, p. 41; *14.3.2*, p. *177*; 22.1.2, p. 344.

R. P. Khare and R. Nath
*On the Analysis of Complex TSC Patterns
(a) Vol. **44**, No. 2, p. 627–631 (1977).
See also 13.4, p. 140; 22.6, p. 392.

N. L. P. Sinha, A. K. Chaudhuri, and H. N. Bose
Structur Analysis of Lead Telluride Films and Their Sensitization
(a) Vol. **44**, No. 2, p. K127–K129 (1977).
See also *1.5*, p. *30*; 16, p. 205; 22, p. 331.

14.3.2 Surfaces and Interfaces

Original Papers and Short Notes

E. M. Itskovich, A. A. Kornyshev, and M. A. Vorotyntsev
*Electric Current across the Metal–Solid Electrolyte Interface (I): Direct Current, Current–Voltage Characteristic
(a) Vol. **39**, No. 1, p. 229—238 (1977).
See also 14.3.4, p. 183; 22.5.1, p. 373.

A. A. Kornyshev and M. A. Vorotyntsev
*Electric Current across the Metal–Solid Electrolyte Interface (II): Low-Amplitude Alternating Current
(a) Vol. **39**, No. 2, p. 573—582 (1977).
See also 14.3.4, p. 183.

H. Kuzmany and Z. Szalmassy
*Systematic Study of the Acoustoelectric Interaction of Surface Waves with Inhomogeneously Distributed Free Carriers in a Photoconductor
(a) Vol. **40**, No. 1, p. 75—82 (1977).
See also *7*, p. *62*; 14.3.3, p. 178; 16, p. 201; 22.4.1, p. 362; 22.8.1, p. 403.

K. Takasaki and S. Tanaka
*Tunnelling Study of $Pb_{1-x}Sn_xTe$
(a) Vol. **40**, No. 1, p. 173—180 (1977).
See also 6, p. 54; 13.1, p. 116; 22.8, p. 397.

J. Toušková and R. Kužel
*Some Properties of Ag–CdTe Diodes
(a) Vol. **40**, No. 1, p. 309—313 (1977).
See also 14.3.3, p. 178; 22.4.3, p. 367.

E. W. Kreutz
*Magneto-Surface Experiments on the III–V Compound Semiconductor InSb
(a) Vol. **40**, No. 2, p. 415—423 (1977).
See also *13.3*, p. *126*; 22.2.3, p. 356.

Th. Englert, K. von Klitzing, R. Silbermann, and G. Landwehr
Influence of the Surface on the Galvanomagnetic Properties of Tellurium
(b) Vol. **81**, No. 1, p. 119—127 (1977).
See also *13.3*, p. *126*; 22.1.3, p. 345.

F.-W. Breitbarth and H.-J. Tiller
Tunnelling in Metal–Glow-Polymer–Metal Sandwich Structures
(a) Vol. **41**, No. 1, p. 151—156 (1977).
See also 13.3, p. 126; *14.3.4*, p. *185*; 22.9, p. 415.

D. Goldschmidt, T. Bernstein, and P. S. Rudman
The Kinetics of Photodissolution of Silver in Amorphous As_2S_3 Films
(a) Vol. **41**, No. 1, p. 283—287 (1977).
See also *11*, p. *95*; 22, p. 324.

S. N. Kozlov
Slow Relaxation Kinetics on a Heterogeneous Semiconductor Surface
(a) Vol. **42**, No. 1, p. 115—124 (1977).
See also *13.3*, p. *127*; 22, p. 326.

J. Buxo, G. Sarrabayrouse, and D. Esteve
The Optically Biased MIS Tunneling Schottky Barrier
(a) Vol. **42**, No. 1, p. 173—179 (1977).
See also *14.3.3*, p. *180*.

H. Fujiyasu, T. Hatano, S. Itoh, M. Sekinobu, and O. Ohtsuki
*n-Type GaAs MIS Structures
(a) Vol. **42**, No. 1, p. K25—K27 (1977).
See also *14.3.3*, p. *180*; 22.2.1, p. 352.

V. F. Kiselev, V. V. Kurylev, and N. L. Levshin
*The Effect of Adsorption on the Electroconductivity of PcFe and PcCu Films
(a) Vol. **42**, No. 1, p. K61—K64 (1977).
See also *14.3.1*, p. *174*; 22.9, p. 417.

C.-P. Czaya, K. Keck, and H. Rüthlein
Influence of Surface Carriers upon the Galvanomagnetic Effects in Slightly Doped Tellurium Samples at Low Temperatures
(b) Vol. **82**, No. 2, p. 595—602 (1977).
See also *13.4*, p. *136*; 22.1.3, p. 346.

A. T. Åberg
*Low-Temperature Annealing of Al-Implanted Si–SiO_2 Structures
(a) Vol. **42**, No. 2, p. 639—645 (1977).
See also *13.3*, p. *127*; 22.1.2, p. 341.

H. Kato, T. Kanda, K. Yasuda, A. Yoshida, and T. Arizumi
*Electron Beam Irradiation Effects on SnO_2–Si Photodiodes
(a) Vol. **42**, No. 2, p. 769—774 (1977).
See also *11*, p. *97*; 16, p. 203; 22.1.2, p. 341.

V. I. Chaikin, K. U. Guga, V. K. Malyutenko, and A. V. Sachenko
*Photoconductivity of Indium Antimonide with Different Surface Treatments
(a) Vol. **43**, No. 1, p. 311—316 (1977).
See also *16*, p. *204*; 22.2.3, p. 356.

I. Gabaš and J. Toušek
*Surface Photovoltage on CdTe
(a) Vol. **43**, No. 1, p. 351—356 (1977).
See also *16*, p. *204*; 22.4.3, p. 367.

J. Hüller and M. T. Pham
*Characterization of Ion Implanted Silicon by the Electrolytic Reverse Current
(a) Vol. **43**, No. 2, p. 505—512 (1977).
See also *11*, p. *98*; 22.1.2, p. 342.

V. Kunigelis, A. Jucys, E. Garška, and V. Vitkauskas
*Influence of the Surface Potential on the Acoustoelectric Interaction in Semiconductor–Piezoelectric Structures
(a) Vol. **43**, No. 2, p. 699—703 (1977).
See also *7*, p. *63*; 22.4.1, p. 364.

D. R. Franceschetti and J. R. Macdonald
*Comment on the Small-Signal AC Response of Solid Electrolytes
(a) Vol. **43**, No. 2, p. K169—K173 (1977).
See also 14.3.4, p. 187.

D. L. Smith
*Recombination Currents in a p–n Heterojunction Diode
(a) Vol. **44**, No. 1, p. 381—390 (1977).
See also *14.3.3*, p. *181*; 22.2.1, p. 353.

D. R. Choudhury, A. K. Chowdhury, and A. N. Chakravarti
*On the Modification of the Einstein Relation for Semiconductor Inversion Layers in the Presence of a Quantizing Magnetic Field
(a) Vol. **44**, No. 1, p. K111—K115 (1977).
See also 22.1.2, p. 343.

G. Buchheim, B. Junghans, and H. Lippmann
*Application of the Kelvin Method for Oxide Charge Evaluation in Si–SiO_2 Structures
(a) Vol. **44**, No. 2, p. 585—591 (1977).
See also 2, p. 41; *14.3.1*, p. *175*; 22.1.2, p. 344.

14.3.3 Devices. Junctions (Contact Problems See 14.3.4)

Review Articles

K. W. Böer
The CdS/Cu_2S Solar Cell (I): Minority Carrier Generation and Transport in the Cu_2S Emitter
(a) Vol. **40**, No. 2, p. 355–384 (1977).
See also 14.3.4, p. 183; *16*, p. *201*; 20.1, p. 247; 22, p. 320; 22.4.1, p. 362.

T. D. Dzhafarov
Atomic Diffusion in Semiconductor Epitaxial Structures
(a) Vol. **42**, No. 1, p. 11–45 (1977).
See also *9*, p. *69*; 10.2, p. 83; 14.3.1, p. 173; 20.2, p. 265; 22.1, p. 332; 22.2, p. 348; 22.4, p. 361.

Original Papers and Short Notes

H. Runge
*Device Geometry Dependent Effects of Deep Levels on $C(U)$ Measurements of Ion-Implanted MOS Diodes
(a) Vol. **39**, No. 1, p. 141–146 (1977).
See also *11*, p. *94*; 22.1.2, p. 337.

M. Chybicki
*Electrical Properties of Metal–Polymer–Metal (MPM) and Metal–Polymer–Silicon (MPS) Structures with Thin Polymer Films
(a) Vol. **39**, No. 1, p. 271–279 (1977).
See also 14.3.1, p. 173; 22.9, p. 414.

I. Munteanu
*The Effect of Effective Lifetime Modulation in Semiconductors
(a) Vol. **39**, No. 1, p. 291–300 (1977).
See also *13.3*, p. *126*; 22, p. 321.

Vijender Sharma and P. C. Mathur
*Effect of Junction Inhomogeneity on AC Voltage of n^+pp^+ GaAs IMPATT Diodes
(a) Vol. **39**, No. 1, p. K59–K61 (1977).
See also 22.2.1, p. 350.

I. Mojzes, B. Pödör, and I. Balogh
*On the Temperature Dependence of Peak Electron Velocity and Threshold Field Measured on GaAs Gunn Diodes
(a) Vol. **39**, No. 2, p. K123–K125 (1977).
See also *14.3.4*, p. *184*; 22.2.1, p. 350.

B. Rheinländer, A. Zehe, and G. Kretzschmar
*Near Field Topography of High Frequency Modulated GaAs LED's
(a) Vol. **39**, No. 2, p. K139–K142 (1977).
See also *20.3*, p. *268*; 22.2.1, p. 351.

H. Kuzmany and Z. Szalmassy
Systematic Study of the Acoustoelectric Interaction of Surface Waves with Inhomogeneously Distributed Free Carriers in a Photoconductor
(a) Vol. **40**, No. 1, p. 75–82 (1977).
See also 7, p. *62*; *14.3.2*, p. *176*; 16, p. 201; 22.4.1, p. 362; 22.8.1, p. 403.

J. Toušková and R. Kužel
Some Properties of Ag–CdTe Diodes
(a) Vol. **40**, No. 1, p. 309–313 (1977).
See also *14.3.2*, p. *176*; 22.4.3, p. 367.

AMITABH JAIN and B. J. SMITH
A New Method for Demonstrating the Bourgoin Mechanism in Semiconductors
(a) Vol. **40**, No. 1, p. K31—K33 (1977).
See also *10.2*, p. *86*; 22.1.2, p. 338.

P. KRISPIN
*Impurity Detection by the Admittance–Frequency Dispersion
(a) Vol. **40**, No. 1, p. K81—K84 (1977).
See also *13.4*, p. *132*; 22.2.2, p. 354.

K. KURUMADA, T. MIZUTANI, and A. ISHIDA
*Bi-Stable Switching Phenomena in GaAs Planar Devices with Schottky Gate
(a) Vol. **40**, No. 2, p. K93—K96 (1977).
See also 22.2.1, p. 351.

N. M. HOSSEINI and B. K. JONES
*Noise in Spinel Ceramics
(a) Vol. **40**, No. 2, p. K185—K189 (1977).
See also 22.8.2, p. 410.

V. RAMAMURTHY and P. K. CHATURVEDI
*Measurement of Heat-Flow Resistance
(a) Vol. **40**, No. 2, p. K191—K193 (1977).

W. SIEGEL, H. BECHERER, G. KÜHNEL, and E. ZIEGLER
*Electroluminescence in $ZnSiP_2$ Schottky Diodes
(a) Vol. **41**, No. 1, p. 75—80 (1977).
See also *20.3*, p. *270*; 22.8, p. 398.

A. SCHLACHETZKI
*Switching in Miniaturized Planar Gunn Devices
(a) Vol. **41**, No. 1, p. 103—111 (1977).
See also 22.2.1, p. 351.

M. J. HOWES, D. V. MORGAN, and W. J. DEVLIN
*Applications of Magnetoresistance Measurements in the Evaluation of Transferred Electron Device Performance
(a) Vol. **41**, No. 1, p. 117—124 (1977).
See also 22.1.1, p. 334; 22.1.2, p. 339; 22.2.1, p. 351; 22.2.2, p. 354.

N. B. LUKYANCHIKOVA, G. S. PEKAR, N. N. TKACHENKO, HOANG MI SHIN, and M. K. SHEINKMAN
*Injection Electroluminescence in Forward-Biased ZnS Metal–Semiconductor Diodes
(a) Vol. **41**, No. 1, p. 299—305 (1977).
See also *20.3*, p. *270*; 22.4.1, p. 363.

K. BHATTACHARYA, P. BASU, and H. SAHA
*Back Wall Schottky Barrier Solar Cell with an Interfacial Layer
(a) Vol. **41**, No. 1, p. 317—321 (1977).
See also 13.3, p. 126; *16*, p. *202*; 22.1.2, p. 339; 22.2.1, p. 351.

YU. I. GORKUN, V. L. KOROBANOV, V. S. LYSENKO, V. A. NOVOMINSKII, and V. I. TURCHANIKOV
*Effect of Bombardment with Low-Energy Ar^+ Ions on Electron Scattering in Accumulation Layers of n-Ge
(a) Vol. **41**, No. 1, p. 339—346 (1977).
See also 11, p. 95; *13.4*, p. *134*; 22.1.1, p. 334.

A. S. LOGGINOV and V. E. SOLOVEV
*Analysis of the Current Carrier Concentration Distribution and Threshold Properties of Stripe-Geometry $Ga_xAl_{1-x}As$ Lasers
(a) Vol. **41**, No. 2, p. 371—378 (1977).
See also *20.2*, p. *265*; 22.2.4, p. 357.

R. D. Gould and C. A. Hogarth
Angular Distribution Measurements of Electrons Emitted from Thin Film Au–SiO_x–Au Diode and Triode Structures
(a) Vol. **41**, No. 2, p. 439–442 (1977).
See also *17*, p. *206*; 22.6, p. 388.

M. Vodenicharova and C. Vodenicharov
*Electrode- and Bulk-Limited Currents in Thin Films of Aliphatic Unsaturated Poly-(1,3,4)-Oxadiazoles
(a) Vol. **41**, No. 2, p. 487–491 (1977).
See also *14.3.1*, p. *174*; 22.9, p. 416.

H. Krause and R. Grünler
*DC Current–Voltage Behaviour of SiO_2 Layers
(a) Vol. **42**, No. 1, p. 149–155 (1977).
See also *13.4*, p. *136*; 22.6, p. 389.

J. Buxo, G. Sarrabayrouse, and D. Esteve
*The Optically Biased MIS Tunneling Schottky Barrier
(a) Vol. **42**, No. 1, p. 173–179 (1977).
See also 14.3.2, p. 176.

I. M. Kotina, S. R. Novikov, and T. I. Pirozhkova
*Determination of the Parameters of Impurity Centres from the Thermally Stimulated Current of the p–n Junction Space Charge Region
(a) Vol. **42**, No. 1, p. 397–401 (1977).
See also *13.4*, p. *136*; 22, p. 326.

K. Asai, T. Sugeta, M. Ida, and M. Fujimoto
*T-Shaped Schottky Barrier Gate GaAs FET
(a) Vol. **42**, No. 1, p. K7–K9 (1977).
See also 22.2.1, p. 352.

H. Fujiyasu, T. Hatano, S. Itoh, M. Sekinobu, and O. Ohtsuki
*n-Type GaAs MIS Structures
(a) Vol. **42**, No. 1, p. K25–K27 (1977).
See also *14.3.2*, p. *177*; 22.2.1, p. 352.

M. Hirose, S. Yokoyama, and Y. Osaka
Surface States in GaAs Tunnel MIS Structures
(a) Vol. **42**, No. 2, p. 483–488 (1977).
See also *13.3*, p. *127*; 22.2.1, p. 352.

U. Traxlmayr, W. F. Fallmann, and F. Paschke
*The Influence of Magnetic Fields on Gunn Effect Planar Devices in Corbino Geometry
(a) Vol. **42**, No. 2, p. 525–528 (1977).
See also *14.3.4*, p. *185*; 18, p. 212; 22.2.1, p. 352.

A. E. Sapega
*Determination of Surface States in Selenium Diodes by Conductance and Capacitance Studies
(a) Vol. **42**, No. 2, p. 585–594 (1977).
See also *13.3*, p. *127*; 22.1.3, p. 346.

K. P. Pande
*Rectification Properties of CdS Schottky Barrier Diodes with Evaporated Au/Ti as Blocking Contact
(a) Vol. **42**, No. 2, p. 615–619 (1977).
See also *14.3.4*, p. *185*; 22.4.1, p. 363.

D. Jäger, R. Heidemann, and R. Kassing
Influence of Carrier Transit Time on the Small-Signal Admittance of SCLC Diodes
(a) Vol. **42**, No. 2, p. 657–665 (1977).
See also *14.3.4*, p. *186*; 22.1.2, p. 341.

U. Hartmann and P. Schley
*The MIS HF Capacitance and Its Minimum
(a) Vol. **42**, No. 2, p. 667—673 (1977).
See also 14.3.4, p. 186; 22, p. 327.

K. Kliefoth and B. Petzel
*Velocity–Field Characteristics of GaAs Gunn Diodes with Different Impurity Concentration
(a) Vol. **42**, No. 2, p. K133—K135 (1977).
See also 22.2.1, p. 352.

G. H. Schwuttke
Low Cost Silicon for Solar Energy Application
(a) Vol. **43**, No. 1, p. 43—51 (1977).
See also 3, p. 42; *16*, p. *204*; 20.1, p. 260; 22.1.2, p. 341.

U. Hartmann
*Charge Measurements in Thin Insulating Layers of MIS and MIM Structures by Means of Current–Voltage Characteristics and Application to the MOS System
(a) Vol. **43**, No. 1, p. 103—110 (1977).
See also *14.3.1*, p. *175*; 22.6, p. 390.

N. Stojadinović
*A New Analytical Expression for the Impurity Profile of Hyperabrupt Varicap Diodes
(a) Vol. **43**, No. 1, p. K91—K94 (1977).

D. D. Nedeoglo, Dang Huy Lam, and A. V. Simashkevich
*Electrical Properties of the Metal–ZnSe Contact
(a) Vol. **44**, No. 1, p. 83—89 (1977).
See also *14.3.4*, p. *187*; 22.4.2, p. 366.

H.-J. Anklam, W. Brodkorb, and W. Haubenreisser
*A Regional Approximation for a Real p^+–i–n^+ Double-Injection Structure by Including Both Diffusion and Space Charge Effects
(a) Vol. **44**, No. 1, p. 161—171 (1977).
See also *14.3.4*, p. *187*; 22, p. 330.

R. Heidemann, D. Jäger, and R. Kassing
*Minority and Majority Carrier Mobilities Determined by Microwave Measurements
(a) Vol. **44**, No. 1, p. 191—195 (1977).
See also *14.3.4*, p. *187*; 22.1.2, p. 343.

H. Knauer, J. Richter, and P. Seidel
*A Direct Calculation of the Resonance Tunneling in Metal–Insulator–Metal Tunnel Junctions
(a) Vol. **44**, No. 1, p. 303—312 (1977).
See also 13.4, p. 139; 22, p. 330.

D. L. Smith
*Recombination Currents in a p–n Heterojunction Diode
(a) Vol. **44**, No. 1, p. 381—390 (1977).
See also *14.3.2*, p. *177*; 22.2.1, p. 353.

Vijender Sharma and P. C. Mathur
*Large-Signal Impedance Measurements and Power-Efficiency Consideration of n^+pp^+ GaAs IMPATT Diodes
(a) Vol. **44**, No. 1, p. 401—410 (1977).
See also *14.3.4*, p. *187*; 22.2.1, p. 353.

B. G. Caswell and J. Woods
*Some Effects of Oxygen on As-Plated CdS–Cu_2S Photovoltaic Cells
(a) Vol. **44**, No. 1, p. K47—K51 (1977).
See also *16*, p. *205*; 22, p. 330; 22.4.1, p. 364.

T. L. Tansley
*Thermally-Assisted Tunnelling in Composite Barrier Heterojunctions
(a) Vol. **44**, No. 2, p. 449—455 (1977).
See also 13.4, p. 140; 22.2.1, p. 353; 22.2.4, p. 357.

G. Kaden and H. Reimer
*Bestimmung der Dichte und des Einfangquerschnittes von Oberflächenzuständen mit dem Oberflächenladungspumpeffekt
(Determination of the Density and Capture Cross Section of Surface States by the Surface Charge Pumping Effect)
(a) Vol. **44**, No. 2, p. K117—K121 (1977).
See also *13.3*, p. *128*; 22, p. 330.

K. Woodbridge and M. E. Ertl
*Pulse Refrigeration in Bi_2Te_3
(a) Vol. **44**, No. 2, p. K123—K125 (1977).
See also 15, p. 200; 22.7, p. 395.

14.3.4 High-Field Phenomena, Space-Charge Effects, Inhomogeneities, Injected Carriers

(Electroluminescence See 20.3; Junctions See 14.3.3)

Review Article

K. W. Böer
The CdS/Cu_2S Solar Cell (I): Minority Carrier Generation and Transport in the Cu_2S Emitter
(a) Vol. **40**, No. 2, p. 355–384 (1977).
See also 14.3.3, p. 178; *16*, p. *201*; 20.1, p. 247; 22, p. 320; 22.4.1, p. 362.

Original Papers and Short Notes

D. P. Bhattacharya and Z. S. Kachlishvili
*On the Theory of the Impurity Breakdown in a Magnetic Field
(a) Vol. **39**, No. 1, p. 49–56 (1977).
See also *13*, p. *107*; 22, p. 321.

E. M. Itskovich, A. A. Kornyshev, and M. A. Vorotyntsev
Electric Current across the Metal–Solid Electrolyte Interface (I): Direct Current, Current–Voltage Characteristic
(a) Vol. **39**, No. 1, p. 229–238 (1977).
See also *14.3.2*, p. *176*; 22.5.1, p. 373.

A. Samoć, M. Samoć, and J. Sworakowski
Interpretation of One-Carrier Thermally Stimulated Currents and Isothermal Decay Currents (III): Influence of Blocking Contacts
(a) Vol. **39**, No. 1, p. 337–344 (1977).
See also *13.4*, p. *130*; 14.4, p. 189; 22.9, p. 414.

C. M. Vodenicharov, A. A. Stanchev, and M. Lozev
*Thermionic Emission in the System M–GeS–M
(a) Vol. **39**, No. 1, p. K1–K5 (1977).
See also 14.3.1, p. 173; 22, p. 321.

J. von Borzeszkowski and J. Scholz
*On the Electric Field Dependence of the Conductivity of Holes in Si at Low Temperatures (II): The Influence of the Variation of Steady-State Hole Density with Electric Field on the Conductivity
(b) Vol. **79**, No. 2, p. 525–531 (1977).
See also 6, p. 52; 13.1, p. 115; 22.1.2, p. 337.

V. M. Fomin and E. P. Pokatilov
*Multiphoton Absorption by Charge Carriers in Semiconductors in a Magnetic Field
(b) Vol. **79**, No. 2, p. 595–604 (1977).
See also *20.1*, p. *249*; 22, p. 321.

A. A. Kornyshev and M. A. Vorotyntsev
Electric Current across the Metal–Solid Electrolyte Interface (II): Low-Amplitude Alternating Current
(a) Vol. **39**, No. 2, p. 573–582 (1977).
See also *14.3.2*, p. *176*.

B.-P. Kietis, R. Reksnys, and A. Sakalas
*Ohmic and Acoustoelectric Currents in the Hall Effect for the Investigation of Bipolar Conductivity
(a) Vol. **39**, No. 2, p. 601–605 (1977).
See also 7, p. *62*; 22.4.1, p. 362.

I. MOJZES, B. PÖDÖR, and I. BALOGH
*On the Temperature Dependence of Peak Electron Velocity and Threshold Field Measured on GaAs Gunn Diodes
(a) Vol. **39**, No. 2, p. K123—K125 (1977).
See also *14.3.3*, p. *178*; 22.2.1, p. 350.

H. S. REEHAL
*Poole-Frenkel Effect with a Screened Coulomb Potential
(b) Vol. **80**, No. 1, p. K63—K65 (1977).
See also *13.4*, p. *132*.

V. CAPOZZI, G. MARIOTTO, M. MONTAGNA, A. CINGOLANI, and A. MINAFRA
*Space-Charge-Limited Currents in GaSe at Different Temperatures
(a) Vol. **40**, No. 1, p. 93—100 (1977).
See also *13.4*, p. *132*; 22.3, p. 358.

A. F. KRAVCHENKO, B. S. LISENKER, and YU. E. MARONCHUK
*Influence of High Electric Field on Capture and Recombination of Carriers in the GaAs Epitaxial Layers
(a) Vol. **40**, No. 1, p. 321—332 (1977).
See also *16*, p. *202*; 22.2.1, p. 351.

H. L. OCZKOWSKI, N. ROMEO, and L. TARRICONE
*Impact Ionization Electroluminescence in ZnSe Single Crystals
(a) Vol. **40**, No. 1, p. K49—K52 (1977).
See also *20.3*, p. *269*; 22.4.2, p. 365.

B. I. FUCHS, M. S. KAGAN, R. A. SURIS, and N. G. ZHDANOVA
*Capacitance Oscillations in Compensated Germanium
(a) Vol. **40**, No. 1, p. K61—K63 (1977).
See also 22.1.1, p. 334.

S. S. KUBAKADDI and B. S. KRISHNAMURTHY
*The Electron–Two Short-Wavelength Phonon Scattering in Non-Polar Semiconductors
(b) Vol. **80**, No. 2, p. 603—609 (1977).
See also *6*, p. *54*; 22.1.2, p. 338.

YU. E. PERLIN, B. S. TSUKERBLAT, and T. SINGH DOD
*Moments of Electroabsorption of Jahn-Teller Impurity Centres
(b) Vol. **80**, No. 2, p. 703—707 (1977).
See also 13.4, p. 132; *20.1*, p. *252*; 22, p. 323.

P. TISCHER and L. TREITINGER
*Threshold and Memory Switching in Single Crystals of the Magnetic Semiconductor $CdCr_2Se_4$
(a) Vol. **40**, No. 2, p. 551—557 (1977).
See also 22.8.2, p. 410.

J. M. PERANSIN, E. GROUBERT, and B. GOUTIERS
*Time Investigation of Parametric Interactions in Acoustoelectric Oscillators
(a) Vol. **40**, No. 2, p. 583—588 (1977).
See also *7*, p. *62*; 22.4.1, p. 362.

J. M. PAWLIKOWSKI
*Schematic Energy Band Diagram of Metal–$Cd_xHg_{1-x}Te$ Contacts
(a) Vol. **40**, No. 2, p. 613—620 (1977).
See also 13.3, p. 126; 22.4.4, p. 368.

H.-G. HOLZHÜTTER and M. MOCKER
*Generalized Boltzmann Equation with Field-Dependent Collison Terms Inclusive of Interband Transitions
(b) Vol. **81**, No. 1, p. 139—146 (1977).
See also *13*, p. *109*.

M. Brauer
*Dynamical Response of Electrons in GaAs at 300 K
(b) Vol. **81**, No. 1, p. 147—152 (1977).
See also *13*, p. *109*; 22.2.1, p. 351.

G. D. Guseinov, S. G. Abdullaeva, E. M. Godzhaev, A. Sh. Mekhtiev, L. A. Alieva, and M. Z. Ismailov
*Electroabsorption of $TlInS_2$ Single Crystals
(b) Vol. **81**, No. 1, p. K47—K50 (1977).
See also 13.5.1, p. 145; *20.1*, p. *254*; 22.8, p. 398.

M. L. Young and C. P. Fender
*Sub-Threshold Velocity–Field Characteristics of $InAs_{1-x}P_x$ Alloys
(a) Vol. **41**, No. 1, p. 85—94 (1977).
See also 13.1, p. 117; 22.2.4, p. 357.

F.-W. Breitbarth and H.-J. Tiller
*Tunnelling in Metal–Glow-Polymer–Metal Sandwich Structures
(a) Vol. **41**, No. 1, p. 151—156 (1977).
See also 13.3, p. 126; 14.3.2, p. 176; 22.9, p. 415.

L. Eglinski, A. F. Rudolph, T. Banys, and A. Dargys
*Microwave Breakdown in Tellurium
(a) Vol. **41**, No. 1, p. 177—181 (1977).
See also 13.4, p. 134; 22.1.3, p. 345.

S. Guha and S. Ghosh
*Convective Instability in n-InSb in the Presence of Crossed Electric and Magnetic Fields
(a) Vol. **41**, No. 1, p. 249—254 (1977).
See also 22.2.3, p. 356.

W. Bohmeyer and W. Hoerstel
*Hall Effect in p-Tellurium at High Electric Fields
(a) Vol. **41**, No. 1, p. K27—K30 (1977).
See also 22.1.3, p. 345.

W. Teubner and B. Fromm
*Some Remarks on the Poole-Frenkel Effect for Centres with Screened Potential
(a) Vol. **41**, No. 1, p. K47—K49 (1977).
See also 22, p. 324.

S. Nešpurek and J. Sworakowski
*A Differential Method of Analysis of Steady-State Space-Charge-Limited Current–Voltage Characteristics
(a) Vol. **41**, No. 2, p. 619—627 (1977).
See also 13.4, p. 135; 22, p. 325.

U. Traxlmayr, W. F. Fallmann, and F. Paschke
*The Influence of Magnetic Fields on Gunn Effect Planar Devices in Corbino Geometry
(a) Vol. **42**, No. 2, p. 525—528 (1977).
See also *14.3.3*, p. *180*; 18, p. 212; 22.2.1, p. 352.

K. P. Pande
*Rectification Properties of CdS Schottky Barrier Diodes with Evaporated Au/Ti as Blocking Contact
(a) Vol. **42**, No. 2, p. 615—619 (1977).
See also *14.3.3*, p. *180*; 22.4.1, p. 363.

R. Wolf, W. Kühn, R. Enderlein, and H. Lange
*Determination of Charge Carrier Concentration from Low-Field Electroreflectance Spectra
(a) Vol. **42**, No. 2, p. 629—637 (1977).
See also *20.1*, p. *258*; 22.2.4, p. 357.

D. JÄGER, R. HEIDEMANN, and R. KASSING
*Influence of Carrier Transit Time on the Small-Signal Admittance of SCLC Diodes
(a) Vol. **42**, No. 2, p. 657–665 (1977).
See also 14.3.3, p. 180; 22.1.2, p. 341.

U. HARTMANN and P. SCHLEY
The MIS HF Capacitance and Its Minimum
(a) Vol. **42**, No. 2, p. 667–673 (1977).
See also *14.3.3*, p. *181*; 22, p. 327.

L. M. BASKIN, O. I. LVOV, and G. N. FURSEY
*On the Theory of Field Emission from p-Type Semiconductors
(a) Vol. **42**, No. 2, p. 757–767 (1977).
See also 13.3, p. 127; 22, p. 327.

C. M. VODENICHAROV
A Simple Method for the Determination of the Effective Electron Mass from Current–Voltage Characteristics in the Thermionic Region
(a) Vol. **42**, No. 2, p. 785–789 (1977).
See also *13.1*, p. *121*; *14.3.1*, p. *174*; 22, p. 327; 22.6, p. 390; 22.9, p. 417.

R. MACH and E. U. MESSERSCHMIDT
*Physical Properties of the Au–CdF_2 Metal–Semiconductor Contact
(a) Vol. **42**, No. 2, p. K187–K190 (1977).
See also 16, p. 204; 22.5, p. 371.

H. BEHRENDT and W. HOERSTEL
*Electric Field Dependence of the Electron Mobility in Tellurium at 77 K
(b) Vol. **83**, No. 1, p. 223–228 (1977).
See also 22.1.3, p. 346.

A. KISIEL and B. OLEŚ
*The Influence of a Light Beam on the Electroreflectance Spectra of Germanium
(b) Vol. **83**, No. 1, p. K35–K39 (1977).
See also *20.1*, p. *259*; 22.1.1, p. 335.

S. GUHA and P. K. SEN
*Two-Stream Instability in n-Type Gallium Arsenide
(a) Vol. **43**, No. 1, p. 89–97 (1977).
See also 6, p. 59; 13, p. 111; 22.2.1, p. 353.

A. N. ZYUGANOV, S. V. SVECHNIKOV, and P. S. SMERTENKO
*Effect of Integrally Weak Diffusion on the Parameters of Current–Voltage Characteristics of Quasi-Monopolar Semiconductors
(a) Vol. **43**, No. 1, p. 333–341 (1977).
See also *14.3*, p. *170*; 22, p. 328.

G. U. REINSPERGER and W. LUDWIG
*The Contribution of Once Scattered Electrons to the Transmission Coefficient of Hot Electrons in Tunnel Triodes
(a) Vol. **43**, No. 1, p. K77–K80 (1977).
See also *13*, p. *111*; 22.6, p. 391.

E. M. BASKIN, M. D. BLOKH, M. V. ENTIN, and L. I. MAGARILL
*Current Quadratic in Field and Photogalvanic Effect in Crystals without Inversion Centre
(b) Vol. **83**, No. 2, p. K97–K100 (1977).
See also *16*, p. *204*; 22.2, p. 348.

A. REKLAITIS
*High Field Electron Diffusion in Indium Antimonide
(b) Vol. **83**, No. 2, p. K121–K124 (1977).
See also 22.1.3, p. 356.

E. Nebauer, P. Süptitz, and I. Willert
Determination of the Charge of Migrating Impurities from Electrodiffusion Profiles: the System a-As_2Se_3:Cu
(a) Vol. **43**, No. 2, p. 451—458 (1977).
See also *2*, p. *39*; *9*, p. *71*; 22, p. 329.

P. Lilienkamp and H. Pfleiderer
*An ESFI SOS Magnetodiode
(a) Vol. **43**, No. 2, p. 479—486 (1977).
See also *14.3.1*, p. *175*; 22.1.2, p. 342.

T. O. Gegechkori and Z. S. Kachlishvili
*Hot Electron Current–Voltage Curves in a Transverse Magnetic Field with Account of Phonon Heating
(a) Vol. **43**, No. 2, p. 513—518 (1977).
See also 6, p. 59; *13*, p. *111*.

K. O. Lee and T. T. Gan
*Space-Charge-Limited Currents in Evaporated Films of Pentacene
(a) Vol. **43**, No. 2, p. 565—571 (1977).
See also *14.3.1*, p. *175*; 22.9, p. 418.

V. Zhelev and J. Malinowski
*Ionic Field Effect in Epitaxial AgBr Layers
(a) Vol. **43**, No. 2, p. 675—683 (1977).
See also 13.3, p. 128; 22.5.1, p. 374.

D. R. Franceschetti and J. R. Macdonald
*Comment on the Small-Signal AC Response of Solid Electrolytes
(a) Vol. **43**, No. 2, p. K169—K173 (1977).
See also *14.3.2*, p. *177*.

H. S. Reehal and C. B. Thomas
Poole-Frenkel Conduction in Amorphous Solids for Centres with a Screened Coulomb Potential
(b) Vol. **84**, No. 1, p. K13—K16 (1977).
See also 2, p. 40; *13.4*, p. *138*; 22.8, p. 400.

D. D. Nedeoglo, Dang Huy Lam, and A. V. Simashkevich
*Electrical Properties of the Metal–ZnSe Contact
(a) Vol. **44**, No. 1, p. 83—89 (1977).
See also *14.3.3*, p. *181*; 22.4.2, p. 366.

H.-J. Anklam, W. Brodkorb, and W. Haubenreisser
*A Regional Approximation for a Real p^+–i–n^+ Double-Injection Structure by Including Both Diffusion and Space Charge Effects
(a) Vol. **44**, No. 1, p. 161—171 (1977).
See also *14.3.3*, p. *181*; 22, p. 330.

R. Heidemann, D. Jäger, and R. Kassing
*Minority and Majority Carrier Mobilities Determined by Microwave Measurements
(a) Vol. **44**, No. 1, p. 191—195 (1977).
See also *14.3.3*, p. *181*; 22.1.2, p. 343.

Vijender Sharma and P. C. Mathur
*Large-Signal Impedance Measurements and Power-Efficiency Consideration of n^+pp^+ GaAs IMPATT Diodes
(a) Vol. **44**, No. 1, p. 401—410 (1977).
See also *14.3.3*, p. *181*; 22.2.1, p. 353.

A. I. Popov, I. Kh. Geller, and V. K. Shemetova
*Memory and Threshold Switching Effects in Amorphous Selenium
(a) Vol. **44**, No. 1, p. K71—K73 (1977).
See also *2*, p. *40*; 22.1.3, p. 347.

NGUYEN VAN HIEU and NGUYEN AI VIET
*Interband Electronic Raman Scattering by a Three-Band Semiconductor in the Presence of a Resonant Electromagnetic Field
(b) Vol. **84**, No. 2, p. 417–422 (1977).
See also 13.1, p. 122; *20.1*, p. *262*; 22.2, p. 349; 22.4, p. 361.

M. M. BABAEV and T. M. GASYMOV
*Thermopower and the Transverse Nernst-Ettingshausen Effect in Semiconductors in a Strong Electric Field
(b) Vol. **84**, No. 2, p. 473–483 (1977).
See also 6, p. 60; 13.4, p. 139; *15*, p. *199*; 22, p. 330.

F. LUKEŠ
*Electroreflectance Spectrum of InAs in the Range of E_0 and $E_0 + \Delta_0$ Transitions
(b) Vol. **84**, No. 2, p. K113–K117 (1977).
See also 13.1, p. 123; *20.1*, p. *263*; 22.2.1, p. 353.

K. JARAŠIŪNAS and J. VAITKUS
*Investigation of Non-Equilibrium Processes in Semiconductors by the Method of Transient Holograms
(a) Vol. **44**, No. 2, p. 793–800 (1977).
See also 20, p. 246; 22.1.2, p. 344; 22.4.2, p. 366.

V. A. GRITSENKO, E. E. MEERSON, K. P. MOGILNIKOV, and S. P. SINITSA
*On the Conductivity of Amorphous Silicon Nitride in High Electric Fields
(a) Vol. **44**, No. 2, p. K167–K170 (1977).
See also *2*, p. *41*; 22, p. 331.

14.4 Dielectrics

Review Article

K. Shahi
*Transport Studies on Superionic Conductors
(a) Vol. **41**, No. 1, p. 11–44 (1977).
See also 9, p. 69; 10.2, p. 83; 15, p. 198; 22.5.1, p. 373; 22.8, p. 396; 22.9, p. 414.

Original Papers and Short Notes

I. J. R. Baumvol, A. Vasquez, J. Martinez, and F. C. Zawislak
Temperature Dependence of the Electric Field Gradient at Ta Nuclei in $CaHfO_3$
(b) Vol. **79**, No. 1, p. K65–K69 (1977).
See also *13.4*, p. *129*; 22.8.1, p. 402.

M. E. van Hulle, and W. Maenhout-van der Vorst
*Ionic Conductivity and Space Charge Region in Silver Bromide (111) Microcrystals
(a) Vol. **39**, No. 1, p. 253–258 (1977).
See also *10.2*, p. *83*; 22.5.1, p. 373.

A. Samoć, M. Samoć, and J. Sworakowski
Interpretation of One-Carrier Thermally Stimulated Currents and Isothermal Decay Currents (III): Influence of Blocking Contacts
(a) Vol. **39**, No. 1, p. 337–344 (1977).
See also *13.4*, p. *130*; 14.3.4, p. 183; 22.9, p. 414.

J. Albers and H. Küppers
Dielectric and Thermal Properties of Ammonium Hydrogen Oxalate Hemihydrate
(a) Vol. **39**, No. 1, p. K49–K51 (1977).
See also *1.2*, p. *14*; 8, p. 64; 22.9, p. 414.

E. Mohler and B. Thomas
*Electro-Optic Properties of CuCl below the Absorption Edge
(b) Vol. **79**, No. 2, p. 509–517 (1977).
See also *20.1*, p. *249*; 22.5, p. 370.

F. J. Owens
*Evidence for Zero-Field Fluctuations in Cr^{3+} near the Phase Transition in $NH_4Al(SO_4)_2 \cdot$ $\cdot$ $12\,H_2O$
(b) Vol. **79**, No. 2, p. 623–628 (1977).
See also 19, p. 235; 22.8, p. 396.

A. Scalabrin, A. S. Chaves, D. S. Shim, and S. P. S. Porto
Temperature Dependence of the A_1 and E Optical Phonons in $BaTiO_3$
(b) Vol. **79**, No. 2, p. 731–742 (1977).
See also *6*, p. *52*; 20.1, p. 249; 22.8.1, p. 402.

J. Prakash and F. Fischer
*Ionic Thermocurrent and Optical Absorption Studies of S^{--}–Vacancy Dipoles in KI
(a) Vol. **39**, No. 2, p. 499–507 (1977).
See also *10.2*, p. *84*; 20.1, p. 250; 22.5.2, p. 375.

K. V. Rao and K. Suryanarayana Rao
*Variation of the Dielectric Constant with Temperature in Ionic Crystals
(a) Vol. **39**, No. 2, p. K119–K121 (1977).
See also 22.5.2, p. 376.

K. Hübner
Chemical Bond and Related Properties of SiO_2 (I): Character of the Chemical Bond
(a) Vol. **40**, No. 1, p. 133–140 (1977).
See also 2, p. 35; *4*, p. *44*; *13.1*, p. *116*; 20.1, p. 251; 22.6, p. 387.

M. E. van Hulle
*Dielectric Loss Measurements of AgBr (100) Emulsions
(a) Vol. **40**, No. 1, p. K57—K60 (1977).
See also *10.2*, p. *86*; 22.5.1, p. 373.

K. Bärner, R. Braunstein, and E. Chock
*The Optical Properties of MnAs and $MnAs_{0.85}P_{0.15}$
(b) Vol. **80**, No. 2, p. 451—460 (1977).
See also 13.1, p. 116; *20.1*, p. *252*; 22.8, p. 397.

S. C. Zílio and M. de Souza
The Dielectric Relaxation of Impurity–Vacancy Complexes in KI Doped with Cl^- Anion
(b) Vol. **80**, No. 2, p. 597—601 (1977).
See also *10.2*, p. *86*; 22.5.2, p. 377.

U. Büchner and W. Guse
*The Dielectric Tensor Components of Czochralski-Grown 2:1 Mullite Single Crystals
(b) Vol. **80**, No. 2, p. 615—618 (1977).
See also 13.1, p. 116; 22.6.1, p. 393.

G. Oelgart, H. Krüger, and R. Herrmann
*Lattice Dielectric Constant of $Bi_{1-x}Sb_x$
(b) Vol. **80**, No. 2, p. K101—K104 (1977).
See also 22.7, p. 394.

W. Bollmann
Evaluation of the Association Parameters of Quadripoles (Dimers of Dipoles) in Ionic Crystals from Conductivity Investigations (II): Association Parameters of (O^{--}–F_V^-) Dipoles and Their Dimers in CaF_2 Crystals
(a) Vol. **40**, No. 2, p. 409—414 (1977).
See also *10.2*, p. *86*; 22.5.3, p. 383.

F. Walz
*Completely Automated Method for the Measurement of Dielectric Aftereffects in Ferroelectric Materials
(a) Vol. **40**, No. 2, p. 455—461 (1977).
See also 14.4.1, p. 195; 22.8.1, p. 403.

K. Hübner
Chemical Bond and Related Properties of SiO_2 (II): Structural Trends
(a) Vol. **40**, No. 2, p. 487—495 (1977).
See also 2, p. 36; *4*, p. *45*; *13.1*, p. *117*; 20.1, p. 252; 22.6, p. 387.

M. E. van Hulle and W. Maenhout-van der Vorst
*Ionic Conductivity of AgBr and AgBr:Pb^{2+} Single Crystals
(a) Vol. **40**, No. 2, p. K173—K177 (1977).
See also *10.2*, p. *87*; 22.5.1, p. 373.

J. Schreiber
*Pseudo-Spin Motion in Solid Solutions of the Type $K(H_{1-x}D_x)_2PO_4$
(b) Vol. **81**, No. 1, p. 371—378 (1977).
See also *14.4.1*, p. *195*; 22.8.1, p. 404.

B. A. Lukiyanets and K. D. Tovstyuk
*The Exciton Contribution to the Dielectric Susceptibility of a Crystal with Extended Defects
(b) Vol. **81**, No. 1, p. K1—K3 (1977).
See also *13.5.1*, p. *145*.

K. K. Srivastava, D. R. Goyal, A. Kumar, K. N. Lakshminarayan, O. S. Panwar, and I. Krishan
*Dielectric Properties of Glassy $As_{10}Te_{75}Ge_{15}$
(a) Vol. **41**, No. 1, p. 323—329 (1977).
See also 2, p. 36; 22.8, p. 398.

M. Malinkowski, A. Pietraszko, and M. Połomska
*Thermal Expansion and Dielectric Permittivity in the Vicinity of Transition Points of Lead Germanate
(a) Vol. **41**, No. 1, p. K55—K58 (1977).
See also *8*, p. *65*; 22.8.1, p. 404.

H. Kresse, D. Demus, and S. König
*Dielectric Relaxation of Liquid Crystals with Nematic and Smectic Modifications
(a) Vol. **41**, No. 1, p. K67—K69 (1977).
See also *2*, p. *36*; 22.9, p. 416.

F. Brosens, L. F. Lemmens, and J. T. Devreese
*Comments on the Static Dielectric Function of Jellium
(b) Vol. **81**, No. 2, p. 551—556 (1977).

V. B. Campos, O. Hipólito, and R. Lobo
Correlations and Plasmons in Quasi-One-Dimensional Metals
(b) Vol. **81**, No. 2, p. 657—663 (1977).
See also *13.5.2*, p. *149*; 21, p. 281.

M. Lannoo and G. Allan
Contribution of the d-Electron Screening to the Properties of Impurities in Transition Metals
(b) Vol. **81**, No. 2, p. 681—690 (1977).
See also *13.4*, p. *134*; 21.1, p. 293.

S. Radhakrishna and S. Haridoss
*Methods of Analysis of Piezo-Stimulated Current Spectra
(a) Vol. **41**, No. 2, p. 649—652 (1977).
See also *10.2*, p. *88*; 22.5.2, p. 378.

Yu. V. Bogatyrev and F. P. Korshunov
*The Effect of High-Speed Electrons of Different Energy on the Charge Storage in SiO_2 and Si_3N_4 Dielectric Films
(a) Vol. **41**, No. 2, p. K151—K154 (1977).
See also *11*, p. *96*; 22, p. 325; 22.6, p. 389.

M. Zieliński and M. Kryszewski
Thermal Sampling Technique for the Thermally Stimulated Discharge in Polymers. Model Calculations
(a) Vol. **42**, No. 1, p. 305—314 (1977).
See also *2*, p. *37*; 22.9, p. 417.

P. C. Mehendru, N. L. Pathak, K. Jain, and P. Mehendru
*Electrical Conduction in Pure and Iodine-Doped Polyvinyl Acetate Thick Films
(a) Vol. **42**, No. 1, p. 403—407 (1977).
See also 20.1, p. 257; 22.9, p. 417.

H. D. Riccius, D. S. Smith, B. G. Withford, and L. H. Jones
*Difference-Frequency Mixing and Second-Harmonic Generation of 3.39 μm Laser Radiation in $LiNbO_3$
(a) Vol. **42**, No. 1, p. K53—K55 (1977).
See also *20*, p. *245*; 22.8.1, p. 406.

T. Ohgaku and N. Takeuchi
*Dielectric Loss of KCl–KBr Mixed Crystals Doped with Sr^{2+}
(a) Vol. **42**, No. 1, p. K83—K85 (1977).
See also 10.2, p. 89; 22.5.4, p. 385.

K. Hübner
Chemical Bond and Related Properties of SiO_2 (III): Core-Level Shifts in SiO_x
(a) Vol. **42**, No. 2, p. 501—509 (1977).
See also 2, p. 38; *4*, p. *45*; *13.1*, p. *120*; 20.1, p. 258; 22.6, p. 390.

C. H. Chen
Electron Energy Loss Studies of Direct Nonvertical Interband Transitions in Silicon
(b) Vol. **83**, No. 1, p. 347—351 (1977).
See also *13.1*, p. *121*; 22.1.2, p. 341.

N. Saleh and M. Gawish
*Effect of Heat Treatment on the Electronic Properties of Calcium Borate Glasses Containing Iron Oxide
(a) Vol. **43**, No. 1, p. 99—102 (1977).
See also 2, p. 39; 22.8.1, p. 406.

J. Eisner
*The Physical Properties of TGS Single Crystals, Grown from Aqueous TGS Solutions Containing Aniline
(a) Vol. **43**, No. 1, p. K1—K4 (1977).
See also 3, p. 43; 22.9, p. 418.

U. Büchner
Wave-Vector Dependence of the Electron Energy Losses of PbS, PbSe, and PbTe
(b) Vol. **83**, No. 2, p. 493—500 (1977).
See also *13.1*, p. *122*; 22, p. 328.

G. Paasch and P. Rennert
*Quadratic Electronic Polarizability of the Interacting Electron Gas
(b) Vol. **83**, No. 2, p. 501—508 (1977).
See also *13*, p. *111*.

G. Oelgart, R. Herrmann, and U. Meschter
*The Lattice Dielectric Constant in $Bi_{100-x}Sb_x$ as a Function of x
(b) Vol. **83**, No. 2, p. 521—528 (1977).
See also 22.7, p. 395.

G. Duggan, G. J. Morgan, N. Shahtahmasebi, and A. Lettington
The Effective Screening Function in Insulating Solids
(b) Vol. **83**, No. 2, p. 543—553 (1977).
See also *13.4*, p. *137*; 22.1.1, p. 335.

A. K. Jonscher
*A Many-Body Model of Dielectric Polarisation in Solids (I): Experimental Data and Existing Interpretations
(b) Vol. **83**, No. 2, p. 585—597 (1977).
See also 2, p. 39; 14.4.1, p. 196; 22, p. 329; 22.9, p. 418.

F. G. dos Reis, Hilda A. Cerdeira, and R. Luzzi
*Dielectric Response of Saturated Semiconductors
(b) Vol. **83**, No. 2, p. 607—614 (1977).
See also 13.1, p. 122; *20*, p. *246*; 22, p. 329.

A. A. Kornyshev, A. I. Rubinshtein, and M. A. Vorotyntsev
*Image Potential near a Dielectric–Plasma-Like Medium Interface
(b) Vol. **84**, No. 1, p. 125—132 (1977).
See also *13*, p. *112*.

A. K. Jonscher
*A Many-Body Model of Dielectric Polarisation in Solids (II): The Universal Model
(b) Vol. **84**, No. 1, p. 159—166 (1977).
See also 2, p. 39; 14.4.1, p. 196; 22, p. 329; 22.9, p. 419.

M. E. van Hulle and Maenhout-van der Vorst
*Space Charge Characteristics of Silver Halide Microcrystals
(a) Vol. **44**, No. 1, p. 229—236 (1977).
See also *10.2*, p. *92*; 22.5.1, p. 374.

D. L. Kirk and S. E. Ormrod
*The Mechanism of Charge Compensation Occurring in Polycrystalline β″- Alumina
(a) Vol. **44**, No. 2, p. 643—650 (1977).
See also *9*, p. *71*; 10.2, p. 93; 22.6, p. 392.

G. de Mey and H. J. Pauwels
*Time Constant of the Ion Current in Thin Film Capacitors
(a) Vol. **44**, No. 2, p. K195—K199 (1977).
See also 22, p. 331.

14.4.1 Ferroelectrics

Original Papers and Short Notes

V. M. Fridkin, B. N. Popov, and K. A. Verkhovskaya
*Effect of Anomalous Bulk Photovoltage in Ferroelectrics
(a) Vol. **39**, No. 1, p. 193—201 (1977).
See also *16*, p. *201*; 22.8.1, p. 402.

K. Irie
*Temperature Dependence of the Shift of the Acceptor Level in Ferroelectric and Photoconductive SbSI
(a) Vol. **39**, No. 1, p. 313—320 (1977).
See also *13.4*, p. *130*; 22.5.4, p. 385.

H. B. Haanstra and H. Ihrig
*Voltage Contrast Imaging of PTC-Type $BaTiO_3$ Ceramics Having Low and High Titanium Excess
(a) Vol. **39**, No. 1, p. K7—K10 (1977).
See also *1.4*, p. *24*; 22.8.1, p. 402.

H. Happ and G. Rother
*Investigation on the Far-Infrared Response of DKDP by Asymmetric Fourier Spectroscopy
(b) Vol. **79**, No. 2, p. 473—477 (1977).
See also 6, p. 52; 20.1, p. 248; 22.8.1, p. 402.

V. S. Vikhnin and A. Ya. Gordon
*Paraelectric Resonance of Hydrogen Bonds in Ferroelectrics and Related Phenomena
(b) Vol. **79**, No. 2, p. 495—499 (1977).
See also 22.8.1, p. 402.

D. Kapor
*On the Mean-Field Approach to the Transverse Ising Model
(b) Vol. **79**, No. 2, p. K111—K114 (1977).

W. Müller-Lierheim, T. Suski, and H. H. Otto
Factor Group Analysis of the Raman Spectrum of $Pb_5Ge_3O_{11}$
(b) Vol. **80**, No. 1, p. 31—41 (1977).
See also *6*, p. *53*; 13.5.3, p. 151; 20.1, p. 250; 22.8.1, p. 403.

A. Pierrefeu, E. F. Steigmeier, and B. Dorner
Inelastic Neutron Scattering in SbSI near the Ferroelectric Phase Transformation
(b) Vol. **80**, No. 1, p. 167—171 (1977).
See also *6*, p. *53*; 22.8, p. 397.

M. Levalois and G. Allais
*Study of Short-Range Interactions Ti–O and Nb–O in Ferroelectrics
(a) Vol. **40**, No. 1, p. 181—187 (1977).

G. L. Mailyan and N. M. Plakida
*Fluctuations of Order Parameter and Anharmonic Interaction in the Vibronic Model of Ferroelectrics
(b) Vol. **80**, No. 2, p. 543—547 (1977).
See also *6*, p. *54*.

V. Winterfeldt and G. Schaak
Phonon Contributions to the Anomalous Specific Heat of Triglycine Sulfate near T_c
(b) Vol. **80**, No. 2, p. 559—567 (1977).
See also 1.2, p. 15; *6*, p. *54*; *8*, p. *65*; 22.9, p. 415.

S. Stamenković, S. D. Stojanović, and R. B. Žakula
*The Influence of Lattice Vibrations on the Parametric Resonance in Hydrogen-Bonded Ferroelectrics by Means of Parallel Electric Microwave Pumping
(b) Vol. **80**, No. 2, p. 661—668 (1977).
See also *6*, p. *55*.

S. Tsunekawa, M. Suezawa, and H. Takei
*New Type Ferroelastic Behaviour of $NdNbO_4$ Crystals
(a) Vol. **40**, No. 2, p. 437—446 (1977).
See also *10.2*, p. *86*; 22.8.1, p. 403.

F. Walz
Completely Automated Method for the Measurement of Dielectric Aftereffects in Ferroelectric Materials
(a) Vol. **40**, No. 2, p. 455—461 (1977).
See also *14.4*, p. *190*; 22.8.1, p. 403.

R. Blinc, J. Slak, and M. Luzar
Deuteron NMR Study of the Role of Ammonium Ions in the Antiferroelectric Transition in $ND_4D_2AsO_4$
(a) Vol. **40**, No. 2, p. K121—K125 (1977).
See also *19*, p. *238*; 22.8, p. 398.

G. Schmidt, J. V. Cieminski, W. Zimmermann, and V. A. Isupov
*Electrostriction in $Sr_{0.75}Ba_{0.25}Nb_2O_6$ Crystals
(a) Vol. **40**, No. 2, p. K153—K156 (1977).
See also 22.8.1, p. 404.

J. Azoulay, D. Gerlich, E. Wiener-Avnear, and I. Pelah
*Piezoelectric Coupled Soft Mode in CsH_2AsO_4 and CsD_2AsO_4 Studied by Brillouin Scattering
(b) Vol. **81**, No. 1, p. 295—306 (1977).
See also 6, p. 55; 22.8, p. 398.

J. Schreiber
*Pseudo-Spin Motion in Solid Solutions of the Type $K(H_{1-x}D_x)_2PO_4$
(b) Vol. **81**, No. 1, p. 371—378 (1977).
See also *14.4*, p. *190*; 22.8.1, p. 404.

N. Schmidt and R. Send
*The Role of 90° Domain Processes for the Determination of the Dielectric Anisotropy in Ferroelectric PLZT Ceramics
(a) Vol. **41**, No. 1, p. 157—161 (1977).
See also 22.8, p. 398.

S. T. Liu and L. E. Cross
*Primary Pyroelectricity in Strontium Barium Niobate ($Sr_{0.5}Ba_{0.5}Nb_2O_6$) Single Crystals
(a) Vol. **41**, No. 1, p. K83—K87 (1977).
See also 22.8.1, p. 404.

N. Yamamoto, K. Yagi, and G. Honjo
*Electron Microscopic Studies of Ferroelectric and Ferroelastic $Gd_2(MoO_4)_3$ (I): General Features of Ferroelectric Domain Wall, Antiphase Boundary, and Crystal Defects
(a) Vol. **41**, No. 2, p. 523—534 (1977).
See also 22.8.1, p. 405.

H. Schmid and L. A. Pétermann
*Dielectric Constant and Electric Resistivity of Copper Chlorine Boracite, $Cu_3B_7O_{13}Cl$ (Cu–Cl–B)
(a) Vol. **41**, No. 2, p. K147—K150 (1977).
See also 22.8, p. 399.

N. YAMAMOTO, K. YAGI, and G. HONJO
*Electron Microscopic Studies of Ferroelectric and Ferroelastic $Gd_2(MoO_4)_3$ (II): Interactions between Ferroelectric Domain Walls
(a) Vol. **42**, No. 1. p. 257—265 (1977).
See also 22.8.1, p. 405.

F. MILIA, G. LAHAJNAR, M. ROŽMARIN, J. SLAK, and R. BLINC
*Proton Magnetic Resonance Study of Molecular Motion in $(NH_4)_2HAsO_4$ and $(NH_4)_2HPO_4$
(a) Vol. **42**, No. 1, p. 315—318 (1977).
See also *19*, p. *240*; 22.8.1, p. 405.

D. VAJDA
*The Dependence of the Landau Kinetic Coefficient L on the Degree of Deuteration in $K(H_{1-n}D_n)_2PO_4$ Crystals
(a) Vol. **42**, No. 1, p. K37—K39 (1977).
See also 22.8.1, p. 406.

I. SAVATINOVA and E. ANACHKOVA
Raman Study of Dynamic Disorder and the Ferroelectric Transition in $K_4Fe(CN)_6 \cdot 3\ H_2O$
(b) Vol. **82**, No. 2, p. 677—683 (1977).
See also *6*, p. *57*; *20.1*, p. *258*; 22.8, p. 399.

S. TRIMPER
Phase Transition in Ferroelectric Liquid Crystals
(b) Vol. **82**, No. 2, p. K75—K79 (1977).
See also *1.2*, p. *17*; 2, p. 38.

S. K. HOFFMANN, L. S. SZCZEPANIAK, and J. STANKOWSKI
*Microscopic Mechanism of the Temperature Dependence of the Spontaneous Polarization in TGS Crystals
(b) Vol. **82**, No. 2, p. K85—K87 (1977).
See also *1.2*, p. *17*; 19, p. 241; 22.9, p. 417.

D. BÄUERLE, W. B. HOLZAPFEL, A. PINCZUK, and Y. YACOBY
*Temperature and Hydrostatic Pressure Dependence of Vibrational Modes in $PbTi_{1-x}Zr_xO_3$
(b) Vol. **83**, No. 1, p. 99—107 (1977).
See also *1.2*, p. *18*; 6, p. 58; 20.1, p. 259; 22.8.1, p. 406.

B. KHELIFA, A. DELAHAIGUE, and P. JOUVE
*Soft Mode and Debye Relaxations in Thiourea
(b) Vol. **83**, No. 1, p. 139—145 (1977).
See also *6*, p. *58*; 20.1, p. 259; 22.9, p. 417.

R. MAŃKA
The Thermoelectric Power for a Small Polaron in Ferroelectrics
(b) Vol. **83**, No. 1, p. K69—K73 (1977).
See also *13.5.3*, p. *152*; *15*, p. *199*; 22.8.1, p. 406.

A. BLUMEN and C. MERKEL
Energy Band Calculations on Helical Systems
(b) Vol. **83**, No. 2, p. 425—431 (1977).
See also *13.1*, p. *121*; 22.5, p. 371.

A. K. JONSCHER
A Many-Body Model of Dielectric Polarisation in Solids (I): Experimental Data and Existing Interpretations
(b) Vol. **83**, No. 2, p. 585—597 (1977).
See also 2, p. 39; *14.4*, p. *192*; 22, p. 329; 22.9, p. 418.

A. K. JONSCHER
A Many-Body Model of Dielectric Polarisation in Solids (II): The Universal Model
(b) Vol. **84**, No. 1, p. 159—166 (1977).
See also 2, p. 39; *14.4*, p. *192*; 22, p. 329; 22.9, p. 419.

I. SAVATINOVA and E. ANACHKOVA
*Raman Spectrum and Dielectric Relaxation in $K_4Fe(CN)_6 \cdot 3\,H_2O$
(b) Vol. 84, No. 1, p. 401–406 (1977).
See also 1.2, p. 20; 6, p. 60; *20.1*, p. *262*; 22.8, p. 400.

N. YAMAMOTO, K. YAGI, and G. HONJO
*Electron Microscopic Studies of Ferroelectric and Ferroelastic $Gd_2(MoO_4)_3$ (III): Image Analysis of Antiphase Boundary and Observation of Phase Transformation
(a) Vol. 44, No. 1, p. 147–160 (1977).
See also 22.8.1, p. 407.

V. V. GAGULIN, N. V. FADEEVA, A. G. BELOUS, L. A. SEVASTIANOVA, A. V. TITOV, M. V. PLOTNIKOVA, K. P. MITROFANOV, E. V. ZUBOVA, S. P. SOLOVEV, and YU. N. VENEVTSEV
*New Rhenium Containing Seignettomagnets and Ferroelectrics
(a) Vol. 44, No. 1, p. 247–257 (1977).
See also 5, p. 49; 18.2, p. 223; *18.4*, p. *234*; 22.8, p. 400.

A. KÜHNEL, S. WENDT, and J. WESSELINOWA
*Dynamic Behaviour of the Ising Model in a Transverse Field
(b) Vol. 84, No. 2, p. 653–664 (1977).
See also 18.2, p. 224.

H. HUSSINGER and H.-G. UNRUH
*Phase Boundaries in Ammonium Sulfate
(a) Vol. 44, No. 2, p. 525–529 (1977).
See also 22.8.1, p. 408.

C. V. MANJUNATH and R. SRINIVASAN
Electron Spin Resonance Study of Copper(II) Doped Ferroelectric Ammonium Sulphate
(a) Vol. 44, No. 2, p. 687–692 (1977).
See also *19*, p. *243*; 22.8.1, p. 408.

A. M. VARAPRASAD, A. L. SHASHIMOHAN, A. B. BISWAS, and D. K. CHAKRABARTY
*Dielectric and Infrared Spectral Studies of the $(Na_{1/2}La_{1/2})_xBa_{1-x}TiO_3$ System
(a) Vol. 44, No. 2, p. 787–791 (1977).
See also 20.1, p. 264; 22.8.1, p. 408.

T. I. CHEKMASOVA, I. S. KABANOV, and V. I. YUZVAK
*The Pressure–Temperature Phase Diagram of $LiNH_4SO_4$
(a) Vol. 44, No. 2, p. K155–K157 (1977).
See also *1.2*, p. *20*; 19, p. 243; 22.8.1, p. 408.

15. Thermoelectric and Thermomagnetic Properties

Review Articles

M. I. Klinger and A. A. Samokhvalov
Electron Conduction in Magnetite and Ferrites
(b) Vol. **79**, No. 1, p. 9—48 (1977).
Erratum
(b) Vol. **83**, No. 1, p. 353 (1977).
See also 13.1, p. 114; *13.5.3*, p. *151*; *14.3*, p. *165*; 18.3, p. 228; 22.8.2, p. 409.

K. Shahi
Transport Studies on Superionic Conductors
(a) Vol. **41**, No. 1, p. 11—44 (1977).
See also 9, p. 69; 10.2, p. 83; *14.4*, p. *189*; 22.5.1, p. 373; 22.8, p. 396; 22.9, p. 414.

Original Papers and Short Notes

K. A. Long
A Theoretical Investigation of the Specific Heats and Transport Properties of the Actinides
(b) Vol. **79**, No. 1, p. 155—165 (1977).
See also *6*, p. *51*; 8, p. 64; *14.1*, p. *157*; 21.5, p. 314.

S. Cristoloveanu and P. Viktorovitch
*Influence of Surfaces on the Ettingshausen Effect. Application to the Characterization of Narrow-Gap Semiconductors
(a) Vol. **39**, No. 1, p. 109—116 (1977).
See also 14.3, p. 165; 22.2.3, p. 356.

G. J. Ashwell, D. D. Eley, M. R. Willis, and J. Woodward
Electrical and Magnetic Properties of Two Crystallographic Forms of $(DEPE)^{2+}(TCNQ)_4^{2-}$
(b) Vol. **79**, No. 2, p. 629—636 (1977).
See also *14.1*, p. *157*; *14.3*, p. *165*; 18.1, p. 213; 22.9, p. 414.

J. F. Baumard and E. Tani
*Thermoelectric Power in Reduced Pure and Nb-Doped TiO_2 Rutile at High Temperature
(a) Vol. **39**, No. 2, p. 373—382 (1977).
See also 10.2, p. 84; 22.6, p. 386.

J. M. Pelletier, R. Borrelly et E. Pernoux
*Influence des traitements thermiques sur les pouvoir thermoélectrique des alliages cuivre–fer
(Influence of Thermal Treatment on the Thermoelectric Power of Copper–Iron Alloys)
(a) Vol. **39**, No. 2, p. 525—532 (1977).
See also 21.1, p. 289; 21.1.1, p. 302.

K. L. Chopra, A. P. Thakoor, S. K. Barthwal, and P. Nath
Electron Transport Properties of Amorphous Cu–Ag Films
(a) Vol. **40**, No. 1, p. 247—255 (1977).
See also *2*, p. *35*; *14.3.1*, p. *173*; 21.1, p. 291; 21.6, p. 315.

M. G. Townsend, J. R. Gosselin, J. L. Horwood, L. G. Ripley, and R. J. Tremblay
Violarite, a Metallic Natural Spinel
(a) Vol. **40**, No. 1, p. K25—K29 (1977).
See also *5*, p. *47*; *18.1*, p. *213*; 22.8.2, p. 410.

R. B. Somoano, V. Hadek, S. P. S. Yen, A. Rembaum, C. H. Hsu, R. J. Deck, T. Datta, and A. M. Hermann
Electrical Properties of (DEPE) $(TCNQ)_4$
(b) Vol. **81**, No. 1, p. 281—286 (1977).
See also *14*, p. *155*; 22.9, p. 415.

P. I. Baranskii, I. S. Buda, and L. A. Shcherbina
Structure of the Thermomagnetic Tensor in Semiconductors of n-Ge Type
(b) Vol. **81**, No. 2, p. 709—718 (1977).
See also *14.3*, p. *168*; 22.1.1, p. 334.

M. Zoaeter, A. Conan, and D. Delaunay
*Transport Measurements and Band Structure of Bulk $MoSe_{2-x}$ Compounds
(a) Vol. **41**, No. 2, p. 629—635 (1977).
See also 13.1, p. 118; *14.3*, p. *169*; 22, p. 325.

K. M. Demchuk and I. M. Tsidilkovskii
Scattering of Electrons by the Deformation Potential in Doped InSb
(b) Vol. **82**, No. 1, p. 59—67 (1977).
See also *13.1*, p. *118*; 14.3, p. 169; 22.2.3, p. 356.

O. P. Hansen
*Weak-Field Thermomagnetic Coefficients of a Trigonal Bismuth Sample
(b) Vol. **82**, No. 1, p. 133—138 (1977).
See also 21.7, p. 319.

E. Shatkovskii and A. Tamaševičius
Sign Inversion of Photomagnetoelectric Effect in InAs at Intense Photoexcitation
(a) Vol. **42**, No. 1, p. 111—114 (1977).
See also *16*, p. *203*; 22.2.1, p. 352.

T. Suzuki, M. Hirose, S. Ogose, and Y. Osaka
Electrical Transport and Photoconductivity in Amorphous Silicon
(a) Vol. **42**, No. 1, p. 337—344 (1977).
See also 2, p. 37; 13.1, p. 120; *14.3.1*, p. *174*; *16*, p. *203*; 22.1.2, p. 340.

A. Jędrzejczak and T. Dietl
*Erratum to: Thermomagnetic Properties of n-Type and p-Type HgTe
(b) Vol. **82**, No. 2, p. 717 (1977).
See also *13.1*, p. *120*; 14.3, p. 170; 22.4.3, p. 367.

W. Eichler
*Thermomagnetic Effects in Pure Imperfect Tellurium Single Crystals
(a) Vol. **42**, No. 2, p. K107—K110 (1977).
See also 13.4, p. 137; 22.1.3, p. 346.

I. P. Zvyagin
Correlation Effects in Hopping Conduction
(b) Vol. **83**, No. 1, p. 63—70 (1977).
See also 2, p. 38; 13, p. 110; *14.3*, p. *170*; 22, p. 327.

R. Mańka
*The Thermoelectric Power for a Small Polaron in Ferroelectrics
(b) Vol. **83**, No. 1, p. K69—K73 (1977).
See also *13.5.3*, p. *152*; 14.4.1, p. 196; 22.8.1, p. 406.

M. Goodarz and R. D. Barnard
Deviations from Matthiessen's Rule in Silver Containing Low-Density Defects
(b) Vol. **83**, No. 2, p. 555—564 (1977).
See also *13.2*, p. *125*; 13.4, p. 137; 14.1, p. 159; 21.6, p. 317.

S. I. Radautsan, E. K. Arushanov, L. N. Lukyanova, and A. N. Nateprov
*The Thermoelectric Power of $Cd_3(As_xP_{1-x})_2$
(a) Vol. **43**, No. 2, p. K159—K164 (1977).
See also 13.1, p. 122; 22.3.1, p. 360.

M. M. Babaev and T. M. Gasymov
*Thermopower and the Transverse Nernst-Ettingshausen Effect in Semiconductors in a Strong Electric Field
(b) Vol. **84**, No. 2, p. 473—483 (1977).
See also 6, p. 60; 13.4, p. 139; *14.3.4*, p. *188*; 22, p. 330.

N. R. Mitra, J. Thakur, and P. L. Srivastava
*Resistivity and Thermoelectric Power of Liquid Lithium and Sodium from Harrison's First Principle
(b) Vol. **84**, No. 2, p. 797—801 (1977).
See also 2, p. 40; *14.1*, p. *160*; 21, p. 287.

K. Woodbridge and M. E. Ertl
Pulse Refrigeration in Bi_2Te_3
(a) Vol. **44**, No. 2, p. K123—K125 (1977).
See also *14.3.3*, p. *182*; 22.7, p. 395.

16. Photoconductivity. Photovoltaic Effects

Review Articles

SH. M. KOGAN and T. M. LIFSHITS
*Photoelectric Spectroscopy — A New Method of Analysis of Impurities in Semiconductors
(a) Vol. **39**, No. 1, p. 11—39 (1977).
See also *13.4*, p. *129*; 22, p. 320; 22.1.1, p. 333; 22.1.2, p. 337; 22.2.1, p. 350.

K. W. BÖER
*The CdS/Cu_2S Solar Cell (I): Minority Carrier Generation and Transport in the Cu_2S Emitter
(a) Vol. **40**, No. 2, p. 355—384 (1977).
See also 14.3.3, p. 178; 14.3.4, p. 183; 20.1, p. 247; 22, p. 320; 22.4.1, p. 362.

Original Papers and Short Notes

V. M. FRIDKIN, B. N. POPOV, and K. A. VERKHOVSKAYA
Effect of Anomalous Bulk Photovoltage in Ferroelectrics
(a) Vol. **39**, No. 1, p. 193—201 (1977).
See also *14.4.1*, p. *194*; 22.8.1, p. 402.

W. BARTH, K. ELSAESSER, W. GÜTH, and E. KAMIENECKI
Photoconductivity in Correlation to Absorption in Plastically Deformed Germanium
(a) Vol. **39**, No. 1, p. 249—252 (1977).
See also *13.4*, p. *130*; 20.1, p. 248; 22.1.1, p. 333.

J. SARAIE, H. KATO, N. YAMADA, SH. KAIDA, and T. TANAKA
*Preparation and Photoconductive Properties of Sintered Films of CdS–CdTe Mixed Crystals
(a) Vol. **39**, No. 1, p. 331—336 (1977).
See also 22.4.4, p. 368.

W. REHM, R.FISCHER, J . STUKE, and H. WAGNER
*Photo and Dark Conductivity of Doped Amorphous Silicon
(b) Vol. **79**, No. 2, p. 539—547 (1977).
See also *2*, p. *33*; 14.3.1, p. 173; 22.1.2, p. 337.

V. A. STUKAN, S. F. MARENKIN, V. I. TRIFONOV, and V. YA. SHEVCHENKO
*Opto-Electrical Properties of p-CdSb in the Submillimetre Wave Region
(a) Vol. **39**, No. 2, p. 445—452 (1977).
See also *20.1*, p. *250*; 22.3, p. 358.

K. ZIEGLER and U. BIRKHOLZ
Photoelectric Properties of Si_2Te_3 Single Crystals
(a) Vol. **39**, No. 2, p. 467—475 (1977).
See also *13.4*, p. *131*; 14.3, p. 166; 20.3, p. 268; 22, p. 322.

H. KUZMANY and Z. SZALMASSY
Systematic Study of the Acoustoelectric Interaction of Surface Waves with Inhomogeneously Distributed Free Carriers in a Photoconductor
(a) Vol. **40**, No. 1, p. 75—82 (1977).
See also *7*, p. *62*; *14.3.2*, p. *176*; 14.3.3, p. 178; 22.4.1, p. 362; 22.8.1, p. 403.

W. BOLLMANN
*The Origin of Photoelectrons and the Concentration of Point Defects in $LiNbO_3$ Crystals
(a) Vol. **40**, No. 1, p. 83—91 (1977).
See also *10.2*, p. *85*; 22.8.1, p. 403.

HJ. BERNHARDT
Near-Edge Absorption of Lead Molybdate Crystals
(a) Vol. **40**, No. 1, p. 257—264 (1977).
See also *20.1*, p. *252*; *20.3*, p. *269*; 22.8.1, p. 403.

A. F. Kravchenko, B. S. Lisenker, and Yu. E. Maronchuk
*Influence of High Electric Field on Capture and Recombination of Carriers in the GaAs Epitaxial Layers
(a) Vol. **40**, No. 1, p. 321—332 (1977).
See also *14.3.4*, p. *184*; 22.2.1, p. 351.

H. Häusler and T. Pusch
*Photoconductivity Measurements on Amorphous GeSe Films
(a) Vol. **40**, No. 1, p. K15—K17 (1977).
See also *2*, p. *35*; 22.7, p. 394.

M. A. Salehi Manshadi and J. Woods
*Photoconductive Effects in Copper-Doped Cadmium Sulphide
(a) Vol. **40**, No. 1, p. K43—K47 (1977).
See also 22.4.1, p. 362.

A. N. Georgobiani, Yu. V. Ozerov, E. E. Strumban, and V. S. Vinogradov
Influence of Random Electric Fields on the Polarization of Deep Centre Electroabsorption
(b) Vol. **80**, No. 2, p. 589—596 (1977).
See also *13.4*, p. *132*; 20.1, p. 252; 20.3, p. 269; 22.4.1, p. 362.

A. F. Halff and J. Schoonman
*An Experimental Investigation of the Photoconductivity of Lead Halides
(a) Vol. **40**, No. 2, p. 511—518 (1977).
See also 22.5, p. 370; 22.5.4, p. 385.

E. L. Frankevich, M. M. Tribel, I. A. Sokolik, and B. V. Kotov
*Photoconductivity of Charge Transfer Complex Crystals. Anthracene–Dimethylpyromellitimide
(a) Vol. **40**, No. 2, p. 655—662 (1977).
See also 13.5.1, p. 144; 22.9, p. 415.

K. Bhattacharya, P. Basu, and H. Saha
*Back Wall Schottky Barrier Solar Cell with an Interfacial Layer
(a) Vol. **41**, No. 1, p. 317—321 (1977).
See also 13.3, p. 126; *14.3.3*, p. *179*; 22.1.2, p. 339; 22.2.1, p. 351.

P. Schlotter, J. Kalinowski, and H. Bässler
Photoionization of Triplet Excitons in Tetracene Crystals
(b) Vol. **81**, No. 2, p. 521—526 (1977).
See also *13.5.1*, p. *145*; 22.9, p. 416.

K. Kocot and J. M. Baranowski
The Resonant States of the Ti(d^2) Impurity in CdSe
(b) Vol. **81**, No. 2, p. 629—635 (1977).
See also *13.4*, p. *134*; 20.1, p. 255; 22.4.2, p. 366.

W. Plesiewicz
*Photoelectronic Investigations of Multi-Level Gold Centres in Ge:Sb:Au
(a) Vol. **41**, No. 2, p. 403—409 (1977).
See also *13.4*, p. *135*; 14.3, p. 168; 22.1.1, p. 334.

D. Mergel and R. Labusch
*Optical Excitation of Dislocation States in Germanium (I): Experiments
(a) Vol. **41**, No. 2, p. 431—438 (1977).
See also *13.4*, p. *135*; 20.1, p. 256; 22.1.1, p. 334.

M. Z. Iqbal
*Deep Level Photocapacitance Spectra of Green-Emitting GaP Lamps
(a) Vol. **41**, No. 2, p. K155—K156 (1977).
See also *13.4*, p. *135*; 22.2.2, p. 354.

G. N. Galkin, K. H. Herrmann, and E. A. Bobrova
*Recombination in Tellurium at High Optical Excitation
(b) Vol. 82, No. 1, p. 237—242 (1977).
See also *14.3*, p. *169*; 22.1.3, p. 345.

J. P. Zielinger, C. Noguet, and M. Tapiero
*Photomemory Effect in Cu_2O Single Crystals. Kinetics of the Excitation and De-Excitation Processes
(a) Vol. 42, No. 1, p. 91—100 (1977).
See also 13.4, p. 136; 22.6, p. 389.

E. Shatkovskii and A. Tamaševičius
*Sign Inversion of Photomagnetoelectric Effect in InAs at Intense Photoexcitation
(a) Vol. 42, No. 1, p. 111—114 (1977).
See also 15, p. 199; 22.2.1, p. 352.

W. Ecke
*The Photomagnetoelectric Effect in Heteroepitaxial Silicon Films
(a) Vol. 42, No. 1, p. 157—163 (1977).
See also *14.3.1*, p. *174*; 22.1.2, p. 340.

D. Mergel and R. Labusch
*Optical Excitation of Dislocation States in Germanium (II): Analysis of the Experimental Results
(a) Vol. 42, No. 1, p. 165—171 (1977).
See also 13.1, p. 119; *13.4*, p. *136*; 22.1.1, p. 335.

T. Suzuki, M. Hirose, S. Ogose, and Y. Osaka
*Electrical Transport and Photoconductivity in Amorphous Silicon
(a) Vol. 42, No. 1, p. 337—344 (1977).
See also 2, p. 37; 13.1, p. 120; *14.3.1*, p. *174*; 15, p. 199; 22.1.2, p. 340.

V. B. Anzin, M. I. Eremets, Yu. V. Kosichkin, A. I. Nadezhdinskii, and A. M. Shirokov
*Measurement of the Energy Gap in Tellurium under Pressure
(a) Vol. 42, No. 1, p. 385—390 (1977).
See also 12.2, p. 105; 13.1, p. 120; *20.1*, p. *257*; 22.1.3, p. 346.

A. F. Kravchenko, B. S. Lisenker, Yu. E. Maronckuk, A. P. Sherstyakov, and V. N. Sherstyakova
Investigation of Recombination of Non-Equilibrium Carriers in GaAs Epitaxial Layers
(a) Vol. 42, No. 2, p. 647—655 (1977).
See also *13.4*, p. *137*; *14.3.1*, p. *174*; 22.2.1, p. 352.

D. D. Nedeoglo, Dang Huy Lam, and A. V. Simashkevich
The Effect of Thermal Treatment and Doping on the Electrical Properties of Zinc Selenide
(a) Vol. 42, No. 2, p. 675—679 (1977).
See also *14.3*, p. *170*; 22.4.2, p. 366.

M. Janai and P. S. Rudman
*The Kinetics of the Light-Enhanced Vaporization of Amorphous As_2S_3 Thin Films
(a) Vol. 42, No. 2, p. 729—737 (1977).
See also *2*, p. *38*; 14.3.1, p. 174; 22, p. 327.

H. Kato, T. Kanda, K. Yasuda, A. Yoshida, and T. Arizumi
Electron Beam Irradiation Effects on SnO_2–Si Photodiodes
(a) Vol. 42, No. 2, p. 769—774 (1977).
See also *11*, p. *97*; *14.3.2*, p. *177*; 22.1.2, p. 341.

G. Giro and P. di Marco
*Origin of Photocarriers in $Ru(2,2'\text{-Bipyridine})_3Cl_2 \cdot 6\,H_2O$ Single Crystals
(a) Vol. 42, No. 2, p. K179—K182 (1977).
See also *20.1*, p. *259*; 20.3, p. 272; 22.9, p. 417.

R. Mach and E. U. Messerschmidt
Physical Properties of the Au–CdF_2 Metal–Semiconductor Contact
(a) Vol. **42**, No. 2, p. K187–K190 (1977).
See also *14.3.4*, p. *186*; 22.5, p. 371.

T. Wosiński and T. Figielski
*Spin-Dependent Recombination at Exchange-Coupled Dislocation Centres in Silicon
(b) Vol. **83**, No. 1, p. 93–98 (1977).
See also *10.2*, p. *90*; 13.4, p. 137; 22.1.2, p. 341.

G. H. Schwuttke
*Low Cost Silicon for Solar Energy Application
(a) Vol. **43**, No. 1, p. 43–51 (1977).
See also 3, p. 42; 14.3.3, p. 181; 20.1, p. 260; 22.1.2, p. 341.

V. I. Chaikin, K. U. Guga, V. K. Malyutenko, and A. V. Sachenko
*Photoconductivity of Indium Antimonide with Different Surface Treatments
(a) Vol. **43**, No. 1, p. 311–316 (1977).
See also *14.3.2*, p. *177*; 22.2.3, p. 356.

I. Gabaš and J. Toušek
*Surface Photovoltage on CdTe
(a) Vol. **43**, No. 1, p. 351–356 (1977).
See also *14.3.2*, p. *177*; 22.4.3, p. 367.

T. X. Hoai and K. H. Herrmann
*Recombination in $Pb_{0.83}Sn_{0.17}Te$ at High Levels of Optical Excitation
(b) Vol. **83**, No. 2, p. 465–470 (1977).
See also 22.8, p. 400.

A. Fort and A. Coret
*Triplet Absorption of Crystalline Anthracene by High Resolution Photoconductivity Spectroscopy
(b) Vol. **83**, No. 2, p. 599–606 (1977).
See also *13.5.1*, p. *147*; 22.9, p. 418.

E. M. Baskin, M. D. Blokh, M. V. Entin, and L. I. Magarill
*Current Quadratic in Field and Photogalvanic Effect in Crystals without Inversion Centre
(b) Vol. **83**, No. 2, p. K97–K100 (1977).
See also *14.3.4*, p. *186*; 22.2, p. 348.

A. Zaręba, A. Lubomirska-Wittlin, A. Szadkowski, and B. Krukowska-Fulde
Photoconversion from the Insulating to the Semiconducting State of $CdF_2:MF_3$
(b) Vol. **83**, No. 2, p. K155–K157 (1977).
See also *13.4*, p. *138*; 22.5.3, p. 384.

B.-P. Kietis, R. Reksnys, A. Sakalas, and P. Höschl
Investigation of the Bipolar Photoconductivity in CdTe
(a) Vol. **43**, No. 2, p. 705–709 (1977).
See also 7, p. *63*; *14.3*, p. *171*; 22.4.3, p. 367.

P. di Marco and G. Giro
*Photoconductive Properties of Fluorene Single Crystals
(a) Vol. **43**, No. 2, p. K193–K197 (1977).
See also 20.1, p. 261; 20.3, p. 274; 22.9, p. 419.

C. Viger, C. Vautier, and D. Carles
*Recombination and Trapping Processes in Amorphous Materials in Presence of an Exponential Distribution of Traps
(a) Vol. **44**, No. 1, p. 97–106 (1977).
See also 2, p. 40; 13.4, p. 139; 14.3.1, p. 175; 22.1.3, p. 347.

H.-J. Kos and D. Neubert
*Two-Step Photoconductivity by Dislocations in Silicon
(a) Vol. **44**, No. 1, p. 259–264 (1977).
See also 10.2, p. 92; 22.1.2, p. **343**.

B. G. Caswell and J. Woods
*Some Effects of Oxygen on As-Plated CdS–Cu_2S Photovoltaic Cells
(a) Vol. **44**, No. 1, p. K47–K51 (1977).
See also *14.3.3*, p. *181*; 22, p. 330; 22.4.1, p. **364**.

G. Jacobsen
*Photovoltaic Spectra as a Function of the Light Intensity in ZnS Single Crystals
(a) Vol. **44**, No. 1, p. K67–K70 (1977).
See also 22.4.1, p. 364.

A. P. Radlinski
On the Properties of Cobalt Impurities in Zinc Selenide Crystals (I): Effect of In-Band Positron of States on Photoelectric Properties
(b) Vol. **84**, No. 2, p. 503–511 (1977).
See also *13.4*, p. *139*; 20.1, p. 262; 22.4.2, p. **366**.

A. Lyubchenko, I. Potykievich, A. Fedorov, and M. Sheinkman
*On the Mechanism of Photoluminescence of Cadmium Diphosphide Doped by Bi and Zn
(a) Vol. **44**, No. 2, p. 747–753 (1977).
See also *20.3*, p. *275*; 22.3, p. **359**.

N. L. P. Sinha, A. K. Chaudhuri, and H. N. Bose
Structural Analysis of Lead Telluride Films and Their Sensitization
(a) Vol. **44**, No. 2, p. K127–K129 (1977).
See also *1.5*, p. *30*; 14.3.1, p. 175; 22, p. **331**.

W.-A. Martin and G. Hartung
*Anisotropic Photoinduced ESR Signals from the Si–SiO_2 Interface on Silicon Single Crystals
(a) Vol. **44**, No. 2, p. K159–K161 (1977).
See also *19*, p. *243*; 22.1.2, p. **344**.

17. Emission of Electrons and Ions

Original Papers and Short Notes

GY. FARKAS, Z. GY. HORVÁTH, L.-A. LOMPRE, and G. PETITE
*Anomalous Behaviour of the Surface Multiphoton Photoeffect Induced by Single Selected Ultrashort Laser Pulses
(a) Vol. **39**, No. 1, p. K25—K27 (1977).
See also 21.6, p. 315.

V. LANTTO
*A Theoretical Study of the Absorption of Photoelectrons Created by X-Rays in Semi-Infinite Targets
(b) Vol. **79**, No. 2, p. 699—711 (1977).
See also *9*, p. *69*; 21, p. 277; 21.1, p. 289; 21.6, p. 315.

H. DÜSTERHÖFT and A. IHLENFELD
*The Energy Distribution of Positive Secondary Ions Emitted from Metal and Semiconductor Targets Bombarded with 12 keV Ar^+ Ions
(a) Vol. **39**, No. 2, p. K147—K150 (1977).
See also 21, p. 278; 21.1, p. 290; 22.1.2, p. 338.

F. L. BATTYE, A. GOLDMANN, and L. KASPER
Ultraviolet Photoelectron Valence Band Studies on Phthalocyanine Compounds
(b) Vol. **80**, No. 2, p. 425—432 (1977).
See also *13.1*, p. *116*; 22.9, p. 415.

F. WERFEL, G. DRÄGER, O. BRÜMMER, and M. JURISCH
XS and XPS Investigation of the V_3Si Electron Structure
(b) Vol. **80**, No. 2, p. K95—K99 (1977).
See also *13.1*, p. *117*; 20, p. 245; 22.7, p. 394.

H. OVEREIJNDER, M. SZYMOŃSKI, A. HARING, and A. E. DE VRIES
*The Mechanism of the Electron Sputtering Process of Alkali Halides
(b) Vol. **81**, No. 1, p. K11—K14 (1977).
See also *10.2*, p. *87*; 22.5.2, p. 378.

R. NÜRNBERGER, F.-J. HIMPSEL, E. E. KOCH, and N. SCHWENTNER
Band Formation in Xenon–Argon Alloys Studied by Photoelectron Spectroscopy
(b) Vol. **81**, No. 2, p. 503—509 (1977).
See also *13.1*, p. *118*; 23, p. 420.

R. D. GOULD and C. A. HOGARTH
*Angular Distribution Measurements of Electrons Emitted from Thin Film Au–SiO_x–Au Diode and Triode Structures
(a) Vol. **41**, No. 2, p. 439—442 (1977).
See also 14.3.3, p. 180; 22.6, p. 388.

R. TARANKO
*The Effect of the Model Hybridized Energy Bands on the Field Emitted Electrons from Metals
(b) Vol. **82**, No. 1, p. K5—K7 (1977).
See also *13.1*, p. *119*; 21, p. 282.

H.-J. FITTING, H. GLAEFEKE, W. WILD, and J. LANGE
*Energy and Angular Distribution of Exoelectrons
(a) Vol. **42**, No. 1, p. K75—K77 (1977).
See also 22.6, p. 389.

U. BRUNSMANN and A. SCHARMANN
*A Simulation of the Maxwellian Exoelectron Energy Distributions with Regard to Patch Fields
(a) Vol. **42**, No. 1, p. K79—K81 (1977).
See also 22.5.2, p. 379.

V. Siegel, W. Rasp, and H.-H. Kirchner
*The Influence of Desorption Processes on the Exoelectron Yield of BeO
(a) Vol. **42**, No. 2, p. K147—K150 (1977).
See also 22.6, p. 390.

L. I. Johansson, A. L. Hagström, A. Platau, and S.-E. Karlsson
Adsorption of Oxygen at Room Temperature on Polycrystalline Titanium Studied by Ultraviolet and X-Ray Photoelectron Spectroscopy
(b) Vol. **83**, No. 1, p. 77—84 (1977).
See also *13.3*, p. *127*; 21.1, p. 297.

F. Werfel, G. Dräger, and O. Brümmer
Electronic Structure Investigation of Metallic Vanadium by XS and XPS
(b) Vol. **83**, No. 1, p. 257—262 (1977).
See also *13.1*, p. *121*; 20, p. 245; 21.1, p. 297.

A. M. Brodskii
*Specific Features of Many-Photon Photoemission from Metals and Semiconductors in Strong Fields
(b) Vol. **83**, No. 1, p. 331—340 (1977).
See also 21, p. 284; 22, p. 328.

G. Oertel and E.-H. Weber
*Growth of Thin SiO_2 Films on Clean Si (111) Surfaces by Low-Pressure Oxidation and Their Evaporation
(a) Vol. **43**, No. 1, p. 141—150 (1977).
See also 1.5, p. 30; *20.1*, p. *260*; 22.6, p. 390.

V. G. Aleshin, T. I. Serebryakova und A. I. Kharlamov
Charge State of Boron in Transition Metal Borides
(b) Vol. **83**, No. 2, p. 537—542 (1977).
See also *13*, p. *111*; 22, p. 328.

Yu. G. Borodko, M. G. Kaplunov, T. M. Moravskaya, and K. I. Pokhodnya
X-Ray Photoelectron Spectra of Single Crystals of Tetracyanoquinodimethane and Its Salts
(b) Vol. **83**, No. 2, p. K141—K144 (1977).
See also 13, p. 111; *20*, p. *246*; 22.9, p. 418.

U. Brunsmann and A. Scharmann
*Thermally Stimulated Exoelectron Emission of NaF. Electron Bombardment and Exoelectron Energy Investigations
(a) Vol. **43**, No. 2, p. 519—528 (1977).
See also 22.5.2, p. 380.

N. I. Konyushkina, I. V. Krylova, V. N. Opekunov, and A. A. Predvoditelev
*Thermostimulated Exoemission from LiF
(a) Vol. **43**, No. 2, p. 639—644 (1977).
See also 22.5.2, p. 380.

V. Siegel and W. Rasp
*The TSEE of Lapped and Unlapped BeO Ceramics
(a) Vol. **43**, No. 2, p. K117—K121 (1977).
See also 1.6, p. 32; 13.3, p. 128; 22.6, p. 391.

17.1 Field Emission Microscope Investigations

Original Papers and Short Notes

J.-M. Derochette and J. Marien
*Field Emission Study of Hydrogen Adsorption on Smooth and Perfectly Ordered Iridium Surfaces (111) and (100) — Influence of Steps and Iridium Clusters on the Surface Potential of the Hydrogen Film
(a) Vol. **39**, No. 1, p. 281—289 (1977).
See also *1.6*, p. *31*; 21, p. 277.

C. V. Dharmadhikari, D. S. Joag, and P. I. Kanitkar
*Surface Migration and Overgrowth of LaB_6 on Tungsten Field Emitter
(a) Vol. **42**, No. 2, p. K99—K101 (1977).
See also 22, p. 327.

18. Magnetic Properties

Original Papers and Short Notes

F. K. Schulte
Electronic Density of States, Heat Capacity, and Spin Susceptibility of Thin Metal Films
(b) Vol. **79**, No. 1, p. K5—K8 (1977).
See also 8, p. 64; *13.3*, p. *126*; 21, p. 276.

A. N. Chakravarti and A. K. Chowdhury
Influence of Band Non-Parabolicity on the Oscillatory Dependence of the Diffusivity–Mobility Ratio in Degenerate Semiconductors on Magnetic Quantization
(b) Vol. **79**, No. 1, p. K31—K35 (1977).
See also 13.1, p. 114; *14.3*, p. *165*; 22, p. 320.

H. U. Habermeier
The Penetration Field and the Magnetization Curve of Type-I Superconductors Containing Lattice Defects
(a) Vol. **39**, No. 1, p. 203—212 (1977).
See also 10.1, p. 75; *14.2*, p. *161*; 21, p. 276.

H. Oesterreicher
*Possible Disorder of Magnetic Components (Magnetic Scatter Order) in Compounds on the Basis of Transition Metals
(a) Vol. **39**, No. 1, p. K41—K43 (1977).
See also 21, p. 277; 21.1, p. 289.

R. M. Catchings
*Magnetic Susceptibility Studies of Iron-Dysprosium Sodium Borate Glasses
(a) Vol. **39**, No. 1, p. K71—K74 (1977).
See also *2*, p. *33*; 22.8, p. 396.

H. Oesterreicher
*Magnetic Studies on GdCuGe, $Gd_6Cu_8Ge_8$, $GdCu_2Ge_2$, and $PrCu_2Ge_2$
(a) Vol. **39**, No. 1, p. K75—K77 (1977).
See also *18.4*, p. *231*; 22.7, p. 394.

G. Bastard, C. Rigaux, and A. Mycielski
Giant Spin Splitting Induced by Exchange Interactions in $Hg_{1-k}Mn_kTe$ Mixed Crystals
(b) Vol. **79**, No. 2, p. 585—593 (1977).
See also *13.1*, p. *115*; 20.1, p. 249; 22.4.4, p. 368.

B. A. Aronzon and E. Z. Meilikhov
Helicon Damping in Semiconductors in the Quantum Limit
(b) Vol. **79**, No. 2, p. 753—762 (1977).
See also *14.3*, p. *166*; *20*, p. *244*; 22.2.3, p. 356.

I. N. Douglas
*Magnetic Circular Dichroism of Co^{2+} Ions in CdF_2 and CaF_2
(b) Vol. **79**, No. 2, p. K99—K101 (1977).
See also 13.4, p. 131; *20.1*, p. *249*; 22.5, p. 370; 22.5.3, p. 383.

G. N. Belozerskii, Yu. P. Khimich, and V. N. Gitsovich
Determination of the Parameters of Hyperfine Interaction from Mössbauer Spectra of Polycrystals in an Externally Applied Magnetic Field
(b) Vol. **79**, No. 2, p. K125—K130 (1977).
See also *5*, p. *47*; 22.8.2, p. 409.

M. Horobiowski, T. Skośkiewicz, and E. Trojnar
Magnetization of the Superconducting Palladium Hydrides
(b) Vol. **79**, No. 2, p. K147—K150 (1977).
See also *14.2*, p. *161*; 21, p. 277.

C. K. Sarkar and P. K. Basu
Inelastic Acoustic Phonon Scattering in Nonparabolic Semiconductors under Longitudinal Quantizing Magnetic Field
(b) Vol. **79**, No. 2, p. K157–K160 (1977).
See also *14.3*, p. *166*; 22, p. 322.

H. Oesterreicher
*Magnetic Properties of Gd_2Al
(a) Vol. **39**, No. 2, p. K91–K93 (1977).
See also 22.7, p. 394.

C. A. S. Lima and L. C. M. Miranda
Plasmon-Enhanced Sound Amplification in Piezoelectric Semiconductors under a Strong Magnetic Field
(b) Vol. **80**, No. 1, p. 57–62 (1977).
See also 7, p. 62; *13.5.2*, p. *149*; 22.2.1, p. 351.

D. Kostopoulos and J. Grammatikakis
*Changes in the State of Magnetite due to Magnetic Fields Applied at 78 K Observed i n Magnetoresistance Measurements
(a) Vol. **40**, No. 1, p. 141–145 (1977).
See also *14.3*, p. *167*; 22.8.2, p. 410.

M. S. Sodha, V. K. Tripathi, and D. P. Singh
*Effect of a Strong Magnetic Field on the Self-Focusing of a Laser Beam in InSb
(a) Vol. **40**, No. 1, p. 315–320 (1977).
See also *20.2*, p. *265*; 22.2.3, p. 356.

E. Wieser and S. Mager
*Investigation of the Localized Magnetic Moments on Fe and Mn in Disordered Fe–Mn–Al Alloys by Magnetic Diffuse Neutron Scattering
(a) Vol. **40**, No. 2, p. 497–501 (1977).
See also 21.1, p. 292; 21.1.1, p. 303.

H. Oesterreicher
*Magnetic Properties of Scatter Order Compounds RFeAl (R = Gd, Tb, Dy, Ho, Er, Tm, Lu and Y)
(a) Vol. **40**, No. 2, p. K139–K143 (1977).
See also 21, p. 280; 21.1.1, p. 303; 21.4, p. 312.

J. Donecker and J. Kluge
*Magneto-Optical Investigation of the Ground State of the Indirect Free Exciton in Si and GaP
(b) Vol. **81**, No. 1, p. 199–203 (1977).
See also 13.5.1, p. 145; *20.1*, p. *253*; 22.1.2, p. 339; 22.2.2, p. 354.

Yu. V. Rakitin, V. T. Kalinnikov, and W. E. Hatfield
*Magnetic and Thermodynamic Properties of Heisenberg Open Finite Chains (I): Susceptibility in Zero Field
(b) Vol. **81**, No. 1, p. 379–385 (1977).

J. Lakner
*The Effect of Non-Transition Metals on the Magnetic Susceptibility of Aluminium Solid Solutions
(b) Vol. **81**, No. 2, p. 593–601 (1977).
See also 13.2, p. 124; 21, p. 280.

Y. Guldner, C. Rigaux, A. Mycielski, and Y. Couder
*Magnetooptical Investigation of $Hg_{1-x}Cd_xTe$ Mixed Crystals (I): Semimetallic Configuration
(b) Vol. **81**, No. 2, p. 615–627 (1977).
See also 13.1, p. 118; *20.1*, p. *255*; 22.4.4, p. 368.

W. THIELEMANN and F. BERGAUER
*Test of the Random Fluctuations of the Conduction Band Edge in Heavily Doped n-Type GaAs by Means of the Magnetic Circular Dichroism
(b) Vol. **81**, No. 2, p. K91–K94 (1977).
See also 13.1, p. 118; *20.1*, p. *255*; 22.2.1, p. 352.

G. FROMMEYER and K. F. SCHNEIDER
*Size Effect and Interface Contribution to Diamagnetism of Silver–Copper Composite Wires with Fibre Structure
(a) Vol. **41**, No. 2, p. 653–658 (1977).
See also 21.1, p. 294; 21.6, p. 316.

YU. V. RAKITIN, V. T. KALINNIKOV, and W. E. HATFIELD
*Magnetic and Thermodynamic Properties of Heisenberg Open Finite Chains (II): Specific Heat
(b) Vol. **82**, No. 1, p. 69–73 (1977).
See also *8*, p. *66*.

Y. GULDNER, C. RIGAUX, A. MYCIELSKI, and Y. COUDER
*Magnetooptical Investigation of $Hg_{1-x}Cd_xTe$ Mixed Crystals (II): Semiconducting Configuration and Semimetal → Semiconductor Transition
(b) Vol. **82**, No. 1, p. 149–158 (1977).
See also 13.1, p. 118; *20.1*, p. *256*; 22.4.4, p. 368.

M. MIJATOVIĆ and S. MILOŠEVIĆ
*The Linear Ising Chain in the Three-Dimensional Phonon Field
(b) Vol. **82**, No. 1, p. 193–203 (1977).
See also *6*, p. *57*.

B. SZPUNAR and B. KOZARZEWSKI
*The Application of CPA to Calculations of the Mean Magnetic Moment in the $Gd_{1-x}Ni_x$, $Gd_{1-x}Fe_x$, $Gd_{1-x}Co_x$, and $Y_{1-x}Co_x$ Intermetallic Compounds
(b) Vol. **82**, No. 1, p. 205–211 (1977).
See also 22.7, p. 394.

J. MERTSCHING
*Transformations of the Hubbard Model
(b) Vol. **82**, No. 1, p. 289–296 (1977).
See also *13.1*, p. *119*.

Y. D. YAO, S. ARAJS, and E. E. ANDERSON
Phenomenological "Dilution Model" for Kondo-Type Effects in Concentrated Alloy Systems
(b) Vol. **82**, No. 1, p. 343–347 (1977).
See also *13.4*, p. *136*; *14.1*, p. *158*; 21.1, p. 295.

W. SCHNITKER and F. R. KESSLER
*Piezoreflectance at the E_1 and $E_1 + \Delta_1$ Edges of Ge in High Magnetic Fields
(b) Vol. **82**, No. 1, p. K35–K37 (1977).
See also 13.1, p. 119; *20.1*, p. *257*; 22.1.1, p. 335.

Y. MURAOKA, M. SHIGA, and Y. NAKAMURA
*Magnetic Properties and Mössbauer Effect of $A(Fe_{1-x}B_x)_2$ (A = Y or Zr, B = Al or Ni) Laves Phase Intermetallic Compounds
(a) Vol. **42**, No. 1, p. 369–374 (1977).
See also 1, p. 10; *5*, p. *48*; 21.1.1, p. 304.

E. R. VANCE and N. B. MANSON
*Magnetic Circular Dichroism of Neutron-Irradiated MgO and CaO
(b) Vol. **82**, No. 2, p. 473–479 (1977).
See also 13.4, p. 136; *20*, p. *245*; 22.6, p. 390.

U. Traxlmayr, W. F. Fallmann, and F. Paschke
The Influence of Magnetic Fields on Gunn Effect Planar Devices in Corbino Geometry
(a) Vol. **42**, No. 2, p. 525—528 (1977).
See also *14.3.3*, p. *180*; *14.3.4*, p. *185*; 22.2.1, p. 352.

L. G. Ferreira, S. R. Salinas, and M. J. Oliveira
*On a Variational Procedure for Obtaining the Thermodynamic Properties of Statistical Models
(b) Vol. **83**, No. 1, p. 229—237 (1977).
See also *13*, p. *110*.

N. F. Berk, J. Rosenthal, L. Yarmus, and C. E. Swenberg
*Magnetic Field Dependence of Prompt Fluorescence from Crystalline Tetracene
(b) Vol. **83**, No. 1, p. K1—K6 (1977).
See also *20.3*, p. *272*; 22.9, p. 418.

B. Rheinländer, S. Haase, and H. Weinert
*On the Determination of the Free Carrier Relaxation Time by Transverse Magneto-Plasma Reflectivity Measurements on Sn-Doped n-Type GaAs
(b) Vol. **83**, No. 1, p. K83—K86 (1977).
See also 13.5.2, p. 150; *20.1*, p. *260*; 22.2.1, p. 353.

I. V. Aleksandrov and R. G. Gabrielyan
*On the Theory of Spin-Lattice Relaxation Due to Diffusion to Impurities in Solids and Liquids
(b) Vol. **83**, No. 2, p. K173—K177 (1977).
See also 9, p. 70; *13*, p. *111*.

R. Karcher, U. Köbler, K. Lüders, and Z. Szücs
*Temperature Dependence of Knight Shift and Susceptibility of Single Crystal Molybdenum
(b) Vol. **84**, No. 1, p. 189—193 (1977).
See also *19*, p. *242*; 21, p. 286.

O. Uemura and T. Satow
*The Magnetic Susceptibility of Antimony and Bismuth Chalcogenides in the Liquid State
(b) Vol. **84**, No. 1, p. 353—360 (1977).
See also *2*, p. *39*; 22, p. 330.

Z. Biegański and M. Drulis
Low-Temperature Phase Transitions in Lanthanum and Neodymium Trihydrides. Specific Heat Measurements
(a) Vol. **44**, No. 1, p. 91—96 (1977).
See also *8*, p. *67*; 21.4, p. 313.

N. E. Alekseevskii, G. Wolf, N. M. Dobrovolksii, and C. Hohlfeld
Thermal and Magnetic Properties of the Superconducting Compound $YbMo_6S_8$ at Low Temperatures
(a) Vol. **44**, No. 1, p. K79—K82 (1977).
See also *8*, p. *68*; *14.2*, p. *164*; 22.8, p. 400.

V. S. Lubimov, E. F. Makarov, and S. P. Ionov
*The Magnetic Properties of Charge-Ordered Crystals
(b) Vol. **84**, No. 2, p. 485—492 (1977).
See also *13.5*, p. *142*.

L. N. Dwivedi and C. M. Bhandari
*Effects of Random Local Magnetic Fields in the Ising Model ($S = 1, \frac{3}{2}$)
(b) Vol. **84**, No. 2, p. K125—K129 (1977).

A. Delapalme et M. Wintenberger
*Etude par neutrons polarisés de l'aimantation spontanée de la cubanite
(Study of the Spontaneous Magnetization of Cubanite by Polarized Neutrons)
(a) Vol. **44**, No. 2, p. 713—716 (1977).
See also 22.8, p. 401.

18.1 Paramagnetic Properties

Original Papers and Short Notes

J. C. Phillips
*Paramagnetic Domain Tips in Quasi-One-Dimensional Conductors
(b) Vol. **79**, No. 1, p. 111—114 (1977).
See also 22.9, p. 414.

H. Ortbauer, W. Steiner, and R. Haferl
Mössbauer Studies of $Y(Fe_xCo_{1-x})_2$ in the Region of Small Fe Concentration
(a) Vol. **39**, No. 1, p. 157—165 (1977).
See also *5*, p. *47*; 18.2, p. 217; 21, p. 276; 21.1, p. 288; 21.1.1, p. 302.

R. A. Tahir-Kheli and K. Kawasaki
*Phenomenological Description of Dynamics in a Diluted Anisotropic Heisenberg Paramagnet at High Temperatures
(b) Vol. **79**, No. 2, p. 413—422 (1977).
See also 18.2, p. 217; 18.4, p. 231.

G. J. Ashwell, D. D. Eley, M. R. Willis, and J. Woodward
Electrical and Magnetic Properties of Two Crystallographic Forms of $(DEPE)^{2+}(TCNQ)_4^{-2}$
(b) Vol. **79**, No. 2, p. 629—636 (1977).
See also *14.1*, p. *157*; *14.3*, p. *165*; 15, p. 198; 22.9, p. 414.

K. de Strooper
*Magnetic Properties of Some Co^{2+} Containing Spinels
(a) Vol. **39**, No. 2, p. 431—435 (1977).
See also 18.3, p. 228; 18.4, p. 231; 22.8.2, p. 409.

P. Delhaes, G. Keryer, S. Flandrois and J. P. Manceau
Study of the Specific Heat of Double Salts of TCNQ with Non-Radical Cations
(b) Vol. **80**, No. 1, p. 125—136 (1977).
See also *8*, p. *64*; 22.9, p. 414.

A. R. Vasconcellos and R. Luzzi
Light Scattering and Spin Diffusion in Doped Semiconductors
(b) Vol. **80**, No. 1, p. 181—191 (1977).
See also *13.5*, p. *141*; 20.1, p. 251; 22.2.1, p. 351.

C. Scherer and Y. Barjhoux
*Reaction Field and Dipolar Interaction in Quasi-Onedimensional Heisenberg Systems. Application to $CsNiF_3$
(b) Vol. **80**, No. 1, p. 313—320 (1977).
See also 22.5.4, p. 385.

F. G. Brady Moreira, I. P. Fittipaldi, S. M. Rezende, R. A. Tahir-Kheli, and B. Žekš
*Field-Induced Phase Transitions in a Randomly Diluted Antiferromagnet. Mean-Field Description of Classical Spin Model
(b) Vol. **80**, No. 1, p. 385—394 (1977).
See also 1.2, p. 15; *18.4*, p. *231*.

O. Uemura and T. Satow
The Amorphous Structure of $Cd(Ge_{1-x}Sn_x)As_2$ with $x = 0.0$, 0.1, and 0.2
(a) Vol. **40**, No. 1, p. 303—308 (1977).
See also *2*, p. *35*; 22.8, p. 397.

M. G. Townsend, J. R. Gosselin, J. L. Horwood, L. G. Ripley, and R. J. Tremblay
*Violarite, a Metallic Natural Spinel
(a) Vol. **40**, No. 1, p. K25—K29 (1977).
See also *5*, p. *47*; 15, p. 198; 22.8.2, p. 410.

G. G. KHALYULLIN
*Concentration Dependence of Exchange Field in Dilute Metals
(b) Vol. **80**, No. 2, p. 709–711 (1977).
See also 21, p. 279.

J. K. YAKINTHOS, H. GAMARI-SEALE, and J. LAFOREST
*Magnetic Properties of R_3Pd_4 (R = Pr, Nd, Sm, Gd, Tb, Dy, Ho, Er, Tm) Intermetallic Compounds
(a) Vol. **40**, No. 2, p. K105–K108 (1977).
See also *18.4*, p. *232*; 21.4, p. 311.

T. ANAGNOSTOPOULOS, P. IKONOMOU et J. K. YAKINTHOS
*Propriétés magnétiques des composés $SmAu_6$, $TbAu_6$ et $DyAu_6$
(Magnetic Properties of the Compounds $SmAu_6$, $TbAu_6$, and $DyAu_6$)
(a) Vol. **40**, No. 2, p. K109–K112 (1977).
See also *18.4*, p. *232*; 21.4, p. 312; 21.6, p. 315.

W. BRÜCKNER, U. GERLACH, and B. THUSS
*Phase Diagram of $V_{1-x}Al_xO_2$
(a) Vol. **40**, No. 2, p. K131–K134 (1977).
See also *1.2*, p. *16*; 22.6.1, p. 393.

K. SUGAWARA
*Effect of Crystal Field on the Inelastic Neutron Scattering Cross-Section
(b) Vol. **81**, No. 1, p. 313–321 (1977).
See also 19, p. 238; 21.4, p. 312.

V. V. GANN, A. I. SKVORTSOV, and A. VAN DEN BOSCH
The Superconducting Transition Temperature of $U_6Fe_{0.98}$ Reduced by Neutron Irradiation
(a) Vol. **41**, No. 1, p. 225–229 (1977).
See also *11*, p. *95*; *14.2*, p. *162*; 21.1.1, p. 303; 21.5, p. 314.

M. F. MOSTAFA, M. SEMARY, and M. A. AHMED
Magnetic Susceptibility Studies of $((CH_3)_2CHNH_3)_2CuCl_4$
(a) Vol. **41**, No. 1, p. 295–297 (1977).
See also *18.2*, p. *220*; 22.9, p. 416.

K. KACZMARSKA and A. CHEŁKOWSKI
*The ESR Linewidth in $CdCr_2S_4$ Single Crystals near Their Curie Point
(b) Vol. **81**, No. 2, p. K95–K97 (1977).
See also *19*, p. *239*; 22.8.2, p. 411.

H. SPIERING, H. VOGEL, and W. IRLER
The Phase Transformation in Ferreous Fluosilicate Hexahydrate. Mössbauer Study
(b) Vol. **82**, No. 1, p. 243–250 (1977).
See also *1.2*, p. *17*; 5, p. 48; 22.8, p. 399.

E. S. GRINBERG and R. R. NIGMATULLIN
*Kinetics of Dilute Solid Paramagnets under "Bottleneck" Conditions
(b) Vol. **82**, No. 1, p. 397–402 (1977).
See also *19*, p. *240*.

W. BRÜCKNER, U. GERLACH, H.-P. BRÜCKNER, W. MOLDENHAUER, and H. OPPERMANN
Influence of Nonstoichiometry on the Phase Transitions in Ga-, Al-, and Fe-Doped VO_2
(a) Vol. **42**, No. 1, p. 295–303 (1977).
See also *1.2*, p. *17*; *3*, p. *42*; 14.3, p. 170; 22.6.1, p. 393.

B. SZPUNAR and P.-A. LINDGÅRD
*Theory of Temperature Dependence of the Magnetization in Rare-Earth–Transition-Metal Alloys
(b) Vol. **82**, No. 2, p. 449–456 (1977).
See also 13, p. 110; *18.4*, p. *233*; 21.1, p. 296; 21.4, p. 312.

K. Hattori
Magnetic Properties of a Random Binary System
(b) Vol. **82**, No. 2, p. 489—494 (1977).
See also *2*, p. *38*; 18.2, p. 221; 18.3, p. 229.

V. Kumar and K. C. Sharma
*Phase Transitions in Spin-One Ferromagnetic Systems
(b) Vol. **82**, No. 2, p. 577—580 (1977).
See also *18.2*, p. *221*.

J. P. Wolfe, R. S. Markiewicz, J. E. Furneaux, S. M. Kelso, and C. D. Jeffries
Magnetostriction of an Electron–Hole Drop in Ge
(b) Vol. **83**, No. 1, p. 305—312 (1977).
See also *13.5.1*, p. *147*; 22.1.1, p. 335.

S. A. Vorontson, D. G. Dolgopolov, V. I. Poltoratskii, and I. V. Svechkarev
*Magnetic Susceptibility Anisotropy of Holes in Tellurium
(b) Vol. **83**, No. 1, p. 341—346 (1977).
See also *13*, p. *111*; 22.1.3, p. 346.

S. Chaudhuri and A. K. Pal
*Thickness Dependence of Curie Temperature of Cu–Ni Films
(a) Vol. **43**, No. 1, p. K9—K13 (1977).
See also 14.1, p. 159; *18.2*, p. *222*; 21.1, p. 297.

J. J. Bara, B. Brandt, A. Gutsze, A. T. Pędziwiatr, Z. M. Stadnik, and A. Wronkowski
*The Mössbauer Effect and Magnetometric Investigations of Iron-Doped Sodium A-Type Zeolite
(a) Vol. **43**, No. 1, p. K41—K44 (1977).
See also *5*, p. *49*; 22.8.1, p. 407.

A. Zygmunt
Magnetic Properties of UGeY Compounds (Y = S, Se, Te)
(a) Vol. **43**, No. 2, p. 573—577 (1977).
See also *18.4*, p. *233*; 22.8, p. 400.

Ö. Sävborg and M. Nygren
Magnetic, Electrical, and Thermal Studies of the $V_{1-x}Re_xO_2$ System with $0 \leqq x \leqq 0.15$
(a) Vol. **43**, No. 2, p. 645—652 (1977).
See also *1.2*, p. *19*; 8, p. 67; 14.3, p. 171; 22.6.1, p. 393.

M. Giordano, M. Martinelli, L. Pardi, and S. Santucci
Electron Spin–Lattice Relaxation Measurements in a Two-Dimensional Crystal
(b) Vol. **84**, No. 1, p. 305—310 (1977).
See also *19*, p. *242*; 22.8, p. 400.

V. F. Kaminskii, M. L. Khidekel, R. B. Lyubovskii, I. F. Shchegolev, R. P. Shibaeva, E. B. Yagubskii, A. V. Zvarykina, and G. L. Zvereva
Metal–Insulator Phase Transition in TTT_2I_3 Affected by Iodine Concentration
(a) Vol. **44**, No. 1, p. 77—82 (1977).
See also *1*, p. *10*; *14*, p. *156*; 22.9, p. 419.

K. de Strooper, A. van Alboom, C. Henriet-Iserentant, R. Vanleerberghe, and R. Vandenberghe
On the Cation Distribution and Magnetic Properties of Cobalt–Zinc Titanates
(a) Vol. **44**, No. 1, p. 215—220 (1977).
See also *18.3*, p. *230*; 22.8.2, p. 412.

S. W. Biernacki
The Multimode Jahn-Teller Coupling for an Orbital Triplet in Cubic Crystals
(b) Vol. **84**, No. 2, p. 699—708 (1977).
See also *13.4*, p. *139*; *20.1*, p. *263*.

A. Mauger
Indirect Exchange in Europium Chalcogenides
(b) Vol. **84**, No. 2, p. 761—771 (1977).
See also *13.1*, p. *123*; 18.2, p. 224; 22, p. 330.

V. I. Muravev and N. B. Yunusov
Covalency and Hyperfine Interaction on the Impurity Ion in Crystals
(b) Vol. **84**, No. 2, p. 791—796 (1977).
See also *13.4*, p. *140*; 19, p. 243; 21.1, p. 300.

E. Kolley and W. Kolley
On the Spin Susceptibility for Itinerant Electrons with Energy-Dependent Effective Interaction
(b) Vol. **84**, No. 2, p. K147—K150 (1977).
See also *13*, p. *113*; 18.2, p. 224.

R. Goblirsch, H. W. Weber, G. Hilscher, R. Grössinger, and W. Steiner
*Search for Magnetic Order in Selected Pseudobinary Systems down to Millikelvin Temperatures
(a) Vol. **44**, No. 2, p. 593—600 (1977).
See also *18.2*, p. *224*; 21.1, p. 300.

T. Satow, O. Uemura, and S. Watanabe
The Amorphous Structure of $CdGe_{1-x}M_xAs_2$ (M = Si, Sn, and Pb) with $x = 0.01$, 0.1, and 0.2
(a) Vol. **44**, No. 2, p. 731—736 (1977).
See also *2*, p. *41*; 22.8, p. 401.

18.2 Ferromagnetic Properties

Review Articles

V. A. Buravikhin, V. V. Litvintsev, and A. I. Ushakov
Polymorphic Transformations and Physical Properties of Thin Films
(a) Vol. **40**, No. 1, p. 11—27 (1977).
See also *1.2*, p. *14*; 1.4, p. 24; *1.5*, p. *29*; 21.1, p. 288; 21.1.1, p. 302.

G. N. Rao and A. K. Singhvi
*Dilute Impurity Hyperfine Fields. Systematics and Comparison with Models
(b) Vol. **84**, No. 1, p. 9—21 (1977).
See also *13.4*, p. *129*; 21.1, p. 288; 21.1.1, p. 302; 21.4, p. 311.

Original Papers and Short Notes

H. Kronmüller, N. König, and F. Walz
Isotope Effects of Diatomic C–H and C–D Complexes in Ni Single Crystals
(b) Vol. **79**, No. 1, p. 237—247 (1977).
See also *9*, p. *69*; 10.1, p. 75; 21.1, p. 288.

L. Billard
*Self-Consistent Cellular Methods for the Ising Problem
(b) Vol. **79**, No. 1, p. 307—311 (1977).

M. V. Medvedev and A. V. Zaborov
*Random Collinear Structures of a Dilute Magnetic Alloy
(b) Vol. **79**, No. 1, p. 379—390 (1977).
See also 21, p. 276.

P. R. R. Pereira and L. C. M. Miranda
Magnon Contribution to the Phonon Damping in Ferromagnetic Semiconductors
(b) Vol. **79**, No. 1, p. K49—K52 (1977).
See also *6*, p. *52*; *13.5.4*, p. *153*; 22.8, p. 396.

H. Ortbauer, W. Steiner, and R. Haferl
Mössbauer Studies of $Y(Fe_xCo_{1-x})_2$ in the Region of Small Fe Concentrations
(a) Vol. **39**, No. 1, p. 157—165 (1977).
See also *5*, p. *47*; 18.1, p. 213; 21, p. 276; 21.1, p. 288; 21.1.1, p. 302.

J. Barton, E. Wieser, and M. Müller
*Investigation of Magnetic Textures in an Fe–Mn Base Alloy by Means of Mössbauer Spectroscopy
(a) Vol. **39**, No. 1, p. 259—264 (1977).
See also 5, p. 47; 21.1, p. 288; 21.1.1, p. 302.

J. Kaczér and P. Novotný
*High-Field Magnetization of U_3As_4
(a) Vol. **39**, No. 1, p. 351—355 (1977).
See also 1.2, p. 14; 22, p. 321.

E. Klugmann, Yu. F. Bydin, and V. A. Volpyas
*Effect of Ion Implantation on the Magnetic Properties of Ferromagnetic Metals
(a) Vol. **39**, No. 1, p. K53—K57 (1977).
See also *11*, p. *94*; 21.1, p. 289.

R. A. Tahir-Kheli and K. Kawasaki
Phenomenological Description of Dynamics in a Diluted Anisotropic Heisenberg Paramagnet at High Temperatures
(b) Vol. **79**, No. 2, p. 413—422 (1977).
See also *18.1*, p. *213*; 18.4, p. 231.

W. Nolting
Influence of Electron–Ion Spin Coupling on the Electronic Properties of a Ferromagnetic Semiconductor
(b) Vol. **79**, No. 2, p. 573–584 (1977).
See also *13.5.4*, p. *153*; 22, p. 321.

I. Ya. Korenblit
*Phase Transition and Instability of the Homogeneous State in Ferromagnetic Semiconductors
(b) Vol. **79**, No. 2, p. K95–K98 (1977).
See also 1.2, p. 14; *13*, p. *107*; 22, p. 321.

J. Richter
*Influence of Off-Diagonal Disorder on Density of States and Itinerant Ferromagnetism
(b) Vol. **79**, No. 2, p. K139–K142 (1977).
See also *2*, p. *34*; 13, p. 108.

J. D. Patterson
*The Curie Temperature of Ferromagnets with Random Anisotropy
(b) Vol. **79**, No. 2, p. K165–K168 (1977).

M. Wautelet, A. Gérard, F. Grandjean, K. de Strooper, and G. Robbrecht
*Magnetic Properties of Fe_2P
(a) Vol. **39**, No. 2, p. 425–430 (1977).
See also 5, p. 47; 22, p. 322.

T. Rajasekharan and K. V. S. Rama Rao
*Nuclear Magnetic Resonance Investigation of Ferromagnetic $MnSb_{1-x}Sn_x$
(a) Vol. **39**, No. 2, p. 485–489 (1977).
See also *19*, p. *236*; 22.8, p. 396.

A. S. Lileev und W. Steiner
*Untersuchung des Ummagnetisierungsvorganges in gesinterten $SmCo_5$ Dauermagneten
(Studies of Remagnetization on Sintered $SmCo_2$ Permanent Magnets)
(a) Vol. **39**, No. 2, p. K105–K109 (1977).
See also 21.1, p. 289; 21.4, p. 311.

L. Kraus and J. Schneider
*Magnetostriction of Amorphous $(Fe_{1-x}Ni_x)_{80}P_{10}B_{10}$ Alloys
(a) Vol. **39**, No. 2, p. K161–K163 (1977).
See also *2*, p. *34*; 19, p. 236; 22.8, p. 397.

J. Klamut and E. Sobula
*The Dependence of Magnetization on Temperature for Ferromagnets with Magnetic Axial Symmetry
(b) Vol. **80**, No. 1, p. K71–K74 (1977).

J. Mizia and F. Kajzar
*Electron Correlations in Strong Ferromagnetic 3d Alloys
(b) Vol. **80**, No. 1, p. K75–K79 (1977).
See also 21.1, p. 290.

A. S. Lileev and W. Steiner
*Untersuchung der magnetischen Eigenschaften von gesintertem $SmCo_5$
(Studies of the Magnetic Properties of Sintered $SmCo_5$)
(a) Vol. **40**, No. 1 p. 125–131 (1977).
See also 21.1, p. 290; 21.4, p. 311.

B. K. Ponomarev, V. G. Thiessen, and G. A. Ivanov-Smolenskii
*Band Magnetism in a Cu + 7at% Mn Alloy
(a) Vol. **40**, No. 1, p. 189–196 (1977).
See also 18.4, p. 232; 21.1, p. 291.

B. Boucher
*Bulk Magnetic Properties of Amorphous $Tb_{52}Ag_{48}$. Evidence of a Strong Local Anisotropy
(a) Vol. **40**, No. 1, p. 197–203 (1977).
See also *2*, p. *35*; 21.4, p. 311; 21.6, p. 315.

H. J. Bauer and B. Baranowski
*In-situ Observation of the Disappearance of the Ferromagnetism of Nickel under High Pressure Gaseous Hydrogen
(a) Vol. **40**, No. 1, p. K35–K38 (1977).
See also 21.1, p. 291.

E. M. Gray and J. H. Smith
*Effect of Ageing on the Magnetic Moment of MnCu Micromagnet
(a) Vol. **40**, No. 1, p. K73–K76 (1977).
See also 14.1, p. 158; 21.1, p. 291.

S. Krzemiński
*Random One-Dimensional Classical Magnet
(b) Vol. **80**, No. 2, p. 651–656 (1977).
See also 8, p. 65; *13*, p. *108*; 18.4, p. 232.

M. Crişan and P. Pîrvu
*Intersection Lines in the Magnetic–Superconducting Phase Transition
(b) Vol. **80**, No. 2, p. K121–K125 (1977).
See also 1.2, p. 16; *14.2*, p. *162*.

H. Berg
*The Magnetic Anisotropy of Orthorhombic $MnAs_{0.97}P_{0.03}$
(a) Vol. **40**, No. 2, p. 559–566 (1977).
See also 18.4, p. 232; 22.8, p. 397.

K. Schröder and T. Joy
*Magnetization of BiMn Particles in a Bi-Rich Bi–Mn Matrix
(a) Vol. **40**, No. 2, p. K145–K148 (1977).
See also 21.1, p. 292; 21.7, p. 319.

G. M. Carneiro and P. R. Rodrigues Pereira
Fermi Liquid Effects in Ferromagnetic Metals
(b) Vol. **81**, No. 1, p. 79–88 (1977).
See also *13.5.4*, p. *153*; 21, p. 280.

F. Kammler and K. Elk
*Numerical Results of the Hubbard III Approximation for Arbitrary Values of the Correlation Energy and the Electron Density (with Application to Disordered Alloys)
(b) Vol. **81**, No. 1, p. 179–184 (1977).
See also 2, p. 36; 13, p. 109.

B. S. Tošić, M. M. Marinković, and S. Berar
The Boson Field Analysis of the Critical Temperature Region in Ferromagnets
(b) Vol. **81**, No. 1, p. 245–252 (1977).
See also *13*, p. *109*.

P. R. R. Pereira, M. A. Amato, and L. C. M. Miranda
Carrier-Assisted Radiation Pumping of Spin Waves in Ferromagnetic Semiconductors under a Strong Magnetic Field
(b) Vol. **81**, No. 1, p. 395–401 (1977).
See also *13.5.4*, p. *153*; 14.3, p. 167; 22, p. 324.

J. Suwalski, L. Dąbrowski, J. Piekoszewski, and Z. Tucholski
Study of Order–Disorder Transformation in Fe–Al–Si Alloys by Mössbauer Spectroscopy
(a) Vol. **41**, No. 1, p. 191–195 (1977).
See also *5*, p. *48*; 21.1.1, p. 303.

M. F. Mostafa, M. Semary, and M. A. Ahmed
*Magnetic Susceptibility Studies of $((CH_3)_2CHNH_3)_2CuCl_4$
(a) Vol. **41**, No. 1, p. 295—297 (1977).
See also 18.1, p. 214; 22.9, p. 416.

H. E. Khodenkov and V. K. Raev
*Statics of Magnetic Bubble Domain Interaction with Infinite Unsaturated Permalloy Sheet
(a) Vol. **41**, No. 1, p. K39—K41 (1977).

G. T. Dubovka
*On the Problem of Magnetic Phase Transitions of Fe–Ni Invar Alloys
(a) Vol. **41**, No. 1, p. K63—K66 (1977).
See also 21.1, p. 293; 21.1.1, p. 303.

W. Blau, S. Mager, and E. Wieser
*Determination of Localized Magnetic Moments in Fe–Cr–Al Alloys and the Electron Structure
(b) Vol. **81**, No. 2, p. 535—544 (1977).
See also 13.1, p. 118; *18.4*, p. *233*; 21.1.1, p. 303.

F. M. Galperin
Hyperfine Field on the Nucleus of an Atom of a Transition d Element (Metal), Magnetic Moment and Electron Structure of a Ferromagnetic Metal Atom
(b) Vol. **81**, No. 2, p. 545—550 (1977).
See also *13*, p. *109*; 21.1, p. 293.

D. I. Pushkarov and Kh. I. Pushkarov
*Solitons in One-Dimensional Ferromagnetic Systems
(b) Vol. **81**, No. 2, p. 703—707 (1977).
See also *13.5.4*, p. *153*.

R. Tahara, Y. Nakamura, M. Inagaki, and Y. Iwama
Mössbauer Study of Spinodal Decomposition in Fe–Cr–Co Alloy
(a) Vol. **41**, No. 2, p. 451—458 (1977).
See also *5*, p. *48*; 21.1, p. 294; 21.1.1, p. 303.

S. K. Arif, I. Sigalas, and D. St. P. Bunbury
Mössbauer Effect of Fe^{57} Situated in (Rare-Earth) Ni_2 Laves Phases
(a) Vol. **41**, No. 2, p. 585—589 (1977).
See also *5*, p. *48*; 21.1, p. 294; 21.4, p. 312.

H. Oesterreicher and H. Bittner
*Hydriding Characteristics and Magnetic Properties of Hydrides in $Y_6(Mn_{1-x}Fe_x)_{23}$
(a) Vol. **41**, No. 2, p. K101—K105 (1977).
See also 21.1, p. 294; 21.1.1, p. 304.

K. P. Belov and L. I. Koroleva
*Magnetic Microregions, Caused by s–d Exchange, in Ferromagnetic Chalcogenide Spinels
(a) Vol. **41**, No. 2, p. K123—K126 (1977).
See also 22.8.2, p. 411.

L. Bauriegel and G. Otto
The Influence of Magnetization of a Monocrystalline Target on the Channeling and Blocking Effect
(a) Vol. **41**, No. 2, p. K193—K195 (1977).
See also *1*, p. *9*; 18.3, p. 229; 21.1, p. 294; 22.8.2, p. 411.

J. Spałek, M. Lubecka, and A. Węgrzyn
*Carrier Concentration and Magnetic Susceptibility of Doped Ferromagnetic Semiconductors with Application to EuO:Eu
(b) Vol. **82**, No. 1, p. 107—116 (1977).
See also *14.3*, p. *169*; 22.6, p. 389.

Yu. I. Balkarei and V. A. Nikitov
On Magnon–Acoustic Phonon Interaction in Magnetic Semiconductors
(b) Vol. **82**, No. 1, p. 127—131 (1977).
See also *6*, p. *57*; *13.5.4*, p. *153*; 22, p. 326.

E. Gratz and C. A. Poldy
*The Influence of Antiferromagnetic Order on the Evaluation of Spin Disorder Resistivity
(b) Vol. **82**, No. 1, p. 159—167 (1977).
See also 13.1, p. 118; 14.1, p. 158; *18.4*, p. *233*; 21.1, p. 295; 21.4, p. 312.

V. V. Vorobev, V. S. Belovol, and V. A. Finkel
*X-Ray Diffraction Study of Magnetostriction and Phase Transitions in a Magnetic Field in Rare Earth Metals and Some Alloys
(b) Vol. **82**, No. 1, p. 213—219 (1977).
See also *1.2*, p. *17*; 21.4, p. 312.

L. Billard, P. Villemain, and A. Chamberod
*Interaction of Atomic Order with Localized Ferromagnetism
(b) Vol. **82**, No. 1, p. 269—278 (1977).
See also *1.2*, p. *17*; 21.1, p. 295; 21.1.1, p. 304.

J. Zieliński
*Delocalization of Electrons in a Double d-Band Model of Transition Metals
(b) Vol. **82**, No. 1, p. K17—K19 (1977).
See also *13.1*, p. *119*; 21.1, p. 295.

S. Takahashi
*Compression-Induced Magnetic Anisotropy and Superlattice Dislocations in Ni_3Fe (I): Calculated Results
(a) Vol. **42**, No. 1, p. 201—207 (1977).
See also *10.1*, p. *79*; 21.1, p. 295; 21.1.1, p. 304.

A. Reich
*On the Theory of Skew Bubble Domains
(a) Vol. **42**, No. 1, p. 345—350 (1977).

J. Drabecki and B. Wysłocki
*Anisotropic Model of the Influence of Nitrides $Fe_{16}N_2$ on the Coercive Force of a FeN Alloy
(a) Vol. **42**, No. 1, p. 359—367 (1977).
See also 21.1.1, p. 304.

K. Hattori
*Magnetic Properties of a Random Binary System
(b) Vol. **82**, No. 2, p. 489—494 (1977).
See also *2*, p. *38*; 18.1, p. 215; 18.3, p. 229.

V. Kumar and K. C. Sharma
*Phase Transitions in Spin-One Ferromagnetic Systems
(b) Vol. **82**, No. 2, p. 577—580 (1977).
See also *18.1*, p. *215*.

I. Tomáš and Z. Klem
*Bistable Bubble Domains in Annular Plates
(a) Vol. **42**, No. 2, p. 459—467 (1977).

S. Takahashi
*Compression-Induced Magnetic Anisotropy and Superlattice Dislocations in Ni_3Fe (II): Experimental Results
(a) Vol. **42**, No. 2, p. 529—536 (1977).
See also *10.1*, p. *79*; 21.1, p. 296; 21.1.1, p. 304.

F. Söffge, E. Steichele, and K. Stierstadt
Thermal Expansion Anomaly of Nickel near the Curie Point
(a) Vol. **42**, No. 2, p. 621—627 (1977).
See also *8*, p. *66*; 12.1, p. 102; 21.1, p. 296.

W. Steiner and R. Haferl
*Hyperfine Interactions in Y_2Fe_{17}
(a) Vol. **42**, No. 2, p. 739–750 (1977).
See also *5*, p. *49*; 21.1.1, p. 305; 21.4, p. 312.

J. F. Dufresne, I. G. Ritchie, and P. Moser
*On the Question of Low-Temperature Magnetic Relaxation in Pure Iron
(a) Vol. **42**, No. 2, p. 751–756 (1977).
See also 21.1.1, p. 305.

M. A. Sigal
*Ferromagnetic Resonance in $BaFe_{12}O_{19}$ with Cylindrical Domain Structure at Arbitrary Magnetization
(a) Vol. **42**, No. 2, p. 775–784 (1977).
See also *19*, p. *241*; 22.8.2, p. 412.

J. A. Baldwin, Jr.
*Stacey's Theory of Thermally Activated Magnetic Domain Wall Motion
(a) Vol. **42**, No. 2, p. K123–K127 (1977).
See also 21.1.1, p. 305.

B. Boucher
*Propriétés magnétiques de l'alliage TbAu Amorphe et massif
(Magnetic Properties of Amorphous and Bulk TbAu Alloy)
(a) Vol. **42**, No. 2, p. K165–K169 (1977).
See also *2*, p. *38*; 21.4, p. 313; 21.6, p. 316.

S. Krzemiński
*Susceptibility of a Random Mixture of Classical Spins on the Infinite Cayley Tree
(b) Vol. **83**, No. 1, p. K63–K67 (1977).
See also *18.4*, p. *233*.

P. Novotný, V. Roskovec, and J. Šternberk
*High-Field Susceptibility of U_3As_4 at Helium Temperature
(b) Vol. **83**, No. 1, p. K87–K90 (1977).
See also 22, p. 328.

W. Steiner, R. Haferl, and B. K. Choi
*Hyperfine Interactions in $RE_2(Fe, Co)_{17}$ (RE = Y, Gd, Dy)
(a) Vol. **43**, No. 1, p. 111–117 (1977).
See also *5*, p. *49*; 21.1.1, p. 305; 21.4, p. 313.

A. G. Moldovan, S. K. Malik, and W. E. Wallace
*Structural and Magnetization Studies on $ThCo_{5-x}Mn_x$ and $ThNi_{5-x}Mn_x$ Ternaries
(a) Vol. **43**, No. 1, p. 317–320 (1977).
See also 1.1, p. 12; 21.1, p. 297; 21.5, p. 314.

S. Chaudhuri and A. K. Pal
*Thickness Dependence of Curie Temperature of Cu–Ni Films
(a) Vol. **43**, No. 1, p. K9–K13 (1977).
See also 14.1, p. 159; *18.1*, p. *215*; 21.1, p. 297.

F. Walz, H. Schreyer, and H. Kronmüller
*The Magnetic After-Effect Spectrum of α-Iron after Low-Temperature Electron Irradiation
(a) Vol. **43**, No. 1, p. K81–K85 (1977).
See also 10.1, p. 80; *11*, p. *97*; 21.1.1, p. 306.

L. H. Dunleavy and H. J. Blythe
*Investigation of the Magnetic After-Effect in α-Iron after Neutron Irradiation at 77 K
(a) Vol. **43**, No. 1, p. K87–K90 (1977).
See also 10.1, p. 80; *11*, p. *97*; 21.1.1, p. 306.

Y. Khwaja and M. Nauciel-Bloch
*Off-Stoichiometry Effects in the Antiferromagnetic FeRh Alloy
(b) Vol. **83**, No. 2, p. 413—424 (1977).
See also 13.1, p. 121; *18.4*, p. *233*; 21.1.1, p. 306.

A. Bobák
*Influence of an Inhomogeneity of the Effective Field on the Properties of a Disordered Heisenberg Ferromagnet
(b) Vol. **83**, No. 2, p. K169—K171 (1977).
See also *2*, p. *39*; 21, p. 285.

V. V. Zolotarev
Erratum to: Magnon Damping Caused by Spin–Spin Exchange Interaction
(b) Vol. **83**, No. 2, p. K179 (1977).
See also *13.5.4*, p. *153*.

I. Tomáš
*Bubble Statics in Some Axially Symmetric Fields
(a) Vol. **43**, No. 2, p. 381—385 (1977).

K. Mukasa and M. Maeda
*Effects of Annealing on Magnetostrictive Vibrations of Co–Fe Alloys
(a) Vol. **43**, No. 2, p. 393—398 (1977).
See also *1.3*, p. *22*; 21.1, p. 297; 21.1.1, p. 306.

W. Steiner, R. Haferl, and H. Welser
*Hyperfine Interactions in $RE_2(Fe, Ni)_{17}$ (RE = Y, Gd, Dy)
(a) Vol. **43**, No. 2, p. 437—442 (1977).
See also *5*, p. *49*; 21.1.1, p. 306; 21.4, p. 313.

W. Mensch and J. Diehl
*The Magnetic After-Effect Spectrum of α-Iron after Low Temperature Neutron Irradiation
(a) Vol. **43**, No. 2, p. K175—K178 (1977).
See also 10.1, p. 81; 11, p. 98; 21.1.1, p. 306.

J. Diehl, W. Mensch, and M. Weller
*Correlations between Magnetic After-Effect and Internal Friction in α-Iron after Low Temperature Neutron Irradiation
(a) Vol. **43**, No. 2, p. K179—K182 (1977).
See also 10.1, p. 81; *11*, p. *98*; 12.1, p. 102; 21.1.1, p. 306.

J. Jelonek and M. Szczygielski
*Frequency Dependence of Initial Magnetic Permeability for $HgCr_2Se_4$ near the Curie Temperature
(a) Vol. **43**, No. 2, p. K199—K200 (1977).
See also 22.8.2, p. 412.

L. Billard, P. Villemain, and A. Chamberod
*Interaction of Atomic Order with Itinerant Ferromagnetism
(b) Vol. **84**, No. 1, p. 335—343 (1977).
See also *1.3*, p. *23*; 21.1.1, p. 306.

R. Kishore
*Disordered Ferromagnetism in Crystalline Metals
(b) Vol. **84**, No. 1, p. K75—K77 (1977).
See also 21, p. 286.

V. V. Gagulin, N. V. Fadeeva, A. G. Belous, L. A. Sevastianova, A. V. Titov, M. V. Plotnikova, K. P. Mitrofanov, E. V. Zubova, S. P. Solovev, and Yu. N. Venevtsev
New Rhenium Containing Signettomagnets and Ferroelectrics
(a) Vol. **44**, No. 1, p. 247—257 (1977).
See also 5, p. 49; *14.4.1*, p. *197*; *18.4*, p. *234*; 22.8, p. 400.

R. Bodenberger und A. Hubert
*Zur Bestimmung der Blochwandenergie von einachsigen Ferromagneten
(On the Determination of the Bloch Wall Energy of Uniaxial Ferromagnets
(a) Vol. **44**, No. 1, p. K7–K11 (1977).
See also 21.1, p. 299; 21.4, p. 313.

J. Degauque, B. Astie, and C. Levade
*A Study on the Interaction between Magnetic Domain Walls and Dislocations in Pure Iron by Measurements at Low Amplitudes of Magnetoelastic Damping
(a) Vol. **44**, No. 1, p. K59–K62 (1977).
See also *10.1*, p. *82*; 21.1.1, p. 306.

V. Tutovan and M. Chiriac
*The Influence of a Magnetic Field, Applied during Annealing, on the Atomic Ordering of the Ni_3Mn Alloy
(a) Vol. **44**, No. 1, p. K63–K66 (1977).
See also 18.4, p. 234; 21.1, p. 299.

H. Ratajczak and I. Łabędzka
*Anomalous Hall Hysteresis Loops in the Vicinity of the Compensation Point in HoCo Amorphous Films
(a) Vol. **44**, No. 1, p. K103–K105 (1977).
See also 2, p. 40; *14.1*, p. *160*; 21.1, p. 299; 21.4, p. 313.

K. Ruebenbauer, J. Fink, H. Schmidt, G. Czjzek, and K. Tomala
*An Investigation of Magnetic Ordering in $Gd_xCe_{1-x}Ru_2$ by ^{155}Gd Mössbauer Spectroscopy
(b) Vol. **84**, No. 2, p. 611–618 (1977).
See also *5*, p. *50*; 14.2, p. 164; 21.4, p. 313.

A. Kühnel, S. Wendt, and J. Wesselinowa
Dynamic Behaviour of the Ising Model in a Transverse Field
(b) Vol. **84**, No. 2, p. 653–664 (1977).
See also *14.4.1*, p. *197*.

K. A. Chao, J. Spałek, and A. M. Oleś
The Kinetic Exchange Interaction in Doubly Degenerate Narrow Bands
(b) Vol. **84**, No. 2, p. 747–759 (1977).
See also *13*, p. *113*; 18.4, p. 234.

A. Mauger
Indirect Exchange in Europium Chalcogenides
(b) Vol. **84**, No. 2, p. 761–771 (1977).
See also *13.1*, p. *123*; 18.1, p. 216; 22, p. 330.

E. Kolley and W. Kolley
On the Spin Susceptibility for Itinerant Electrons with Energy-Dependent Effective Interaction
(b) Vol. **84**, No. 2, p. K147–K150 (1977).
See also *13*, p. *113*; 18.1, p. 216.

N. V. Fedoseeva, I. P. Spevakova, B. V. Beznosikov, E. P. Naiden, and V. A. Khokhlov
*Weak Ferromagnetism and Magnetic Structure of $TlMnCl_3$
(a) Vol. **44**, No. 2, p. 429–434 (1977).
See also *18.4*, p. *234*; 22.8, p. 400.

R. Goblirsch, H. W. Weber, G. Hilscher, R. Grössinger, and W. Steiner
*Search for Magnetic Order in Selected Pseudobinary Systems down to Millikelvin Temperatures
(a) Vol. **44**, No. 2, p. 593–600 (1977).
See also *18.1*, p. *216*; 21.1, p. 300.

V. G. Kleparskii, E. I. Ilyashenko, V. V. Randoshkin, and R. V. Telesnin
*Bubble Conversion to "Anomalous" Behaviour
(a) Vol. **44**, No. 2, p. 771—775 (1977).
See also 22.8.2, p. 413.

S. Ohnuma, K. Watanabe, and T. Masumoto
*Low Coercivity and Zero Magnetostriction of Amorphous Fe–Co–Ni System Alloys
(a) Vol. **44**, No. 2, p. K151—K154 (1977).
See also *2*, p. *41*; 21.1, p. 301; 21.1.1, p. 307.

K. Hoselitz
*Magnetization of Fe–Si–B Metallic Glass
(a) Vol. **44**, No. 2, p. K191—K193 (1977).
See also *2*, p. *41*; 21.1.1, p. 307.

18.2.1 Ferromagnetic Films

Original Papers and Short Notes

O. G. RAMER and C. H. WILTS
*Surface Effects in Spin-Wave Resonance in Thin YIG Films
(b) Vol. **79**, No. 1, p. 313–320 (1977).
See also 19, p. 235; 22.8.2, p. 409.

U. GRADMANN and K. SALEWSKI
*Magnetic "Dead Layers" and Exchange Anisotropy in the Interface of 60Ni/40 Fe(111) and Polycrystalline Mn
(a) Vol. **39**, No. 1, p. 41–48 (1977).
See also 18.4, p. 231; 21.1, p. 288; 21.1.1, p. 302.

O. BOSTANJOGLO and H.-P. GEMÜND
*Curling-Type Magnetic Domains in Narrow Film Strips
(a) Vol. **40**, No. 2, p. K97–K99 (1977).
See also 21.1.1, p. 303.

S. S. NANDRA and P. J. GRUNDY
*Lorentz Microscopy of the Magnetic Domain Structure in Amorphous Rare Earth–Cobalt Films Evaporated at Normal Incidence
(a) Vol. **41**, No. 1, p. 65–74 (1977).
See also 21.1, p. 293; 21.4, p. 312.

W. ANDRÄ and H. DANAN
*Magnetic Interface Anisotropy and Magnetization of Plated Ni and Co Films
(a) Vol. **42**, No. 1, p. 227–233 (1977).
See also 21.1, p. 295.

M. A. CHALABYAN, R. STRAUBEL, and YA. M. POGOSYAN
*A Wall Splitting Effect in Double-Layer Permalloy Films
(a) Vol. **42**, No. 1, p. 273–278 (1977).
See also 21.1, p. 295; 21.1.1, p. 304.

O. BOSTANJOGLO and H. P. GEMÜND
*New Rotation Processes in Magnetic Film Strips
(a) Vol. **42**, No. 2, p. K103–K106 (1977).
See also 21.1.1, p. 305.

U. A. DURASOVA, O. S. KOLOTOV, T. N. LETOVA, and R. V. TELESNIN
*On the Stray Fields Arising during Pulse Reversal in Thin Magnetic Films
(a) Vol. **42**, No. 2, p. K171–K174 (1977).
See also 21.1.1, p. 305.

H. PUSZKARSKI
*Surface-State "Swapping" Effect in Thin Magnetic Films
(b) Vol. **83**, No. 1, p. K25–K27 (1977).
See also *13.3*, p. *127*; 22.8.2, p. 412.

U. DRAEGER, R. STRAUBEL, and W. J. ZIĘTEK
*Calculation of the Wall Curvature in Thin Film Strips with In-Plane Magnetization
(a) Vol. **43**, No. 1, p. K113–K116 (1977).
See also 21.1.1, p. 306.

J. L. VICENT
Conduction Electrons and s–d Exchange Interaction
(b) Vol. **84**, No. 1, p. K71–K74 (1977).
See also 13, p. 112; *14.1*, p. *160*; 21.1, p. 299.

S. R. Herd

*TEM Observation on the Source of Perpendicular Anisotropy in Amorphous GdCo(O_2)
(a) Vol. **44**, No. 1, p. 363–380 (1977).
See also *2*, p. *40*; 21.4, p. 313.

M. A. Rozenblat and S. E. Yurchenko

Hard Bubble Conversion in the Presence of In-Plane Fields
(a) Vol. **44**, No. 1, p. K75–K78 (1977).

18.3 Ferrimagnetic Properties

Review Articles

M. I. Klinger and A. A. Samokhvalov
Electron Conduction in Magnetite and Ferrites
(b) Vol. **79**, No. 1, p. 9—48 (1977).
Erratum
(b) Vol. **83**, No. 1, p. 353 (1977).
See also 13.1, p. 114; *13.5.3*, p. *151*; *14.3*, p. *165*; 15, p. 198; 22.8.2, p. 409.

P. Röschmann and H. Dötsch
*Properties of Magnetostatic Modes in Ferrimagnetic Spheroids
(b) Vol. **82**, No. 1, p. 11—57 (1977).
See also 13.5.4, p. 153; 19, p. 235; 22.8.2, p. 409.

Original Papers and Short Notes

J. A. Baldwin, Jr.
*Force of Interaction between a 180° Domain Wall and a Dislocation in YIG
(a) Vol. **39**, No. 1, p. K17—K19 (1977).
See also *10.2*, p. *84*; 22.8.2, p. 409.

K. de Strooper
Magnetic Properties of Some Co^{2+} Containing Spinels
(a) Vol. **39**, No. 2, p. 431—435 (1977).
See also *18.1*, p. *213*; 18.4, p. 231; 22.8.2, p. 409.

H. Dötsch
*Influence of Spin Precession on the Stability of Bubble Domains
(a) Vol. **39**, No. 2, p. 589—594 (1977).
See also 22.8.2, p. 409.

F. Schneider, K. Melzer, H. Mehner, and G. Dehe
*^{119}Sn Hyperfine Fields in α-Fe_2O_3
(a) Vol. **39**, No. 2, p. K115—K117 (1977).
See also 5, p. 47; 22.6, p. 386.

J. K. Yakinthos et E. Roudaut
*Structure magnétique du composé intermétallique Dy_2Ni_{17}
(Magnetic Structure of the Intermetallic Compound Dy_2Ni_{17})
(a) Vol. **39**, No. 2, p. K151—K155 (1977).
See also 22.7, p. 394.

M. Guillot and M. le Gall
*Superexchange Field and Intense Magnetic Field Dependences of the Faraday Rotation Induced by the Electric Dipole Transitions in YIG
(b) Vol. **80**, No. 1, p. 279—287 (1977).
See also 13, p. 108; *20*, p. *244*; 22.8.2, p. 410.

E. Jäger
*Structure Fluctuations in the Molecular Field Theory of Amorphous Ferrimagnets
(b) Vol. **80**, No. 1, p. K81—K83 (1977).
See also *2*, p. *34*; 21.1, p. 290; 21.4, p. 311.

H.-R. Müller, E. Jäger, and M. Bitterlich
*Influence of Impurities on the Magnetic Properties of Amorphous Gd–Co Thin Films
(b) Vol. **80**, No. 1, p. K85—K87 (1977).
See also *2*, p. *34*; 21.1, p. 290; 21.4, p. 311.

P. Bernstein and T. Merceron
*Model of Photomagnetic Effect in Ni–Zn Ferrite with Small Amount of Cobalt
(a) Vol. **40**, No. 1, p. 159—166 (1977).
See also 22.8.2, p. 410.

A. Ślawska-Waniewska and H. Lipko
*The Effect of High Hydrostatic Pressure on Anisotropy Properties of YIG:Fe^{2+} Single Crystals
(a) Vol. **40**, No. 1, p. K39—K41 (1977).
See also 22.8.2, p. 410.

R. Perthel, W. Keilig, R. Kosicik, U. Röpke, and C.-G. d'Ambly
*Determination of the Magnetic Anisotropy in Amorphous Gd–Co Films by Torque Measurement
(a) Vol. **40**, No. 2, p. K135—K138 (1977).
See also *2*, p. *36*; 21, p. 280; 21.1, p. 292.

K. Ozawa and S. Anzai
*Effect of Pressure on a New Magnetic Transition Temperature in the Ferrimagnetic State of $Mn_{1.77}Sn$
(a) Vol. **40**, No. 2, p. K199—K202 (1977).
See also *14.1*, p. *158*; 21.7, p. 319.

H. Gamari-Seale and J. K. Yakinthos
*Temperature Dependence of the Magnetic Structure of Fe_2MoO_4
(a) Vol. **41**, No. 1, p. 147—150 (1977).
See also 22.8.2, p. 411.

A. Feu
*Dynamic Characteristics of Domain Walls in Polycrystalline YIG. Low-Temperature Behaviour
(a) Vol. **41**, No. 2, p. 361—370 (1977).
See also 22.8.2, p. 411.

L. Bauriegel and G. Otto
The Influence of Magnetization of a Monocrystalline Target on the Channeling and Blocking Effect
(a) Vol. **41**, No. 2, p. K193—K195 (1977).
See also *1*, p. *9*; 18.2, p. 220; 21.1, p. 294; 22.8.2, p. 411.

J. K. Yakinthos
*Determination of Tb_2Ni_{17}, Er_2Ni_{17}, and Tm_2Ni_{17} Magnetic Structure by Neutron Diffraction
(b) Vol. **82**, No. 1, p. 349—356 (1977).
See also 21.1, p. 295; 21.4, p. 312.

Diep-The-Hung
*Effect of the Next-Nearest Neighbour Coupling in a Semi-Infinite Ferrimagnet
(b) Vol. **82**, No. 1, p. 383—389 (1977).
See also 1.6, p. 31; *13.5.4*, p. *153*.

N. L. Pakhomova, M. A. Vinnik, L. M. Kazimenko, L. Jahn und V. A. Kozlov
*Statistisches Modell und Phasenübergang in Nickel–Zink-Ferriten
(Statistical Model and Phase Transition in Nickel–Zinc Ferrites)
(a) Vol. **42**, No. 1, p. 191—199 (1977).
See also 22.8.2, p. 412.

K. Hattori
Magnetic Properties of a Random Binary System
(b) Vol. **82**, No. 2, p. 489—494 (1977).
See also *2*, p. *38*; 18.1, p. 215; 18.2, p. 221.

M. W. Muller
*Site Ordering and Magnetostriction in Garnet
(b) Vol. **83**, No. 1, p. 177–185 (1977).
See also *1*, p. *10*; 22.8.2, p. 412.

D. Linzen, S. Bornmann, P. Görnert, R. Hergt, H. Pfeiffer, W. Schüppel, F. Voigt, and M. Wendt
*Growth-Induced Magnetic Anisotropy of (Sm, Ga)-Substituted YIG Films Grown on {111} Substrates
(a) Vol. **43**, No. 1, p. 277–281 (1977).
See also 3, p. **43**; 22.8.2, p. 412.

N. D. J. Miller
*Non-Reciprocal Propagation of Magnetostatic Volume Waves
(a) Vol. **43**, No. 2, p. 593–600 (1977).
See also 22.8.2, p. 412.

K. de Strooper, A. van Alboom, C. Henriet-Iserentant, R. Vanleerberghe, and R. Vandenberghe
*On the Cation Distribution and Magnetic Properties of Cobalt–Zinc Titanates
(a) Vol. **44**, No. 1, p. 215–220 (1977).
See also 18.1, p. 215; 22.8.2, p. 412.

J. J. Bara, A. T. Pędziwiatr, Z. M. Stadnik, A. Szytuła, J. Todorovič, Z. Tomkowicz, and W. Zarek
*Investigations of Crystal and Magnetic Properties of Nickel Ferrite-Aluminates
(a) Vol. **44**, No. 1, p. 325–331 (1977).
See also *1.1*, p. *13*; 5, p. 49; 22.8.2, p. 413.

J. Loos
*The Mean Field Phase Diagrams of YbIG
(b) Vol. **84**, No. 2, p. 457–463 (1977).
See also 22.8.2, p. 413.

M. Wurlitzer and U. Reinhold
X-Ray Induced Change of Initial Permeability in Polycrystalline YIG(Me^{1+})
(a) Vol. **44**, No. 2, p. 531–538 (1977).
See also *11*, p. *99*; 22.8.2, p. 413.

T. Merceron and J. Loriers
*Magnetic After-Effect Phenomenon in the Iron–Yttrium Praseodymium Garnets
(a) Vol. **44**, No. 2, p. K147–K150 (1977).
See also 22.8.2, p. 413.

18.4 Antiferromagnetic Properties

Original Papers and Short Notes

U. Gradmann and K. Salewski
Magnetic "Dead Layers" and Exchange Anisotropy in the Interface of 60Ni/40 Fe(111) and Polycrystalline Mn
(a) Vol. **39**, No. 1, p. 41–48 (1977).
See also *18.2.1*, p. *226*; 21.1, p. 288; 21.1.1, p. 302.

J. C. Marmeggi, D. Hohlwein, and E. F. Bertaut
*Magnetic Neutron Laue Diffraction Study of the Domain Distribution in α-Fe_2O_3
(a) Vol. **39**, No. 1, p. 57–64 (1977).
See also 22.6, p. 386.

A. Heidemann, K. Kosuge, Y. Ueda, and S. Kachi
*Hyperfine Interaction in V_3O_7
(a) Vol. **39**, No. 1, p. K37–K40 (1977).
See also 22.6, p. 386.

H. Oesterreicher
*Magnetic Studies on GdCuGe, $Gd_6Cu_8Ge_8$, $GdCu_2Ge_2$, and $PrCu_2Ge_2$
(a) Vol. **39**, No. 1, p. K75–K77 (1977).
See also *18*, p. *209*; 22.7, p. 394.

R. A. Tahir-Kheli and K. Kawasaki
Phenomenological Description of Dynamics in a Diluted Anisotropic Heisenberg Paramagnet at High Temperatures
(b) Vol. **79**, No. 2, p. 413–422 (1977).
See also *18.1*, p. *213*; 18.2, p. 217.

D. I. Khomskii and K. I. Kugel
*Degenerate Hubbard Model in a Magnetic Field. Application to Jahn-Teller Systems
(b) Vol. **79**, No. 2, p. 441–450 (1977).
See also 22.8, p. 396.

K. de Strooper
Magnetic Properties of Some Co^{2+} Containing Spinels
(a) Vol. **39**, No. 2, p. 431–435 (1977).
See also *18.1*, p. *213*; 18.3, p. 228; 22.8.2, p. 409.

V. P. Plakhtii, I. V. Golosovskii, M. N. Bedrisova, O. P. Smirnov, V. I. Sokolov, B. V. Mill, and N. N. Parfenova
*Magnetic Ordering and Bond Geometry in Garnets with Fe^{3+} Ions in the Tetrahedral Sublattice
(a) Vol. **39**, No. 2, p. 683–695 (1977).
See also 22.8.2, p. 410.

G. I. Makovetskii, V. M. Ryzhkovskii, and G. M. Shakhlevich
*Neutron Diffraction Studies of Chromium Sulphide with Nickel Arsenide Type Structure
(a) Vol. **39**, No. 2, p. K127–K130 (1977).
See also 22, p. 322.

A. Treindl and K. H. Germann
*Linear Optical Anisotropy Investigations of the Antiferromagnetic Phase Transition in MnO
(b) Vol. **80**, No. 1, p. 159–166 (1977).
See also *20.1*, p. *251*; 22.6, p. 387.

F. G. Brady Moreira, I. P. Fittipaldi, S. M. Rezende, R. A. Tahir-Kheli, and B. Žekš
*Field-Induced Phase Transitions in a Randomyl Diluted Antiferromagnet. Mean-Field Description of Classical Spin Model
(b) Vol. **80**, No. 1, p. 385–394 (1977).
See also 1.2, p. 15; *18.1*, p. *213*.

R. KISHORE
*Semiclassical Theory of Spin Waves in Amorphous Antiferromagnets
(b) Vol. **80**, No. 1, p. K33—K35 (1977).
See also *2*, p. *34*; 13.5, p. 141.

V. G. BARYAKHTAR, A. A. GALKIN, and V. T. TELEPA
*High Hydrostatic Pressure Effect on the Flop of Magnetic Moments of Subla ttices and Néel Temperature in the Antiferromagnetic $CuCl_2 \cdot 2\ H_2O$
(b) Vol. **80**, No. 1, p. K37—K40 (1977).
See also 19, p. 236; 22, p. 323.

G. HAUSCH and E. TÖRÖK
Thermal Expansivity and Elastic Constants of CrFe Alloys
(a) Vol. **40**, No. 1, p. 55—62 (1977).
See also *12.1*, p. *101*; 21.1, p. 290; 21.1.1, p. 303.

B. K. PONOMAREV, V. G. THIESSEN, and G. A. IVANOV-SMOLENSKII
Band Magnetism in a Cu + 7 at% Mn Alloy
(a) Vol. **40**, No. 1, p. 189—196 (1977).
See also *18.2*, p. *218*; 21.1, p. 291.

S. KRZEMIŃSKI
Random One-Dimensional Classical Magnet
(b) Vol. **80**, No. 2, p. 651—656 (1977).
See also 8, p. 65; *13*, p. *108*; *18.2*, p. *219*.

A. KIRFEL, W. SCHÄFER, G. WILL, and K. H. J. BUSCHOW
*The High-Temperature Polymorph β-$FeSO_4$. Stability, Magnetic Structure, and Magnetic Properties
(a) Vol. **40**, No. 2, p. 447—453 (1977).
See also *1.2*, p. *16*; 22.8.1, p. 403.

H. BERG
The Magnetic Anisotropy of Orthorhombic $MnAs_{0.97}P_{0.03}$
(a) Vol. **40**, No. 2, p. 559—566 (1977).
See also *18.2*, p. *219*; 22.8, p. 397.

J. K. YAKINTHOS, H. GAMARI-SEALE, and J. LAFOREST
*Magnetic Properties of R_3Pd_4 (R = Pr, Nd, Sm, Gd, Tb, Dy, Ho, Er, Tm) Intermetallic Compounds
(a) Vol. **40**, No. 2, p. K105—K108 (1977).
See also *18.1*, p. *214*; 21.4, p. 311.

T. ANAGNOSTOPOULOS, P. IKONOMOU et J. K. YAKINTHOS
*Propriétés magnétiques des composés $SmAu_6$, $TbAu_6$ et $DyAu_6$
(Magnetic Properties of the Compounds $SmAu_6$, $TbAu_6$, and $DyAu_6$)
(a) Vol. **40**, No. 2, p. K109—K112 (1977).
See also *18.1*, p. *214*; 21.4, p. 312; 21.6, p. 315.

V. V. EREMENKO, A. I. MASLENNIKOV, and V. M. NAUMENKO
Host and Impurity Excitations in Antiferromagnetic and Paramagnetic $FeCO_3$ in the Far Infrared
(b) Vol. **81**, No. 1, p. 351—357 (1977).
See also *13.4*, p. *133*; 19, p. 238; 22.8.1, p. 404.

G. HAUSCH
*Invar-Like Thermal Expansion Anomaly in Antiferromagnetic Mn–Ni Alloys
(a) Vol. **41**, No. 1, p. K35—K38 (1977).
See also *8*, p. *65*; 21.1, p. 293.

W. BLAU, S. MAGER, and E. WIESER
*Determination of Localized Magnetic Moments in Fe–Cr–Al Alloys and the Electron Structure
(b) Vol. **81**, No. 2, p. 535–544 (1977).
See also 13.1, p. 118; *18.2*, p. *220*; 21.1.1, p. 303.

V. T. KALINNIKOV, YU. V. RAKITIN, VAN H. CRAWFORD, and W. E. HATFIELD
*Near One-Dimensional Antiferromagnetism in Vanadyl Carboxylates
(b) Vol. **81**, No. 2, p. K75–K79 (1977).
See also 19, p. 239; 22.9, p. 416.

A. COLLOMB, D. SAMARAS, B. BOCHU et J. C. JOUBERT
*Propriétés et structure magnétiques de $CaCu_3Ti_4O_{12}$ á structure pérovskite
(Properties and Magnetic Structures of Perovskite $CaCu_3Ti_4O_{12}$)
(a) Vol. **41**, No. 2, p. 459–463 (1977).
See also 22.8.1, p. 405.

N. B. NEŠKOVIĆ, B. BABIĆ, and J. KONSTANTINOVIĆ
High Temperature Anomalous Behaviour of the Crystal Lattice of Hematite
(a) Vol. **41**, No. 2, p. K133–K136 (1977).
See also *1.2*, p. *17*; 22.6, p. 388.

E. GRATZ and C. A. POLDY
*The Influence of Antiferromagnetic Order on the Evaluation of Spin Disorder Resistivity
(b) Vol. **82**, No. 1, p. 159–167 (1977).
See also 13.1, p. 118; 14.1, p. 158; *18.2*, p. *221*; 21.1, p. 295; 21.4, p. 312.

B. SZPUNAR and P.-A. LINDGÅRD
*Theory of Temperature Dependence of the Magnetization in Rare-Earth–Transition-Metal Alloys
(b) Vol. **82**, No. 2, p. 449–456 (1977).
See also 13, p. 110; *18.1*, p. *214*; 21.1, p. 296; 21.4, p. 312.

W. LÖSER
*Thermodynamic Properties of a Finite Disordered Hubbard Cluster
(b) Vol. **83**, No. 1, p. 299–304 (1977).
See also 2, p. 38; *8*, p. *66*.

S. KRZEMIŃSKI
Susceptibility of a Random Mixture of Classical Spins on the Infinite Cayley Tree
(b) Vol. **83**, No. 1, p. K63–K67 (1977).
See also *18.2*, p. *222*.

M. PERNET, C. BERTHET-COLOMINAS, M. A. ALARIO FRANCO et A. N. CHRISTENSEN
*Etude par diffraction neutronique de l'oxyhydroxyde de chrome β-CrOOH
(Study of Chromium Oxyhydroxide (β–CrOOH) by Neutron Diffraction)
(a) Vol. **43**, No. 1, p. 81–88 (1977).
See also *1*, p. *10*; 22.9, p. 418.

Y. KHWAJA and M. NAUCIEL-BLOCH
*Off-Stoichiometry Effects in the Antiferromagnetic FeRh Alloy
(b) Vol. **83**, No. 2, p. 413–424 (1977).
See also 13.1, p. 121; *18.2*, p. *223*; 21.1.1, p. 306.

A. ZYGMUNT
*Magnetic Properties of UGeY Compounds (Y = S, Se, Te)
(a) Vol. **43**, No. 2, p. 573–577 (1977).
See also 18.1, p. 215; 22.8, p. 400.

S. M. BORODACHEV, S. I. MASHAROV, and V. A. VOLKOV
Electrical Resistivity of Antiferromagnetic Alloys. Coherent Potential Approximation
(b) Vol. **84**, No. 1, p. 33–40 (1977).
See also *14.1*, p. *159*; 21, p. 286.

A. Hofstaetter, A. Scharmann, and B. Vitt
Antiferromagnetic Properties of the $Li_2Mn_xTi_{1-x}O_3$ Mixed Crystal System. An EPR Investigation
(b) Vol. 84, No. 1, p. 93—98 (1977).
See also *19*, p. *242*; 22.8.1, p. 407.

M. Schwab and H. G. Kahle
*Specific Heat of $TbPO_4$
(b) Vol. 84, No. 1, p. 167—170 (1977).
See also 1.2, p. 19; *8*, p. *67*; 22.8.1, p. 407.

K. Bärner
Energy Bands and Electron Correlation Effects in Manganese Pnictides
(b) Vol. 84, No. 1, p. 385—392 (1977).
See also *13.1*, p. *122*; 14.1, p. 159; *20.1*, p. *261*; 21.1, p. 299.

V. V. Gagulin, N. V. Fadeeva, A. G. Belous, L. A. Sevastianova, A. V. Titov, M. V. Plotnikova, K. P. Mitrofanov, E. V. Zubova, S. P. Solovev, and Yu. N. Venevtsev
*New Rhenium Containing Seignettomagnets and Ferroelectrics
(a) Vol. 44, No. 1, p. 247—257 (1977).
See also 5, p. 49; *14.4.1*, p. *197*; 18.2, p. 223; 22.8, p. 400.

V. Tutovan and M. Chiriac
*The Influence of a Magnetic Field, Applied during Annealing, on the Atomic Ordering of the Ni_3Mn Alloy
(a) Vol. 44, No. 1, p. K63—K66 (1977).
See also *18.2*, p. *224*; 21.1, p. 299.

J. Mizia
*CPA Treatment of the Extended Hubbard Model. Antiferromagnetic Solution
(b) Vol. 84, No. 2, p. 449—455 (1977).
See also *13*, p. *112*.

A. Kasten and P. J. Becker
*Spectroscopic Investigation of Optically Excited Doublet States of Ho^{3+} in Magnetically Ordered $(Ho_{0.01}, Gd_{0.99})VO_4$
(b) Vol. 84, No. 2, p. 551—560 (1977).
See also 13.4, p. 139; *20.1*, p. *262*; 22.8.1, p. 408.

J. Monecke
One-Particle Spectrum of Correlated Electrons near the Free-Electron Limit
(b) Vol. 84, No. 2, p. 665—672 (1977).
See also *13.1*, p. *123*; 14, p. 156.

K. A. Chao, J. Spałek, and A. M. Oleś
The Kinetic Exchange Interaction in Doubly Degenerate Narrow Bands
(b) Vol. 84, No. 2, p. 747—759 (1977).
See also *13*, p. *113*; 18.2, p. 224.

I. G. Gochev and K. I. Grozdev
*On the Heat Capacity of $Fe(HCOO)_2 \cdot 2\,H_2O$
(b) Vol. 84, No. 2, p. 785—789 (1977).
See also *8*, p. *68*; 22.9, p. 419.

Diep-The-Hung
Effect of Induced Surface Magnetic Moment on Spin-Wave Spectrum
(b) Vol. 84, No. 2, p. K159—K161 (1977).
See also 1.6, p. 32; *13.5.4*, p. *154*.

N. V. Fedoseeva, I. P. Spevakova, B. V. Beznosikov, E. P. Naiden, and V. A. Khokhlov
*Weak Ferromagnetism and Magnetic Structure of $TlMnCl_3$
(a) Vol. 44, No. 2, p. 429—434 (1977).
See also *18.2*, p. *224*; 22.8, p. 400.

19. Magnetic Resonance

Review Article

P. Röschmann and H. Dötsch
Properties of Magnetostatic Modes in Ferrimagnetic Spheroids
(b) Vol. **82**, No. 1, p. 11—57 (1977).
See also 13.5.4, p. 153; *18.3*, p. *228*; 22.8.2, p. 409.

Original Papers and Short Notes

J. M. de Siebenthal and H. Bill
A Perturbed O_2^- Molecule Ion in $SrCl_2$ Single Crystals. EPR Results
(b) Vol. **79**, No. 1, p. 259—267 (1977).
See also *10.2*, p. *83*; 22.5.3, p. 383.

O. G. Ramer and C. H. Wilts
Surface Effects in Spin-Wave Resonance in Thin YIG Films
(b) Vol. **79**, No. 1, p. 313—320 (1977).
See also *18.2.1*, p. *226*; 22.8.2, p. 409.

P. Matthys, M. Vanhaelst, and E. Boesman
*Spin–Lattice Relaxation of the S_2^- Centre in KI
(b) Vol. **79**, No. 1, p. K23—K26 (1977).
See also 22.5.2, p. 375.

P. G. Baranov, Yu. P. Veshchunov, R. A. Zhitnikov, and N. G. Romanov
*EPR, Optical Absorption, and Luminescence of Anisotropic Copper Centres in KCl
(b) Vol. **79**, No. 1, p. K27—K30 (1977).
See also 20.1, p. 248; 20.3, p. 267; 22.5.2, p. 375.

J. Livage, P. Pineau, M. C. Leroy, and M. Michaud
Semiconducting Vanadium Phosphate Glasses
(a) Vol. **39**, No. 1, p. 73—78 (1977).
See also *2*, p. *33*; 22.6.1, p. 393.

G. Berg, M. Dubiel, and F. Fröhlich
T_1-Measurements in Ca-Doped NaCl Crystals
(a) Vol. **39**, No. 1, p. 265—269 (1977).
See also *10.2*, p. *83*; 22.5.2, p. 375.

G. O. Berim and A. R. Kessel
*The Model Hamiltonian and Some Properties of Magnetic Resonance of Triplet Excitons (II): Kubo-Tomita Approximation
(b) Vol. **79**, No. 2, p. 489—494 (1977).
See also *13.5.1*, p. *143*.

R. Karcher, K. Lüders, G. Roth, and Z. Szücs
*Low-Temperature Knight-Shift Investigations of Yttrium Metal
(b) Vol. **79**, No. 2, p. 519—524 (1977).
See also 21.1, p. 289.

F. J. Owens
Evidence for Zero-Field Fluctuations in Cr^{3+} near the Phase Transition in $NH_4Al(SO_4)_2 \cdot 12\,H_2O$
(b) Vol. **79**, No. 2, p. 623—628 (1977).
See also *14.4*, p. *189*; 22.8, p. 396.

J. R. Sambles, G. Sharp-Dent, J. E. Cousins, A. Stesmans, and J. Witters
*Conduction Electron Spin Resonance in Aluminium at 20.98 and 9.27 GHz
(b) Vol. **79**, No. 2, p. 645—654 (1977).
See also 21, p. 277.

M. A. B. Whitaker
*The Influence of Dipolar Energy in Nuclear Double-Resonance
(b) Vol. **79**, No. 2, p. 673—681 (1977).

S. W. Biernacki and M. Hamera
LCAO Approach to Mn^{2+} in ZnSe
(b) Vol. **79**, No. 2, p. 683—689 (1977).
See also *13.4*, p. *130*; 22.4.2, p. 365.

T. Rajasekharan and K. V. S. Rama Rao
*Nuclear Magnetic Resonance Investigation of Ferromagnetic $MnSb_{1-x}Sn_x$
(a) Vol. **39**, No. 2, p. 485—489 (1977).
See also *18.2*, p. *218*; 22.8, p. 396.

K. H. Lee, G. E. Holmberg, and J. H. Crawford, Jr.
Optical and ESR Studies of Holes Centers in γ-Irradiated Al_2O_3
(a) Vol. **39**, No. 2, p. 669—674 (1977).
See also *10.2*, p. *84*; *20.3*, p. *268*; 22.6, p. 386.

P. N. Murty, C. R. K. Murty, and S. Subramanian
*Electron Paramagnetic Resonance Study of X-Irradiated Arsenate-Doped Phosphate: KH_2AsO_4/KH_2PO_4
(a) Vol. **39**, No. 2, p. 675—682 (1977).
See also 11, p. 94; 22.8.1, p. 402.

R. M. Catchings
*EPR Study of the Ferric Oxide–Nickel Oxide–Sodium–Lithium–Borosilicate Glass System
(a) Vol. **39**, No. 2, p. K101—K103 (1977).
See also *2*, p. *34*; 22.8, p. 396.

L. Kraus and J. Schneider
Magnetostriction of Amorphous $(Fe_{1-x}Ni_x)_{80}P_{10}B_{10}$ Alloys
(a) Vol. **39**, No. 2, p. K161—K163 (1977).
See also *2*, p. *34*; *18.2*, p. *218*; 22.8, p. 397.

R. Herrmann and P. Goy
*Investigation of Plasma Effects in Bismuth–Antimony Alloys in the Submillimeter Wave Range
(b) Vol. **80**, No. 1, p. 207—213 (1977).
See also *13.1*, p. *115*; 21.7, p. 319.

H. J. Hackelöer, H. Selbach, O. Kanert, A. W. Sleeswyk, and G. Hut
Determination of the Velocity of Mobile Dislocations by Nuclear Spin Relaxation Measurements
(b) Vol. **80**, No. 1, p. 235—243 (1977).
See also *10.2*, p. *85*; 22.5.2, p. 376.

N. C. Halder
*Calculation of Nuclear Quadrupolar Relaxation in Liquid Metals
(b) Vol. **80**, No. 1, p. 273—278 (1977).
See also *2*, p. *34*; 21, p. 278.

N. Guskos, J. Kuriata, and T. Rewaj
Spin–Lattice Coefficients of Eu^{2+} in BaF_2
(b) Vol. **80**, No. 1, p. K25—K27 (1977).
See also 22.5.3, p. 383.

V. G. Baryakhtar, A. A. Galkin, and V. T. Telepa
High Hydrostatic Pressure Effect on the Flop of Magnetic Moments of Sublattices and Néel Temperature in the Antiferromagnetic $CuCl_2 \cdot 2\,H_2O$
(b) Vol. **80**, No. 1, p. K37—K40 (1977).
See also *18.4*, p. *232*; 22, p. 323.

H. Drulis, I. A. Garifullin, E. G. Kharakhashyan, B. Staliński, N. M. Suleymanov, and M. M. Zaripov
*EPR in the $Pd_{1-y}Gd_yH_x$ System
(a) Vol. **40**, No. 1, p. K11—K14 (1977).
See also 21, p. 279.

G. O. Berim, F. G. Cherkasov, E. G. Kharakhashyan, and Y. I. Talanov
*Line Shape of Conduction Electron Spin Resonance in Spherical Metal Particles
(a) Vol. **40**, No. 1, p. K53—K55 (1977).
See also 21.2, p. 308.

B. Movaghar and L. Schweitzer
*ESR and Conductivity in Amorphous Germanium and Silicon
(b) Vol. **80**, No. 2, p. 491—498 (1977).
See also 2, p. 35; *14.3*, p. *167*; 22.1.1, p. 334; 22.1.2, p. 338.

W. Gehlhoff
*Pseudocubic Angular Dependence of the $M = -1/2$ to $+ 1/2$ EPR Transitions for Ions with $S = 3/2$ by Small Distortions of the Cubic Crystal Field
(b) Vol. **80**, No. 2, p. 549—558 (1977).

V. Müller, G. Schanz, E. Fischer, and E. J. Unlterhorst
*Nuclear Acoustic Resonance Absorption and Dispersion in Vanadium via Coupling to the Magnetic Dipole Moment
(b) Vol. **80**, No. 2, p. 629—639 (1977).
See also 6, p. 54; 21.1, p. 291.

A. Watterich and K. Rakasányi
Optical and ESR Investigations of Mg Centres in X-Irradiated and F-Bleached KCl Crystals at RT
(b) Vol. **80**, No. 2, p. K105—K108 (1977).
See also *10.2*, p. *86*; *20.1*, p. *252*; 22.5.2, p. 377.

T. S. Altshuler, E. G. Kharakhashyan, E. F. Kukovitskii, and M. M. Zaripov
*EPR of Dy^{3+} in Copper Single Crystals
(b) Vol. **80**, No. 2, p. K109—K111 (1977).
See also 21.1, p. 291.

N. Kishimoto, K. Morigaki, K. Murakami, A. Shimizu, and A. Hiraki
*Electron Spin Resonance in the Amorphous Si–Au System
(b) Vol. **80**, No. 2, p. K113—K115 (1977).
Erratum
(b) Vol. **82**, No. 2, p. K123 (1977).
See also *2*, p. *35*; 21.6, p. 315; 22.1.2, p. 339.

M. Friedrich and W. Karthe
*ESR Study of $PbMoO_4$:Fe^{3+}
(b) Vol. **80**, No. 2, p. K127—K130 (1977).
See also 22.8.1, p. 403.

R. L. Havill, J. Marsden, and J. M. Titman
*Nuclear Quadrupole Relaxation in Liquid Mercury
(b) Vol. **80**, No. 2, p. K141—K143 (1977).
See also *2*, p. *36*; 21, p. 279.

H. Röppischer, J. Jacobs, and B. V. Novikov
*Influence of Preparation Conditions on the Optical Band Gap and the Mn^{2+} ESR Spectrum of ZnSe Single Crystals
(a) Vol. **40**, No. 2, p. 575—581 (1977).
See also 13.4, p. 132; *13.5.1*, p. *144*; 22.4.2, p. 365.

H. Röppischer and J. Jacobs
*ESR Method for Investigating Associated Point Defects in ZnSe:Mn Single Crystals
(a) Vol. **40**, No. 2, p. K117—K120 (1977).
See also *10.2*, p. *87*; 22.4.2, p. 365.

R. Blinc, J. Slak, and M. Luzar
*Deutron NMR Study of the Role of Ammonium Ions in the Antiferroelectric Transition in $ND_4D_2AsO_4$
(a) Vol. **40**, No. 2, p. K121—K125 (1977).
See also 14.4.1, p. 195; 22.8, p. 398.

T. Gregorkiewicz and L. Rusiniak
*On the Frequency Dependence of the Classical Cyclotron Resonance Linewidth
(a) Vol. **40**, No. 2, p. K127—K129 (1977).
See also 22.1.1, p. 334.

M. Rotter, B. Sedlák, Z. Šimša, and V. A. M. Brabers
*NMR Study of Valency States in $MnFe_2O_4$
(a) Vol. **40**, No. 2, p. K169—K171 (1977).
See also 22.8.2, p. 410.

A. A. Antipin, A. V. Vinokurov, M. P. Davydova, S. L. Korableva, A. L. Stolov, and A. A. Fedii
Optical Spectra, EPR, and Spin-Lattice Relaxation of Yb^{3+} Ions in Crystals Having Perovskite-Type Structure
(b) Vol. **81**, No. 1, p. 287—293 (1977).
See also *13.4*, p. *133*; 20.3, p. 270; 22.5.4, p. 385.

K. Sugawara
Effect of Crystal Field on the Inelastic Neutron Scattering Cross-Section
(b) Vol. **81**, No. 1, p. 313—321 (1977).
See also *18.1*, p. *214*; 21.4, p. 312.

W. Karthe
*Zero-Field Splitting in Trigonal Symmetry. Cr^{3+} and Fe^{3+} in α-$LiIO_3$
(b) Vol. **81**, No. 1, p. 323—332 (1977).
See also *13.4*, p. *133*; 22.8.1, p. 404.

V. V. Eremenko, A. I. Maslennikov, and V. M. Naumenko
Host and Impurity Excitations in Antiferromagnetic and Paramagnetic $FeCO_3$ in the Far Infrared
(b) Vol. **81**, No. 2, p. 351—357 (1977).
See also *13.4*, p. *133*; 18.4, p. 232; 22.8.1, p. 404.

A. Manoogian and A. Leclerc
*An Explanation to the Temperature Dependent Curves of the Quadrupole Interaction Parameter Q' for $^{27}Al^{3+}$ in Ammonium and Potassium Alums
(b) Vol. **81**, No. 1, p. K15—K18 (1977).
See also 22.8.1, p. 404.

H. S. Maldonado, M. G. Gutierrez, C. Zárate, and G. E. Barberis
*ESR of Mn^{2+} in Lanthanum Magnesium Nitrate Hydrate
(b) Vol. **81**, No. 1, p. K27—K30 (1977).
See also 22.8, p. 398.

M. Kumeda, Y. Jinno, and T. Shimizu
*Electron Spin Resonance of Amorphous Si–Ge Alloys
(b) Vol. **81**, No. 1, p. K71—K74 (1977).
See also *2*, p. *36*; 22.1.1, p. 334; 22.1.2, p. 339.

J. M. MORET and H. BILL
*Experimental Determination of the Spin-Stress Coefficients of Gd^{3+} in the New Host Crystal Na_2S
(a) Vol. **41**, No. 1, p. 163—171 (1977).
See also 13.4, p. 134; 22, p. 324.

G. KÜHN, M. HEIN and S. WARTEWIG
*Manganese-Oxygen Impurities in GaAs
(a) Vol. **41**, No. 1, p. K13—K15 (1977).
See also *10.2*, p. *88*; 22.2.1, p. 351.

T. KIICHI, M. NOMOTO, N. NAKAMURA, and H. CHIHARA
*Phase Transitions in Tetramethylammonium Chlorate
(a) Vol. **41**, No. 1, p. K17—K19 (1977).
See also *1.2*, p. *16*; 22.9, p. 416.

W. GEHLHOFF and K. H. SEGSA
*Anisotropic Broadening of Linewidth in the EPR Spectrum of Fe^0 in Silicon
(a) Vol. **41**, No. 1, p. K21—K23 (1977).
See also 22.1.2, p. 339.

J. PIETRZAK and J. KASPRZAK
*Quadrupole Relaxation of ^{35}Cl Nuclei in Barium and Sodium Chlorates
(a) Vol. **41**, No. 1, p. K25—K26 (1977).
See also 22.8.1, p. 404.

L. BARTELSEN
*The EPR Fine Structure Spectrum of Dislocations in Silicon
(b) Vol. **81**, No. 2, p. 471—478 (1977).
See also 10.2, p. 88; 22.1.2, p. 340.

V. A. SKREBNEV
*Calculation of the Energy Flows between Subsystems by Means of the Non-Equilibrium Statistical Operator
(b) Vol. **81**, No. 2, p. 727—734 (1977).

V. T. KALINNIKOV, YU. V. RAKITIN, VAN H. CRAWFORD, and W. E. HATFIELD
Near One-Dimensional Antiferromagnetism in Vanadyl Carboxylates
(b) Vol. **81**, No. 2, p. K75—K79 (1977).
See also *18.4*, p. *233*; 22.9, p. 416.

K. KACZMARSKA and A. CHEŁKOWSKI
*The ESR Linewidth in $CdCr_2S_4$ Single Crystals near Their Curie Point
(b) Vol. **81**, No. 2, p. K95—K97 (1977).
See also *18.1*, p. *214*; 22.8.2, p. 411.

M. MATYÁŠ
*Magnetic Properties of As_xSe_{100-x} Glasses in the Temperature Range 77 to 300 K
(b) Vol. **81**, No. 2, p. K115—K117 (1977).
See also *2*, p. *37*; 22, p. 325.

I. K. AMANIS et J. G. KLIAVA
*Modèles des centres paramagnétiques formés par les atomes de cuivre et d'argent dans le quartz
(Models of Paramagnetic Centres Formed by Copper and Silver Atoms in Quartz)
(a) Vol. **41**, No. 2, p. 385—392 (1977).
See also 11, p. 95; 22.6, p. 388.

D. GOURIER, J. ANTOINE, D. VIVIEN, J. THERY, J. LIVAGE, and R. COLLONGUES
*EPR Study of Cu^{2+} Ions in the Conduction Planes of Na, Tl, and Ag β-Aluminas
(a) Vol. **41**, No. 2, p. 423—429 (1977).
See also 22.6.1, p. 393.

Y.-H. LEE, J. W. CORBETT, and K. L. BROWER
*EPR of a Carbon-Oxygen–Divacancy Complex in Irradiated Silicon
(a) Vol. **41**, No. 2, p. 637—647 (1977).
See also *10.2*, p. *88*; 11, p. 95; 22.1.2, p. 340.

B. MILSCH and W. WINDSCH
*On the Temperature Dependence of the Zero-Field Splitting Parameter of Cr^{3+} Ions in Guanidinium Aluminium Sulfate Hexahydrate
(a) Vol. **41**, No. 2, p. K141—K146 (1977).
See also 22.9, p. 416.

B. FRICK and D. SIEBERT
*ESR in Semi-Insulating GaAs:Cr and Colorimetric Determination of the Cr Content
(a) Vol. **41**, No. 2, p. K185—K187 (1977).
See also *10.2*, p. *89*; 22.2.1, p. 352.

I. V. CHEPELEVA, E. A. ZHILINSKAYA, V. N. LAZUKIN, A. P. CHERNOV, and V. I. OLKHOVSKII
*Mn^{2+} EPR Study of Cu–As–Se, As–Se–I, and Cu–As–Se–I Glass Systems
(b) Vol. **82**, No. 1, p. 189—192 (1977).
See also 2, p. 37; 22.8, p. 399.

V. D. KOROLEV, S. A. PESKOVATSKII, and V. M. SHULGA
*Spin–Lattice Relaxation in Fe^{3+}-Doped Andalusite Crystals (I): Relaxation Mechanisms
(b) Vol. **82**, No. 1, p. 251—255 (1977).
See also 13.4, p. 135; 22.6, p. 389.

S. A. MARSHALL
Comments on the Magnetic Resonance Absorption Spectrum of the F_2^- Molecule-Ion in Single Crystals of LiF, NaF, and KF
(b) Vol. **82**, No. 1, p. 311—317 (1977).
See also 22.5.2, p. 379.

E. S. GRINBERG and R. R. NIGMATULLIN
*Kinetics of Dilute Solid Paramagnets under "Bottleneck" Conditions
(b) Vol. **82**, No. 1, p. 397—402 (1977).
See also *18.1*, p. *214*.

M. v. ORTENBERG and L. G. RUBIN
*Submillimeter Cyclotron Resonance of Electrons in Stressed Silicon at High Magnetic Fields
(b) Vol. **82**, No. 1, p. K1—K4 (1977).
See also *13.1*, p. *119*; 22.1.2, p. 340.

G. BACQUET, L. YOUDRI, and L. O. ANDERSSON
*A Bond Angle and Wave Function of the CO_3^{3-}–Li^+ Centre in Irradiated Synthetic Single Crystal Calcite
(b) Vol. **82**, No. 1, p. K9—K11 (1977).
See also 22.8.1, p. 405.

H. BIALAS and LI-SHING LI
*Hypersound Excitation at 8.9 GHz by Exploiting Magnetic Anisotropy in Oriented Nickel Films
(a) Vol. **42**, No. 1, p. 125—132 (1977).
See also 1.5, p. 30; *7*, p. *62*; 21.1, p. 295.

F. MILIA, G. LAHAJNAR, M. ROŽMARIN, J. SLAK, and R. BLINC
*Proton Magnetic Resonance Study of Molecular Motion in $(NH_4)_2$ $HAsO_4$ and $(NH_4)_2HPO_4$
(a) Vol. **42**, No. 1, p. 315—318 (1977).
See also *14.4.1*, p. *196*; 22.8.1, p. 405.

M. MORENO
*A Study on the Magnetic Properties of Compressed D_{4h} Clusters
(b) Vol. **82**, No. 2, p. 669—676 (1977).
See also *13.4*, p. *136*; 22.5, p. 371.

M. VANHAELST, P. MATTHYS, and E. BOESMAN
*EPR of Iron Centres in V_2O_5 Single Crystals
(b) Vol. **82**, No. 2, p. K69—K73 (1977).
See also 22.6, p. 390.

S. K. HOFFMANN, L. S. SZCZEPANIAK, and J. STANKOWSKI
Microscopic Mechanism of the Temperature Dependence of the Spontaneous Polarization in TGS Crystals
(b) Vol. **82**, No. 2, p. K85—K87 (1977).
See also *1.2*, p. *17*; *14.4.1*, p. *196*; 22.9, p. 417.

R. C. DUVARNEY and A. K. GARRISON
An EPR-ENDOR Study of the [Li] ° Center in Crystalline and Ceramic BeO
(a) Vol. **42**, No. 2, p. 609—614 (1977).
See also *13.4*, p. *137*; 22.6, p. 390.

M. A. SIGAL
*Ferromagnetic Resonance in $BaFe_{12}O_{19}$ with Cylindrical Domain Structure at Arbitrary Magnetization
(a) Vol. **42**, No. 2, p. 775—784 (1977).
See also *18.2*, p. *222*; 22.8.2, p. 412.

V. LUPEI, A. LUPEI, and S. GEORGESCU
*EPR of U^{5+} in Alkali Fluorides with a New Charge Compensation Mechanism
(b) Vol. **83**, No. 1, p. 71—75 (1977).
See also 22.5.2, p. 380.

A. K. JAIN and G. C. UPRETI
*Low Symmetry Effects in the EPR of Mn^{2+} in $Cd(NO_3)_2 \cdot 4\ H_2O$ Single Crystals
(b) Vol. **83**, No. 1, p. K17—K20 (1977).
See also 22.8.1, p. 406.

D. BRINKMANN, G. FINGER, and H. AREND
*On the Motion of Br^- Ions in $K_2Pt(CN)_4Br_{0.3} \cdot 3.2\ H_2O$ (KCP)
(b) Vol. **83**, No. 1, p. K45—K48 (1977).
See also *10.2*, p. *90*; 22.8, p. 399.

G. N. NEILO, A. D. PROKHOROV, and G. A. TSINTSADZE
*Electron Spin–Lattice Relaxation of Rare-Earth Ions in Lanthanium Magnesium Double Nitrate under Pressure
(b) Vol. **83**, No. 1, p. K59—K61 (1977).
See also 12.2, p. 105; 22.8.1, p. 406.

J. KURIATA, N. GUSKOS, and T. REWAJ
*Temperature Dependence of the Second-Order Spin–Lattice Coefficients for Eu^{2+} in CaF_2 Single Crystals
(b) Vol. **83**, No. 1, p. K79—K82 (1977).
See also *6*, p. *59*; 22.5.3, p. 384.

E. ILDSTAD, I. SVARE, and T. A. FJELDLY
*Diffusion in Lanthanum Fluoride Studied with ^{19}F Relaxation
(a) Vol. **43**, No. 1, p. K65—K68 (1977).
See also *9*, p. *70*; 22.5, p. 371.

M. A. B. WHITAKER
*Modes of Cross-Relaxation in Nuclear Double-Resonance
(b) Vol. **83**, No. 2, p. 573—584 (1977).
See also 22.5.2, p. 380.

H. NAKATA and H. MORISHIMA
*An ESR Study of Annealing Characteristics for Heavily Irradiated Silicon
(a) Vol. **43**, No. 2, p. 417—423 (1977).
See also 11, p. 98; 22.1.2, p. 342.

H. W. den Hartog and H. Benning
Paramagnetic Color Centers in Calciumchlorapatite Doped with F^- Ions
(a) Vol. **43**, No. 2, p. 633—638 (1977).
See also *10.2*, p. *91*; 22.8, p. 400.

W. Bizzaro, J. Rosenthal, N. F. Berk, and L. Yarmus
*The EPR Linewidth of Triplet Excitons in Single-Crystal Anthracene: a Re-Examination
(b) Vol. **84**, No. 1, p. 27—32 (1977).
See also *13.5.1*, p. *147*; 22.9, p. 419.

A. Hofstaetter, A. Scharmann, and B. Vitt
*Antiferromagnetic Properties of the $Li_2Mn_xTi_{1-x}O_3$ Mixed Crystal System. An EPR Investigation
(b) Vol. **84**, No. 1, p. 93—98 (1977).
See also 18.4, p. 234; 22.8.1, p. 407.

M. A. B. Whitaker
*Calculation of Nuclear Double-Resonance Relaxation Rate
(b) Vol. **84**, No. 1, p. 149—158 (1977).
See also 22.5.2, p. 381.

R. Karcher, U. Köbler, K. Lüders, and Z. Scücs
*Temperature Dependence of Knight Shift and Susceptibility of Single Crystal Molybdenum
(b) Vol. **84**, No. 1, p. 189—193 (1977).
See also *18*, p. *212*; 21, p. 286.

E. Schwarzer and H. Haken
Theory of the Influence of Exciton Motion on Proton Relaxation in Organic Solids
(b) Vol. **84**, No. 1, p. 253—267 (1977).
See also *13.5.1*, p. *148*; 22.9, p. 419.

S. Limmer, S. Grande, and A. Lösche
NMR Investigations of the Melting Behaviour of Mesogen Compounds
(b) Vol. **84**, No. 1, p. 283—292 (1977).
See also *1.2*, p. *19*; 22.9, p. 419.

M. Giordano, M. Martinelli, L. Pardi, and S. Santucci
*Electron Spin–Lattice Relaxation Measurements in a Two-Dimensional Crystal
(b) Vol. **84**, No. 1, p. 305—310 (1977).
See also 18.1, p. 215; 22.8, p. 400.

A. Manoogian
*Interpretation of the Temperature Dependent Quadrupole Coupling Constant Q' for $^{27}Al^{3+}$ and $^{71}Ga^{3+}$ in Guanidinium Hexahydrate Crystals
(b) Vol. **84**, No. 1, p. K7—K12 (1977).
See also 22.8.1, p. 407.

R. Garg, A. C. Oliveira, and V. K. Garg
*Quadrupolar Hyperfine Anisotropy in Potassium Dithioferrate
(b) Vol. **84**, No. 1, p. K17—K19 (1977).
See also 22.8.1, p. 407.

M. L. Falin, V. P. Meiklyar, and V. A. Ulanov
*Transferred Hyperfine Interaction of Yb^{3+} in $KMgF_3$
(b) Vol. **84**, No. 1, p. K29—K32 (1977).
See also 22.5, p. 372.

H. R. Khan, K. Lüders, Ch. J. Raub, and Z. Szücs
*Knight Shift of Pt^{195} in A15- and σ-Phases of Nb–Pt
(b) Vol. **84**, No. 1, p. K33—K35 (1977).
See also *14.2*, p. *163*; 21, p. 286; 21.6, p. 317.

E. J. Unterhorst, V. Müller, and G. Schanz
*Nuclear Acoustic Resonance in Liquid Gallium
(b) Vol. **84**, No. 1, p. K53—K56 (1977).
See also 2, p. 40; 21, p. 286.

D. Kropman, A. Sügis, and M. Vinnal
*EPR Investigation of Si–SiO_2 Bonding Layer
(a) Vol. **44**, No. 1, p. K1—K3 (1977).
See also 10.2, p. 92; 22.1.2, p. 343.

V. D. Korolev, S. A. Peskovatskii, and V. M. Shulga
*Spin–Lattice Relaxation in Fe^{3+}-Doped Andalusite Crystals (II): The Effect of Lattice Oscillator Heating
(b) Vol. **84**, No. 2, p. 443—447 (1977).
See also 13.4, p. 139; 22.6, p. 391.

V. I. Lesin, V. P. Sakun, A. I. Pristupa, and E. L. Frankevich
*Reaction Yield Detected Magnetic Resonance Spectra of Intermediate Pairs of Triplet Excitons. Evolution of Spectra during Degeneracy of Energy Levels
(b) Vol. **84**, No. 2, p. 513—520 (1977).
See also *13.5.1*, p. *148*; 22.9, p. 419.

K. Sugawara
*Singly, Doubly, and Multiply Bottlenecked EPR Based on the Linear Response Theory
(b) Vol. **84**, No. 2, p. 709—719 (1977).

V. I. Muravev and N. B. Yunusov
Covalency and Hyperfine Interaction on the Impurity Ion in Crystals
(b) Vol. **84**, No. 2, p. 791—796 (1977).
See also *13.4*, p. *140*; 18.1, p. 216; 21.1, p. 300.

M. V. Eremin and N. I. Silkin
*Theory of the ESR Spectra of Tl^{2+} Ions in K_2SO_4 Single Crystals
(b) Vol. **84**, No. 2, p. 803—811 (1977).
See also 13.4, p. 140; 22.8.1, p. 408.

C. V. Manjunath and R. Srinivasan
*Electron Spin Resonance Study of Copper(II) Doped Ferroelectric Ammonium Sulphate
(a) Vol. **44**, No. 2, p. 687—692 (1977).
See also 14.4.1, p. 197; 22.8.1, p. 408.

C. Korn and D. Teitel
Influence of Vanadium on Hydrogen Diffusion and Phase Constitution in Ti–H
(a) Vol. **44**, No. 2, p. 755—766 (1977).
See also *1.3*, p. *23*; 21.1, p. 300.

T. I. Chekmasova, I. S. Kabanov, and V. I. Yuzvak
The Pressure–Temperature Phase Diagram of $LiNH_4SO_4$
(a) Vol. **44**, No. 2, p. K155—K157 (1977).
See also *1.2*, p. *20*; *14.4.1*, p. *197*; 22.8.1, p. 408.

W.-A. Martin and G. Hartung
*Anisotropic Photoinduced ESR Signals from the Si–SiO_2 Interface on Silicon Single Crystals
(a) Vol. **44**, No. 2, p. K159—K161 (1977).
See also *16*, p. *205*; 22:1.2, p. 344.

R. Durný and D. Barančok
*EPR Investigation of Ag-Doped Ge–S Glasses
(a) Vol. **44**, No. 2, p. K177—K179 (1977).
See also *2*, p. *41*; 22, p. 331.

20. Optical Properties

Original Papers and Short Notes

V. N. Lugovoi
*Electromagnetic Waves in an Electron Plasma in an Alternating External Magnetic Field
(b) Vol. **79**, No. 1, p. 131—138 (1977).
See also *14.3*, p. *165*.

P. O. Nilsson, I. Curelaru, and T. Jarlborg
XPS and APS Studies of V_3Si
(b) Vol. **79**, No. 1, p. 277—281 (1977).
See also *13.1*, p. *114*; 21.1, p. 288.

V. I. Grebennikov, Yu. A. Babanov, and O. B. Sokolov
Extra-Atomic Relaxation and X-Ray Spectra of Narrow-Band Metals (I): Formalism
(b) Vol. **79**, No. 2, p. 423—432 (1977).
See also *13*, p. *107*; 21, p. 277.

B. A. Aronzon and E. Z. Meilikhov
*Helicon Damping in Semiconductors in the Quantum Limit
(b) Vol. **79**, No. 2, p. 753—762 (1977).
See also *14.3*, p. *166*; 18, p. 209; 22.2.3, p. 356.

D. C. Creagh
*Determination of the Mass Attenuation Coefficients and Anomalous Dispersion Corrections for Calcium for X-Ray Wavelengths from $IK\alpha_1$ to $CuK\alpha_1$
(a) Vol. **39**, No 2, p. 705—715 (1977).
See also 21.3, p. 310.

V. I. Grebennikov, Yu. A. Babanov, and O. B. Sokolov
Extra-Atomic Relaxation and X-Ray Spectra of Narrow-Band Metals (II): Results
(b) Vol. **80**, No. 1, p. 73—82 (1977).
See also *13*, p. *108*; 21, p. 278.

V. P. Vijayavargiya, S. N. Gupta, and B. D. Padalia
*Extended X-Ray Absorption Fine Structure (EXAFS) Studies of the Lanthanides
(b) Vol. **80**, No. 1, p. 83—88 (1977).
See also 21.4, p. 311.

M. Guillot and M. Le Gall
*Superexchange Field and Intense Magnetic Field Dependences of the Faraday Rotation Induced by the Electric Dipole Transitions in YIG
(b) Vol. **80**, No. 1, p. 279—287 (1977).
See also 13, p. 108; *18.3*, p. *228*; 22.8.2, p. 410.

J. Bergmann, K. Kneipp, and H. E. Ponath
*Raman-Induced Kerr Effect in $LiIO_3$ Single Crystals
(b) Vol. **80**, No. 1, p. K55—K58 (1977).
See also *6*, p. *54*; 22.8.1, p. 403.

W. Lochmann
Phonon-Assisted Auger Recombination in Semiconductors
(a) Vol. **40**, No. 1, p. 285—292 (1977).
See also *6*, p. *54*; *13.1*, p. *116*; 22.2.1, p. 351.

H. Neddermeyer
Charge Transfer in Aluminium–Magnesium Alloys
(b) Vol. **80**, No. 2, p. 611—614 (1977).
See also *13*, p. *108*; 21, p. 279; 21.3, p. 310.

F. Werfel, G. Dräger, O. Brümmer, and M. Jurisch
XS and XPS Investigation of the V_3Si Electron Structure
(b) Vol. **80**, No. 2, p. K95—K99 (1977).
See also *13.1*, p. *117*; 17, p. 206; 22.7, p. 394.

L. K. Samanta and G. C. Bhar
*Optical Nonlinearity of Some Stannite and Famatinite Crystals
(a) Vol. **41**, No. 1, p. 331—337 (1977).
See also 22.8, p. 398.

W. Schwander and K. Ulmer
*Extension of Crystal Spectroscopy to Small Quantum Energies: The Ni $M_{2,3}$ Spectrum
(a) Vol. **41**, No. 2, p. K167—K170 (1977).
See also 21.1, p. 294.

N. Nhu Dat and N. van Hieu
*On the High-Order Harmonic Generation by Free Charge Carriers in Solids
(b) Vol. **82**, No. 1, p. K39—K40 (1977).
See also *13*, p. *110*.

W. Lochmann
*Scattering Mechanisms in Phonon-Assisted Auger Recombination
(a) Vol. **42**, No. 1, p. 181—185 (1977).
See also *6*, p. *57*; 22.2.1, p. 352.

H. D. Riccius, D. S. Smith, B. G. Withford, and L. H. Jones
*Difference-Frequency Mixing and Second-Harmonic Generation of 3.39 μm Laser Radiation in $LiNbO_3$
(a) Vol. **42**, No. 1, p. K53—K55 (1977).
See also *14.4*, p. *191*; 22.8.1, p. 406.

E. R. Vance and N. B. Manson
*Magnetic Circular Dichroism of Neutron-Irradiated MgO and CaO
(b) Vol. **82**, No. 2, p. 473—479 (1977).
See also 13.4, p. 136; *18*, p. *211*; 22.6, p. 390.

W. Niederlag, W. Blau, and K. Kleinstück
*X-Ray Spectroscopical Investigation of the Electronic Structure of Fe–Cr
(b) Vol. **82**, No. 2, p. 495—499 (1977).
See also 13.1, p. 120; 21.1, p. 296; 21.1.1, p. 304.

M. Taut and A. M. Radwan
*Soft X-Ray Spectra of Al and the Laves Phase $CaAl_2$
(b) Vol. **82**, No. 2, p. 507—511 (1977).
See also *13.1*, p. *120*; 21, p. 283; 22, p. 326.

S. N. Gupta, V. P. Vijayavargiya, B. D. Padalia, B. C. Tripathi, and M. N. Ghatikar
*Extended X-Ray Absorption Fine Structures and Chemical Shifts for the Heavy Rare-Earth Oxides
(b) Vol. **82**, No. 2, p. 603—609 (1977).
See also 22.6, p. 390.

H.-D. Hattendorff
*Transition Radiation from Gold and Silver Films Bombarded with 80 keV Electrons
(a) Vol. **42**, No. 2, p. 489—494 (1977).
See also *11*, p. *96*; 21.6, p. 316.

F. Werfel, G. Dräger, and O. Brümmer
Electronic Structure Investigation of Metallic Vanadium by XS and XPS
(b) Vol. **83**, No. 1, p. 257—262 (1977).
See also *13.1*, p. *121*; 17, p. 207; 21.1, p. 297.

O. B. Sokolov, V. I. Grebennikov, and E. A. Turov
*A New Approach to the Theory of X-Ray Spectra of Metals with Allowance for the Finite Lifetime of the Deep Hole
(b) Vol. 83, No. 1, p. 281—291 (1977).
See also *13*, p. *110*; 21, p. 284.

O. B. Sokolov, V. I. Grebennikov, and E. A. Turov
Resonance Scattering of X-Rays in Metals
(b) Vol. 83, No. 2, p. 383—393 (1977).
See also *13*, p. *111*; 21, p. 285.

Vo Hong Anh and Nguyen Van Trong
On the Theory of Parametric Excitation in Quantum Electron–Phonon Systems
(b) Vol. 83, No. 2, p. 395—404 (1977).
See also *6*, p. *59*; *13.5.2*, p. *150*; 22, p. 328.

F. G. dos Reis, Hilda A. Cerdeira, and R. Luzzi
*Dielectric Response of Saturated Semiconductors
(b) Vol. 83, No. 2, p. 607—614 (1977).
See also 13.1, p. 122; *14.4*, p. *192*; 22, p. 329.

Yu. G. Borodko, M. G. Kaplunov, T. M. Moravskaya, and K. I. Pokhodnya
*X-Ray Photoelectron Spectra of Single Crystals of Tetracyanoquinodimethane and Its Salts
(b) Vol. 83, No. 2, p. K141—K144 (1977).
See also 13, p. 111; 17, p. 207; 22.9, p. 418.

Vo Hong Anh and Nguyen Nhu Dat
Parametric Action of a Strong Radiation Field on Anisotropic Semiconductors
(b) Vol. 84, No. 1, p. 115—123 (1977).
See also *6*, p. *60*; *13.5*, p. *142*; 22.4.1, p. 364.

E. Tegeler, N. Kosuch, G. Wiech, and A. Faessler
*Anisotropic Emission of the X-Ray K-Emission Band of Nitrogen in Hexagonal Boron Nitride
(b) Vol. 84, No. 2, p. 561—567 (1977).
See also 22.2, p. 349.

V. I. Grebennikov, O. B. Sokolov, and E. A. Turov
Response of Metal Electrons to Local Fields of Holes Arising in X-Ray Processes
(b) Vol. 84, No. 2, p. 773—784 (1977).
See also *13*, p. *113*; 21, p. 287.

K. Jarašiūnas and J. Vaitkus
Investigation of Non-Equilibrium Processes in Semiconductors by the Method of Transient Holograms
(a) Vol. 44, No. 2, p. 793—800 (1977).
See also *14.3.4*, p. *188*; 22.1.2, p. 344; 22.4.2, p. 366.

20.1 Spectra. Optical Constants
(X-Ray Spectra See 20)

Review Articles

YU. V. KNYAZEV and M. M. NOSKOV
*The Optical Properties of Rare Earth Metals
(b) Vol. **80**, No. 1, p. 11—29 (1977).
See also 13.1, p. 114; 21.4, p. 311.

K. W. BÖER
The CdS/Cu_2S Solar Cell (I): Minority Carrier Generation and Transport in the Cu_2S Emitter
(a) Vol. **40**, No. 2, p. 355—384 (1977).
See also 14.3.3, p. 178; 14.3.4, p. 183; *16*, p. *201*; 22, p. 320; 22.4.1, p. 362.

A. GOLDMANN
*Band Structure and Optical Properties of Tetrahedrally Coordinated Cu- and Ag-Halides
(b) Vol. **81**, No. 1, p. 9—47 (1977).
See also *13.1*, p. *114*; 13.5.1, p. 143; 22.5, p. 370; 22.5.1, p. 373.

G. BORSTEL and H. J. FALGE
Surface Phonon-Polaritons at Semi-Infinite Crystals
(b) Vol. **83**, No. 1, p. 11—45 (1977).
See also *6*, p. *51*; *13.5*, p. *141*; 22, p. 320; 22.6, p. 386.

V. T. AGEKYAN
Spectroscopic Properties of Semiconductor Crystals with Direct Forbidden Energy Gap
(a) Vol. **43**, No. 1, p. 11—42 (1977).
See also 6, p. 51; *13.5.1*, p. *143*; 20.3, p. 267; 22.6, p. 386.

Original Papers and Short Notes

P. PETELENZ
Mixing of Frenkel Excitons and Ionic Excited States of a Linear Molecular Crystal with Two Molecules in the Unit Cell (II): Physical Consequences
(b) Vol. **79**, No. 1, p. 61—70 (1977).
See also *13.5.1*, p. *143*; 22, p. 320.

D. THEIS
*Wavelength-Modulated Reflectivity Spectra of ZnSe and ZnS from 2.5 to 8 eV
(b) Vol. **79**, No. 1, p. 125—130 (1977).
See also *13.1*, p. *114*; 13.5.1, p. 143; 22.4.1, p. 362; 22.4.2, p. 365.

G. B. SCOTT and J. L. PAGE
*The Absorption Spectra of $Y_3Fe_5O_{12}$ and $Y_3Ga_5O_{12}:Fe^{3+}$ to 5.5 eV
(b) Vol. **79**, No. 1, p. 203—213 (1977).
See also 13.4, p. 129; 22.8.2, p. 409.

I. POLLINI
*Ultraviolet Reflectivity Spectra of $CoCl_2$ and $CoBr_2$ Crystals
(b) Vol. **79**, No. 1, p. 231—236 (1977).
See also *13.1*, p. *114*; 22.5, p. 370.

G. S. AGARWAL and H. D. VOLLMER
Surface Polariton Effects in Spontaneous Emission
(b) Vol. **79**, No. 1, p. 249—257 (1977).
See also *13.3*, p. *126*; 13.5.2, p. 149; 21, p. 276; 22.2.2, p. 354.

S. ONARI, M. CARDONA, E. SCHÖNHERR, and W. STETTER
*Effect of Stress on the Kaman Spectra of Mg_2Si and Mg_2Sn
(b) Vol. **79**, No. 1, p. 269—276 (1977).
See also *6*, p. *51*; 12.2, p. 104; 22, p. 320.

E. Majerníková
*On the Theory of Exciton Absorption in Amorphous Semiconductors
(b) Vol. **79**, No. 1, p. 339—345 (1977).
See also 2, p. **33**; *13.5.1*, p. *143*; 22, p. 320.

E. V. Chisler, V. Yu. Davydov, I. N. Goncharuk, and E. A. Ivanova
*The "Destructive" Fermi Resonance
(b) Vol. **79**, No. 1, p. 347—357 (1977).
See also *6*, p. *51*; 22, p. 320.

T. Iida and R. Monnier
*Erratum to: Static Properties of the Self-Traped Hole in Simple Cubic Cesium Halides
(b) Vol. **79**, No. 1, p. 391 (1977).
See also 6, p. 52; *13.4*, p. *129*; 22.5.2, p. 375.

P. G. Baranov, Yu. P. Veshchunov, R. A. Zhitnikov, and N. G. Romanov
EPR, Optical Absorption, and Luminescence of Anisotropic Copper Centres in KCl
(b) Vol. **79**, No. 1, p. K27—K30 (1977).
See also *19*, p. *235*; 20.3, p. 267; 22.5.2, p. 375.

Č. Koňák, K. Zimmermann, and B. Březina
*Circular Dichroism Spectra and Optical Activity of $LiIO_3$ Crystals
(b) Vol. **79**, No. 1, p. K35—K38 (1977).
See also 22.8.1, p. 402.

J. Szczyrbowski
*The Shape of Absorption Edge of Amorphous Ge, Si, InSb, GaSb, GaAs, and GaP
(b) Vol. **79**, No. 1, p. K43—K48 (1977).
See also *2*, p. *33*; 22.1, p. 332; 22.2, p. 348.

F. Henneberger, K. Henneberger, and J. Voigt
*On the Emission Lineshape of Excitonic Molecules in Wurtzite-Type Semiconductors
(b) Vol. **79**, No. 1, p. K81—K85 (1977).
See also *13.5.1*, p. *143*; 22.4.1, p. 362.

Y. Khan, K. Nahm, M. Rosenberg, and H. Willner
Superconductivity and Semiconductor–Metal Phase Transition in the System $BaPb_{1-x}Bi_xO_3$
(a) Vol. **39**, No. 1, p. 79—88 (1977).
See also 1, p. 9; 13.1, p. 114; *14.2*, p. *161*; 22.6.1, p. 393.

J. Bauer, L. Biste, and D. Bolze
Optical Properties of Aluminium Nitride Prepared by Chemical and Plasmachemical Vapour Deposition
(a) Vol. **39**, No. 1, p. 173—181 (1977).
See also *1.5*, p. *29*; 13.1, p. 114; 22.2, p. 348.

W. Barth, K. Elsaesser, W. Güth, and E. Kamieniecki
Photoconductivity in Correlation to Absorption in Plastically Deformed Germanium
(a) Vol. **39**, No. 1, p. 249—252 (1977).
See also *13.4*, p. *130*; 16, p. 201; 22.1.1, p. 333.

S. I. Subbotin, V. V. Panfilov, and R. T. Molchanova
*Pressure-Induced Polytypic Transformations and Edge Absorption of GaSe
(a) Vol. **39**, No. 1, p. 357—362 (1977).
See also *1.2*, p. *14*; 12.2, p. 104; 13.5.1, p. 143; 22.3, p. 358.

H. Happ and G. Rother
Investigation on the Far-Infrared Response of DKDP by Asymmetric Fourier Spectroscopy
(b) Vol. **79**, No. 2, p. 473—477 (1977).
See also 6, p. 52; *14.4.1*, p. *194*; 22.8.1, p. 402.

E. Mohler and B. Thomas
*Electro-Optic Properties of CuCl below the Absorption Edge
(b) Vol. **79**, No. 2, p. 509—517 (1977).
See also *14.4*, p. *189*; 22.5, p. 370.

G. Bastard, C. Rigaux, and A. Mycielski
Giant Spin Splitting Induced by Exchange Interactions in $Hg_{1-k}Mn_kTe$ Mixed Crystals
(b) Vol. **79**, No. 2, p. 585—593 (1977).
See also *13.1*, p. *115*; 18, p. 209; 22.4.4, p. 368.

V. M. Fomin and E. P. Pokatilov
*Multiphoton Absorption by Charge Carriers in Semiconductors in a Magnetic Field
(b) Vol. **79**, No. 2, p. 595—604 (1977).
See also *14.3.4*, p. *183*; 22, p. 321.

B. N. Ganguly and M. Nicol
Effect of Hydrostatic Pressure on the Vibrational Properties and the Structure of $SrWO_4$ and $PbMoO_4$
(b) Vol. **79**, No. 2, p. 617—622 (1977).
See also 1.2, p. 14; *6*, p. *52*; 12.2, p. 104; 22.8.1, p. 402.

W. Szuskiewicz
*Optical Absorption in HgTe
(b) Vol. **79**, No. 2, p. 691—698 (1977).
See also 13.1, p. 115; 22.4.3, p. 367.

A. Scalabrin, A. S. Chaves, D. S. Shim, and S. P. S. Porto
Temperature Dependence of the A_1 and E Optical Phonons in $BaTiO_3$
(b) Vol. **79**, No. 2, p. 731—742 (1977).
See also *6*, p. *52*; 14.4, p. 189; 22.8.1, p. 402.

J. L. Farvacque, J. C. Doukhan, U. von Alpen, and E. Gmelin
Screw-Dislocation-Induced Scattering Processes and Acceptor States in Te
(b) Vol. **79**, No. 2, p. 763—773 (1977).
See also *13.4*, p. *131*; *14.3*, p. *166*; 22.1.3, p. 345.

M. V. Eremin and A. A. A. Kornienko
The Superposition Model in Crystal Field Theory
(b) Vol. **79**, No. 2, p. 775—785 (1977).
See also 4, p. 44; *13.4*, p. *131*; 20.1, p. 383.

E. Loh, G. Samoggia, E. Reguzzoni, and L. Nosenzo
*Thermally Modulated Ultraviolet Absorption Spectra of $CaF_2:Ce^{3+}$ Crystals
(b) Vol. **79**, No. 2, p. 795—799 (1977).
Addendum
(b) Vol. **81**, No. 1, p. 411 (1977).
See also *13.4*, p. *131*; 22.5.3, p. 383.

I. N. Douglas
*Magnetic Circular Dichroism of Co^{2+} Ions in CdF_2 and CaF_2
(b) Vol. **79**, No. 2, p. K99—K101 (1977).
See also 13.4, p. 131; *18*, p. *209*; 22.5, p. 370; 22.5.3, p. 383.

G. A. Gledhill and J. F. Angress
*Temperature Dependence of the Infrared Reflection Spectra of $RbCl_yBr_{1-y}$ Alloys
(b) Vol. **79**, No. 2, p. K107—K110 (1977).
See also *6*, p. *52*; 22.5.4, p. 385.

J. Bauer
Optical Properties, Band Gap, and Surface Roughness of Si_3N_4
(a) Vol. **39**, No. 2, p. 411—418 (1977).
See also *2*, p. *34*; 13.1, p. 115; 22, p. 322.

V. A. STUKAN, S. F. MARENKIN, V. I. TRIFONOV, and V. YA. SHEVCHENKO
*Opto-Electrical Properties of p-CdSb in the Submillimetre Wave Region
(a) Vol. **39**, No. 2, p. 445–452 (1977).
See also *16*, p. *201*; 22.3, p. 358.

W. BOLLMANN and H.-J. STÖHR
Incorporation and Mobility of OH^- Ions in $LiNbO_3$ Crystals
(a) Vol. **39**, No. 2, p. 477–484 (1977).
See also *10.2*, p. *84*; 14.3, p. 166; 22.8.1, p. 402.

J. PRAKASH and F. FISCHER
Ionic Thermocurrent and Optical Absorption Studies of S^{--}–Vacancy Dipoles in KI
(a) Vol. **39**, No. 2, p. 499–507 (1977).
See also *10.2*, p. *84*; *14.4*, p. *189*; 22.5.2, p. 375.

G. HOROWITZ
*Nondestructive Determination of Free Carrier Parameters of Implanted Gallium Arsenide by Infrared Reflectivity Measurements
(a) Vol. **39**, No. 2, p. 533–540 (1977).
See also 22.2.1, p. 350.

D. BLOOR and F. H. PRESTON
*Optical Propreties of Bis (p-Toluene Sulphonate) Diacetylene (II): Low-Temperature Spectra
(a) Vol. **39**, No. 2, p. 607–614 (1977).
See also 1.1, p. 11; 22.9, p. 414.

K. R. ALLAKHVERDIEV and M. M. TAGYEV
*Infrared Investigation of the GaS_xSe_{1-x} System
(a) Vol. **39**, No. 2, p. K111–K113 (1977).
See also 1.2, p. 15; *6*, p. *53*; 22.3.1, p. 360.

E. L. SILL, J. J. MARTIN, and Z. AL SHAIEB
*Optical Determination of the Eu Concentration in Eu Doped KCl Crystals
(a) Vol. **39**, No. 2, p. K131–K133 (1977).
See also 22.5.2, p. 376.

W. MÜLLER-LIERHEIM, T. SUSKI, and H. H. OTTO
Factor Group Analysis of the Raman Spectrum of $Pb_5Ge_3O_{11}$
(b) Vol. **80**, No. 1, p. 31–41 (1977).
See also *6*, p. *53*; 13.5.3, p. 151; 14.4.1, p. 194; 22.8.1, p. 403.

A. A. ADBUMALIKOV and A. A. KLOCHIKHIN
*The Double Resonance in Two-Phonon Raman Scattering
(b) Vol. **80**, No. 1, p. 43–50 (1977).
See also *6*, p. *53*; 13.1, p. 115.

W. EKARDT and M. I. SHEBOUL
LO-Phonon-Assisted Two-Photon Absorption of the First Excited State of the Biexciton in CuCl
(b) Vol. **80**, No. 1, p. 51–56 (1977).
See also *13.5.1*, p. *144*; 20.3, p. 268; 22.5, p. 370.

E. SCHULLER, H. J. FALGE, and G. BORSTEL
*Dispersion of Bulk Phonon Polaritons inside Reststrahlen Bands
(b) Vol. **80**, No. 1, p. 109–117 (1977).
See also *13.5*, p. *141*; 22.6, p. 386.

H. P. GESERICH, W. MÖLLER, G. SCHEIBER, and L. PINTSCHOVIUS
*Optical Investigation of the Electrical Anisotropy of $(SN)_x$ Single Crystals
(b) Vol. **80**, No. 1, p. 119–123 (1977).
See also 1.4, p. 24; *14.1*, p. *157*; 22, p. 322.

A. TREINDL and K. H. GERMANN
*Linear Optical Anisotropy Investigations of the Antiferromagnetic Phase Transition in MnO
(b) Vol. **80**, No. 1, p. 159—166 (1977).
See also *18.4*, p. *231*; 22.6, p. 387.

A. R. VASCONCELLOS and R. LUZZI
Light Scattering and Spin Diffusion in Doped Semiconductors
(b) Vol. **80**, No. 1, p. 181—191 (1977).
See also *13.5*, p. *141*; 18.1, p. 213; 22.2.1, p. 351.

M. RUEFF and E. SIGMUND
*E–e Jahn-Teller System: Optical Response Studies
(b) Vol. **80**, No. 1, p. 215—223 (1977).
See also *13.4*, p. *131*.

R. ENDERLEIN and F. BECHSTEDT
*Theory of Allowed Resonance Raman Scattering of First and Second Order in a Magnetic Field
(b) Vol. **80**, No. 1, p. 225—234 (1977).
See also *6*, p. *53*; 13.1, p. 115; 13.5.1, p. 144.

H. J. STOLZ, E. PETRI, and A. OTTO
Optical Properties and Plasmon Dispersion in Epitaxial $(SN)_x$ by Electron Energy Loss Spectroscopy
(b) Vol. **80**, No. 1, p. 289—294 (1977).
See also *13.5.2*, p. *149*; 21, p. 278.

W. GĘBICKI and W. NAZAREWICZ
*Long-Wavelength Optical Phonons in $Mn_xHg_{1-x}Te$ Mixed Crystals
(b) Vol. **80**, No. 1, p. 307—311 (1977).
See also *6*, p. *53*; 22.8, p. 397.

H. J. VAN DER VALK and C. HAAS
*Anisotropy of Lattice Vibrations of Layered Compounds
(b) Vol. **80**, No. 1, p. 321—332 (1977).
See also *6*, p. *53*; 22, p. 322; 22.5, p. 370.

G. S. ZAVT
Dynamical Properties of NO_2^- Impurity Molecules in Alkali Halides (I): Theory
(b) Vol. **80**, No. 1, p. 399—408 (1977).
See also *10.2*, p. *85*; 22.5.2, p. 376.

V. K. BASHENOV, KH. S. BAGDASAROV, and V. V. TIMOFEENKO
Universal Valence Bands for Fluorite-Type Crystals
(b) Vol. **80**, No. 1, p. K17—K20 (1977).
See also *13.1*, p. *116*; 22.5.3, p. 383.

W. HÖRIG, H. NEUMANN, H.-J. HÖBLER, and G. KÜHN
*Temperature Dependence of the Absorption Edge in $CuInSe_2$
(b) Vol. **80**, No. 1, p. K21—K23 (1977).
See also 13.1, p. 116; 22.8, p. 397.

H. KÖHLER
Investigation of the Magnetoresistance of n-Type Bi_2Te_3 under Uniaxial Stress
(b) Vol. **80**, No. 1, p. K29—K32 (1977).
See also 13.1, p. 116; *14.3*, p. *166*; 22.7, p. 394.

K. HÜBNER
Chemical Bond and Related Properties of SiO_2 (I): Character of the Chemical Bond
(a) Vol. **40**, No. 1, p. 133—140 (1977).
See also 2, p. 35; *4*, p. *44*; *13.1*, p. *116*; 14.4, p. 189; 22.6, p. 387.

Hj. Bernhardt
*Near-Edge Absorption of Lead Molybdate Crystals
(a) Vol. **40**, No. 1, p. 257–264 (1977).
See also 16, p. 201; *20.3*, p. *269*; 22.8.1, p. 403.

D. Bloor, D. N. Batchelder, and F. H. Preston
*Optical Properties of Bis (p-Toluene Sulphonate) Diacetylene (III): Defect-Induced Spectra
(a) Vol. **40**, No. 1, p. 279–284 (1977).
See also 10.2, p. 85; 20.3, p. 269; 22.9, p. 415.

M. V. Belousov, D. E. Pogarev, and A. A. Shultin
Vibrations of Isotopic Impurities in Crystals Built up by Complex Ions
(b) Vol. **80**, No. 2, p. 417–424 (1977).
See also *6*, p. *54*; 22.8.1, p. 403.

K. Bärner, R. Braunstein, and E. Chock
*The Optical Properties of MnAs and $MnAs_{0.85}P_{0.15}$
(b) Vol. **80**, No. 2, p. 451–460 (1977).
See also 13.1, p. 116; *14.4*, p. *190*; 22.8, p. 397.

C. J. Delbecq, J. P. Stokes, and P. H. Yuster
Thermal Reorientation Kinetics of I_2^- in RbI
(b) Vol. **80**, No. 2, p. 471–475 (1977).
See also *10.2*, p. *86*; 22.5.2, p. 377.

A. N. Georgobiani, Yu. V. Ozerov, E. E. Strumban, and V. S. Vinogradov
Influence of Random Electric Fields on the Polarization of Deep Centre Electroabsorption
(b) Vol. **80**, No. 2, p. 589–596 (1977).
See also *13.4*, p. *132*; 16, p. 202; 20.3, p. 269; 22.4.1, p. 362.

F. P. Koffyberg
*Optical and Modulated Absorption Spectra in the Band Gap Region of Amorphous Vanadium Zinc Borate Semiconductors
(b) Vol. **80**, No. 2, p. 669–673 (1977).
See also 2, p. 35; 13.1, p. 116; 22.8, p. 397.

T. Nasu
*The Temperature Dependence of the Absorption Spectrum of NiO
(b) Vol. **80**, No. 2, p. 675–679 (1977).
See also 13.5.4, p. 153; 22.6, p. 387.

Yu. E. Perlin, B. S. Tsukerblat, and T. Singh Dod
*Moments of Electroabsorption of Jahn-Teller Impurity Centres
(b) Vol. **80**, No. 2, p. 703–707 (1977).
See also 13.4, p. 132; *14.3.4*, p. *184*; 22, p. 323.

A. Watterich and K. Raksányi
*Optical and ESR Investigations of Mg Centres in X-Irradiated and F-Bleached KCl Crystals at RT
(b) Vol. **80**, No. 2, p. K105–K108 (1977).
See also *10.2*, p. *86*; 19, p. 237; 22.5.2, p. 377.

A. Murasik, J. Leciejiwicz, and Z. Żołnierek
*An Observation of the Crystal-Field Transition in UCl_4 by Neutron Spectroscopy
(b) Vol. **80**, No. 2, p. K137–K140 (1977).
See also *13*, p. *108*; 22.5, p. 370.

K. Hübner
Chemical Bond and Related Properties of SiO_2 (II): Structural Trends
(a) Vol. **40**, No. 2, p. 487–495 (1977).
See also 2, p. 36; *4*, p. *45*; *13.1*, p. *117*; 14.4, p. 190; 22.6, p. 387.

B. Prévot, J. Biellmann, M. F. Meftah, and M. Sieskind
*Infra-Red Reflectivity of Non-Stoichiometric Ferrous Oxide
(a) Vol. **40**, No. 2, p. 503—510 (1977).
See also 22.6, p. 387.

G. Chassagne, D. Durand, J. Serughetti, and L. W. Hobbs
The Behaviour of Alkali Metal Colloids in Additively Coloured Potassium Chloride Crystals (I): Study by Optical Absorption and Surface Replicas — Colloid Size Distribution
(a) Vol. **40**, No. 2, p. 629—642 (1977).
See also *10.2*, p. *87*; 22.5.2, p. 377.

M. Wardzyńska and B. M. Wanklyn
*Optical Absorption Spectra and Crystal Field Parameters of Dysprosium and Erbium Germanates
(a) Vol. **40**, No. 2, p. 663—668 (1977).
See also 22.8.1, p. 404.

V. G. Kvachadze, M. F. Zhvaniya, and T. Sh. Davitashvili
*On the Nature of the Radiative Centres of the Halogen Impurity
(a) Vol. **40**, No. 2, p. K113—K115 (1977).
See also 8, p. 65; *10.2*, p. *87*; 22.5.2, p. 377.

R. Katzschmann, R. Kranold, and A. Rehfeld
*Optical Anomaly of Small Particles in Glasses
(a) Vol. **40**, No. 2, p. K161 (1977).
See also *13.1*, p. *117*; 22.4.2, p. 365.

L. A. Rebane, G. S. Zavt, and K. E. Haller
Dynamical Properties of NO_2^- Impurity Molecules in Alkali Halides (II): Vibration Spectra
(b) Vol. **81**, No. 1, p. 57—68 (1977).
See also *10.2*, p. *87*; 22.5.2, p. 377.

J. Laugsch and G. A. Gehring
*Optical Side Bands in a Cooperative Jahn-Teller Material: $TmAsO_4$
(b) Vol. **81**, No. 1, p. 167—178 (1977).
See also 6, p. 55; 13.5.1, p. 145; 22.8.1, p. 404.

T. Arai and T. P. Martin
*Absorption Spectra of S_2 Molecules Embedded in van der Waals Solids
(b) Vol. **81**, No. 1, p. 185—189 (1977).
See also 6, p. 55; 23, p. 420.

J. Donecker and J. Kluge
*Magneto-Optical Investigation of the Ground State of the Indirect Free Exciton in Si and GaP
(b) Vol. **81**, No. 1, p. 199—203 (1977).
See also 13.5.1, p. 145; *18*, p. *210*; 22.1.2, p. 253; 22.2.2, p. 354.

H. Venghaus
*Infrared Reflectance and Dielectric Properties of Pyrolytic Graphite for $\boldsymbol{E} \parallel \boldsymbol{c}$ Polarization
(b) Vol. **81**, No. 1, p. 221—225 (1977).
See also 13.1, p. 117; 13.5.2, p. 149; 22.1, p. 332.

U. Büchner
*Wave-Vector Dependence of the Electron Energy Losses of Boron Nitride and Graphite
(b) Vol. **81**, No. 1, p. 227—234 (1977).
See also 13.1, p. 117; 22.1, p. 332; 22.2, p. 348.

O. Goede
*Deformation Effects of Biexcitons in Wurtzite-Type Crystals
(b) Vol. **81**, No. 1, p. 235—244 (1977).
See also 13.1, p. 117; *13.5.1*, p. *145*; 22.4.1, p. 363.

M. SONDERGELD
*Two-Photon Absorption by Envelope–Hole Coupled Exciton States in Cubic ZnSe (I): Energy Spectrum of the H_d Term and Valence-Band Parameters
(b) Vol. **81**, No. 1, p. 253—262 (1977).
See also *13.5.1*, p. *145*; 22.4.2, p. 365.

J. S. NKOMA
*Theory of Two-Photon Absorption by Surface Polaritons
(b) Vol. **81**, No. 1, p. 307—312 (1977).
See also *13.5*, p. *141*; 22.2.2, p. 354.

P. J. ALONSO and R. ALCALÁ
*On the Optical Absorption Spectrum of Co^{2+} in CaF_2
(b) Vol. **81**, No. 1, p. 333—339 (1977).
See also *13.4*, p. *133*; 22.5.3, p. 384.

V. S. KOROLKOV and A. G. MAKHANEK
The Effects of Coordination Polyhedron Structure on Rare-Earth Ion Spectra
(b) Vol. **81**, No. 1, p. 387—394 (1977).
See also *13.4*, p. *133*; 22.8.2, p. 411.

T. P. MARTIN and H. SCHABER
*Infrared Absorption of Aggregated and Isolated Microcrystals
(b) Vol. **81**, No. 1, p. K41—K45 (1977).
See also 6, p. 55; 22.5.2, p. 378.

G. D. GUSEINOV, S. G. ABDULLAEVA, E. M. GODZHAEV, A. SH. MEKHTIEV, L. A. ALIEVA, and M. Z. ISMAILOV
*Electroabsorption of $TlInS_2$ Single Crystals
(b) Vol. **81**, No. 1, p. K47—K50 (1977).
See also 13.5.1, p. 145; *14.3.4*, p. *185*; 22.8, p. 398.

H. WEINERT, H. NEUMANN, H.-J. HÖBLER, G. KÜHN, and NGUYEN VAN NAM
*Infrared Faraday Effect in n-Type $CuInSe_2$
(b) Vol. **81**, No. 1, p. K59—K61 (1977).
See also *13.1*, p. *117*; 22.8, p. 398.

V. A. ANTONOV, P. A. ARSENEV, SH. A. VAKHIDOV, E. M. IBRAGIMOVA, and D. S. PETROVA
*Spectroscopic Properties of Neodymium Ions in $LiYO_2$ Monocrystals
(a) Vol. **41**, No. 1, p. 45—50 (1977).
See also 10.2, p. 87; 13.4, p. 133; *20.3*, p. *270*; 22.8.1, p. 404.

D. I. M. KNOTTNERUS, A. LANJOUW, and H. W. DEN HARTOG
Electron-Excess Centres in Calciumchlorapatite
(a) Vol. **41**, No. 1, p. 95—101 (1977).
See also *10.2*, p. *87*; 22.8, p. 398.

M. COUZI, A. DAOUD, and R. PERRET
Low-Frequency Raman Spectra and Phase Transitions in Perovskite-Type Layer Compounds. $(CH_3NH_3)_2MnCl_4$ and $(CH_3NH_3)_2CdCl_4$
(a) Vol. **41**, No. 1, p. 271—282 (1977).
See also *1.2*, p. *16*; 6, p. 56; 22.9, p. 416.

K. OZAWA and A. KIKUCHI
*Induced Flow Stress and Optical Absorption in NaCl Crystals Irradiated with Neutrons at Low Temperatures
(a) Vol. **41**, No. 1, p. 289—294 (1977).
See also *10.2*, p. *88*; 11, p. 95; 22.5.2, p. 378.

D. DURAND and F. LUTY
*Raman Scattering from Tunnelling and Hindered Rotational States of CN^- Molecules in KCl
(b) Vol. **81**, No. 2, p. 443—450 (1977).
See also *6*, p. *56*; 22.5.2, p. 378.

M. Sondergeld
*Two-Photon Absorption by Envelope–Hole Coupled Exciton States in Cubic ZnSe (II): Polarization Dependence in the Case of Degenerate Valence Bands
(b) Vol. **81**, No. 2, p. 451–461 (1977).
See also *13.5.1*, p. *145*; 22.4.2, p. 366.

C. Guttel, E. Antic, and P. E. Caro
Crystal Field Parameters for Gd^{3+} in $Gd_2(MoO_4)_3$
(b) Vol. **81**, No. 2, p. 463–470 (1977).
See also *13.4*, p. *134*; 22.8.2, p. 411.

J. Hoshen and R. Kopelman
*Molecular Exciton Cluster States and Spectra with Application to Benzene
(b) Vol. **81**, No. 2, p. 479–491 (1977).
See also *13.5.1*, p. *145*; 22.9, p. 411.

V. Čápek
Localized States on Imperfections and Red Shift of the Absorption Edge in Some Ferromagnetic Semiconductors
(b) Vol. **81**, No. 2, p. 571–577 (1977).
See also *13.4*, p. *134*; 22.8.2, p. 41*6*.

W. J. Huybrechts
*Infrared Absorption by an Optical Polaron in a Magnetic Field
(b) Vol. **81**, No. 2, p. 585–592 (1977).
See also *13.5.3*, p. *151*; 22, p. 325.

S. Benci, M. P. Fontana, and M. Manfredi
On the Excited States of the Tl^+ Monomer Centre in NaI
(b) Vol. **81**, No. 2, p. 603–608 (1977).
See also *13.4*, p. *134*; 20.3, p. 270; 22.5.2, p. 378.

Y. Guldner, C. Rigaux, A. Mycielski, and Y. Couder
*Magnetooptical Investigation of $Hg_{1-x}Cd_xTe$ Mixed Crystals (I): Semimetallic Configuration
(b) Vol. **81**, No. 2, p. 615–627 (1977).
See also 13.1, p. 118; *18*, p. *210*; 22.4.4, p. 368.

K. Kocot and J. M. Baranowski
The Resonant States of the $Ti(d^2)$ Impurity in CdSe
(b) Vol. **81**, No. 2, p. 629–635 (1977).
See also *13.4*, p. *134*; 16, p. 202; 22.4.2, p. 366.

G. Antonioli, D. Bianchi, V. Canevari, U. Emiliani, and P. Podini
*Optical Constants of GaSe at the Fundamental Absorption Edge
(b) Vol. **81**, No. 2, p. 665–670 (1977).
See also 13.5.1, p. 146; 22.3, p. 358.

W. Thielemann and F. Bergauer
*Test of the Random Fluctuations of the Conduction Band Edge in Heavily Doped n-Type GaAs by Means of the Magnetic Circular Dichroism
(b) Vol. **81**, No. 2, p. K91–K94 (1977).
See also 13.1, p. 118; *18*, p. *211*; 22.2.1, p. 352.

L. T. Ho
*Zeeman Effect of the Excitation Spectrum of Neutral Magnesium Donors in Silicon
(b) Vol. **81**, No. 2, p. K99–K101 (1977).
See also *13.4*, p. *134*; 22.1.2, p. 340.

W. Szuszkiewicz
Direct Determination of the Negative Energy Gap E_0 for Semiconductors with Inverted Band Structure
(b) Vol. **81**, No. 2, p. K119–K123 (1977).
See also *13.1*, p. *118*; 22.4.3, p. 367.

E. F. Kustov, V. P. Petrov, D. S. Petrova, and J. P. Udalov
*Absorption and Luminescence Spectra of Nd^{3+} and Er^{3+} Ions in Monocrystals of $CaYAlO_4$
(a) Vol. 41, No. 2, p. 379–383 (1977).
See also *20.3*, p. *270*; 22.8.1, p. 405.

D. Mergel and R. Labusch
Optical Excitation of Dislocation States in Germanium (I): Experiments
(a) Vol. 41, No. 2, p. 431–438 (1977).
See also *13.4*, p. *135*; *16*, p. *202*; 22.1.1, p. 334.

N. A. Blum, C. Feldman, and F. G. Satkiewicz
*Infrared Absorption of Amorphous Boron Films Containing Carbon and Hydrogen
(a) Vol. 41, No. 2, p. 481–486 (1977).
See also 2, p. 37; 22.1, p. 332.

R. D. Newton and W. A. Sibley
Deformation Induced Damage in MgO
(a) Vol. 41, No. 2, p. 569–575 (1977).
See also *10.2*, p. *88*; 20.3, p. 270; 22.6, p. 388.

N. N. Sirota, I. V. Bodnar, and G. F. Smirnova
*Lattice Reflection Spectra of $(GaP)_x$ $(InAs)_{1-x}$ Solid Solutions
(a) Vol. 41, No. 2, p. 669–672 (1977).
See also *13.1*, p. *118*; 22.2.2.4, p. 357.

R. Rauch
*Stark Effect on Sodium-Disturbed F_2 Centres in CaF_2
(a) Vol. 41, No. 2, p. K97–K99 (1977).
See also *10.2*, p. *89*; 20.3, p. 271; 22.5.3, p. 384.

J. Peña and J. M. Cabrera
About Additive Colouration of KCl:Pb
(a) Vol. 41, No. 2, p. K119–K121 (1977).
See also *10.2*, p. *89*; 22.5.2, p. 378.

V. A. Antonov, P. A. Arsenev, and D. S. Petrova
*Spectroscopic Properties of the Nd^{3+} Ion in $Y_2Ti_2O_7$ and $Gd_2Ti_2O_7$ Monocrystals
(a) Vol. 41, No. 2, p. K127–K131 (1977).
See also 20.2, p. 265; *20.3*, p. *271*; 22.8.1, p. 405.

H.-J. Kos and R. Nink
*Z_1 Centres in Mg-Doped Lithium Fluoride
(a) Vol. 41, No. 2, p. K157–K161 (1977).
See also *10.2*, p. *89*; 13.4, p. 135; 22.5.2, p. 378.

S. I. Pekar
*On the Theory of Additional Light Waves and Boundary Conditions in Anisotropic Crystals
(b) Vol. 82, No. 1, p. 83–89 (1977).
See also *13.5.1*, p. *146*; 22, p. 325.

J. Pelzl, P. Engels, and R. Florian
Raman Spectroscopic Study of the Structural Phase Transitions in K_2SnCl_6
(b) Vol. 82, No. 1, p. 145–148 (1977).
See also *1.2*, p. *17*; 6, p. 57; 22.8, p. 399.

Y. Guldner, C. Rigaux, A. Mycielski, and Y. Couder
*Magnetooptical Investigation of $Hg_{1-x}Cd_xTe$ Mixed Crystals (II): Semiconducting Configuration and Semimetal → Semiconductor Transition
(b) Vol. 82, No. 1, p. 149–158 (1977).
See also 13.1, p. 118; *18*, p. *211*; 22.4.4, p. 368.

M. Bacci
*Quadratic Jahn-Teller Effect and Tunneling Splitting in Octahedral Clusters
(b) Vol. 82, No. 1, p. 169—177 (1977).
See also *13.4*, p. *135*; 22.6, p. 389.

W. Schnitker and F. R. Kessler
*Piezoreflectance at the E_1 and $E_1 + \Delta_1$ Edges of Ge in High Magnetic Fields
(b) Vol. 82, No. 1, p. K35—K37 (1977).
See also 13.1, p. 119; *18*, p. *211*; 22.1.1, p. 335.

G. Ambrazevičius, G. Babonas, and A. Šileika
*Thermoreflectance Spectra of $CdSiP_2$
(b) Vol. 82, No. 1, p. K45—K48 (1977).
See also *13.1*, p. *119*; 22.8, p. 399.

M. Kh. Ashurov, Yu. K. Voronko, V. V. Osiko, A. A. Sobol, and M. I. Timoshechkin
Spectroscopic Study of Stoichiometry Deviation in Crystals with Garnet Structure
(a) Vol. 42, No. 1, p. 101—110 (1977).
See also *1.3*, p. *22*; 13.4, p. 136; 20.3, p. 271; 22.8.2, p. 411.

H. Mutka and O. J. A. Tiainen
Annealing Experiments on Fast Neutron Irradiated Selenium Single Crystals
(a) Vol. 42, No. 1, p. 241—245 (1977).
See also *11*, p. *96*; 14.3, p. 169; 22.1.3, p. 346.

G. Jungk
*Determination of Optical Constants. Matching Methods
(a) Vol. 42, No. 1, p. 267—272 (1977).

K. Mori
Transient Colour Centres by UV Light Irradiation in Yttrium Aluminium Garnet Crystals
(a) Vol. 42, No. 1, p. 375—384 (1977).
See also *10.2*, p. *89*; 20.2, p. 265; 20.3, p. 271; 22.8.2, p. 412.

V. B. Anzin, M. I. Eremets, Yu. V. Kosichkin, A. I. Nadezhdinskii, and A. M. Shirokov
*Measurement of the Energy Gap in Tellurium under Pressure
(a) Vol. 42, No. 1, p. 385—390 (1977).
See also 12.2, p. 105; 13.1, p. 120; *16*, p. *203*; 22.1.3, p. 346.

P. C. Mehendru, N. L. Pathak, K. Jain, and P. Mehendru
Electrical Conduction in Pure and Iodine-Doped Polyvinyl Acetate Thick Films
(a) Vol. 42, No. 1, p. 403—407 (1977).
See also *14.4*, p. *191*; 22.9, p. 417.

P. A. Arsenev, A. V. Potemkin, and V. V. Fenin
*Spectroscopic Properties of Erbium Ions in Mixed Aluminates of Rare-Earth Elements
(a) Vol. 42, No. 1, p. K19—K24 (1977).
See also 20.2, p. 266; *20.3*, p. *271*; 22.8.1, p. 405.

I. Pelant and P. Sladký
*The Ratio of Photoelastic Constants p_{11}/p_{12} of AgCl
(a) Vol. 42, No. 1, p. K57—K59 (1977).
See also 7, p. 63; *12.2*, p. *105*; 22.5.1, p. 373.

T. Hata and M. Suzuki
*Mechanism of Optical Modulation Caused by Acoustic Domains in CdS
(a) Vol. 42, No. 1, p. K71—K74 (1977).
See also 7, p. *63*; 22.4.1, p. 363.

H. Stolze, M. Lutz, and P. Grosse
*The Optical Activity of Tellurium
(b) Vol. 82, No. 2, p. 457—466 (1977).
See also *13.1*, p. *120*; 22.1.3, p. 346.

H. Herrmann
*Infrared Studies of the Layered Compound $ZnIn_2S_4$
(b) Vol. 82, No. 2, p. 513—521 (1977).
See also *6*, p. *57*; 22.8, p. 399.

T. Hollstein, U. Kreibig, and F. Leis
*Optical Properties of Cu and Ag in the Intermediate Region between Pure Drude and Interband Absorption
(b) Vol. 82, No. 2, p. 545—556 (1977).
See also 21.1, p. 296; 21.6, p. 316.

W. J. Huybrechts
*Influence of a DC Field on the Free Polaron Infrared Absorption
(b) Vol. 82, No. 2, p. 589—594 (1977).
See also *13.5.3*, p. *152*; 22, p. 327.

H. Paerschke, K.-E. Süsse, D.-G. Welsch, and B. Wilhelmi
*Theory of Absorption and Dispersion Profiles of Molecular Transitions in Dense Media
(b) Vol. 82, No. 2, p. 633—641 (1977).
See also *6*, p. *57*.

I. Savatinova and E. Anachkova
*Raman Study of Dynamic Disorder and the Ferroelectric Transition in $K_4Fe(CN)_6 \cdot 3\,H_2O$
(b) Vol. 82, No. 2, p. 677—683 (1977).
See also *6*, p. *57*; 14.4.1, p. 196; 22.8, p. 399.

V. A. Voloshin and M. G. Krimnus
*Effect of Pressure on the Spectrum of Praseodymium-Doped Single Crystals
(b) Vol. 82, No. 2, p. 695—700 (1977).
See also *12.2*, p. *105*; 22.5, p. 371; 22.8.1, p. 406.

G. L. Belenkii, L. N. Alieva, R. Kh. Nani, and E. Yu. Salaev
*Vibrational Spectrum of InSe
(b) Vol. 82, No. 2, p. 705—709 (1977).
See also *6*, p. *58*; 22.3, p. 358.

M. T. Carlos, M. Siu Li, and M. de Souza
*The F Centre Half-Width in KCN–KCl Mixed Crystals
(b) Vol. 82, No. 2, p. K103—K105 (1977).
See also *10.2*, p. *90*; 22.5.4, p. 385.

I. S. Osadko
Electron–Vibrational Transitions in the Presence of Several Minima of Adiabatic Potentials
(b) Vol. 82, No. 2, p. K107—K110 (1977).
See also *13.4*, p. *136*; 20.3, p. 272.

G. Jungk
Critical Points in the E_0'-Region of Ge from Ellipsometric Data
(b) Vol. 82, No. 2, p. K113—K116 (1977).
See also *13.1*, p. *120*; 22.1.1, p. 335.

K. Hübner
Chemical Bond and Related Properties of SiO_2 (III): Core — Level Shifts in SiO_x
(a) Vol. 42, No. 2, p. 501—509 (1977).
See also 2, p. 38; *4*, p. *45*; *13.1*, p. *120*; 14.4, p. 191; 22.6, p. 390.

R. Wolf, W. Kühn, R. Enderlein, and H. Lange
*Determination of Charge Carrier Concentration from Low-Field Electroreflectance Spectra
(a) Vol. 42, No. 2, p. 629—637 (1977).
See also *14.3.4*, p. *185*; 22.2.4, p. 357.

D. Bäuerle, K. Betzler, H. Hesse, S. Kapphan, and P. Loose
*Phase-Matched Second Harmonic Generation in Urea
(a) Vol. **42**, No. 2, p. K119—K121 (1977).
See also 22.9, p. 417.

G. S. White, K. H. Lee, and J. H. Crawford, Jr.
*Effects of γ-Irradiation upon the Optical Behavior of Spinel
(a) Vol. **42**, No. 2, p. K137—K141 (1977).
See also 11, p. 97; *20.3*, p. *272*; 22.8.2, p. 412.

G. Giro and P. di Marco
*Origin of Photocarriers in $Ru(2,2'\text{-Bipyridine})_3Cl_2 \cdot 6\,H_2O$ Single Crystals
(a) Vol. **42**, No. 2, p. K179—K182 (1977).
See also *16*, p. *203*; 20.3, p. 272; 22.9, p. 417.

P. A. Arsenev, A. V. Potemkin, V. V. Fenin, and I. Senff
Investigation of Stimulated Emission of Ho^{3+} Ions in Mixed Crystals with Perovskite Structure
(a) Vol. **42**, No. 2, p. K183—K185 (1977).
See also *20.2*, p. *266*; 20.3, p. 272; 22.8.1, p. 406.

A. Selkin
Energy Transfer by the Normal Modes in the Exciton Resonance Region
(b) Vol. **83**, No. 1, p. 47—53 (1977).
See also *13.5.1*, p. *146*; 22.4.1, p. 363.

D. Bäuerle, W. B. Holzapfel, A. Pinczuk, and Y. Yacoby
Temperature and Hydrostatic Pressure Dependence of Vibrational Modes in $PbTi_{1-x}Zr_xO_3$
(b) Vol. **83**, No. 1, p. 99—107 (1977).
See also *1.2*, p. *18*; 6, p. 58; *14.4.1*, p. *196*; 22.8.1, p. 406.

B. Khelifa, A. Delahaigue, and P. Jouve
Soft Mode and Debye Relaxations in Thiourea
(b) Vol. **83**, No. 1, p. 139—145 (1977).
See also *6*, p. *58*; *14.4.1*, p. *196*; 22.9, p. 417.

M. Kumar, P. S. Mahesh, and T. P. Sharma
*Second-Order Raman Spectra and Phonon Spectrum of CsF
(b) Vol. **83**, No. 1, p. 155—159 (1977).
See also *6*, p. *58*; 22.5.2, p. 380.

S. Lewonczuk, J. G. Gross, M. A. Khan, and J. Ringeissen
*Optical Properties of Cuprous Halides in the 4 to 30 eV Range
(b) Vol. **83**, No. 1, p. 161—168 (1977).
See also *13.1*, p. *121*; 22.5, p. 371.

R. S. Tripathi, L. L. Moseley, and T. Lukes
*Optical Absorption in Non-Crystalline Semiconductors
(b) Vol. **83**, No. 1, p. 197—207 (1977).
See also 2, p. 38; *13*, p. *110*; 22, p. 327.

F. Bechstedt and R. Enderlein
*Theory of Resonance Raman Scattering in Disordered Solids. Smooth Random Field Model
(b) Vol. **83**, No. 1, p. 239—247 (1977).
See also 2, p. 38; *6*, p. *58*; 13.1, p. 121; 22, p. 328.

J. P. Nérou, S. Jandl, and P. E. Girard
*Thermoreflectance Spectrum of $SnSe_2$
(b) Vol. **83**, No. 1, p. K7—K9 (1977).
See also 22.7, p. 395.

A. Kisiel and B. Oleś
*The Influence of a Light Beam on the Electroreflectance Spectra of Germanium
(b) Vol. **83**, No. 1, p. K35—K39 (1977).
See also *14.3.4*, p. *186*; 22.1.1, p. 335.

TH. HOLLSTEIN, U. KREIBIG, and F. LEIS
*Optical Properties of Au and Al in the Visible, Determined at 300 and 1.5 K
(b) Vol. **83**, No. 1, p. K49—K53 (1977).
See also 21, p. 284; 21.6, p. 316.

B. RHEINLÄNDER, S. HAASE, and H. WEINERT
*On the Determination of the Free Carrier Relaxation Time by Transverse Magneto-Plasma Reflectivity Measurements on Sn-Doped n-Type GaAs
(b) Vol. **83**, No. 1, p. K83—K86 (1977).
See also 13.5.2, p. 150; *18*, p. *212*; 22.2.1, p. 353.

G. H. SCHWUTTKE
Low Cost Silicon for Solar Energy Application
(a) Vol. **43**, No. 1, p. 43—51 (1977).
See also 3, p. 42; 14.3.3, p. 181; *16*, p. *204*; 22.1.2, p. 341.

A. A. KAMINSKII, S. E. SARKISOV, J. BOHM, P. REICHE, D. SCHULTZE, and R. UECKER
Growth, Spectroscopic and Laser Properties of Crystals in the $K_5Bi_{1-x}Nd_x(MoO_4)_4$ System
(a) Vol. **43**, No. 1, p. 71—79 (1977).
See also 3, p. 42; 13.4, p. 137; *20.2*, p. *266*; *20.3*, p. *272*; 22.8, p. 399.

G. OERTEL and E.-H. WEBER
*Growth of Thin SiO_2 Films on Clean Si (111) Surfaces by Low-Pressure Oxidation and Their Evaporation
(a) Vol. **43**, No. 1, p. 141—150 (1977).
See also 1.5, p. 30; *17*, p. *207*; 22.6, p. 390.

M. G. KAPLUNOV, K. I. POKHODNYA, A. I. KOTOV, E. B. YAGUBSKII, T. A. KITAEVA, and YU. G. BORODKO
*Optical Properties of the New Quasi-One-Dimensional Conductor $(TSeT)Br_{0.5}$
(a) Vol. **43**, No. 1, p. K73—K76 (1977).
See also 22.9, p. 418.

G. BUSSE and A. PFALLER
*Stress Spectroscopy of a Resonant Bandmode with a Far Infrared Laser
(a) Vol. **43**, No. 1, p. K103—K107 (1977).
See also 12.2, p. 106; 22.5.2, p. 380.

M. ZÁVĚTOVÁ and I. GREGORA
*Refractive Index of Amorphous Arsenic
(a) Vol. **43**, No. 1, p. K109—K111 (1977).
See also *2*, p. *39*; 22.1, p. 332.

V. V. MAKSIMENKO, A. J. SIMONOV, and A. A. LUSHNIKOV
*Far-Infrared Absorption in Small Metallic Particles
(b) Vol. **83**, No. 2, p. 377—382 (1977).
See also *13.3*, p. *127*; 13.5.2, p. 150.

D. HADŽIAHMETOVIĆ, M. PIRIĆ, and B. S. TOŠIĆ
Kinematical Exciton Levels and Adsorption of Light
(b) Vol. **83**, No. 2, p. 479—483 (1977).
See also *13.5.1*, p. *147*; 20.3, p. 273.

A. M. BRODSKII and M. I. URBAKH
On the Dependence of Light Reflection from Metals on Adatom Characteristics
(b) Vol. **83**, No. 2, p. 633—644 (1977).
See also *13.3*, p. *128*; 21, p. 285.

B. N. GANGULY, M. NICOL, and W. D. ELLENSON
Raman Spectroscopic Studies of the High Pressure Phase Transformation of $NaClO_3$
(b) Vol. **83**, No. 2, p. K115—K120 (1977).
See also *1.2*, p. *19*; *6*, p. *59*; 12.2, p. 106; 22.8.1, p. 407.

NGUYEN THE KHOI and J. A. GAJ
*Fundamental Absorption Edge of $Cd_{1-x}Mn_xTe$ Mixed Crystals
(b) Vol. **83**, No. 2, p. K133—K135 (1977).
See also 13.5.1, p. 147; 22.4.4, p. 369.

V. P. MUSHINSKII, L. I. PALAKI, and V. V. CHEBOTARU
*Optical Absorption in α-Ga_2S_3 Single Crystals
(b) Vol. **83**, No. 2, p. K149—K154 (1977).
See also 13.5.1, p. 147; 22.3, p. 358.

J. HÄRTWIG, B. DIETRICH, R. BÖTTGER, and G. LÜBKE
*Connection between Lattice Deformation and Coloration Pattern in Synthetic Quartz Monocrystals
(a) Vol. **43**, No. 2, p. K131—K133 (1977).
See also *10.2*, p. *91*; 22.6, p. 391.

T. A. MCMATH, G. JACKLE, and J. C. IRWIN
*On Refractive Index Measurements in GaS and GaSe
(a) Vol. **43**, No. 2, p. K151—K154 (1977).
See also 22.3, p. 359.

P. DI MARCO and G. GIRO
Photoconductive Properties of Fluorene Single Crystals
(a) Vol. **43**, No. 2, p. K193—K197 (1977).
See also *16*, p. *204*; 20.3, p. 274; 22.9, p. 419.

A. YA. KORETS, V. P. FEDOROV, A. V. KORSHUNOV, and V. A. NAUMOV
*Spectroscopic Investigation of Orientational Disorder in an Acenaphthylene Molecular Crystal
(b) Vol. **84**, No. 1, p. 107—113 (1977).
See also 13.4, p. 138; *13.5.1*, p. *148*; 22.9, p. 419.

V. A. KRYLOV and W. ULRICI
*Influence of an Electric Field on the Jahn-Teller Split Absorption Band $^4A_{2g} \rightarrow {}^4T_{1g}(P)$ of Co^{2+} in CdF_2 and CaF_2
(b) Vol. **84**, No. 1, p. 215—225 (1977).
See also *13.4*, p. *138*; 22.5.3, p. 384.

W. EKARDT
Theory of Excitons in Zincblende-Structure Semiconductors in Intermediate Magnetic Fields
(b) Vol. **84**, No. 1, p. 293—304 (1977).
See also *13.5.1*, p. *148*; 22.2.1, p. 353; 22.2.2, p. 355.

LE MANH HOANG and J. M. BARANOWSKI
*Absorption and Luminescence of $V(d^3)$ in II—VI Compound Semiconductors
(b) Vol. **84**, No. 1, p. 361—365 (1977).
See also 13.4, p. 138; *20.3*, p. *274*; 22.4, p. 361.

K. BÄRNER
*Energy Bands and Electron Correlation Effects in Manganese Pnictides
(b) Vol. **84**, No. 1, p. 385—392 (1977).
See also *13.1*, p. *122*; 14.1, p. 159; 18.4, p. 234; 21.1, p. 299.

A. RANFAGNI and G. VILIANI
Spin–Orbit-Allowed Tunneling Splitting in the $T \times \varepsilon_g$ Jahn-Teller Problem. Splitting of the Zero-Phonon Line
(b) Vol. **84**, No. 1, p. 393—399 (1977).
See also *13.4*, p. *138*; 22.5.4, p. 385; 22.6, p. 391.

I. SAVATINOVA and E. ANACHKOVA
*Raman Spectrum and Dielectric Relaxation in $K_4Fe(CN)_6 \cdot 3\,H_2O$
(b) Vol. 84, No. 1, p. 401–406 (1977).
See also 1.2, p. 20; 6, p. 60; *14.4.1*, p. *197*; 22.8, p. 400.

I. V. BODNAR, A. G. KAROZA, and G. F. SMIRNOVA
*Lattice Reflectivity Spectra of $CuGaS_2$–$CuGaSe_2$ Solid Solutions
(b) Vol. 84, No. 1, p. K65–K69 (1977).
See also *6*, p. *60*; 22.8, p. 400.

P. A. ARSENEV, KH. S. BAGDASAROV, A. V. POTEMKIN, and V. V. FENIN
*Spectra of Absorption, Luminescence, and Stimulated Emission of Neodymium in Crystals of Mixed Yttrium and Gadolinium Scandates
(a) Vol. 44, No. 1, p. 51–57 (1977).
See also 20.2, p. 266; *20.3*, p. *274*; 22.8.1, p. 407.

S. RADHAKRISHNA and V. S. SIVASANKAR
F_{Z_1} Centres in Cesium Halides
(a) Vol. 44, No. 1, p. 345–351 (1977).
See also *10.2*, p. *92*; 22.5.2, p. 381.

V. V. BORSHCH, V. S. KOVAL, I. V. POTYKEVICH, and I. V. FEKESHGAZI
*Birefringence and Optical Activity of CdP_2
(a) Vol. 44, No. 1, p. K15–K19 (1977).
See also 22.3, p. 359.

A. SZADKOWSKI, A. LUBOMIRSKA-WITTLIN, A. ZARĘBA, and B. KRUKOWSKA-FULDE
*On Conversion Processes of $CdF_2:MF_3$
(a) Vol. 44, No. 1, p. K43–K46 (1977).
See also *14.3*, p. *172*; 22.5.3, p. 384.

NGUYEN VAN HIEU and NGUYEN AI VIET
*Interband Electronic Raman Scattering by a Three-Band Semiconductor in the Presence of a Resonant Electromagnetic Field
(b) Vol. 84, No. 2, p. 417–422 (1977).
See also 13.1, p. 122; *14.3.4*, p. *188*; 22.2, p. 349; 22.4, p. 361.

M. N. SHARMA and A. KANT
A New Approach to Lattice Properties of Alkali Halide Crystals
(b) Vol. 84, No. 2, p. 427–434 (1977).
See also *4*, p. *46*; *8*, p. *68*; 22.5.2, p. 381.

A. P. RADLINSKI
On the Properties of Cobalt Impurities in Zinc Selenide Crystals (I): Effect of In-Band Position of States on Photoelectric Properties
(b) Vol. 84, No. 2, p. 503–511 (1977).
See also *13.4*, p. *139*; 16, p. 205; 22.4.2, p. 366.

A. KASTEN and P. J. BECKER
*Spectroscopic Investigation of Optically Excited Doublet States of Ho^{3+} in Magnetically Ordered $(Ho_{0.01}, Gd_{0.99})\,VO_4$
(b) Vol. 84, No. 2, p. 551–560 (1977).
See also 13.4, p. 139; *18.4*, p. *234*; 22.8.1, p. 408.

H. LANGE and I. A. ILIEV
*Evidence of Coulomb Interaction at the E_1 Interband Transition of Hexagonal CdSe
(b) Vol. 84, No. 2, p. 569–574 (1977).
See also 13.1, p. 122; *13.5.1*, p. *148*; 22.4.2, p. 366.

W. RICHTER, H. KÖHLER, and C. R. BECKER
*A Raman and Far-Infrared Investigation of Phonons in the Rhombohedral V_2–VI_3 Compounds: Bi_2Te_3, Bi_2Se_3, Sb_2Te_3 and $Bi_2(Te_{1-x}Se_x)_3$ $(0 < x < 1)$, $(Bi_{1-y}Sb_y)_2Te_3$ $(0 < y < 1)$
(b) Vol. 84, No. 2, p. 619–628 (1977).
See also *6*, p. *60*; 22.7, p. 395.

K. HEIDRICH, W. SCHÄFER, M. SCHREIBER, G. TRENDEL, and J. TREUSCH
*Optical Properties of Polymorphous Thallous Halide Mixed Crystal Layers
(b) Vol. 84, No. 2, p. 629–635 (1977).
See also *13.1*, p. *122*; 22.5.4, p. 385.

J. GARDAVSKÝ, I. BARVÍK, and M. ZVÁRA
*Jahn-Teller Effect on the $^3T_1(^3F)$ State and Resonance of the $^3T_1(^3P)$ State with the Conduction Band in CdTe:Ti^{2+}
(b) Vol. 84, No. 2, p. 691–698 (1977).
See also *13.4*, p. *139*; 22.4.3, p. 367.

S. W. BIERNACKI
*The Multimode Jahn-Teller Coupling for an Orbital Triplet in Cubic Crystals
(b) Vol. 84, No. 2, p. 699–708 (1977).
See also *13.4*, p. *139*; 18.1, p. 215.

VU DUY PHACH, A. BIVAS, B. HÖNERLAGE, and J. B. GRUN
*Biexciton Resonant Two-Photon Absorption in CuCl
(b) Vol. 84, No. 2, p. 731–740 (1977).
See also *13.5.1*, p. *148*; 22.5, p. 372.

A. TRUŢIA, F. IOVA, and M. VODĂ
*Concentration of "Impurity" Ion Clusters in a Two-Dimensional Host Lattice
(b) Vol. 84, No. 2, p. 817–824 (1977).
See also *10.2*, p. *92*; 22.5, p. 372.

A. ANEDDA and E. FORTIN
Excitons in HgI_2
(b) Vol. 84, No. 2, p. K87–K90 (1977).
See also *13.5.1*, p. *148*; 22.5, p. 372.

E. SCHUBERTH
First Order Raman Scattering from Ar Crystals: Evidence for the H.C.P. Phase
(b) Vol. 84, No. 2, p. K91–K94 (1977).
See also *1.2*, p. *20*; 6, p. 60; 23, p. 420.

A. M. LEJUS and D. MICHEL
*Raman Spectrum of Er_2O_3 Sesquioxide
(b) Vol. 84, No. 2, p. K105–K108 (1977).
See also *6*, p. *61*; 22.6, p. 391.

F. LUKEŠ
*Electroreflectance Spectrum of InAs in the Range of E_0 and $E_0 + \Delta_0$ Transitions
(b) Vol. 84, No. 2, p. K113–K117 (1977).
See also 13.1, p. 123; *14.3.4*, p. *188*; 22.2.1, p. 353.

P. HLÍDEK, M. ZVÁRA, and V. PROSSER
*Photoinduced Infra-Red Optical Absorption in $CdCr_2Se_4$ Magnetic Semiconductor
(b) Vol. 84, No. 2, p. K119–K123 (1977).
See also 22.8.2, p. 413.

P. HENNL
*Low Temperature Photochemistry on Chalcogen–Vacancy Centres in Alkali Halide Crystals
(b) Vol. 84, No. 2, p. K151–K154 (1977).
See also *10.2*, p. *92*; 22.5.2, p. 381.

W. ULRÍCI
CdF_2:Cr^{2+} — an Example for an Orthorhombic Jahn-Teller Distortion
(b) Vol. 84, No. 2, p. K155–K157 (1977).
See also *13.4*, p. *140*; 22.5.3, p. 384.

H.-J. KOS and R. NINK
*Mechanism of Restored Thermoluminescence in Lithium Fluoride
(a) Vol. 44, No. 2, p. 505–510 (1977).
See also *10.2*, p. *93*; *20.3*, p. *275*; 22.5.2, p. 381.

S. K. MEHTA, J. F. MERKLIN, and H. J. DONNERT
*Thermoluminescence Related Z Centers in LiF:Mg, Ti
(a) Vol. **44**, No. 2, p. 679—685 (1977).
See also *10.2*, p. *93*; 22.5.2, p. 382.

V. M. ORERA and R. ALCALÁ
*Optical Properties of Cation Colloidal Particles in CaF_2 and SrF_2
(a) Vol. **44**, No. 2, p. 717—723 (1977).
See also *10.2*, p. *93*; 22.5.3, p. 384.

A. M. VARAPRASAD, A. L. SHASHIMOHAN, A. B. BISWAS, and D. K. CHAKRABARTY
Dielectric and Infrared Spectral Studies of the $(Na_{1/2}La_{1/2})_xBa_{1-x}TiO_3$ System
(a) Vol. **44**, No. 2, p. 787—791 (1977).
See also *14.4.1*, p. *197*; 22.8.1, p. 408.

V. R. DUMKE
Image Effect in Previously Electrolytically Coloured KCl–OH^-
(a) Vol. **44**, No. 2, p. K133—K134 (1977).
See also *10.2*, p. *93*; 22.5.2, p. 382.

E. I. TERUKOV, H. WAGNER, W. REICHELT und H. OPPERMANN
*IR-spektroskopische Untersuchungen zum Halbleiter–Metall-Phasenübergang des V_3O_5
(IR Spectroscopic Investigations on the Semiconductor–Metal-Phase Transition of V_3O_5)
(a) Vol. **44**, No. 2, p. K187—K190 (1977).
See also *1.2*, p. *20*; 6, p. 61; 22.6, p. 392.

20.2 Lasers

Review Article

T. D. Dzhafarov
Atomic Diffusion in Semiconductor Epitaxial Structures
(a) Vol. **42**, No. 1, p. 11–45 (1977).
See also *9*, p. *69*; 10.2, p. 83; 14.3.1, p. 173; 14.3.3, p. 178; 22.1, p. 332; 22.2, p. 348; 22.4, p. 361.

Original Papers and Short Notes

A. A. Kaminskii, T. I. Butaeva, V. A. Fedorov, Kh. S. Bagdasarov, and A. G. Petrosyan
Absorption, Luminescence, and Stimulated Emission Investigations in $Lu_3Al_5O_{12}$–Er^{3+} Crystals
(a) Vol. **39**, No. 2, p. 541–548 (1977).
See also *13.4*, p. *131*; 20.3, p. 268; 22.8, p. 396.

H. Hora
*Corrections of the Long Beating Length in Quantum Modulated Electron Beams
(b) Vol. **80**, No. 1, p. 143–147 (1977).
Erratum
(b) Vol. **83**, No. 1, p. 355 (1977).
See also *13.5.2*, p. *149*.

W. A. Szilva and L. E. Murr
*Variation of Laser Mirror Metal Microstructure and Its Effect on Reflectivity at 10.6 μm
(a) Vol. **40**, No. 1, p. 211–224 (1977).
See also *1.4*, p. *24*; 21.1, p. 291.

M. S. Sodha, V. K. Tripathi, and D. P. Singh
*Effect of a Strong Magnetic Field on the Self-Focusing of a Laser Beam in InSb
(a) Vol. **40**, No. 1, p. 315–320 (1977).
See also *18*, p. *210*; 22.2.3, p. 356.

A. M. Prokhorov, A. A. Kaminskii, V. V. Osiko, M. I. Timoshechkin, E. V. Zharikov, T. I. Butaeva, S. E. Sarkisov, A. G. Petrosyan, and V. A. Fedorov
*Investigations of the 3 μm Stimulated Emission from Er^{3+} Ions in Aluminium Garnets at Room Temperature
(a) Vol. **40**, No. 1, p. K69–K72 (1977).
Erratum
(a) Vol. **41**, No. 2, p. K199 (1977).
See also 22.8.2, p. 410.

A. S. Logginov and V. E. Solovev
*Analysis of the Current Carrier Concentration Distribution and Threshold Properties of Stripe-Geometry $Ga_xAl_{1-x}As$ Lasers
(a) Vol. **41**, No. 2, p. 371–378 (1977).
See also *14.3.3*, p. *179*; 22.2.4, p. 357.

V. A. Antonov, P. A. Arsenev, and D. S. Petrova
Spectroscopic Properties of the Nd^{3+} Ion in $Y_2Ti_2O_7$ and $Gd_2Ti_2O_7$ Monocrystals
(a) Vol. **41**, No. 2, p. K127–K131 (1977).
See also *20.1*, p. *256*; *20.3*, p. *271*; 22.8.1, p. 405.

K. Mori
Transient Colour Centres by UV Light Irradiation in Yttrium Aluminium Garnet Crystals
(a) Vol. **42**, No. 1, p. 375–384 (1977).
See also *10.2*, p. *89*; 20.1, p. 257; 20.3, p. 271; 22.8.2, p. 412.

P. A. Arsenev, A. V. Potemkin, and V. V. Fenin
Spectroscopic Properties of Erbium Ions in Mixed Aluminates of Rare-Earth Elements
(a) Vol. **42**, No. 1, p. K19–K24 (1977).
See also *20.1*, p. *257*; *20.3*, p. *271*; 22.8.1, p. 405.

H. Haug and S. Koch
*On the Theory of Laser Action in Dense Exciton Systems
(b) Vol. **82**, No. 2, p. 531–543 (1977).
See also *13.5.1*, p. *146*; 22.4.1, p. 363.

H. Rüfer and W. Ruppel
*Luminescence of Graded $Zn_xCd_{1-x}S$ Single Crystals
(a) Vol. **42**, No. 2, p. 519–524 (1977).
See also 13.5.1, p. 146; *20.3*, p. *272*; 22.4.4, p. 368.

P. A. Arsenev, A. V. Potemkin, V. V. Fenin, and I. Senff
*Investigation of Stimulated Emission of Ho^{3+} Ions in Mixed Crystals with Perovskite Structure
(a) Vol. **42**, No. 2, p. K183–K185 (1977).
See also 20.1, p. 259; 20.3, p. 272; 22.8.1, p. 406.

A. A. Kaminskii, S. E. Sarkisov, J. Bohm, P. Reiche, D. Schultze, and R. Uecker
*Growth, Spectroscopic and Laser Properties of Crystals in the $K_5Bi_{1-x}Nd_x(MoO_4)_4$ System
(a) Vol. **43**, No. 1, p. 71–79 (1977).
See also 3, p. 42; 13.4, p. 137; 20.1, p. 260; *20.3*, p. *272*; 22.8, p. 399.

P. A. Arsenev, A. V. Potemkin, V. V. Fenin, and I. Senff
*Investigation of Stimulated Emission of Er^{3+} Ions in Mixed Crystals with Perovskite Structure
(a) Vol. **43**, No. 1, p. K15–K18 (1977).
See also 13.4, p. 137; 22.8, p. 399.

R. Moncorgé, D. Pacheco, and B. Di Bartolo
*Thermal Behavior of the Shape and Width of the 9145 Å Laser Line of Nd^{3+} in $CaWO_4$
(a) Vol. **43**, No. 1, p. K45–K48 (1977).
See also 22.8.1, p. 407.

P. A. Arsenev, Kh. S. Bagdasarov, A. V. Potemkin, and V. V. Fenin
Spectra of Absorption, Luminescence, and Stimulated Emission of Neodymium in Crystals of Mixed Yttrium and Gadolinium Scandates
(a) Vol. **44**, No. 1, p. 51–57 (1977).
See also *20.1*, p. *262*; *20.3*, p. *274*; 22.8.1, p. 407.

20.3 Luminescence (X-Ray Spectra See 20)

Review Article

V. T. Agekyan
Spectroscopic Properties of Semiconductor Crystals with Direct Forbidden Energy Gap
(a) Vol. **43**, No. 1, p. 11—42 (1977).
See also 6, p. 51; *13.5.1*, p. *143*; 20.1, p. 247; 22.6, p. 386.

Original Papers and Short Notes

M. F. Trinkler and I. S. Zolovkina
*A-Luminescence of Alkali Halides Activated by Monovalent Mercury-Like Ions
(b) Vol. **79**, No. 1, p. 49—59 (1977).
See also *13.4*, p. *129*; 22.5.2, p. 375.

S. I. Golubov and Yu. V. Konobeev
Energy Transfer by Excitons in a Tetracene-Doped Anthracene Crystal
(b) Vol. **79**, No. 1, p. 79—84 (1977).
See also 13.4, p. 129; *13.5.1*, p. *143*; 22.9, p. 414.

D. Engemann and R. Fischer
*Photoluminescence in Amorphous Silicon
(b) Vol. **79**, No. 1, p. 195—202 (1977).
See also *2*, p. *33*; 22.1.2, p. 337.

P. G. Baranov, Yu. P. Veshchunov, R. A. Zhitnikov, and N. G. Romanov
EPR, Optical Absorption, and Luminescence of Anisotropic Copper Centres in KCl
(b) Vol. **79**, No. 1, p. K27—K30 (1977).
See also *19*, p. *235*; 20.1, p. 248; 22.5.2, p. 375.

A. Vanderhaeghe and G. Jacobs
*The Lifetime of the F_H Centre in CsBr:Cl and CsBr:F Crystals
(b) Vol. **79**, No. 1, p. K75—K79 (1977).
See also 10.2, p. 83; 22.5.3, p. 383.

A. J. H. Macke
*Luminescence and Energy Transfer in the Ordered Perovskite System $La_2MgSn_{1-x}Ti_xO_6$
(a) Vol. **39**, No. 1, p. 117—123 (1977).
See also 22.8, p. 396.

J.-P. Aicardi, J.-P. Leyris, and G. Masse
*Cathodoluminescence of Cinnabar. Evidence of a Donor–Acceptor-Type Transition
(a) Vol. **39**, No. 1, p. 125—131 (1977).
See also 13.4, p. 130; 22.4.1, p. 362.

J. C. Gâcon, J. C. Souillat, J. Sériot, and B. Di Bartolo
*Radiative and Radiationless Transition Rate Measurements Using a Tunable Pulsed Dye Laser. Application to Sm^{2+} Fluorescence in BaClF
(a) Vol. **39**, No. 1, p. 147—156 (1977).
See also 13.4, p. 130; 22.5.3, p. 383.

A. Suzuki, H. Matsunami, and T. Tanaka
*Erratum to: Very Slow Decays of the Luminescence from Boron-Doped 4 H–SiC
(a) Vol. **39**, No. 1, p. K79 (1977).
See also 22.3, p. 358.

M. Ikeya and J. H. Crawford, Jr.
*Recombination Luminescence in Alkali Halides Doped with Monovalent Cations. $(V_K + e)_A$-Type Emission
(b) Vol. **79**, No. 2, p. 559—566 (1977).
See also 10.2, p. 84; 22.5.2, p. 375.

A. Y. S. Kung and J. M. Vail
A Molecular Model for the Emission Process of F_A(II)-Centres in Alkali Halides
(b) Vol. **79**, No. 2, p. 663—672 (1977).
See also *10.2*, p. *84*; 13.4, p. 130; 22.5.2, p. 375.

G. K. Averkieva, V. S. Grigoreva, I. A. Maltseva, V. D. Prochukhan, and Yu. V. Rud
*Photoluminescence of p-Type $ZnGeP_2$ Crystals
(a) Vol. **39**, No. 2, p. 453—457 (1977).
See also 13.4, p. 131; 22.8, p. 396.

K. Ziegler and U. Birkholz
Photoelectric Properties of Si_2Te_3 Single Crystals
(a) Vol. **39**, No. 2, p. 467—475 (1977).
See also *13.4*, p. *131*; 14.3, p. 166; 16, p. 201; 22, p. 322.

A. A. Kaminskii, T. I. Butaeva, V. A. Fedorov, Kh. S. Bagdasarov, and A. G. Petrosyan
Absorption, Luminescence, and Stimulated Emission Investigations in $Lu_3Al_5O_{12}$–Er^{3+} Crystals
(a) Vol. **39**, No. 2, p. 541—548 (1977).
See also *13.4*, p. *131*; 20.2, p. 265; 22.8, p. 396.

K. H. Lee, G. E. Holmberg, and J. H. Crawford, Jr.
*Optical and ESR Studies of Holes Centers in γ-Irradiated Al_2O_3
(a) Vol. **39**, No. 2, p. 669—674 (1977).
See also *10.2*, p. *84*; 19, p. 236 ; 22.6, p. 386.

B. Rheinländer, A. Zehe, and G. Kretzschmar
*Near Field Topography of High Frequency Modulated GaAs LED's
(a) Vol. **39**, No. 2, p. K139—K142 (1977).
See also *14.3.3*, p. *178*; 22.2.1, p. 351.

G. Blasse and G. P. M. van den Heuvel
*Luminescence of U^{6+}-Activated Li_2WO_4
(a) Vol. **39**, No. 2, p. K157—K160 (1977).
See also 22.8.1, p. 403.

W. Ekardt and M. I. Sheboul
LO-Phonon-Assisted Two-Photon Absorption of the First Excited State of the Biexciton in CuCl
(b) Vol. **80**, No. 1, p. 51—56 (1977).
See also *13.5.1*, p. *144*; 20.1, p. 250; 22.5, p. 370.

L. Mărculescu
*Luminescence of Lead-Activated Sodium Chloride Single Crystals
(b) Vol. **80**, No. 1, p. 265—272 (1977).
See also 13.4, p. 131; 22.5.2, p. 376.

J. B. Aladekomo, S. Arnold, and M. Pope
*Triplet Exciton Diffusion and Double Photon Absorption in Tetracene
(b) Vol. **80**, No. 1, p. 333—340 (1977).
See also *13.5.1*, p. *144*; 22.9, p. 414.

E. Mugeński and J. Z. Damm
*On the 420 nm Emission Band in Pb^{2+} Doped KCl Crystals
(b) Vol. **80**, No. 1, p. K59—K62 (1977).
See also 22.5.2, p. 376.

E. Grilli and M. Guzzi
*Recombination Process of Photoexcited Carriers in $ZnIn_2S_4$
(a) Vol. **40**, No. 1, p. 69—74 (1977).
See also 22.8, p. 397.

Hj. Bernhardt
*Near-Edge Absorption of Lead Molybdate Crystals
(a) Vol. **40**, No. 1, p. 257—264 (1977).
See also **16**, p. 201; *20.1*, p. *252*; 22.8.1, p. 403.

D. Bloor, D. N. Batchelder, and F. H. Preston
Optical Properties of Bis (p-Toluene Sulphonate) Diacetylene (III): Defect-Induced Spectra
(a) Vol. **40**, No. 1, p. 279—284 (1977).
See also 10.2, p. 85; *20.1*, p. *252*; 22.9, p. 415.

H. L. Oczkowski, N. Romeo, and L. Tarricone
*Impact Ionization Electroluminescence in ZnSe Single Crystals
(a) Vol. **40**, No. 1, p. K49—K52 (1977).
See also *14.3.4*, p. *184*; 22.4.2, p. 365.

M. Krystek
*On the Evaluation Method of Balarin and Zetzsche
(a) Vol. **40**, No. 1, p. K65—K68 (1977).
See also *13.4*, p. *132*; 22, p. 323.

J. Collet, J. Barrau, M. Brousseau, and H. Maaref
Simultaneous Kinetics of Drops and Excitons in Silicon during the Drop Formation
(b) Vol. **80**, No. 2, p. 461—470 (1977).
See also *13.5.1*, p. *144*; 22.1.2, p. 338.

F. Thuselt
*Spin-Dependent Luminescence of Donor–Acceptor Pairs in Semiconductors (I): General Theory
(b) Vol. **80**, No. 2, p. 499—507 (1977).
See also *13.4*, p. *132*; 22, p. 323.

A. N. Georgobiani, Yu. V. Ozerov, E. E. Strumban, and V. S. Vinogradov
Influence of Random Electric Fields on the Polarization of Deep Centre Electroabsorption
(b) Vol. **80**, No. 2, p. 589—596 (1977).
See also *13.4*, p. *132*; 16, p. 202; 20.1, p. 252; 22.4.1, p. 362.

B. P. Chandra
*Stress Dependence of Triboluminescence in Crystals
(a) Vol. **40**, No. 2, p. 403—408 (1977).
See also *12.2*, p. *104*; 22.9, p. 415.

K. A. Shoaib, F. H. Hashmi, M. Ali, S. J. H. Bukhari, and C. A. Majid
*Thermoluminescence from X-Ray-Irradiated Stabilized ZrO_2 Single Crystals
(a) Vol. **40**, No. 2, p. 605—612 (1977).
See also 10.2, p. 86; 13.4, p. 133; 22.6, p. 387.

M. Binder, N. E. Geacintov, M. Pope, and C. E. Swenberg
*Magnetic Field Modulated Scintillation from Tetracene-Doped Anthracene Crystals under α-Particle Irradiation
(a) Vol. **40**, No. 2, p. 621—628 (1977).
See also 11, p. 95; *13.5.1*, p. *144*; 22.9, p. 415.

Y. E. Perlin, A. A. Kaminskii, S. I. Klokishner, V. N. Enakii, Kh. S. Bagdasarov, G. A. Bogomolova, and D. N. Vylegzhanin
Thermal Broadening and Shift of the Energy Levels of TR^{3+} Ions in Crystals
(a) Vol. **40**, No. 2, p. 643—653 (1977).
See also 6, p. 55; *13.4*, p. *133*; 22.8.1, p. 403.

F. Thuselt and K. Unger
*Spin-Dependent Luminescence of Donor–Acceptor Pairs in Semiconductors (II): Application to Inner Strains
(b) Vol. **81**, No. 1, p. 89—95 (1977).
See also *13.4*, p. *133*; 22, p. 324.

F. Bechstedt and F. Henneberger
*Two-Photon Resonance Raman Scattering by Excitons via an Intermediate Excitonic Molecule (I): Bare Exciton Approach
(b) Vol. **81**, No. 1, p. 211—220 (1977).
See also *13.5.1*, p. *145*; 22.4.1, p. 363; 22.5, p. 370.

A. A. Antipin, A. V. Vinokurov, M. P. Davydova, S. L. Korableva, A. L. Stolov, and A. A. Fedii
Optical Spectra, EPR, and Spin–Lattice Relaxation of Yb^{3+} Ions in Crystals Having Perovskite-Type Structure
(b) Vol. **81**, No. 1, p. 287—293 (1977).
See also *13.4*, p. *133*; 19, p. 238; 22.5.4, p. 385.

A. Freiberg and L. A. Rebane
*Peculiarities of the Electron–Phonon Interaction in the Spectra of the Molecular Ions O_2^-, S_2^-, and NO_2^- in Alkali Halide Crystals
(b) Vol. **81**, No. 1, p. 359—369 (1977).
See also *6*, p. *55*; 22.5.2, p. 377.

V. A. Antonov, P. A. Arsenev, Sh. A. Vakhidov, E. M. Ibragimova, and D. S. Petrova
*Spectroscopic Properties of Neodymium Ions in $LiYO_2$ Monocrystals
(a) Vol. **41**, No. 1, p. 45—50 (1977).
See also 10.2, p. 87; 13.4, p. 133; *20.1*, p. *254*; 22.8.1, p. 404.

W. Siegel, H. Becherer, G. Kühnel, and E. Ziegler
*Electroluminescence in $ZnSiP_2$ Schottky Diodes
(a) Vol. **41**, No. 1, p. 75—80 (1977).
See also *14.3.3*, p. *179*; 22.8, p. 398.

N. B. Lukyanchikova, G. S. Pekar, N. N. Tkachenko, Hoang Mi Shin, and M. K. Sweinkman
*Injection Electroluminescence in Forward-Biased ZnS Metal–Semiconductor Diodes
(a) Vol. **41**, No. 1, p. 299—305 (1977).
See also *14.3.3.*, p. *179*; 22.4.1, p. 363.

S. Benci, M. P. Fontana, and M. Manfredi
On the Excited States of the Tl^+ Monomer Centre in NaI
(b) Vol. **81**, No. 2, p. 603—608 (1977).
See also *13.4*, p. *134*; 20.1, p. 255; 22.5.2, p. 378.

E. F. Kustov, V. P. Petrov, D. S. Petrova, and J. P. Udalov
*Absorption and Luminescence Spectra of Nd^{3+} and Er^{3+} Ions in Monocrystals of $CaYAlO_4$
(a) Vol. **41**, No. 2, p. 379—383 (1977).
See also *20.1*, p. *256*; 22.8.1, p. 405.

R. Visocekas and A. Geoffroy
*Tunnelling Afterglows and Retrapping in Calcite
(a) Vol. **41**, No. 2, p. 499—503 (1977).
See also *10.2*, p. *88*; 22.8.1, p. 405.

W. Stręk, C. Szafrański, E. Łukowiak, Z. Mazurak, and B. Jeżowska-Trzebiatowska
*Fluorescence Quenching in Neodymium Pentaphosphate
(a) Vol. **41**, No. 2, p. 547—553 (1977).
See also 13.4, p. 135; 22.8.1, p. 405.

R. D. Newton and W. A. Sibley
Deformation Induced Damage in MgO
(a) Vol. **41**, No. 2, p. 569—575 (1977).
See also *10.2*, p. *88*; 20.1, p. 256; 22.6, p. 388.

A. B. van Oosterhout
*Tungstate Luminescence in Ordered Perovskites
(a) Vol. **41**, No. 2, p. 607—617 (1977).
See also 13.4, p. 135; 22.8.1, p. 405.

K. D. Glinchuk, A. V. Prokhorovich, V. E. Rodionov, and V. I. Vovnenko
*The Scheme of Electronic Transitions via 0.94, 1.0, 1.2, and 1.3 eV Radiative Centres in n-GaAs
(a) Vol. **41**, No. 2, p. 659—668 (1977).
See also *13.4*, p. *135*; 22.2.1, p. 352.

R. Rauch
Stark Effect on Sodium-Disturbed F_2 Centres in CaF_2
(a) Vol. **41**, No. 2, p. K97—K99 (1977).
See also *10.2*, p. *89*; *20.1*, p. *256*; 22.5.3, p. 384.

V. A. Antonov, P. A. Arsenev, and D. S. Petrova
*Spectroscopic Properties of the Nd^{3+} Ion in $Y_2Ti_2O_7$ and $Gd_2Ti_2O_7$ Monocrystals
(a) Vol. **41**, No. 2, p. K127—K131 (1977).
See also *20.1*, p. *256*; 20.2, p. 265; 22.8.1, p. 405.

A. F. Ellervee
*Luminescence of Pb^{2+} and Bi^{3+} Centres in Alkali-Earth Sulphides and Oxides
(b) Vol. **82**, No. 1, p. 91—98 (1977).
See also 13.4, p. 135; 22.4.1, p. 363; 22.6, p. 389.

E. Ostertag and J. B. Grun
*Subnanosecond Time-Resolved Spectroscopy of the Biexciton Luminescence in CuCl
(b) Vol. **82**, No. 1, p. 335—342 (1977).
See also *13.5.1*, p. *146*; 22.5, p. 371.

L. Bosi and M. Nimis
On the Decay of the M_A Centre in KCl:Li and KCl:Na
(b) Vol. **82**, No. 1, p. K49—K51 (1977).
See also *10.2*, p. *89*; *13.4*, p. *136*; 22.5.2, p. 379.

M. Kh. Ashurov, Yu. K. Voronko, V. V. Osiko, A. A. Sobol, and M. I. Timoshechkin
Spectroscopic Study of Stoichiometry Deviation in Crystals with Garnet Structure
(a) Vol. **42**, No. 1, p. 101—110 (1977).
See also *1.3*, p. *22*; 13.4, p. 136; 20.1, p. 257; 22.8.2, p. 411.

K. Mori
Transient Colour Centres by UV Light Irradiation in Yttrium Aluminium Garnet Crystals
(a) Vol. **42**, No. 1, p. 375—384 (1977).
See also *10.2*, p. *89*; 20.1, p. 257; 20.2, p. 265; 22.8.2, p. 412.

P. A. Arsenev, A. V. Potemkin, and V. V. Fenin
*Spectroscopic Properties of Erbium Ions in Mixed Aluminates of Rare-Earth Elements
(a) Vol. **42**, No. 1, p. K19—K24 (1977).
See also *20.1*, p. *257*; 20.2, p. 266; 22.8.1, p. 405.

A. L. Lelyakov
*Luminescence Band Shape of Free Excitons in AgBr Crystals at High Exciton Density
(b) Vol. **82**, No. 2, p. 711—716 (1977).
See also 6, p. 58; *13.5.1*, p. *146*; 22.5.1, p. 373.

V. Hizhnyakov and I. Tehver
*On the Theory of Hot Luminescence and Resonant Raman Effect of Impurity Centres
(b) Vol. **82**, No. 2, p. K89—K92 (1977).
See also *6*, p. *58*; 10.2, p. 89.

I. S. OSADKO
Electron–Vibrational Transitions in the Presence of Several Minima of Adiabatic Potentials
(b) Vol. **82**, No. 2, p. K107–K110 (1977).
See also *13.4*, p. *136*; 20.1, p. 258.

V. A. LISOVENKO, L. A. KHUTORNAYA, M. T. SHPAK, and E. N. VELIKAYA
*Basal Dislocations in Anthracene Crystals
(a) Vol. **42**, No. 2, p. 433–437 (1977).
See also *10.2*, p. *90*; 22.9, p. 417.

H. RÜFER and W. RUPPEL
*Luminescence of Graded $Zn_xCd_{1-x}S$ Single Crystals
(a) Vol. **42**, No. 2, p. 519–524 (1977).
See also 13.5.1, p. 146; *20.2*, p. *266*; 22.4.4, p. 368.

G. MASSÉ, J.-P. AICARDI, and C. BUTTI
*Study of Band-Gap Shrinkage in Cinnabar (α-HgS) by Cathodoluminescence under Very Strong Excitation
(a) Vol. **42**, No. 2, p. 545–552 (1977).
See also *13.1*, p. *121*; 22.4.1, p. 363.

G. S. WHITE, K. H. LEE, and J. H. CRAWFORD, JR.
*Effects of γ-Irradiation upon the Optical Behavior of Spinel
(a) Vol. **42**, No. 2, p. K137–K141 (1977).
See also 11, p. 97; *20.1*, p. *259*; 22.8.2, p. 412.

R. K. GARTIA
*Optical Bleaching — a Technique to Clean Thermoluminescence Peaks
(a) Vol. **42**, No. 2, p. K155–K158 (1977).
See also *10.2*, p. *90*; 22.5.2, p. 380.

G. GIRO and P. DI MARCO
Origin of Photocarriers in $Ru(2,2'\text{-Bipyridine})_3Cl_2 \cdot 6\,H_2O$ Single Crystals
(a) Vol. **42**, No. 2, p. K179–K182 (1977).
See also *16*, p. *203*; *20.1*, p. *259*; 22.9, p. 417.

P. A. ARSENEV, A. V. POTEMKIN, V. V. FENIN, and I. SENFF
Investigation of Stimulated Emission of Ho^{3+} Ions in Mixed Crystals with Perovskite Structure
(a) Vol. **42**, No. 2, p. K183–K185 (1977).
See also 20.1, p. 259; *20.2*, p. *266*; 22.8.1, p. 406.

J. BURGOS, M. POPE, CH. E. SWENBERG, and R. R. ALFANO
*Heterofission in Pentacene-Doped Tetracene Single Crystals
(b) Vol. **83**, No. 1, p. 249–256 (1977).
See also *13.5.1*, p. *146*; 22.9, p. 418.

N. F. BERK, J. ROSENTHAL, L. YARMUS, and C. E. SWENBERG
*Magnetic Field Dependence of Prompt Fluorescence from Crystalline Tetracene
(b) Vol. **83**, No. 1, p. K1–K6 (1977).
See also *18*, p. *212*; 22.9, p. 418.

A. A. KAMINSKII, S. E. SARKISOV, J. BOHM, P. REICHE, D. SCHULTZE, and R. UECKER
*Growth, Spectroscopic and Laser Properties of Crystals in the $K_5Bi_{1-x}Nd_x(MoO_4)_4$ System
(a) Vol. **43**, No. 1, p. 71–79 (1977).
See also 3, p. 42; 13.4, p. 137; 20.1, p. 260; *20.2*, p. *266*; 22.8, p. 399.

F. R. KESSLER und E. BREDNER
*Kathodolumineszenz von Selen
(Cathodoluminescence of Selenium)
(a) Vol. **43**, No. 1, p. 159–166 (1977).
See also 13.5.1, p. 147; 22.1.3, p. 346.

H. Raidt, G. Kästner, M. Menninger, and G. Voigt
On Real Structure Investigations of $Ga_xIn_{1-x}P$ by Electron Microscopic Methods
(a) Vol. **43**, No. 1, p. 167–173 (1977).
See also *1.4*, p. *27*; *3*, p. *43*; 22.2.4, p. 357.

G. Massé, J. P. Aicardi, and J. P. Leyris
*Green Emission of Mercury Sulphide (α-HgS)
(a) Vol. **43**, No. 1, p. 191–201 (1977).
See also 6, p. 59; 13.4, p. 137; 22.4.1, p. 363.

R. Rentzsch and I. S. Shlimak
*Photoluminescence of Heavily Doped and Compensated Germanium
(a) Vol. **43**, No. 1, p. 231–238 (1977).
See also 22.1.1, p. 335.

A. L. Kartuzhanskii, A. G. Kavetskii, L. D. Kekhva, T. E. Kekhva, A. P. Kiselev, and B. T. Plachenov
*Recombination Processes during Excitation of $AgCl:Me^{2+}$ Crystals by Ionizing Radiation
(a) Vol. **43**, No. 1, p. 321–327 (1977).
See also *13.4*, p. *137*; 22.5.1, p. 373.

F. Henneberger, K. Henneberger, and J. Voigt
*Two-Photon Resonance Raman Scattering by Excitons via an Intermediate Excitonic Molecule (II): Polarization Properties and Polariton Effects
(b) Vol. **83**, No. 2, p. 439–450 (1977).
See also *13.5.1*, p. *147*; 22.4.1, p. 364; 22.5, p. 371.

A. Vanderhaeghe and G. Jacobs
On F- and F_H-Centres in Cesium Halides
(b) Vol. **83**, No. 2, p. 471–477 (1977).
See also *10.2*, p. *90*; 22.5.2, p. 380.

D. Hadžiahmetović, M. Pirić, and B. S. Tošić
Kinematical Exciton Levels and Absorption of Light
(b) Vol. **83**, No. 2, p. 479–483 (1977).
See also *13.5.1*, p. *147*; 20.1, p. 260.

B. Hönerlage, Vu Duy Phach, A. Bivas, and E. Ostertag
*Polariton Effects in Resonant Raman Scattering in CuCl
(b) Vol. **83**, No. 2, p. K101–K103 (1977).
See also *13.5*, p. *141*; 22.5, p. 372.

N. A. Drozdov, A. A. Patrin, and V. D. Tkachev
*On the Nature of the Dislocation Luminescence in Silicon
(b) Vol. **83**, No. 2, p. K137–K139 (1977).
See also 10.2, p. 91; 22.1.2, p. 342.

H. Eichele and M. Schwoerer
*Fluorescence of Localized States in a Partially Polymerized Diacetylene Single Crystal
(a) Vol. **43**, No. 2, p. 465–472 (1977).
See also 13.4, p. 138; 22.9, p. 418.

R. Bindemann, H. Fischer, and K. Kreher
*Determination of the Concentration of Substitutional O in GaP by Means of Donor–Acceptor Pair Luminescence
(a) Vol. **43**, No. 2, p. 529–533 (1977).
See also *13.4*, p. *138*; 22.2.2, p. 354.

T. Matsumura, M. Sudo, C. Tatsuyama, and S. Ichimura
*Photoluminescence in GaSe
(a) Vol. **43**, No. 2, p. 685–693 (1977).
See also 13.5.1, p. 147; 22.3, p. 359.

Z. LUDMER
*A Metastable Excimer-Emitting Crystal Form of 9-Cyanoanthracene
(a) Vol. **43**, No. 2, p. 695—698 (1977).
See also 22.9, p. 419.

P. DI MARCO and G. GIRO
Photoconductive Properties of Fluorene Single Crystals
(a) Vol. **43**, No. 2, p. K193—K197 (1977).
See also *16*, p. *204*; 20.1, p. 261; 22.9, p. 419.

M. CASALBONI, U. M. GRASSANO, A. SCACCO, and A. TANGA
*Emission Spectra of NaBr:Tl^+ Excited in the A-Band
(b) Vol. **84**, No. 1, p. 23—26 (1977).
See also *13.4*, p. *138*; 22.5.2, p. 381.

A. BIVAS, VU DUY PHACH, B. HÖNERLAGE, and J. B. GRUN
*Induced Absorption Due to Exciton–Biexciton Transitions in CuCl
(b) Vol. **84**, No. 1, p. 235—243 (1977).
See also *13.5.1*, p. *148*; 22.5, p. 372.

LE MANH HOANG and J. M. BARANOWSKI
*Absorption and Luminescence of V(d^3) in II–VI Compound Semiconductors
(b) Vol. **84**, No. 1, p. 361—365 (1977).
See also 13.4, p. 138; *20.1*, p. *261*; 22.4, p. 361.

R. SCHWABE, F. THUSELT, and T. HÄNSEL
Exciton–Phonon Energy Transfer in Semiconductors
(b) Vol. **84**, No. 1, p. K43—K47 (1977).
See also *6*, p. *60*; *13.5.1*, p. *148*; 22.3, p. 359.

R. KINK, A. LÔHMUS, M. SELG, and T. SOOVIK
*Luminescence Lifetimes of Crystalline Xenon
(b) Vol. **84**, No. 1, p. K61—K63 (1977).
See also 23, p. 420.

P. A. ARSENEV, KH. S. BAGDASAROV, A. V. POTEMKIN, and V. V. FENIN
*Spectra of Absorption, Luminescence, and Stimulated Emission of Neodymium in Crystals of Mixed Yttrium and Gadolinium Scandates
(a) Vol. **44**, No. 1, p. 51—57 (1977).
See also *20.1*, p. *262*; 20.2, p. 266; 22.8.1, p. 407.

C. LAWTHER and K. V. ANAND
*Physical Mechanism of Current Conduction and Light Emission in High-Resistivity ZnS:Mn Thin Films
(a) Vol. **44**, No. 1, p. 313—323 (1977).
See also *14.3.1*, p. *175*; 22.4.1, p. 364.

R. K. GARTIA
*Thermoluminescence of Z_3 Centres in X-Irradiated Mg-Doped LiF Crystals
(a) Vol. **44**, No. 1, p. K21—K24 (1977).
See also *10.2*, p. *92*; 22.5.2, p. 381.

A. ZARĘBA, A. J. NADOLNY, and B. KRUKOWSKA-FULDE
*Thermoluminescence of Mn Centres in CdF_2 Crystals
(a) Vol. **44**, No. 1, p. K83—K85 (1977).
See also 22.5.3, p. 384.

A .D. AVEZOV, S. A. VAKHIDOV, E. M. GASANOV, A. I. NOVOZHILOV, and M. I. SAMOILOVICH
*Some Luminescence Properties of Synthetic Mica Fluorphlogopite
(a) Vol. **44**, No. 1, p. K87—K89 (1977).
See also 22.8.1, p. 407.

W. Schmid
*Auger Lifetimes for Excitons Bound to Neutral Donors and Acceptors in Si
(b) Vol. **84**, No. 2, p. 529—540 (1977).
See also *13.5.1*, p. *148*; 22.1.2, p. 344.

M. Aguilar, F. Jaque, and F. Agulló-López
*X-Ray Induced Luminescence in NaCl:Ca^{++}
(b) Vol. **84**, No. 2, p. 595—601 (1977).
See also 10.2, p. 92; 22.5.2, p. 381.

E. Ostertag, A. Bivas, and J. B. Grun
*Polariton Effects on the Recombination of Cold Biexcitons
(b) Vol. **84**, No. 2, p. 673—680 (1977).
See also *13.5.1*, p. *148*; 22.5, p. 372.

H.-J. Kos and R. Nink
*Mechanism of Restored Thermoluminescence in Lithium Fluoride
(a) Vol. **44**, No. 2, p. 505—510 (1977).
See also *10.2*, p. *93*; 20.1, p. 263; 22.5.2, p. 381.

C. Lawther and J. Woods
*Forward-Bias Electroluminescence in ZnSe Diodes
(a) Vol. **44**, No. 2, p. 693—702 (1977).
See also 22.4.2, p. 366.

A. Lyubchenko, I. Potykievich, A. Fedorov, and M. Sheinkman
*On the Mechanism of Photoluminescence of Cadmium Diphosphide Doped by Bi and Zn
(a) Vol. **44**, No. 2, p. 747—753 (1977).
See also *16*, p. *205*; 22.3, p. 359.

K. D. Glinchuk and A. V. Prokhorovich
*Effect of the 0.94, 1.0, 1.2, and 1.3 eV Radiative Centres on the Intrinsic Luminescence Intensity in n-GaAs
(a) Vol. **44**, No. 2, p. 777—785 (1977).
See also *13.4*, p. *140*; 22.2.1, p. 353.

G. P. M. van den Heuvel and G. Blasse
*The Blue Emission of $PbCl_2$
(a) Vol. **44**, No. 2, p. K163—K166 (1977).
See also 22.5, p. 372.

21. Metals. Alloys

Original Papers and Short Notes

M. Fähnle and H. Kronmüller
On the Domain Structure near the Penetration Field of the Intermediate State in Type-I Superconductors
(b) Vol. **79**, No. 1, p. 139–147 (1977).
See also *14.2*, p. *161*.

F. K. Schulte
Energies and Fermi Level of Electrons in Thin Size-Quantized Metal Films
(b) Vol. **79**, No. 1, p. 149–153 (1977).
See also *13.3*, p. *126*.

G. S. Agarwal and H. D. Vollmer
Surface Polariton Effects in Spontaneous Emission
(b) Vol. **79**, No. 1, p. 249–257 (1977).
See also *13.3*, p. *126*; 13.5.2, p. 149; 20.1, p. 247; 22.2.2, p. 354.

E. Kolley and W. Kolley
On the Theory of Phonon-Modulated Electronic Transport in Off-Diagonal Random Alloys
(b) Vol. **79**, No. 1, p. 325–337 (1977).
See also *2*, p. *33*; 6, p. 51; *14*, p. *155*.

M. V. Medvedev and A. V. Zaborov
Random Collinear Structures of a Dilute Magnetic Alloy
(b) Vol. **79**, No. 1, p. 379–390 (1977).
See also *18.2*, p. *217*.

F. K. Schulte
Electronic Density of States, Heat Capacity, and Spin Susceptibility of Thin Metal Films
(b) Vol. **79**, No. 1, p. K5–K8 (1977).
See also 8, p. 64; *13.3*, p. *126*; 18, p. 209.

M. A. Belogolovskii, Yu. M. Ivanchenko, E. V. Kochergin, A. A. Lisyanskii, and Yu. V. Medvedev
Interaction of Electrons with Phonons and Magnetic Impurities in Tunneling Junctions
(b) Vol. **79**, No. 1, p. K17–K21 (1977).
See also 6, p. 52; *14.1*, p. *157*; *14.2*, p. *161*.

T. Soma
Electronic Screening Effect on the Structure Stability of Face-Centered Tetragonal In
(b) Vol. **79**, No. 1, p. K53–K56 (1977).
See also *4*, p. *44*; *13*, p. *107*.

H. Ortbauer, W. Steiner, and R. Haferl
Mössbauer Studies of $Y(Fe_x Co_{1-x})_2$ in the Region of Small Fe Concentrations
(a) Vol. **39**, No. 1, p. 157–165 (1977).
See also *5*, p. *47*; 18.1, p. 213; 18.2, p. 217; 21.1, p. 288; 21.1.1, p. 302.

H. U. Habermeier
The Penetration Field and the Magnetization Curve of Type-I Superconductors Containing Lattice Defects
(a) Vol. **39**, No. 1, p. 203–212 (1977).
See also 10.1, p. 75; *14.2*, p. *161*; 18, p. 209.

E. Ronander and S. Kritzinger
The Structure of Multiple-Faulted Loop Configurations in Quenched Aluminium and Aluminium Alloys
(a) Vol. **39**, No. 1, p. 239–248 (1977).
See also *10.1*, p. *75*.

J.-M. Derochette and J. Marien
Field Emission Study of Hydrogen Adsorption on Smooth and Perfectly Ordered Iridium Surfaces (111) and (100) — Influence of Steps and Iridium Clusters on the Surface Potential of the Hydrogen Film
(a) Vol. **39**, No. 1, p. 281—289 (1977).
See also *1.6*, p. *31*; *17.1*, p. *208*.

V. S. Postnikov, V. V. Postnikov, and V. S. Zheleznyi
Superconductivity in Mo–Re System Alloy Films Produced by Electron Beam Evaporation in High Vacuum
(a) Vol. **39**, No. 1, p. K21—K23 (1977).
See also *14.2*, p. *161*.

H. Oesterreicher
Possible Disorder of Magnetic Components (Magnetic Scatter Order) in Compounds on the Basis of Transition Metals
(a) Vol. **39**, No. 1, p. K41—K43 (1977).
See also *18*, p. *209*; 21.1, p. 289.

A. Feher, M. Reiffers, P. Petrovič, and Š. Jánoš
The Electrical Resistivity of Nb–Ti and Nb–Ti–Zr Alloys at Low Temperatures
(a) Vol. **39**, No. 1, p. K67—K69 (1977).
See also *14.1*, p. *157*; 14.2, p. 161.

V. I. Grebennikov, Yu. A. Babanov, and O. B. Sokolov
Extra-Atomic Relaxation and X-Ray Spectra of Narrow-Band Metals (I): Formalism
(b) Vol. **79**, No. 2, p. 423—432 (1977).
See also *13*, p. *107*; 20, p. 244.

E. Kolley and W. Kolley
Calculation of the Temperature-Dependent DC Conductivity of Disordered Alloys in Off-Diagonal CPA
(b) Vol. **79**, No. 2, p. 461—472 (1977).
See also *2*, p. *33*; 6, p. 52; 13, p. 107; *14.1*, p. *157*.

J. R. Sambles, G. Sharp-Dent, J. E. Cousins, A. Stesmans, and J. Witters
Conduction Electron Spin Resonance in Aluminium at 20.98 and 9.27 GHz
(b) Vol. **79**, No. 2, p. 645—654 (1977).
See also *19*, p. *235*.

V. Lantto
A Theoretical Study of the Absorption of Photoelectrons Created by X-Rays in Semi-Infinite Targets
(b) Vol. **79**, No. 2, p. 699—711 (1977).
See also *9*, p. *69*; *17*, p. *206*; 21.1, p. 289; 21.6, p. 315.

S. C. Goel, Y. R. Midha, and V. S. Nanda
Applications of a Simple Equation of State for Metals
(b) Vol. **79**, No. 2, p. K103—K105 (1977).
See also *8*, p. *64*; 21.1, p. 289; 21.6, p. 315.

M. Horobiowski, T. Skośkiewicz, and E. Trojnar
Magnetization of the Superconducting Palladium Hydrides
(b) Vol. **79**, No. 2, p. K147—K150 (1977).
See also *14.2*, p. *161*; 18, p. 209.

C. P. Chang and C. W. Chen
Observations on Interaction of Glissile Dislocations in Niobium by Transmission Electron Microscopy
(a) Vol. **39**, No. 2, p. 549—560 (1977).
See also *10.1*, p. *75*.

A. Vincent, S. M. Seyed Reihani, and J. Perez
The Nature of Dislocation-Pinning Defects in Cold-Worked Aluminium Studied by Ultrasonic Waves
(a) Vol. **39**, No. 2, p. 651—660 (1977).
See also 7, p. 62; *10.1*, p. *76*.

E. Bonetti, E. Evangelista, P. Gondi, and R. Tognato
Deformation and Annealing Effects on High-Temperature Internal Friction Peaks of Aluminium Single and Polycrystals
(a) Vol. **39**, No. 2, p. 661—667 (1977).
See also *10.1*, p. *76*; *12.1*, p. *101*.

R. Wang and M. D. Merz
Polymorphic Bonding and Thermal Stability of Elemental Noncrystalline Solids
(a) Vol. **39**, No. 2, p. 697—703 (1977).
See also 1.2, p. 15; *2*, p. *34*; *4*, p. *44*; 22.1, p. 332.

G. Dlubek, O. Brümmer, and N. Meyendorf
A Study of the Influence of Mono-and Multivacancies on the Peak Height of the Angular Correlation Curve of the Positron Annihilation Radiation in Al, Cu, and Ni
(a) Vol. **39**, No. 2, p. K95—K99 (1977).
See also *10.1*, p. *76*; 21.1, p. 289.

H. Düsterhöft and A. Ihlenfeld
The Energy Distribution of Positive Secondary Ions Emitted from Metal and Semiconductor Targets Bombarded with 12 keV Ar^+ Ions
(a) Vol. **39**, No. 2, p. K147—K150 (1977).
See also *17*, p. *206*; 21.1, p. 290; 22.1.2, p. 338.

V. I. Grebennikov, Yu. A. Babanov, and O. B. Sokolov
Extra-Atomic Relaxation and X-Ray Spectra of Narrow-Band Metals (II): Results
(b) Vol. **80**, No. 1, p. 73—82 (1977).
See also *13*, p. *108*; 20, p. 244.

O. Takai, R. Yamamoto, and M. Doyama
The Formation Energies of a Vacancy and a Divacancy and the Binding Energy of a Divacancy in Simple Metals
(b) Vol. **80**, No. 1, p. 193—199 (1977).
See also *10.1*, p. *76*.

G. Iadonisi, V. Marigliano Ramaglia, and G. P. Zucchelli
Atomic Hydrogen Levels near a Metallic Surface
(b) Vol. **80**, No. 1, p. 245—254 (1977).
See also *13.3*, p. *126*.

N. C. Halder
Calculation of Nuclear Quadrupolar Relaxation in Liquid Metals
(b) Vol. **80**, No. 1, p. 273—278 (1977).
See also *2*, p. *34*; *19*, p. *236*.

H. J. Stolz, E. Petri, and A. Otto
Optical Properties and Plasmon Dispersion in Epitaxial $(SN)_x$ by Electron Energy Loss Spectroscopy
(b) Vol. **80**, No. 1, p. 289—294 (1977).
See also *13.5.2*, p. *149*; 20.1, p. 251.

A. B. Makhnovetskii and G. L. Krasko
The Pseudopotential Method and the Problem of Ordering in Alloys (III): Intermetallic Compounds LiAl, LiTl, and NaTl
(b) Vol. **80**, No. 1, p. 341—351 (1977).
See also 1.3, p. 21; *4*, p. *44*.

V. Geist, R. Flagmeyer, D. Stephan, and H.-J. Ullrich
Kossel Interferences of Proton-Induced AlKα Radiation
(a) Vol. **40**, No. 1, p. 113–117 (1977).
See also *1.1*, p. *11*; 21.1, p. 290.

D. Segers, C. Dauwe, M. Dorikens, and L. Dorikens-Vanpraet
On the Effect of Thermal Expansion on Positron Annihilation Lineshape Factors
(a) Vol. **40**, No. 1, p. 153–157 (1977).
See also *8*, p. *65*.

A. van Veen, L. M. Caspers, E. V. Kornelsen, R. Fastenau, A. van Gorkum, and A. Warnaar
Vacancy Creation by Helium Trapping at Substitutional Krypton in Tungsten
(a) Vol. **40**, No. 1, p. 235–246 (1977).
See also *10.1*, p. *76*.

B. Wielke, W. Tikvic, and G. Schoeck
Cottrell-Stokes Law in Cadmium and Zinc
(a) Vol. **40**, No. 1, p. 271–278 (1977).
See also *10.1*, p. *76*; 21.1, p. 291.

H. Drulis, I. A. Garifullin, E. G. Kharakhashyan, B. Staliński, N. M. Suleymanov, and M. M. Zaripov
EPR in the $Pd_{1-y}Gd_yH_x$ System
(a) Vol. **40**, No. 1, p. K11–K14 (1977).
See also *19*, p. *237*.

J. P. Badiali, A. Defrain, M. Dupont, and C. Regnaut
Effective Pair Potential and Thermodynamical Properties of Liquid Gallium
(b) Vol. **80**, No. 2, p. 533–541 (1977).
See also *2*, p. *35*; 8, p. 65.

H. Neddermeyer
Charge Transfer in Aluminium–Magnesium Alloys
(b) Vol. **80**, No. 2, p. 611–614 (1977).
See also *13*, p. *108*; 20, p. 244; 21.3, p. 310.

G. G. Khalyullin
Concentration Dependence of Exchange Field in Dilute Metals
(b) Vol. **80**, No. 2, p. 709–711 (1977).
See also *18.1*, p. *214*.

R. Yamamoto, N. Owada, O. Takai, and M. Doyama
Calculations of the Heat of Solution of the Impurities in Aluminium and the Cohesive Energies of Simple Metals
(b) Vol. **80**, No. 2, p. K131–K135 (1977).
See also *4*, p. *45*; *10.1*, p. *77*.

R. L. Havill, J. Marsden, and J. M. Titman
Nuclear Quadrupole Relaxation in Liquid Mercury
(b) Vol. **80**, No. 2, p. K141–K143 (1977).
See also *2*, p. *36*; *19*, p. *237*.

G. Athanassiadis, G. Le Caer, J. Foct, and L. Rimlinger
Study of Ternay Ordered Solid Solutions Derived from Fe_3Al by Substitution
(a) Vol. **40**, No. 2, p. 425–435 (1977).
See also *1.3*, p. *21*; 5, p. 48; 21.1.1, p. 303.

U. Gahn
Short- and Long-Range Order of B.C.C. Solid Solutions
(a) Vol. **40**, No. 2, p. 463–470 (1977).
See also *1.2*, p. *16*; *1.3*, p. *21*.

Yu. V. Konobeev, V. I. Tscherbak, V. N. Bykov, and A. V. Subbotin
Determination of the Vacancy Diffusivity in Molybdenum and Nickel from Data on the Width of the Grain Boundary Void Denuded Zone
(a) Vol. **40**, No.2, p. K89—K92 (1977).
See also *9*, p. *69*; *11*, p. *95*; 21.1, p. 292.

R. Perthel, W. Keilig, R. Kosicik, U. Röpke, and C.-G. d'Ambly
Determination of the Magnetic Anisotropy in Amorphous Gd–Co Films by Torque Measurement
(a) Vol. **40**, No. 2, p. K135—K138 (1977).
See also *2*, p. *36*; *18.3*, p. *229*; 21.1, p. 292.

H. Oesterreicher
Magnetic Properties of Scatter Order Compounds RFeAl (R = Gd, Tb, Dy, Ho, Er, Tm, Lu, and Y)
(a) Vol. **40**, No. 2, p. K139—K143 (1977).
See also *18*, p. *210*; 21.1.1, p. 303; 21.4, p. 312.

R. Kroggel, H. Löffler, and O. Kabisch
Experimental Investigations on the Position of the Coherency Spinodal Curve in Al–Zn (15, 16, and 18 at%) Alloys
(a) Vol. **40**, No. 2, p. K163—K164 (1977).
See also 1.2, p. 16; *10.1*, p. *77*; 21.1, p. 292.

G. M. Carneiro and P. R. Rodrigues Pereira
Fermi Liquid Effects in Ferromagnetic Metals
(b) Vol. **81**, No. 1, p. 79—88 (1977).
See also *13.5.4*, p. *153*; 18.2, p. 219.

A. Naudon and A. M. Flank
Spinodal Curve and Metastable Miscibility Gap in Low-Concentrated Al–Ag Alloys
(a) Vol. **41**, No. 1, p. 207—211 (1977).
See also *1.2*, p. *16*; *1.3*, p. *21*; 21.6, p. 316.

M. Condat and M. Fayard
On the Formation of Prismatic Loops in Quenched F.C.C. Metals and Alloys
(a) Vol. **41**, No. 1, p. K5—K8 (1977).
See also *1.4*, p. *26*; *10.1*, p. *78*.

J. Sołtys
Order–Disorder Phase Transitions in the Heusler Alloy Ni_2MnIn
(a) Vol. **41**, No. 1, p. K51—K54 (1977).
See also *1.3*, p. *21*; 21.1, p. 293.

S. Saterlie, M. L. Johnson, P. Alexopoulos, F. Seppi, R. Ure, and J. G. Byrne
Positron Annihilation Observations of Shot Peened Aluminium Alloys
(a) Vol. **41**, No. 1, p. K75—K78 (1977).
See also *12.1*, p. *101*.

S. S. Bhandari and J. Varma
Anharmonicity Effect in the Oxides of Pb, Sn, and Bi
(a) Vol. **41**, No. 1, p. K89—K92 (1977).
See also *5*, p. *48*; 21.7, p. 319.

F. Brosens, J. Cornelis, D. C. Wallace, and J. T. Devreese
A Method to Derive Interatomic Potentials in Metals from Experimental Phonon Spectra
(b) Vol. **81**, No. 2, p. 557—564 (1977).
See also *4*, p. *45*; 6, p. 56; 21.2, p. 308.

J. Lakner
The Effect of Non-Transition Metals on the Magnetic Susceptibility of Aluminium Solid Solutions
(b) Vol. **81**, No. 2, p. 593—601 (1977).
See also 13.2, p. 124; *18*, p. *210*.

V. B. Campos, O. Hipólito, and R. Lobo
Correlations and Plasmons in Quasi-One-Dimensional Metals
(b) Vol. **81**, No. 2, p. 657–663 (1977).
See also *13.5.2*, p. *149*; 14.4, p. 191.

E. Kolley and W. Kolley
Coherent Ladder Approximation for the Random Hubbard Model
(b) Vol. **81**, No. 2, p. 735–745 (1977).
See also *13*, p. *109*.

A. Balogh and I. Dézsi
Further Positron Lifetimes of Some Elements
(b) Vol. **81**, No. 2, p. K81–K82 (1977).
See also *13*, p. *110*; 22.1.3, p. 345.

V. Christoph and K. Elk
Electrical Conductivity of Disordered Substitutional Alloys with Strong Electron Correlation
(b) Vol. **81**, No. 2, p. K125–K130 (1977).
See also *14.1*, p. *158*.

H. D. Wiesinger
Possibilities for the Increase of the Transition Temperature to Superconductivity from the Metal-Scientific Point of View
(a) Vol. **41**, No. 2, p. 465–479 (1977).
See also *1.3*, p. *21*; *14.2*, p. *162*.

S. H. van den Brink, A. van den Beukel, and P. G. McCormick
The Influence of Specimen Dimensions and Machine Stiffness on the Portevin-Le Chatelier Effect in an Al–Cu Alloy
(a) Vol. **41**, No. 2, p. 513–521 (1977).
See also *10.1*, p. *78*; 21.1, p. 294.

G. A. Hassan
Superplastic Behaviour of an Aluminium–Magnesium Alloy Effected by Bidirectional Twisting
(a) Vol. **41**, No. 2, p. 541–546 (1977).
See also *10.1*, p. *78*; 21.3, p. 310.

G. Laslaz, G. Kostorz, M. Roth, P. Guyot, and R. J. Stewart
Interparticle Interference in Neutron Small-Angle Scattering of Al–6.8 at% Zn during Ageing at 133 °C
(a) Vol. **41**, No. 2, p. 577–583 (1977).
See also *1.4*, p. *26*; 10.1, p. 78; 21.1, p. 294.

J. J. Couderc, J. Bras and M. Fagot
Precipitation in the Iron-Gallium System in the Vicinity of 25 at % Ga
(a) Vol. **41**, No. 2, p. 595–605 (1977).
See also *1.2*, p. *17*; *1.3*, p. *22*; 21.1.1, p. 304.

J. P. Simon
Annealing Study of the Vacancy-Impurity Interaction in the Diluted Alloys $\underline{\mathrm{Al}}$Zr and $\underline{\mathrm{Al}}$Cr
(a) Vol. **41**, No. 2, p. K107–K112 (1977).
See also *10.1*, p. *78*; 21.1, p. 294.

D. Wagner and F. Stangler
Activation Energy and Primary Characteristics of the Bordoni Relaxation Peak of Highest Purity Polycrystalline Lead
(a) Vol. **41**, No. 2, p. K137–K140 (1977).
See also *10.1*, p. *78*; *12.1*, p. *101*.

S. Takács and A. Pevala
The Angular Dependence of Critical Currents in Superconductors with Long Narrow Defects
(a) Vol. **41**, No. 2, p. K175–K179 (1977).
See also 10.1, p. 78; *14.2*, p. *162*.

I. Stulíková, P. Lukáč, and G. I. Shklarevskaya
The Critical Resolved Shear Stress of Cd–Ag Single Crystals at Low Temperatures
(a) Vol. **41**, No. 2, p. K197—K198 (1977).
See also *10.1*, p. *78*; 21.6, p. 316.

O. Singh
Ultrasonic Absorption in the Intermediate State of Gallium and Indium
(b) Vol. **82**, No. 1, p. 179—188 (1977).
See also *7*, p. *62*; 14.2, p. 162.

S. Tanigawa, K. Hinode, and M. Doyama
Phenomenological Interpretation of Positron Lifetimes in Metals
(b) Vol. **82**, No. 1, p. 369—376 (1977).
See also *11*, p. *96*.

R. Taranko
The Effect of the Model Hybridized Energy Bands on the Field Emitted Electrons from Metals
(b) Vol. **82**, No. 1, p. K5—K7 (1977).
See also *13.1*, p. *119*; *17*, p. *206*.

A. Lodder and J. C. C. de Ruiter
The Anisotropy of the Lattice Distortion Structure Factor in the Electron–Impurity Scattering Rate
(b) Vol. **82**, No. 1, p. K23—K28 (1977).
See also *10.1*, p. *78*.

Mahendra Kumar and M. P. Hemkar
Density of States and Fermi Energy of Simple Metals
(b) Vol. **82**, No. 1, p. K29—K34 (1977).
See also *13.2*, p. *124*; 21.2, p. 309.

V. B. Vykhodets, A. Ya. Fishman, and V. B. Demin
An Analysis of Deviations from Sievert's Law for Hydrogen in Metals
(b) Vol. **82**, No. 1, p. K59—K61 (1977).
See also *10.1*, p. *78*.

G. Dlubek, O. Brümmer, N. Meyendorf, I. Ya. Dekhtyar and R. G. Fedchenko
Positron Annihilation Studies of Annealing of Cystal Defects in Plastically Deformed PdFe Alloys
(a) Vol. **42**, No. 1, p. K15—K18 (1977).
See also *10.1*, p. *79*.

F. Antonangeli, A. Balzarotti, A. Bianconi, E. Burattini, P. Perfetti, and A. Kisiel
Influence of the Hydrogenation on the Electrical Resistance of Palladium Thin Films
(a) Vol. **42**, No. 1, p. K41—K45 (1977).
See also *14.1*, p. *159*.

C. L. Reynolds, Jr. and P. R. Couchman
Tait's Equation of State at High Pressures and the Isothermal Bulk Modulus
(a) Vol. **42**, No. 1, p. K47—K52 (1977).
See also *12.1*, p. *102*; *12.2*, p. *105*; 22.5.2, p. 379; 23, p. 420.

L. C. R. Alfred
Charge Screening in an Electron Gas of Arbitrary Degeneracy
(b) Vol. **82**, No. 2, p. 467—471 (1977).
See also *13.1*, p. *120*; *14*, p. *156*; 22, p. 326.

W. F. W. M. van Heugten
An Interatomic Potential for Molybdenum from Phonon Spectra
(b) Vol. **82**, No. 2, p. 501—506 (1977).
See also 6, p. 57; *10.1*, p. *79*.

M. Taut and A. M. Radwan
Soft X-Ray Spectra of Al and the Laves Phase $CaAl_2$
(b) Vol. 82, No. 2, p. 507—511 (1977).
See also *13.1*, p. *120*; *20*, p. *245*; 22, p. 326.

H. J. de Bruin, H. Bakker, and L. P. van der Mey
Diffusion Correlation in Random Primitive Cubic Alloys
(b) Vol. 82, No. 2, p. 581—588 (1977).
See also *9*, p. *70*; 10.1, p. 79.

K. Suzuki, M. Kitagawa, and Y. H. Ohtsuki
Theory of Surf-Riding Electrons in the Wake Potential of Ion Beams in a Metal
(b) Vol. 82, No. 2, p. 643—650 (1977).
See also *13.5.2*, p. *149*.

C. Sułkowski
Electronic Structure and Superconductivity in Solid Solutions Based on In, Pb, and Tl
(b) Vol. 82, No. 2, p. 657—662 (1977).
See also *14.2*, p. *162*.

V. V. Maksimenko, A. J. Simonov, and A. A. Lushnikov
Surface Plasmons in Thin Metal Films, Circular Cylinders, and Spherical Voids
(b) Vol. 82, No. 2, p. 685—693 (1977).
See also *13.5.2*, p. *149*.

G. K. Bristow, T. R. Finlayson, and T. F. Smith
Hall Coefficient Sign Reversals in 2H Transition Metal Dichalcogenides
(b) Vol. 82, No. 2, p. K81—K84 (1977).
See also *13.2*, p. *124*; *14.1*, p. *159*.

M. Rohr and P. Müller
Superconducting Properties of Cold-Worked Core-Wires on the Basis of A15-Phases
(a) Vol. 42, No. 2, p. 477—482 (1977).
See also *1.2*, p. *18*; *14.2*, p. *162*.

N. A. Stolwijk, T. Spruijt, M. A. Hoetjes-Eijkil, and H. Bakker
Measurements of Diffusion Coefficients of ^{60}Co in the Intermetallic Compound CoGa
(a) Vol. 42, No. 2, p. 537—544 (1977).
See also *9*, p. *70*; 21.1, p. 296.

M. J. Witcomb
Microstructure and Superconducting Properties of Niobium and Tantalum Films Sputtered onto Copper Substrates
(a) Vol. 42, No. 2, p. 595—607 (1977).
See also *1.4*, p. *27*; *14.2*, p. *162*.

P. Čapková and J. Šedivý
X-Ray Debye Characteristic Temperatures of Magnesium and α-Titanium
(a) Vol. 42, No. 2, p. 681—686 (1977).
See also *6*, p. *58*; *8*, p. *66*; 21.3, p. 310.

M. Fumeron, J. P. Guillot and A. Dauger
Model X-Ray Scattering by Ellipsoidal Giunier-Preston Zones in the Case of Al–Zn
(a) Vol. 42, No. 2, p. 687—697 (1977).
See also *1.3*, p. *22*; 12.1, p. 102; 21.1, p. 296.

W. D. Schmidt and G. Albrecht
Enhancement of the Instability Current of Superconducting Thin Film Bridges in Perpendicular Magnetic Fields
(a) Vol. 42, No. 2, p. K143—K145 (1977).
See also *14.2*, p. *163*.

J. Sosnowski, M. Dryś, and J. Szymaszek
Properties of Superconducting Nb–Ga Alloys with A15 Type Structure
(a) Vol. **42**, No. 2, p. K159—K164 (1977).
See also *14.2*, p. *163*.

M. L. Johnson, S. Panchanadeeswaran, S. Saterlie, and J. G. Byrne
Positron Measurements of Aging in Complex Aluminium Alloys
(a) Vol. **42**, No. 2, p. K175—K177 (1977).
See also *10.1*, p. *80*.

K. Saermark, J. Lebech, and H. Johansen
On the Concept of Cyclotron Resonance in Metals
(b) Vol. **83**, No. 1, p. 187—195 (1977).
See also *13.2*, p. *124*; *14.1*, p. *159*.

G. Paasch and M. Hietschold
A Layer Summation for Electrostatic Surface Problems. Application to the Classical Cleavage Energy
(b) Vol. **83**, No. 1, p. 209—222 (1977).
See also *13.3*, p. *127*.

O. B. Sokolov, V. I. Grebennikov, and E. A. Turov
A New Approach to the Theory of X-Ray Spectra of Metals with Allowance for the Finite Lifetime of the Deep Hole
(b) Vol. **83**, No. 1, p. 281—291 (1977).
See also *13*, p. *110*; 20, p. 246.

R. K. Puri and L. R. Gupta
Computation of Force Constant Change of Dilute Mössbauer Impurities in Metallic Hosts from Mössbauer Parameters
(b) Vol. **83**, No. 1, p. 293—298 (1977).
See also *5*, p. *49*; 6, p. 58.

I. Ya. Dekhtyar and V. I. Shevchenko
Positron Annihilation in Hydrides of Transition Metals
(b) Vol. **83**, No. 1, p. 323—330 (1977).
See also *13*, p. *110*.

A. M. Brodskii
Specific Features of Many-Photon Photoemission from Metals and Semiconductors in Strong Fields
(b) Vol. **83**, No. 1, p. 331—340 (1977).
See also *17*, p. *207*; 22, p. 328.

Th. Hollstein, U. Kreibig, and F. Leis
Optical Properties of Au and Al in the Visible, Determined at 300 and 1.5 K
(b) Vol. **83**, No. 1, p. K49—K53 (1977).
See also *20.1*, p. *260*; 21.6, p. 316.

V. I. Roslyakov, A. S. Rudnev, E. I. Sirotinin, A. F. Tulinov, and V. A. Khodyrev
Analysis of Backscattering Energy Spectra and Determination of Energy Loss of Channelling Particles
(a) Vol. **43**, No. 1, p. 59—70 (1977).
See also *11*, p. *97*.

R. de Ridder, G. van Tendeloo, and S. Amelinckx
The Ordering of Vacancies in $Ni_{1-x}Al$. Study of the Transition State by Electron Diffraction
(a) Vol. **43**, No. 1, p. 133—139 (1977).
See also *1.3*, p. *22*; 10.1, p. 80; 21.1, p. 297.

U. Essmann, W. Wiethaup, and H. U. Habermeier
The Intermediate State of Superconducting Pure Tantalum
(a) Vol. **43**, No. 1, p. 151—158 (1977).
See also 1.4, p. 27; *14.2*, p. *163*.

H.-J. Fitting, H. Glaefeke, and W. Wild
Electron Penetration and Energy Transfer in Solid Targets
(a) Vol. **43**, No. 1, p. 185—190 (1977).
See also *11*, p. *97*; 21.1, p. 297; 21.3, p. 310; 21.6, p. 317.

Y. Uchida and F. Fujimoto
Electron Channeling in ⟨111⟩ Direction of Tungsten Crystals (I): Energy Range of 200 to 1000 keV
(a) Vol. **43**, No. 1, p. 269—275 (1977).
See also *11*, p. *97*; 13, p. 111.

S. Takeuchi, K. Maeda, and T. Suzuki
Temperature and Strain-Rate Dependence of Flow Stress of Tantalum in the Superconducting and Normal States
(a) Vol. **43**, No. 1, p. 289—294 (1977).
See also *10.1*, p. *80*; *14.2*, p. *163*.

B. K. Jap
On the Dependence of the Diffracted Beam Intensity Features with Electron Accelerating Voltage
(a) Vol. **43**, No. 1, p. K37—K40 (1977).
See also *1.1*, p. *12*.

O. B. Sokolov, V. I. Grebennikov, and E. A. Turov
Resonance Scattering of X-Rays in Metals
(b) Vol. **83**, No. 2, p. 383—393 (1977).
See also *13*, p. *111*; 20, p. 246.

S. K. Sarkar, S. K. Das, D. Roy, and S. Sengupta
Three-Body Interaction and Lattice Dynamics of Metals
(b) Vol. **83**, No. 2, p. 615—624 (1977).
See also *6*, p. *59*; 21.1, p. 297; 21.6, p. 317.

A. M. Brodskii and M. I. Urbakh
On the Dependence of Light Reflection from Metals on Adatom Characteristics
(b) Vol. **83**, No. 2, p. 633—644 (1977).
See also *13.3*, p. *128*; 20.1, p. 260.

A. F. Akkerman
Correlations between the Elastic Energy Transferred in Solids by Implanted Hydrogen Ions and Hydrogen Trapping
(b) Vol. **83**, No. 2, p. K145—K147 (1977).
See also *11*, p. *98*; 22.1.2, p. 342.

J. C. Ashley and R. H. Ritchie
The Influence of Damping on the Mean Free Path of Low Energy Electrons in an Electron Gas
(b) Vol. **83**, No. 2, p. K159—K163 (1977).
See also *13.5*, p. *142*; 13.5.2, p. 150.

M. P. Das
An Effective Periodic Potential for Disordered Binary Alloys
(b) Vol. **83**, No. 2, p. K165—K168 (1977).
See also 2, p. 39; *13*, p. *111*.

A. Bobák
Influence of an Inhomogeneity of the Effective Field on the Properties of a Disordered Heisenberg Ferromagnet
(b) Vol. **83**, No. 2, p. K169—K171 (1977).
See also *2*, p. *39*; *18.2*, p. *223*.

C. GAULT, A. DAUGER, and P. BOCH
Variations of the Elastic Constants of Aluminium–Magnesium Single Crystals with Guinier-Preston Zones
(a) Vol. **43**, No. 2, p. 625—632 (1977).
See also 1.3, p. 23; *12.1*, p. *102*; 21.3, p. 310.

C. B. SATTERTHWAITE
On Isotope Effects in $PdH(D)_x$
(a) Vol. **43**, No. 2, p. K147—K149 (1977).
See also 8, p. 67; 13, p. 112; *14.2*, p. *163*.

S. M. BORODACHEV, S. I. MASHAROV, and V. A. VOLKOV
*Electrical Resistivity of Antiferromagnetic Alloys. Coherent Potential Approximation
(b) Vol. **84**, No. 1, p. 33—40 (1977).
See also *14.1*, p. *159*; 18.4, p. 233.

R. KARCHER, U. KÖBLER, K. LÜDERS, and Z. SCÜCS
Temperature Dependence of Knight Shift and Susceptibility of Single Crystal Molybdenum
(b) Vol. **84**, No. 1, p. 189—193 (1977).
See also *18*, p. *212*; *19*, p. *242*.

W. WASSERBÄCH
Low-Temperature Thermal Conductivity of Pure and Impure Niobium and Tantalum Single Crystals
(b) Vol. **84**, No. 1, p. 205—214 (1977).
See also *8*, p. *67*; 10.1, p. 81.

M. FÄHNLE
Influence of Fast Neutron Irradiation on the Critical Current Densities in Nb_3Sn (I): Low Neutron Doses ($\Phi < 2 \times 10^{18}$ neutrons/cm^2)
(b) Vol. **84**, No. 1, p. 245—251 (1977).
See also *11*, p. *98*; *14.2*, p. *163*.

J. KUDRNOVSKÝ and J. PETRU
Electrical Conductivity of Two-Band Models for Discordered Binary Substitutional Alloys: Theory
(b) Vol. **84**, No. 1, p. 325—333 (1977).
See also 13.1, p. 122; *14.1*, p. *159*.

J. PETRŮ and J. KUDRNOVSKÝ
Electrical Conductivity of Two-Band Metals for Disordered Binary Substitutional Alloys: Numerical Investigation
(b) Vol. **84**, No. 1, p. K21—K23 (1977).
See also *14.1*, p. *159*.

H. R. KHAN, K. LÜDERS, CH. J. RAUB, and Z. SZÜCS
Knight Shift of Pt^{195} in A15- and σ-Phases of Nb–Pt
(b) Vol. **84**, No. 1, p. K33—K35 (1977).
See also *14.2*, p. *163*; *19*, p. *242*; 21.6, p. 317.

E. J. UNTERHORST, V. MÜLLER, and G. SCHANZ
Nuclear Acoustic Resonance in Liquid Gallium
(b) Vol. **84**, No. 1, p. K53—K56 (1977).
See also 2, p. 40; *19*, p. *243*.

R. KISHORE
Disordered Ferromagnetism in Crystalline Metals
(b) Vol. **84**, No. 1, p. K75—K77 (1977).
See also *18.2*, p. *223*.

R. Z. VALIEV and O. A. KAIBYSHEV
Mechanism of Superplastic Deformation in a Magnesium Alloy (I): Structural Changes and Operative Deformation Mechanisms
(a) Vol. **44**, No. 1, p. 65—76 (1977).
See also *10.1*, p. *81*; 21.3, p. 310.

T. Yamamoto
Experimental Aspects of Electron Channeling Patterns in Scanning Electron Microscopy (I): Measurement of Contrast
(a) Vol. **44**, No. 1, p. 137–146 (1977).
See also *1.4*, p. *28*; *10.1*, p. *81*; 21.1, p. 299; 21.6, p. 317; 22.1.2, p. 343.

E. B. Leiko, A. Luft, and E. M. Nadgornii
*Slip Rosettes and Dislocation Structure Formed by Microindentations in Molybdenum Single Crystals
(a) Vol. **44**, No. 1, p. 285–293 (1977).
See also *10.1*, p. *82*.

J. D. Livingston
Grain Size in A-15 Reaction Layers
(a) Vol. **44**, No. 1, p. 295–301 (1977).
See also *1.4*, p. *28*; 14.2, p. 164.

P. Gondi, C. Patuelli, and R. Tognato
Dislocation Source Activation by Load Impulses
(a) Vol. **44**, No. 1, p. 333–338 (1977).
See also *10.1*, p. *82*; 12.1, p. 103.

E. Bonetti, E. Evangelista, and P. Gondi
Kè Peak Analysis by Means of Internal Friction Measurements during Deformation
(a) Vol. **44**, No. 1, p. K31–K33 (1977).
See also *10.1*, p. *82*; *12.1*, p. *103*.

W. Richter, K. Blüthner, H.-J. Köhler, and G. Albrecht
Thin Film DC SQUID with Transformer
(a) Vol. **44**, No. 1, p. K99–K101 (1977).
See also *14.2*, p. *164*.

M. Fähnle
The Influence of Fast Neutron Irradiation on the Critical Current Densities in Nb_3Sn (II): High Neutron Doses ($\Phi < 2 \times 10^{18}$ neutrons/cm^2)
(b) Vol. **84**, No. 2, p. 541–549 (1977).
See also *11*, p. *99*; *14.2*, p. *164*.

V. I. Grebennikov, O. B. Sokolov, and E. A. Turov
Response of Metal Electrons to Local Fields of Holes Arising in X-Ray Processes
(b) Vol. **84**, No. 2, p. 773–784 (1977).
See also *13*, p. *113*; 20, p. 246.

N. R. Mitra, J. Thakur, and P. L. Srivastava
Resistivity and Thermoelectric Power of Liquid Lithium and Sodium from Harrison's First Principle
(b) Vol. **84**, No. 2, p. 797–801 (1977).
See also 2, p. 40; *14.1*, p. *160*; *15*, p. *200*.

K. Takanaka and R. Yamamoto
Temperature-Dependent Multi-Ion Interactions in Simple Metals
(b) Vol. **84**, No. 2, p. 813–816 (1977).
See also *4*, p. *46*.

T. Yamamoto
Experimental Aspects of Electron Channeling Patterns in Scanning Electron Microscopy (II): Estimation of Contrast Depth
(a) Vol. **44**, No. 2, p. 467–476 (1977).
See also *1.4*, p. *28*; *10.1*, p. *82*; 21.1, p. 300; 21.6, p. 318; 22.1.2, p. 344.

R. Z. Valiev and O. A. Kaibyshev
Mechanism of Superplastic Deformation in a Magnesium Alloy (II): The Role of Grain Boundaries
(a) Vol. **44**, No. 2, p. 477–484 (1977).
See also 1.4, p. 28; *10.1*, p. *82*; 21.3, p. 310.

21.1 Transition Metals (Group Sc to Zn) and Their Alloys

Review Articles

V. A. Buravikhin, V. V. Litvintsev, and A. I. Ushakov
Polymorphic Transformations and Physical Properties of Thin Films
(a) Vol. **40**, No. 1, p. 11—27 (1977).
See also *1.2*, p. *14*; 1.4, p. 24; *1.5*, p. *29*; 18.2, p. 217; 21.1.1, p. 302.

G. N. Rao and A. K. Singhvi
Dilute Impurity Hyperfine Fields. Systematics and Comparison with Models
(b) Vol. **84**, No. 1, p. 9—21 (1977).
See also *13.4*, p. *129*; *18.2*, p. *217*; 21.1.1, p. 302; 21.4, p. 311.

Original Papers and Short Notes

G. F. Syrykh, A. P. Zhernov, M. G. Zemlyanov, N. A. Chernoplekov, and G. I. Solovev
Deformation of the Vanadium Phonon Spectrum Due to Additions of Uranium Atoms
(b) Vol. **79**, No. 1, p. 105—109 (1977).
See also *6*, p. *51*.

P. Rennert and A. M. Radwan
Structural Investigation of the Laves Phase $MgZn_2$ with Model Potential Calculations
(b) Vol. **79**, No. 1, p. 167—173 (1977).
See also *1.3*, p. *21*; 10.1, p. 75; 13, p. 107; 21.3, p. 310.

H. Kronmüller, N. König, and F. Walz
Isotope Effects of Diatomic C–H and C–D Complexes in Ni Single Crystals
(b) Vol. **79**, No. 1, p. 237—247 (1977).
See also *9*, p. *69*; 10.1, p. 75; 18.2, p. 217.

P. O. Nilsson, I. Curelaru, and T. Jarlborg
XPS and APS Studies of V_3Si
(b) Vol. **79**, No. 1, p. 277—281 (1977).
See also *13.1*, p. *114*; 20, p. 244.

P. Parikh and D. R. Hay
Crystallographic Aspects of Elastic Dispersion in Body-Centered Cubic Transition Metals
(b) Vol. **79**, No. 1, p. 299—305 (1977).
See also *6*, p. *51*.

U. Gradmann and K. Salewski
Magnetic "Dead Layers" and Exchange Anisotropy in the Interface of 60 Ni/40 Fe(111) and Polycrystalline Mn
(a) Vol. **39**, No. 1, p. 41—48 (1977).
See also *18.2.1*, p. *226*; 18.4, p. 231; 21.1.1, p. 302.

M. Wilkens, M. L. Jenkins, and K.-H. Katerbau
TEM Diffraction Contrast of Lattice Defects Causing Strain Contrast and Structure Factor Contrast Simultaneously
(a) Vol. **39**, No. 1, p. 103—107 (1977).
See also *10.1*, p. *75*.

H. Ortbauer, W. Steiner, and R. Haferl
Mössbauer Studies of $Y(Fe_xCo_{1-x})_2$ in the Region of Small Fe Concentrations
(a) Vol. **39**, No. 1, p. 157—165 (1977).
See also *5*, p. *47*; 18.1, p. 213; 18.2, p. 217; 21, p. 276; 21.1.1, p. 302.

J. Barton, E. Wieser, and M. Müller
Investigation of Magnetic Textures in an Fe–Mn Base Alloy by Means of Mössbauer Spectroscopy
(a) Vol. **39**, No. 1, p. 259—264 (1977).
See also 5, p. 47; *18.2*, p. *217*; 21.1.1, p. 302.

H. Oesterreicher
Possible Disorder of Magnetic Components (Magnetic Scatter Order) in Compounds on the Basis of Transition Metals
(a) Vol. **39**, No. 1, p. K41—K43 (1977).
See also *18*, p. *209*; 21, p. 277.

E. Klugmann, Yu. F. Bydin, and V. A. Volpyas
Effect of Ion Implantation on the Magnetic Properties of Ferromagnetic Metals
(a) Vol. **39**, No. 1, p. K53—K57 (1977).
See also *11*, p. *94*; *18.2*, p. *217*.

R. Karcher, K. Lüders, G. Roth, and Z. Szücs
Low-Temperature Knight-Shift Investigations of Yttrium Metal
(b) Vol. **79**, No. 2, p. 519—524 (1977).
See also *19*, p. *235*.

V. Lantto
A Theoretical Study of the Absorption of Photoelectrons Created by X-Rays in Semi-Infinite Targets
(b) Vol. **79**, No. 2, p. 699—711 (1977).
See also *9*, p. *69*; *17*, p. *206*; 21, p. 277; 21.6, p. 315.

S. C. Goel, Y. R. Midha, and V. S. Nanda
Applications of a Simple Equation of State for Metals
(b) Vol. **79**, No. 2, p. K103—K105 (1977).
See also *8*, p. *64*; 21, p. 277; 21.6, p. 315.

G. Dlubek, O. Brümmer, and P. Sickert
Study of the Nature of the Processes Occurring in the Recovery Stages III and IV of Plastically Deformed Ni of Different Purity by Means of the Positron Annihilation Method
(a) Vol. **39**, No. 2, p. 401—410 (1977).
See also *10.1*, p. *75*.

J. M. Pelletier, R. Borrelly and E. Pernoux
Influence of Thermal Treatment on the Thermoelectric Power of Copper–Iron Alloys
(a) Vol. **39**, No. 2, p. 525—532 (1977).
See also *15*, p. *198*; 21.1.1, p. 302.

H. Föll and M. Wilkens
Transmission Electron Microscope Studies of Dislocation Loops in Heavy-Ion Irradiated H.C.P. Cobalt
(a) Vol. **39**, No. 2, p. 561—571 (1977).
See also *10.1*, p. *76*; 11, p. 94.

G. Dlubek, O. Brümmer, and N. Meyendorf
A Study of the Influence of Mono- and Multivacancies on the Peak Height of the Angular Correlation Curve of the Positron Annihilation Radiation in Al, Cu, and Ni
(a) Vol. **39**, No. 2, p. K95—K99 (1977).
See also *10.1*, p. *76*; 21, p. 278.

A. S. Lileev and W. Steiner
Studies of Remagnetization on Sintered $SmCo_2$ Permanent Magnets
(a) Vol. **39**, No. 2, p. K105—K109 (1977).
See also *18.2*, p. *218*; 21.4, p. 311.

I. M. Chapnik
Compounds of Transition Metals with Possible Closed Electron Shells outside the Inert Gas Shells
(a) Vol. **39**, No. 2, p. K135—K138 (1977).
See also *13*, p. *108*; 21.1.1, p. 302; 22.7, p. 394.

E. V. Kornelsen, A. van Veen, L. M. Caspers, and H. Bakker
Thermal Release of Injected Helium from $Co_{56}Ga_{44}$
(a) Vol. **39**, No. 2, p. K143—K146 (1977).
See also *10.1*, p. *76*; *11*, p. *94*.

H. Düsterhöft and A. Ihlenfeld
The Energy Distribution of Positive Secondary Ions Emitted from Metal and Semiconductor Targets Bombarded with 12 keV Ar^+ Ions
(a) Vol. **39**, No. 2, p. K147—K150 (1977).
See also *17*, p. *206*; 21, p. 278; 22.1.2, p. 338.

W. Dander and H. E. Schaefer
Electrical Resistivity Recovery of Cobalt after Low-Temperature Electron Irradiation
(b) Vol. **80**, No. 1, p. 173—179 (1977).
See also *11*, p. *94*; *14.1*, p. *157*.

E. Mrosan
Residual Resistivity for Electrons Scattered by Spin-Dependent Potentials
(b) Vol. **80**, No. 1, p. K49—K53 (1977).
See also *14.1*, p. *158*.

J. Mizia and F. Kajzar
Electron Correlation in Strong Ferromagnetic 3d Alloys
(b) Vol. **80**, No. 1, p. K75—K79 (1977).
See also *18.2*, p. *218*.

E. Jäger
Structure Fluctuations in the Molecular Field Theory of Amorphous Ferrimagnets
(b) Vol. **80**, No. 1, p. K81—K83 (1977).
See also *2*, p. *34*; *18.3*, p. *228*; 21.4, p. 311.

H.-R. Müller, E. Jäger, and M. Bitterlich
Influence of Impurities on the Magnetic Properties of Amorphous Gd–Co Thin Films
(b) Vol. **80**, No. 1, p. K85—K87 (1977).
See also *2*, p. *34*; *18.3*, p. *228*; 21.4, p. 311.

G. A. Hassan
Very High Ductility of Technically Pure Copper by Bidirectional Twisting at Room Temperature
(a) Vol. **40**, No. 1, p. 43—47 (1977).
See also *10.1*, p. *76*; *12.1*, p. *101*.

G. Hausch and E. Török
Thermal Expansivity and Elastic Constants of CrFe Alloys
(a) Vol. **40**, No. 1, p. 55—62 (1977).
See also *12.1*, p. *101*; 18.4, p. 232; 21.1.1, p. 303.

V. Geist, R. Flagmeyer, D. Stephan, and H.-J. Ullrich
Kossel Interferences of Proton-Induced AlKα Radiation
(a) Vol. **40**, No. 1, p. 113—117 (1977).
See also *1.1*, p. *11*; 21, p. 279.

A. S. Lileev and W. Steiner
Studies of the Magnetic Properties of Sintered $SmCo_5$
(a) Vol. **40**, No. 1, p. 125—131 (1977).
See also *18.2*, p. *218*; 21.4, p. 311.

J. Victoria, J. Piqueras, R. González, and L. Brú
Temperature Dependence of Flow Stress in Fatigued High-Purity Copper Polycrystals
(a) Vol. **40**, No. 1, p. 167—171 (1977).
See also *10.1*, p. *76*.

B. K. Ponomarev, V. G. Thiessen, and G. A. Ivanov-Smolenskii
Band Magnetism in a Cu + 7 at% Mn Alloy
(a) Vol. **40**, No. 1, p. 189–196 (1977).
See also *18.2*, p. *218*; 18.4, p. 232.

W. A. Szilva and L. E. Murr
Variation of Laser Mirror Metal Microstructure and Its Effect on Reflectivity at 10.6 μm
(a) Vol. **40**, No. 1, p. 211–224 (1977).
See also *1.4*, p. *24*; *20.2*, p. *265*.

K. L. Chopra, A. P. Thakoor, S. K. Barthwal, and P. Nath
Electron Transport Properties of Amorphous Cu–Ag Films
(a) Vol. **40**, No. 1, p. 247–255 (1977).
See also *2*, p. *35*; *14.3.1*, p. *173*; 15, p. 198; 21.6, p. 315.

B. Wielke, W. Tikvic, and G. Schoeck
Cottrell-Stokes Law in Cadmium and Zinc
(a) Vol. **40**, No. 1, p. 271–278 (1977).
See also *10.1*, p. *76*; 21, p. 279.

J. Th. M. de Hosson
On the Formation of Argon–Vacancy Clusters in Copper Irradiated with 4 to 6 kV Argon Ions
(a) Vol. **40**, No. 1, p. 293–301 (1977).
See also *10.1*, p. *77*.

J. Victoria, E. González, and J. Piqueras
Influence of Purity on the Temperature Dependence of Flow Stress in Fatigued Copper Polycrystals
(a) Vol. **40**, No. 1, p. K19–K23 (1977).
See also *10.1*, p. *77*.

H. J. Bauer, and B. Baranowski
In-situ Observation of the Disappearance of the Ferromagnetism of Nickel under High Pressure Gaseous Hydrogen
(a) Vol. **40**, No. 1, p. K35–K38 (1977).
See also *18.2*, p. *219*.

E. M. Gray and J. H. Smith
Effect of Ageing on the Magnetic Moment of MnCu Micromagnet
(a) Vol. **40**, No. 1, p. K73–K76 (1977).
See also 14.1, p. 158; *18.2*, p. *219*.

M. Kolář
Surface Density of States in a Simple Model of Nickel
(b) Vol. **80**, No. 2, p. 521–531 (1977).
See also *13.3*, p. *126*.

V. Müller, G. Schanz, E. Fischer, and E. J. Unterhorst
Nuclear Acoustic Resonance Absorption and Dispersion in Vanadium via Coupling to the Magnetic Dipole Moment
(b) Vol. **80**, No. 2, p. 629–639 (1977).
See also 6, p. 54; *19*, p. *237*.

T. S. Altshuler, E. G. Kharakhashyan, E. F. Kukovitskii, and M. M. Zaripov
EPR of Dy^{3+} in Copper Single Crystals
(b) Vol. **80**, No. 2, p. K109–K111 (1977).
See also *19*, p. *237*.

Th. Müller and P. Paufler
Yield Strength of the Monocrystalline Intermetallic Compound $MgZn_2$
(a) Vol. **40**, No. 2, p. 471–477 (1977).
See also *10.1*, p. *77*; 21.3, p. 310.

T. F. M. KORTEKAAS and J. J. M. FRANSE
Thermal Expansion of Some Ni_3Al and Ni–Pt Alloys
(a) Vol. **40**, No. 2, p. 479—485 (1977).
See also *8*, p. *65*; 10.1, p. 77.

E. WIESER and S. MAGER
Investigation of the Localized Magnetic Moments on Fe and Mn in Disordered Fe–Mn–Al Alloys by Magnetic Diffuse Neutron Scattering
(a) Vol. **40**, No. 2, p. 497—501 (1977).
See also *18*, p. *210*; 21.1.1, p. 303.

J. MERLIN, M. KOHEN, and P. F. GOBIN
Conditions for the Appeareance of Striations in the Stress-Strain Curves of Substitutional Alloys
(a) Vol. **40**, No. 2, p. 519—530 (1977).
See also *10.1*, p. *77*.

G. VAN TENDERLOO, R. WOLF, and S. AMELINCKX
The Microstructure of the Alloy Au_5Mn_2: A Domain Structure with 84 Variants
(a) Vol. **40**, No. 2, p. 531—550 (1977).
See also *1.3*, p. *21*; *1.4*, p. *25*; 21.6, p. 315.

R. URAO
On {211} Twins in B.C.C. Cr Films Electrodeposited onto {100} Cu Substrates
(a) Vol. **40**, No. 2, p. 567—574 (1977).
See also *1.4*, p. *25*; *1.5*, p. *29*.

S. HANADA and O. IZUMI
Orientation Dependence of Yield Stress and Operative Slip Systems of β-CuZn Single Crystals at Low Temperatures
(a) Vol. **40**, No. 2, p. 589—598 (1977).
See also *10.1*, p. *77*; *12.1*, p. *101*.

R. BONNET and E. E. LAUFER
Precise Determination of the Relative Orientation of Two Crystals from the Analysis of Spot Diffraction Patterns
(a) Vol. **40**, No. 2, p. 599—603 (1977).
See also *1*, p. *9*; *1.4*, p. *25*.

YU. V. KONOBEEV, V. I. TSCHERBAK, V. N. BYKOV, and A. V. SUBBOTIN
Determination of the Vacancy Diffusivity in Molybdenum and Nickel from Data on the Width of the Grain Boundary Void Denuded Zone
(a) Vol. **40**, No. 2, p. K89—K92 (1977).
See also *9*, p. *69*; *11*, p. *95*; 21, p. 280.

R. PERTHEL, W. KEILIG, R. KOSICIK, U. RÖPKE, and C.-G. D'AMBLY
Determination of the Magnetic Anisotropy in Amorphous Gd–Co Films by Torque Measurement
(a) Vol. **40**, No. 1, p. K135—K138 (1977).
See also *2*, p. *36*; *18.3*, p. *229*; 21, p. 280.

K. SCHRÖDER and T. JOY
Magnetization of BiMn Particles in a Bi-Rich Bi-Mn Matrix
(a) Vol. **40**, No. 2, p. K145—K148 (1977).
See also *18.2*, p. *219*; 21.7, p. 319.

R. KROGGEL, H. LÖFFLER, and O. KABISCH
Experimental Investigations on the Position of the Coherency Spinodal Curve in Al–Zn (15, 16, and 18 at%) Alloys
(a) Vol. **40**, No. 2, p. K163—K164 (1977).
See also 1.2, p. 16; *10.1*, p. *77*; 21, p. 280.

P. Kratochvíl
The Stress Dependence of the Activation Volume in Copper Base Alloy Single Crystals
(a) Vol. **40**, No. 2 p. K165—K167 (1977).
See also *10.1*, p. *77*; 12.1, p. 101; 21.6, p. 315.

Y. Waseda and J. G. Wright
The Resistivity of Amorphous Transition Elements
(b) Vol. **81**, No. 1, p. K37—K40 (1977).
See also *2*, p. *36*; *14.1*, p. *158*.

S. S. Nandra and P. J. Grundy
Lorentz Microscopy of the Magnetic Domain Structure in Amorphous Rare Earth–Cobalt Films Evaporated at Normal Incidence
(a) Vol. **41**, No. 1, p. 65—74 (1977).
See also *18.2.1*, p. *226*; 21.4, p. 312.

K. Salama and G. A. Alers
The Composition Dependence of the Third-Order Elastic Constants of the Cu–Ni System
(a) Vol. **41**, No. 1, p. 241—247 (1977).
See also *12.1*, p. *101*.

B. Hertel, K.-H. Katerbau, and R. Bair
A Possible Source of Error in the Determination of the Type of Small Lattice Defects from Electron Micrographs
(a) Vol. **41**, No. 1, p. K1—K4 (1977).
See also *10.1*, p. *77*; 11, p. 95.

G. Hausch
Invar-Like Thermal Expansion Anomaly in Antiferromagnetic Mn–Ni Alloys
(a) Vol. **41**, No. 1, p. K35—K38 (1977).
See also *8*, p. *65*; *18.4*, p. *232*.

F. N. Tavadze, G. M. Surmava, and M. M. Gvichiya
Investigation of the Diffusion Process of Zinc in Nickel Whiskers
(a) Vol. **41**, No. 1, p. K43—K45 (1977).
See also *9*, p. *70*.

J. Sołtys
Order–Disorder Phase Transitions in the Heusler Alloy Ni_2MnIn
(a) Vol. **41**, No. 1, p. K51—K54 (1977).
See also *1.3*, p. *21*; 21, p. 280.

G. T. Dubovka
On the Problem of Magnetic Phase Transitions of Fe–Ni Invar Alloys
(a) Vol. **41**, No. 1, p. K63—K66 (1977).
See also *18.2*, p. *220*; 21.1.1, p. 303.

F. M. Galperin
Hyperfine Field on the Nucleus of an Atom of a Transition d Element (Metal), Magnetic Moment and Electron Structure of a Ferromagnetic Metal Atom
(b) Vol. **81**, No. 2, p. 545—550 (1977).
See also *13*, p. *109*; 18.2, p. 220.

M. Lannoo and G. Allan
Contribution of the d-Electron Screening to the Properties of Impurities in Transition Metals
(b) Vol. **81**, No. 2, p. 681—690 (1977).
See also *13.4*, p. *134*; 14.4, p. 191.

E. V. Kozlov, V. N. Emelyanov, and V. A. Antonov
Quantum Theory of the Order–Disorder Transition Energy in $AuCu_3$
(b) Vol. **81**, No. 2, p. 719—725 (1977).
See also *1.2*, p. *16*; *1.3*, p. *21*; 21.6, p. 316.

A. Holas
One-Phonon Scattering of Ultra Cold Neutrons in Copper
(b) Vol. **81**, No. 2, p. K107—K110 (1977).
See also *6*, p. *56*.

R. Tahara, Y. Nakamura, M. Inagaki, and Y. Iwama
Mössbauer Study of Spinodal Decomposition in Fe–Cr–Co Alloy
(a) Vol. **41**, No. 2, p. 451—458 (1977).
See also *5*, p. *48*; 18.2, p. 220; 21.1.1, p. 303.

Amar Nath Nigam and Ratna Rani
Surface States of Chemisorbed Hydrogen and Electrical Conduction in Thin Hydrogenated Nickel Foils
(a) Vol. **41**, No. 2, p. 493—498 (1977).
See also *13.3*, p. *127*; *14.1*, p. *158*.

S. H. van den Brink, A. van den Beukel, and P. G. McCormick
The Influence of Specimen Dimensions and Machine Stiffness on the Portevin-Le Chatelier Effect in an Al–Cu Alloy
(a) Vol. **41**, No. 2, p. 513—521 (1977).
See also *10.1*, p. *78*; 21, p. 281.

G. Laslaz, G. Kostorz, M. Roth, P. Guyot, and R. J. Stewart
Interparticle Interference in Neutron Small-Angle Scattering of Al–6.8 at% Zn during Ageing at 133 °C
(a) Vol. **41**, No. 2, p. 577—583 (1977).
See also *1.4*, p. *26*; 10.1, p. 78; 21, p. 281.

S. K. Arif, I. Sigalas, and D. St. P. Bunbury
Mössbauer Effect of Fe^{57} Situated in (Rare-Earth) Ni_2 Laves Phases
(a) Vol. **41**, No. 2, p. 585—589 (1977).
See also *5*, p. *48*; 18.2, p. 220; 21.4, p. 312.

G. Frommeyer and K. F. Schneider
Size Effect and Interface Contribution to Diamagnetism of Silver–Copper Composite Wires with Fibre Structure
(a) Vol. **41**, No. 2, p. 653—658 (1977).
See also *18*, p. *211*; 21.6, p. 316.

H. Oesterreicher, and H. Bittner
Hydriding Characteristics and Magnetic Properties of Hydrides in $Y_6(Mn_{1-x}Fe_x)_{23}$
(a) Vol. **41**, No. 2, p. K101—K105 (1977).
See also *18.2*, p. *220*; 21.1.1, p. 304.

J. P. Simon
Annealing Study of the Vacancy–Impurity Interaction in the Diluted Alloys $\underline{Al}$Zr and $\underline{Al}$Cr
(a) Vol. **41**, No. 2, p. K107—K112 (1977).
See also *10.1*, p. *78*; 21, p. 281.

W. Schwander and K. Ulmer
Extension of Crystal Spectroscopy to Small Quantum Energies: The Ni $M_{2,3}$ Spectrum
(a) Vol. **41**, No. 2, p. K167—K170 (1977).
See also *20*, p. *245*.

L. Bauriegel and G. Otto
The Influence of Magnetization of a Monocrystalline Target on the Channeling and Blocking Effect
(a) Vol. **41**, No. 2, p. K193—K195 (1977).
See also *1*, p. *9* ; 18.2, p. 220; 18.3, p. 229; 22.8.2, p. 411.

E. GRATZ and C. A. POLDY
The Influence of Antiferromagnetic Order on the Evaluation of Spin Disorder Resistivity
(b) Vol. **82**, No. 1, p. 159–167 (1977).
See also 13.1, p. 118; 14.1, p. 158; *18.2*, p. *221*; *18.4*, p. *233*; 21.4, p. 312.

L. BILLARD, P. VILLEMAIN, and A. CHAMBEROD
Interaction of Atomic Order with Localized Ferromagnetism
(b) Vol. **82**, No. 1, p. 269–278 (1977).
See also *1.2*, p. *17*; *18.2*, p. *221*; 21.1.1, p. 304.

Y. D. YAO, S. ARAJS, and E. E. ANDERSON
Phenomenological "Dilution Model" for Kondo-Type Effects in Concentrated Alloy Systems
(b) Vol. **82**, No. 1, p. 343–347 (1977).
See also *13.4*, p. *136*; *14.1*, p. *158*; 18, p. 211.

J. K. YAKINTHOS
Determination of Tb_2Ni_{17}, Er_2Ni_{17}, and Tm_2Ni_{17} Magnetic Structure by Neutron Diffraction
(b) Vol. **82**, No. 1, p. 349–356 (1977).
See also *18.3*, p. *229*; 21.4, p. 312.

J. ZIELIŃSKI
Delocalization of Electrons in a Double d-Band Model of Transition Metals
(b) Vol. **82**, No. 1, p. K17–K19 (1977).
See also *13.1*, p. *119*; *18.2*, p. *221*.

H. BIALAS and LI-SHING LI
Hypersound Excitation at 8.9 GHz by Exploiting Magnetic Anisotropy in Oriented Nickel Films
(a) Vol. **42**, No. 1, p. 125–132 (1977).
See also 1.5, p. 30; *7*, p. *62*; *19*, p. *240*.

S. TAKAHASHI
Compression-Induced Magnetic Anisotropy and Superlattice Dislocations in Ni_3Fe (I): Calculated Results
(a) Vol. **42**, No. 1, p. 201–207 (1977).
See also *10.1*, p. *79*; *18.2*, p. *221*; 21.1.1, p. 304.

W. ANDRÄ and H. DANAN
Magnetic Interface Anisotropy and Magnetization of Plated Ni and Co Films
(a) Vol. **42**, No. 1, p. 227–233 (1977).
See also *18.2.1*, p. *226*.

L. T. MIYADA, S. WATANABE, K. TANAKA, and A. ISORĒ
Internal Friction Peaks in Cold-Worked and Irradiated Titanium
(a) Vol. **42**, No. 1, p. 235–239 (1977).
See also *10.1*, p. *79*; 11, p. 96; *12.1*, p. *101*.

M. A. CHALABYAN, R. STRAUBEL, and YA. M. POGOSYAN
A Wall Splitting Effect in Double-Layer Permalloy Films
(a) Vol. **42**, No. 1, p. 273–278 (1977).
See also *18.2.1*, p. *226*; 21.1.1, p. 304.

O. BUCK and A. SEEGER
Self-Interstitial Migration and Crowdion Conversion in Nickel Studied by Dislocation Pinning Experiments
(a) Vol. **42**, No. 1, p. 285–293 (1977).
See also *10.1*, p. *79*; 11, p. 96; 12.1, p. 102.

A. A. ADAMENKO and A. M. SHALAEV
On the Application of the Positron Annihilation Method to the Dislocation Density Evaluation in Metals
(a) Vol. **42**, No. 1, p. K29–K31 (1977).
See also *10.1*, p. *79*.

A. NARAYANASAMY, T. ERICSSON, T. NAGARAJAN, and P. MUTHUKUMARASAMY
Mössbauer Studies on the $Fe_{1-x}Ti_{1+x}$ System
(a) Vol. **42**, No. 1, p. K65—K69 (1977).
See also *5*, p. *48*; 21.1.1, p. 304.

YA. M. SOIFER and V. G. SHTEINBERG
Low Temperature Internal Friction and Microplastic Strain in Copper
(a) Vol. **42**, No. 1, p. K87—K90 (1977).
See also *10.1*, p. *79*; *12.1*, p. *102*.

B. SZPUNAR and P.-A. LINDGÅRD
Theory of Temperature Dependence of the Magnetization in Rare-Earth–Transition-Metal Alloys
(b) Vol. **82**, No. 2, p. 449—456 (1977).
See also 13, p. 110; *18.1*, p. *214*; *18.4*, p. *233*; 21.4, p. 312.

W. NIEDERLAG, W. BLAU, and K. KLEINSTÜCK
X-Ray Spectroscopical Investigation of the Electronic Structure of Fe–Cr
(b) Vol. **82**, No. 2, p. 495—499 (1977).
See also 13.1, p. 120; *20*, p. *245*; 21.1.1, p. 304.

T. HOLLSTEIN, U. KREIBIG, and F. LEIS
Optical Properties of Cu and Ag in the Intermediate Region between Pure Drude and Interband Absorption
(b) Vol. **82**, No. 2, p. 545—556 (1977).
See also *20.1*, p. *258*; 21.6, p. 316.

F. A. KHVAYA, A. A. KATSNELSON, V. M. SILONOV, and M. M. KHRUSHCHOV
A Pseudopotential Approach to the Electronic Theory of Short-Range Order
(b) Vol. **82**, No. 2, p. 701—704 (1977).
See also *1.3*, p. *22*.

Y. AOKI and Y. GOTOH
Electron-Microscopic Study on the Low-Temperature Phase of the Co-Rich Co–V Alloy System
(a) Vol. **42**, No. 2, p. 469—475 (1977).
See also *1.3*, p. *22*; *1.4*, p. *27*.

S. TAKAHASHI
Compression-Induced Magnetic Anisotropy and Superlattice Dislocations in Ni_3Fe (II): Experimental Results
(a) Vol. **42**, No. 2, p. 529—536 (1977).
See also *10.1*, p. *79*; *18.2*, p. *221*; 21.1.1, p. 304.

N. A. STOLWIJK, T. SPRUIJT, M. A. HOETJES-EIJKIL, and H. BAKKER
Measurements of Diffusion Coefficients of ^{60}Co in the Intermetallic Compound CoGa
(a) Vol. **42**, No. 2, p. 537—544 (1977).
See also *9*, p. *70*; 21, p. 283.

F. SÖFFGE, E. STEICHELE, and K. STIERSTADT
Thermal Expansion Anomaly of Nickel near the Curie Point
(a) Vol. **42**, No. 2, p. 621—627 (1977).
See also *8*, p. *66*; 12.1, p. 102; 18.2, p. 221.

M. FUMERON, J. P. GUILLOT and A. DAUGER
Model X-Ray Scattering by Ellipsoidal Giunier-Preston Zones in the Case of Al–Zn
(a) Vol. **42**, No. 2, p. 687—697 (1977).
See also *1.3*, p. *22*; 12.1, p. 102; 21, p. 283.

M. HALBWACHS
The Mobilities of Self-Interstitials and Vacancies in a Concentrated Ag–Zn Alloy
(a) Vol. **42**, No. 2, p. 707—712 (1977).
See also *9*, p. *70*; *10.1*, p. *80*; 21.6, p. 316.

G. A. Hassan
Effect of Pre-Torsional Deformation at Different Temperatures on the Creep Characteristics of Copper
(a) Vol. **42**, No. 2, p. K115—K117 (1977).
See also *12.1*, p. *102*.

L. I. Johansson, A. L. Hagström, A. Platau, and S.-E. Karlsson
Adsorption of Oxygen at Room Temperature on Polycrystalline Titanium Studied by Ultraviolet and X-Ray Photoelectron Spectroscopy
(b) Vol. **83**, No. 1, p. 77—84 (1977).
See also *13.3*, p. *127*; 17, p. 207.

F. Werfel, G. Dräger, and O. Brümmer
Electronic Structure Investigation of Metallic Vanadium by XS and XPS
(b) Vol. **83**, No. 1, p. 257—262 (1977).
See also *13.1*, p. *121*; 17, p. 207; 20, p. 245.

R. de Ridder, G. van Tendeloo, and S. Amelinckx
The Ordering of Vacancies in $Ni_{1-x}Al$. Study of the Transition State by Electron Diffraction
(a) Vol. **43**, No. 1, p. 133—139 (1977).
See also *1.3*, p. *22*; 10.1, p. 80; 21, p. 284.

H.-J. Fitting, H. Glaefeke, and W. Wild
Electron Penetration and Energy Transfer in Solid Targets
(a) Vol. **43**, No. 1, p. 185—190 (1977).
See also *11*, p. *97*; 21, p. 285; 21.3, p. 310; 21.6, p. 317.

A. G. Moldovan, S. K. Malik, and W. E. Wallace
Structural and Magnetization Studies on $ThCo_{5-x}Mn_x$ and $ThNi_{5-x}Mn_x$ Ternaries
(a) Vol. **43**, No. 1, p. 317—320 (1977).
See also 1.1, p. 12; *18.2*, p. *222*; 21.5, p. 314.

H. M. Eissa
Hexagonal Networks in the Basal Plane of Zn Single Crystals
(a) Vol. **43**, No. 1, p. 343—349 (1977).
See also *1.4*, p. *27*; *10.1*, p. *80*.

S. Chaudhuri and A. K. Pal
Thickness Dependence of Curie Temperature of Cu–Ni Films
(a) Vol. **43**, No. 1, p. K9—K13 (1977).
See also 14.1, p. 159; *18.1*, p. *215*; *18.2*, p. *222*.

S. K. Sarkar, S. K. Das, D. Roy, and S. Sengupta
Three-Body Interaction and Lattice Dynamics of Metals
(b) Vol. **83**, No. 2, p. 615—624 (1977).
See also *6*, p. *59*; 21, p. 285; 21.6, p. 317.

K. Mukasa and M. Maeda
Effects of Annealing on Magnetostrictive Vibrations of Co–Fe Alloys
(a) Vol. **43**, No. 2, p. 393—398 (1977).
See also *1.3*, p. *22*; *18.2*, p. *223*; 21.1.1, p. 306.

C. Jourdan and J. Gastaldi
X-Ray Topographic Investigation of Dislocations in Titanium Single Crystals Grown by Recrystallization
(a) Vol. **43**, No. 2, p. 425—435 (1977).
See also 1.2, p. 19; *1.4*, p. *27*; *10.1*, p. *80*.

T. Schober and A. Carl
A Differential Thermal Analysis Study of the Vanadium–Hydrogen System
(a) Vol. **43**, No. 2, p. 443—449 (1977).
See also *1.2*, p. *19*; 1.3, p. 23; 8, p. 67.

G. Vanderschaeve and T. Sarrazin
On the Nature of Deformation Twins in Ordered Ni_3V
(a) Vol. **43**, No. 2, p. 459–464 (1977).
See also *1.4*, p. *27*; 10.1, p. 80.

H. Neuhäuser and G. Schulz
Etch Pit Study of Dislocation Breakaway in Pre-Yield Deformation of Copper Single Crystals
(a) Vol. **43**, No. 2, p. 493–503 (1977).
See also *10.1*, p. *81*.

G. van Tendeloo, and S. Amelinckx
Long-Period Antiphase Boundary Structures in the Ordered Alloy Au_4Zn (I): One-Dimensional Long-Period Antiphase Boundary Structure
(a) Vol. **43**, No. 2, p. 553–564 (1977).
See also *1.3*, p. *23*; *1.4*, p. *28*; 21.6, p. 317.

S. Myhra
Defect Recovery in Zn after Low Temperature Electron Irradiation
(a) Vol. **43**, No. 2, p. 579–587 (1977).
See also *10.1*, p. *81*; *11*, p. *98*.

H. P. Aubauer
Morphological Stability Study in Binary Alloys (II): Stability of Coherent Concentration Modulations
(a) Vol. **43**, No. 2, p. 601–610 (1977).
See also *1.2*, p. *19*; 4, p. 46.

M. Yamaguchi and Y. Umakoshi
Slip Geometry and Anomalous Strengthening in β-Cu(ZnMn) Single Crystals
(a) Vol. **43**, No. 2, p. 667–674 (1977).
See also *10.1*, p. *81*; 12.1, p. 102.

A. J. Morton
The Displacement Vector Associated with Inversion Boundaries in γ-Brass Type Phases — A Reassessment
(a) Vol. **43**, No. 2, p. K123–K125 (1977).
See also 1.3, p. 23; *1.4*, p. *28*.

N. Ōtsuka, Y. Hanawa, and S. Nagakura
Crystal Structure and Phase Transition of Mn_6N_5 Studied by Electron Diffraction
(a) Vol. **43**, No. 2, p. K127–K129 (1977).
See also *1.2*, p. *19*.

Y. Kimura
On the Relation between the Density of States in the Cubic State and Phase Transition Temperatures (T_e, T_m) of V–Zr–Nb Ternary Laves Compounds
(a) Vol. **43**, No. 2, p. K141–K146 (1977).
See also *1.2*, p. *19*; 8, p. 67; 13, p. 112; *14.2*, p. *163*.

S. D. Mamchich
Phenomenological Dependence of the Deformation Activation Volume upon the Deformation Value
(a) Vol. **43**, No. 2, p. K155–K158 (1977).
See also *10.1*, p. *81*.

J. Victoria, R. Gonzáles, J. Piqueras, and L. Brú
Relationship between Point Defect Clusters and the Temperature Dependence of Yield Stress in Fatigued Copper Polycrystals
(a) Vol. **43**, No. 2, p. K187–K191 (1977).
See also *10.1*, p. *81*.

L. Dagens
Calculation of a Resonant Model Potential for Copper, Silver, and Gold
(b) Vol. **84**, No. 1, p. 311—324 (1977).
See also *4*, p. *46*; *13*, p. *112*; 21.6, p. 317.

K. Bärner
Energy Bands and Electron Correlation Effects in Manganese Pnictides
(b) Vol. **84**, No. 1, p. 385—392 (1977).
See also *13.1*, p. *122*; 14.1, p. 159; 18.4, p. 234; *20.1*, p. *261*.

J. L. Vicent
Conduction Electrons and s-d Exchange Interaction
(b) Vol. **84**, No. 1, p. K71—K74 (1977).
See also 13, p. 112; *14.1*, p. *160*; *18.2.1*, p. *226*.

J. Buchar and Z. Bílek
Amplitude and Frequency Response of Copper Single Crystals to One-Dimensional Transient Stress Waves
(a) Vol. **44**, No. 1, p. 117—125 (1977).
See also *10.1*, p. *81*; 12.1, p. 103.

T. Yamamoto
Experimental Aspects of Electron Channeling Patterns in Scanning Electron Microscopy (I): Measurement of Contrast
(a) Vol. **44**, No. 1, p. 137—146 (1977).
See also *1.4*, p. *28*; *10.1*, p. *81*; 21, p. 287; 21.6, p. 317; 22.1.2, p. 343.

A. J. Morton
Inversion Domains in γ-Brass Type Phases. Stabilisation Mechanism–The Role of Electron Concentration
(a) Vol. **44**, No. 1, p. 205—214 (1977).
See also *1.3*, p. *23*; *1.4*, p. *28*.

M. Jurisch, K.-H. Berthel, and H.-J. Ullrich
Influence of Chemical Composition within the Range of Homogeneity on Phase Transition and Transition Temperature of V_3Si Single Crystals (I):Preparation and Characterization of Single Crystals
(a) Vol. **44**, No. 1, p. 277—284 (1977).
See also 1.3, p. 23; *3*, p. *43*; 14.2, p. 164.

M. Stubičar, E. Babić, D. Subašić, D. Pavuna, and Ž. Marohnić
Phase Transformations during Isochronal Annealing of $Fe_{40}Ni_{40}B_{20}$ Glass
(a) Vol. **44**, No. 1, p. 339—344 (1977).
See also *1.2*, p. *20*; *2*, p. *40*; 21.1.1, p. 306.

R. Bodenberger and A. Hubert
On the Determination of the Bloch Wall Energy of Uniaxial Ferromagnets
(a) Vol. **44**, No. 1, p. K7—K11 (1977).
See also *18.2*, p. *224*; 21.4, p. 313.

V. Tutovan and M. Chiriac
The Influence of a Magnetic Field, Applied during Annealing, on the Atomic Ordering of the Ni_3Mn Alloy
(a) Vol. **44**, No. 1, p. K63—K66 (1977).
See also *18.2*, p. *224*; 18.4, p. 234.

H. Ratajczak and I. Łabędzka
Anomalous Hall Hysteresis Loops in the Vicinity of the Compensation Point in HoCo Amorphous Films
(a) Vol. **44**, No. 1, p. K103—K105 (1977).
See also 2, p. 40; *14.1*, p. *160*; *18.2*, p. *224*; 21.4, p. 313.

S. C. GOEL, Y. R. MIDHA, and V. S. NANDA
A Low Temperature Equation of State for Monovalent Metals
(b) Vol. 84, No. 2, p. 423—426 (1977).
See also *8*, p. *68*.

V. I. MURAVEV and N. B. YUNUSOV
Covalency and Hyperfine Interaction on the Impurity Ion in Crystals
(b) Vol. 84, No. 2, p. 791—796 (1977).
See also *13.4*, p. *140*; 18.1, p. 216; 19, p. 243.

P. FRANZOSI
The Conduction Electron *g*-Factor on the Fermi Surface of Copper
(b) Vol. 84, No. 2, p. K131—K135 (1977).
See also *13.2*, p. *125*.

P. FRANZOSI
The Geometrical Curvature Factor on the Fermi Surface of the Noble Metals
(b) Vol. 84, No. 2, p. K137—K141 (1977).
See also *13.2*, p. *125*; 21.6, p. 318.

T. YAMAMOTO
Experimental Aspects of Electron Channeling Patterns in Scanning Electron Microscopy (II): Estimation of Contrast Depth
(a) Vol. 44, No. 2, p. 467—476 (1977).
See also *1.4*, p. *28*; *10.1*, p. *82*; 21, p. 287; 21.6, p. 318; 22.1.2, p. 344.

P. PAUFLER, E. ZEDLER, H.-J. ULLRICH, K.-H. BERTHEL, U. KRÄMER, M. JURISCH, K. RICHTER, and K. EICHLER
Influence of Chemical Composition within the Range of Homogeneity on Phase Transition and Transition Temperature of V_3Si Single Crystals (II): On the Defect Structure of V_3Si Single Crystals
(a) Vol. 44, No. 2, p. 499—504 (1977).
See also *10.1*, p. *82*; 14.2, p. 164.

K. MAIER
Self-Diffusion in Copper at "Low" Temperatures
(a) Vol. 44, No. 2, p. 567—576 (1977).
See also *9*, p. *71*; 10.1, p. 82.

R. GOBLIRSCH, H. W. WEBER, G. HILSCHER, R. GRÖSSINGER, and W. STEINER
Search for Magnetic Order in Selected Pseudobinary Systems down to Millikelvin Temperatures
(a) Vol. 44, No. 2, p. 593—600 (1977).
See also *18.1*, p. *216*; *18.2*, p. *224*.

N. RUSOVIĆ and H. WARLIMONT
The Elastic Behaviour of β_2-NiAl Alloys
(a) Vol. 44, No. 2, p. 609—619 (1977).
See also *12.1*, p. *103*.

H. P. FISCHER, K.-H. STEINMETZ, and E. NEMBACH
A New High-Resolution, High-Intensity Apparatus for the Study of Fermi Surfaces of Alloys by Positron Annihilation
(a) Vol. 44, No. 2, p. 669—673 (1977).
See also *13.2*, p. *125*.

C. KORN and D. TEITEL
Influence of Vanadium on Hydrogen Diffusion and Phase Constitution in Ti–H
(a) Vol. 44, No. 2, p. 755—766 (1977).
See also *1.3*, p. *23*; 19, p. 243.

S. Ohnuma, K. Watanabe, and T. Masumoto
Low Coercivity and Zero Magnetostriction of Amorphous Fe–Co–Ni System Alloys
(a) Vol. **44**, No. 2, p. K151—K154 (1977).
See also *2*, p. *41*; *18.2*, p. *225*; 21.1.1, p. 307.

A. Wolfenden
Mechanical Damping at 40 kHz in Mn–Cu Alloys
(a) Vol. **44**, No. 2, p. K171—K175 (1977).
See also *12.1*, p. *103*.

21.1.1 Fe and Fe Alloys

Review Articles

V. A. Buravikhin, V. V. Litvintsev, and A. I. Ushakov
Polymorphic Transformations and Physical Properties of Thin Films
(a) Vol. **40**, No. 1, p. 11–27 (1977).
See also *1.2*, p. *14*; 1.4, p. 24; *1.5*, p. *29*; 18.2, p. 217; 21.1, p. 288.

G. N. Rao and A. K. Singhvi
Dilute Impurity Hyperfine Fields. Systematics and Comparison with Models
(b) Vol. **84**, No. 1, p. 9–21 (1977).
See also *13.4*, p. *129*; *18.2*, p. *217*; 21.1, p. 288; 21.4, p. 311.

Original Papers and Short Notes

U. Gradmann and K. Salewski
Magnetic "Dead Layers" and Exchange Anisotropy in the Interface of 60 Ni/40 Fe(111) and Polycrystalline Mn
(a) Vol. **39**, No. 1, p. 41–48 (1977).
See also *18.2.1*, p. *226*; 18.4, p. 231; 21.1, p. 288.

J. D. Stefhenson
Some Applications of Characteristic Square Etch Pits Formed on {100} Surfaces of Fe–3wt% Si Crystals
(a) Vol. **39**, No. 1, p. 89–101 (1977).
See also *1.4*, p. *24*; 10.1, p. 75.

H. Ortbauer, W. Steiner, and R. Haferl
Mössbauer Studies of $Y(Fe_xCo_{1-x})_2$ in the Region of Small Fe Concentrations
(a) Vol. **39**, No. 1, p. 157–165 (1977).
See also *5*, p. *47*; 18.1, p. 213; 18.2, p. 217; 21, p. 276; 21.1, p. 288.

J. Barton, E. Wieser, and M. Müller
Investigation of Magnetic Textures in an Fe–Mn Base Alloy by Means of Mössbauer Spectroscopy
(a) Vol. **39**, No. 1, p. 259–264 (1977).
See also 5, p. 47; *18.2*, p. *217*; 21.1, p. 288.

K. Kojima and M. Meshii
Surface Film Softening of Iron Single Crystals Oriented for Single Slip
(a) Vol. **39**, No. 2, p. 491–498 (1977).
See also *1.6*, p. *31*; *10.1*, p. *75*.

J. M. Pelletier, R. Borrelly and E. Pernoux
Influence of Thermal Treatment on the Thermoelectric Power of Copper–Iron Alloys
(a) Vol. **39**, No. 2, p. 525–532 (1977).
See also *15*, p. *198*; 21.1, p. 289.

I. M. Chapnik
Compounds of Transition Metals with Possible Closed Electron Shells outside the Inert Gas Shells
(a) Vol. **39**, No. 2, p. K135–K138 (1977).
See also *13*, p. *108*; 21.1, p. 289; 22.7, p. 394.

J. A. Sawicki and B. D. Sawicka
Comparability of ^{57}Fe Mössbauer Data in Ion-Implanted and Amorphous FeGe
(b) Vol. **80**, No. 1, p. K41–K44 (1977).
See also 2, p. 34; *5*, p. *47*.

G. Hausch and E. Török
Thermal Expansivity and Elastic Constants of CrFe Alloys
(a) Vol. **40**, No. 1, p. 55—62 (1977).
See also *12.1*, p. *101*; 18.4, p. 232; 21.1, p. 290.

A. L. Roitburd
Predominant Effect of the Internal Stress Relaxation on Microstructural and Kinetic Features of Martensitic Transformation
(a) Vol. **40**, No. 1, p. 333—342 (1977).
See also *1.2*, p. *15*; 1.4, p. 25.

G. Athanassiadis, G. Le Caer, J. Foct, and L. Rimlinger
Study of Ternary Ordered Solid Solutions Derived from Fe_3Al by Substitution
(a) Vol. **40**, No. 2, p. 425—435 (1977).
See also *1.3*, p. *21*; 5, p. 48; 21, p. 279.

E. Wieser and S. Mager
Investigation of the Localized Magnetic Moments on Fe and Mn in Disordered Fe–Mn–Al Alloys by Magnetic Diffuse Neutron Scattering
(a) Vol. **40**, No. 2, p. 497—501 (1977).
See also *18*, p. *210*; 21.1, p. 292.

O. Bostanjoglo and H.-P. Gemünd
Curling-Type Magnetic Domains in Narrow Film Strips
(a) Vol. **40**, No. 2, p. K97—K99 (1977).
See also *18.2.1*, p. *226*.

H. Oesterreicher
Magnetic Properties of Scatter Order Compounds RFeAl (R = Gd, Tb, Dy, Ho, Er, Tm, Lu, and Y)
(a) Vol. **40**, No. 2, p. K139—K143 (1977).
See also *18*, p. *210*; 21, p. 280; 21.4, p. 312.

J. Suwalski, L. Dąbrowski, J. Piekoszewski, and Z. Tucholski
Study of Order–Disorder Transformation in Fe–Al–Si Alloys by Mössbauer Spectroscopy
(a) Vol. **41**, No. 1, p. 191—195 (1977).
See also *5*, p. *48*; 18.2, p. 219.

V. V. Gann, A. I. Skvortsov, and A. van den Bosch
The Superconducting Transition Temperature of $U_6Fe_{0.98}$ Reduced by Neutron Irradiation
(a) Vol. **41**, No. 1, p. 225—229 (1977).
See also *11*, p. *95*; *14.2*, p. *162*; 18.1, p. 214; 21.5, p. 314.

G. T. Dubovka
On the Problem of Magnetic Phase Transitions of Fe–Ni Invar Alloys
(a) Vol. **41**, No. 1, p. K63—K66 (1977).
See also *18.2*, p. *220*; 21.1, p. 293.

W. Blau, S. Mager, and E. Wieser
Determination of Localized Magnetic Moments in Fe–Cr–Al Alloys and the Electron Structure
(b) Vol. **81**, No. 2, p. 535—544 (1977).
See also 13.1, p. 118; *18.2*, p. *220*; *18.4*, p. *233*.

O. Uemura, Y. Suzuki, and T. Satow
The Amorphous Structure of the Fe–Ge System
(a) Vol. **41**, No. 2, p. 417—421 (1977).
See also *2*, p. *37*; 22.1.1, p. 334.

R. Tahara, Y. Nakamura, M. Inagaki, and Y. Iwama
Mössbauer Study of Spinodal Decomposition in Fe–Cr–Co Alloy
(a) Vol. **41**, No. 2, p. 451—458 (1977).
See also *5*, p. *48*; 18.2, p. 220; 21.1, p. 294.

J. J. Couderc, J. Bras and M. Fagot
Precipitation in the Iron–Gallium System in the Vicinity of 25 at % Ga
(a) Vol. **41**, No. 2, p. 595–605 (1977).
See also *1.2*, p. *17*; *1.3*, p. *22*; 21, p. 281.

H. Oesterreicher and H. Bittner
Hydriding Characteristics and Magnetic Properties of Hydrides in $Y_6(Mn_{1-x}Fe_x)_{23}$
(a) Vol. **41**, No. 2, p. K101–K105 (1977).
See also *18.2*, p. *220*; 21.1, p. 294.

R. Kamel, S. A. Mahmoud, and F. M. Mansy
A Dislocation–Carbon Atom Relaxation in 4% Carbon Iron
(a) Vol. **41**, No. 2, p. K113–K117 (1977).
See also *10.1*, p. *78*; *12.1*, p. *101*.

L. Billard, P. Villemain, and A. Chamberod
Interaction of Atomic Order with Localized Ferromagnetism
(b) Vol. **82**, No. 1, p. 269–278 (1977).
See also *1.2*, p. *17*; *18.2*, p. *221*; 21.1, p. 295.

S. Takahashi
Compression-Induced Magnetic Anisotropy and Superlattice Dislocations in Ni_3Fe (I): Calculated Results
(a) Vol. **42**, No. 1, p. 201–207 (1977).
See also *10.1*, p. *79*; *18.2*, p. *221*; 21.1, p. 295.

M. A. Chalabyan, R. Straubel, and Ya. M. Pogosyan
A Wall Splitting Effect in Double-Layer Permalloy Films
(a) Vol. **42**, No. 1, p. 273–278 (1977).
See also *18.2.1*, p. *226*; 21.1, p. 295.

J. Drabecki and B. Wysłocki
Anisotropic Model of the Influence of Nitrides $Fe_{16}N_2$ on the Coercive Force of a FeN Alloy
(a) Vol. **42**, No. 1, p. 359–367 (1977).
See also *18.2*, p. *221*.

Y. Muraoka, M. Shiga, and Y. Nakamura
Magnetic Properties and Mössbauer Effect of $A(Fe_{1-x}B_x)_2$ (A = Y or Zr, B = Al or Ni) Laves Phase Intermetallic Compounds
(a) Vol. **42**, No. 1, p. 369–374 (1977).
See also 1, p. 10; *5*, p. *48*; *18*, p. *211*.

A. Narayanasamy, T. Ericsson, T. Nagarajan, and P. Muthukumarasamy
Mössbauer Studies on the $Fe_{1-x}Ti_{1+x}$ System
(a) Vol. **42**, No. 1, p. K65–K69 (1977).
See also *5*, p. *48*; 21.1, p. 296.

W. Niederlag, W. Blau, and K. Kleinstück
X-Ray Spectroscopical Investigation of the Electronic Structure of Fe–Cr
(b) Vol. **82**, No. 2, p. 495–499 (1977).
See also 13.1, p. 120; *20*, p. *245*; 21.1, p. 296.

V. Hivert, P. Groh, P. Moser, and W. Frank
Internal Friction Peaks Due to Dislocation Relaxations in Plastically Deformed and/or Irradiated High-Purity α-Iron (I): Experiments
(a) Vol. **42**, No. 2, p. 511–518 (1977).
See also *10.1*, p. *79*; 11, p. 97; *12.1*, p. *102*.

S. Takahashi
Compression-Induced Magnetic Anisotropy and Superlattice Dislocations in Ni_3Fe (II): Experimental Results
(a) Vol. **42**, No. 2, p. 529–536 (1977).
See also *10.1*, p. *79*; *18.2*, p. *221*; 21.1, p. 296.

A. Machová, V. Paidar, and F. Kroupa
Theory of Elastic Constants of Fe–Si Alloys at 0 K
(a) Vol. **42**, No. 2, p. 713—720 (1977).
See also *12.1*, p. *102*.

W. Steiner and R. Haferl
Hyperfine Interactions in Y_2Fe_{17}
(a) Vol. **42**, No. 2, p. 739—750 (1977).
See also *5*, p. *49*; *18.2*, p. *222*; 21.4, p. 312.

J. F. Dufresne, I. G. Ritschie, and P. Moser
On the Question of Low-Temperature Magnetic Relaxation in Pure Iron
(a) Vol. **42**, No. 2, p. 751—756 (1977).
See also *18.2*, p. *222*.

H. H. Mende and W. Hagen
X-Ray Topographic Investigations of Deformed Iron Whiskers
(a) Vol. **42**, No. 2, p. K95—K98 (1977).
See also *1.4*, p. *27*; *10.1*, p. *80*.

O. Bostanjoglo and H. P. Gemünd
New Rotation Processes in Magnetic Film Strips
(a) Vol. **42**, No. 2, p. K103—K106 (1977).
See also *18.2.1*, p. *226*.

J. A. Baldwin, Jr.
Stacey's Theory of Thermally Activated Magnetic Domain Wall Motion
(a) Vol. **42**, No. 2, p. K123—K127 (1977).
See also *18.2*, p. *222*.

U. A. Durasova, O. S. Kolotov, T. N. Letova, and R. V. Telesnin
On the Stray Fields Arising during Pulse Reversal in Thin Magnetic Films
(a) Vol. **42**, No. 2, p. K171—K174 (1977).
See also *18.2.1*, p. *226*.

W. Steiner, R. Haferl, and B. K. Choi
Hyperfine Interactions in $RE_2(Fe, Co)_{17}$ (RE = Y, Gd, Dy)
(a) Vol. **43**, No.1, p. 111—117 (1977).
See also *5*, p. *49*; *18.2*, p. *222*; 21.4, p. 313.

B. K. Jain, A. K. Singh, and K. Chandra
Effect of Alloying Elements in Plain Carbon Steels
(a) Vol. **43**, No. 1, p. 357—361 (1977).
See also *1.3*, p. *22*; 5, p. 49.

V. A. Pavlov and N. L. Pecherkina
Cold-Rolling Cavity Formation in Armco-Fe
(a) Vol. **43**, No. 1, p. K5—K8 (1977).
See also *10.1*, p. *80*.

E. Ya. Mikhlin and V. V. Nelaev
The Effect of Uniaxial Compression upon Dimensions and Shape of the Frenkel Defect Recombination Zone in α-Iron
(a) Vol. **43**, No. 1, p. K23—K27 (1977).
See also *10.1*, p. *80*.

Hideo Nakajima and Ken-ichi Hirano
Influence of the Addition of Nickel and Chromium on the Electromigration of Carbon in γ-Iron
(a) Vol. **43**, No. 1, p. K57—K60 (1977).
See also *9*, p. *70*.

F. Walz, H. Schreyer, and H. Kronmüller
The Magnetic After-Effect Spectrum of α-Iron after Low-Temperature Electron Irradiation
(a) Vol. **43**, No. 1, p. K81—K85 (1977).
See also 10.1, p. 80; *11*, p. *97*; *18.2*, p. *222*.

L. H. Dunleavy and H. J. Blythe
Investigation of the Magnetic After-Effect in α-Iron after Neutron Irradiation at 77 K
(a) Vol. **43**, No. 1, p. K87—K90 (1977).
See also 10.1, p. 80; *11*, p. *97*; *18.2*, p. *222*.

U. Draeger, R. Straubel, and W. J. Ziętek
Calculation of the Wall Curvature in Thin Film Strips with In-Plane Magnetization
(a) Vol. **43**, No. 1, p. K113—K116 (1977).
See also *18.2.1*, p. *226*.

Y. Khwaja and M. Nauciel-Bloch
Off-Stoichiometry Effects in the Antiferromagnetic FeRh Alloy
(b) Vol. **83**, No. 2, p. 413—424 (1977).
See also 13.1, p. 121; *18.2*, p. *223*; *18.4*, p. *233*.

K. Mukasa and M. Maeda
Effects of Annealing on Magnetostrictive Vibrations of Co–Fe Alloys
(a) Vol. **43**, No. 2, p. 393—398 (1977).
See also *1.3*, p. *22*; *18.2*, p. *223*; **21.1**, p. **297**.

W. Steiner, R. Haferl, and H. Welser
Hyperfine Interactions in $RE_2(Fe, Ni)_{17}$ (RE = Y, Gd, Dy)
(a) Vol. **43**, No. 2, p. 437—442 (1977).
See also *5*, p. *49*; *18.2*, p. *223*; **21.4**. p. **313**.

V. Paidar
Diffusion in the Force Field in Solid Solutions
(a) Vol. **43**, No. 2, p. K165—K168 (1977).
See also *9*, p. *71*.

W. Mensch and J. Diehl
The Magnetic After-Effect Spectrum of α-Iron after Low Temperature Neutron Irradiation
(a) Vol. **43**, No. 2, p. K175—K178 (1977).
See also 10.1, p. 81; 11, p. 98; *18.2*, p. *223*.

J. Diehl, W. Mensch, and M. Weller
Correlations between Magnetic After-Effect and Internal Friction in α-Iron after Low Temperature Neutron Irradiation
(a) Vol. **43**, No. 2, p. K179—K182 (1977).
See also 10.1, p. 81; *11*, p. *98*; 12.1, p. 102; *18.2*, p. *223*.

L. Billard, P. Villemain, and A. Chamberod
Interaction of Atomic Order with Itinerant Ferromagnetism
(b) Vol. **84**, No.1, p. 335—343 (1977).
See also *1.3*, p. *23*; *18.2*, p. *223*.

M. Stubičar, E. Babić, D. Subašić, D. Pavuna, and Ž. Marohnić
Phase Transformations during Isochronal Annealing of $Fe_{40}Ni_{40}B_{20}$ Glass
(a) Vol. **44**, No. 1, p. 339—344 (1977).
See also *1.2*, p. *20*; *2*, p. *40*; 21.1, p. 299.

J. Degauque, B. Astie, and C. Levade
A Study on the Interaction between Magnetic Domain Walls and Dislocations in Pure Iron by Measurements at Low Amplitudes of Magnetoelastic Damping
(a) Vol. **44**, No. 1, p. K59—K62 (1977).
See also *10.1*, p. *82*; *18.2*, p. *224*.

F. Kroupa
Vibration Modes of Split Screw Dislocations in B.C.C. Metals
(b) Vol. **84**, No. 2, p. 725–730 (1977).
See also *10.1*, p. *82*.

M. Sakoh and M. Shimizu
Thermodynamics of Phase Transitions in Iron
(a) Vol. **44**, No. 2, p. 511–516 (1977).
See also *1.2*, p. *20*; *8*, p. *68*.

U. Gradmann and P. Tillmanns
Supersaturation and Mode of Growth for Fe Films on Cu(111) — An Experimental Study Using LEED and AES
(a) Vol. **44**, No. 2, p. 539–547 (1977).
See also *1.5*, p. *30*; *3*, p. *43*.

J. Parisot and G. Ferron
A Calculation of Dislocation Densities in Quenched α-Iron from Internal Friction Experiments
(a) Vol. **44**, No. 2, p. 549–555 (1977).
See also *10.1*, p. *82*; 12.1, p. 103.

S. Ohnuma, K. Watanabe, and T. Masumoto
Low Coercivity and Zero Magnetostriction of Amorphous Fe–Co–Ni System Alloys
(a) Vol. **44**, No. 2, p. K151–K154 (1977).
See also *2*, p. *41*; *18.2*, p. *225*; 21.1, p. 301.

K. Hoselitz
Magnetization of Fe–Si–B Metallic Glass
(a) Vol. **44**, No. 2, p. K191–K193 (1977).
See also *2*, p. *41*; *18.2*, p. *225*.

21.2 Alkali Metals

Original Papers and Short Notes

R. C. Rai and M. P. Hemkar
A Simple Phenomenological Model for the Lattice Dynamics of B.C.C. Metals
(b) Vol. **79**, No. 1, p. 289–298 (1977).
See also *6*, p. *51*; 12.1, p. 101.

M. Yahagi and K. Iwamura
Temperature Dependences of Electrical Resistivity and Thermal Expansion in the Intermetallic Compound LiIn
(a) Vol. **39**, No. 1, p. 189–192 (1977).
See also *8*, p. *64*; 10.1, p. 75; *14.1*, p. *157*.

K. Saermark and J. Lebech
Dispersion of Cyclotron Waves in Metals and Observability of Many-Body Effects
(b) Vol. **79**, No. 2, p. 723–730 (1977).
See also *13.2*, p. *124*; 14.1, p. 157; 21.6, p. 315.

K. H. Weilacher, H. Roth-Seefrid, and H. Bross
Lattice Dynamics of Lithium. A Theoretical Microscopic Calculation
(b) Vol. **80**, No. 1, p. 137–142 (1977).
See also *6*, p. *53*.

B. Prasad and R. S. Srivastava
Lattice Dynamics and Thermal Properties of Simple Metals Based on Model Pseudopotential Calculations
(b) Vol. **80**, No. 1, p. 379–383 (1977).
See also *6*, p. *54*; *8*, p. *64*.

W. Schülke
Fermi Surface Mapping from the One-Dimensional Fourier Transform of Compton Profiles
(b) Vol. **80**, No. 1, p. K67–K70 (1977).
See also *13.2*, p. *124*.

G. O. Berim, F. G. Cherkasov, E. G. Kharakhashyan, and Y. I. Talanov
Line Shape of Conduction Electron Spin Resonance in Spherical Metal Particles
(a) Vol. **40**, No. 1, p. K53–K55 (1977).
See also *19*, p. *237*.

F. Perrot
A Model for Vacancies in Simple Metals Including an Ionic Potential in the Self-Consistent Density Calculation
(b) Vol. **81**, No. 1, p. 205–210 (1977).
See also *10.1*, p. *77*.

O. P. Gupta, H. L. Kharoo, and M. P. Hemkar
Ideal Thermal Resistivity of Lithium, Potassium, and Rubidium
(b) Vol. **81**, No. 1, p. 275–280 (1977).
See also *6*, p. *55*; *8*, p. *65*.

F. Brosens, J. Cornelis, D. C. Wallace, and J. T. Devreese
A Method to Derive Interatomic Potentials in Metals from Experimental Phonon Spectra
(b) Vol. **81**, No. 2, p. 557–564 (1977).
See also *4*, p. *45*; 6, p. 56; 21, p. 280.

W. Schülke
The One-Dimensional Fourier Transform of Compton Profiles. Theorems and Applications
(b) Vol. **82**, No. 1, p. 229–235 (1977).
See also *13.1*, p. *119*; *13.2*, p. *124*; 22.1.2, p. 340.

MAHENDRA KUMAR and M. P. HEMKAR
Density of States and Fermi Energy of Simple Metals
(b) Vol. **82**, No. 1, p. K29–K34 (1977).
See also *13.2*, p. *124*; 21, p. 282.

I. GYÉMÁNT and G. SOLT
Formation Energy of Spherical Cavities in Metals
(b) Vol. **82**, No. 2, p. 651–655 (1977).
See also *4*, p. *45*; *10.1*, p. *79*.

E. N. MGBENU
The Lithium Band Structure
(b) Vol. **83**, No. 1, p. K75–K77 (1977).
See also *13.1*, p. *121*; 13.2, p. 125.

R. N. SINGH, P. L. SRIVASTAVA, and N. R. MITRA
Phonon Frequency, Electrical Resistivity, and Inter-Ionic Pair Potential of Li
(b) Vol. **83**, No. 2, p. 651–657 (1977).
See also 4, p. 46; *6*, p. *59*; 14.1, p. 159.

21.3 Alkali-Earth Metals

Original Papers and Short Notes

P. Rennert and A. M. Radwan
Structure Investigation of the Laves Phase $MgZn_2$ with Model Potential Calculations
(b) Vol. **79**, No. 1, p. 167–173 (1977).
See also *1.3*, p. *21*; 10.1, p. 75; 13, p. 107; 21.1, p. 288.

H. Eschrig, K. Feldmann, K. Hennig, W. Matz, and P. Paufler
Phonon Spectra of the Laves Phase Intermetallic Compound $CaMg_2$
(b) Vol. **79**, No. 1, p. 283–288 (1977).
See also *6*, p. *51*.

D. C. Creagh
Determination of the Mass Attenuation Coefficients and Anomalous Dispersion Corrections for Calcium for X-Ray Wavelengths from $IK\alpha_1$ to $CuK\alpha_1$
(a) Vol. **39**, No. 2, p. 705–715 (1977).
See also *20*, p. *244*.

H. Neddermeyer
Charge Transfer in Aluminium–Magnesium Alloys
(b) Vol. **80**, No. 2, p. 611–614 (1977).
See also *13*, p. *108*; 20, p. 244; 21, p. 279.

Th. Müller and P. Paufler
Yield Strength of the Monocrystalline Intermetallic Compound $MgZn_2$
(a) Vol. **40**, No. 2, p. 471–477 (1977).
See also *10.1*, p. *77*; 21.1, p. 291.

G. A. Hassan
Superplastic Behaviour of an Aluminium–Magnesium Alloy Effected by Bidirectional Twisting
(a) Vol. **41**, No. 2, p. 541–546 (1977).
See also *10.1*, p. *78*; 21, p. 281.

P. Čapková and J. Šedivý
X-Ray Debye Characteristic Temperatures of Magnesium and α-Titanium
(a) Vol. **42**, No. 2, p. 681–686 (1977).
See also *6*, p. *58*; *8*, p. *66*; 21, p. 283.

H.-J. Fitting, H. Glaefeke, and W. Wild
Electron Penetration and Energy Transfer in Solid Targets
(a) Vol. **43**, No. 1, p. 185–190 (1977).
See also *11*, p. *97*; 21, p. 285; 21.1, p. 297; 21.6, p. 317.

C. Gault, A. Dauger, and P. Boch
Variations of the Elastic Constants of Aluminium–Magnesium Single Crystals with Guinier-Preston Zones
(a) Vol. **43**, No. 2, p. 625–632 (1977).
See also 1.3, p. 23; *12.1*, p. *102*; 21, p. 286.

R. Z. Valiev and O. A. Kaibyshev
Mechanism of Superplastic Deformation in a Magnesium Alloy (I): Structural Changes and Operative Deformation Mechanisms
(a) Vol. **44**, No. 1, p. 65–76 (1977).
See also *10.1*, p. *81*; 21, p. 286.

R. Z. Valiev and O. A. Kaibyshev
Mechanism of Superplastic Deformation in a Magnesium Alloy (II): The Role of Grain Boundaries
(a) Vol. **44**, No. 2, p. 477–484 (1977).
See also 1.4, p. 28; *10.1*, p. *82*; 21, p. 287.

21.4 Rare-Earth Metals

Review Articles

Yu. V. Knyazev and M. M. Noskov
The Optical Properties of Rare-Earth Metals
(b) Vol. **80**, No. 1, p. 11—29 (1977).
See also 13.1, p. 114; *20.1*, p. *247*.

G. N. Rao and A. K. Singhvi
Dilute Impurity Hyperfine Fields. Systematics and Comparison with Models
(b) Vol. **84**, No. 1, p. 9—21 (1977).
See also *13.4*, p. *129*; *18.2*, p. *217*; 21.1, p. 288; 21.1.1, p. 302.

Original Papers and Short Notes

D. Glötzel and L. Fritsche
Volume Dependence of the Electronic Structure of Lanthanum and Cerium
(b) Vol. **79**, No. 1, p. 85—92 (1977).
See also *13.1*, p. *114*; 13.2, p. 124.

A. S. Lileev und W. Steiner
Studies of Remagnetization on Sintered $SmCO_2$ Permanent Magnets
(a) Vol. **39**, No. 2, p. K105—K109 (1977).
See also *18.2*, p. *218*; 21.1, p. 289.

V. P. Vijayavargiya, S. N. Gupta, and B. D. Padalia
Extended X-Ray Absorption Fine Structure (EXAFS) Studies of the Lanthanides
(b) Vol. **80**, No. 1, p. 83—88 (1977).
See also *20*, p. *244*.

E. Jäger
Structure Fluctuations in the Molecular Field Theory of Amorphous Ferrimagnets
(b) Vol. **80**, No. 1, p. K81—K83 (1977).
See also *2*, p. *34*; *18.3*, p. *228*; 21.1, p. 290.

H.-R. Müller, E. Jäger, and M. Bitterlich
Influence of Impurities on the Magnetic Properties of Amorphous Gd–Co Thin Films
(b) Vol. **80**, No. 1, p. K85—K87 (1977).
See also *2*, p. *34*; *18.3*, p. *228*; 21.1, p. 290.

J. N. Daou, J. E. Bonnet, P. Vajda, M. Biget, A. Lucasson, and P. Lucasson
The Effect of Low-Temperature Electron Irradiation on Lutetium and Its Hydrogen Solid Solutions
(a) Vol. **40**, No. 1, p. 101—108 (1977).
See also *11*, p. *95*.

A. S. Lileev and W. Steiner
Studies of the Magnetic Properties of Sintered $SmCo_5$
(a) Vol. **40**, No. 1, p. 125—131 (1977).
See also *18.2*, p. *218*; 21.1, p. 290.

B. Boucher
Bulk Magnetic Properties of Amorphous $Tb_{52}Ag_{48}$. Evidence of a Strong Local Anisotropy
(a) Vol. **40**, No. 1, p. 197—203 (1977).
See also *2*, p. *35*; *18.2*, p. *219*; 21.6, p. 315.

J. K. Yakinthos, H. Gamari-Seale, and J. Laforest
Magnetic Properties of R_3Pd_4 (R = Pr, Nd, Sm, Gd, Tb, Dy, Ho, Er, Tm) Intermetallic Compounds
(a) Vol. **40**, No. 2, p. K105—K108 (1977).
See also *18.1*, p. *214*; *18.4*, p. *232*.

T. Anagnostopoulos, P. Ikonomou and J. K. Yakinthos
Magnetic Properties of the Compounds $SmAu_6$, $TbAu_6$, and $DyAu_6$
(a) Vol. **40**, No. 2, p. K109—K112 (1977).
See also *18.1*, p. *214*; *18.4*, p. *232*; 21.6, p. 315.

H. Oesterreicher
Magnetic Properties of Scatter Order Compounds RFeAl (R = Gd, Tb, Dy, Ho, Er, Tm, Lu, and Y)
(a) Vol. **40**, No. 2, p. K139—K143 (1977).
See also *18*, p. *210*; 21, p. 280; 21.1.1, p. 303.

K. Sugawara
Effect of Crystal Field on the Inelastic Neutron Scattering Cross-Section
(b) Vol. **81**, No. 1, p. 313—321 (1977).
See also *18.1*, p. *214*; 19, p. 238.

Š. Jánoš, A. Feher, and K. Flachbart
Magnetoresistance of Pr, Nd, Sm, and Eu at Low Temperatures
(b) Vol. **81**, No. 1, p. K19—K22 (1977).
See also *14.1*, p. *158*.

S. S. Nandra and P. J. Grundy
Lorentz Microscopy of the Magnetic Domain Structure in Amorphous Rare Earth–Cobalt Films Evaporated at Normal Incidence
(a) Vol. **41**, No. 1, p. 65—74 (1977).
See also *18.2.1*, p. *226*; 21.1, p. 293.

S. K. Arif, I. Sigalas, and D. St. P. Bunbury
Mössbauer Effect of Fe^{57} Situated in (Rare-Earth) Ni_2 Laves Phases
(a) Vol. **41**, No. 2, p. 585—589 (1977).
See also *5*, p. *48*; 18.2, p. 220; 21.1, p. 294.

E. Gratz and C. A. Poldy
The Influence of Antiferromagnetic Order on the Evaluation of Spin Disorder Resistivity
(b) Vol. **82**, No. 1, p. 159—167 (1977).
See also 13.1, p. 118; 14.1, p. 158; *18.2*, p. *221*; *18.4*, p. *233*; 21.1, p. 295.

V. V. Vorobev, V. S. Belovol, and V. A. Finkel
X-Ray Diffraction Study of Magnetostriction and Phase Transitions in a Magnetic Field in Rare Earth Metals and Some Alloys
(b) Vol. **82**, No. 1, p. 213—219 (1977).
See also *1.2*, p. *17*; *18.2*, p. *221*.

J. K. Yakinthos
Determination of Tb_2Ni_{17}, Er_2Ni_{17}, and Tm_2Ni_{17} Magnetic Structure by Neutron Diffraction
(b) Vol. **82**, No. 1, p. 349—356 (1977).
See also *18.3*, p. *229*; 21.1, p. 295.

R. Ramji Rao and A. Ramanand
Third-Order Elastic Constants, Thermal Expansion, and Bulk Modulus of Holmium
(a) Vol. **42**, No. 1, p. 247—255 (1977).
See also 6, p. 57; *8*, p. *66*; *12.1*, p. *102*.

B. Szpunar and P.-A. Lindgård
Theory of Temperature Dependence of the Magnetization in Rare-Earth–Transition-Metal Alloys
(b) Vol. **82**, No. 2, p. 449—456 (1977).
See also 13, p. 110; *18.1*, p. *214*; *18.4*, p. *233*; 21.1, p. 296.

W. Steiner and R. Haferl
Hyperfine Interactions in Y_2Fe_{17}
(a) Vol. **42**, No. 2, p. 739—750 (1977).
See also *5*, p. *49*; *18.2*, p. *222*; 21.1.1, p. 305.

B. Boucher
Magnetic Properties of Amorphous and Bulk TbAu Alloy
(a) Vol. **42**, No. 2, p. K165—K169 (1977).
See also *2*, p. *38*; *18.2*, p. *222*; 21.6, p. 316.

W. Steiner, R. Haferl, and B. K. Choi
Hyperfine Interactions in $RE_2(Fe, Co)_{17}$ (RE = Y, Gd, Dy)
(a) Vol. **43**, No. 1, p. 111—117 (1977).
See also *5*, p. *49*; *18.2*, p. *222*; 21.1.1, p. 305.

H. Krizek, N. N. R. Taylor, and W. D. Corner
Resistivity of Some Light–Heavy Rare-Earth Alloys (I): Resistivity Variation in the Temperature Range 4.2 to 300 K
(a) Vol. **43**, No. 1, p. 251—260 (1977).
See also *14.1*, p. *159*.

I. O. Bashkin, E. G. Ponyatovskii, and M. E. Kost
Phase Transformations in Hydrides of Rare-Earth Metals under Pressure (I): Structural Transition in $NdH_{2.82}$
(b) Vol. **83**, No. 2, p. 517—520 (1977).
See also *1.2*, p. *19*; 14.1, p. 159.

W. Steiner, R. Haferl, and H. Welser
Hyperfine Interactions in $RE_2(Fe, Ni)_{17}$ (RE = Y, Gd, Dy)
(a) Vol. **43**, No. 2, p. 437—442 (1977).
See also *5*, p. *49*; *18.2*, p. *223*; 21.1.1, p. 306.

Z. Biegański and M. Drulis
Low-Temperature Phase Transitions in Lanthanum and Neodymium Trihydrides. Specific Heat Measurements
(a) Vol. **44**, No. 1, p. 91—96 (1977).
See also *8*, p. *67*; 18, p. 212.

S. R. Herd
TEM Observation on the Source of Perpendicular Anisotropy in Amorphous $GdCo(O_2)$
(a) Vol. **44**, No. 1, p. 363—380 (1977).
See also *2*, p. *40*; *18.2.1*, p. *227*.

R. Bodenberger und A. Hubert
On the Determination of the Bloch Wall Energy of Uniaxial Ferromagnets
(a) Vol. **44**, No. 1, p. K7—K11 (1977).
See also *18.2*, p. *224*; 21.1, p. 299.

H. Ratajczak and I. Łabędzka
Anomalous Hall Hysteresis Loops in the Vicinity of the Compensation Point in HoCo Amorphous Films
(a) Vol. **44**, No. 1, p. K103—K105 (1977).
See also 2, p. 40; *14.1*, p. *160*; *18.2*, p. *224*; 21.1, p. 299.

K. Ruebenbauer, J. Fink, H. Schmidt, G. Czjzek, and K. Tomala
An Investigation of Magnetic Ordering in $Gd_xCe_{1-x}Ru_2$ by ^{155}Gd Mössbauer Spectroscopy
(b) Vol. **84**, No. 2, p. 611—618 (1977).
See also *5*, p. *50*; 14.2, p. 164; *18.2*, p. *224*.

21.5 Actinides

Original Papers and Short Notes

K. A. Long
A Theoretical Investigation of the Specific Heats and Transport Properties of the Actinides
(b) Vol. **79**, No. 1, p. 155–165 (1977).
See also *6*, p. *51*; 8, p. 64; *14.1*, p. *157*; 15, p. 198.

V. V. Gann, A. I. Skvortsov, and A. van den Bosch
The Superconducting Transition Temperature of $U_6Fe_{0.98}$ Reduced by Neutron Irradiation
(a) Vol. **41**, No. 1, p. 225–229 (1977).
See also *11*, p. *95*; *14.2*, p. *162*; 18.1, p. 214; 21.1.1, p. 303.

A. G. Moldovan, S. K. Malik, and W. E. Wallace
Structural and Magnetization Studies on $ThCo_{5-x}Mn_x$ and $ThNi_{5-x}Mn_x$ Ternaries
(a) Vol. **43**, No. 1, p. 317–320 (1977).
See also 1.1, p. 12; *18.2*, p. *222*; 21.1, p. 297.

21.6 Noble Metals

Original Papers and Short Notes

GY. FARKAS, Z. GY. HORVÁTH, L.-A. LOMPRE, and G. PETITE
Anomalous Behaviour of the Surface Multiphoton Photoeffect Induced by Single Selected Ultrashort Laser Pulses
(a) Vol. **39**, No. 1, p. K25—K27 (1977).
See also *17*, p. *206*.

V. LANTTO
A Theoretical Study of the Absorption of Photoelectrons Created by X-Rays in Semi-Infinite Targets
(b) Vol. **79**, No. 2, p. 699—711 (1977).
See also *9*, p. *69*; *17*, p. *206*; 21, p. 277; 21.1, p. 289.

K. SAERMARK and J. LEBECH
Dispersion of Cyclotron Waves in Metals and Observability of Many-Body Effects
(b) Vol. **79**, No. 2, p. 723—730 (1977).
See also *13.2*, p. *124*; 14.1, p. 157; 21.2, p. 308.

N. SINGH, J. SINGH, and S. PRAKASH
Residual Resistivity of Noble Metals
(b) Vol. **79**, No. 2, p. 787—794 (1977).
See also 13.4, p. 131; *14.1*, p. *157*.

S. C. GOEL, Y. R. MIDHA, and V. S. NANDA
Applications of a Simple Equation of State for Metals
(b) Vol. **79**, No. 2, p. K103—K105 (1977).
See also *8*, p. *64*; 21, p. 277; 21.1, p. 289.

B. BOUCHER
Bulk Magnetic Properties of Amorphous $Tb_{52}Ag_{48}$. Evidence of a Strong Local Anisotropy
(a) Vol. **40**, No. 1, p. 197—203 (1977).
See also *2*, p. *35*; *18.2*, p. *219*; 21.4, p. 311.

K. L. CHOPRA, A. P. THAKOOR, S. K. BARTHWAL, and P. NATH
Electron Transport Properties of Amorphous Cu–Ag Films
(a) Vol. **40**, No. 1, p. 247—255 (1977).
See also *2*, p. *35*; *14.3.1*, p. *173*; 15, p. 198; 21.1, p. 291.

N. KISHIMOTO, K. MORIGAKI, K. MURAKAMI, A. SHIMIZU, and A. HIRAKI
Electron Spin Resonance in the Amorphous Si–Au System
(b) Vol. **80**, No. 2, p. K113—K115 (1977).
Erratum
(b) Vol. **82**, No. 2, p. K123 (1977).
See also *2*, p. *35*; *19*, p. *237*; 22.1.2, p. 339.

G. VAN TENDELOO, R. WOLF, and S. AMELINCKX
The Microstructure of the Alloy Au_5Mn_2: A Domain Structure with 84 Variants
(a) Vol. **40**, No. 2, p. 531—550 (1977).
See also *1.3*, p. *21*; *1.4*, p. *25*; 21.1, p. 292.

T. ANAGNOSTOPOULOS, P. IKONOMOU and J. K. YAKINTHOS
Magnetic Properties of the Compounds $SmAu_6$, $TbAu_6$, and $DyAu_6$
(a) Vol. **40**, No. 2, p. K109—K112 (1977).
See also *18.1*, p. *214*; *18.4*, p. *232*; 21.4, p. 312.

P. KRATOCHVÍL
The Stress Dependence of the Activation Volume in Copper Base Alloy Single Crystals
(a) Vol. **40**, No. 2, p. K165—K167 (1977).
See also *10.1*, p. *77*; 12.1, p. 101; 21.1, p. 293.

A. Naudon and A. M. Flank
Spinodal Curve and Metastable Miscibility Gap in Low-Concentrated Al–Ag Alloys
(a) Vol. **41**, No. 1, p. 207–211 (1977).
See also *1.2*, p. *16*; *1.3*, p. *21*; 21, p. 280.

E. V. Kozlov, V. N. Emelyanov, and V. A. Antonov
Quantum Theory of the Order–Disorder Transition Energy in $AuCu_3$
(b) Vol. **81**, No. 2, p. 719–725 (1977).
See also *1.2*, p. *16*; *1.3*, p. *21*; 21.1, p. 293.

G. Frommeyer and K. F. Schneider
Size Effect and Interface Contribution to Diamagnetism of Silver–Copper Composite Wires with Fibre Structure
(a) Vol. **41**, No. 2, p. 653–658 (1977).
See also *18*, p. *211*; 21.1, p. 294.

I. Stulíková, P. Lukáč, and G. I. Shklarevskaya
The Critical Resolved Shear Stress of Cd–Ag Single Crystals at Low Temperatures
(a) Vol. **41**, No. 2, p. K197–K198 (1977).
See also *10.1*, p. *78*; 21, p. 282.

M. Galtier, B. Pardo and D. Renard
Study of the Validity of the Real and Imaginary Terms of the Lattice Potential by an Electron Microscopic Method
(a) Vol. **42**, No. 1, p. 141–148 (1977).
See also *1.1*, p. *12*.

E. W. Kreutz and N. Sotnik
Kikuchi Patterns of Spherical Ag Crystals in the Bragg Case
(a) Vol. **42**, No. 1, p. K11–K14 (1977).
See also *1.1*, p. *12*; *1.6*, p. *31*.

T. Hollstein, U. Kreibig, and F. Leis
Optical Properties of Cu and Ag in the Intermediate Region between Pure Drude and Interband Absorption
(b) Vol. **82**, No. 2, p. 545–556 (1977).
See also *20.1*, p. *258*; 21.1, p. 296.

H.-D. Hattendorff
Transition Radiation from Gold and Silver Films Bombarded with 80 keV Electrons
(a) Vol. **42**, No. 2, p. 489–494 (1977).
See also *11*, p. *96*; *20*, p. *245*.

M. J. Yacamán and T. Ocaña Z.
High-Resolution Dark-Field Electron Microscopy of Small Metal Particles
(a) Vol. **42**, No. 2, p. 571–577 (1977).
See also *1.5*, p. *30*.

M. Halbwachs
The Mobilities of Self-Interstitials and Vacancies in a Concentrated Ag–Zn Alloy
(a) Vol. **42**, No. 2, p. 707–712 (1977).
See also *9*, p. *70*; *10.1*, p. *80*; 21.1, p. 296.

B. Boucher
Magnetic Properties of Amophous and Bulk TbAu Alloy
(a) Vol. **42**, No. 2, p. K165–K169 (1977).
See also *2*, p. *38*; *18.2*, p. *222*; 21.4, p. 313.

Th. Hollstein, U. Kreibig, and F. Leis
Optical Properties of Au and Al in the Visible, Determined at 300 and 1.5 K
(b) Vol. **83**, No. 1, p. K49–K53 (1977).
See also *20.1*, p. *260*; 21, p. 284.

H.-J. Fitting, H. Glaefeke, and W. Wild
Electron Penetration and Energy Transfer in Solid Targets
(a) Vol. **43**, No. 1, p. 185–190 (1977).
See also *11*, p. *97*; 21, p. 285; 21.1, p. 297; 21.3, p. 310.

K. Koyama and T. Kaneko
Radiative Electrons Capture by Backscattered Ions from Solid Surfaces
(b) Vol. **83**, No. 2, p. 485–492 (1977).
See also *11*, p. *98*; 13.3, p. 127.

M. Goodarz and R. D. Barnard
Deviations from Matthiessen's Rule in Silver Containing Low-Density Defects
(b) Vol. **83**, No. 2, p. 555–564 (1977).
See also *13.2*, p. *125*; 13.4, p. 137; 14.1, p. 159; 15, p. 199.

S. K. Sarkar, S. K. Das, D. Roy, and S. Sengupta
Three-Body Interaction and Lattice Dynamics of Metals
(b) Vol. **83**, No. 2, p. 615–624 (1977).
See also *6*, p. *59*; 21, p. 285; 21.1, p. 297.

G. van Tendeloo and S. Amelinckx
Long-Period Antiphase Boundary Structures in the Ordered Alloy Au_4Zn (I): One-Dimensional Long-Period Antiphase Boundary Structure
(a) Vol. **43**, No. 2, p. 553–564 (1977).
See also *1.3*, p. *23*; *1.4*, p. *28*; 21.1, p. 298.

S. Bharati, S. Badrinarayanan, and A. P. B. Sinha
On the Diffusion of Iron in Silver
(a) Vol. **43**, No. 2, p. 653–657 (1977).
See also *9*, p. *71*.

L. Dagens
Calculation of a Resonant Model Potential for Copper, Silver, and Gold
(b) Vol. **84**, No. 1, p. 311–324 (1977).
See also *4*, p. *46*; *13*, p. *112*; 21.1, p. 299.

H. R. Khan, K. Lüders, Ch. J. Raub, and Z. Szücs
Knight Shift of Pt^{195} in A15- and σ-Phases of Nb–Pt
(b) Vol. **84**, No. 1, p. K33–K35 (1977).
See also *14.2*, p. *163*; *19*, p. *242*; 21, p. 286.

T. Yamamoto
Experimental Aspects of Electron Channeling Patterns in Scanning Electron Microscopy (I): Measurement of Contrast
(a) Vol. **44**, No. 1, p. 137–146 (1977).
See also *1.4*, p. *28*; *10.1*, p. *81*; 21, p. 287; 21.1, p. 299; 22.1.2, p. 343.

J. M. Corbett and S. S. Sheinin
A Determination of the Reflections Required to Describe Electron Diffraction Phenomena in Gold
(a) Vol. **44**, No. 1, p. 353–361 (1977).
See also *1.1*, p. *13*.

S. Bharati and A. P. B. Sinha
Isotope Effect and Correlation Factor for the Diffusion of Cadmium in Silver
(a) Vol. **44**, No. 1, p. 391–399 (1977).
See also *9*, p. *71*.

R. Kamel, E. Aziz, and K. H. Georgy
Strain Hardening of Pre-Fatigued Ag–5 at% Cu Solid Solution at High Temperatures
(a) Vol. **44**, No. 1, p. K53–K57 (1977).
See also *10.1*, p. *82*.

P. Franzosi
The Geometrical Curvature Factor on the Fermi Surface of the Noble Metals
(b) Vol. **84**, No. 2, p. K137—K141 (1977).
See also *13.2*, p. *125*; 21.1, p. 300.

T. Yamamoto
Experimental Aspects of Electron Channeling Patterns in Scanning Electron Microscopy (II): Estimation of Contrast Depth
(a) Vol. **44**, No. 2, p. 467—476 (1977).
See also *1.4*, p. *28*; *10.1*, p. *82*; 21, p. 287; 21.1, p. 300; 22.1.2, p. 344.

21.7 Semi-Metals (Semiconducting Alloys See 22.7)

Original Papers and Short Notes

W. Braune, G. Kuka, S. Hess, H.-U. Müller, and T. Jung
Cyclotron Resonance and Quantum Oscillations in Semimetallic Bismuth–Antimony Alloys with $E_g = 0$
(b) Vol. **79**, No. 2, p. 501—508 (1977).
See also *13.2*, p. *124*; 14.1, p. 157.

E. Antončík
Calibration of the Isomer Shift for ^{119}Sn
(b) Vol. **79**, No. 2, p. 605—611 (1977).
See also *5*, p. *47*; 13.4, p. 130.

R. Herrmann and P. Goy
Investigation of Plasma Effects in Bismuth–Antimony Alloys in the Submillimeter Wave Range
(b) Vol. **80**, No. 1, p. 207—213 (1977).
See also *13.1*, p. *115*; *19*, p. *236*.

V. A. Kovarskii, N. F. Perelman, I. A. Chaikovskii, and V. N. Chebotar
Conductivity in Thin (Quantizing) Wires
(b) Vol. **80**, No. 1, p. K5—K8 (1977).
See also *14.1*, p. *158*.

K. Schröder and T. Joy
Magnetization of BiMn Particles in a Bi-Rich Bi–Mn Matrix
(a) Vol. **40**, No. 2, p. K145—K148 (1977).
See also *18.2*, p. *219*; 21.1, p. 292.

K. Ozawa and S. Anzai
Effect of Pressure on a New Magnetic Transition Temperature in the Ferrimagnetic State of $Mn_{1.77}Sn$
(a) Vol. **40**, No. 2, p. K199—K202 (1977).
See also *14.1*, p. *158*; *18.3*, p. *229*.

A. B. Makhnovetskii and G. L. Krasko
Pseudopotential Theory and the Problem of Covalent Bonding (II): Crystals of Si, Ge, and α-Sn as-Ordered Systems
(b) Vol. **81**, No. 1, p. 263—273 (1977).
See also *1*, p. *9*; *4*, p. *45*; 22.1.1, p. 334; 22.1.2, p. 339.

S. S. Bhandari and J. Varma
Anharmonicity Effect in the Oxides of Pb, Sn, and Bi
(a) Vol. **41**, No. 1, p. K89—K92 (1977).
See also *5*, p. *48*; 21, p. 280.

O. P. Hansen
Weak-Field Thermomagnetic Coefficients of a Trigonal Bismuth Sample
(b) Vol. **82**, No. 1, p. 133—138 (1977).
See also *15*, p. *199*.

K. Bartkowski, D. Włosewicz, and J. Rafałowicz
Anisotropy of Thermal Conductivity of Tin Monocrystals Depending on Chemical Admixtures in the Temperature Range 5 to 70 K
(a) Vol. **42**, No. 1, p. 351—357 (1977).
See also *8*, p. *66*.

W. Schnelle and U. Dillner
Thickness Dependence of Resistivity and Its Temperature Coefficient of Polycrystalline Bismuth Films
(a) Vol. **44**, No. 1, p. 197—204 (1977).
See also *14.1*, p. *160*; *14.3.1*, p. *175*.

22. Semiconductors and Ionic Crystals

Review Articles

SH. M. KOGAN and T. M. LIFSHITS
Photoelectric Spectroscopy — A New Method of Analysis of Impurities in Semiconductors
(a) Vol. **39**, No. 1, p. 11—39 (1977).
See also *13.4*, p. *129*; *16*, p. *201*; 22.1.1, p. 333; 22.1.2, p. 337; 22.2.1, p. 350.

K. W. BÖER
The CdS/Cu_2S Solar Cell (I): Minority Carrier Generation and Transport in the Cu_2S Emitter
(a) Vol. **40**, No. 2, p. 355—384 (1977).
See also 14.3.3, p. 178; 14.3.4, p. 183; *16*, p. *201*; 20.1, p. 247; 22.4.1, p. 362.

G. BORSTEL and H. J. FALGE
Surface Phonon-Polaritons at Semi-Infinite Crystals
(b) Vol. **83**, No. 1, p. 11—45 (1977).
See also *6*, p. *51*; *13.5*, p. *141*; 20.1, p. 247; 22.6, p. 386.

Original Papers and Short Notes

P. PETELENZ
Mixing of Frenkel Excitons and Ionic Excited States of a Linear Molecular Crystal with Two Molecules in the Unit Cell (II): Physical Consequences
(b) Vol. **79**, No. 1, p. 61—70 (1977).
See also *13.5.1*, p. *143*; 20.1, p. 247.

P. GOSAR
Quantum Mechanics of Dislocation Kinks in Covalent Semiconductors
(b) Vol. **79**, No. 1, p. 175—183 (1977).
See also *10.2*, p. *83*.

S. ONARI, M. CARDONA, E. SCHÖNHERR, and W. STETTER
Effect of Stress on the Raman Spectra of Mg_2Si and Mg_2Sn
(b) Vol. **79**, No. 1, p. 269—276 (1977).
See also *6*, p. *51*; 12.2, p. 104; *20.1*, p. *247*.

E. MAJERNÍKOVÁ
On the Theory of Exciton Absorption in Amorphous Semiconductors
(b) Vol. **79**, No. 1, p. 339—345 (1977).
See also 2, p. 33; *13.5.1*, p. *143*; *20.1*, p. *248*.

E. V. CHISLER, V. YU. DAVYDOV, I. N. GONCHARUK, and E. A. IVANOVA
The "Destructive" Fermi Resonance
(b) Vol. **79**, No. 1, p. 347—357 (1977).
See also *6*, p. *51*; *20.1*, p. *248*.

L. BÁNYAI and A. ALDEA
Semi-Classical and Quantum-Mechanical Theory of Hopping Conduction
(b) Vol. **79**, No. 1, p. 365—377 (1977).
Erratum
(b) Vol. **81**, No. 1, p. 409 (1977).
See also 6, p. 51; 13.4, p. 129; *14.3*, p. *165*.

R. KISHORE
Exactly Solvable Off-Diagonal Disorder
(b) Vol. **79**, No. 1, p. K9—K11 (1977).
See also *2*, p. *33*; *13*, p. *107*.

A. N. CHAKRAVARTI and A. K. CHOWDHURY
Influence of Band Non-Parabolicity on the Oscillatory Dependence of the Diffusivity–Mobility Ratio in Degenerate Semiconductors on Magnetic Quantization
(b) Vol. **79**, No. 1, p. K31—K34 (1977).
See also 13.1, p. 114; *14.3*, p. *165*; 18, p. 209.

S. P. Reifman
The Influence of a Strong Magnetic Field on the Self-Trapping of Electron in Homopolar Crystals
(b) Vol. **79**, No. 1, p. K57—K59 (1977).
See also *13.5.3*, p. *151*.

D. P. Bhattacharya and Z. S. Kachlishvili
On the Theory of the Impurity Breakdown in a Magnetic Field
(a) Vol. **39**, No. 1, p. 49—56 (1977).
See also *13*, p. *107*; *14.3.4*, p. *183*.

I. Munteanu
The Effect of Effective Lifetime Modulation in Semiconductors
(a) Vol. **39**, No. 1, p. 291—300 (1977).
See also *13.3*, p. *126*; *14.3.3*, p. *178*.

J. Kaczér and P. Novotný
High-Field Magnetization of U_3As_4
(a) Vol. **39**, No. 1, p. 351—355 (1977).
See also 1.2, p. 14; *18.2*, p. *217*.

C. M. Vodenicharov, A. A. Stanchev, and M. Lozev
Thermionic Emission in the System M–GeS–M
(a) Vol. **39**, No. 1, p. K1—K5 (1977).
See also 14.3.1, p. 173; *14.3.4*, p. *183*.

W. Nolting
Influence of Electron–Ion Spin Coupling on the Electronic Properties of a Ferromagnetic Semiconductor
(b) Vol. **79**, No. 2, p. 573—584 (1977).
See also *13.5.4*, p. *153*; 18.2, p. 218.

V. M. Fomin and E. P. Pokatilov
Multiphoton Absorption by Charge Carriers in Semiconductors in a Magnetic Field
(b) Vol. **79**, No. 2, p. 595—604 (1977).
See also *14.3.4*, p. *183*; *20.1*, p. *249*.

B. Gerlach
On the Binding Energy of Wannier Excitons in Strong Anisotropic Crystals
(b) Vol. **79**, No. 2, p. 613—616 (1977).
See also *13.5.1*, p. *144*.

G. Cubiotti, E. Donato, E. S. Giuliano, and R. Ruggeri
On the Influence of Disorder on the Electron Density of States in a Simplified Model for a Quasi-Onedimensional Conductor
(b) Vol. **79**, No. 2, p. 655—661 (1977).
See also *2*, p. *33*; *13*, p. *107*; 14, p. 155.

S. H. El-Mahalawy and B. L. Evans
Temperature Dependence of the Electrical Conductivity and Hall Coefficient in 2H-MoS_2, $MoSe_2$, WSe_2, and $MoTe_2$
(b) Vol. **79**, No. 2, p. 713—722 (1977).
See also 13.1, p. 115; *14.3*, p. *166*.

R. A. Evarestov and V. A. Lovchikov
Large Unit Cell Calculations of Solids in the CNDO Approximation
(b) Vol. **79**, No. 2, p. 743—751 (1977).
See also *4*, p. *44*; *13*, p. *107*.

I. Ya. Korenblit
Phase Transition and Instability of the Homogeneous State in Ferromagnetic Semiconductors
(b) Vol. **79**, No. 2, p. K95—K98 (1977).
See also 1.2, p. 14; *13*, p. *107*; *18.2*, p. *218*.

A. P. Dzhotyan
Bose-Einstein Condensation of Excitons in Thin Semiconductors
(b) Vol. **79**, No. 2, p. K143—K146 (1977).
See also *13.5.1*, p. *144*.

W. Gorzkowski
Excitons Bound to Ionized Donors in Polar Semiconductors
(b) Vol. **79**, No. 2, p. K151—K152 (1977).
See also *13.5.1*, p. *144*; 22.4.1, p. 362; 22.4.2, p. 365.

C. K. Sarkar and P. K. Basu
Inelastic Acoustic Phonon Scattering in Nonparabolic Semiconductors under Longitudinal Quantizing Magnetic Field
(b) Vol. **79**, No. 2, p. K157—K160 (1977).
See also *14.3*, p. *166*; 18, p. 210.

R. de Ridder, D. van Dyck, G. van Tendeloo, J. van Landuyt, and S. Amelinckx
Microstructural Study of Copper-Intercalated Niobium Disulphide and Tantalum Disulphide Cu_xNbS_2 and Cu_xTaS_2 (III): Study of the Diffuse Scattering: the Transition State
(a) Vol. **39**, No. 2, p. 383—399 (1977).
See also *1.2*, p. *15*; 1.4, p. 24.

J. Bauer
Optical Properties, Band Gap, and Surface Roughness of Si_3N_4
(a) Vol. **39**, No. 2, p. 411—418 (1977).
See also *2*, p. *34*; 13.1, p. 115; 20.1, p. 249.

M. Wautelet, A. Gérard, F. Grandjean, K. de Strooper, and G. Robbrecht
Magnetic Properties of Fe_2P
(a) Vol. **39**, No. 2, p. 425—430 (1977).
See also 5, p. 47; *18.2*, p. *218*.

P. P. Seregin, V. P. Sivkov, F. S. Nasredinov, L. N. Vasilev, Yu. V. Krylnikov, and Yu. P. Kostikov
The Influence of the Crystal-to-Glass Transition on the Local Structure of Semiconductors
(a) Vol. **39**, No. 2, p. 437—444 (1977).
See also *1.2*, p. *15*; 2, p. 34; 5, p. 47.

K. Ziegler and U. Birkholz
Photoelectric Properties of Si_2Te_3 Single Crystals
(a) Vol. **39**, No. 2, p. 467—475 (1977).
See also *13.4*, p. *131*; 14.3, p. 166; 16, p. 201; 20.3, p. 268.

Meera Rao and O. N. Srivastava
Interpolytypic Transformations and the Electrical Behaviour of PbI_2 Crystals
(a) Vol. **39**, No. 2, p. K81—K84 (1977).
See also *1.1*, p. *11*; *1.2*, p. *15*; 14.3, p. 166.

G. I. Makovetskii, V. M. Ryzhkovskii, and G. M. Shakhlevich
Neutron Diffraction Studies of Chromium Sulphide with Nickel Arsenide Type Structure
(a) Vol. **39**, No. 2, p. K127—K130 (1977).
See also *18.4*, p. *231*.

H. P. Geserich, W. Möller, G. Scheiber, and L. Pintschovius
Optical Investigation of the Electrical Anisotropy of $(SN)_x$ Single Crystals
(b) Vol. **80**, No. 1, p. 119—123 (1977).
See also 1.4, p. 24; *14.1*, p. *157*; *20.1*, p. *250*.

H. J. L. van der Valk and C. Haas
Anisotropy of Lattice Vibrations of Layered Compounds
(b) Vol. **80**, No. 1, p. 321—332 (1977).
See also *6*, p. *53*; *20.1*, p. *251*; 22.5, p. 370.

G. V. Ionova, S. P. Ionov, O. M. Pachev, and E. F. Makarov
Orbital-Charge Ordering in Crystals
(b) Vol. **80**, No. 1, p. K9—K12 (1977).
See also *13*, p. *108*.

S. P. Ionov, V. S. Lubimov, L. A. Manakova, and E. F. Makarov
The Charge-Ordered State and Magnetic Collective Excitations
(b) Vol. **80**, No. 1, p. K13—K15 (1977).
See also *13*, p. *108*; 13.5, p. 141.

V. G. Baryakhtar, A. A. Galkin, and V. T. Telepa
High Hydrostatic Pressure Effect on the Flop of Magnetic Moments of Sublattices and Néel Temperature in the Antiferromagnetic $CuCl_2 \cdot 2\,H_2O$
(b) Vol. **80**, No. 1, p. K37—K40 (1977).
See also *18.4*, p. *232*; 19, p. 236.

B. Fromm and W. Teubner
Screening Effects in Amorphous Semiconductors Influencing the Poole-Frenkel Conduction
(a) Vol. **40**, No. 1, p. 63—68 (1977).
See also *2*, p. *35*; *14.3.1*, p. *173*.

R. de Ridder, G. van Tendeloo, J. van Landuyt, D. van Dyck, and S. Amelinckx
Erratum to: Microstructural Study of Copper-Intercalated Niobium Disulphide and Tantalum Disulphide Cu_xNbS_2 and Cu_xTaS_2 (I)
(a) Vol. **40**, No. 1, p. 343 (1977).
See also *1.2*, p. *15*; 1.4, p. 25.

M. Krystek
On the Evaluation Method of Balarin and Zetzsche
(a) Vol. **40**, No. 1, p. K65—K68 (1977).
See also *13.4*, p. *132*; *20.3*, p. *269*.

H. R. Bilger and J. P. Nougier
Perturbation of a Homogeneous Current Path by a Stub
(a) Vol. **40**, No. 1, p. K85—K88 (1977).
See also *14.3*, p. *167*.

J. L. Farvacque and P. Lenglart
Inhomogeneous Strain Field Scattering Deformation Potential (II): Free-Carrier Mobility Associated With Dislocation Strain Fields in Semiconductors
(b) Vol. **80**, No. 2, p. 433—439 (1977).
See also *10.2*, p. *86*; *14.3*, p. *167*; 22.1.3, p. 345.

F. Thuselt
Spin-Dependent Luminescence of Donor–Acceptor Pairs in Semiconductors (I): General Theory
(b) Vol. **80**, No. 2, p. 499—507 (1977).
See also *13.4*, p. *132*; *20.3*, p. *269*.

H. Böttger and V. V. Bryksin
A Theory of the Hall Effect in the Hopping Region in Disordered Systems (I): Rate Equation in Presence of Electrical and Magnetic Fields
(b) Vol. **80**, No. 2, p. 569—578 (1977).
See also 2, p. 35; 6, p. 54; 13, p. 108; 13.5.3, p. 151; *14.3*, p. *167*.

G. M. Shmelev, I. A. Chaikovskii, V. V. Pavlovich, and E. M. Epshtein
Electron–Phonon Interaction in a Superlattice
(b) Vol. **80**, No. 2, p. 697—701 (1977).
See also *6*, p. *55*.

Yu. E. Perlin, B. S. Tsukerblat, and T. Singh Dod
Moments of Electroabsorption of Jahn-Teller Impurity Centres
(b) Vol. **80**, No. 2, p. 703—707 (1977).
See also 13.4, p. 132; *14.3.4*, p. *184*; *20.1*, p. *252*.

V. K. Bashenov, M. Bardashova, and A. M. Mutal
Baldereschi Points for Some Noncubic Lattices
(b) Vol. **80**, No. 2, p. K89—K93 (1977).
See also *13.1*, p. *116*.

R. de Ridder, D. van Dyck, G. van Tendeloo, and S. Amelinckx
A Cluster Model for the Transition State and Its Study by Means of Electron Diffraction (II): Application to Some Particular Systems
(a) Vol. **40**, No. 2, p. 669—683 (1977).
See also *1.4*, p. *25*; 10.2, p. 87.

J. J. Legendre, M. Huber, and M. Sauvage
Observation of Linear and Planar Defects in $TiS_{1.7}$ Polytypic Crystals
(a) Vol. **40**, No. 2, p. K101—K103 (1977).
See also *1.4*, p. *25*.

A. Buroff, E. Nebauer, P. Süptitz, and I. Willert
Migration of Silver and Gold in Amorphous As_2S_3
(a) Vol. **40**, No. 2, p. K195—K198 (1977).
See also *2*, p. *36*; *9*, p. *70*.

F. Thuselt and K. Unger
Spin-Dependent Luminescence of Donor–Acceptor Pairs in Semiconductors (II): Application to Inner Strains
(b) Vol. **81**, No. 1, p. 89—95 (1977).
See also *13.4*, p. *133*; *20.3*, p. *269*.

H. Böttger and V. V. Bryksin
A Theory of the Hall Effect in the Hopping Region in Disordered Systems (II): Random Resistor Network and Hall Effect
(b) Vol. **81**, No. 1, p. 97—106 (1977).
See also 2, p. 36; 6, p. 55; 13, p. 109; 13.5.3, p. 151; *14.3*, p. *167*.

P. R. R. Pereira, M. A. Amato, and L. C. M. Miranda
Carrier-Assisted Radiation Pumping of Spin Waves in Ferromagnetic Semiconductors under a Strong Magnetic Field
(b) Vol. **81**, No. 1, p. 395—401 (1977).
See also *13.5.4*, p. *153*; 14.3, p. 167; 18.2, p. 219.

D. Chattopadhyay
Average Energy of Conduction Electrons in Non-Parabolic Semiconductors
(b) Vol. **81**, No. 1, p. K51—K53 (1977).
See also 13.1, p. 117; *14.3*, p. *168*.

A. K. Goyal and S. C. Goyal
Dispersion of Phonons in NbC
(b) Vol. **81**, No. 1, p. K63—K65 (1977).
See also *6*, p. *56*; 12.2, p. 104.

J. M. Moret and H. Bill
Experimental Determination of the Spin-Stress Coefficients of Gd^{3+} in the New Host Crystal Na_2S
(a) Vol. **41**, No. 1, p. 163—171 (1977).
See also 13.4, p. 134; *19*, p. *239*.

D. Goldschmidt, T. Bernstein, and P. S. Rudman
The Kinetics of Photodissolution of Silver in Amorphous As_2S_3 Films
(a) Vol. **41**, No. 1, p. 283—287 (1977).
See also *11*, p. *95*; 14.3.2, p. 176.

W. Teubner and B. Fromm
Some Remarks on the Poole-Frenkel Effect for Centres with Screened Potential
(a) Vol. **41**, No. 1, p. K47—K49 (1977).
See also *14.3.4*, p. *185*.

I. M. Chapnik
On the Possibility of Semiconducting Properties in Diborides ReB_2, TcB_2, OsB_2, and RuB_2
(a) Vol. **41**, No. 1, p. K71—K73 (1977).
See also *14.3*, p. *168*.

H. Böttger and V. V. Bryksin
A Theory of the Hall Effect in the Hopping Region in Disordered Systems (III): DC Hall Effect
(b) Vol. **81**, No. 2, p. 433—441 (1977).
See also 2, p. 37; 6, p. 56; 13, p. 109; 13.5.3, p. 151; *14.3*, p. *168*.

W. J. Huybrechts
Infrared Absorption by an Optical Polaron in a Magnetic Field
(b) Vol. **81**, No. 2, p. 585—592 (1977).
See also *13.5.3*, p. *151*; *20.1*, p. *255*.

S. P. Ionov, V. S. Lubimov, L. A. Manakova, and E. F. Makarov
The Electronic Spectra of Charge-Ordered Crystals
(b) Vol. **81**, No. 2, p. 691—701 (1977).
See also *13*, p. *109*; 13.5, p. 141.

K. S. Dubey
Contribution to the Lattice Thermal Conductivity Due to the Correction Term in the Callaway Integral
(b) Vol. **81**, No. 2, p. K83—K86 (1977).
See also *6*, p. *56*; *8*, p. *65*.

M. Matyáš
Magnetic Properties of As_xSe_{100-x} Glasses in the Temperature Range 77 to 300 K
(b) Vol. **81**, No. 2, p. K115—K117 (1977).
See also *2*, p. *37*; *19*, p. *239*.

S. Nešpurek and J. Sworakowski
A Differential Method of Analysis of Steady-State Space-Charge-Limited Current–Voltage Characteristics
(a) Vol. **41**, No. 2, p. 619—627 (1977).
See also 13.4, p. 135; *14.3.4*, p. *185*.

M. Zoaeter, A. Conan, and D. Delaunay
Transport Measurements and Band Structure of Bulk $MoSe_{2-x}$ Compounds
(a) Vol. **41**, No. 2, p. 629—635 (1977).
See also 13.1, p. 118; *14.3*, p. *169*; *15*, p. *199*.

Yu. V. Bogatyrev and F. P. Korshunov
The Effect of High-Speed Electrons of Different Energy on the Charge Storage in SiO_2 and Si_3N_4 Dielectric Films
(a) Vol. **41**, No. 2, p. K151—K154 (1977).
See also *11*, p. *96*; *14.4*, p. *191*; 22.6, p. 389.

R. Erlandsson, U. Birkholz, and S.-E. Karlsson
Surface Investigation of Si_2Te_3 with Auger Spectroscopy
(a) Vol. **41**, No. 2, p. K163—K165 (1977).
See also *1.6*, p. *31*.

S. I. Pekar
On the Theory of Additional Light Waves and Boundary Conditions in Anisotropic Crystals
(b) Vol. **82**, No. 1, p. 83—89 (1977).
See also *13.5.1*, p. *146*; *20.1*, p. *256*.

I. I. Tupizin and I. V. Abarenkov
The Crystal Potential and Summation and Diverging Series
(b) Vol. **82**, No. 1, p. 99—105 (1977).
See also *1.1*, p. *11*.

YU. I. BALKAREI and V. A. NIKITOV
On Magnon–Acoustic Phonon Interaction in Magnetic Semiconductors
(b) Vol. **82**, No. 1, p. 127—131 (1977).
See also *6*, p. *57*; *13.5.4*, p. *153*; 18.2, p. 221.

J. R. CHELIKOWSKY, M. SCHLÜTER, and M. L. COHEN
Semi-Empirical Pseudopotential Calculation for the Electronic Structure of Polymeric Sulfur Nitride
(b) Vol. **82**, No. 1, p. 357—367 (1977).
See also *13.1*, p. *119*; 14, p. 155.

G. M. SHMELEV, I. A. CHAIKOVSKII, V. V. PAVLOVICH, and E. M. EPSHTEIN
Plasma Oscillations in a Superlattice
(b) Vol. **82**, No. 1, p. 391—395 (1977).
See also *13.5.2*, p. *149*.

S. LARA, R. MOREIRA XAVIER, and C. A. TAFT
On the Indirect Exchange in Intrinsic Semiconductors
(b) Vol. **82**, No. 1, p. K21—K22 (1977).
See also *13.1*, p. *119*.

K. S. DUBEY
Role of Point-Defect Scattering in the Lattice Thermal Conductivity of an Insulator at Low Temperatures
(b) Vol. **82**, No. 1, p. K63—K67 (1977).
See also 6, p. 57; *8*, p. *66*.

S. N. KOZLOV
Slow Relaxation Kinetics on a Heterogeneous Semiconductor Surface
(a) Vol. **42**, No. 1, p. 115—124 (1977).
See also *13.3*, p. *127*; 14.3.2, p. 176.

A. H. MARSHAK and K. M. VAN VLIET
Current in Semiconductors with Position-Dependent Band Gap and Electron Affinity
(a) Vol. **42**, No. 1, p. 279—283 (1977).
See also 13.4, p. 136; *14.3*, p. *169*.

G. VAN TENDELOO, R. DE RIDDER, L. VAN GOETHEM, D. VAN DYCK, J. VAN LANDUYT, and S. AMELINCKX
Microstructural Study of Octahedral Intercalation Compounds of Tantalum and Niobium Sulphide
(a) Vol. **42**, No. 1, p. 319—335 (1977).
Erratum
(a) Vol. **44**, No. 1, p. 411(1977).
See also 1.3, p. 22; *1.4*, p. *26*.

I. M. KOTINA, S. R. NOVIKOV, and T. I. PIROZHKOVA
Determination of the Parameters of Impurity Centres from the Thermally Stimulated Current of the p–n Junction Space Charge Region
(a) Vol. **42**, No. 1, p. 397—401 (1977).
See also *13.4*, p. *136*; *14.3.3*, p. *180*.

L. C. R. ALFRED
Charge Screening in an Electron Gas of Arbitrary Degeneracy
(b) Vol. **82**, No. 2, p. 467—471 (1977).
See also *13.1*, p. *120*; *14*, p. *156*; 21, p. 282.

M. TAUT and A. M. RADWAN
Soft X-Ray Spectra of Al and the Laves Phase $CaAl_2$
(b) Vol. **82**, No. 2, p. 507—511 (1977).
See also *13.1*, p. *120*; *20*, p. *245*; 21, p. 283.

K. A. Agyei and J. L. Birman
A New Group-Theoretical Approach to Cell-Preserving Phase Transitions in Crystals
(b) Vol. **82**, No. 2, p. 565—575 (1977).
See also *1.2*, p. *17*.

W. J. Huybrechts
Influence of a DC Field on the Free Polaron Infrared Absorption
(b) Vol. **82**, No. 2, p. 589—594 (1977).
See also *13.5.3*, p. *152*; *20.1*, p. *258*.

W. John
Comment on the Paper "Exactly Solvable Off-Diagonal Disorder"
(b) Vol. **82**, No. 2, p. K111—K112 (1977).
See also *2*, p. *38*; *13*, p. *110*.

J. van Landuyt, G. van Tendeloo, and S. Amelinckx
Electron Diffraction Study of Inter- and Intrapolytypic Phase Transitions in Transition Metal Dichalcogenides (IV): Cinematographic Observations of the $1T_\beta \rightarrow 1T_\gamma$-Phase Transition in TaS_2
(a) Vol. **42**, No. 2, p. 565—570 (1977).
See also *1.1*, p. *12*; *1.2*, p. *18*.

U. Hartmann and P. Schley
The MIS HF Capacitance and Its Minimum
(a) Vol. **42**, No. 2, p. 667—673 (1977).
See also *14.3.3*, p. *181*; 14.3.4, p. 186.

W. Rehwald, J. R. Sandercock, and M. Rossinelli
Elastic Properties of KCN and $K(CN)_{1-x}Cl_x$
(a) Vol. **42**, No. 2, p. 699—705 (1977).
See also *12.2*, p. *105*.

M. Janai and P. S. Rudman
The Kinetics of the Light-Enhanced Vaporization of Amorphous As_2S_3 Thin Films
(a) Vol. **42**, No. 2, p. 729—737 (1977).
See also *2*, p. *38*; 14.3.1, p. 174; *16*, p. *203*.

L. M. Baskin, O. I. Lvov, and G. N. Fursey
On the Theory of Field Emission from p-Type Semiconductors
(a) Vol. **42**, No. 2, p. 757—767 (1977).
See also 13.3, p. 127; *14.3.4*, p. *186*.

C. M. Vodenicharov
A Simple Method for the Determination of the Effective Electron Mass from Current-Voltage Characteristics in the Thermionic Region
(a) Vol. **42**, No. 2, p. 785—789 (1977).
See also *13.1*, p. *121*; *14.3.1*, p. *174*; 14.3.4, p. 186; 22.6, p. 390; 22.9, p. 417.

C. V. Dharmadhikari, D. S. Joag, and P. I. Kanitkar
Surface Migration and Overgrowth of LaB_6 on Tungsten Field Emitter
(a) Vol. **42**, No. 2, p. K99—K101 (1977).
See also *17.1*, p. *208*.

I. P. Zvyagin
Correlation Effects in Hopping Conduction
(b) Vol. **83**, No. 1, p. 63—70 (1977).
See also 2, p. 38; 13, p. 110; *14.3*, p. *170*; 15, p. 199.

R. S. Tripathi, L. L. Moseley, and T. Lukes
Optical Absorption in Non-Crystalline Semiconductors
(b) Vol. **83**, No. 1, p. 197—207 (1977).
See also 2, p. 38; *13*, p. *110*; *20.1*, p. *259*.

F. Bechstedt and R. Enderlein
Theory of Resonance Raman Scattering in Disordered Solids. Smooth Random Field Model
(b) Vol. **83**, No. 1, p. 239—247 (1977).
See also 2, p. 38; *6*, p. *58*; 13.1, p. 121; *20.1*, p. *259*.

A. M. Brodskii
Specific Features of Many-Photon Photoemission from Metals and Semiconductors in Strong Fields
(b) Vol. **83**, No. 1, p. 331—340 (1977).
See also *17*, p. *207*; 21, p. 284.

U. V. Desnica and N. B. Urli
An Approach to the Problem of the Displacement Energy Threshold in Semiconductors
(b) Vol. **83**, No. 1, p. K41—K44 (1977).
See also *10.2*, p. *90*; 11, p. 97.

R. Pässler
High-Temperature Limit of the Non-Radiative Multiphonon Carrier-Capture Cross-Sections of Deep Traps in Semiconductors
(b) Vol. **83**, No. 1, p. K55—K58 (1977).
See also *13.4*, p. *137*.

P. Novotný, V. Roskovec, and J. Šternberk
High-Field Susceptibility of U_3As_4 at Helium Temperature
(b) Vol. **83**, No. 1, p. K87—K90 (1977).
See also *18.2*, p. *222*.

G. R. Umarov
Isoelectronic Series and Trends in Band Structures
(b) Vol. **83**, No. 1, p. K93—K95 (1977).
See also *4*, p. *45*; *13.1*, p. *121*.

A. N. Zyuganov, S. V. Svechnikov, and P. S. Smertenko
Effect of Integrally Weak Diffusion on the Parameters of Current–Voltage Characteristics of Quasi-Monopolar Semiconductors
(a) Vol. **43**, No. 1, p. 333—341 (1977).
See also *14.3*, p. *170*; 14.3.4, p. 186.

Vo Hong Anh and Nguyen Van Trong
On the Theory of Parametric Excitation in Quantum Electron–Phonon Systems
(b) Vol. **83**, No. 2, p. 395—404 (1977).
See also *6*, p. *59*; *13.5.2*, p. *150*; 20, p. 246.

U. Büchner
Wave-Vector Dependence of the Electron Energy Losses of PbS, PbSe, and PbTe
(b) Vol. **83**, No. 2, p. 493—500 (1977).
See also *13.1*, p. *122*; 14.4, p. 192.

S. C. Goyal and R. Prakash
Many-Body Effects on the Second-Order Elastic Constants of Solids with ZnS Structure
(b) Vol. **83**, No. 2, p. 509—515 (1977).
See also *4*, p. *46*; *12.2*, p. *106*; 22.1, p. 332; 22.1.1, p. 335; 22.1.2, p. 342; 22.2, p. 348; 22.3, p. 358; 22.4, p. 361.

V. G. Aleshin, T. I. Serebryakova and A. I. Kharlamov
Charge State of Boron in Transition Metal Borides
(b) Vol. **83**, No. 2, p. 537—542 (1977).
See also *13*, p. *111*; 17, p. 207.

K. Pesz and T. Luty
On the Charge Carrier Mobilities in Molecular Crystals
(b) Vol. **83**, No. 2, p. 565—571 (1977).
See also *13.5.3*, p. *152*; 14.3, p. 171.

A. K. Jonscher
A Many-Body Model of Dielectric Polarisation in Solids (I): Experimental Data and Existing Interpretations
(b) Vol. **83**, No. 2, p. 585—597 (1977).
See also 2, p. 39; *14.4*, p. *192*; 14.4.1, p. 196; 22.9, p. 418.

F. G. dos Reis, Hilda A. Cerdeira, and R. Luzzi
Dielectric Response of Saturated Semiconductors
(b) Vol. **83**, No. 2, p. 607—614 (1977).
See also 13.1, p. 122; *14.4*, p. *192*; *20*, p. *246*.

K. O. Kechechyan and A. A. Kirakossyan
Wannier-Mott Exciton in the Field of Screw Dislocations
(b) Vol. **83**, No. 2, p. K105—K109 (1977).
See also 10.2, p. 91; *13.5.1*, p. *147*.

R. Pässler
Low-Temperature Dependences of the Nonradiative Multiphonon Carrier-Capture Cross-Sections of Deep Traps in Semiconductors
(b) Vol. **83**, No. 2, p. K111—K114 (1977).
Erratum
(b) Vol. **84**, No. 2, p. K163 (1977).
See also *13.4*, p. *137*.

E. Nebauer, P. Süptitz, and I. Willert
Determination of the Charge of Migrating Impurities from Electrodiffusion Profiles: the System a-As_2Se_3:Cu
(a) Vol. **43**, No. 2, p. 451—458 (1977).
See also *2*, p. *39*; *9*, p. *71*; 14.3.4, p. 187.

D. van Dyck, R. de Ridder, G. van Tendeloo, and S. Amelinckx
A Cluster Model for the Transition State and Its Study by Means of Electron Diffraction (III): Generalisations of the Theory and Relation to the SRO Parameters
(a) Vol. **43**, No. 2, p. 541—552 (1977).
See also *1.4*, p. *27*; 10.2, p. 91.

A. S. Vishnevskii, I. Ya. Dekhtyar, E. G. Madatova, V. S. Mikhalenkov, and A. Čížek
Positron Annihilation in Boron Nitride Crystals
(a) Vol. **43**, No. 2, p. 589—592 (1977).
See also *10.2*, p. *91*.

G. van Tendeloo, J. van Landuyt, and S. Amelinckx
Electron Diffraction Evidence for a One-Dimensional Peierls Distortion in a Transition Metal Trichalcogenide: TaS_3
(a) Vol. **43**, No. 2, p. K137—K140 (1977).
See also *1.1*, p. *13*; 1.4, p. 28.

L. A. Manakova, E. F. Makarov, and S. P. Ionov
The Electronic Spectra and Electromagnetic Properties of Charge-Ordered Systems
(b) Vol. **84**, No. 1, p. 55—67 (1977).
See also *13*, p. *112*; 13.5, p. 142.

V. L. Bonch-Bruevich
Coulomb Gap in a Disordered Semiconductor
(b) Vol. **84**, No. 1, p. 99—105 (1977).
See also *13*, p. *112*; 14.3, p. 171.

A. K. Jonscher
A Many-Body Model of Dielectric Polarisation in Solids (II): The Universal Model
(b) Vol. **84**, No. 1, p. 159—166 (1977).
See also 2, p. 39; *14.4*, p. *192*; 14.4.1, p. 196; 22.9, p. 419.

O. UEMURA and T. SATOW
The Magnetic Susceptibility of Antimony and Bismuth Chalcogenides in the Liquid State
(b) Vol. **84**, No. 1, p. 353—360 (1977).
See also *2*, p. *39*; *18*, p. *212*.

L. A. MANAKOVA, E. F. MAKAROV, V. S. LUBIMOV, and S. P. IONOV
Some Peculiarities of Charge-Ordered Systems at Constant Electric Field
(b) Vol. **84**, No. 1, p. K1—K6 (1977).
See also *13*, p. *112*; 13.5, p. 142.

B. MESTNIK and Z. OGORELEC
Madelung Constants of Stoichiometric and Non-Stoichiometric Average Crystals
(b) Vol. **84**, No. 1, p. K37—K41 (1977).
See also *4*, p. *46*.

H.-J. ANKLAM, W. BRODKORB, and W. HAUBENREISSER
A Regional Approximation for a Real p^+–i–n^+ Double-Injection Structure by Including Both Diffusion and Space Charge Effects
(a) Vol. **44**, No. 1, p. 161—171 (1977).
See also *14.3.3*, p. *181*; *14.3.4*, p. *187*.

H. KNAUER, J. RICHTER, and P. SEIDEL
A Direct Calculation of the Resonance Tunneling in Metal–Insulator–Metal Tunnel Junctions
(a) Vol. **44**, No. 1, p. 303—312 (1977).
See also 13.4, p. 139; *14.3.3*, p. *181*.

B. G. CASWELL and J. WOODS
Some Effects of Oxygen on As-Plated CdS–Cu_2S Photovoltaic Cells
(a) Vol. **44**, No. 1, p. K47—K51 (1977).
See also *14.3.3*, p. *181*; *16*, p. *205*; 22.4.1, p. 364.

M. M. BABAEV and T. M. GASYMOV
Thermopower and the Transverse Nernst-Ettingshausen Effect in Semiconductors in a Strong Electric Field
(b) Vol. **84**, No. 2, p. 473—483 (1977).
See also 6, p. 60; 13.4, p. 139; *14.3.4*, p. *188*; *15*, p. *199*.

A. MAUGER
Indirect Exchange in Europium Chalcogenides
(b) Vol. **84**, No. 2, p. 761—771 (1977).
See also *13.1*, p. *123*; 18.1, p. 216; 18.2, p. 224.

E. MAJERNÍKOVÁ
Quantum Shift of the Electron Density of States in a Gaussian Random Potential
(b) Vol. **84**, No. 2, p. K109—K111 (1977).
See also *2*, p. *40*; *13*, p. *113*.

S. LARA, R. MOREIRA XAVIER, and C. A. TAFT
Remarks on the Effects of Pressure on the Energy Gaps in EuO and EuS
(b) Vol. **84**, No. 2, p. K143—K145 (1977).
See also *12.2*, p. *106*; *13.1*, p. *123*; 22.6, p. 391.

A. TONEVA and A. ALAZHAZHIAN
Mobility in p-$Pb_{1-x}Sn_xTe$
(a) Vol. **44**, No. 2, p. 621—626 (1977).
See also 13.4, p. 140; *14.3*, p. *172*.

G. KADEN und H. REIMER
Determination of the Density and Capture Cross Section of Surface States by the Surface Charge Pumping Effect
(a) Vol. **44**, No. 2, p. K117—K121 (1977).
See also *13.3*, p. *128*; *14.3.3*, p. *182*.

N. L. P. Sinha, A. K. Chaudhuri, and H. N. Bose
Structural Analysis of Lead Telluride Films and Their Sensitization
(a) Vol. **44**, No. 2, p. K127—K129 (1977).
See also *1.5*, p. *30*; 14.3.1, p. 175; 16, p. 205.

V. A. Gritsenko, E. E. Meerson, K. P. Mogilnikov, and S. P. Sinitsa
On the Conductivity of Amorphous Silicon Nitride in High Electric Fields
(a) Vol. **44**, No. 2, p. K167—K170 (1977).
See also *2*, p. *41*; *14.3.4*, p. *188*.

R. Durný and D. Barančok
EPR Investigation of Ag-Doped Ge–S Glasses
(a) Vol. **44**, No. 2, p. K177—K179 (1977).
See also *2*, p. *41*; *19*, p. *243*.

G. De Mey and H. J. Pauwels
Time Constant of the Ion Current in Thin Film Capacitors
(a) Vol. **44**, No. 2, p. K195—K199 (1977).
See also *14.4*, p. *193*.

22.1 Elements

Review Article

T. D. DZHAFAROV
Atomic Diffusion in Semiconductor Epitaxial Structures
(a) Vol. **42**, No. 1, p. 11–45 (1977).
See also *9*, p. *69*; 10.2, p. 83; 14.3.1, p. 173; 14.3.3, p. 178; 20.2, p. 265; 22.2, p. 348; 22.4, p. 361.

Original Papers and Short Notes

J. SZCZYRBOWSKI
The Shape of Absorption Edge of Amorphous Ge, Si, InSb, GaSb, GaAs, and GaP
(b) Vol. **79**, No. 1, p. K43–K48 (1977).
See also *2*, p. *33*; *20.1*, p. *248*; 22.2, p. 348.

A. MILCHEV and R. PICKENHAIN
The Quadratic Response of a Fermi Gas. Building-Up of the Covalent Bonding in Zincblende Semiconductors. A Nondiagonal Density Matrix Treatment
(b) Vol. **79**, No. 2, p. 549–558 (1977).
See also *4*, p. *44*; 13, p. 107; 22.1.1, p. 333; 22.1.2, p. 338; 22.2.1, p. 350.

R. WANG and M. D. MERZ
Polymorphic Bonding and Thermal Stability of Elemental Noncrystalline Solids
(a) Vol. **39**, No. 2, p. 697–703 (1977).
See also 1.2, p. 15; *2*, p. *34*; *4*, p. *44*; 21, p. 278.

H. VENGHAUS
Infrared Reflectance and Dielectric Properties of Pyrolytic Graphite for $\boldsymbol{E} \parallel \boldsymbol{c}$ Polarization
(b) Vol. **81**, No. 1, p. 221–225 (1977).
See also 13.1, p. 117; 13.5.2, p. 149; *20.1*, p. *253*.

U. BÜCHNER
Wave-Vector Dependence of the Electron Energy Losses of Boron Nitride and Graphite
(b) Vol. **81**, No. 1, p. 227–234 (1977).
See also 13.1, p. 117; *20.1*, p. *253*; 22.2, p. 348.

N. A. BLUM, C. FELDMAN, and F. G. SATKIEWICZ
Infrared Absorption of Amorphous Boron Films Containing Carbon and Hydrogen
(a) Vol. **41**, No. 2, p. 481–486 (1977).
See also 2, p. 37; *20.1*, p. *256*.

B. I. SHKLOVSKII
Thickness Dependence of the Hopping Conduction in Amorphous Films. Experiment versus Theory
(b) Vol. **83**, No. 1, p. K11–K15 (1977).
See also *2*, p. *38*; *14.3.1*, p. *175*.

M. ZÁVĚTOVÁ and I. GREGORA
Refractive Index of Amorphous Arsenic
(a) Vol. **43**, No. 1, p. K109–K111 (1977).
See also *2*, p. *39*; *20.1*, p. *260*.

S. C. GOYAL and R. PRAKASH
Many-Body Effects on the Second-Order Elastic Constants of Solids with ZnS Structure
(b) Vol. **83**, No. 2, p. 509–515 (1977).
See also *4*, p. *46*; *12.2*, p. *106*; 22, p. 328; 22.1.1, p. 335; 22.1.2, p. 342; 22.2, p. 348; 22.3, p. 358; 22.4, p. 361.

A. A. BEDNYAKOV, F. BERNHARD, YU. V. BULGAKOV, V. L. CHERNOV, and U. MÜLLER-JAHREIS
Energy Loss Straggling of Protons and Helium Ions in Carbon
(b) Vol. **84**, No. 1, p. K79–K81 (1977).
See also *11*, p. *98*.

22.1.1 Germanium

Review Articles

SH. M. KOGAN and T. M. LIFSHITS
Photoelectric Spectroscopy — A New Method of Analysis of Impurities in Semiconductors
(a) Vol. **39**, No. 1, p. 11—39 (1977).
See also *13.4*, p. *129*; *16*, p. *201*; 22, p. 320; 22.1.2, p. 337; 22.2.1, p. 350.

L. N. ALEKSANDROV
Transition Regions in Epitaxial Films
(a) Vol. **44**, No. 1, p. 11—43 (1977).
See also *1.5*, p. *29*; *3*, p. *42*; 14.3.1, p. 173; 22.1.2, p. 337; 22.2.1, p. 350.

Original Papers and Short Notes

N. C. HALDER, J. M. MONTANTE, and W. E. SWARTZ, JR.
Evidence of a Polycrystalline Phase in Amorphous Ge Films Evaporated at Room Temperature
(a) Vol. **39**, No. 1, p. 213—222 (1977).
See also *1.5*, p. *29*; *2*, p. *33*; 13.1, p. 115; 14.3.1, p. 173.

W. BARTH, K. ELSAESSER, W. GÜTH, and E. KAMIENIECKI
Photoconductivity in Correlation to Absorption in Plastically Deformed Germanium
(a) Vol. **39**, No. 1, p. 249—252 (1977).
See also *13.4*, p. *130*; 16, p. 201; 20.1, p. 248.

L. N. ALEKSANDROV and I. A. ENTIN
On the Kinetics of Thin-Film Growth under High Adatom Concentrations
(a) Vol. **39**, No. 1, p. 327—330 (1977).
See also *1.5*, p. *29*; *3*, p. *42*.

U. HEILIG
Determination of the Concentration of Slow States at Real Germanium Surfaces by Means of Field Effect Measurements
(a) Vol. **39**, No. 1, p. K29—K31 (1977).
See also *13.3*, p. *126*.

A. MILCHEV and R. PICKENHAIN
The Quadratic Response of a Fermi Gas. Building-Up of the Covalent Bonding in Zincblende Semiconductors. A Nondiagonal Density Matrix Treatment
(b) Vol. **79**, No. 2, p. 549—558 (1977).
See also *4*, p. *44*; 13, p. 107; 22.1, p. 332; 22.1.2, p. 338; 22.2.1, p. 350.

S. WINTER
Bound Electron States Close to the Conduction Band in Germanium Due to 60 ° Dislocations
(b) Vol. **79**, No. 2, p. 637—644 (1977).
See also *13.4*, p. *130*.

K. S. DUBEY
Contribution to the Lattice Thermal Conductivity Due to the Correction Term in the Frame of the Generalized Callaway Integral: Application to Ce
(b) Vol. **79**, No. 2, p. K119—K124 (1977).
See also *6*, p. *53*; *8*, p. *64*.

W. I. KINNEY and G. S. CARGILL III
Angle-of-Deposition Effects in Evaporated Amorphous Germanium
(a) Vol. **40**, No. 1, p. 37—42 (1977).
See also *2*, p. *35*.

B. I. Fuchs, M. S. Kagan, R. A. Suris, and N. G. Zhdanova
Capacitance Oscillations in Compensated Germanium
(a) Vol. **40**, No. 1, p. K61–K63 (1977).
See also *14.3.4*, p. *184*.

B. Movaghar and L. Schweitzer
ESR and Conductivity in Amorphous Germanium and Silicon
(b) Vol. **80**, No. 2, p. 491–498 (1977).
See also 2, p. 35; *14.3*, p. *167*; *19*, p. *237*; 22.1.2, p. 338.

T. Gregorkiewicz and L. Rusiniak
On the Frequency Dependence of the Classical Cyclotron Resonance Linewidth
(a) Vol. **40**, No. 2, p. K127–K129 (1977).
See also *19*, p. *238*.

A. B. Makhnovetskii and G. L. Krasko
Pseudopotential Theory and the Problem of Covalent Bonding (II): Crystals of Si, Ge, and α-Sn as-Ordered Systems
(b) Vol. **81**, No. 1, p. 263–273 (1977).
See also *1*, p. *9*; *4*, p. *45*; 21.7, p. 319; 22.1.2, p. 339.

M. Kumeda, Y. Jinno, and T. Shimizu
Electron Spin Resonance of Amorphous Si–Ge Alloys
(b) Vol. **81**, No. 1, p. K71–K74 (1977).
See also *2*, p. *36*; *19*, p. *238*; 22.1.2, p. 339.

M. J. Howes, D. V. Morgan, and W. J. Devlin
Applications of Magnetoresistance Measurements in the Evaluation of Transferred Electron Device Performance
(a) Vol. **41**, No. 1, p. 117–124 (1977).
See also *14.3.3*, p. *179*; 22.1.2, p. 339; 22.2.1, p. 351; 22.1.2, p. 354.

Yu. I. Gorkun, V. L. Korobanov, V. S. Lysenko, V. A. Novominskii, and V. I. Turchanikov
Effect of Bombardment with Low-Energy Ar^+ Ions on Electron Scattering in Accumulation Layers of n-Ge
(a) Vol. **41**, No. 1, p. 339–346 (1977).
See also 11, p. 95; *13.4*, p. *134*; *14.3.3*, p. *179*.

P. I. Baranskii, I. S. Buda, and L. A. Shcherbina
Structure of the Thermomagnetic Tensor in Semiconductors of n-Ge Type
(b) Vol. **81**, No. 2, p. 709–718 (1977).
See also *14.3*, p. *168*; 15, p. 199.

W. Plesiewicz
Photoelectronic Investigations of Multi-Level Gold Centres in Ge:Sb:Au
(a) Vol. **41**, No. 2, p. 403–409 (1977).
See also *13.4*, p. *135*; 14.3, p. 168; *16*, p. *202*.

O. Uemura, Y. Suzuki, and T. Satow
The Amorphous Structure of the Fe–Ge System
(a) Vol. **41**, No. 2, p. 417–421 (1977).
See also *2*, p. *37*; 21.1.1, p. 303.

D. Mergel and R. Labusch
Optical Excitation of Dislocation States in Germanium (I): Experiments
(a) Vol. **41**, No. 2, p. 431–438 (1977).
See also *13.4*, p. *135*; *16*, p. *202*; 20.1, p. 256.

K. Yasuda, A. Yoshida, and T. Arizumi
The Effects of Annealing on Mott's Parameters for Hopping Conduction in Amorphous Ge
(a) Vol. **41**, No. 2, p. K181–K184 (1977).
See also *2*, p. *37*; *14.3.1*, p. *174*.

T. Soma
Temperature Dependence of the Grüneisen Constant of Si and Ge
(b) Vol. **82**, No. 1, p. 319—324 (1977).
See also *6*, p. *57*; 8, p. 66; 22.1.2, p. 340.

W. Schnitker and F. R. Kessler
Piezoreflectance at the E_1 and $E_1 + \Delta_1$ Edges of Ge in High Magnetic Fields
(b) Vol. **82**, No. 1, p. K35—K37 (1977).
See also 13.1, p. 119; *18*, p. *211*; *20.1*, p. *257*.

D. Mergel and R. Labusch
Optical Excitation of Dislocation States in Germanium (II): Analysis of the Experimental Results
(a) Vol. **42**, No. 1, p. 165—171 (1977).
See also 13.1, p. 119; *13.4*, p. *136*; *16*, p. *203*.

R. K. Jain
Calculation of the Fermi Level, Minority Carrier Concentration, Effective Intrinsic Concentration, and Einstein Relation in n- and p-Type Germanium and Silicon
(a) Vol. **42**, No. 1, p. 221—226 (1977).
See also *13.1*, p. *120*; *14.3*, p. *169*; 22.1.2, p. 340.

G. Jungk
Critical Points in the E_0'-Region of Ge from Ellipsometric Data
(b) Vol. **82**, No. 2, p. K113—K116 (1977).
See also *13.1*, p. *120*; 20.1, p. 258.

J. P. Wolfe, R. S. Markiewicz, J. E. Furneaux, S. M. Kelso, and C. D. Jefries
Magnetostriction of an Electron–Hole Drop in Ge
(b) Vol. **83**, No. 1, p. 305—312 (1977).
See also *13.5.1*, p. *147*; 18.1, p. 215.

A. Kisiel and B. Oles
The Influence of a Light Beam on the Electroreflectance Spectra of Germanium
(b) Vol. **83**, No. 1, p. K35—K39 (1977).
See also *14.3.4*, p. *186*; *20.1*, p. *259*.

S. Lakeou, S. Cristoloveanu, and A. Chovet
Magnetoresistance Effect in Near Intrinsic Semiconductors. Influence of Sample Geometry. A New Method for the Determination of Carrier Densities and Mobilities
(a) Vol. **43**, No. 1, p. 213—222 (1977).
See also *14.3*, p. *170*; 22.1.2, p. 341; 22.2.3, p. 356.

R. Rentzsch and I. S. Shlimak
Photoluminescence of Heavily Doped and Compensated Germanium
(a) Vol. **43**, No. 1, p. 231—238 (1977).
See also *20.3*, p. *273*.

S. C. Goyal and R. Prakash
Many-Body Effects on the Second-Order Elastic Constants of Solids with ZnS Structure
(b) Vol. **83**, No. 2, p. 509—515 (1977).
See also *4*, p. *46*; *12.2*, p. *106*; 22, p. 328; 22.1, p. 332; 22.1.2, p. 342; 22.2, p. 348; 22.3, p. 358; 22.4, p. 361.

G. Duggan, G. J. Morgan, N. Shahtahmasebi, and A. Lettington
The Effective Screening Function in Insulating Solids
(b) Vol. **83**, No. 2, p. 543—553 (1977).
See also *13.4*, p. *137*; 14.4, p. 192.

R. S. Markiewicz
Acoustic Phonon Scattering by Electron–Hole Droplets in Ge: Deformation Potentials
(b) Vol. **83**, No. 2, p. 659—668 (1977).
See also *13.5.1*, p. *147*; 14.3, p. 171.

O. P. Pchelyakov, E. B. Gorokhov, S. V. Pokrovskaya, L. V. Sokolov, and S. I. Stenin
Structure of SiO_2 Pyrolitic Films on Ge Substrates
(a) Vol. **43**, No. 2, p. 659—666 (1977).
See also *1.5*, p. *30*; 2, p. 39; 22.6, p. 391.

A. Cavallini, P. Gondi, and A. Castaldini
Dislocation Energy Levels in Ge
(a) Vol. **43**, No. 2, p. K205—K208 (1977).
See also 10.2, p. 91; *13.4*, p. *138*; 14.3, p. 171.

E. Palaniyandi and S. Balasubramanian
Polarizabilities of Donors in Elemental Semiconductors
(b) Vol. **84**, No. 1, p. 171—174 (1977).
See also *13.4*, p. *138*; *14.3*, p. *171*; 22.1.2, p. 343.

M. Dhwaj Singh
The Effect of Internal Strain and Electrical Field of an Impurity on Electron–Phonon Interaction in n-Type Germanium and Silicon
(b) Vol. **84**, No. 1, p. 375—384 (1977).
See also *6*, p. *60*; 8, p. 67; 22.1.2, p. 343.

P. Kocevar
Drag Effects in a Heated Electron–Phonon System. n-Germanium and n-Silicon
(b) Vol. **84**, No. 2, p. 681—690 (1977).
See also *6*, p. *60*; 22.1.2, p. 344.

V. P. Dobrego, O. P. Ermolaev, and V. D. Tkachev
Investigation of Shallow Radiation Defect Levels Produced in Ge by Fast-Neutron Irradiation
(a) Vol. **44**, No. 2, p. 435—442 (1977).
See also *11*, p. *99*; *13.4*, p. *140*.

22.1.2 Silicon

Review Articles

SH. M. KOGAN and T. M. LIFSHITS
Photoelectric Spectroscopy — A New Method of Analysis of Impurities in Semiconductors
(a) Vol. **39**, No. 1, p. 11—39 (1977).
See also *13.4*, p. *129*; *16*, p. *201*; 22, p. 320; 22.1.1, p. **333**; 22.2.1, p. 350.

L. N. ALEKSANDROV
Transition Regions in Epitaxial Films
(a) Vol. **44**, No. 1, p. 11—43 (1977).
See also *1.5*, p. *29*; *3*, p. *42*; 14.3.1, p. 173; 22.1.1, p. **333**; 22.2.1, p. 350.

Original Papers and Short Notes

R. VON BALTZ and B. URBAN
Interaction of Holes with Local Modes and One and Two Optical Phonons as Revealed in M–I–Silicon Tunnelling Spectra
(b) Vol. **79**, No. 1, p. 185—194 (1977).
See also *13.5.3*, p. *151*.

D. ENGEMANN and R. FISCHER
Photoluminescence in Amorphous Silicon
(b) Vol. **79**, No. 1, p. 195—202 (1977).
See also *2*, p. *33*; *20.3*, p. *267*.

T. ŌMURA, M. KITAGAWA, and Y. H. OHTSUKI
Theory of Axial Dechanneling
(b) Vol. **79**, No. 1, p. 321—324 (1977).
See also *9*, p. *69*; 11, p. **94**.

J. GIBER, A. HÜBLER, and D. MARTON
On the Local Energy Levels of a Silicon Vacancy. A Model of the Defect-Molecule in the Wannier Approximation
(b) Vol. **79**, No. 1, p. K71—K74 (1977).
See also *10.2*, p. *83*; *13.4*, p. *129*.

H. RUNGE
Device Geometry Dependent Effects of Deep Levels on $C(U)$ Measurements of Ion-Implanted MOS Diodes
(a) Vol. **39**, No. 1, p. 141—146 (1977).
See also *11*, p. *94*; *14.3.3*, p. *178*.

F. BURKHARDT and C. WAGNER
Depth Distribution of Phosphorus Implanted into Silicon
(a) Vol. **39**, No. 1, p. K63—K65 (1977).
See also *11*, p. *94*.

J. VON BORZESZKOWSKI and J. SCHOLZ
On the Electric Field Dependence of the Conductivity of Holes in Si at Low Temperatures (II): The Influence of the Variation of Steady-State Hole Density with Electric Field on the Conductivity
(b) Vol. **79**, No. 2, p. 525—531 (1977).
See also 6, p. 52; 13.1, p. 115; *14.3.4*, p. *183*.

W. REHM, R. FISCHER, J. STUKE, and H. WAGNER
Photo and Dark Conductivity of Doped Amorphous Silicon
(b) Vol. **79**, No. 2, p. 539—547 (1977).
See also *2*, p. *33*; 14.3.1, p. 173; *16*, p. *201*.

A. Milchev and R. Pickenhain
The Quadratic Response of a Fermi Gas. Building-Up of the Covalent Bonding in Zincblende Semiconductors. A Nondiagonal Density Matrix Treatment
(b) Vol. **79**, No. 2, p. 549–558 (1977).
See also *4*, p. *44*; 13, p. 107; 22.1, p. 332; 22.1.1, p. 333; 22.2.1, p. 350.

A. Q. Sarker
Spherical Effective Impurity Electron Mass in Si
(b) Vol. **79**, No. 2, p. 567–571 (1977).
See also *13.4*, p. *130*; 14.3, p. 165.

J. Giber, P. Deák, and D. Marton
The Mechanical Stress Deriving from Molecular Fitting on the Si–SiO_2 Interface and Its Effect upon the Development and Distribution of Vacancies
(b) Vol. **79**, No. 2, p. K89–K93 (1977).
See also *10.2*, p. *84*.

S. K. Maksimov and D. I. Piskunov
Correct Calculation of the Stress Relaxation on the Foil Surface in Constructing Theoretical Electron Micrographs
(a) Vol. **39**, No. 2, p. 459–466 (1977).
See also *10.2*, p. *84*.

H. Runge
Distribution of Implanted Ions under Arbitrarily Shaped Mask Edges
(a) Vol. **39**, No. 2, p. 595–599 (1977).
See also *11*, p. *94*.

T. Bedyńska, R. Bubáková, and Z. Šourek
Erratum to: Comparison between Experimental and Theoretical Dislocation Image for the Bragg Case
(a) Vol. **39**, No. 2, p. 725 (1977).
See also *10.2*, p. *85*.

H. Düsterhöft and A. Ihlenfeld
The Energy Distribution of Positive Secondary Ions Emitted from Metal and Semiconductor Targets Bombarded with 12 keV Ar^+ Ions
(a) Vol. **39**, No. 2, p. K147–K150 (1977).
See also *17*, p. *206*; 21, p. 278; 21.1, p. 290.

I. Alstrup and S. Marklund
The Electron States Associated with the Core Region of the 60° Dislocation in Silicon
(b) Vol. **80**, No. 1, p. 301–306 (1977).
See also *13.4*, p. *132*.

Amitabh Jain and B. Smith
A New Method for Demonstrating the Bourgoin Mechanism in Semiconductors
(a) Vol. **40**, No. 1, p. K31–K33 (1977).
See also *10.2*, p. *86*; 14.3.3, p. 179.

J. Collet, J. Barrau, M. Brousseau, and H. Maaref
Simultaneous Kinetics of Drops and Excitons in Silicon during the Drop Formation
(b) Vol. **80**, No. 2, p. 461–470 (1977).
See also *13.5.1*, p. *144*; 20.3, p. 269.

B. Movaghar and L. Schweitzer
ESR and Conductivity in Amorphous Germanium and Silicon
(b) Vol. **80**, No. 2, p. 491–498 (1977).
See also 2, p. 35; *14.3*, p. *167*; *19*, p. *237*; 22.1.1, p. 334.

S. S. Kubakaddi and B. S. Krishnamurthy
The Electron–Two Short-Wavelength Phonon Scattering in Non-Polar Semiconductors
(b) Vol. **80**, No. 2, p. 603–609 (1977).
See also *6*, p. *54*; *14.3.4*, p. *184*.

P. N. Dyachkov and A. A. Levin
Bond Orbital Model for Tetrahedral Semiconductors on the Base of Molecular Parameters. Covalent Crystals and Perturbative Treatment of Partially Ionic Semiconductors
(b) Vol. **80**, No. 2, p. 681–690 (1977).
See also *4*, p. *44*; *13.1*, p. *116*; 22.1.3, p. 345; 22.2, p. 348; 22.4, p. 361.

N. Kishimoto, K. Morigaki, K. Murakami, A. Shimizu, and A. Hiraki
Electron Spin Resonance in the Amorphous Si–Au System
(b) Vol. **80**, No. 2, p. K113–K115 (1977).
Erratum
(b) Vol. **82**, No. 2, p. K123 (1977).
See also *2*, p. *35*; *19*, p. *237*; 21.6, p. 315.

J. Donecker and J. Kluge
Magneto-Optical Investigation of the Ground State of the Indirect Free Exciton in Si and GaP
(b) Vol. **81**, No. 1, p. 199–203 (1977).
See also 13.5.1, p. 145; *18*, p. *210*; *20.1*, p. *253*; 22.2.2, p. 354.

A. B. Makhnovetskii and G. L. Krasko
Pseudopotential Theory and the Problem of Covalent Bonding (II): Crystals of Si, Ge, and α-Sn as-Ordered Systems
(b) Vol. **81**, No. 1, p. 263–273 (1977).
See also *1*, p. *9*; *4*, p. *45*; 21.7, p. 319; 22.1.1, p. 334.

M. Kumeda, Y. Jinno, and T. Shimizu
Electron Spin Resonance of Amorphous Si–Ge Alloys
(b) Vol. **81**, No. 1, p. K71–K74 (1977).
See also *2*, p. *36*; *19*, p. *238*; 22.1.1, p. 334.

Á. Barna, P. B. Barna, G. Radnóczi, L. Tóth, and P. Thomas
A Comparative Study of the Structure of Evaporated and Glow Discharge Silicon
(a) Vol. **41**, No. 1, p. 81–84 (1977).
See also *1.5*, p. *29*; 2, p. 36.

M. J. Howes, D. V. Morgan, and W. J. Devlin
Applications of Magnetoresistance Measurements in the Evaluation of Transferred Electron Device Performance
(a) Vol. **41**, No. 1, p. 117–124 (1977).
See also *14.3.3*, p. *179*; 22.1.1, p. 334; 22.2.1, p. 351; 22.2.2, p. 354.

W. Pietrenko
Influence of Temperature and Impurity Concentration on the Piezoresistance in n-Silicon
(a) Vol. **41**, No. 1, p. 197–205 (1977).
See also *14.3*, p. *168*.

K. Bhattacharya, P. Basu, and H. Saha
Back Wall Schottky Barrier Solar Cell with an Interfacial Layer
(a) Vol. **41**, No. 1, p. 317–321 (1977).
See also 13.3, p. 126; *14.3.3*, p. *179*; *16*, p. *202*; 22.2.1, p. 351.

W. Gehlhoff and K. H. Segsa
Anisotropic Broadening of Linewidth in the EPR Spectrum of Fe^0 in Silicon
(a) Vol. **41**, No. 1, p. K21–K23 (1977).
See also *19*, p. *239*.

E. A. Matson and S. A. Polyakov
On the "Evolutionary Selection" Principle in Relation to the Growth of Polycrystalline Silicon Films
(a) Vol. **41**, No. 1, p. K93–K95 (1977).
See also *1.5*, p. *29*; *3*, p. *42*.

L. Bartelsen
The EPR Fine Structure Spectrum of Dislocations in Silicon
(b) Vol. **81**, No. 2, p. 471–478 (1977).
See also 10.2, p. 88; *19*, p. *239*.

Vijay A. Singh, C. Weigel, J. W. Corbett, and L. M. Roth
Vibrational and Electronic Structure of Hydrogen-Related Defects in Silicon Calculated by the Extended Hückel Theory
(b) Vol. **81**, No. 2, p. 637–646 (1977).
See also 6, p. 56; *13.4*, p. *134*.

L. T. Ho
Zeeman Effect of the Excitation Spectrum of Neutral Magnesium Donors in Silicon
(b) Vol. **81**, No. 2, p. K99–K101 (1977).
See also *13.4*, p. *134*; *20.1*, p. *255*.

K. B. Tolpygo and V. A. Telezhkin
Role of Interelectronic Correlation in the Explanation of Paramagnetic Properties of the Impurity–Vacancy Complex in Silicon
(b) Vol. **81**, No. 2, p. K139–K140 (1977).
See also *13.4*, p. *135*.

Y.-H. Lee, J. W. Corbett, and K. L. Brower
EPR of a Carbon-Oxygen–Divacancy Complex in Irradiated Silicon
(a) Vol. **41**, No. 2, p. 637–647 (1977).
See also *10.2*, p. *88*; 11, p. 95; *19*, p. *240*.

H.-D. Geiler, G. Götz, K.-D. Klinge, and N. Triem
Investigation of Laser Induced Diffusion and Annealing Processes of Arsenic-Implanted Silicon Crystals
(a) Vol. **41**, No. 2, p. K171–K173 (1977).
See also 9, p. 70; *11*, p. *96*.

W. Schülke
The One-Dimensional Fourier Transform of Compton Profiles. Theorems and Applications
(b) Vol. **82**, No. 1, p. 229–235 (1977).
See also *13.1*, p. *119*; *13.2*, p. *124*; 21.2, p. 308.

T. Soma
Temperature Dependence of the Grüneisen Constant of Si and Ge
(b) Vol. **82**, No. 1, p. 319–324 (1977).
See also *6*, p. *57*; 8, p. 66; 22.1.1, p. 335.

M. v. Ortenberg and L. G. Rubin
Submillimeter Cyclotron Resonance of Electrons in Stressed Silicon at High Magnetic Fields
(b) Vol. **82**, No. 1, p. K1–K4 (1977).
See also *13.1*, p. *119*; *19*, p. *240*.

W. Ecke
The Photomagnetoelectric Effect in Heteroepitaxial Silicon Films
(a) Vol. **42**, No. 1, p. 157–163 (1977).
See also *14.3.1*, p. *174*; *16*, p. *203*.

R. K. Jain
Calculation of the Fermi Level, Minority Carrier Concentration, Effective Intrinsic Concentration, and Einstein Relation in n- and p-Type Germanium and Silicon
(a) Vol. **42**, No. 1, p. 221–226 (1977).
See also *13.1*, p. *120*; *14.3*, p. *169*; 22.1.1, p. 335.

T. Suzuki, M. Hirose, S. Ogose, and Y. Osaka
Electrical Transport and Photoconductivity in Amorphous Silicon
(a) Vol. **42**, No. 1, p. 337–344 (1977).
See also 2, p. 37; 13.1, p. 120; *14.3.1*, p. *174*; 15, p. 199; *16*, p. *203*.

A. M. AFANASEV, M. V. KOVALCHUK, E. K. KOVEV, and V. G. KOHN
*X-Ray Diffraction in a Perfect Crystal with Disturbed Surface Layer
(a) Vol. **42**, No. 1, p. 415–422 (1977).
See also *1*, p. *10*; *1.6*, p. *31*.

J. HÄRTWIG
X-Ray Diffraction in the Extremely Asymmetric Laue Case with a Small Angle between the Crystal Surface and the Incident Beam
(a) Vol. **42**, No. 2, p. 495–500 (1977).
See also *1.1*, p. *12*.

G. H. SCHWUTTKE, K. YANG, and H. KAPPERT
Lifetime Control in Silicon through Impact Sound Stressing
(a) Vol. **42**, No. 2, p. 553–564 (1977).
See also *14.3*, p. *170*; 14.3.1, p. 174.

A. T. ÅBERG
Low-Temperature Annealing of Al-Implanted Si–SiO_2 Structures
(a) Vol. **42**, No. 2, p. 639–645 (1977).
See also *13.3*, p. *127*; *14.3.2*, p. *177*.

D. JÄGER, R. HEIDEMANN, and R. KASSING
Influence of Carrier Transit Time on the Small-Signal Admittance of SCLC Diodes
(a) Vol. **42**, No. 2, p. 657–665 (1977).
See also 14.3.3, p. 180; *14.3.4*, p. *186*.

H. KATO, T. KANDA, K. YASUDA, A. YOSHIDA, and T. ARIZUMI
Electron Beam Irradiation Effects on SnO_2–Si Photodiodes
(a) Vol. **42**, No. 2, p. 769–774 (1977).
See also *11*, p. *97*; *14.3.2*, p. *177*; 16, p. 203.

H. OPPOLZER
Reflection X-Ray Topography of Oxidation-Induced Stacking Faults in Silicon
(a) Vol. **42**, No. 2, p. K91–K93 (1977).
See also *10.2*, p. *90*.

T. WOSIŃSKI and T. FIGIELSKI
Spin-Dependent Recombination at Exchange-Coupled Dislocation Centres in Silicon
(b) Vol. **83**, No. 1, p. 93–98 (1977).
See also *10.2*, p. *90*; 13.4, p. 137; *16*, p. *204*.

C. H. CHEN
Electron Energy Loss Studies of Direct Nonvertical Interband Transitions in Silicon
(b) Vol. **83**, No. 1, p. 347–351 (1977).
See also *13.1*, p. *121*; 14.4, p. 192.

G. H. SCHWUTTKE
Low Cost Silicon for Solar Energy Application
(a) Vol. **43**, No. 1, p. 43–51 (1977).
See also 3, p. 42; 14.3.3, p. 181; *16*, p. *204*; 20.1, p. 260.

G. SIXT, H. KAPPERT, and G. H. SCHWUTTKE
Minority Carrier Lifetime and Defect Structure in Silicon after Cesium Implantation
(a) Vol. **43**, No. 1, p. 119–131 (1977).
See also *11*, p. *97*; *14.3*, p. *170*.

S. LAKEOU, S. CRISTOLOVEANU, and A. CHOVET
Magnetoresistance Effect in Near Intrinsic Semiconductors. Influence of Sample Geometry. A New Method for the Determination of Carrier Densities and Mobilities
(a) Vol. **43**, No. 1, p. 213–222 (1977).
See also *14.3*, p. *170*; 22.1.1, p. 335; 22.2.3, p. 356.

H. FLIETNER, W. FÜSSEL, and N. D. SINH
Interface States of the Si/SiO_2-System and Their Separation in Groups
(a) Vol. **43**, No. 1, p. K99–K101 (1977).
See also *13.3*, p. *127*.

S. C. GOYAL and R. PRAKASH
Many-Body Effects on the Second-Order Elastic Constants of Solids with ZnS Structure
(b) Vol. **83**, No. 2, p. 509–515 (1977).
See also *4*, p. *46*; *12.2*, p. *106*; 22, p. 328; 22.1, p. 332; 22.1.1, p. 335; 22.2, p. 348; 22.3, p. 358; 22.4, p. 361.

N. A. DROZDOV, A. A. PATRIN, and V. D. TKACHEV
On the Nature of the Dislocation Luminescence in Silicon
(b) Vol. **83**, No. 2, p. K137–K139 (1977).
See also 10.2, p. 91; *20.3*, p. *273*.

A. F. AKKERMAN
Correlations between the Elastic Energy Transferred in Solids by Implanted Hydrogen Ions and Hydrogen Trapping
(b) Vol. **83**, No. 2, p. K145–K147 (1977).
See also *11*, p. *98*; 21, p. 285.

V. A. GRAZHULIS, V. V. KVEDER, and V. YU. MUKHINA
Investigation of the Energy Spectrum and Kinetic Phenomena in Dislocated Si Crystals (I)
(a) Vol. **43**, No. 2, p. 407–415 (1977).
See also 10.2, p. 91; *13.4*, p. *138*; *14.3*, p. *171*.

H. NAKATA and H. MORISHIMA
An ESR Study of Annealing Characteristics for Heavily Irradiated Silicon
(a) Vol. **43**, No. 2, p. 417–423 (1977).
See also 11, p. 98; *19*, p. *241*.

M. SAUVAGE, J. F. PETROFF, and P. SKALICKY
Synchroton Radiation Topographic Evidence for the Interaction of Plane-Polarized X-Ray Photons with Perfect Crystals
(a) Vol. **43**, No. 2, p. 473–477 (1977).
See also *1.1*, p. *13*.

P. LILIENKAMP and H. PFLEIDERER
An ESFI SOS Magnetodiode
(a) Vol. **43**, No. 2, p. 479–486 (1977).
See also *14.3.1*, p. *175*; *14.3.4*, p. *187*.

U. BONSE, W. GRAEFF, R. TEWORTE, and H. RAUCH
Oscillatory Structure of Laue Case Rocking Curves
(a) Vol. **43**, No. 2, p. 487–492 (1977).
See also *1.1*, p. *13*.

J. HÜLLER and M. T. PHAM
Characterization of Ion Implanted Silicon by the Electrolytic Reverse Current
(a) Vol. **43**, No. 2, p. 505–512 (1977).
See also *11*, p. *98*; *14.3.2*, p. *177*.

I. CHERNOV, YU. KRYUCHKOV, A. MAMONTOV, N. SLAVIN, and YU. TIMOSHNIKOV
Lattice Location of ^{16}O Implanted in Si Crystals by Backscattering
(a) Vol. **43**, No. 2, p. K201–K203 (1977).
See also *10.2*, p. *91*; *11*, p. *98*.

M. A. KUMAKHOV
Theory of Radiation of Charged Particles Channeled in a Crystal
(b) Vol. **84**, No. 1, p. 41–54 (1977).
See also *11*, p. *98*; 13, p. 112.

E. PALANIYANDI and S. BALASUBRAMANIAN
Polarizabilities of Donors in Elemental Semiconductors
(b) Vol. 84, No. 1, p. 171—174 (1977).
See also *13.4*, p. *138*; *14.3*, p. *171*; 22.1.1, p. 336.

L. PASEMANN
A Non-Diagonally Screened Model Potential for Semiconductors Applied to the Band Structure Calculation of Si
(b) Vol. 84, No. 1, p. 175—188 (1977).
See also *13.1*, p. *122*.

M. DHWAJ SINGH
The Effect of Internal Strain and Electrical Field of an Impurity on Electron–Phonon Interaction in n-Type Germanium and Silicon
(b) Vol. 84, No. 1, p. 375—384 (1977).
See also *6*, p. *60*; 8, p. 67; 22.1.1, p. 336.

E. ZIELIŃSKA-ROHOZIŃSKA
X-Ray Diffraction Investigations of Annealing Characteristics in P^+-Implanted Silicon Crystals
(a) Vol. 44, No. 1, p. 59—64 (1977).
See also *1.4*, p. *28*; 11, p. 99.

V. A. GRAZHULIS, V. V. KVEDER, and V. YU. MUKHINA
Investigation of the Energy Spectrum and Kinetic Phenomena in Dislocated Si Crystals (II): Microwave Conductivity
(a) Vol. 44, No. 1, p. 107—115 (1977).
See also 10.2, p. 91; *13.4*, p. *139*; *14.3*, p. *171*.

G. ALBERTINI, A. BOEUF, B. KLAR, S. LAGOMARSINO, S. MAZKEDIAN, S. MELONE, P. PULITI, and F. RUSTICHELLI
Dynamical Neutron Diffraction by Curved Crystals in the Laue Geometry
(a) Vol. 44, No. 1, p. 127—136 (1977).
See also *1.1*, p. *13*.

T. YAMAMOTO
Experimental Aspects of Electron Channeling Patterns in Scanning Electron Microscopy (I): Measurement of Contrast
(a) Vol. 44, No. 1, p. 137—146 (1977).
See also *1.4*, p. *28*; *10.1*, p. *81*; 21, p. 287; 21.1, p. 299; 21.6, p. 317.

R. HEIDEMANN, D. JÄGER, and R. KASSING
Minority and Majority Carrier Mobilities Determined by Microwave Measurements
(a) Vol. 44, No. 1, p. 191—195 (1977).
See also *14.3.3*, p. *181*; *14.3.4*, p. *187*.

H.-J. KOS and D. NEUBERT
Two-Step Photoconductivity by Dislocations in Silicon
(a) Vol. 44, No. 1, p. 259—264 (1977).
See also 10.2, p. 92; *16*, p. *205*.

D. KROPMAN, A. SÜGIS, and M. VINNAL
EPR Investigation of Si–SiO_2 Bonding Layer
(a) Vol. 44, No. 1, p. K1—K3 (1977).
See also 10.2, p. 92; *19*, p. *243*.

D. R. CHOUDHURY, A. K. CHOWDHURY, and A. N. CHAKRAVARTI
On the Modification of the Einstein Relation for Semiconductor Inversion Layers in the Presence of a Quantizing Magnetic Field
(a) Vol. 44, No. 1, p. K111—K115 (1977).
See also *14.3.2*, p. *177*.

W. Schmid
Auger Lifetimes for Excitons Bound to Neutral Donors and Acceptors in Si
(b) Vol. **84**, No. 2, p. 529–540 (1977).
See also *13.5.1*, p. *148*; *20.3*, p. *275*.

M. A. Kumakhov and R. Wedell
Theory of Radiation of Relativistic Channelled Particles
(b) Vol. **84**, No. 2, p. 581–593 (1977).
See also *11*, p. *99*; 13, p. 112.

P. Kocevar
Drag Effects in a Heated Electron–Phonon System. n-Germanium and n-Silicon
(b) Vol. **84**, No. 2, p. 681–690 (1977).
See also *6*, p. *60*; 22.1.1, p. 336.

T. Yamamoto
Experimental Aspects of Electron Channeling Patterns in Scanning Electron Microscopy (II): Estimation of Contrast Depth
(a) Vol. **44**, No. 2, p. 467–476 (1977).
See also *1.4*, p. *28*; *10.1*, p. *82*; 21, p. 287; 21.1, p. 300; 21.6, p. 318.

G. Buchheim, B. Junghans, and H. Lippmann
Application of the Kelvin Method for Oxide Charge Evaluation in Si–SiO_2 Structures
(a) Vol. **44**, No. 2, p. 585–591 (1977).
See also 2, p. 41; *14.3.1*, p. *175*; *14.3.2*, p. *177*.

K. Jarašiūnas and J. Vaitkus
Investigation of Non-Equilibrium Processes in Semiconductors by the Method of Transient Holograms
(a) Vol. **44**, No. 2, p. 793–800 (1977).
See also *14.3.4*, p. *188*; 20, p. 246; 22.4.2, p. 366.

F. K. Naehring, A. Schmidt, and J. Schöneich
Ion Induced Carbon Contamination and Recoil Implantation
(a) Vol. **44**, No. 2, p. K141–K145 (1977).
See also 1.5, p. 30; *11*, p. *99*.

W.-A. Martin and G. Hartung
Anisotropic Photoinduced ESR Signals from the Si–SiO_2 Interface on Silicon Single Crystals
(a) Vol. **44**, No. 2, p. K159–K161 (1977).
See also *16*, p. *205*; *19*, p. *243*.

22.1.3 Group VI Elements

Original Papers and Short Notes

J. L. Farvacque, J. C. Doukhan, U. von Alpen, and E. Gmelin
Screw-Dislocation-Induced Scattering Processes and Acceptor States in Te
(b) Vol. **79**, No. 2, p. 763—773 (1977).
See also *13.4*, p. *131*; *14.3*, p. *166*; 20.1, p. 249.

A. Kubový and M. Janda
Influence of Tunneling through Barriers on Grain Boundaries on Hole Mobility in Polycrystalline Tellurium Films
(a) Vol. **40**, No. 1, p. 225—233 (1977).
See also *14.3.1*, p. *173*.

J. L. Farvacque and P. Lenglart
Inhomogeneous Strain Field Scattering Deformation Potential (II): Free-Carrier Mobility Associated with Dislocation Strain Fields in Semiconductors
(b) Vol. **80**, No. 2, p. 433—439 (1977).
See also *10.2*, p. *86*; *14.3*, p. *167*; 22, p. 323.

P. N. Dyachkov and A. A. Levin
Bond Orbital Model for Tetrahedral Semiconductors on the Base of Molecular Parameters. Covalent Crystals and Perturbative Treatment of Partially Ionic Semiconductors
(b) Vol. **80**, No. 2, p. 681—690 (1977).
See also *4*, p. *44*; *13.1*, p. *116*; 22.1.2, p. 339; 22.2, p. 348; 22.4, p. 361.

Th. Englert, K. von Klitzing, R. Silbermann, and G. Landwehr
Influence of the Surface on the Galvanomagnetic Properties of Tellurium
(b) Vol. **81**, No. 1, p. 119—127 (1977).
See also *13.3*, p. *126*; 14.3.2, p. 176.

K. Hamilton
Theory of the Nuclear Electric Field Gradient in Trigonal Tellurium
(b) Vol. **81**, No. 1, p. K23—K26 (1977).
See also *5*, p. *48*.

L. Eglinski, A. F. Rudolph, T. Banys, and A. Dargys
Microwave Breakdown in Tellurium
(a) Vol. **41**, No. 1, p. 177—181 (1977).
See also 13.4, p. 134; *14.3.4*, p. *185*.

W. Bohmeyer and W. Hoerstel
Hall Effect in p-Tellurium at High Electric Fields
(a) Vol. **41**, No. 1, p. K27—K30 (1977).
See also *14.3.4*, p. *185*.

A. Balogh and I. Dézsi
Further Positron Lifetimes of Some Elements
(b) Vol. **81**, No. 2, p. K81—K82 (1977).
See also *13*, p. *110*; 21, p. 281.

R. M. Mehra, P. C. Mathur, A. K. Kathuria, and Radhey Shyam
Dependence of Conductivity and Activation Energy on Tellurium Concentration in Amorphous Selenium
(a) Vol. **41**, No. 2, p. K189—K192 (1977).
See also *2*, p. *37*; *14.3*, p. *169*.

G. N. Galkin, K. H. Herrmann, and E. A. Bobrova
Recombination in Tellurium at High Optical Excitation
(b) Vol. **82**, No. 1, p. 237—242 (1977).
See also *14.3*, p. *169*; *16*, p. *203*.

H. Mutka and O. J. A. Tiainen
Annealing Experiments on Fast Neutron Irradiated Selenium Single Crystals
(a) Vol. **42**, No. 1, p. 241—245 (1977).
See also *11*, p. *96*; 14.3, p. 169; 20.1, p. 257.

V. B. Anzin, M. I. Eremets, Yu. V. Kosichkin, A. I. Nadezhdinskii, and A. M. Shirokov
Measurement of the Energy Gap in Tellurium under Pressure
(a) Vol. **42**, No. 1, p. 385—390 (1977).
See also 12.2, p. 105; 13.1, p. 120; *16*, p. *203*; *20.1*, p. *257*.

E. Kittinger
Temperature Dependence of Sound Velocity in Vitreous Selenium
(a) Vol. **42**, No. 1, p. K33—K35 (1977).
See also *2*, p. *37*; *7*, p. *62*.

H. Stolze, M. Lutz, and P. Grosse
The Optical Activity of Tellurium
(b) Vol. **82**, No. 2, p. 457—466 (1977).
See als o *13.1*, p. *120*; *20.1*, p. *257*.

C.-P. Czaya, K. Keck, and H. Rüthlein
Influence of Surface Carriers upon the Galvanomagnetic Effects in Slightly Doped Tellurium Samples at Low Temperatures
(b) Vol. **82**, No. 2, p. 595—602 (1977).
See also *13.4*, p. *136*; 14.3.2, p. 177.

A. E. Sapega
Determination of Surface States in Selenium Diodes by Conductance and Capacitance Studies
(a) Vol. **42**, No. 2, p. 585—594 (1977).
See also *13.3*, p. *127*; *14.3.3*, p. *180*.

W. Eichler
Thermomagnetic Effects in Pure Imperfect Tellurium Single Crystals
(a) Vol. **42**, No. 2, p. K107—K110 (1977).
See also 13.4, p. 137; *15*, p. *199*.

H. Behrendt and W. Hoerstel
Electric Field Dependence of the Electron Mobility in Tellurium at 77 K
(b) Vol. **83**, No. 1, p. 223—228 (1977).
See also *14.3.4*, p. *186*.

S. A. Vorontsov, D. G. Dolgopolov, V. I. Poltoratskii, and I. V. Svechkarev
Magnetic Susceptibility Anisotropy of Holes in Tellurium
(b) Vol. **83**, No. 1, p. 341—346 (1977).
See also *13*, p. *111*; *18.1*, p. *215*.

F. R. Kessler and E. Bredner
Cathodoluminescence of Selenium
(a) Vol. **43**, No. 1, p. 159—166 (1977).
See also 13.5.1, p. 147; *20.3*, p. *272*.

P. J. Walsh, G. C. Vezzoli, and L. W. Doremus
Thermodynamic Basis for Generalization of Group VI-A Melting Curves
(a) Vol. **43**, No. 1, p. 305—310 (1977).
See also *1.1*, p. *12*; *1.2*, p. *18*; 8, p. 67.

J. C. Malaurent and J. Dixmier
Random Chain Assembly Model for Amorphous Sulfur, Selenium, and Tellurium
(a) Vol. **43**, No. 1, p. K61—K64 (1977).
See also *2*, p. *39*.

C. Viger, C. Vautier, and D. Carles
Recombination and Trapping Processes in Amorphous Materials in Presence of an Exponential Distribution of Traps
(a) Vol. **44**, No. 1, p. 97–106 (1977).
See also 2, p. 40; 13.4, p. 139; 14.3.1, p. 175; *16*, p. *204*.

E. Kittinger
Time Dependent Thermal Expansion of Vitreous Selenium
(a) Vol. **44**, No. 1, p. K35–K37 (1977).
See also *2*, p. *40*; *8*, p. *68*.

A. I. Popov, I. Kh. Geller, and V. K. Shemetova
Memory and Threshold Switching Effects in Amorphous Selenium
(a) Vol. **44**, No. 1, p. K71–K73 (1977).
See also *2*, p. *40*; *14.3.4*, p. *187*.

V. Valvoda and J. Ječný
X-Ray Investigation of the Anisotropic Thermal Motion of Atoms in Tellurium
(a) Vol. **44**, No. 2, p. 557–565 (1977).
See also *6*, p. *61*.

22.2 III–V Compounds

Review Article

T. D. Dzhafarov
Atomic Diffusion in Semiconductor Epitaxial Structures
(a) Vol. **42**, No. 1, p. 11–45 (1977).
See also *9*, p. *69*; 10.2, p. 83; 14.3.1, p. 173; 14.3.3, p. 178; 20.2, p. 265; 22.1, p. 332; 22.4, p. 361.

Original Papers and Short Notes

F. Thuselt and K. Unger
Exchange Splitting of Donor–Acceptor Pairs
(b) Vol. **79**, No. 1, p. K39–K42 (1977).
See also *13.4*, p. *129*; 13.5.1, p. 143.

J. Szczyrbowski
The Shape of Absorption Edge of Amorphous Ge, Si, InSb, GaSb, GaAs, and GaP
(b) Vol. **79**, No. 1, p. K43–K48 (1977).
See also *2*, p. *33*; *20.1*, p. *248*; 22.1, p. 332.

J. Bauer, L. Biste, and D. Bolze
Optical Properties of Aluminium Nitride Prepared by Chemical and Plasmachemical Vapour Deposition
(a) Vol. **39**, No. 1, p. 173–181 (1977).
See also *1.5*, p. *29*; 13.1, p. 114; 20.1, p. 248.

B. Tuck and M. D. Zahari
"Up-Hill" Diffusion of Zinc in the III–V Semiconductors
(a) Vol. **39**, No. 2, p. 509–515 (1977).
See also *9*, p. *69*.

P. N. Dyachkov and A. A. Levin
Bond Orbital Model for Tetrahedral Semiconductors on the Base of Molecular Parameters Covalent Crystals and Perturbative Treatment of Partially Ionic Semiconductors
(b) Vol. **80**, No. 2, p. 681–690 (1977).
See also *4*, p. *44*; *13.1*, p. *116*; 22.1.2, p. 339; 22.1.3, p. 345; 22.4, p. 361.

U. Büchner
Wave-Vector Dependence of the Electron Energy Losses of Boron Nitride and Graphite
(b) Vol. **81**, No. 1, p. 227–234 (1977).
See also 13.1, p. 117; *20.1*, p. *253*; 22.1, p. 332.

H.-R. Trebin
Polarons in the Degenerate-Band Case (II): Weak to Intermediate Hole–Phonon Coupling
(b) Vol. **81**, No. 2, p. 527–534 (1977).
See also 13.1, p. 118; *13.5.3*, p. *151*; 22.4, p. 361.

S. C. Goyal and R. Prakash
Many-Body Effects on the Second-Order Elastic Constants of Solids with ZnS Structure
(b) Vol. **83**, No. 2, p. 509–515 (1977).
See also *4*, p. *46*; *12.2*, p. *106*; 22, p. 328; 22.1, p. 332; 22.1.1, p. 335; 22.1.2, p. 342; 22.3, p. 358; 22.4, p. 361.

E. M. Baskin, M. D. Blokh, M. V. Entin, and L. I. Magarill
Current Quadratic in Field and Photogalvanic Effect in Crystals without Inversion Centre
(b) Vol. **83**, No. 2, p. K97–K100 (1977).
See also *14.3.4*, p. *186*; *16*, p. *204*.

Nguyen Van Hieu and Nguyen Ai Viet
Interband Electronic Raman Scattering by a Three-Band Semiconductor in the Presence of a Resonant Electromagnetic Field
(b) Vol. **84**, No. 2, p. 417–422 (1977).
See also 13.1, p. 122; *14.3.4*, p. *188*; *20.1*, p. *262*; 22.4, p. 361.

E. Tegeler, N. Kosuch, G. Wiech, and A. Faessler
Anisotropic Emission of the X-Ray K-Emission Band of Nitrogen in Hexagonal Boron Nitride
(b) Vol. **84**, No. 2, p. 561–567 (1977).
See also *20*, p. *246*.

22.2.1 Arsenides

Review Articles

SH. M. KOGAN and T. M. LIFSHITS
Photoelectric Spectroscopy — A New Method of Analysis of Impurities in Semiconductors
(a) Vol. **39**, No. 1, p. 11—39 (1977).
See also *13.4*, p. *129*; *16*, p. *201*; 22, p. 320; 22.1.1, p. 333; 22.1.2, p. 337.

I. N. ALEKANDROV
Transition Regions in Epitaxial Films
(a) Vol. **44**, No. 1, p. 11—43 (1977).
See also *1.5*, p. *29*; *3*, p. *42*; 14.3.1, p. 173; 22.1.1, p. 333; 22.1.2, p. 337.

Original Papers and Short Notes

A. PHILIPP, F. KUCHAR, and K. SEEGER
Submillimeter Conductivity of Thermal and Warm Electrons in GaAs and InP
(b) Vol. **79**, No. 1, p. 115—124 (1977).
See also *14.3*, p. *165*; 22.2.2, p. 354.

J. P. STRØMSHEIM, T. OLSEN, and D. J. RUZICKA
Analyses of Surface Preparation, Anodic Oxidation, and Epitaxial Layers of GaAs Using Nuclear Backscattering
(a) Vol. **39**, No. 1, p. 167—172 (1977).
See also *1.5*, p. *29*; *1.6*, p. *31*.

VIJENDER SHARMA and P. C. MATHUR
Effect of Junction Inhomogeneity on AC Voltage of n^+pp^+ GaAs IMPATT Diodes
(a) Vol. **39**, No. 1, p. K59—K61 (1977).
See also *14.3.3*, p. *178*.

MARTIN SCHMIDT
Acceptor Ground States in Nearly Cubic Semiconductors
(b) Vol. **79**, No. 2, p. 533—538 (1977).
See also *13.4*, p. *130*.

A. MILCHEV and R. PICKENHAIN
The Quadratic Response of a Fermi Gas. Building-Up of the Covalent Bonding in Zincblende Semiconductors. A Nondiagonal Density Matrix Treatment
(b) Vol. **79**, No. 2, p. 549—558 (1977).
See also *4*, p. *44*; 13, p. 107; 22.1, p. 332; 22.1.1, p. 333; 22.1.2, p. 338.

G. HOROWITZ
Nondestructive Determination of Free Carrier Parameters of Implanted Gallium Arsenide by Infrared Reflectivity Measurements
(a) Vol. **39**, No. 2, p. 533—540 (1977).
See also *20.1*, p. *250*.

L. I. FILINA, A. G. ILYASHENKO, YU. V. VOROBEV, YU. I. KARKHANIN, and M. K. SHEINKMAN
The Thermally Stimulated Conductivity and the Features of the Process of Recombination of Charge Carriers in Semi-Insulating GaAs(Cr)
(a) Vol. **39**, No. 2, p. 717—724 (1977).
See also *13.4*, p. *131*; *14.3*, p. *166*.

I. MOJZES, B. PÖDÖR, and I. BALOGH
On the Temperature Dependence of Peak Electron Velocity and Threshold Field Measured on GaAs Gunn Diodes
(a) Vol. **39**, No. 2, p. K123—K125 (1977).
See also *14.3.3*, p. *178*; *14.3.4*, p. *184*.

B. Rheinländer, A. Zehe, and G. Kretzschmar
Near Field Topography of High Frequency Modulated GaAs LED's
(a) Vol. **39**, No. 2, p. K139—K142 (1977).
See also *14.3.3*, p. *178*; *20.3*, p. *268*.

C. A. S. Lima and L. C. M. Miranda
Plasmon-Enhanced Sound Amplification in Piezoelectric Semiconductors under a Strong Magnetic Field
(b) Vol. **80**, No. 1, p. 57—62 (1977).
See also 7, p. 62; *13.5.2*, p. *149*; 18, p. 210.

A. R. Vasconcellos and R. Luzzi
Light Scattering and Spin Diffusion in Doped Semiconductors
(b) Vol. **80**, No. 1, p. 181—191 (1977).
See also *13.5*, p. *141*; 18.1, p. 213; 20.1, p. 251.

W. Lochmann
Phonon-Assisted Auger Recombination in Semiconductors
(a) Vol. **40**, No. 1, p. 285—292 (1977).
See also *6*, p. *54*; *13.1*, p. *116*; 20, p. 244.

A. F. Kravchenko, B. S. Lisenker, and Yu. E. Maronchuk
Influence of High Electric Field on Capture and Recombination of Carriers in the GaAs Epitaxial Layers
(a) Vol. **40**, No. 1, p. 321—332 (1977).
See also *14.3.4*, p. *184*; *16*, p. *202*.

K. Kurumada, T. Mizutani, and A. Ishida
Bi-Stable Switching Phenomena in GaAs Planar Devices with Schottky Gate
(a) Vol. **40**, No. 2, p. K93—K96 (1977).
See also *14.3.3*, p. *179*.

M. Brauer
Dynamical Response of Electrons in GaAs at 300 K
(b) Vol. **81**, No. 1, p. 147—152 (1977).
See also *13*, p. *109*; *14.3.4*, p. *185*.

A. Schlachetzki
Switching in Miniaturized Planar Gunn Devices
(a) Vol. **41**, No. 1, p. 103—111 (1977).
See also *14.3.3*, p. *179*.

M. J. Howes, D. V. Morgan, and W. J. Devlin
Applications of Magnetoresistance Measurements in the Evaluation of Transferred Electron Device Performance
(a) Vol. **41**, No. 1, p. 117—124 (1977).
See also *14.3.3*, p. *179*; 22.1.1, p. 334; 22.1.2, p. 339; 22.2.2, p. 354.

T. Matsushita and J. Hayashi
X-Ray Structure Factors and the Debye-Waller Factor of Gallium Arsenide Crystals Determined from Full Widths at Half Maximum of Bragg Case Diffraction Curves
(a) Vol. **41**, No. 1, p. 139—145 (1977).
See also *1.1*, p. *11*; 6, p. 56.

K. Bhattacharya, P. Basu, and H. Saha
Back Wall Schottky Barrier Solar Cell with an Interfacial Layer
(a) Vol. **41**, No. 1, p. 317—321 (1977).
See also 13.3, p. 126; *14.3.3*, p. *179*; *16*, p. *202*; 22.1.2, p. 339.

G. Kühn, M. Hein and S. Wartewig
Manganese–Oxygen Impurities in GaAs
(a) Vol. **41**, No. 1, p. K13—K15 (1977).
See also *10.2*, p. *88*; *19*, p. *239*.

W. Thielemann and F. Bergauer
Test of the Random Fluctuations of the Conduction Band Edge in Heavily Doped n-Type GaAs by Means of the Magnetic Circular Dichroism
(b) Vol. **81**, No. 2, p. K91—K94 (1977).
See also 13.1, p. 118; *18*, p. *211*; *20.1*, p. *255*.

P. L. Vuillermoz, J. Jouglar, A. Laugier, and H. R. Winteler
Phonon–Defect Interaction in Dislocation-Free GaAs
(a) Vol. **41**, No. 2, p. 561—567 (1977).
See also 6, p. 56; 8, p. 66; *10.2*, p. *88*.

K. D. Glinchuk, A. V. Prokhorovich, V. E. Rodionov, and V. I. Vovnenko
The Scheme of Electronic Transitions via 0.94, 1.0, 1.2, and 1.3 eV Radiative Centres in n-GaAs
(a) Vol. **41**, No. 2, p. 659—668 (1977).
See also *13.4*, p. *135*; *20.3*, p. *271*.

B. Frick and D. Siebert
ESR in Semi-Insulating GaAs:Cr and Colorimetric Determination of the Cr Content
(a) Vol. **41**, No. 2, p. K185—K187 (1977).
See also *10.2*, p. *89*; *19*, p. *240*.

E. Shatkovskii and A. Tamaševičius
Sign Inversion of Photomagnetoelectric Effect in InAs at Intense Photoexcitation
(a) Vol. **42**, No. 1, p. 111—114 (1977).
See also 15, p. 199; *16*, p. *203*.

W. Lochmann
Scattering Mechanisms in Phonon-Assisted Auger Recombination
(a) Vol. **42**, No. 1, p. 181—185 (1977).
See also *6*, p. *57*; *20*, p. *245*.

K. Asai, T. Sugeta, M. Ida, and M. Fujimoto
T-Shaped Schottky Barrier Gate GaAs FET
(a) Vol. **42**, No. 1, p. K7—K9 (1977).
See also *14.3.3*, p. *180*.

H. Fujiyasu, T. Hatano, S. Itoh, M. Sekinobu, and O. Ohtsuki
n-Type GaAs MIS Structures
(a) Vol. **42**, No. 1, p. K25—K27 (1977).
See also *14.3.2*, p. *177*; *14.3.3*, p. *180*.

M. Hirose, S. Yokoyama, and Y. Osaka
Surface States in GaAs Tunnel MIS Structures
(a) Vol. **42**, No. 2, p. 483—488 (1977).
See also *13.3*, p. *127*; 14.3.3, p. 180.

U. Traxlmayr, W. F. Fallmann, and F. Paschke
The Influence of Magnetic Fields on Gunn Effect Planar Devices in Corbino Geometry
(a) Vol. **42**, No. 2, p. 525—528 (1977).
See also *14.3.3*, p. *180*; *14.3.4*, p. *185*; 18, p. 212.

A. F. Kravchenko, B. S. Lisenker, Yu. E. Maronchuk, A. P. Sherstyakov, and V. N. Sherstyakova
Investigation of Recombination of Non-Equilibrium Carriers in GaAs Epitaxial Layers
(a) Vol. **42**, No. 2, p. 647—655 (1977).
See also *13.4*, p. *137*; *14.3.1*, p. *174*; 16, p. 203.

K. Kliefoth and B. Petzel
Velocity–Field Characteristics of GaAs Gunn Diodes with Different Impurity Concentration
(a) Vol. **42**, No. 2, p. K133—K135 (1977).
See also *14.3.3*, p. *181*.

M. Rösler and R. Zimmermann
Phase Diagram of Electron–Hole Liquids in Polar Semiconductors
(b) Vol. **83**, No. 1, p. 85–91 (1977).
See also *13.5*, p. *141*; 13.5.1, p. 146; 13.5.2, p. 149; 13.5.3, p. 152; 22.4.1, p. 363; 22.4.2, p. 366; 22.4.3, p. 367.

B. Rheinländer, S. Haase, and H. Weinert
On the Determination of the Free Carrier Relaxation Time by Transverse Magneto-Plasma Reflectivity Measurements on Sn-Doped n-Type GaAs
(b) Vol. **83**, No. 1, p. K83–K86 (1977).
See also 13.5.2, p. 150; *18*, p. *212*; *20.1*, p. *260*.

S. Guha and P. K. Sen
Two-Stream Instability in n-Type Gallium Arsenide
(a) Vol. **43**, No. 1, p. 89–97 (1977).
See also 6, p. 59; 13, p. 111; *14.3.4*, p. *186*.

W. Ekardt
Theory of Excitons in Zincblende-Structure Semiconductors in Intermediate Magnetic Fields
(b) Vol. **84**, No. 1, p. 293–304 (1977).
See also *13.5.1*, p. *148*; 20.1, p. 261; 22.2.2, p. 355.

D. L. Smith
Recombination Currents in a p–n Heterojunction Diode
(a) Vol. **44**, No. 1, p. 381–390 (1977).
See also *14.3.2*, p. *177*; *14.3.3*, p. *181*.

Vijender Sharma and P. C. Mathur
Large-Signal Impedance Measurements and Power-Efficiency Consideration of n^+pp^+ GaAs IMPATT Diodes
(a) Vol. **44**, No. 1, p. 401–410 (1977).
See also *14.3.3*, p. *181*; *14.3.4*, p. *187*.

F. Lukeš
Electroreflectance Spectrum of InAs in the Range of E_0 and $E_0 + \Delta_0$ Transitions
(b) Vol. **84**, No. 2, p. K113–K117 (1977).
See also 13.1, p. 123; *14.3.4*, p. *188*; *20.1*, p. *263*.

T. L. Tansley
Thermally-Assisted Tunnelling in Composite Barrier Heterojunctions
(a) Vol. **44**, No. 2, p. 449–455 (1977).
See also 13.4, p. 140; *14.3.3*, p. *182*; 22.2.4, p. 357.

H. Neumann, K. Jacobs, Nguyen Van Nam, W. Koj, and C. Krause
Properties of Liquid Phase Epitaxial Ge-Doped GaAs
(a) Vol. **44**, No. 2, p. 675–678 (1977).
See also *3*, p. *43*; *10.2*, p. *93*.

K. D. Glinchuk and A. V. Prokhorovich
Effect of the 0.94, 1.0, 1.2, and 1.3 eV Radiative Centres on the Intrinsic Luminescence Intensity in n-GaAs
(a) Vol. **44**, No. 2, p. 777–785 (1977).
See also *13.4*, p. *140*; *20.3*, p. *275*.

22.2.2 Phosphides

Original Papers and Short Notes

A. PHILIPP, F. KUCHAR, and K. SEEGER
Submillimeter Conductivity of Thermal and Warm Electrons in GaAs and InP
(b) Vol. **79**, No. 1, p. 115–124 (1977).
See also *14.3*, p. *165*; 22.2.1, p. 350.

G. S. AGARWAL and H. D. VOLLMER
Surface Polariton Effects in Spontaneous Emission
(b) Vol. **79**, No. 1, p. 249–257 (1977).
See also *13.3*, p. *126*; 13.5.2, p. 149; 20.1, p. 247; 21, p. 276.

F. S. SHISHIYANU, V. GH. GHEORGHIU, and S. K. PALAZOV
Diffusion, Solubility, and Electrical Activity of Co and Fe in InP
(a) Vol. **40**, No. 1, p. 29–35 (1977).
See also *9*, p. *69*; 10.2, p. 85; 14.3, p. 167.

P. KRISPIN
Impurity Detection by the Admittance–Frequency Dispersion
(a) Vol. **40**, No. 1, p. K81–K84 (1977).
See also *13.4*, p. *132*; *14.3.3*, p. *179*.

J. DONECKER and J. KLUGE
Magneto-Optical Investigation of the Ground State of the Indirect Free Exciton in Si and GaP
(b) Vol. **81**, No. 1, p. 199–203 (1977).
See also 13.5.1, p. 145; *18*, p. *210*; *20.1*, p. *253*; 22.1.2, p. 339.

J. S. NKOMA
Theory of Two-Photon Absorption by Surface Polaritons
(b) Vol. **81**, No. 1, p. 307–312 (1977).
See also *13.5*, p. *141*; *20.1*, p. *254*.

M. J. HOWES, D. V. MORGAN, and W. J. DEVLIN
Applications of Magnetoresistance Measurements in the Evaluation of Transfered Electron Device Performance
(a) Vol. **41**, No. 1, p. 117–124 (1977).
See also *14.3.3*, p. *179*; 22.1.1, p. 334; 22.1.2, p. 339; 22.2.1, p. 351.

M. Z. IQBAL
Deep Level Photocapacitance Spectra of Green-Emitting GaP Lamps
(a) Vol. **41**, No. 2, p. K155–K156 (1977).
See also *13.4*, p. *135*; *16*, p. *202*.

H. NEUMANN, H.-G. ERNST, E. BUTTER, P. KRAMER, and W. KOHL
Electrical Properties of Liquid Phase Epitaxial GaP Grown from Tin Solution
(a) Vol. **43**, No. 1, p. K95–K98 (1977).
See also *14.3*, p. *171*.

R. BINDEMANN, H. FISCHER, and K. KREHER
Determination of the Concentration of Substitutional O in GaP by Means of Donor–Acceptor Pair Luminescence
(a) Vol. **43**, No. 2, p. 529–533 (1977).
See also *13.4*, p. *138*; *20.3*, p. *273*.

W. Ekardt
Theory of Excitons in Zincblende-Structure Semiconductors in Intermediate Magnetic Fields
(b) Vol. **84**, No. 1, p. 293–304 (1977).
See also *13.5.1*, p. *148*; 20.1, p. 261; 22.2.1, p. 353.

H. Rosin, G. Freydank, R. Klein, M. Bruchholz, and V. Gottschalch
Reveal of Dislocation Etch Pits on {001} GaP with Hot Phosphoric Acid
(a) Vol. **44**, No. 1, p. K13–K14 (1977).
See also *10.2*, p. *92*.

22.2.3 Antimonides

Original Papers and Short Notes

S. Cristoloveanu and P. Viktorovitch
Influence of Surfaces on the Ettingshausen Effect. Application to the Characterization of Narrow-Gap Semiconductors
(a) Vol. **39**, No. 1, p. 109–116 (1977).
See also 14.3, p. 165; *15*, p. *198*.

B. A. Aronzon and E. Z. Meilikhov
Helicon Damping in Semiconductors in the Quantum Limit
(b) Vol. **79**, No. 2, p. 753–762 (1977).
See also *14.3*, p. *166*; 18, p. 209; *20*, p. *244*.

H. van Cong
Activity Coefficients of Electrons with Small Effective Mass in Heavily Doped Semiconductors
(b) Vol. **80**, No. 1, p. K45–K47 (1977).
See also *14.3*, p. *166*.

M. S. Sodha, V. K. Tripathi, and D. P. Singh
Effect of a Strong Magnetic Field on the Self-Focusing of a Laser Beam in InSb
(a) Vol. **40**, No. 1, p. 315–320 (1977).
See also *18*, p. *210*; *20.2*, p. *265*.

E. W. Kreutz
Magneto-Surface Experiments on the III–V Compound Semiconductor InSb
(a) Vol. **40**, No. 2, p. 415–423 (1977).
See also *13.3*, p. *126*; *14.3.2*, p. *176*.

S. Guha and S. Ghosh
Convective Instability in n-InSb in the Presence of Crossed Electric and Magnetic Fields
(a) Vol. **41**, No. 1, p. 249–254 (1977).
See also *14.3.4*, p. *185*.

K. M. Demchuk and I. M. Tsidilkovskii
Scattering of Electrons by the Deformation Potential in Doped InSb
(b) Vol. **82**, No. 1, p. 59–67 (1977).
See also *13.1*, p. *118*; 14.3, p. 169; 15, p. 199.

I. Gorczyca
Spin-Dependent Wave Functions and Band Parameters at the Γ Point of InSb
(b) Vol. **82**, No. 1, p. 279–288 (1977).
See also *13.1*, p. *119*.

S. Lakeou, S. Cristoloveanu, and A. Chovet
Magnetoresistance Effect in Near Intrinsic Semiconductors. Influence of Sample Geometry. A New Method for the Determination of Carrier Densities and Mobilities
(a) Vol. **43**, No. 1, p. 213–222 (1977).
See also *14.3*, p. *170*; 22.1.1, p. 335; 22.1.2, p. 341.

V. I. Chaikin, K. U. Guga, V. K. Malyutenko, and A. V. Sachenko
Photoconductivity of Indium Antimonide with Different Surface Treatments
(a) Vol. **43**, No. 1, p. 311–316 (1977).
See also *14.3.2*, p. *177*; *16*, p. *204*.

A. Reklaitis
High Field Electron Diffusion in Indium Antimonide
(b) Vol. **83**, No. 2, p. K121–K124 (1977).
See also *14.3.4*, p. *186*.

M. M. Kaila
Effect of Magnetic Field on Electrical Effects of Dislocations in InSb
(a) Vol. **44**, No. 2, p. K135–K139 (1977).
See also *10.2*, p. *93*; *14.3*, p. *172*.

22.2.4 Mixed Crystals

Original Papers and Short Notes

H. G. Brühl, L. Hildisch, H. Morwinski, W. Schmidt, and E. Schubert
Diffuse X-Ray Scattering and Compositional Disorder in $Ga_xIn_{1-x}P$
(a) Vol. **39**, No. 1, p. 133–139 (1977).
See also *1.1*, p. *11*; 1.2, p. 14.

M. L. Young and C. P. Fender
Sub-Threshold Velocity–Field Characteristics of $InAs_{1-x}P_x$ Alloys
(a) Vol. **41**, No. 1, p. 85–94 (1977).
See also 13.1, p. 117; *14.3.4*, p. *185*.

A. S. Logginov and V. E. Solovev
Analysis of the Current Carrier Concentration Distribution and Threshold Properties of Stripe-Geometry $Ga_xAl_{1-x}As$ Lasers
(a) Vol. **41**, No. 2, p. 371–378 (1977).
See also *14.3.3*, p. *179*; *20.2*, p. *265*.

N. N. Sirota, I. V. Bodnar, and G. F. Smirnova
Lattice Reflection Spectra of $(GaP)_x(InAs)_{1-x}$ Solid Solutions
(a) Vol. **41**, No. 2, p. 669–672 (1977).
See also *13.1*, p. *118*; *20.1*, p. *256*.

A. S. Popov and I. Y. Yanchev
On the Disorder Scattering in $GaAs_{1-x}P_x$ Solid Solutions
(a) Vol. **41**, No. 2, p. 687–691 (1977).
See also *14.3*, p. *169*.

G. Müller, M. Haubold, R. Schimko, C.-E. Richter, and G. Schwarz
The Diffusion of Ion-Implanted Zinc in $GaAs_{1-x}P_x$
(a) Vol. **42**, No. 2, p. 579–583 (1977).
See also *9*, p. *70*; *11*, p. *97*.

R. Wolf, W. Kühn, R. Enderlein, and H. Lange
Determination of Charge Carrier Concentration from Low-Field Electroreflectance Spectra
(a) Vol. **42**, No. 2, p. 629–637 (1977).
See also *14.3.4*, p. *185*; *20.1*, p. *258*.

E. E. Matyas
Phase Diagram of the InAs–AlAs Pseudobinary System
(a) Vol. **42**, No. 2, p. K129–K131 (1977).
See also *1.2*, p. *18*.

H. Raidt, G. Kästner, H. Menninger, and G. Voigt
On Real Structure Investigations of $Ga_xIn_{1-x}P$ by Electron Microscopic Methods
(a) Vol. **43**, No. 1, p. 167–173 (1977).
See also *1.4*, p. *27*; *3*, p. *43*; 20.3, p. 273.

T. L. Tansley
Thermally-Assisted Tunnelling in Composite Barrier Heterojunctions
(a) Vol. **44**, No. 2, p. 449–455 (1977).
See also 13.4, p. 140; *14.3.3*, p. *182*; 22.2.1, p. 353.

22.3 IV—IV, II—V, and III—VI Compounds

Original Papers and Short Notes

S. I. Subbotin, V. V. Panfilov, and R. T. Molchanova
Pressure-Induced Polytypic Transformations and Edge Absorption of GaSe
(a) Vol. **39**, No. 1, p. 357—362 (1977).
See also *1.2*, p. *14*; 12.2, p. 104; 13.5.1, p. 143; *20.1*, p. *248*.

A. Suzuki, H. Matsunami, and T. Tanaka
Erratum to: Very Slow Decays of the Luminescence from Boron-Doped 4H–SiC
(a) Vol. **39**, No. 1, p. K79 (1977).
See also *20.3*, p. *267*.

V. A. Stukan, S. F. Marenkin, V. I. Trifonov, and V. Ya. Shevchenko
Opto-Electrical Properties of p-CdSb in the Submillimetre Wave Region
(a) Vol. **39**, No. 2, p. 445—452 (1977).
See also *16*, p. *201*; *20.1*, p. *250*.

V. Capozzi, G. Mariotto, M. Montagna, A. Cingolani, and A. Minafra
Space-Charge-Limited Currents in GaSe at Different Temperatures
(a) Vol. **40**, No. 1, p. 93—100 (1977).
See also *13.4*, p. *132*; *14.3.4*, p. *184*.

S. Popović, B. Čelustka, Ž. Ružić-Toroš, and D. Broz
X-Ray Diffraction Study and Semiconducting Properties of the System Ga_2Se_3–In_2Se_3
(a) Vol. **41**, No. 1, p. 255—262 (1977).
See also *1.2*, p. *16*; 14.3, p. 168.

V. A. Stukan, V. I. Trifonov, V. Ya. Shevchenko, and G. I. Goncharenko
Electrical Properties of p-CdSb at Low Temperatures
(a) Vol. **41**, No. 1, p. 307—315 (1977).
See also 13.1, p. 117; 13.4, p. 134; *14.3*, p. *168*.

G. Antonioli, D. Bianchi, V. Canevari, U. Emiliani, and P. Podini
Optical Constants of GaSe at the Fundamental Absorption Edge
(b) Vol. **81**, No. 2, p. 665—670 (1977).
See also 13.5.1, p. 146; *20.1*, p. *255*.

G. L. Belenkii, L. N. Alieva, R. Kh. Nani, and E. Yu. Salaev
Vibrational Spectrum of InSe
(b) Vol. **82**, No. 2, p. 705—709 (1977).
See also *6*, p. *58*; *20.1*, p. *258*.

V. K. Bashenov, S. A. Garyainov, and D. I. Marvakov
Band Structure of Layer Gallium Selenide
(b) Vol. **82**, No. 2, p. K95—K97 (1977).
See also *13.1*, p. *120*.

Yu. M. Tairov, V. F. Tsvetkov, M. A. Chernov, and V. A. Taranets
Investigation of Phase Transformations and Polytype Stability of β-SiC
(a) Vol. **43**, No. 1, p. 363—369 (1977).
See also *1*, p. *10*; 1.2, p. 18.

S. C. Goyal and R. Prakash
Many-Body Effects on the Second-Order Elastic Constants of Solids with ZnS Structure
(b) Vol. **83**, No. 2, p. 509—515 (1977).
See also *4*, p. *46*; *12.2*, p. *106*; 22, p. 328; 22.1, p. 332; 22.1.1, p. 335; 22.1.2, p. 342; 22.2, p. 348; 22.4, p. 361.

V. P. Mushinskii, L. I. Palaki, and V. V. Chebotaru
Optical Absorption in α-Ga_2S_3 Single Crystals
(b) Vol. **83**, No. 2, p. K149—K154 (1977).
See also 13.5.1, p. 147; *20.1*, p. *261*.

T. Matsumura, M. Sudo, C. Tatsuyama, and S. Ichimura
Photoluminescence in GaSe
(a) Vol. **43**, No. 2, p. 685—693 (1977).
See also 13.5.1, p. 147; *20.3*, p. *273*.

T. A. McMath, G. Jackle, and J. C. Irwin
On Refractive Index Measurements in GaS and GaSe
(a) Vol. **43**, No. 2, p. K151—K154 (1977).
See also *20.1*, p. *261*.

R. Schwabe, F. Thuselt, and T. Hänsel
Exciton–Phonon Energy Transfer in Semiconductors
(b) Vol. **84**, No. 1, p. K43—K47 (1977).
See also *6*, p. *60*; *13.5.1*, p. *148*; 20.3, p. 274.

M. K. Anis and A. R. Piercy
Growth of Single Crystals of GaSe with Natural Facets at Large Angles to the Layers
(a) Vol. **44**, No. 1, p. K5—K6 (1977).
See also *3*, p. *43*.

V. V. Borshch, V. S. Koval, I. V. Potykevich, and I. V. Fekeshgazi
Birefringence and Optical Activity of CdP_2
(a) Vol. **44**, No. 1, p. K15—K19 (1977).
See also *20.1*, p. *262*.

L. P. Mokhracheva, V. A. Tskhai, and P. V. Geld
Chemical Binding and Energy Band Structure of Vanadium Monocarbide
(b) Vol. **84**, No. 2, p. 465—471 (1977).
See also 4, p. 46; *13.1*, p. *122*.

A. Lyubchenko, I. Potykievich, A. Fedorov, and M. Sheinkman
On the Mechanism of Photoluminescence of Cadmium Diphosphide Doped by Bi and Zn
(a) Vol. **44**, No. 2, p. 747—753 (1977).
See also *16*, p. *205*; *20.3*, p. *275*.

22.3.1 Mixed Crystals

Original Papers and Short Notes

H.-G. Holzhütter
On the Theory of Disorder Scattering in Mixed Crystal Systems
(b) Vol. **79**, No. 1, p. K13–K15 (1977).
See also *13*, p. *107*; 22.4.4, p. 368.

L. Gouskov, A. Gouskov, V. Lemos, W. May, and H. Sampaio
Electrical Properties of $GaTe_xSe_{1-x}$ Crystals
(a) Vol. **39**, No. 1, p. 65–71 (1977).
See also 1.2, p. 14; *13.4*, p. *129*; *14.3*, p. *165*.

K. R. Allakhverdiev and M. M. Tagyev
Infrared Investigation of the GaS_xSe_{1-x} System
(a) Vol. **39**, No. 2, p. K111–K113 (1977).
See also 1.2, p. 15; *6*, p. *53*; *20.1*, p. *250*.

S. G. Abdullaeva and H. S. Mamedov
X-Ray Investigation of Single Crystals of $GaSe_xS_{1-x}$ ($0 \leqq x \leqq 1$)
(a) Vol. **40**, No. 1, p. K7–K9 (1977).
See also *1*, p. *9*.

J. Cisowski and W. Żdanowicz
Pressure and Temperature Dependence of the Hall Coefficient in $Cd_{3-x}Zn_xAs_2$ Solid Solutions
(a) Vol. **41**, No. 1, p. K59–K62 (1977).
See also 12.2, p. 105; *14.3*, p. *168*.

S. I. Radautsan, E. A. Arushanov, L. N. Lukyanova, and A. N. Nateprov
The Thermoelectric Power of $Cd_3(As_xP_{1-x})_2$
(a) Vol. **43**, No. 2, p. K159–K164 (1977).
See also 13.1, p. 122; *15*, p. *199*.

B. Etlinger, R. Trojko, and U. V. Desnica
Thermal Decomposition of $Al_2In_{12}S_{21}$
(a) Vol. **44**, No. 1, p. K29–K30 (1977).
See also *3*, p. *43*.

22.4 II–VI Compounds

Review Article

T. D. Dzhafarov
Atomic Diffusion in Semiconductor Epitaxial Structures
(a) Vol. **42**, No. 1, p. 11—45 (1977).
See also *9*, p. *69*; 10.2, p. 83; 14.3.1, p. 173; 14.3.3, p. 178; 20.2, p. 265; 22.1, p. 332; 22.2, p. 348.

Original Papers and Short Notes

P. N. Dyachkov and A. A. Levin
Bond Orbital Model for Tetrahedral Semiconductors on the Base of Molecular Parameters. Covalent Crystals and Perturbative Treatment of Partially Ionic Semiconductors
(b) Vol. **80**, No. 2, p. 681—690 (1977).
See also *4*, p. *44*; *13.1*, p. *116*; 22.1.2, p. 339; 22.1.3, p. 345; 22.2, p. 348.

H.-R. Trebin
Polarons in the Degenerate-Band Case (II): Weak to Intermediate Hole–Phonon Coupling
(b) Vol. **81**, No. 2, p. 527—534 (1977).
See also 13.1, p. 118; *13.5.3*, p. *151*; 22.2, p. 348.

S. C. Goyal and R. Prakash
Many-Body Effects on the Second-Order Elastic Constants of Solids with ZnS Structure
(b) Vol. **83**, No. 2, p. 509—515 (1977).
See also *4*, p. *46*; *12.2*, p. *106*; 22, p. 328; 22.1, p. 332; 22.1.1, p. 335; 22.1.2, p. 342; 22.2, p. 348; 22.3, p. 358.

Le Manh Hoang and J. M. Baranowski
Absorption and Luminescence of V(d^3) in II–VI Compound Semiconductors
(b) Vol. **84**, No. 1, p. 361—365 (1977).
See also 13.4, p. 138; *20.1*, p. *261*; *20.3*, p. *274*.

Nguyen Van Hieu and Nguyen Ai Viet
Interband Electronic Raman Scattering by a Three-Band Semiconductor in the Presence of a Resonant Electromagnetic Field
(b) Vol. **84**, No. 2, p. 417—422 (1977).
See also 13.1, p. 122; *14.3.4*, p. *188*; *20.1*, p. *262*; 22.2, p. 349.

22.4.1 Sulphides

Review Article

K. W. Böer
The CdS/Cu_2S Solar Cell (I): Minority Carrier Generation and Transport in the Cu_2S Emitter
(a) Vol. **40**, No. 2, p. 355—384 (1977).
See also 14.3.3, p. 178; 14.3.4, p. 183; *16*, p. *201*; 20.1, p. 247; 22, p. 320.

Original Papers and Short Notes

D. Theis
Wavelength-Modulated Reflectivity Spectra of ZnSe and ZnS from 2.5 to 8 eV
(b) Vol. **79**, No. 1, p. 125—130 (1977).
See also *13.1*, p. *114*; 13.5.1, p. 143; *20.1*, p. *247*; 22.4.2, p. 365.

F. Henneberger, K. Henneberger, and J. Voigt
On the Emission Lineshape of Excitonic Molecules in Wurtzite-Type Semiconductors
(b) Vol. **79**, No. 1, p. K81—K85 (1977).
See also *13.5.1*, p. *143*; *20.1*, p. *248*.

J.-P. Aicardi, J.-P. Leyris, and G. Masse
Cathodoluminescence of Cinnabar. Evidence of a Donor–Acceptor-Type Transition
(a) Vol. **39**, No. 1, p. 125—131 (1977).
See also 13.4, p. 130; *20.3*, p. *267*.

U. V. Desnica
Indirect Doping as a Possibility for Obtaining p-Type Conductivity CdS
(a) Vol. **39**, No. 1, p. K33—K36 (1977).
See also *11*, p. *94*; *14.3*, p. *165*.

W. Gorzkowski
Excitons Bound to Ionized Donors in Polar Semiconductors
(b) Vol. **79**, No. 2, p. K151—K152 (1977).
See also *13.5.1*, p. *144*; 22, p. 322; 22.4.2, p. 365.

B.-P. Kietis, R. Reksnys, and A. Sakalas
Ohmic and Acoustoelectric Currents in the Hall Effect for the Investigation of Bipolar Conductivity
(a) Vol. **39**, No. 2, p. 601—605 (1977).
See also 7, p. *62*; *14.3.4*, p. *183*.

H. Kuzmany and Z. Szalmassy
Systematic Study of the Acoustoelectric Interaction of Surface Waves with Inhomogeneously Distributed Free Carriers in a Photoconductor
(a) Vol. **40**, No. 1, p. 75—82 (1977).
See also 7, p. *62*; *14.3.2*, p. *176*; 14.3.3, p. 178; 16, p. 201; 22.8.1, p. 403.

M. A. Salehi Manshadi and J. Woods
Photoconductive Effects in Copper-Doped Cadmium Sulphide
(a) Vol. **40**, No. 1, p. K43—K47 (1977).
See also *16*, p. *202*.

A. N. Georgobiani, Yu. V. Ozerov, E. E. Strumban, and V. S. Vinogradov
Influence of Random Electric Fields on the Polarization of Deep Centre Electroabsorption
(b) Vol. **80**, No. 2, p. 589—596 (1977).
See also *13.4*, p. *132*; 16, p. 202; 20.1, p. 252; 20.3, p. 269.

J. M. Peransin, E. Groubert, and B. Goutiers
Time Investigation of Parametric Interactions in Acoustoelectric Oscillators
(a) Vol. **40**, No. 2, p. 583—588 (1977).
See also 7, p. *62*; *14.3.4*, p. *184*.

F. Bechstedt and F. Henneberger
Two-Photon Resonance Raman Scattering by Excitons via an Intermediate Excitonic Molecule (I): Bare-Exciton Approach
(b) Vol. **81**, No. 1, p. 211—220 (1977).
See also *13.5.1*, p. *145*; *20.3*, p. *270*; 22.5, p. 370.

O. Goede
Deformation Effects of Biexcitons in Wurtzite-Type Crystals
(b) Vol. **81**, No. 1, p. 235—244 (1977).
See also 13.1, p. 117; *13.5.1*, p. *145*; *20.1*, p. *253*.

N. B. Lukyanchikova, G. S. Pekar, N. N. Tkachenko, Hoang Mi Shin, and M. K. Sheinkman
Injection Electroluminescence in Forward-Biased ZnS Metal–Semiconductor Diodes
(a) Vol. **41**, No. 1, p. 299—305 (1977).
See also *14.3.3*, p. *179*; *20.3*, p. *270*.

S. Chatterjee and P. Sinha
Orthogonalized Augmented Plane-Wave Method for the Solution of the Energy Band Problem in Solids
(b) Vol. **81**, No. 2, p. 493—502 (1977).
See also *13.1*, p. *118*.

A. F. Ellervee
Luminescence of Pb^{2+} and Bi^{3+} Centres in Alkali-Earth Sulphides and Oxides
(b) Vol. **82**, No. 1, p. 91—98 (1977).
See also 13.4, p. 135; *20.3*, p. *271*; 22.6, p. 389.

T. Hata and M. Suzuki
Mechanism of Optical Modulation Caused by Acoustic Domains in CdS
(a) Vol. **42**, No. 1, p. K71—K74 (1977).
See also *7*, p. *63*; *20.1*, p. *257*.

H. Haug and S. Koch
On the Theory of Laser Action in Dense Exciton Systems
(b) Vol. **82**, No. 2, p. 531—543 (1977).
See also *13.5.1*, p. *146*; *20.2*, p. *266*.

G. Massé, J.-P. Aicardi, and C. Butti
Study of Band-Gap Shrinkage in Cinnabar (α-HgS) by Cathodoluminescence under Very Strong Excitation
(a) Vol. **42**, No. 2, p. 545—552 (1977).
See also *13.1*, p. *121*; *20.3*, p. *272*.

K. P. Pande
Rectification Properties of CdS Schottky Barrier Diodes with Evaporated Au/Ti as Blocking Contact
(a) Vol. **42**, No. 2, p. 615—619 (1977).
See also *14.3.3*, p. *180*; *14.3.4*, p. *185*.

A. Selkin
Energy Transfer by the Normal Modes in the Exciton Resonance Region
(b) Vol. **83**, No. 1, p. 47—53 (1977).
See also *13.5.1*, p. *146*; 20.1, p. 259.

M. Rösler and R. Zimmermann
Phase Diagram of Electron–Hole Liquids in Polar Semiconductors
(b) Vol. **83**, No. 1, p. 85—91 (1977).
See also *13.5*, p. *141*; 13.5.1, p. 146; 13.5.2, p. 149; 13.5.3, p. 152; 22.2.1, p. 353; 22.4.2, p. 366; 22.4.3, p. 367.

G. Massé, J. P. Aicardi, and J. P. Leyris
Green Emission of Mercury Sulphide (α-HgS)
(a) Vol. **43**, No. 1, p. 191—201 (1977).
See also 6, p. 59; 13.4, p. 137; *20.3*, p. *273*.

F. Henneberger, K. Henneberger, and J. Voigt
Two-Photon Resonance Raman Scattering by Excitons via an Intermediate Excitonic Molecule (II): Polarization Properties and Polariton Effects
(b) Vol. **83**, No. 2, p. 439–450 (1977).
See also *13.5.1*, p. *147*; *20.3*, p. *273*; 22.5, p. 371.

V. Kunigėlis, A. Jucys, E. Garška, and V. Vitkauskas
Influence of the Surface Potential on the Acoustoelectric Interaction in Semiconductor–Piezoelectric Structures
(a) Vol. **43**, No. 2, p. 699–703 (1977).
See also *7*, p. *63*; *14.3.2*, p. *177*.

Vo Hong Anh and Nguyen Nhu Dat
Parametric Action of a Strong Radiation Field on Anisotropic Semiconductors
(b) Vol. **84**, No. 1, p. 115–123 (1977).
See also *6*, p. *60*; *13.5*, p. *142*; 20, p. 246.

H. J. Hoffmann and W. Sohn
Analysis of Localized Levels in Semiconducting CdS from Generation–Recombination Noise Spectra
(a) Vol. **44**, No. 1, p. 237–246 (1977).
See also *13.4*, p. *139*; *14.3*, p. *171*.

C. Lawther and K. V. Anand
Physical Mechanism of Current Conduction and Light Emission in High-Resistivity ZnS: Mn Thin Films
(a) Vol. **44**, No. 1, p. 313–323 (1977).
See also *14.3.1*, p. *175*; *20.3*, p. *274*.

B. G. Caswell and J. Woods
Some Effects of Oxygen on As-Plated CdS–Su_2S Photovoltaic Cells
(a) Vol. **44**, No. 1, p. K47–K51 (1977).
See also *14.3.3*, p. *181*; *16*, p. *205*; 22, p. 330.

G. Jacobsen
Photovoltaic Spectra as a Function of the Light Intensity in ZnS Single Crystals
(a) Vol. **44**, No. 1, p. K67–K70 (1977).
See also *16*, p. *205*.

22.4.2 Selenides

Original Papers and Short Notes

D. Theis
Wavelength-Modulated Reflectivity Spectra of ZnSe and ZnS from 2.5 to 8 eV
(b) Vol. **79**, No. 1, p. 125—130 (1977).
See also *13.1*, p. *114*; 13.5.1, p. 143; *20.1*, p. *247*; 22.4.1, p. 362.

S. W. Biernacki and M. Hamera
LCAO Approach to Mn^{2+} in ZnSe
(b) Vol. **79**, No. 2, p. 683—689 (1977).
See also *13.4*, p. *130*; 19, p. 236.

W. Gorzkowski
Excitons Bound to Ionized Donors in Polar Semiconductors
(b) Vol. **79**, No. 2, p. K151—K152 (1977).
See also *13.5.1*, p. *144*; 22, p. 322; 22.4.1, p. 362.

D. D. Nedeoglo
Formation and Properties of the Impurity Band in n-ZnSe
(b) Vol. **80**, No. 1, p. 369—377 (1977).
See also *13.4*, p. *132*; *14.3*, p. *166*.

H. L. Oczkowski, N. Romeo, and L. Tarricone
Impact Ionization Electroluminescence in ZnSe Single Crystals
(a) Vol. **40**, No. 1, p. K49—K52 (1977).
See also *14.3.4*, p. *184*; *20.3*, p. *269*.

H. Röppischer, J. Jacobs, and B. V. Novikov
Influence of Preparation Conditions on the Optical Band Gap and the Mn^{2+} ESR Spectrum of ZnSe Single Crystals
(a) Vol. **40**, No. 2, p. 575—581 (1977).
See also 13.4, p. 132; *13.5.1*, p. *144*; *19*, p. *237*.

H. Röppischer and J. Jacobs
ESR Method for Investigating Associated Point Defects in ZnSe:Mn Single Crystals
(a) Vol. **40**, No. 2, p. K117—K120 (1977).
See also *10.2*, p. *87*; *19*, p. *238*.

R. Katzschmann, R. Kranold, and A. Rehfeld
Optical Anomaly of Small Particles in Glasses
(a) Vol. **40**, No. 2, p. K161 (1977).
See also *13.1*, p. *117*; *20.1*, p. *253*.

M. Sondergeld
Two-Photon Absorption by Envelope–Hole Coupled Exciton States in Cubic ZnSe (I): Energy Spectrum of the H_d Term and Valence-Band Parameters
(b) Vol. **81**, No. 1, p. 253—262 (1977).
See also *13.5.1*, p. *145*; *20.1*, p. *254*.

S. W. Biernacki, M. Hamera, and S. Wartewig
Ligand Hyperfine Structure of Mn^{2+} in ZnSe as Calculated by EHT–SCCC, Scattered-Wave, and LCAO Methods
(b) Vol. **81**, No. 1, p. K5—K9 (1977).
See also *13.4*, p. *133*.

M. K. Rao and S. R. Jawalekar
Hall Voltage Measurements on CdSe Films
(a) Vol. **41**, No. 1, p. K31—K33 (1977).
See also *14.3.1*, p. *174*.

M. Sondergeld
Two-Photon Absorption by Envelope–Hole Coupled Exciton States in Cubic ZnSe (II): Polarization Dependence in the Case of Degenerate Valence Bands
(b) Vol. **81**, No. 2, p. 451–461 (1977).
See also *13.5.1*, p. *145*; *20.1*, p. *255*.

K. Kocot and J. M. Baranowski
The Resonant States of the Ti(d^2) Impurity in CdSe
(b) Vol. **81**, No. 2, p. 629–635 (1977).
See also *13.4*, p. *134*; 16, p. 202; 20.1, p. 255.

D. D. Nedeoglo, Dang Huy Lam, and A. V. Simashkevich
The Effect of Thermal Treatment and Doping on the Electrical Properties of Zinc Selenide
(a) Vol. **42**, No. 2, p. 675–679 (1977).
See also *14.3*, p. *170*; 16, p. 203.

B. R. Sethi, P. L. Talwar, O. P. Sharma, and P. C. Mathur
The Electrical Transport Properties of a Zinc Selenide Crystal Highly Annealed in Molten Zinc
(a) Vol. **42**, No. 2, p. 791–795 (1977).
See also 13.5.3, p. 152; *14.3*, p. *170*.

M. Rösler and R. Zimmermann
Phase Diagram of Electron–Hole Liquids in Polar Semiconductors
(b) Vol. **83**, No. 1, p. 85–91 (1977).
See also *13.5*, p. *141*; 13.5.1, p. 146; 13.5.2, p. 149; 13.5.3, p. 152; 22.2.1, p. 353; 22.4.1, p. 363; 22.4.3, p. 367.

D. D. Nedeoglo, Dang Huy Lam, and A. V. Simashkevich
Electrical Properties of the Metal–ZnSe Contact
(a) Vol. **44**, No. 1, p. 83–89 (1977).
See also *14.3.3*, p. *181*; *14.3.4*, p. *187*.

S. Al-Dallal
Epitaxial Growth of CdSe on ZnTe (111)
(a) Vol. **44**, No. 1, p. 183–189 (1977).
See also *1.5*, p. *30*; *3*, p. *43*.

A. P. Radlinski
On the Properties of Cobalt Impurities in Zinc Selenide Crystals (I): Effect of In-Band Position of States on Photoelectric Properties
(b) Vol. **84**, No. 2, p. 503–511 (1977).
See also *13.4*, p. *139*; 16, p. 205; 20.1, p. 262.

H. Lange and I. A. Iliev
Evidence of Coulomb Interaction at the E_1 Interband Transition of Hexagonal CdSe
(b) Vol. **84**, No. 2, p. 569–574 (1977).
See also 13.1, p. 122; *13.5.1*, p. *148*; *20.1*, p. *262*.

C. Lawther and J. Woods
Forward-Bias Electroluminescence in ZnSe Diodes
(a) Vol. **44**, No. 2, p. 693–702 (1977).
See also *20.3*, p. *275*.

K. Jarašiūnas and J. Vaitkus
Investigation of Non-Equilibrium Processes in Semiconductors by the Method of Transient Holograms
(a) Vol. **44**, No. 2, p. 793–800 (1977).
See also *14.3.4*, p. *188*; 20, p. 246; 22.1.2, p. 344.

22.4.3 Tellurides

Original Papers and Short Notes

W. Szuszkiewicz
Optical Absorption in HgTe
(b) Vol. **79**, No. 2, p. 691—698 (1977).
See also 13.1, p. 115; *20.1*, p. *249*.

J. Toušková and R. Kužel
Some Properties of Ag–CdTe Diodes
(a) Vol. **40**, No. 1, p. 309—313 (1977).
See also *14.3.2*, p. *176*; 14.3.3, p. 178.

W. Szuszkiewicz
Direct Determination of the Negative Energy Gap E_0 for Semiconductors with Inverted Band Structure
(b) Vol. **81**, No. 2, p. K119—K123 (1977).
See also *13.1*, p. *118*; 20.1, p. 255.

A. Jędrzejczak and T. Dietl
Erratum to: Thermomagnetic Properties of n-Type and p-Type HgTe
(b) Vol. **82**, No. 2, p. 717 (1977).
See also *13.1*, p. *120*; 14.3, p. 170; *15*, p. *199*.

M. Rössler and R. Zimmermann
Phase Diagram of Electron–Hole Liquids in Polar Semiconductors
(b) Vol. **83**, No. 1, p. 85—91 (1977).
See also *13.5*, p. *141*; 13.5.1, p. 146; 13.5.2, p. 149; 13.5.3, p. 152; 22.2.1, p. 353; 22.4.1, p. 363; 22.4.2, p. 366.

I. Gabaš and J. Toušek
Surface Photovoltage on CdTe
(a) Vol. **43**, No. 1, p. 351—356 (1977).
See also *14.3.2*, p. *177*; *16*, p. *204*.

B.-P. Kietis, R. Reksnys, A. Sakalas, and P. Höschl
Investigation of the Bipolar Photoconductivity in CdTe
(a) Vol. **43**, No. 2, p. 705—709 (1977).
See also *7*, p. *63*; *14.3*, p. *171*; 16, p. 204.

J. Gardavský, I. Barvík, and M. Zvára
Jahn-Teller Effect on the $^3T_1(^3F)$ State and Resonance of the $^3T_1(^3P)$ State with the Conduction Band in CdTe:Ti^{2+}
(b) Vol. **84**, No. 2, p. 691—698 (1977).
See also *13.4*, p. *139*; *20.1*, p. *263*.

P. Byszewski and R. R. Gałązka
Resonance Acceptor Level in $Hg_{0.98}Mn_{0.02}Te$
(b) Vol. **84**, No. 2, p. K83—K86 (1977).
See also *13.4*, p. *140*; *14.3*, p. *172*.

22.4.4 Mixed Crystals

Original Papers and Short Notes

H.-G. Holzhütter
On the Theory of Disorder Scattering in Mixed Crystal Systems
(b) Vol. **79**, No. 1, p. K13–K15 (1977).
See also *13*, p. *107*; 22.3.1, p. 360.

J. Saraie, H. Kato, N. Yamada, Sh. Kaida, and T. Tanaka
Preparation and Photoconductive Properties of Sintered Films of CdS–CdTe Mixed Crystals
(a) Vol. **39**, No. 1, p. 331–336 (1977).
See also *16*, p. *201*.

V. T. Bublik and A. A. Zaitsev
Determination of the Energy of Mixing of HgTe–CdTe Solid Solutions by Diffuse X-Ray Scattering
(a) Vol. **39**, No. 1, p. 345–349 (1977).
See also 1.2, p. 14; *1.3*, p. *21*.

G. Bastard, C. Rigaux, and A. Mycielski
Giant Spin Splitting Induced by Exchange Interactions in $Hg_{1-k}Mn_kTe$ Mixed Crystals
(b) Vol. **79**, No. 2, p. 585–593 (1977).
See also *13.1*, p. *115*; 18, p. 209; 20.1, p. 249.

G. Bastard
Resonant Acceptor Level in Zero-Gap Semiconductors. Single and Multiple Scattering Effects
(b) Vol. **80**, No. 2, p. 641–650 (1977).
See also *13.4*, p. *132*; 14.3, p. 167.

J. M. Pawlikowski
Schematic Energy Band Diagram of Metal-$Cd_xHg_{1-x}Te$ Contacts
(a) Vol. **40**, No. 2, p. 613–620 (1977).
See also 13.3, p. 126; *14.3.4*, p. *184*.

K. Sierański, J. Szatkowski, J. M. Pawlikowski, E. Majchrowska, and E. Płaczek-Popko
Fermi Level in $Cd_xHg_{1-x}Te$
(b) Vol. **81**, No. 1, p. K67–K70 (1977).
See also *13.1*, p. *117*; 14.3, p. 168.

Y. Guldner, C. Rigaux, A. Mycielski, and Y. Couder
Magnetooptical Investigation of $Hg_{1-x}Cd_xTe$ Mixed Crystals (I): Semimetallic Configuration
(b) Vol. **81**, No. 2, p. 615–627 (1977).
See also 13.1, p. 118; *18*, p. *210*; *20.1*, p. *255*.

Y. Guldner, C. Rigaux, A. Mycielski, and Y. Couder
Magnetooptical Investigation of $Hg_{1-x}Cd_xTe$ Mixed Crystals (II): Semiconducting Configuration and Semimetal → Semiconductor Transition
(b) Vol. **82**, No. 1, p. 149–158 (1977).
See also 13.1, p. 118; *18*, p. *211*; *20.1*, p. *256*.

H. Rüfer and W. Ruppel
Luminescence of Graded $Zn_xCd_{1-x}S$ Single Crystals
(a) Vol. **42**, No. 2, p. 519–524 (1977).
See also 13.5.1, p. 146; *20.2*, p. *266*; *20.3*, p. *272*.

J. Szatkowski, K. Sierański, J. M. Pawlikowski, E. Płaczek-Popko, P. Becla, and E. Dudziak

Electrical Transport Properties of Epitaxial Graded-Gap $Cd_xHg_{1-x}Te$ Layers
(a) Vol. **42**, No. 2, p. 721–727 (1977).
See also 13.1, p. 121; *14.3.1*, p. *174*.

Nguyen The Khoi and J. A. Gaj

Fundamental Absorption Edge of $Cd_{1-x}Mn_xTe$ Mixed Crystals
(b) Vol. **83**, No. 2, p. K133–K135 (1977).
See also 13.5.1, p. 147; *20.1*, p. *261*.

22.5 Halides

Review Article

A. Goldmann
Band Structure and Optical Properties of Tetrahedrally Coordinated Cu- and Ag-Halides
(b) Vol. **81**, No. 1, p. 9–47 (1977).
See also *13.1*, p. *114*; 13.5.1, p. 143; *20.1*, p. *247*; 22.5.1, p. 373.

Original Papers and Short Notes

I. Pollini
Ultraviolet Reflectivity Spectra of $CoCl_2$ and $CoBr_2$ Crystals
(b) Vol. **79**, No. 1, p. 231–236 (1977).
See also *13.1*, p. *114*; *20.1*, p. *247*.

E. Mohler and B. Thomas
Electro-Optic Properties of CuCl below the Absorption Edge
(b) Vol. **79**, No. 2, p. 509–517 (1977).
See also *14.4*, p. *189*; *20.1*, p. *249*.

I. N. Douglas
Magnetic Circular Dichroism of Co^{2+} Ions in CdF_2 and CaF_2
(b) Vol. **79**, No. 2, p. K99–K101 (1977).
See also 13.4, p. 131; *18*, p. *209*; *20.1*, p. *249*; 22.5.3, p. 383.

W. Ekardt and M. I. Sheboul
LO-Phonon-Assisted Two-Photon Absorption of the First Excited State of the Biexciton in CuCl
(b) Vol. **80**, No. 1, p. 51–56 (1977).
See also *13.5.1*, p. *144*; 20.1, p. 250; 20.3, p. 268.

A. D. Brothers and D. M. Wieliczka
Temperature and Hydrostatic Pressure Measurements of the Exciton Peaks in BiI_3 and SbI_3
(b) Vol. **80**, No. 1, p. 201–206 (1977).
See also 12.2, p. 104; 13.1, p. 115; *13.5.1*, p. *144*.

A. A. Haav, E. Peljo, and P. Suortti
Thermal Vibration Amplitudes of TlCl between 296 and 534 K
(b) Vol. **80**, No. 1, p. 255–264 (1977).
See also 4, p. 44; *6*, p. *53*.

H. J. L. van der Valk and C. Haas
Anisotropy of Lattice Vibrations of Layered Compounds
(b) Vol. **80**, No. 1, p. 321–332 (1977).
See also *6*, p. *53*; *20.1*, p. *251*; 22, p. 322.

A. Murasik, J. Leciejewicz, and Z. Żołnierek
An Observation of the Crystal-Field Transition in UCl_4 by Neutron Spectroscopy
(b) Vol. **80**, No. 2, p. K137–K140 (1977).
See also *13*, p. *108*; *20.1*, p. *252*.

A. F. Halff and J. Schoonman
An Experimental Investigation of the Photoconductivity of Lead Halides
(a) Vol. **40**, No. 2, p. 511–518 (1977).
See also *16*, p. *202*; 22.5.4, p. 385.

F. Bechstedt and F. Henneberger
Two-Photon Resonance Raman Scattering by Excitons via an Intermediate Excitonic Molecule (I): Bare-Exciton Approach
(b) Vol. **81**, No. 1, p. 211–220 (1977).
See also *13.5.1*, p. *145*; *20.3*, p. *270*; 22.4.1, p. 363.

V. K. Bashenov, I. Baumann, and V. V. Timofeenko
Upper Valence Bands of Rutile-Type Fluorides
(b) Vol. **81**, No. 1, p. K55—K58 (1977).
See also *13.1*, p. *117*.

E. Ostertag and J. B. Grun
Subnanosecond Time-Resolved Spectroscopy of the Biexciton Luminescence in CuCl
(b) Vol. **82**, No. 1, p. 335—342 (1977).
See also *13.5.1*, p. *146*; *20.3*, p. *271*.

K. K. Bajaj and C. Aldrich
Effective Electron–Hole Interaction for Intermediate and Strong Electron–Phonon Coupling — A Comment
(b) Vol. **82**, No. 2, p. 663—668 (1977).
See also *13.5.1*, p. *146*; 13.5.3, p. 152; 22.6, p. 390.

M. Moreno
A Study on the Magnetic Properties of Compressed D_{4h} Clusters
(b) Vol. **82**, No. 2, p. 669—676 (1977).
See also *13.4*, p. *136*; *19*, p. *240*.

V. A. Voloshin and M. G. Krimnus
Effect of Pressure on the Spectrum of Praseodymium-Doped Single Crystals
(b) Vol. **82**, No. 2, p. 695—700 (1977).
See also *12.2*, p. *105*; *20.1*, p. *258*; 22.8.1, p. 406.

V. Ya. Altshul, V. K. Bashenov, and V. V. Timofeenko
Upper Valence Bands of Perobskite- and Tysonite-Type Fluorides
(b) Vol. **82**, No. 2, p. K99—K101 (1977).
See also *13.1*, p. *120*.

R. Mach and E. U. Messerschmidt
Physical Properties of the Au–CdF_2 Metal–Semiconductor Contact
(a) Vol. **42**, No. 2, p. K187—K190 (1977).
See also *14.3.4*, p. *186*; 16, p. 204.

M. B. Haxton
The Phonon Spectra of β-CuI (I): The Model
(b) Vol. **83**, No. 1, p. 147—154 (1977).
See also *6*, p. *58*.

S. Lewonczuk, J. G. Gross, M. A. Khan, and J. Ringeissen
Optical Properties of Cuprous Halides in the 4 to 30 eV Range
(b) Vol. **83**, No. 1, p. 161—168 (1977).
See also *13.1*, p. *121*; *20.1*, p. *259*.

E. Ildstad, I. Svare, and T. A. Fjeldly
Diffusion in Lanthanum Fluoride Studied with ^{19}F Relaxation
(a) Vol. **43**, No. 1, p. K65—K68 (1977).
See also *9*, p. *70*; *19*, p. *241*.

A. Blumen and C. Merkel
Energy Band Calculations on Helical Systems
(b) Vol. **83**, No. 2, p. 425—431 (1977).
See also *13.1*, p. *121*; 14.4.1, p. 196.

F. Henneberger, K. Henneberger, and J. Voigt
Two-Photon Resonance Raman Scattering by Excitons via an Intermediate Excitonic Molecule (II): Polarization Properties and Polariton Effects
(b) Vol. **83**, No. 2, p. 439—450 (1977).
See also *13.5.1*, p. *147*; *20.3*, p. *273*; 22.4.1, p. 364.

M. B. Haxton
The Phonon Spectra of β-CuI (II): The Numerical Evaluation
(b) Vol. **83**, No. 2, p. 529—536 (1977).
See also *6*, p. *59*.

B. Hönerlage, Yu Duy Phach, A. Bivas, and E. Ostertag
Polariton Effects in Resonant Raman Scattering in CuCl
(b) Vol. **83**, No. 2, p. K101—K103 (1977).
See also *13.5*, p. *141*; *20.3*, p. *273*.

A. Bivas, Vu Duy Phach, B. Hönerlage, and J. B. Grun
Induced Absorption Due to Exciton–Biexcitons Transitions in CuCl
(b) Vol. **48**, No. 1, p. 235—243 (1977).
See also *13.5.1*, p. *148*; *20.3*, p. *274*.

M. L. Falin, V. P. Meiklyar, and V. A. Ulanov
Transferred Hyperfine Interaction of Yb^{3+} in $KMgF_3$
(b) Vol. **84**, No. 1, p. K29—K32 (1977).
See also *19*, p. *242*.

E. Ostertag, A. Bivas, and J. B. Grun
Polariton Effects on the Recombination of Cold Biexcitons
(b) Vol. **84**, No. 2, p. 673—680 (1977).
See also *13.5.1*, p. *148*; *20.3*, p. *275*.

Vu Duy Phach, A. Bivas, B. Hönerlage, and J. B. Grun
Biexciton Resonant Two-Photon Absorption in CuCl
(b) Vol. **84**, No. 2, p. 731—740 (1977).
See also *13.5.1*, p. *148*; *20.1*, p. *263*.

A. Truţia, F. Iova, and M. Vodă
Concentration of "Impurity" Ion Clusters in a Two-Dimensional Host Lattice
(b) Vol. **84**, No. 2, p. 817—824 (1977).
See also *10.2*, p. *92*; *20.1*, p. *263*.

A. Anedda and E. Fortin
Excitons in HgI_2
(b) Vol. **84**, No. 2, p. K87—K90 (1977).
See also *13.5.1*, p. *148*; 20.1, p. 263.

G. P. M. van den Heuvel and G. Blasse
The Blue Emission of $PbCl_2$
(a) Vol. **44**, No. 2, p. K163—K166 (1977).
See also *20.3*, p. *275*.

22.5.1 Silver Halides

Review Articles

A. Goldmann
Band Structure and Optical Properties of Tetrahedrally Coordinated Cu- and Ag-Halides
(b) Vol. **81**, No. 1, p. 9—47 (1977).
See also *13.1*, p. *114*; 13.5.1, p. 143; *20.1*, p. *247*; 22.5, p. 370.

K. Shahi
Transport Studies on Superionic Conductors
(a) Vol. **41**, No. 1, p. 11—44 (1977).
See also 9, p. 69; 10.2, p. 83; *14.4*, p. *189*; 15, p. 198; 22.8, p. 396; 22.9, p. 414.

Original Papers and Short Notes

E. M. Itskovich, A. A. Kornyshev, and M. A. Vorotyntsev
Electric Current across the Metal–Solid Electrolyte Interface (I): Direct Current, Current–Voltage Characteristic
(a) Vol. **39**, No. 1, p. 229—238 (1977).
See also *14.3.2*, p. *176*; 14.3.4, p. 183.

M. E. van Hulle and W. Maenhout-van der Vorst
Ionic Conductivity and Space Charge Region in Silver Bromide (111) Microcrystals
(a) Vol. **39**, No. 1, p. 253—258 (1977).
See also *10.2*, p. *83*; *14.4*, p. *189*.

W. Bollmann
Evaluation of the Association Parameters of Quadripoles (Dimers of Dipoles) in Ionic Crystals from Conductivity Investigations (I): General
(a) Vol. **40**, No. 1, p. 49—53 (1977).
See also *10.2*, p. *85*; 22.5.2, p. 376; 22.5.3, p. 383.

M. E. van Hulle
Dielectric Loss Measurements of AgBr (100) Emulsions
(a) Vol. **40**, No. 1, p. K57—K60 (1977).
See also *10.2*, p. *86*; *14.4*, p. *190*.

M. E. van Hulle and W. Maenhout-van der Vorst
Ionic Conductivity of AgBr and AgBr:Pb^{2+} Single Crystals
(a) Vol. **40**, No. 2, p. K173—K177 (1977).
See also *10.2*, p. *87*; *14.4*, p. *190*.

M. Suzuki and H. Okazaki
The Structure of α-AgI
(a) Vol. **42**, No. 1, p. 133—140 (1977).
See also *1.1*, p. *12*.

I. Pelant and P. Sladký
The Ratio of Photoelastic Constants p_{11}/p_{12} of AgCl
(a) Vol. **42**, No. 1, p. K57—K59 (1977).
See also 7, p. 63; *12.2*, p. *105*; *20.1*, p. *257*.

A. V. Lelyakov
Luminescence Band Shape of Free Excitons in AgBr Crystals at High Exciton Density
(b) Vol. **82**, No. 2, p. 711—716 (1977).
See also 6, p. 58; *13.5.1*, p. *146*; *20.3*, p. *271*.

A. L. Kartuzhanskii, A. G. Kavetskii, L. D. Kekhva, T. E. Kekhva, A. P. Kiselev, and B. T. Plachenov
Recombination Processes during Excitation of AgCl:Me^{2+} Crystals by Ionizing Radiation
(a) Vol. **43**, No. 1, p. 321—327 (1977).
See also *13.4*, p. *137*; *20.3*, p. *273*.

V. Zhelev and J. Malinowski
Ionic Field Effect in Epitaxial AgBr Layers
(a) Vol. **43**, No. 2, p. 675—683 (1977).
See also 13.3, p. 128; *14.3.4*, p. *187*.

M. E. van Hulle and W. Maenhout-van der Vorst
Space Charge Characteristics of Silver Halide Microcrystals
(a) Vol. **44**, No. 1, p. 229—236 (1977).
See also *10.2*, p. *92*; *14.4*, p. *193*.

22.5.2 Alkali Halides

Original Papers and Short Notes

M. F. TRINKLER and I. S. ZOLOVKINA
A-Luminescence of Alkali Halides Activated by Monovalent Mercury-Like Ions
(b) Vol. **79**, No. 1, p. 49–59 (1977).
See also *13.4*, p. *129*; *20.3*, p. *267*.

M. N. SHARMA, A. KANT, and K. D. MISRA
A New Approach to Overlap Repulsion
(b) Vol. **79**, No. 1, p. 359–364 (1977).
See also *4*, p. *44*; 12.2, p. 104.

T. IIDA and R. MONNIER
Erratum to: Static Properties of the Self-Traped Hole in Simple Cubic Cesium Halides
(b) Vol. **79**, No. 1, p. 391 (1977).
See also 6, p. 52; *13.4*, p. *129*; 20.1, p. 248.

P. MATTHYS, M. VANHAELST, and E. BOESMAN
Spin–Lattice Relaxation of the S_2^- Centre in KI
(b) Vol. **79**, No. 1, p. K23–K26 (1977).
See also *19*, p. *235*.

P. G. BARANOV, YU. P. VESHCHUNOV, R. A. ZHITNIKOV, and N. G. ROMANOV
EPR, Optical Absorption, and Luminescence of Anisotropic Copper Centres in KCl
(b) Vol. **79**, No. 1, p. K27–K30 (1977).
See also *19*, p. *235*; 20.1, p. 248; 20.3, p. 267.

V. BELZNER and F. GRANZER
Atomistic Calculations on the Dissociation of an Edge Dislocation in NaCl
(a) Vol. **39**, No. 1, p. 183–187 (1977).
See also *10.2*, p. *83*.

G. BERG, M. DUBIEL, and F. FRÖHLICH
T_1-Measurements in Ca-Doped NaCl Crystals
(a) Vol. **39**, No. 1, p. 265–269 (1977).
See also *10.2*, p. *83*; 19, p. 235.

V. I. SAVENKO, L. A. KACHANOVA, and E. D. SHCHUKIN
The Generation of Dislocations near the Surfaces of Crystals under Microindentation
(a) Vol. **39**, No. 1, p. K11–K15 (1977).
See also *1.6*, p. *31*; *10.2*, p. *83*; 12.2, p. 104.

M. IKEYA and J. H. CRAWFORD, JR.
Recombination Luminescence in Alkali Halides Doped with Monovalent Cations. $(V_K + e)_A$-Type Emission
(b) Vol. **79**, No. 2, p. 559–566 (1977).
See also 10.2, p. 84; *20.3*, p. *267*.

A. Y. S. KUNG and J. M. VAIL
A Molecular Model for the Emission Process of F_A(II)-Centres in Alkali Halides
(b) Vol. **79**, No. 2, p. 663–672 (1977).
See also *10.2*, p. *84*; 13.4, p. 130; 20.3, p. 268.

A. F. BURENKOV and F. F. KOMAROV
The Effect of Atomic Chemical Bonding under Fast Ion Stopping in Solids
(b) Vol. **79**, No. 2, p. K161–K164 (1977).
See also *11*, p. *94*.

J. PRAKASH and F. FISCHER
Ionic Thermocurrent and Optical Absorption Studies of S^{--}–Vacancy Dipoles in KI
(a) Vol. **39**, No. 2, p. 499–507 (1977).
See also *10.2*, p. *84*; *14.4*, p. *189*; 20.1, p. 250.

R. Albrecht, V. Schmidt, and V. I. Betekhtin
The Damage Process Preceding Semi-Brittle Fracture in Dependence on Deformation (I): Scanning Microscopic In-Situ Studies of Microcracks in NaCl Crystals
(a) Vol. **39**, No. 2, p. 621–630 (1977).
See also *10.2*, p. *84*; *12.2*, p. *104*.

M. Yacamán, L. W. Hobbs, and M. J. Goringe
The Direct Observation of Suzuki-Phase Precipitates in NaCl:$MnCl_2$ by Transmission Electron Microscopy
(a) Vol. **39**, No. 2, p. K85–K90 (1977).
See also *1.4*, p. *24*; *10.2*, p. *85*.

K. V. Rao and K. Suryanarayana Rao
Variation of the Dielectric Constant with Temperature in Ionic Crystals
(a) Vol. **39**, No. 2, p. K119–K121 (1977).
See also *14.4*, p. *189*.

E. L. Sill, J. J. Martin, and Z. Al Shaieb
Optical Determination of the Eu Concentration in Eu Doped KCl Crystals
(a) Vol. **39**, No. 2, p. K131–K133 (1977).
See also *20.1*, p. *250*.

V. K. Garg, D. S. Puri, and M. P. Verma
On the Third-Order Elastic Constants of Ionic Solids with NaCl Structure
(b) Vol. **80**, No. 1, p. 63–72 (1977).
See also *10.2*, p. *85*; *12.2*, p. *104*.

H. J. Hackelöer, H. Selbach, O. Kanert, A. W. Sleeswyk, and G. Hut
Determination of the Velocity of Mobile Dislocations by Nuclear Spin Relaxation Measurements
(b) Vol. **80**, No. 1, p. 235–243 (1977).
See also *10.2*, p. *85*; 19, p. 236.

L. Mărculescu
Luminescence of Lead-Activated Sodium Chloride Single Crystals
(b) Vol. **80**, No. 1, p. 265–272 (1977).
See also 13.4, p. 131; *20.3*, p. *268*.

G. S. Zavt
Dynamical Properties of NO_2^- Impurity Molecules in Alkali Halides (I): Theory
(b) Vol. **80**, No. 1, p. 399–408 (1977).
See also *10.2*, p. *85*; 20.1, p. 251.

E. Mugeński and J. Z. Damm
On the 420 nm Emission Band in Pb^{2+} Doped KCl Crystals
(b) Vol. **80**, No. 1, p. K59–K62 (1977).
See also *20.3*, p. *268*.

W. Bollmann
Evaluation of the Association Parameters of Quadripoles (Dimers of Dipoles) in Ionic Crystals from Conductivity Investigations (I): General
(a) Vol. **40**, No. 1, p. 49–53 (1977).
See also *10.2*, p. *85*; 22.5.1, p. 373; 22.5.3, p. 383.

J. A. Jordão, A. Sakamoto, and M. de Souza
The Ultrasonic Attenuation due to Cu^+ Substitutional Ions in KCl
(a) Vol. **40**, No. 1, p. 109–112 (1977).
See also 7, p. 62; *10.2*, p. *85*.

R. Albrecht, V. Schmidt, and V. I. Betekhtin
The Damage Process Preceding Semi-Brittle Fracture in Dependence on Deformation (II): A Discussion on Results of Microcrack Development in NaCl Crystals
(a) Vol. **40**, No. 1, p. 147–152 (1977).
See also *10.2*, p. *85*; *12.2*, p. *104*.

A. A. Urusovskaya, N. L. Sizova, and I. A. Rachkov
Nature of the Yield Point in CsI Crystals
(a) Vol. **40**, No. 1, p. K1–K6 (1977).
See also *10.2*, p. *85*.

C. J. Delbecq, J. P. Stokes, and P. H. Yuster
Thermal Reorientation Kinetics of I_2^- in RbI
(b) Vol. **80**, No. 2, p. 471–475 (1977).
See also *10.2*, p. *86*; 20.1, p. 252.

S. C. Zílio and M. de Souza
The Dielectric Relaxation of Impurity–Vacancy Complexes in KI Doped with Cl^- Anions
(b) Vol. **80**, No. 2, p. 597–601 (1977).
See also *10.2*, p. *86*; 14.4, p. 190.

A. I. Anselm and I. G. Lang
Theory of the Excitonic Mechanism of Crystal Defect Formation
(b) Vol. **80**, No. 2, p. 691–695 (1977).
See also *10.2*, p. *86*; *13.5.1*, p. *144*.

A. Watterich and K. Raksányi
Optical and ESR Investigations of Mg Centres in X-Irradiated and F-Bleached KCl Crystals at RT
(b) Vol. **80**, No. 2, p. K105–K108 (1977).
See also *10.2*, p. *86*; 19, p. 237; *20.1*, p. *252*.

L. Kemter and H. Strunk
Dislocation Densities in Deformed NaCl Single Crystals Determined by Transmission Electron Microscopy
(a) Vol. **40**, No. 2, p. 385–391 (1977).
See also *10.2*, p. *86*.

G. Chassagne, D. Durand, J. Serughetti, and L. W. Hobbs
The Behaviour of Alkali Metal Colloids in Additively Coloured Potassium Chloride Crystals (I): Study by Optical Absorption and Surface Replicas — Colloid Size Distribution
(a) Vol. **40**, No. 2, p. 629–642 (1977).
See also *10.2*, p. *87*; 20.1, p. 253.

V. G. Kvachadze, M. F. Zhvaniya, and T. Sh. Davitashvili
On the Nature of the Radiative Centres of the Halogen Impurity
(a) Vol. **40**, No. 2, p. K113–K115 (1977).
See also 8, p. 65; *10.2*, p. *87*; *20.1*, p. *253*.

B. Jarosz
Dislocation Charge in X-Irradiated NaCl
(a) Vol. **40**, No. 2, p. K149–K151 (1977).
See also *10.2*, p. *87*.

L. A. Rebane, G. S. Zavt, and K. E. Haller
Dynamical Properties of NO_2^- Impurity Molecules in Alkali Halides (II): Vibrational Spectra
(b) Vol. **81**, No. 1, p. 57–68 (1977).
See also *10.2*, p. *87*; 20.1, p. 253.

A. Freiberg and L. A. Rebane
Peculiarities of the Electron–Phonon Interaction in the Spectra of the Molecular Ions O_2^-, S_2^- and NO_2^- in Alkali Halide Crystals
(b) Vol. **81**, No. 1, p. 359–369 (1977).
See also *6*, p. *55*; *20.3*, p. *270*.

C. M. Goel and B. Dayal
Lattice Dynamics of Rubidium Halide Crystals
(b) Vol. **81**, No. 1, p. 403–407 (1977).
See also *6*, p. *55*.

H. Overeijnder, M. Szymoński, A. Haring, and A. E. de Vries
The Mechanism of the Electron Sputtering Process of Alkali Halides
(b) Vol. **81**, No. 1, p. K11–K14 (1977).
See also *10.2*, p. *87*; *17*, p. *206*.

T. P. Martin and H. Schaber
Infrared Absorption of Aggregated and Isolated Microcrystals
(b) Vol. **81**, No. 1, p. K41–K45 (1977).
See also 6, p. 55; *20.1*, p. *254*.

G. Chassagne, D. Durand, J. Serughetti, and L. W. Hobbs
The Behaviour of Alkali Metal Colloids in Additively Coloured Potassium Chloride Crystals (II): Study by Transmission Electron Microscopy — Colloid Nucleation and Growth
(a) Vol. **41**, No. 1, p. 183–190 (1977).
See also *10.2*, p. *87*.

K. Ozawa and A. Kikuchi
Induced Flow Stress and Optical Absorption in NaCl Crystals Irradiated with Neutrons at Low Temperatures
(a) Vol. **41**, No. 1, p. 289–294 (1977).
See also *10.2*, p. *88*; 11, p. 95; *20.1*, p. *254*.

D. Durand and F. Luty
Raman Scattering from Tunnelling and Hindered Rotational States of CN^- Molecules in KCl
(b) Vol. **81**, No. 2, p. 443–450 (1977).
See also *6*, p. *56*; *20.1*, p. *254*.

S. Benci, M. P. Fontana, and M. Manfredi
On the Excited States of the Tl^+ Monomer Centre in NaI
(b) Vol. **81**, No. 2, p. 603–608 (1977).
See also *13.4*, p. *134*; 20.1, p. 255; 20.3, p. 270.

T. Chiba, G. B. Dürr, and W. Brandt
Enhancement Effect on Positron Annihilation with Tightly Bound Electrons
(b) Vol. **81**, No. 2, p. 609–614 (1977).
See also *13*, p. *109*; 22.6, p. 388.

A. A. Urusovskaya, N. L. Sizova, I. A. Rachkov, Ya. A. Zakharin, and V. M. Dobryak
Influence of Impurities on the Mechanical Properties of CsI Crystals
(a) Vol. **41**, No. 2, p. 443–450 (1977).
See also *10.2*, p. *88*.

S. Radhakrishna and S. Haridoss
Methods of Analysis of Piezo-Stimulated Current Spectra
(a) Vol. **41**, No. 2, p. 649–652 (1977).
See also *10.2*, p. *88*; *14.4*, p. *191*.

G. A. Andreev, M. Hartmanová, and V. A. Klimov
Precipitation in Nickel-Doped Sodium Chloride Crystals
(a) Vol. **41**, No. 2, p. 679–685 (1977).
See also *10.2*, p. *89*.

J. Peña and J. M. Cabrera
About Additive Colouration of KCl:Pb
(a) Vol. **41**, No. 2, p. K119–K121 (1977).
See also *10.2*, p. *89*; 20.1, p. 256.

H.-J. Kos and R. Nink
Z_1 Centres in Mg-Doped Lithium Fluoride
(a) Vol. **41**, No. 2, p. K157–K161 (1977).
See also *10.2*, p. *89*; 13.4, p. 135; *20.1*, p. *256*.

T. Nagarajan, P. Ramasamy, and S. Ramasamy
Positron Annihilation in Colour Centres
(b) Vol. **82**, No. 1, p. 75—81 (1977).
See also *10.2*, p. *89*.

V. K. Garg and M. P. Verma
Evaluation of Some Important Lattice Sums on NaCl and CsCl Structures
(b) Vol. **82**, No. 1, p. 139—143 (1977).
See also *1.1*, p. *11*; 12.2, p. 105.

S. A. Marshall
Comments on the Magnetic Resonance Absorption Spectrum of the F_2^- Molecule-Ion in Single Crystals of LiF, NaF, and KF
(b) Vol. **82**, No. 1, p. 311—317 (1977).
See also *19*, p. *240*.

O. P. Sharma, A. P. Gupta, and Jai Shanker
Effect of the Three Body Interactions on the Cohesive Energies of Alkali Halides
(b) Vol. **82**, No. 1, p. K13—K16 (1977).
See also *4*, p. *45*; 12.2, p. 105.

L. Bosi and M. Nimis
On the Decay of the M_A Centre in KCl:Li and KCl:Na
(b) Vol. **82**, No. 1, p. K49—K51 (1977).
See also *10.2*, p. *89*; *13.4*, p. *136*; 20.3, p. 271.

M. N. Sharma
A Simple Expression for the Anderson-Grüncisen Parameter
(b) Vol. **82**, No. 1, p. K53—K58 (1977).
See also *8*, p. *66*.

G. Berg, F. Fröhlich, and D. Schneider
On the Existence of Charged Aggregates of Divalent Impurities in NaCl-Type Crystals (Point Charge Approximation)
(a) Vol. **42**, No. 1, p. 73—77 (1977).
See also *10.2*, p. *89*.

C. L. Reynolds, Jr. and P. R. Couchman
Tait's Equation of State at High Pressure and the Isothermal Bulk Modulus
(a) Vol. **42**, No. 1, p. K47—K52 (1977).
See also *12.1*, p. *102*; *12.2*, p. *105*; 21, p. 282; 23, p. 420.

U. Brunsmann and A. Scharmann
A Simulation of the Maxwellian Exoelectron Energy Distributions with Regard to Patch Fields
(a) Vol. **42**, No. 1, p. K79—K81 (1977).
See also *17*, p. *206*.

V. K. Garg, D. S. Puri, and M. P. Verma
Many-Body Effects on the Fourth-Order Elastic Constants of Ionic Solids (II): Applications
(b) Vol. **82**, No. 2, p. 481—488 (1977).
See also *4*, p. *45*; *12.2*, p. *105*.

S. K. Chakrabarti
On the Polymorphic Phase Transition in Alkali Halides
(b) Vol. **82**, No. 2, p. 557—564 (1977).
See also *1.2*, p. *17*; *12.2*, p. *105*.

C. L. Reynolds, Jr.
Influence of Pressure on the Melting Temperature of Alkali Halides
(a) Vol. **42**, No. 2, p. K151—K153 (1977).
See also *8*, p. *66*.

R. K. Gartia
Optical Bleaching — a Technique to Clean Thermoluminescence Peaks
(a) Vol. **42**, No. 2, p. K155—K158 (1977).
See also *10.2*, p. *90*; *20.3*, p. *272*.

G. R. Augst
Distortion of Ionic Crystals by Vacancies
(b) Vol. **83**, No. 1, p. 55—61 (1977).
See also *10.2*, p. *90*.

V. Lupei, A. Lupei, and S. Georgescu
EPR of U^{5+} in Alkali Fluorides with a New Charge Compensation Mechanism
(b) Vol. **83**, No. 1, p. 71—75 (1977).
See also *19*, p. *241*.

M. Kumar, P. S. Mahesh, and T. P. Sharma
Second-Order Raman Spectra and Phonon Spectrum of CsF
(b) Vol. **83**, No. 1, p. 155—159 (1977).
See also *6*, p. *58*; *20.1*, p. *259*.

J. M. Herreros and F. Jaque
F-Thermal Bleaching of X-Irradiated NaCl
(b) Vol. **83**, No. 1, p. K21—K23 (1977).
See also *10.2*, p. *90*.

W. Bollmann
Extrema of the Temperature-Dependent Degree of Association of Point Defects in Ionic Crystals
(a) Vol. **43**, No. 1, p. 175—184 (1977).
See also *10.2*, p. *90*; 22.5.3, p. 384.

G. Busse and A. Pfaller
Stress Spectroscopy of a Resonant Bandmode with a Far Infrared Laser
(a) Vol. **43**, No. 1, p. K103—K107 (1977).
See also 12.2, p. 106; *20.1*, p. *260*.

A. Vanderhaeghe and G. Jacobs
On F- and F_H-Centres in Cesium Halides
(b) Vol. **83**, No. 2, p. 471—477 (1977).
See also *10.2*, p. *90*; 20.3, p. 273.

M. A. B. Whitaker
Modes of Cross-Relaxation in Nuclear Double-Resonance
(b) Vol. **83**, No. 2, p. 573—584 (1977).
See also *19*, p. *241*.

S. Paul and S. Sengupta
A Simple Model for the Calculation of Heat of Formation for High-Concentration Substitutional Defects in Alkali Halide Crystals
(b) Vol. **83**, No. 2, p. 645—650 (1977).
See also *10.2*, p. *91*.

U. Brunsmann and A. Scharmann
Thermally Stimulated Exoelectron Emission of NaF. Electron Bombardment and Exoelectron Energy Investigations
(a) Vol. **43**, No. 2, p. 519—528 (1977).
See also *17*, p. *207*.

N. I. Konyushkina, I. V. Krylova, V. N. Opekunov, and A. A. Predvoditelev
Thermostimulated Exoemission from LiF
(a) Vol. **43**, No. 2, p. 639—644 (1977).
See also *17*, p. *207*.

M. Casalboni, U. M. Grassano, A. Scacco, and A. Tanga
Emission Spectra of NaBr:Tl^+ Excited in the A-Band
(b) Vol. **84**, No. 1, p. 23–26 (1977).
See also *13.4*, p. *138*; *20.3*, p. *274*.

M. A. B. Whitaker
Calculation of Nuclear Double-Resonance Relaxation Rate
(b) Vol. **84**, No. 1, p. 149–158 (1977).
See also *19*, p. *242*.

H. R. Anand
Mössbauer Study of Dispersed Iron Ions in Alkali Halides
(b) Vol. **84**, No. 1, p. 227–233 (1977).
See also *5*, p. *49*.

S. Radhakrishna and V. S. Sivasankar
F_{Z_1} Centres in Cesium Halides
(a) Vol. **44**, No. 1, p. 345–351 (1977).
See also *10.2*, p. *92*; 20.1, p. 262.

R. K. Gartia
Thermoluminescence of Z_3 Centres in X-Irradiated Mg-Doped LiF Crystals
(a) Vol. **44**, No. 1, p. K21–K24 (1977).
See also *10.2*, p. *92*; *20.3*, p. *274*.

V. I. Zaitsev, V. I. Barbashov, and Yu. B. Tkachenko
Effect of High Hydrostatic Pressure on the Dislocation Mobility in NaCl Crystals
(a) Vol. **44**, No. 1, p. K39–K41 (1977).
See also *10.2*, p. *92*; *12.2*, p. *106*.

M. N. Sharma and A. Kant
A New Approach to Lattice Properties of Alkali Halide Crystals
(b) Vol. **84**, No. 2, p. 427–434 (1977).
See also *4*, p. *46*; *8*, p. *68*; 20.1, p. 262.

M. Aguilar, F. Jaque, and F. Agulló-López
X-Ray Induced Luminescence in NaCl:Ca^{++}
(b) Vol. **84**, No. 2, p. 595–601 (1977).
See also 10.2, p. 92; *20.3*, p. *275*.

P. Hennl
Low Temperature Photochemistry on Chalcogen–Vacancy Centres in Alkali Halide Crystals
(b) Vol. **84**, No. 2, p. K151–K154 (1977).
See also *10.2*, p. *92*; *20.1*, p. *263*.

E. V. Shvedov, V. S. Postnikov, and V. M. Ievlev
Nucleus Saturation Density and Epitaxy in the Condensation of Metals on Alkali-Halide Crystals (I): Saturation Density
(a) Vol. **44**, No. 2, p. 423–427 (1977).
See also *3*, p. *43*; *10.2*, p. *92*.

V. Z. Bengus and T. P. Kovalenko
Mean Mobile Dislocation Velocity in LiF Crystals under Strain
(a) Vol. **44**, No. 2, p. 457–465 (1977).
See also *10.2*, p. *93*.

H.-J. Kos and R. Nink
Mechanism of Restored Thermoluminescence in Lithium Fluoride
(a) Vol. **44**, No. 2, p. 505–510 (1977).
See also *10.2*, p. *93*; 20.1, p. 263; *20.3*, p. *275*.

E. A. van den Bergen, R. Delhez, and E. J. Mittemeijer
An X-Ray Diffraction Study of Diffusional Homogenization in RbCl–KCl Powder Blends
(a) Vol. **44**, No. 2, p. 517—524 (1977).
See also 1, p. 10; *9*, p. *71*.

S. K. Mehta, J. F. Merklin, and H. J. Donnert
Thermoluminescence Related Z Centers in LiF:Mg, Ti
(a) Vol. **44**, No. 2, p. 679—685 (1977).
See also *10.2*, p. *93*; *20.1*, p. *264*.

Ah Mee Hor and P. W. M. Jacobs
Thermal Depolarization in Crystals of Rubidium Bromide Doped with Ba^{2+} and Sn^{2+}
(a) Vol. **44**, No. 2, p. 725—729 (1977).
See also *10.2*, p. *93*.

V. S. Sivasankar and S. Radhakrishna
Electric Field Gradient at the F_{Z_1} Electron Site in Cesium Halides
(a) Vol. **44**, No. 2, p. 741—745 (1977).
See also *10.2*, p. *93*; 13.4, p. 140.

V. R. Dumke
Image Effect in Previously Electrolytically Coloured KCl–OH^-
(a) Vol. **44**, No. 2, p. K133—K134 (1977).
See also *10.2*, p. *93*; 20.1, p. 264.

22.5.3 Alkali-Earth Halides

Original Papers and Short Notes

J. M. DE SIEBENTHAL and H. BILL
A Perturbed O_2^- Molecule Ion in $SrCl_2$ Single Crystals. EPR Results
(b) Vol. **79**, No. 1, p. 259—267 (1977).
See also *10.2*, p. *83*; 19, p. 235.

A. VANDERHAEGHE and G. JACOBS
The Lifetime of the F_H Centre in CsBr:Cl and CsBr:F Crystals
(b) Vol. **79**, No. 1, p. K75—K79 (1977).
See also 10.2, p. 83; *20.3*, p. *267*.

J. C. GÂCON, J. C. SOUILLAT, J. SÉRIOT, and B. DI BARTOLO
Radiative and Radiationless Transition Rate Measurements Using a Tunable Pulsed Dye Laser. Application to Sm^{2+} Fluorescence in BaClF
(a) Vol. **39**, No. 1, p. 147—156 (1977).
See also 13.4, p. 130; *20.3*, p. *267*.

M. V. EREMIN and A. A. KORNIENKO
The Superposition Model in Crystal Field Theory
(b) Vol. **79**, No. 2, p. 775—785 (1977).
See also 4, p. 44; *13.4*, p. *131*; 20.1, p. 249.

E. LOH, G. SAMOGGIA, E. REGUZZONI, and L. NOSENZO
Thermally Modulated Ultraviolet Absorption Spectra of $CaF_2:Ce^{3+}$ Crystals
(b) Vol. **79**, No. 2, p. 795—799 (1977).
Addendum
(b) Vol. **81**, No. 1, p. 411 (1977).
See also *13.4*, p. *131*; *20.1*, p. *249*.

I. N. DOUGLAS
Magnetic Circular Dichroism of Co^{2+} Ions in CdF_2 and CaF_2
(b) Vol. **79**, No. 2, p. K99—K101 (1977).
See also 13.4, p. 131; *18*, p. *209*; *20.1*, p. *249*; 22.5, p. 370.

V. KUMAR, N. DASS, and K. CHANDRA
On the Investigations of the Crystalline Field of Nd^{3+} and Yb^{3+} Ions at O_h Sites in Calcium Fluoride
(b) Vol. **80**, No. 1, p. K1—K4 (1977).
See also *13.4*, p. *132*.

V. K. BASHENOV, KH. S. BAGDASAROV, and V. V. TIMOFEENKO
Universal Valence Bands for Fluorite-Type Crystals
(b) Vol. **80**, No. 1, p. K17—K20 (1977).
See also *13.1*, p. *116*; 20.1, p. 251.

N. GUSKOS, J. KURIATA, and T. REWAJ
Spin-Lattice Coefficients of Eu^{2+} in BaF_2
(b) Vol. **80**, No. 1, p. K25—K27 (1977).
See also *19*, p. *236*.

W. BOLLMANN
Evaluation of the Association Parameters of Quadripoles (Dimers of Dipoles) in Ionic Crystals from Conductivity Investigations (I): General
(a) Vol. **40**, No. 1, p. 49—53 (1977).
See also *10.2*, p. *85*; 22.5.1, p. 373; 22.5.2, p. 376.

W. BOLLMANN
Evaluation of the Association Parameters of Quadripoles (Dimers of Dipoles) in Ionic Crystals from Conductivity Investigations (II): Association Parameters of $[O^{--}\text{–}F_v^-]$ Dipoles and Their Dimers in CaF_2 Crystals
(a) Vol. **40**, No. 2, p. 409—414 (1977).
See also *10.2*, p. *86*; 14.4, p. 190.

P. J. Alonso and R. Alcalá
On the Optical Absorption Spectrum of Co^{2+} in CaF_2
(b) Vol. **81**, No. 1, p. 333–339 (1977).
See also *13.4*, p. *133*; *20.1*, p. *254*.

R. Rauch
Stark Effect on Sodium-Disturbed F_2 Centres in CaF_2
(a) Vol. **41**, No. 2, p. K97–K99 (1977).
See also *10.2*, p. *89*; *20.1*, p. *256*; 20.3, p. 271.

Y. Sato
Mössbauer Effect of γ-Irradiated CaF_2 Doped with ^{57}Co
(b) Vol. **82**, No. 2, p. 611–616 (1977).
See also *5*, p. *48*; 11, p. 96.

J. Kurjata, N. Guskos, and T. Rewaj
Temperature Dependence of the Second-Order Spin–Lattice Coefficients for Eu^{2+} in CaF_2 Single Crystals
(b) Vol. **83**, No. 1, p. K79–K82 (1977).
See also *6*, p. *59*; *19*, p. *241*.

W. Bollmann
Extrema of the Temperature-Dependent Degree of Association of Point Defects in Ionic Crystals
(a) Vol. **43**, No. 1, p. 175–184 (1977).
See also *10.2*, p. *90*; 22.5.2, p. 380.

Cao-Xuân An
Determination of the Elastic Constants of $SrCl_2$ by Brillouin Scattering and Study of Their Variations with Temperature
(a) Vol. **43**, No. 1, p. K69–K72 (1977).
See also 6, p. 59; *12.2*, p. *106*.

A. Zaręba, A. Lubomirska-Wittlin, A. Szadkowski, and B. Krukowska-Fulde
Photoconversion from the Insulating to the Semiconducting State of $CdF_2:MF_3$
(b) Vol. **83**, No. 2, p. K155–K157 (1977).
See also *13.4*, p. *138*; 16, p. 204.

V. A. Krylov and W. Ulrici
Influence of an Electric Field on the Jahn-Teller Split Absorption Band $^4A_{2g} \rightarrow {}^4T_{1g}(P)$ of Co^{+2} in CdF_2 and CaF_2
(b) Vol. **84**, No. 1, p. 215–225 (1977).
See also *13.4*, p. *138*; *20.1*, p. *261*.

A. Szadkowski, A. Lubomirska-Wittlin, A. Zaręba, and B. Krukowska-Fulde
On Conversion Processes of $CdF_2:MF_3$
(a) Vol. **44**, No. 1, p. K43–K46 (1977).
See also *14.3*, p. *172*; *20.1*, p. *262*.

A. Zaręba, A. J. Nadolny, and B. Krukowska-Fulde
Thermoluminescence of Mn Centres in CdF_2 Crystals
(a) Vol. **44**, No. 1, p. K83–K85 (1977).
See also *20.3*, p. *274*.

W. Ulríci
$CdF_2:Cr^{2+}$ — an Example for an Orthorhombic Jahn-Teller Distortion
(b) Vol. **84**, No. 2, p. K155–K157 (1977).
See also *13.4*, p. *140*; 20.1, p. 263.

V. M. Orera and R. Alcalá
Optical Properties of Cation Colloidal Particles in CaF_2 and SrF_2
(a) Vol. **44**, No. 2, p. 717–723 (1977).
See also *10.2*, p. *93*; *20.1*, p. *264*.

22.5.4 Mixed Crystals

Original Papers and Short Notes

K. Irie
Temperature Dependence of the Shift of the Acceptor Level in Ferroelectric and Photoconductive SbSI
(a) Vol. **39**, No. 1, p. 313–320 (1977).
See also *13.4*, p. *130*; *14.4.1*, p. *194*.

G. A. Gledhill and J. F. Angress
Temperature Dependence of the Infrared Reflection Spectra of $RbCl_yBr_{1-y}$ Alloys
(b) Vol. **79**, No. 2, p. K107–K110 (1977).
See also *6*, p. *52*; *20.1*, p. *249*.

C. Scherer and Y. Barjhoux
Reaction Field and Dipolar Interaction in Quasi-Onedimensional Heisenberg Systems. Application to $CsNiF_3$
(b) Vol. **80**, No. 1, p. 313–320 (1977).
See also *18.1*, p. *213*.

A. F. Halff and J. Schoonman
An Experimental Investigation of the Photoconductivity of Lead Halides
(a) Vol. **40**, No. 2, p. 511–518 (1977).
See also *16*, p. *202*; 22.5, p. 370.

A. A. Antipin, A. V. Vinokurov, M. P. Davydova, S. L. Korableva, A. L. Stolov, and A. A. Fedii
Optical Spectra, EPR, and Spin-Lattice Relaxation of Yb^{3+} Ions in Crystals Having Perovskite-Type Structure
(b) Vol. **81**, No. 1, p. 287–293 (1977).
See also *13.4*, p. *133*; 19, p. 238; 20.3, p. 270.

M. V. Eremin and Yu. V. Rakitin
Channel Model in Isotropic Exchange Theory (II)
(b) Vol. **82**, No. 1, p. 221–228 (1977).
See also 1, p. 10; *13*, p. *110*; 22.6, p. 389.

T. Ohgaku and N. Takeuchi
Dielectric Loss of KCl–KBr Mixed Crystals Doped with Sr^{2+}
(a) Vol. **42**, No. 1, p. K83–K85 (1977).
See also 10.2, p. 89; *14.4*, p. *191*.

M. T. Carlos, M. Siu Li, and M. de Souza
The F Centre Half-Width in KCN–KCl Mixed Crystals
(b) Vol. **82**, No. 2, p. K103–K105 (1977).
See also *10.2*, p. *90*; *20.1*, p. *258*.

A. Ranfagni and G. Viliani
Spin-Orbit-Allowed Tunneling Splitting in the $T \times \varepsilon_g$ Jahn-Teller Problem. Splitting of the Zero-Phonon Line
(b) Vol. **84**, No. 1, p. 393–399 (1977).
See also *13.4*, p. *138*; 20.1, p. 261; 22.6, p. 391.

K. Heidrich, W. Schäfer, M. Schreiber, G. Trendel, and J. Treusch
Optical Properties of Polymorphous Thallous Halide Mixed Crystal Layers
(b) Vol. **84**, No. 2, p. 629–635 (1977).
See also *13.1*, p. *122*; *20.1*, p. *263*.

J. B. Hartmann and J. J. Martin
Thermal Conductivity of Irradiated $RbMgF_3$
(b) Vol. **84**, No. 2, p. 721–724 (1977).
See also *8*, p. *68*; 11, p. 99.

22.6 Simple Oxides

Review Articles

G. Borstel and H. J. Falge
Surface Phonon-Polaritons at Semi-Infinite Crystals
(b) Vol. **83**, No. 1, p. 11—45 (1977).
See also *6*, p. *51*; *13.5*, p. *141*; 20.1, p. 247; 22, p. 320.

V. T. Agekyan
Spectroscopic Properties of Semiconductor Crystals with Direct Forbidden Energy Gap
(a) Vol. **43**, No. 1, p. 11—42 (1977).
See also 6, p. 51; *13.5.1*, p. *143*; 20.1, p. 247; 20.3, p. 267.

Original Papers and Short Notes

J. C. Marmeggi, D. Hohlwein, and E. F. Bertaut
Magnetic Neutron Laue Diffraction Study of the Domain Distribution in α-Fe_2O_3
(a) Vol. **39**, No. 1, p. 57—64 (1977).
See also *18.4*, p. *231*.

B. J. Pletka, T. E. Mitchell, and A. H. Heuer
Solid Solution Hardening of Sapphire (α-Al_2O_3)
(a) Vol. **39**, No. 1, p. 301—311 (1977).
See also *10.2*, p. *83*.

A. Heidemann, K. Kosuge, Y. Ueda, and S. Kachi
Hyperfine Interaction in V_3O_7
(a) Vol. **39**, No. 1, p. K37—K40 (1977).
See also *18.4*, p. *231*.

W. Klement, Jr.
Phase Transitions in Ag_2O to 39 kbar
(a) Vol. **39**, No. 1, p. K45—K48 (1977).
See also *1.2*, p. *14*.

R. P. Goyal and S. C. Goyal
Dispersion of Phonons in Transition Metal Oxides (CoO and NiO)
(b) Vol. **79**, No. 2, p. K115—K117 (1977).
See also *6*, p. *52*.

J. F. Baumard and E. Tani
Thermoelectric Power in Reduced Pure and Nb-Doped TiO_2 Rutile at High Temperature
(a) Vol. **39**, No. 2, p. 373—382 (1977).
See also 10.2, p. 84; *15*, p. *198*.

K. H. Lee, G. E. Holmberg, and J. H. Crawford, Jr.
Optical and ESR Studies of Holes Centers in γ-Irradiated Al_2O_3
(a) Vol. **39**, No. 2, p. 669—674 (1977).
See also *10.2*, p. *84*; 19, p. 236; *20.3*, p. *268*.

F. Schneider, K. Melzer, H. Mehner, and G. Dehe
^{119}Sn Hyperfine Fields in α-Fe_2O_3
(a) Vol. **39**, No. 2, p. K115—K117 (1977).
See also 5, p. 47; *18.3*, p. *228*.

E. Schuller, H. J. Falge, and G. Borstel
Dispersion of Bulk Phonon Polaritons inside Reststrahlen Bands
(b) Vol. **80**, No. 1, p. 109—117 (1977).
See also *13.5*, p. *141*; *20.1*, p. *250*.

A. Treindl and K. H. Germann
Linear Optical Anisotropy Investigations of the Antiferromagnetic Phase Transition in MnO
(b) Vol. **80**, No. 1, p. 159–166 (1977).
See also *18.4*, p. *231*; *20.1*, p. *251*.

S. Mochizuki
Microwave Conductivity of NiO at High Temperatures
(a) Vol. **40**, No. 1, p. 119–123 (1977).
See also *14.3*, p. *167*.

K. Hübner
Chemical Bond and Related Properties of SiO_2(I): Character of the Chemical Bond
(a) Vol. **40**, No. 1, p. 133–140 (1977).
See also 2, p. 35; *4*, p. *44*; *13.1*, p. *116*; 14.4, p. 189; 20.1, p. 251.

F. Eichhorn, K. Hennig, B. Lippold, and W. Matz
Diffraction of a Pulsed Thermal Neutron Beam on an Elastically Bent Quartz Crystal Plate
(a) Vol. **40**, No. 1, p. 205–210 (1977).
See also *1*, p. *9*; *11*, p. *95*.

S. E. Ormrod and D. L. Kirk
The Anomalous Behaviour of Na^+ Ions in Polycrystalline β''-Alumina
(a) Vol. **40**, No. 1, p. 265–270 (1977).
See also *9*, p. *69*.

T. Nasu
The Temperature Dependence of the Absorption Spectrum of NiO
(b) Vol. **80**, No. 2, p. 675–679 (1977).
See also 13.5.4, p. 153; *20.1*, p. *252*.

K. Hübner
Chemical Bond and Related Properties of SiO_2 (II): Structural Trends
(a) Vol. **40**, No. 2, p. 487–495 (1977).
See also 2, p. 36; *4*, p. *45*; *13.1*, p. *117*; 14.4, p. 190; 20.1, p. 252.

B. Prévot, J. Biellmann, M. F. Meftah, and M. Sieskind
Infra-Red Reflectivity of Non-Stoichiometric Ferrous Oxide
(a) Vol. **40**, No. 2, p. 503–510 (1977).
See also *20.1*, p. *253*.

K. A. Shoaib, F. H. Hashmi, M. Ali, S. J. H. Bukhari, and C. A. Majid
Thermoluminescence from X-Ray-Irradiated Stabilized ZrO_2 Single Crystals
(a) Vol. **40**, No. 2, p. 605–612 (1977).
See also 10.2, p. 86; 13.4, p. 133; *20.3*, p. *269*.

K.-D. Ufert and E. I. Terukov
On the Problem of the Phase Transition in Disordered VO_2 Thin Films
(a) Vol. **40**, No. 2, p. K157–K159 (1977).
See also *14.3.1*, p. *173*.

L. Resca and R. Resta
Energy Bands in Cubic Ice. Ab Initio Calculation Using the Method of Linear Combination of Molecular Orbitals
(b) Vol. **81**, No. 1, p. 129–138 (1977).
See also *13.1*, p. *117*.

R. P. Goyal, K. K. Sarkar, and S. C. Goyal
Erratum to: Dispersion of Phonons in Transition Metal Oxides (MnO)
(b) Vol. **81**, No. 1, p. 409 (1977).
See also *6*, p. *55*.

E. I. Terukov, W. Reichelt, M. P. Volotzkoi, F. A. Chudnovskii, and H. Oppermann
Hydrostatic Pressure Dependence of the Phase Transition of V_3O_5
(a) Vol. **41**, No. 1, p. 125–128 (1977).
See also *1.2*, p. *16*; 14.3, p. 168.

L. van Goethem, J. van Landuyt, and S. Amelinckx
The $\alpha \rightarrow \beta$ Transition in Amethyst Quartz as Studied by Electron Microscopy and Diffraction. The Interaction of Dauphiné with Brazil Twins
(a) Vol. **41**, No. 1, p. 129–137 (1977).
See also *1.2*, p. *16*; *1.4*, p. *25*.

T. Chiba, G. B. Dürr, and W. Brandt
Enhancement Effect on Positron Annihilation with Tightly Bound Electrons
(b) Vol. **81**, No. 2, p. 609–614 (1977).
See also *13*, p. *109*; 22.5.2, p. 378.

I. K. Amanis and J. G. Kliava
Models of Paramagnetic Centres Formed by Copper and Silver Atoms in Quartz
(a) Vol. **41**, No. 2, p. 385–392 (1977).
See also 11, p. 95; *19*, p. *239*.

S. Mochizuki and T. Sakurai
Electrical Conduction of MgO Doped with Cr at High Temperatures in Oxidizing Atmosphere
(a) Vol. **41**, No. 2, p. 411–415 (1977).
See also *14.3*, p. *168*.

R. D. Gould and C. A. Hogarth
Angular Distribution Measurements of Electrons Emitted from Thin Film Au–SiO_x–Au Diode and Triode Structures
(a) Vol. **41**, No. 2, p. 439–442 (1977).
See also 14.3.3, p. 180; *17*, p. *206*.

R. de Ridder, G. van Tendeloo, D. van Dyck, and S. Amelinckx
Application of the Cluster Model to the Study of Ordering of Atom Strings in Ternary W–Nb Oxides
(a) Vol. **41**, No. 2, p. 555–560 (1977).
See also *1.4*, p. *26*; *10.2*, p. *88*.

R. D. Newton and W. A. Sibley
Deformation Induced Damage in MgO
(a) Vol. **41**, No. 2, p. 569–575 (1977).
See also *10.2*, p. *88*; 20.1, p. 256; 20.3, p. 270.

S. Mochizuki
Electrical Conductivity of α-Fe_2O_3
(a) Vol. **41**, No. 2, p. 591–594 (1977).
See also *14.3*, p. *169*.

G. Dolino and J. P. Bachheimer
Fundamental and Second-Harmonic Light Scattering by the α–β Coexistence State of Quartz
(a) Vol. **41**, No. 2, p. 673–677 (1977).
See also *1.2*, p. *17*.

N. B. Nešković, B. Babić, and J. Konstantinović
High Temperature Anomalous Behaviour of the Crystal Lattice of Hematite
(a) Vol. **41**, No. 2, p. K133–K136 (1977).
See also *1.2*, p. *17*; 18.4, p. 233.

YU. V. BOGATYREV and F. P. KORSHUNOV
The Effect of High-Speed Electrons of Different Energy on the Charge Storage in SiO_2 and Si_3N_4 Dielectric Films
(a) Vol. **41**, No. 2, p. K151—K154 (1977).
See also *11*, p. *96*; *14.4*, p. *191*; 22, p. 325.

A. F. ELLERVEE
Luminescence of Pb^{2+} and Bi^{3+} Centres in Alkali-Earth Sulphides and Oxides
(b) Vol. **82**, No. 1, p. 91—98 (1977).
See also 13.4, p. 135; *20.3*, p. *271*; 22.4.1, p. 363.

J. SPAŁEK, M. LUBECKA, and A. WĘGRZYN
Carrier Concentration and Magnetic Susceptibility of Doped Ferromagnetic Semiconductors with Application to EuO:Eu
(b) Vol. **82**, No. 1, p. 107—116 (1977).
See also *14.3*, p. *169*; *18.2*, p. *220*.

M. BACCI
Quadratic Jahn-Teller Effect and Tunneling Splitting in Octahedral Clusters
(b) Vol. **82**, No. 1, p. 169—177 (1977).
See also *13.4*, p. *135*; *20.1*, p. *257*.

M. V. EREMIN and YU. V. RAKITIN
Channel Model in Isotropic Exchange Theory (II)
(b) Vol. **82**, No. 1, p. 221—228 (1977).
See also 1, p. 10; *13*, p. *110*; 22.5.4, p. 385.

V. D. KOROLEV, S. A. PESKOVATSKII, and V. M. SHULGA
Spin–Lattice Relaxation in Fe^{3+}-Doped Andalusite Crystals (I): Relaxation Mechanisms
(b) Vol. **82**, No. 1, p. 251—255 (1977).
See also 13.4, p. 135; *19*, p. *240*.

F. APPEL, H. BETHGE, and U. MESSERSCHMIDT
Dislocation Motion and Multiplication at the Deformation of MgO Single Crystals in the High Voltage Electron Microscope
(a) Vol. **42**, No. 1, p. 61—71 (1977).
See also *1.4*, p. *26*; 10.2, p. 89.

J. P. ZIELINGER, C. NOGUET, and M. TAPIERO
Photomemory Effect in Cu_2O Single Crystals. Kinetics of the Excitation and De-Excitation Processes
(a) Vol. **42**, No. 1, p. 91—100 (1977).
See also 13.4, p. 136; *16*, p. *203*.

H. KRAUSE and R. GRÜNLER
DC Current–Voltage Behaviour of SiO_2 Layers
(a) Vol. **42**, No. 1, p. 149—155 (1977).
See also *13.4*, p. *136*; *14.3.3*, p. *180*.

K.-D. UFERT
Doping of VO_2 Thin Films by Ion Implantation
(a) Vol. **42**, No. 1, p. 187—190 (1977).
See also *11*, p. *96*; *14.3.1*, p. *174*.

R. BARRIO, D. RÍOS, and G. TORRES
Radiation Damage in Cu_2O
(a) Vol. **42**, No. 1, p. K1—K5 (1977).
See also *11*, p. *96*.

H.-J. FITTING, H. GLAEFEKE, W. WILD, and J. LANGE
Energy and Angular Distribution of Exoelectrons
(a) Vol. **42**, No. 1, p. K75—K77 (1977).
See also *17*, p. *206*.

E. R. Vance and N. B. Manson
Magnetic Circular Dichroism of Neutron-Irradiated MgO and CaO
(b) Vol. **82**, No. 2, p. 473–479 (1977).
See also 13.4, p. 136; *18*, p. *211*; *20*, p. *245*.

S. N. Gupta, V. P. Vijayavargiya, B. D. Padalia, B. C. Tripathi, and M. N. Ghatikar
Extended X-Ray Absorption Fine Structures and Chemical Shifts for the Heavy Rare-Earth Oxides
(b) Vol. **82**, No. 2, p. 603–609 (1977).
See also *20*, p. *245*.

K. K. Bajaj and C. Aldrich
Effective Electron–Hole Interaction for Intermediate and Strong Electron–Phonon Coupling — A Comment
(b) Vol. **82**, No. 2, p. 663–668 (1977).
See also *13.5.1*, p. *146*; 13.5.3, p. 152; 22.5, p. 371.

M. Vanhaelst, P. Matthys, and E. Boesman
EPR of Iron Centres in V_2O_5 Single Crystals
(b) Vol. **82**, No. 2, p. K69–K73 (1977).
See also *19*, p. *241*.

K. Hübner
Chemical Bond and Related Properties of SiO_2 (III): Core — Level Shifts in SiO_x
(a) Vol. **42**, No. 2, p. 501–509 (1977).
See also 2, p. 38; *4*, p. *45*; *13.1*, p. *120*; 14.4, p. 191; 20.1, p. 258.

R. C. DuVarney and A. K. Garrison
An EPR-ENDOR Study of the [Li]° Center in Crystalline and Ceramic BeO
(a) Vol. **42**, No. 2, p. 609–614 (1977).
See also *13.4*, p. *137*; 19, p. 241.

C. M. Vodenicharov
A Simple Method for the Determination of the Effective Electron Mass from Current–Voltage Characteristics in the Thermionic Region
(a) Vol. **42**, No. 2, p. 785–789 (1977).
See also *13.1*, p. *121*; *14.3.1*, p. *174*; 14.3.4, p. 186; 22, p. 327; 22.9, p. 417.

V. Siegel, W. Rasp, and H.-H. Kirchner
The Influence of Desorption Processes on the Exoelectron Yield of BeO
(a) Vol. **42**, No. 2, p. K147–K150 (1977).
See also *17*, p. *207*.

U. Hartmann
Charge Measurements in Thin Insulating Layers of MIS and MIM Structures by Means of Current–Voltage Characteristics and Application to the MOS System
(a) Vol. **43**, No. 1, p. 103–110 (1977).
See also *14.3.1*, p. *175*; *14.3.3*, p. *181*.

G. Oertel and E.-H. Weber
Growth of Thin SiO_2 Films on Clean Si (111) Surfaces by Low-Pressure Oxidation and Their Evaporation
(a) Vol. **43**, No. 1, p. 141–150 (1977).
See also 1.5, p. 30; *17*, p. *207*; *20.1*, p. *260*.

A. Mosser, S. Shindo, and Y. Adachi
Angular and Temperature Dependence of the LEED Kikuchi Pattern Intensity
(a) Vol. **43**, No. 1, p. 295–304 (1977).
See also *1.6*, p. *32*.

N. N. Voinova and V. G. Govorkov
Temperature–Time Dependence of the Yield Point of Ruby by Compression and Extension
(a) Vol. **43**, No. 1, p. K49–K52 (1977).
See also *10.2*, p. *90*.

G. U. Reinsperger and W. Ludwig
The Contribution of Once Scattered Electrons to the Transmission Coefficient of Hot Electrons in Tunnel Triodes
(a) Vol. **43**, No. 1, p. K77—K80 (1977).
See also *13*, p. *111*; *14.3.4*, p. *186*.

O. P. Pchelyakov, E. B. Gorokhov, S. V. Pokrovskaya, L. V. Sokolov, and S. I. Stenin
Structure of SiO_2 Pyrolitic Films on Ge Substrates
(a) Vol. **43**, No. 2, p. 659—666 (1977).
See also *1.5*, p. *30*; 2, p. 39; 22.1.1, p. 336.

V. Siegel and W. Rasp
The TSEE of Lapped and Unlapped BeO Ceramics
(a) Vol. **43**, No. 2, p. K117—K121 (1977).
See also 1.6, p. 32; 13.3, p. 128; *17*, p. *207*.

J. Härtwig, B. Dietrich, R. Böttger, and G. Lübke
Connection between Lattice Deformation and Coloration Pattern in Synthetic Quartz Monocrystals
(a) Vol. **43**, No. 2, p. K131—K133 (1977).
See also *10.2*, p. *91*; *20.1*, p. *261*.

M. Németh-Sallay, Á. Barna, A. Lörinczy, and I. C. Szép
Induced Crystallization of Amorphous Silicon Oxide
(a) Vol. **43**, No. 2, p. K135—K136 (1977).
See also 1.5, p. 30; *1.6*, p. *32*; 2, p. 39.

A. Ranfagni and G. Viliani
Spin–Orbit-Allowed Tunneling Splitting in the $T \times \varepsilon_g$ Jahn-Teller Problem. Splitting of the Zero-Phonon Line
(b) Vol. **84**, No. 1, p. 393—399 (1977).
See also *13.4*, p. *138*; 20.1, p. 261; 22.5.4, p. 385.

W. Nowak
Scattered Light Investigations on Amorphous Systems
(a) Vol. **44**, No. 1, p. 265—275 (1977).
See also *2*, p. *40*; 10.2, p. 92.

W. Brückner, E. Wieser, W. Moldenhauer, and W. Reichelt
On the Conductivity Transition in V_3O_5
(a) Vol. **44**, No. 1, p. K95—K98 (1977)).
See also *1.2*, p. *20*; 5, p. 49; 14.3, p. 172.

V. D. Korolev, S. A. Peskovatskii, and V. M. Shulga
Spin–Lattice Relaxation in Fe^{3+}-Doped Andalusite Crystals (II): The Effect of Lattice Oscillator Heating
(b) Vol. **84**, No. 2, p. 443—447 (1977).
See also 13.4, p. 139; *19*, p. *243*.

A. M. Lejus and D. Michel
Raman Spectrum of Er_2O_3 Sesquioxide
(b) Vol. **84**, No. 2, p. K105—K108 (1977).
See also *6*, p. *61*; *20.1*, p. *263*.

S. Lara, R. Moreira Xavier, and C. A. Taft
Remarks on the Effects of Pressure on the Energy Gaps in EuO and EuS
(b) Vol. **84**, No. 2, p. K143—K145 (1977).
See also *12.2*, p. *106*; *13.1*, p. *123*; 22, p. 330.

F. Dettmann, K. Blüthner, P. Pertsch, P. Weber, and G. Albrecht
A Study of V–VO_x–Pb Josephson Tunnel Junctions
(a) Vol. **44**, No. 2, p. 577—584 (1977).
See also *14.2*, p. *164*.

R. P. Khare and R. Nath
On the Analysis of Complex TSC Patterns
(a) Vol. **44**, No. 2, p. 627—631 (1977).
See also 13.4, p. 140; *14.3.1*, p. *175*.

D. L. Kirk and S. E. Ormrod
The Mechanism of Charge Compensation Occurring in Polycrystalline β″-Alumina
(a) Vol. **44**, No. 2, p. 643—650 (1977).
See also *9*, p. *71*; 10.2, p. 93; *14.4*, p. *193*.

E. I. Terukov, H. Wagner, W. Reichelt und H. Oppermann
IR Spectroscopic Investigations on the Semiconductor–Metal Phase Transition of V_3O_5
(a) Vol. **44**, No. 2, p. K187—K190 (1977).
See also *1.2*, p. *20*; 6, p. 61; *20.1*, p. *264*.

22.6.1 Mixed Crystals

Original Papers and Short Notes

J. Livage, P. Pineau, M. C. Leroy, and M. Michaud
Semiconducting Vanadium Phosphate Glasses
(a) Vol. **39**, No. 1, p. 73—78 (1977).
See also *2*, p. *33*; 19, p. 235.

Y. Khan, K. Nahm, M. Rosenberg, and H. Willner
Superconductivity and Semiconductor–Metal Phase Transition in the System $BaPb_{1-x}Bi_xO_3$
(a) Vol. **39**, No. 1, p. 79—88 (1977).
See also 1, p. 9; 13.1, p. 114; *14.2*, p. *161*; 20.1, p. 248.

U. Büchner and W. Guse
The Dielectric Tensor Components of Czochralski-Grown 2:1 Mullite Single Crystals
(b) Vol. **80**, No. 2, p. 615—618 (1977).
See also 13.1, p. 116; *14.4*, p. *190*.

W. Brückner, U. Gerlach, and B. Thuss
Phase Diagram of $V_{1-x}Al_xO_2$
(a) Vol. **40**, No. 2, p. K131—K134 (1977).
See also *1.2*, p. *16*; *18.1*, p. *214*.

R. Nagarajan and J. K. Srivastava
Anomalous Mössbauer Lineshapes in the Magnetically Ordered Ga_2O_3–Fe_2O_3 System
(b) Vol. **81**, No. 1, p. 107—117 (1977).
See also *5*, p. *48*.

D. Gourier, J. Antoine, D. Vivien, J. Thery, J. Livage, and R. Collongues
EPR Study of Cu^{2+} Ions in the Conduction Planes of Na, Tl and Ag β-Aluminas
(a) Vol. **41**, No. 2, p. 423—429 (1977).
See also *19*, p. *239*.

W. Brückner, U. Gerlach, H.-P. Brückner, W. Moldenhauer, and H. Oppermann
Influence of Nonstoichiometry on the Phase Transitions in Ga-, Al-, and Fe-Doped VO_2
(a) Vol. **42**, No. 1, p. 295—303 (1977).
See also *1.2*, p. *17*; *3*, p. *42*; 14.3, p. 170; 18.1, p. 214.

Y. Li and D. Y. Chung
A Method for the Determination of the Elastic Constants for Crystals with Low Trigonal Symmetry
(a) Vol. **43**, No. 1, p. 329—332 (1977).
See also *7*, p. *63*; *12.2*, p. *106*.

Ö. Sävborg and M. Nygren
Magnetic, Electrical, and Thermal Studies of the $V_{1-x}Re_xO_2$ System with $0 \leqq x \leqq 0.15$
(a) Vol. **43**, No. 2, p. 645—652 (1977).
See also *1.2*, p. *19*; 8, p. 67; 14.3, p. 171; 18.1, p. 215.

22.7 Semiconducting Intermetallic Compounds

Original Papers and Short Notes

H. Oesterreicher
Magnetic Studies on GdCuGe, $Gd_6Cu_8Ge_8$, $GdCu_2Ge_2$, and $PrCu_2Ge_2$
(a) Vol. **39**, No. 1, p. K75—K77 (1977).
See also *18*, p. *209*; *18.4*, p. *231*.

H. Oesterreicher
Magnetic Properties of Gd_2Al
(a) Vol. **39**, No. 2, p. K91—K93 (1977).
See also *18*, p. *210*.

I. M. Chapnik
Compounds of Transition Metals with Possible Closed Electron Shells outside the Inert Gas Shells
(a) Vol. **39**, No. 2, p. K135—K138 (1977).
See also *13*, p. *108*; 21.1, p. 289; 21.1.1, p. 302.

J. K. Yakinthos and E. Roudaut
Magnetic Structure of the Intermetallic Compound Dy_2Ni_{17}
(a) Vol. **39**, No. 2, p. K151—K155 (1977).
See also *18.3*, p. *228*.

H. Köhler
Investigation of the Magnetoresistance of n-Type Bi_2Te_3 under Uniaxial Stress
(b) Vol. **80**, No. 1, p. K29—K32 (1977).
See also 13.1, p. 116; *14.3*, p. *166*; 20.1, p. 251.

H. Häusler and T. Pusch
Photoconductivity Measurements on Amorphous GeSe Films
(a) Vol. **40**, No. 1, p. K15—K17 (1977).
See also *2*, p. *35*; *16*, p. *202*.

F. Werfel, G. Dräger, O. Brümmer, and M. Jurisch
XS and XPS Investigation of the V_3Si Electron Structure
(b) Vol. **80**, No. 2, p. K95—K99 (1977).
See also *13.1*, p. *117*; 17, p. 206; 20, p. 245.

G. Oelgart, H. Krüger, and R. Herrmann
Lattice Dielectric Constant of $Bi_{1-x}Sb_x$
(b) Vol. **80**, No. 2, p. K101—K104 (1977).
See also *14.4*, p. *190*.

M. Snykers, R. Serneels, P. Delavignette, R. Gevers, J. van Landuyt, and S. Amelinckx
Diffraction Contrast between Inversion Domains and at Inversion Domain Boundaries
(a) Vol. **41**, No. 1, p. 51—63 (1977).
See also *1.4*, p. *25*.

G. Oelgart and R. Herrmann
Non-Ellipsoidal Fermi Surface in Semiconducting $Bi_{1-x}Sb_x$
(b) Vol. **81**, No. 2, p. K135—K138 (1977).
See also *13.2*, p. *124*.

B. Szpunar and B. Kozarzewski
The Application of CPA to Calculations of the Mean Magnetic Moment in the $Gd_{1-x}Ni_x$ $Gd_{1-x}Fe_x$, $Gd_{1-x}Co_x$, and $Y_{1-x}Co_x$ Intermetallic Compounds
(b) Vol. **82**, No. 1, p. 205—211 (1977).
See also *18*, p. *211*.

J. P. Nérou, S. Jandl, and P. E. Girard
Thermoreflectance Spectrum of SnS_2
(b) Vol. **83**, No. 1, p. K7–K9 (1977).
See also *20.1*, p. *259*.

G. Oelgart, R. Herrmann, and U. Meschter
The Lattice Dielectric Constant in $Bi_{100-x}Sb_x$ as a Function of x
(b) Vol. **83**, No. 2, p. 521–528 (1977).
See also *14.4*, p. *192*.

K. S. Dubey and R. H. Misho
Three-Phonon Scattering Relaxation Rate and Phonon Conductivity. Application to Mg_2Ge
(b) Vol. **84**, No. 1, p. 69–81 (1977).
See also *6*, p. *59*; 8, p. 67.

H. Köhler and A. Freudenberger
Investigation of the Highest Valence Band in $(Bi_{1-x}Sb_x)_2Te_3$ Crystals
(b) Vol. **84**, No. 1, p. 195–203 (1977).
See also *13.1*, p. *122*; *13.2*, p. *125*; 14.3, p. 171.

W. Richter, H. Köhler, and C. R. Becker
A Raman and Far-Infrared Investigation of Phonons in the Rhombohedral V_2–VI_3 Compounds: Bi_2Te_3, Bi_2Se_3, Sb_2Te_3 and $Bi_2(Te_{1-y}Se_y)_3$ $(0 < y < 1)$, $(Bi_{1-x}Sb_x)_2Te_3$ $(0 < x < 1)$
(b) Vol. **84**, No. 2, p. 619–628 (1977).
See also *6*, p. *60*; *20.1*, p. *262*.

K. Woodbridge and M. E. Ertl
Pulse Refrigeration in Bi_2Te_3
(a) Vol. **44**, No. 2, p. K123–K125 (1977).
See also *14.3.3*, p. *182*; 15, p. 200.

22.8 Tertiary and Higher Compounds

Review Article

K. Shahi
Transport Studies on Superionic Conductors
(a) Vol. **41**, No. 1, p. 11–44 (1977).
See also 9, p. 69; 10.2, p. 83; *14.4*, p. *189*; 15, p. 198; 22.5.1, p. 373; 22.9, p. 414

Original Papers and Short Notes

P. R. R. Pereira and L. C. M. Miranda
Magnon Contribution to the Phonon Damping in Ferromagnetic Semiconductors
(b) Vol. **79**, No. 1, p. K49–K52 (1977).
See also *6*, p. *52*; *13.5.4*, p. *153*; 18.2, p. 217.

A. J. H. Macke
Luminescence and Energy Transfer in the Ordered Perovskite System $La_2MgSn_{1-x}Ti_xO_6$
(a) Vol. **39**, No. 1, p. 117–123 (1977).
See also *20.3*, p. *267*.

R. M. Catchings
Magnetic Susceptibility Studies of Iron-Dysprosium Sodium Borate Glasses
(a) Vol. **39**, No. 1, p. K71–K74 (1977).
See also *2*, p. *33*; *18*, p. *209*.

D. I. Khomskii and K. I. Kugel
Degenerate Hubbard Model in a Magnetic Field. Application to Jahn-Teller Systems
(b) Vol. **79**, No. 2, p. 441–450 (1977).
See also *18.4*, p. *231*.

F. J. Owens
Evidence for Zero-Field Fluctuations in Cr^{3+} near the Phase Transition in $NH_4Al(SO_4)_2 \cdot$ $\cdot\ 12\ H_2O$
(b) Vol. **79**, No. 2, p. 623–628 (1977).
See also *14.4*, p. *189*; 19, p. 235.

G. K. Averkieva, V. S. Grigoreva, I. A. Maltseva, V. D. Prochukhan, and Yu. V. Rud
Photoluminescence of p-Type $ZnGeP_2$ Crystals
(a) Vol. **39**, No. 2, p. 453–457 (1977).
See also 13.4, p. 131; *20.3*, p. *268*.

T. Rajasekharan and K. V. S. Rama Rao
Nuclear Magnetic Resonance Investigation of Ferromagnetic $MnSb_{1-x}Sn_x$
(a) Vol. **39**, No. 2, p. 485–489 (1977).
See also *18.2*, p. *218*; *19*, p. *236*.

A. A. Kaminskii, T. I. Butaeva, V. A. Fedorov, Kh. S. Bagdasarov, and A. G. Petrosyan
Absorption, Luminescence, and Stimulated Emission Investigations in $Lu_3Al_5O_{12}$–Er^{3+} Crystals
(a) Vol. **39**, No. 2, p. 541–548 (1977).
See also *13.4*, p. *131*; 20.2, p. 265; 20.3, p. 268.

J. N. Boland, B. E. Hobbs, and A. C. McLaren
The Defect Structure in Natural and Experimentally Deformed Kyanite
(a) Vol. **39**, No. 2, p. 631–641 (1977).
See also *1.4*, p. *24*; *10.2*, p. *84*.

R. M. Catchings
EPR Study of the Ferric Oxide–Nickel Oxide–Sodium–Lithium–Borosilicate Glass System
(a) Vol. **39**, No. 2, p. K101–K103 (1977).
See also *2*, p. *34*; *19*, p. *236*.

L. Kraus and J. Schneider
Magnetostriction of Amorphous $(Fe_{1-x}Ni_x)_{80}P_{10}B_{10}$ Alloys
(a) Vol. **39**, No. 2, p. K161 – K163 (1977).
See also *2*, p. *34*; *18.2*, p. *218*; **19**, p. **236**.

K. Takita, Truong Dinh Thanh, and S. Tanaka
Galvanomagnetic Effects and Alloy Scattering in n-Type $Pb_{1-x}Sn_xTe$ at Low Temperatures
(b) Vol. **80**, No. 1, p. 149 – 158 (1977).
See also *14.3*, p. *166*.

A. Pierrefeu, E. F. Steigmeier, and B. Dorner
Inelastic Neutron Scattering in SbSI near the Ferroelectric Phase Transformation
(b) Vol. **80**, No. 1, p. 167 – 171 (1977).
See also *6*, p. *53*; 14.4.1, p. 194.

W. Gębicki and W. Nazarewicz
Long-Wavelength Optical Phonons in $Mn_xHg_{1-x}Te$ Mixed Crystals
(b) Vol. **80**, No. 1, p. 307 – 311 (1977).
See also *6*, p. *53*; *20.1*, p. *251*.

W. Hörig, H. Neumann, H.-J. Höbler, and G. Kühn
Temperature Dependence of the Absorption Edge in $CuInSe_2$
(b) Vol. **80**, No. 1, p. K21 – K24 (1977).
See also 13.1, p. 116; *20.1*, p. *251*.

E. Grilli and M. Guzzi
Recombination Process of Photoexcited Carriers in $ZnIn_2S_4$
(a) Vol. **40**, No. 1, p. 69 – 74 (1977).
See also *20.3*, p. *268*.

K. Takasaki and S. Tanaka
Tunnelling Study of $Pb_{1-x}Sn_xTe$
(a) Vol. **40**, No. 1, p. 173 – 180 (1977).
See also 6, p. 54; 13.1, p. 116; *14.3.2*, p. *176*.

O. Uemura and T. Satow
The Amorphous Structure of $Cd(Ge_{1-x}Sn_x)As_2$ with $x = 0.0$, 0.1, and 0.2
(a) Vol. **40**, No. 1, p. 303 – 308 (1977).
See also *2*, p. *35*; 18.1, p. 213.

K. Bärner, R. Braunstein, and E. Chock
The Optical Properties of MnAs and $MnAs_{0.85}P_{0.15}$
(b) Vol. **80**, No. 2, p. 451 – 460 (1977).
See also 13.1, p. 116; *14.4*, p. *190*; *20.1*, p. *252*.

F. P. Koffyberg
Optical and Modulated Absorption Spectra in the Band Gap Region of Amorphous Vanadium Zinc Borate Semiconductors
(b) Vol. **80**, No. 2, p. 669 – 673 (1977).
See also 2, p. 35; 13.1, p. 116; *20.1*, p. *252*.

H. Berg
The Magnetic Anisotropy of Orthorhombic $MnAs_{0.97}P_{0.03}$
(a) Vol. **40**, No. 2, p. 559 – 566 (1977).
See also *18.2*, p. *219*; 18.4, p. 232.

B. B. Sharma, R. Ayyar, and H. Singh
Stability of the Tetrahedral Phase in the $A_2^{I}B^{IV}C_3^{VI}$ Group of Compounds
(a) Vol. **40**, No. 2, p. 691 – 696 (1977).
See also *1*, p. *9*; 1.4, p. 25.

R. Blinc, J. Slak, and M. Luzar
Deuteron NMR Study of the Role of Ammonium Ions in the Antiferroelectric Transition in $ND_4D_2AsO_4$
(a) Vol. **40**, No. 2, p. K121—K125 (1977).
See also 14.4.1, p. 195; *19*, p. *238*.

J. Azoulay, D. Gerlich, E. Wiener-Avnear, and I. Pelah
Piezoelectric Coupled Soft Mode in CsH_2AsO_4 and CsD_2AsO_4 Studied by Brillouin Scattering
(b) Vol. **81**, No. 1, p. 295—306 (1977).
See also 6, p. 55; *14.4.1*, p. *195*.

H. S. Maldonado, M. G. Gutierrez, C. Zárate, and G. E. Barberis
ESR of Mn^{2+} in Lanthanum Magnesium Nitrate Hydrate
(b) Vol. **81**, No. 1, p. K27—K30 (1977).
See also *19*, p. *238*.

G. D. Guseinov, S. G. Abdullaeva, E. M. Godzhaev, A. Sh. Mekhtiev, L. A. Alieva, and M. Z. Ismailov
Electroabsorption of $TlInS_2$ Single Crystals
(b) Vol. **81**, No. 1, p. K47—K50 (1977).
See also 13.5.1, p. 145; *14.3.4*, p. *185*; *20.1*, p. *254*.

H. Weinert, H. Neumann, H.-J. Höbler, G. Kühn, and Nguyen Van Nam
Infrared Faraday Effect in n-Type $CuInSe_2$
(b) Vol. **81**, No. 1, p. K59—K61 (1977).
See also *13.1*, p. *117*; *20.1*, p. *254*.

W. Siegel, H. Becherer, G. Kühnel, and E. Ziegler
Electroluminescence in $ZnSiP_2$ Schottky Diodes
(a) Vol. **41**, No. 1, p. 75—80 (1977).
See also *14.3.3*, p. *179*; *20.3*, p. *270*.

D. I. M. Knottnerus, A. Lanjouw, and H. W. den Hartog
Electron-Excess Centres in Calciumchlorapatite
(a) Vol. **41**, No. 1, p. 95—101 (1977).
See also *10.2*, p. *87*; 20.1, p. 254.

H. Föll and K. Ulmer
A New Approach for an Absolute Determination of $2d$ Values of Spectrometer Crystals
(a) Vol. **41**, No. 1, p. 113—116 (1977).
See also *1.1*, p. *11*.

N. Schmidt and R. Send
The Role of 90° Domain Processes for the Determination of the Dielectric Anisotropy in Ferroelectric PLZT Ceramics
(a) Vol. **41**, No. 1, p. 157—161 (1977).
See also *14.4.1*, p. *195*.

D. B. Marshall and A. C. McLaren
Elastic Twinning in Experimentally Deformed Plagioclase Feldspars
(a) Vol. **41**, No. 1, p. 231—240 (1977).
See also *1.4*, p. *26*; 10.2, p. 87.

K. K. Srivastava, D. R. Goyal, A. Kumar, K. N. Lakshminarayan, O. S. Panwar, and I. Krishan
Dielectric Properties of Glassy $As_{10}Te_{75}Ge_{15}$
(a) Vol. **41**, No. 1, p. 323—329 (1977).
See also 2, p. 36; *14.4*, p. *190*.

L. K. Samanta and G. C. Bhar
Optical Nonlinearity of Some Stannite and Famatinite Crystals
(a) Vol. **41**, No. 1, p. 331—337 (1977).
See also *20*, p. *245*.

H. Schmid and L. A. Pétermann
Dielectric Constant and Electric Resistivity of Copper Chlorine Boracite, $Cu_3B_7O_{13}Cl$ (Cu–Cl–B)
(a) Vol. **41**, No. 2, p. K147—K150 (1977).
See also *14.4.1*, p. *195*.

J. Pelzl, P. Engels, and R. Florian
Raman Spectroscopic Study of the Structural Phase Transitions in K_2SnCl_6
(b) Vol. **82**, No. 1, p. 145—148 (1977).
See also *1.2*, p. *17*; 6, p. 57; 20.1, p. 256.

I. V. Chepeleva, E. A. Zhilinskaya, V. N. Lazukin, A. P. Chernov, and V. I. Olkhovskii
Mn^{2+} EPR Study of Cu–As–Se, As–Se–I, and Cu–As–Se–I Glass Systems
(b) Vol. **82**, No. 1, p. 189—192 (1977).
See also 2, p. 37; *19*, p. *240*.

H. Spiering, H. Vogel, and W. Irler
The Phase Transformation in Ferreous Fluosilicate Hexahydrate. Mössbauer Study
(b) Vol. **82**, No. 1, p. 243—250 (1977).
See also *1.2*, p. *17*; 5, p. 48; 18.1, p. 214.

G. Ambrazevičius, G. Babonas, and A. Šileika
Thermoreflectance Spectra of $CdSiP_2$
(b) Vol. **82**, No. 1, p. K45—K48 (1977).
See also *13.1*, p. *119*; *20.1*, p. *257*.

H. Herrmann
Infrared Studies of the Layered Compound $ZnIn_2S_4$
(b) Vol. **82**, No. 2, p. 513—521 (1977).
See also *6*, p. *57*; *20.1*, p. *258*.

I. Savatinova and E. Anachkova
Raman Study of Dynamic Disorder and the Ferroelectric Transition in $K_4Fe(CN)_6 \cdot 3\ H_2O$
(b) Vol. **82**, No. 2, p. 677—683 (1977).
See also *6*, p. *57*; 14.4.1, p. 196; *20.1*, p. *258*.

R. Chandra and S. Lokanathan
Electric Field Gradient in Biotite Mica
(b) Vol. **83**, No. 1, p. 273—280 (1977).
See also *5*, p. *49*.

D. Brinkmann, G. Finger, and H. Arend
On the Motion of Br^- Ions in $K_2Pt(CN)_4Br_{0.3} \cdot 3.2\ H_2O$ (KCP)
(b) Vol. **83**, No. 1, p. K45—K48 (1977).
See also *10.2*, p. *90*; *19*, p. *241*.

S. Hodorowicz, M. Ciechanowicz-Rutkowska, J. M. Janik, and J. A. Janik
The X-Ray Powder Investigation of Phase Transitions in $[Ni(NH_3)_6](ClO_4)_2$
(a) Vol. **43**, No. 1, p. 53—57 (1977).
See also *1.2*, p. *18*.

A. A. Kaminskii, S. E. Sarkisov, J. Bohm, P. Reiche, D. Schultze, and R. Uecker
Growth, Spectroscopic and Laser Properties of Crystals in the $K_5Bi_{1-x}Nd_x(MoO_4)_4$ System
(a) Vol. **43**, No. 1, p. 71—79 (1977).
See also 3, p. 42; 13.4, p. 137; 20.1, p. 260; *20.2*, p. *266*; *20.3*, p. *272*.

P. A. Arsenev, A. V. Potemkin, V. V. Fenin, and I. Senff
Investigation of Stimulated Emission of Er^{3+} Ions in Mixed Crystals with Perovskite Structure
(a) Vol. **43**, No. 1, p. K15—K18 (1977).
See also 13.4, p. 137; *20.2*, p. *266*.

P. Müller and M. Rohr
Lattice Instability of Superconducting Cold-Worked Core-Wires on the Base of PbMo S
(a) Vol. **43**, No. 1, p. K19—K22 (1977).
See also *1.2*, p. *18*; *14.2*, p. *163*.

T. X. Hoai and K. H. Herrmann
Recombination in $Pb_{0.83}Sn_{0.17}Te$ at High Levels of Optical Excitation
(b) Vol. **83**, No. 2, p. 465–470 (1977).
See also *16*, p. *204*.

A. Zygmunt
Magnetic Properties of UGeY Compounds (Y = S, Se, Te)
(a) Vol. **43**, No. 2, p. 573–577 (1977).
See also 18.1, p. 215; *18.4*, p. *233*.

H. W. den Hartog and H. Benning
Paramagnetic Color Centers in Calciumchlorapatite Doped with F^- Ions
(a) Vol. **43**, No. 2, p. 633–638 (1977).
See also *10.2*, p. *91*; 19, p. 242.

M. Giordano, M. Martinelli, L. Pardi, and S. Santucci
Electron Spin-Lattice Relaxation Measurements in a Two-Dimensional Crystal
(b) Vol. **84**, No. 1, p. 305–310 (1977).
See also 18.1, p. 215; *19*, p. *242*.

I. Savatinova and E. Anachkova
Raman Spectrum and Dielectric Relaxation in $K_4Fe(CN)_6 \cdot 3\,H_2O$
(b) Vol. **84**, No. 1, p. 401–406 (1977).
See also 1.2, p. 20; 6, p. 60; *14.4.1*, p. *197*; *20.1*, p. *262*.

H. S. Reehal and C. B. Thomas
Poole-Frenkel Conduction in Amorphous Solids for Centres with a Screened Coulomb Potential
(b) Vol. **84**, No. 1, p. K13–K16 (1977).
See also 2, p. 40; *13.4*, p. *138*; *14.3.4*, p. *187*.

I. V. Bodnar, A. G. Karoza, and G. F. Smirnova
Lattice Reflectivity Spectra of $CuGaS_2$–$CuGaSe_2$ Solid Solutions
(b) Vol. **84**, No. 1, p. K65–K69 (1977).
See also *6*, p. *60*; *20.1*, p. *262*.

V. V. Gagulin, N. V. Fadeeva, A. G. Belous, L. A. Sevastianova, A. V. Titov, M. V. Plotnikova, K. P. Mitrofanov, E. V. Zubova, S. P. Solovev, and Yu. N. Venevtsev
New Rhenium Containing Seignettomagnets and Ferroelectrics
(a) Vol. **44**, No. 1, p. 247–257 (1977).
See also 5, p. 49; *14.4.1*, p. *197*; 18.2, p. 223; *18.4*, p. *234*.

N. E. Alekseevskii, G. Wolf, N. M. Dobrovolskii, and C. Hohlfeld
Thermal and Magnetic Properties of the Superconducting Compound $YbMo_6S_8$ at Low Temperatures
(a) Vol. **44**, No. 1, p. K79–K82 (1977).
See also *8*, p. *68*; *14.2*, p. *164*; 18, p. 212.

N. V. Fedoseeva, I. P. Spevakova, B. V. Beznosikov, E. P. Naiden, and V. A. Khokhlov
Weak Ferromagnetism and Magnetic Structure of $TlMnCl_3$
(a) Vol. **44**, No. 2, p. 429–434 (1977).
See also *18.2*, p. *224*; *18.4*, p. *234*.

Y. Epelboin, M. Gandais, and C. Willaime
Influence of the Elastic Anisotropy on the Contrast of Dislocation Images in Electron Microscopy and X-Ray Topography. Application to Feldspars
(a) Vol. **44**, No. 2, p. 651–659 (1977).
See also *10.2*, p. *93*; *12.2*, p. *106*.

A. Delapalme and M. Wintenberger
Study of the Spontaneous Magnetization of Cubanite by Polarized Neutrons
(a) Vol. **44**, No. 2, p. 713—716 (1977).
See also *18*, p. *212*.

T. Satow, O. Uemura, and S. Watanabe
The Amorphous Structure of $CdGe_{1-x}M_xAs_2$ (M = Si, Sn, and Pb) with x = 0.0, 0.1, and 0.2
(a) Vol. **44**, No. 2, p. 731—736 (1977).
See also *2*, p. *41*; 18.1, p. 216.

22.8.1 Salts of Oxy-Acids (Carbonates, Phosphates, Silicates Including Mica etc.)

Original Papers and Short Notes

Č. Konák, K. Zimmermann, and B. Březina
Circular Dichroism Spectra and Optical Activity of $LiIO_3$ Crystals
(b) Vol. **79**, No. 1, p. K35–K38 (1977).
See also *20.1*, p. *248*.

I. J. R. Baumvol, A. Vasquez, J. Martinez, and F. C. Zawislak
Temperature Dependence of the Electric Field Gradient at Ta Nuclei in $CaHfO_3$
(b) Vol. **79**, No. 1, p. K65–K69 (1977).
See also *13.4*, p. *129*; 14.4, p. 189.

V. M. Fridkin, B. N. Popov, and K. A. Verkhovskaya
Effect of Anomalous Bulk Photovoltage in Ferroelectrics
(a) Vol. **39**, No. 1, p. 193–201 (1977).
See also *14.4.1*, p. *194*; *16*, p. *201*.

H. B. Haanstra and H. Ihrig
Voltage Contrast Imaging of PTC-Type $BaTiO_3$ Ceramics Having Low and High Titanium Excess
(a) Vol. **39**, No. 1, p. K7–K10 (1977).
See also *1.4*, p. *24*; *14.4.1*, p. *194*.

H. Happ and G. Rother
Investigation on the Far-Infrared Response of DKDP by Asymmetric Fourier Spectroscopy
(b) Vol. **79**, No. 2, p. 473–477 (1977).
See also 6, p. 52; *14.4.1*, p. *194*; 20.1, p. 248.

V. S. Vikhnin and A. Ya. Gordon
Paraelectric Resonance of Hydrogen Bonds in Ferroelectrics and Related Phenomena
(b) Vol. **79**, No. 2, p. 495–499 (1977).
See also *14.4.1*, p. *194*.

B. N. Ganguly and M. Nicol
Effect of Hydrostatic Pressure on the Vibrational Properties and the Structure of $SrWO_4$ and $PbMoO_4$
(b) Vol. **79**, No. 2, p. 617–622 (1977).
See also 1.2, p. 14; *6*, p. *52*; 12.2, p. 104; 20.1, p. 249.

A. Scalabrin, A. S. Chaves, D. S. Shim, and S. P. S. Porto
Temperature Dependence of the A_1 and E Optical Phonons in $BaTiO_3$
(b) Vol. **79**, No. 2, p. 731–742 (1977).
See also *6*, p. *52*; 14.4, p. 189; 20.1, p. 249.

R. Kmieć, K. Łątka, and K. Ruebenbauer
Resonant Self-Absorption Correction for the $Ba^{119m}SnO_3$ Mössbauer Sources
(b) Vol. **79**, No. 2, p. K137–K138 (1977).
See also *5*, p. *47*.

W. Bollmann and H.-J. Stöhr
Incorporation and Mobility of OH^- Ions in $LiNbO_3$ Crystals
(a) Vol. **39**, No. 2, p. 477–484 (1977).
See also *10.2*, p. *84*; 14.3, p. 166; 20.1, p. 250.

P. N. Murty, C. R. K. Murty, and S. Subramanian
Electron Paramagnetic Resonance Study of X-Irradiated Arsenate-Doped Phosphate: KH_2AsO_4/KH_2PO_4
(a) Vol. **39**, No. 2, p. 675–682 (1977).
See also 11, p. 94; *19*, p. *236*.

G. BLASSE and G. P. M. VAN DEN HEUVEL
Luminescence of U^{6+}-Activated Li_2WO_4
(a) Vol. **39**, No. 2, p. K157—K160 (1977).
See also *20.3*, p. *268*.

W. MÜLLER-LIERHEIM, T. SUSKI, and H. H. OTTO
Factor Group Analysis of the Raman Spectrum of $Pb_5Ge_3O_{11}$
(b) Vol. **80**, No. 1, p. 31—41 (1977).
See also *6*, p. *53*; 13.5.3, p. 151; 14.4.1, p. 194; 20.1, p. 250.

J. BERGMANN, K. KNEIPP, and H. E. PONATH
Raman-Induced Kerr Effect in $LiIO_3$ Single Crystals
(b) Vol. **80**, No. 1, p. K55—K58 (1977).
See also *6*, p. *54*; *20*, p. *244*.

H. KUZMANY and Z. SZALMASSY
Systematic Study of the Acoustoelectric Interaction of Surface Waves with Inhomogeneously Distributed Free Carriers in a Photoconductor
(a) Vol. **40**, No. 1, p. 75—82 (1977).
See also *7*, p. *62*; *14.3.2*, p. *176*; 14.3.3, p. 178; 16, p. 201; 22.4.1, p. 362.

W. BOLLMANN
The Origin of Photoelectrons and the Concentration of Point Defects in $LiNbO_3$ Crystals
(a) Vol. **40**, No. 1, p. 83—91 (1977).
See also *10.2*, p. *85*; *16*, p. *201*.

HJ. BERNHARDT
Near-Edge Absorption of Lead Molybdate Crystals
(a) Vol. **40**, No. 1, p. 257—264 (1977).
See also 16, p. 201; *20.1*, p. *252*; *20.3*, p. *269*.

M. V. BELOUSOV, D. E. POGAREV, and A. A. SHULTIN
Vibrations of Isotopic Impurities in Crystals Built up by Complex Ions
(b) Vol. **80**, No. 2, p. 417—424 (1977).
See also *6*, p. *54*; 20.1, p. 252.

M. FRIEDRICH and W. KARTHE
ESR Study of $PbMoO_4:Fe^{3+}$
(b) Vol. **80**, No. 2, p. K127—K130 (1977).
See also *19*, p. *237*.

S. TSUNEKAWA, M. SUEZAWA, and H. TAKEI
New Type Ferroelastic Behaviour of $NdNbO_4$ Crystals
(a) Vol. **40**, No. 2, p. 437—446 (1977).
See also *10.2*, p. *86*; *14.4.1*, p. *195*.

A. KIRFEL, W. SCHÄFER, G. WILL, and K. H. J. BUSCHOW
The High-Temperature Polymorph β-$FeSO_4$. Stability, Magnetic Structure, and Magnetic Properties
(a) Vol. **40**, No. 2, p. 447—453 (1977).
See also *1.2*, p. *16*; *18.4*, p. *232*.

F. WALZ
Completely Automated Method for the Measurement of Dielectric Aftereffects in Ferroelectric Materials
(a) Vol. **40**, No. 2, p. 455—461 (1977).
See also *14.4*, p. *190*; 14.4.1, p. 195.

Y. E. PERLIN, A. A. KAMINSKII, S. I. KLOKISHNER, V. N. ENAKII, KH. S. BAGDASAROV, G. A. BOGOMOLOVA, and D. N. VYLEGZHANIN
Thermal Broadening and Shift of the Energy Levels of TR^{3+} Ions in Crystals
(a) Vol. **40**, No. 2, p. 643—653 (1977).
See also 6, p. 55; *13.4*, p. *133*; 20.3, p. 269.

M. WARDZYŃSKA and B. M. WANKLYN
Optical Absorption Spectra and Crystal Field Parameters of Dysprosium and Erbium Germanates
(a) Vol. **40**, No. 2, p. 663—668 (1977).
See also *20.1*, p. *253*.

G. SCHMIDT, J. V. CIEMINSKI, W. ZIMMERMANN, and V. A. ISUPOV
Electrostriction in $Sr_{0.75}Ba_{0.25}Nb_2O_6$ Crystals
(a) Vol. **40**, No. 2, p. K153—K156 (1977).
See also *14.4.1*, p. *195*.

J. LAUGSCH and G. A. GEHRING
Optical Side Bands in a Cooperative Jahn-Teller Material: $TmAsO_4$
(b) Vol. **81**, No. 1, p. 167—178 (1977).
See also 6, p. 55; 13.5.1, p. 145; *20.1*, p. *253*.

W. KARTHE
Zero-Field Splitting in Trigonal Symmetry. Cr^{3+} and Fe^{3+} in α-$LiIO_3$
(b) Vol. **81**, No. 1, p. 323—332 (1977).
See also *13.4*, p. *133*; *19*, p. *238*.

V. V. EREMENKO, A. I. MASLENNIKOV, and V. M. NAUMENKO
Host and Impurity Excitations in Antiferromagnetic and Paramagnetic $FeCO_3$ in the Far Infrared
(b) Vol. **81**, No. 1, p. 351—357 (1977).
See also *13.4*, p. *133*; 18.4, p. 232; 19, p. 238.

J. SCHREIBER
Pseudo-Spin Motion in Solid Solutions of the Type $K(H_{1-x}D_x)_2PO_4$
(b) Vol. **81**, No. 1, p. 371—378 (1977).
See also *14.4*, p. *190*; *14.4.1*, p. *195*.

A. MANOOGIAN and A. LECLERC
An Explanation to the Temperature Dependent Curves of the Quadrupole Interaction Parameter Q' for $^{27}Al^{3+}$ in Ammonium and Potassium Alums
(b) Vol. **81**, No. 1, p. K15—K18 (1977).
See also *19*, p. *238*.

V. A. ANTONOV, P. A. ARSENEV, SH. A. VAKHIDOV, E. M. IBRAGIMOVA, and D. S. PETROVA
Spectroscopic Properties of Neodymium Ions in $LiYO_2$ Monocrystals
(a) Vol. **41**, No. 1, p. 45—50 (1977).
See also 10.2, p. 87; 13.4, p. 133; *20.1*, p. *254*; *20.3*, p. *270*.

J. SAWICKI, B. SAWICKA, and O. GZOWSKI
Conversion Electron Mössbauer Analysis of Glass Surfaces
(a) Vol. **41**, No. 1, p. 173—176 (1977).
See also 2, p. 36; *5*, p. *48*.

J. PIETRZAK and J. KASPRZAK
Quadrupole Relaxation of ^{35}Cl Nuclei in Barium and Sodium Chlorates
(a) Vol. **41**, No. 1, p. K25—K26 (1977).
See also *19*, p. *239*.

M. MALINOWSKI, A. PIETRASZKO, and M. POŁOMSKA
Thermal Expansion and Dielectric Permittivity in the Vicinity of Transition Points of Lead Germanate
(a) Vol. **41**, No. 1, p. K55—K58 (1977).
See also *8*, p. *65*; *14.4*, p. *191*.

S. T. LIU and L. E. CROSS
Primary Pyroelectricity in Strontium Barium Niobate ($Sr_{0.5}Ba_{0.5}Nb_2O_6$) Single Crystals
(a) Vol. **41**, No. 1, p. K83—K87 (1977).
See also *14.4.1*, p. *195*.

E. F. Kustov, V. P. Petrov, D. S. Petrova, and J. P. Udalov
Absorption and Luminescence Spectra of Nd^{3+} and Er^{3+} Ions in Monocrystals of $CaYAlO_4$
(a) Vol. **41**, No. 2, p. 379—383 (1977).
See also *20.1*, p. *256*; *20.3*, p. *270*.

A. Collomb, D. Samaras, B. Bochu, and J. C. Joubert
Properties and Magnetic Structures of Perovskite $CaCu_3Ti_4O_{12}$
(a) Vol. **41**, No. 2, p. 459—463 (1977).
See also *18.4*, p. *233*.

R. Visocekas and A. Geoffroy
Tunnelling Afterglows and Retrapping in Calcite
(a) Vol. **41**, No. 2, p. 499—503 (1977).
See also *10.2*, p. *88*; *20.3*, p. *270*.

N. Yamamoto, K. Yagi, and G. Honjo
Electron Microscopic Studies of Ferroelectric and Ferroelastic $Gd_2(MoO_4)_3$(I): General Features of Ferroelectric Domain Wall, Antiphase Boundary, and Crystal Defects
(a) Vol. **41**, No. 2, p. 523—534 (1977).
See also *14.4.1*, p. *195*.

M. Böhm, O. Erb, and A. Scharmann
Thermally Stimulated Currents in $Ca(NbO_3)_2$
(a) Vol. **41**, No. 2, p. 535—540 (1977).
See also *14.3*, p. *169*.

W. Stręk, C. Szafrański, E. Łukowiak, Z. Mazurak, and B. Jeżowska- Trzebiatowska
Fluorescence Quenching in Neodymium Pentaphosphate
(a) Vol. **41**, No. 2, p. 547—553 (1977).
See also 13.4, p. 135; *20.3*, p. *270*.

A. B. van Oosterhout
Tungstate Luminescence in Ordered Perovskites
(a) Vol. **41**, No. 2, p. 607—617 (1977).
See also 13.4, p. 135; *20.3*, p. *271*.

V. A. Antonov, P. A. Arsenev, and D. S. Petrova
Spectroscopic Properties of the Nd^{3+} Ion in $Y_2Ti_2O_7$ and $Gd_2Ti_2O_7$ Monocrystals
(a) Vol. **41**, No. 2, p. K127—K131 (1977).
See also *20.1*, p. *256*; 20.2, p. 265; *20.3*, p. *271*.

G. Bacquet, L. Youdri, and L. O. Andersson
Bond Angle and Wave Function of the CO_3^{3-}–Li^+ Centre in Irradiated Synthetic Single Crystal Calcite
(b) Vol. **82**, No. 1, p. K9—K11 (1977).
See also *19*, p. *240*.

N. Yamamoto, K. Yagi, and G. Honjo
Electron Microscopic Studies of Ferroelectric and Ferroelastic $Gd_2(MoO_4)_3$ (II): Interactions between Ferroelectric Domain Walls
(a) Vol. **42**, No. 1, p. 257—265 (1977).
See also *14.4.1*, p. *196*.

F. Milia, G. Lahajnar, M. Rožmarin, J. Slak, and R. Blinc
Proton Magnetic Resonance Study of Molecular Motion in $(NH_4)_2HAsO_4$ and $(NH_4)_2HPO_4$
(a) Vol. **42**, No. 1, p. 315—318 (1977).
See also *14.4.1*, p. *196*; *19*, p. *240*.

P. A. Arsenev, A. V. Potemkin, and V. V. Fenin
Spectroscopic Properties of Erbium Ions in Mixed Aluminates of Rare-Earth Elements
(a) Vol. **42**, No. 1, p. K19—K24 (1977).
See also *20.1*, p. *257*; 20.2, p. 266; *20.3*, p. *271*.

D. Vajda
The Dependence of the Landau Kinetic Coefficient L on the Degree of Deuteration in $K(H_{1-n}D_n)_2PO_4$ Crystals
(a) Vol. **42**, No. 1, p. K37—K39 (1977).
See also *14.4.1*, p. *196*.

H. D. Riccius, D. S. Smith, B. G. Withford, and L. H. Jones
Difference-Frequency Mixing and Second-Harmonic Generation of 3.39 μm Laser Radiation in $LiNbO_3$
(a) Vol. **42**, No. 1, p. K53—K55 (1977).
See also *14.4*, p. *191*; *20*, p. *245*.

V. A. Voloshin and M. G. Krimnus
Effect of Pressure on the Spectrum of Praseodymium-Doped Single Crystals
(b) Vol. **82**, No. 2, p. 695—700 (1977).
See also *12.2*, p. *105*; *20.1*, p. *258*; 22.5, p. 371.

P. A. Arsenev, A. V. Potemkin, V. V. Fenin, and I. Senff
Investigation of Stimulated Emission of Ho^{3+} Ions in Mixed Crystals with Perovskite Structure
(a) Vol. **42**, No. 2, p. K183—K185 (1977).
See also 20.1, p. 259; *20.2*, p. *266*; 20.3, p. 272.

D. Bäuerle, W. B. Holzapfel, A. Pinczuk, and Y. Yacoby
Temperature and Hydrostatic Pressure Dependence of Vibrational Modes in $PbTi_{1-x}Zr_xO_3$
(b) Vol. **83**, No. 1, p. 99—107 (1977).
See also *1.2*, p. *18*; 6, p. 58; *14.4.1*, p. *196*; 20.1, p. 259.

A. K. Jain and G. C. Upreti
Low Symmetry Effects in the EPR of Mn^{2+} in $Cd(NO_3)_2 \cdot 4\,H_2O$ Single Crystals
(b) Vol. **83**, No. 1, p. K17—K20 (1977).
See also *19*, p. *241*.

G. N. Neilo, A. D. Prokhorov, and G. A. Tsintsadze
Electron Spin–Lattice Relaxation of Rare-Earth Ions in Lanthanium Magnesium Double Nitrate under Pressure
(b) Vol. **83**, No. 1, p. K59—K61 (1977).
See also 12.2, p. 105; *19*, p. *241*.

R. Mańka
The Thermoelectric Power for a Small Polaron in Ferroelectrics
(b) Vol. **83**, No. 1, p. K69—K73 (1977).
See also *13.5.3*, p. *152*; 14.4.1, p. 196; *15*, p. *199*.

N. Saleh and M. Gawish
Effect of Heat Treatment on the Electronic Properties of Calcium Borate Glasses Containing Iron Oxide
(a) Vol. **43**, No. 1, p. 99—102 (1977).
See also 2, p. 39; *14.4*, p. *192*.

S. A. Hamid
Piezoelectric Properties of $Li_{1-x}H_xIO_3$
(a) Vol. **43**, No. 1, p. K29—K30 (1977).
See also *14.3*, p. *170*.

S. N. Vaidya, D. K. Joshi, M. D. Karkhanavala, and I. K. Gopalakrishnan
High Pressure Phase Transformation in La_2CuO_4
(a) Vol. **43**, No. 1, p. K31—K35 (1977).
See also *1.2*, p. *19*; *14.3*, p. *170*.

J. J. BARA, B. BRANDT, A. GUTSZE, A. T. PĘDZIWIATR, Z. M. STADNIK, and A. WRONKOWSKI
The Mössbauer Effect and Magnetometric Investigations of Iron-Doped Sodium A-Type Zeolite
(a) Vol. **43**, No. 1, p. K41—K44 (1977).
See also *5*, p. *49*; *18.1*, p. *215*.

R. MONCORGÉ, D. PACHECO, and B. DI BARTOLO
Thermal Behavior of the Shape and Width of the 9145 Å Laser Line of Nd^{3+} in $CaWO_4$
(a) Vol. **43**, No. 1, p. K45—K48 (1977).
See also *20.1*, p. *266*.

B. N. GANGULY, M. NICOL, and W. D. ELLENSON
Raman Spectroscopic Studies of the High Pressure Phase Transformation of $NaClO_3$
(b) Vol. **83**, No. 2, p. K115—K120 (1977).
See also *1.2*, p. *19*; *6*, p. *59*; 12.2, p. 106; 20.1, p. 260.

A. HOFSTAETTER, A. SCHARMANN, and B. VITT
Antiferromagnetic Properties of the $Li_2Mn_xTi_{1-x}O_3$ Mixed Crystal System. An EPR Investigation
(b) Vol. **84**, No. 1, p. 93—98 (1977).
See also 18.4, p. 234; *19*, p. *242*.

M. SCHWAB and H. G. KAHLE
Specific Heat of $TbPO_4$
(b) Vol. **84**, No. 1, p. 167—170 (1977).
See also 1.2, p. 19; *8*, p. *67*; *18.4*, p. *234*.

A. MANOOGIAN
Interpretation of the Temperature Dependent Quadrupole Coupling Constant Q' for $^{27}Al^{3+}$ and $^{71}Ga^{3+}$ in Guanidinium Hexahydrate Crystals
(b) Vol. **84**, No. 1, p. K7—K12 (1977).
See also *19*, p. *242*.

R. GARG, A. C. OLIVEIRA, and V. K. GARG
Quadrupolar Hyperfine Anisotropy in Potassium Dithioferrate
(b) Vol. **84**, No. 1, p. K17—K19 (1977).
See also *19*, p. *242*.

K. S. DUBEY
Phonon Conductivity of Sapphire
(b) Vol. **84**, No. 1, p. K57—K60 (1977).
See also *6*, p. *60*; 8, p. 67.

P. A. ARSENEV, KH. S. BAGDASAROV, A. V. POTEMKIN, and V. V. FENIN
Spectra of Absorption, Luminescence, and Stimulated Emission of Neodymium in Crystals of Mixed Yttrium and Gadolinium Scandates
(a) Vol. **44**, No. 1, p. 51—57 (1977).
See also *20.1*, p. *262*; 20.2, p. 266; *20.3*, p. *274*.

N. YAMAMOTO, K. YAGI, and G. HONJO
Electron Microscopic Studies of Ferroelectric and Ferroelastic $Gd_2(MoO_4)_3$ (III): Image Analysis of Antiphase Boundary and Observation of Phase Transformation
(a) Vol. **44**, No. 1, p. 147—160 (1977).
See also *14.4.1*, p. *197*.

U. ROY, K. PETROV, I. TSOLOVSKI, and P. PESHEV
On the Structural Phase Transition and the Thermal Expansion of Spinel Lithium Titanate
(a) Vol. **44**, No. 1, p. K25—K27 (1977).
See also *1.2*, p. *20*; 8, p. 67.

A. D. AVEZOV, S. A. VAKHIDOV, E. M. GASANOV, A. I. NOVOZHILOV, and M. I. SAMOILOVICH
Some Luminescence Properties of Synthetic Mica Fluorphlogopite
(a) Vol. **44**, No. 1, p. K87—K89 (1977).
See also *20.3*, p. *274*.

I. B. Barkan, M. V. Entin, and S. I. Marennikov
Conductivity of Fe-Doped $LiNbO_3$ Crystals
(a) Vol. **44**, No. 1, p. K91—K94 (1977).
See also *14.3*, p. *172*.

J. J. Bara and B. F. Bogacz
Investigations of Thickness Dependences of Mössbauer Absorption and Scattering Lines
(a) Vol. **44**, No. 1, p. K107—K109 (1977).
See also *5*, p. *50*.

A. Kasten and P. J. Becker
Spectroscopic Investigation of Optically Excited Doublet States of Ho^{3+} in Magnetically Ordered $(Ho_{0.01}, Gd_{0.99})VO_4$
(b) Vol. **84**, No. 2, p. 551—560 (1977).
See also 13.4, p. 139; *18.4*, p. *234*; *20.1*, p. *262*.

M. V. Eremin and N. I. Silkin
Theory of the ESR Spectra of Tl^{2+} Ions in K_2SO_4 Single Crystals
(b) Vol. **84**, No. 2, p. 803—811 (1977).
See also 13.4, p. 140; *19*, p. *243*.

H. Hussinger and H.-G. Unruh
Phase Boundaries in Ammonium Sulfate
(a) Vol. **44**, No. 2, p. 525—529 (1977).
See also *14.4.1*, p. *197*.

M. Henry and F. Varret
Mössbauer Polarimetry Using Fluosilicates. Double Motion Drive and Effective Thickness Measurements
(a) Vol. **44**, No. 2, p. 601—607 (1977).
See also *5*, p. *50*.

C. V. Manjunath and R. Srinivasan
Electron Spin Resonance Study of Copper (II) Doped Ferroelectric Ammonium Sulphate
(a) Vol. **44**, No. 2, p. 687—692 (1977).
See also 14.4.1, p. 197; *19*, p. *243*.

A. M. Varaprasad, A. L. Shashimohan, A. B. Biswas, and D. K. Chakrabarty
Dielectric and Infrared Spectral Studies of the $(Na_{1/2}La_{1/2})_xBa_{1-x}TiO_3$ System
(a) Vol. **44**, No. 2, p. 787—791 (1977).
See also *14.4.1*, p. *197*; 20.1, p. 264.

T. I. Chekmasova, I. S. Kabanov, and V. I. Yuzvak
The Pressure-Temperature Phase Diagram of $LiNH_4SO_4$
(a) Vol. **44**, No. 2, p. K155—K157 (1977).
See also *1.2*, p. *20*; *14.4.1*, p. *197*; 19, p. 243.

P. J. Ouseph and H. E. Groskreutz
Mössbauer Study of Iron in Mica
(a) Vol. **44**, No. 2, p. K181—K185 (1977).
See also *5*, p. *50*.

22.8.2 Spinels, Garnets, Ferrites

Review Articles

M. I. Klinger and A. A. Samokhvalov
Electron Conduction in Magnetite and Ferrites
(b) Vol. **79**, No. 1, p. 9—48 (1977).
Erratum
(b) Vol. **83**, No. 1, p. 353 (1977).
See also 13.1, p. 114; *13.5.3*, p. *151*; *14.3*, p. *165*; 15, p. 198; 18.3, p. 228.

P. Röschmann and H. Dötsch
Properties of Magnetostatic Modes in Ferrimagnetic Spheroids
(b) Vol. **82**, No. 1, p. 11—57 (1977).
See also 13.5.4, p. 153; *18.3*, p. *228*; 19, p. 235.

Original Papers and Short Notes

G. B. Scott and J. L. Page
The Absorption Spectra of $Y_3Fe_5O_{12}$ and $Y_3Ga_5O_{12}:Fe^{3+}$ to 5.5 eV
(b) Vol. **79**, No. 1, p. 203—213 (1977).
See also 13.4, p. 129; *20.1*, p. *247*.

O. G. Ramer and C. H. Wilts
Surface Effects in Spin-Wave Resonance in Thin YIG Films
(b) Vol. **79**, No. 1, p. 313—320 (1977).
See also *18.2.1*, p. *226*; 19, p. 235.

J. A. Baldwin, Jr.
Force of Interaction between a 180° Domain Wall and a Dislocation in YIG
(a) Vol. **39**, No. 1, p. K17—K19 (1977).
See also *10.2*, p. *84*; *18.3*, p. *228*.

G. N. Belozerskii, Yu. P. Khimich, and V. N. Gitsovich
Determination of the Parameters of Hyperfine Interaction from Mössbauer Spectra of Polycrystals in an Externally Applied Magnetic Field
(b) Vol. **79**, No. 2, p. K125—K130 (1977).
See also *5*, p. *47*; 18, p. 209.

K. de Strooper
Magnetic Properties of Some Co^{2+} Containing Spinels
(a) Vol. **39**, No. 2, p. 431—435 (1977).
See also *18.1*, p. *213*; 18.3, p. 228; 18.4, p. 231.

J. Bohm, H.-D. Kürsten, P. Reiche and R. Schalge
Morphology of Dislocations and Segregations in Gadolinium Garnet (GGG)
(a) Vol. **39**, No. 2, p. 517—523 (1977).
See also *1.4*, p. *24*; *10.2*, p. *84*.

H. Dötsch
Influence of Spin Precession on the Stability of Bubble Domains
(a) Vol. **39**, No. 2, p. 589—594 (1977).
See also *18.3*, p. *228*.

J. Piekoszewski, L. Dąbrowski, J. Suwalski, and S. Makolągwa
Mössbauer Study of Local Spin Structure in Singly Substituted YIG Ferrites
(a) Vol. **39**, No. 2, p. 643—650 (1977).
See also *5*, p. *47*.

V. P. Plakhtii, I. V. Golosovskii, M. N. Bedrisova, O. P. Smirnov, V. I. Sokolov, B. V. Mill, and N. N. Parfenova
Magnetic Ordering and Bond Geometry in Garnets with Fe^{3+} Ions in the Tetrahedral Sublattice
(a) Vol. **39**, No. 2, p. 683–695 (1977).
See also *18.4*, p. *231*.

M. Guillot and M. Le Gall
Superexchange Field and Intense Magnetic Field Dependences of the Faraday Rotation Induced by the Electric Dipole Transitions in YIG
(b) Vol. **80**, No. 1, p. 279–287 (1977).
See also 13, p. 108; *18.3*, p. *228*; *20*, p. *244*.

D. Kostopoulos and J. Grammatikakis
Changes in the State of Magnetite due to Magnetic Fields Applied at 78 K Observed in Magnetoresistance Measurements
(a) Vol. **40**, No. 1, p. 141–145 (1977).
See also *14.3*, p. *167*; *18*, p. *210*.

P. Bernstein and T. Merceron
Model of Photomagnetic Effect in Ni–Zn Ferrite with Small Amount of Cobalt
(a) Vol. **40**, No. 1, p. 159–166 (1977).
See also *18.3*, p. *229*.

M. G. Townsend, J. R. Gosselin, J. L. Horwood, L. G. Ripley, and R. J. Tremblay
Violarite, a Metallic Natural Spinel
(a) Vol. **40**, No. 1, p. K25–K29 (1977).
See also *5*, p. *47*; 15, p. 198; *18.1*, p. *213*.

A. Ślawska-Waniewska and H. Lipko
The Effect of High Hydrostatic Pressure on Anisotropy Properties of YIG:Fe^{2+} Single Crystals
(a) Vol. **40**, No. 1, p. K39–K41 (1977).
See also *18.3*, p. *229*.

A. M. Prokhorov, A. A. Kaminskii, V. V. Osiko, M. I. Timoshechkin, E. V. Zharikov, T. I. Butaeva, S. E. Sarkisov, A. G. Petrosyan, and V. A. Fedorov
Investigations of the 3 μm Stimulated Emission from Er^{3+} Ions in Aluminium Garnets at Room Temperature
(a) Vol. **40**, No. 1, p. K69–K72 (1977).
Erratum
(a) Vol. **41**, No. 2, p. K199 (1977).
See also *20.2*, p. *265*.

P. Tischer and L. Treitinger
Threshold and Memory Switching in Single Crystals of the Magnetic Semiconductor $CdCr_2Se_4$
(a) Vol. **40**, No. 2, p. 551–557 (1977).
See also *14.3.4*, p. *184*.

M. Rotter, B. Sedlák, Z. Šimša, and V. A. M. Brabers
NMR Study of Valency States in $MnFe_2O_4$
(a) Vol. **40**, No. 2, p. K169–K171 (1977).
See also *19*, p. *238*.

C. Lacour and M. Paulus
The Effect of Equilibrium Segregation at the Boundaries in the Discontinuous Growth Region
(a) Vol. **40**, No. 2, p. K179–K181 (1977).
See also *1.4*, p. *25*; *3*, p. *42*.

N. M. Hosseini and B. K. Jones
Noise in Spinel Ceramics
(a) Vol. **40**, No. 2, p. K185–K189 (1977).
See also *14.3.3*, p. *179*.

V. S. KOROLKOV and A. G. MAKHANEK
The Effects of Coordination Polyhedron Structure on Rare-Earth Ion Spectra
(b) Vol. **81**, No. 1, p. 387–394 (1977).
See also *13.4*, p. *133*; 20.1, p. 254.

H. GAMARI-SEALE and J. K. YAKINTHOS
Temperature Dependence of the Magnetic Structure of Fe_2MoO_4
(a) Vol. **41**, No. 1, p. 147–150 (1977).
See also *18.3*, p. *229*.

G. WENDE and P. GÖRNERT
Study of ACRT Influence on Crystal Growth in High-Temperature Solutions by the "High-Resolution Induced Striation Method"
(a) Vol. **41**, No. 1, p. 263–270 (1977).
See also *3*, p. *42*.

J. W. MATTHEWS and T. S. PLASKETT
Erratum to: Optical Birefringence Images of Edge Dislocations
(a) Vol. **41**, No. 1, p. 347 (1977).
See also *10.2*, p. *88*.

C. GUTTEL, E. ANTIC, and P. E. CARO
Crystal Field Parameters for Gd^{3+} in $Gd_2(MoO_4)_3$
(b) Vol. **81**, No. 2, p. 463–470 (1977).
See also *13.4*, p. *134* ; 20.1, p. 255.

V. ČÁPEK
Localized States on Imperfections and Red Shift of the Absorption Edge in Some Ferromagnetic Semiconductors
(b) Vol. **81**, No. 2, p. 571–577 (1977).
See also *13.4*, p. *134*; 20.1, p. 255.

K. KACZMARSKA and A. CHEŁKOWSKI
The ESR Linewidth in $CdCr_2S_4$ Single Crystals near Their Curie Point
(b) Vol. **81**, No. 2, p. K95–K97 (1977).
See also *18.1*, p. *214*; *19*, p. *239*.

A. FEU
Dynamic Characteristics of Domain Walls in Polycrystalline YIG. Low-Temperature Behaviour
(a) Vol. **41**, No. 2, p. 361–370 (1977).
See also *18.3*, p. *229*.

P. GÖRNERT, S. BORNMANN, F. VOIGT, and M. WENDT
Study of the Liquid Phase Epitaxy Process of Garnet Layers by Induced Striations
(a) Vol. **41**, No. 2, p. 505–511 (1977).
See also *3*, p. *42*.

K. P. BELOV and L. I. KOROLEVA
Magnetic Microregions, Caused by s–d Exchange, in Ferromagnetic Chalcogenide Spinels
(a) Vol. **41**, No. 2, p. K123–K126 (1977).
See also *18.2*, p. *220*.

L. BAURIEGEL and G. OTTO
The Influence of Magnetization of a Monocrystalline Target on the Channeling and Blocking Effect
(a) Vol. **41**, No. 2, p. K193–K195 (1977).
See also *1*, p. *9*; 18.2, p. 220; 18.3, p. 229; 21.1, p. 294.

M. KH. ASHUROV, YU. K. VORONKO, V. V. OSIKO, A. A. SOBOL, and M. I. TIMOSHECHKIN
Spectroscopic Study of Stoichiometry Deviation in Crystals with Garnet Structure
(a) Vol. **42**, No. 1, p. 101–110 (1977).
See also *1.3*, p. *22*; 13.4, p. 136; 20.1, p. 257; 20.3, p. 271.

N. L. PAKHOMOVA, M. A. VINNIK, L. M. KAZIMENKO, L. JAHN and V. A. KOZLOV
Statistical Model and Phase Transition in Nickel-Zinc Ferrites
(a) Vol. **42**, No. 1, p. 191—199 (1977).
See also *18.3*, p. *229*.

K. MORI
Transient Colour Centres by UV Light Irradiation in Yttrium Aluminium Garnet Crystals
(a) Vol. **42**, No. 1, p. 375—384 (1977).
See also *10.2*, p. *89*; 20.1, p. 257; 20.2, p. 265; 20.3, p. 271.

M. A. SIGAL
Ferromagnetic Resonance in $BaF_{12}O_{19}$ with Cylindrical Domain Structure at Arbitrary Magnetization
(a) Vol. **42**, No. 2, p. 775—784 (1977).
See also *18.2*, p. *222*; *19*, p. *241*.

G. S. WHITE, K. H. LEE, and J. H. CRAWFORD, JR.
Effects of γ-Irradiation upon the Optical Behavior of Spinel
(a) Vol. **42**, No. 2, p. K137—K141 (1977).
See also 11, p. 97; *20.1*, p. *259*; *20.3*, p. *272*.

M. W. MULLER
Site Ordering and Magnetostriction in Garnet
(b) Vol. **83**, No. 1, p. 177—185 (1977).
See also *1*, p. *10*; *18.3*, p. *230*.

H. PUSZKARSKI
Surface-State "Swapping" Effect in Thin Magnetic Films
(b) Vol. **83**, No. 1, p. K25—K27 (1977).
See also *13.3*, p. *127*; *18.2.1*, p. *226*.

D. LINZEN, S. BORNMANN, P. GÖRNERT, R. HERGT, H. PFEIFFER, W. SCHÜPPEL, F. VOIGT, and M. WENDT
Growth-Induced Magnetic Anisotropy of (Sm, Ga)-Substituted YIG Films Grown on {111} Substrates
(a) Vol. **43**, No. 1, p. 277—281 (1977).
See also 3, p. 43; *18.3*, p. *230*.

N. D. J. MILLER
Non-Reciprocal Propagation of Magnetostatic Volume Waves
(a) Vol. **43**, No. 2, p. 593—600 (1977).
See also *18.3*, p. *230*.

J. DE SITTER, A. GOVAERT, E. DE GRAVE, D. CHAMBAERE, and G. ROBBRECHT
A Mössbauer Study of Ca^{2+}-Containing Magnetites
(a) Vol. **43**, No. 2, p. 619—624 (1977).
See also *5*, p. *49*.

J. JELONEK and M. SZCZYGIELSKI
Frequency Dependence of Initial Magnetic Permeability for $HgCr_2Se_4$ near the Curie Temperature
(a) Vol. **43**, No. 2, p. K199—K200 (1977).
See also *18.2*, p. *223*.

N. K. RADHAKRISHNAN and A. B. BISWAS
A Neutron Diffraction Study of the Spinel Oxide $CuMn_2O_4$
(a) Vol. **44**, No. 1, p. 45—49 (1977).
See also *1*, p. *10*.

K. DE STROOPER, A. VAN ALBOOM, C. HENRIET-ISERENTANT, R. VANLEERBERGHE, and R. VANDENBERGHE
On the Cation Distribution and Magnetic Properties of Cobalt–Zinc Titanates
(a) Vol. **44**, No. 1, p. 215—220 (1977).
See also 18.1, p. 215; *18.3*, p. *230*.

H. Pfeiffer
Growth Rate Dependence of the Growth-Induced Anisotropy in Garnets. A Simple Theoretical Model
(a) Vol. **44**, No. 1, p. 221—227 (1977).
See also *3*, p. *43*.

J. J. Bara, A. T. Pędziwiatr, Z. M. Stadnik, A. Szytuła, J. Todorovič, Z. Tomkowicz, and W. Zarek
Investigations of Crystal and Magnetic Properties of Nickel Ferrite-Aluminates
(a) Vol. **44**, No. 1, p. 325—331 (1977).
See also *1.1*, p. *13*; 5, p. 49; *18.3*, p. *230*.

J. Loos
The Mean Field Phase Diagrams of YbIG
(b) Vol. **84**, No. 2, p. 457—463 (1977).
See also *18.3*, p. *230*.

P. Hlídek, M. Zvára, and V. Prosser
Photoinduced Infra-Red Optical Absorption in $CdCr_2Se_4$ Magnetic Semiconductor
(b) Vol. **84**, No. 2, p. K119—K123 (1977).
See also *20.1*, p. *263*.

A. K. Singh, B. K. Jain, and K. Chandra
Mössbauer Studies of Natural Goethite and Bog Iron Ore
(a) Vol. **44**, No. 2, p. 443—447 (1977).
See also *5*, p. *50*.

M. Wurlitzer and U. Reinhold
X-Ray Induced Change of Initial Permeability in Polycrystalline YIG(Me^{4+})
(a) Vol. **44**, No. 2, p. 531—538 (1977).
See also *11*, p. *99*; 18.3, p. 230.

J. J. Bara
Cation Distributions in Nickel Ferrite-Aluminates
(a) Vol. **44**, No. 2, p. 737—740 (1977).
See also *1*, p. *10*.

V. G. Kleparskii, E. I. Ilyashenko, V. V. Randoshkin, and R. V. Telesnin
Bubble Conversion to "Anomalous" Behaviour
(a) Vol. **44**, No. 2, p. 771—775 (1977).
See also *18.2*, p. *225*.

T. Merceron and J. Loriers
Magnetic After-Effect Phenomenon in the Iron–Yttrium Praseodymium Garnets
(a) Vol. **44**, No. 2, p. K147—K150 (1977).
See also *18.3*, p. *230*.

22.9 Organic Semiconductors

Review Article

K. Shahi
Transport Studies on Superionic Conductors
(a) Vol. **41**, No. 1, p. 11—44 (1977).
See also 9, p. 69; 10.2, p. 83; *14.4*, p. *189*; 15, p. 198; 22.5.1, p. 373; 22.8, p. 396.

Original Papers and Short Notes

S. I Golubov and Yu. V. Konobeev
Energy Transfer by Excitons in a Tetracene-Doped Anthracene Crystal
(b) Vol. **79**, No. 1, p. 79—84 (1977).
See also 13.4, p. 129; *13.5.1*, p. *143*; 20.3, p. 267.

J. C. Phillips
Paramagnetic Domain Tips in Quasi-One-Dimensional Conductors
(b) Vol. **79**, No. 1, p. 111—114 (1977).
See also *18.1*, p. *213*.

M. Chybicki
Electrical Properties of Metal–Polymer–Metal (MPM) and Metal–Polymer–Silicon (MPS) Structures with Thin Polymer Films
(a) Vol. **39**, No. 1, p. 271—279 (1977).
See also 14.3.1, p. 173; *14.3.3*, p. *178*.

A. Samoć, M. Samoć, and J. Sworakowski
Interpretation of One-Carrier Thermally Stimulated Currents and Isothermal Decay Currents (III): Influence of Blocking Contacts
(a) Vol. **39**, No. 1, p. 337—344 (1977).
See also *13.4*, p. *130*; 14.3.4, p. 183; 14.4, p. 189.

J. Albers and H. Küppers
Dielectric and Thermal Properties of Ammonium Hydrogen Oxalate Hemihydrate
(a) Vol. **39**, No. 1, p. K49—K51 (1977).
See also *1.2*, p. *14*; 8, p. 64; 14.4, p. 189.

G. J. Ashwell, D. D. Eley, M. R. Willis, and J. Woodward
Electrical and Magnetic Properties of Two Crystallographic Forms of $(DEPE)^{2+}(TCNQ)_4^{2-}$
(b) Vol. **79**, No. 2, p. 629—636 (1977).
See also *14.1*, p. *157*; *14.3*, p. *165*; 15, p. 198; 18.1, p. 213.

A. Hadni, R. Thomas, and C. Erhard
An Unusual Type of Epitaxial Growth
(a) Vol. **39**, No. 2, p. 419—424 (1977).
See also *1.5*, p. *29*; *3*, p. *42*.

D. Bloor and F. H. Preston
Optical Properties of Bis (p-Toluene Sulphonate) Diacetylene (II): Low Temperature Spectra
(a) Vol. **39**, No. 2, p. 607—614 (1977).
See also 1.1, p. 11; *20.1*, p. *250*.

P. Delhaes, G. Keryer, S. Flandrois and J. P. Manceau
Study of the Specific Heat of Double Salts of TCNQ with Non-Radical Cations
(b) Vol. **80**, No. 1, p. 125—136 (1977).
See also *8*, p. *64*; 18.1, p. 213.

J. B. Aladekomo, S. Arnold, and M. Pope
Triplet Exciton Diffusion and Double Photon Absorption in Tetracene
(b) Vol. **80**, No. 1, p. 333—340 (1977).
See also *13.5.1*, p. *144*; *20.3*, p. *268*.

J. C. Phillips
A Simple Explanation for the "$4k_F$ Anomaly" in TTF-TCNQ
(b) Vol. **80**, No. 1, p. 395—398 (1977).
See also *1*, p. *9*; *14*, p. *155*.

D. Bloor, D. N. Batchelder, and F. H. Preston
Optical Properties of Bis (p-Toluene Sulphonate) Diacetylene (III): Defect-Induced Spectra
(a) Vol. **40**, No. 1, p. 279—284 (1977).
See also 10.2, p. 85; *20.1*, p. *252*; 20.3, p. 269.

F. L. Battye, A. Goldmann, and L. Kasper
Ultraviolet Photoelectron Valence Band Studies on Phthalocyanine Compounds
(b) Vol. **80**, No. 2, p. 425—432 (1977).
See also *13.1*, p. *116*; 17, p. 206.

V. Winterfeldt and G. Schaack
Phonon Contributions to the Anomalous Specific Heat of Triglycine Sulfate near T_c
(b) Vol. **80**, No. 2, p. 559—567 (1977).
See also 1.2, p. 15; *6*, p. *54*; *8*, p. *65*; 14.4.1, p. 194.

J. Di Persio and B. Escaig
Initiation of Dislocation Slip in the B.C.C. Lattice. The Case of the van der Waals B.C.C. Hexamine
(a) Vol. **40**, No. 2, p. 393—401 (1977).
See also *10.2*, p. *86*.

B. P. Chandra
Stress Dependence of Triboluminescence in Crystals
(a) Vol. **40**, No. 2, p. 403—408 (1977).
See also *12.2*, p. *104*; *20.3*, p. *269*.

M. Binder, N. E. Geacintov, M. Pope, and C. E. Swenberg
Magnetic Field Modulated Scintillation from Tetracene-Doped Anthracene Crystals under α-Particle Irradiation
(a) Vol. **40**, No. 2, p. 621—628 (1977).
See also 11, p. 95; *13.5.1*, p. *144*; *20.3*, p. *269*.

E. L. Frankevich, M. M. Tribel, I. A. Sokolik, and B. V. Kotov
Photoconductivity of Charge Transfer Complex Crystals. Anthracene–Dimethylpyromellitimide
(a) Vol. **40**, No. 2, p. 655—662 (1977).
See also 13.5.1, p. 144; *16*, p. *202*.

P. Petelenz and B. Petelenz
Exciton Dissociation in the Electric Field of the Charged Defect
(b) Vol. **81**, No. 1, p. 49—56 (1977).
See also *13.5.1*, p. *145*.

T. Kondow, U. Mizutani, and T. B. Massalski
Low Temperature Specific Heats of TCNQ Complexes
(b) Vol. **81**, No. 1, p. 157—165 (1977).
See also *8*, p. *65*.

R. B. Somoano, V. Hadek, S. P. S. Yen, A. Rembaum, C. H. Hsu, R. J. Deck, T. Datta, and A. M. Hermann
Electrical Properties of (DEPE) $(TCNQ)_4$
(b) Vol. **81**, No. 1, p. 281—286 (1977).
See also *14*, p. *155*; 15, p. 198.

F.-W. Breitbarth and H.-J. Tiller
Tunnelling in Metal–Glow-Polymer–Metal Sandwich Structures
(a) Vol. **41**, No. 1, p. 151—156 (1977).
See also 13.3, p. 126; 14.3.2, p. 176; *14.3.4*, p. *185*.

M. COUZI, A. DAOUD, and R. PERRET
Low-Frequency Raman Spectra and Phase Transitions in Perovskite-Type Layer Compounds. $(CH_3NH_3)_2MnCl_4$ and $(CH_3NH_3)_2CdCl_4$
(a) Vol. **41**, No. 1, p. 271–282 (1977).
See also *1.2*, p. *16*; 6, p. 56; 20.1, p. 254.

M. F. MOSTAFA, M. SEMARY, and M. A. AHMED
Magnetic Susceptibility Studies of $((CH_3)_2CHNH_3)_2CuCl_4$
(a) Vol. **41**, No. 1, p. 295–297 (1977).
See also 18.1, p. 214; *18.2*, p. *220*.

T. KIICHI, M. NOMOTO, N. NAKAMURA, and H. CHIHARA
Phase Transitions in Tetramethylammonium Chlorate
(a) Vol. **41**, No. 1, p. K17–K19 (1977).
See also *1.2*, p. *16*; *19*, p. *239*.

H. KRESSE, D. DEMUS, and S. KÖNIG
Dielectric Relaxation of Liquid Crystals with Nematic and Smectic Modifications
(a) Vol. **41**, No. 1, p. K67–K69 (1977).
See also *2*, p. *36*; *14.4*, p. *191*.

J. HOSHEN and R. KOPELMAN
Molecular Exciton Cluster States and Spectra with Application to Benzene
(b) Vol. **81**, No. 2, p. 479–491 (1977).
See also *13.5.1*, p. *145*; *20.1*, p. *255*.

P. SCHLOTTER, J. KALINOWSKI, and H. BÄSSLER
Photoionization of Triplet Excitons in Tetracene Crystals
(b) Vol. **81**, No. 2, p. 521–526 (1977).
See also *13.5.1*, p. *145*; 16, p. 202.

T. GOWOREK and C. RYBKA
Impurities and Positronium Formation in Molecular Crystals
(b) Vol. **81**, No. 2, p. 565–569 (1977).
See also *10.2*, p. *88*.

V. T. KALINNIKOV, YU. V. RAKITIN, VAN H. CRAWFORD, and W. E. HATFIELD
Near One-Dimensional Antiferromagnetism in Vanadyl Carboxylates
(b) Vol. **81**, No. 2, p. K75–K79 (1977).
See also *18.4*, p. *233*; 19, p. 239.

M. VODENICHAROVA and V. VODENICHAROV
Electrode- and Bulk-Limited Currents in Thin Films of Aliphatic Unsaturated Poly(1,3,4)-Oxadiazoles
(a) Vol. **41**, No. 2, p. 487–491 (1977).
See also *14.3.1*, p. *174*; *14.3.3*, p. *180*.

B. MILSCH and W. WINDSCH
On the Temperature Dependence of the Zero-Field Splitting Parameter of Cr^{3+} Ions in Guanidinium Aluminium Sulfate Hexahydrate
(a) Vol. **41**, No. 2, p. K141–K146 (1977).
See also *19*, p. *240*.

JAI SINGH
Calculation of the Energy Band Structures and Carrier Mobilities in γ-Platinum Phthalocyanine Single Crystals
(b) Vol. **82**, No. 1, p. 263–268 (1977).
See also *13.1*, p. *119*; 14.3, p. 169.

J. E. LOWTHER
Proton Transfer by Dynamic Interactions and the Quantum Theory of Mutations in DNA
(b) Vol. **82**, No. 1, p. K41–K44 (1977).
See also *2*, p. *37*; *6*, p. *57*.

M. Zieliński and M. Kryszewski
Thermal Sampling Technique for the Thermally Stimulated Discharge in Polymers. Model Calculations
(a) Vol. **42**, No. 1, p. 305–314 (1977).
See also *2*, p. *37*; 14.4, p. 191.

P. C. Mehendru, N. L. Pathak, K. Jain, and P. Mehendru
Electrical Conduction in Pure and Iodine-Doped Polyvinyl Acetate Thick Films
(a) Vol. **42**, No. 1, p. 403–407 (1977).
See also *14.4*, p. *191*; 20.1, p. 257.

V. F. Kiselev, V. V. Kurylev, and N. L. Levshin
*The Effect of Adsorption on the Electroconductivity of PcFe and PcCu Films
(a) Vol. **42**, No. 1, p. K61–K64 (1977).
See also *14.3.1*, p. *174*; *14.3.2*, p. *177*.

S. K. Hoffmann, L. S. Szczepaniak, and J. Stankowski
Microscopic Mechanism of the Temperature Dependence of the Spontaneous Polarization in TGS Crystals
(b) Vol. **82**, No. 2, p. K85–K87 (1977).
See also *1.2*, p. *17*; *14.4.1*, p. *196*; 19, p. 241.

L. B. Coleman, A. M. Hermann, R. Williams, and R. B. Somoano
On the Crystal Phases of (DEPE) $(TCNQ)_4$
(b) Vol. **82**, No. 2, p. K117–K121 (1977).
See also *1.1*, p. *12*.

V. A. Lisovenko, L. A. Khutornaya, M. T. Shpak, and E. N. Velikaya
Basal Dislocations in Anthracene Crystals
(a) Vol. **42**, No. 2, p. 433–437 (1977).
See also *10.2*, p. *90*; *20.3*, p. *272*.

C. M. Vodenicharov
A Simple Method for the Determination of the Effective Electron Mass from Current–Voltage Characteristics in the Thermionic Region
(a) Vol. **42**, No. 2, p. 785–789 (1977).
See also *13.1*, p. *121*; *14.3.1*, p. *174*; 14.3.4, p. 186; 22, p. 327; 22.6, p. 390.

D. Bäuerle, K. Betzler, H. Hesse, S. Kapphan, and P. Loose
Phase-Matched Second Harmonic Generation in Urea
(a) Vol. **42**, No. 2, p. K119–K121 (1977).
See also *20.1*, p. *259*.

G. Giro and P. di Marco
Origin of Photocarriers in Ru(2,2′-Bipyridine)$_3$Cl$_2$ · 6 H_2O Single Crystals
(a) Vol. **42**, No. 2, p. K179–K182 (1977).
See also *16*, p. *203*; *20.1*, p. *259*; 20.3, p. 272.

J. C. Phillips
The "Extra" Phase Transition in TTF-TCNQ
(b) Vol. **83**, No. 1, p. 123–127 (1977).
See also *1.2*, p. *18*; 1.4, p. 27; 12.2, p. 105.

B. Khelifa, A. Delahaigue, and P. Jouve
Soft Mode and Debye Relaxations in Thiourea
(b) Vol. **83**, No. 1, p. 139–145 (1977).
See also *6*, p. *58*; *14.4.1*, p. *196*; 20.1, p. 259.

P. Petelenz
Effect of the Mixing between Frenkel Excitons and Charge Transfer States on Charge Carrier Transfer in Molecular Crystals
(b) Vol. **83**, No. 1, p. 169–175 (1977).
See also *13.5.1*, p. *146*.

J. BURGOS, M. POPE, CH. E. SWENBERG, and R. R. ALFANO
Heterofission in Pentacene-Doped Tetracene Single Crystals
(b) Vol. **83**, No. 1, p. 249–256 (1977).
See also *13.5.1*, p. *146*; *20.3*, p. *272*.

N. F. BERK, J. ROSENTHAL, L. YARMUS, and C. E. SWENBERG
Magnetic Field Dependence of Prompt Fluorescence from Crystalline Tetracene
(b) Vol. **83**, No. 1, p. K1–K6 (1977).
See also *18*, p. *212*; *20.3*, p. *272*.

T. GOWOREK and C. RYBKA
Positronium Formation in p-Terphenyl at Elevated Temperatures
(b) Vol. **83**, No. 1, p. K91–K92 (1977).
See also *10.2*, p. *90*.

M. PERNET, C. BERTHET-COLOMINAS, M. A. ALARIO FRANCO and A. N. CHRISTENSEN
Study of Chromium Oxyhydroxide (β-CrOOH) by Neutron Diffraction
(a) Vol. **43**, No. 1, p. 81–88 (1977).
See also *1*, p. *10*; *18.4*, p. *233*.

G. CHAPUIS
X-Ray Study of the First-Order Phase Transition Pcab–Bmab in $(CH_3CH_2NH_3)_2CdCl_4$
(a) Vol. **43**, No. 1, p. 203–212 (1977).
See also *1.2*, p. *18*.

J. EISNER
The Physical Properties of TGS Single Crystals, Grown from Aqueous TGS Solutions Containing Aniline
(a) Vol. **43**, No. 1, p. K1–K4 (1977).
See also 3, p. 43; *14.4*, p. *192*.

M. G. KAPLUNOV, K. I. POKHODNYA, A. I. KOTOV, E. B. YAGUBSKII, T. A. KITAEVA, and YU. G. BORODKO
Optical Properties of the New Quasi-One-Dimensional Conductor $(TSeT)Br_{0.5}$
(a) Vol. **43**, No. 1, p. K73–K76 (1977).
See also *20.1*, p. *260*.

A. K. JONSCHER
A Many-Body Model of Dielectric Polarisation in Solids (I): Experimental Data and Existing Interpretations
(b) Vol. **83**, No. 2, p. 585–597 (1977).
See also 2, p. 39; *14.4*, p. *192*; 14.4.1, p. 196; 22, p. 329.

A. FORT and A. CORET
Triplet Absorption of Crystalline Anthracene by High Resolution Photoconductivity Spectroscopy
(b) Vol. **83**, No. 2, p. 599–606 (1977).
See also *13.5.1*, p. *147*; *16*, p. *204*.

YU. G. BORODKO, M. G. KAPLUNOV, T. M. MORAVSKAYA, and K. I. POKHADNYA
X-Ray Photoelectron Spectra of Single Crystals of Tetracyanoquinodimethane and Its Salts
(b) Vol. **83**, No. 2, p. K141–K144 (1977).
See also 13, p. 111; 17, p. 207; *20*, p. *246*.

H. EICHELE and M. SCHWOERER
Fluorescence of Localized States in a Partially Polymerized Diacetylene Single Crystal
(a) Vol. **43**, No. 2, p. 465–472 (1977).
See also 13.4, p. 138; *20.3*, p. *273*.

K. O. LEE and T. T. GAN
Space-Charge-Limited Currents in Evaporated Films of Pentacene
(a) Vol. **43**, No. 2, p. 565–571 (1977).
See also *14.3.1*, p. *175*; *14.3.4*, p. *187*.

Z. Ludmer
A Metastable Excimer-Emitting Crystal Form of 9-Cyanoanthracene
(a) Vol. 43, No. 2, p. 695–698 (1977).
See also *20.3*, p. *274*.

P. Di Marco and G. Giro
Photoconductive Properties of Fluorene Single Crystals
(a) Vol. 43, No. 2, p. K193–K197 (1977).
See also *16*, p. *204*; 20.1, p. 261; 20.3, p. 274.

W. Bizzaro, J. Rosenthal, N. F. Berk, and L. Yarmus
The EPR Linewidth of Triplet Excitons in Single-Crystal Anthracene: a Re-Examination
(b) Vol. 84, No. 1, p. 27–32 (1977).
See also *13.5.1*, p. *147*; *19*, p. *242*.

A. Ya. Korets, V. P. Fedorov, A. V. Korshunov, and V. A. Naumov
Spectroscopic Investigation of Orientational Disorder in an Acenaphthylene Molecular Crystal
(b) Vol. 84, No. 1, p. 107–113 (1977).
See also 13.4, p. 138; *13.5.1*, p. *148*; *20.1*, p. *261*.

A. K. Jonscher
A Many-Body Model of Dielectric Polarisation in Solids (II): The Universal Model
(b) Vol. 84, No. 1, p. 159–166 (1977).
See also 2, p. 39; *14.4*, p. *192*; 14.4.1, p. 196; 22, p. 329.

E. Schwarzer and H. Haken
Theory of the Influence of Exciton Motion on Proton Relaxation in Organic Solids
(b) Vol. 84, No. 1, p. 253–267 (1977).
See also *13.5.1*, p. *148*; 19, p. 242.

S. Limmer, S. Grande, and A. Lösche
NMR Investigations of the Melting Behaviour of Mesogen Compounds
(b) Vol. 84, No. 1, p. 283–292 (1977).
See also *1.2*, p. *19*; 19, p. 242.

T. Goworek, C. Rybka, and J. Wawryszczuk
Positronium Lifetimes in Molecular Crystal with Impurities
(b) Vol. 84, No. 1, p. K49–K52 (1977).
See also *10.2*, p. *91*.

V. F. Kaminskii, M. L. Khidekel, R. B. Lyubovskii, I. F. Shchegolev, R. P. Shibaeva, E. B. Yagubskii, A. V. Zvarykina, and G. L. Zvereva
Metal–Insulator Phase Transition in TTT_2I_3 Affected by Iodine Concentration
(a) Vol. 44, No. 1, p. 77–82 (1977).
See also *1*, p. *10*; *14*, p. *156*; 18.1, p. 215.

V. I. Lesin, V. P. Sakun, A. I. Pristupa, and E. L. Frankevich
Reaction Yield Detected Magnetic Resonance Spectra of Intermediate Pairs of Triplet Excitons. Evolution of Spectra during Degeneracy of Energy Levels
(b) Vol. 84, No. 2, p. 513–520 (1977).
See also *13.5.1*, p. *148*; *19*, p. *243*.

I. G. Gochev and K. I. Grozdev
On the Heat Capacity of $Fe(HCOO)_2 \cdot 2\,H_2O$
(b) Vol. 84, No. 2, p. 785–789 (1977).
See also *8*, p. *68*; *18.4*, p. *234*.

R. Kind
On the Soft Modes in $(CH_3NH_3)_2CdCl_4$
(a) Vol. 44, No. 2, p. 661–667 (1977).
See also *1.2*, p. *20*; 6, p. 61.

23. Solidified Gases

Original Papers and Short Notes

I. J. Gupta and S. K. Trikha
Analysis of Lattice Thermal Conductivity of Solid Argon at Low Temperatures
(b) Vol. **80**, No. 1, p. 353–359 (1977).
See also *6*, p. *53*; *8*, p. *64*.

T. Arai and T. P. Martin
Absorption Spectra of S_2 Molecules Embedded in van der Waals Solids
(b) Vol. **81**, No. 1, p. 185–189 (1977).
See also 6, p. 55; *20.1*, p. *253*.

R. Nürnberger, F.-J. Himpsel, E. E. Koch, and N. Schwentner
Band Formation in Xenon–Argon Alloys Studied by Photoelectron Spectroscopy
(b) Vol. **81**, No. 2, p. 503–509 (1977).
See also *13.1*, p. *118*; 17, p. 206.

C. L. Reynolds, Jr. and P. R. Couchman
Tait's Equation of State at High Pressures and the Isothermal Bulk Modulus
(a) Vol. **42**, No. 1, p. K47–K52 (1977).
See also *12.1*, p. *102*; *12.2*, p. *105*; 21, p. 282; 22.5.2, p. 379.

A. T. Macrander and R. K. Crawford
Density of Solid Argon at Melting
(a) Vol. **43**, No. 2, p. 611–617 (1977).
See also *1.2*, p. *19*.

R. Kink, A. Lôhmus, M. Selg, and T. Soovik
Luminescence Lifetimes of Crystalline Xenon
(b) Vol. **84**, No. 1, p. K61–K63 (1977).
See also *20.3*, p. *274*.

E. Schuberth
First Order Raman Scattering from Ar Crystals: Evidence for the H.C.P. Phase
(b) Vol. **84**, No. 1, p. K91–K94 (1977).
See also *1.2*, p. *20*; 6, p. 60; 20.1, p. 263.

I. J. Gupta and S. K. Trikha
On the Low Temperature Lattice Thermal Conductivity of Solid Argon
(b) Vol. **84**, No. 2, p. K95–K99 (1977).
See also *6*, p. *61*; *8*, p. *68*.

Author Index

I. V. Abarenkov	(b) **82**,	99 (1977)	*1.1*, p. *11*; 22, p. 325
S. G. Abdullaeva	(a) **40**,	K7 (1977)	*1*, p. *9*; 22.3.1, p. 360
	(b) **81**,	K47 (1977)	13.5.1, p. 145; *14.3.4*, p. *185*; *20.1*, p. *254*; 22.8, p. 398
A. A. Abdumalikov	(b) **80**,	43 (1977)	*6*, p. *53*; 13.1, p. 115; *20.1*, p. *250*
A. T. Åberg	(a) **42**,	639 (1977)	*13.3*, p. *127*; *14.3.2*, p. *177*; 22.1.2, p. 341
B. S. Abramovich	(b) **81**,	341 (1977)	*13.5.2*, p. *149*
Y. Adachi	(a) **43**,	295 (1977)	*1.6*, p. *32*; 22.6, p. 390
A. A. Adamenko	(a) **42**,	K29 (1977)	*10.1*, p. *79*; 21.1, p. 295
A. M. Afanasev	(a) **42**,	415 (1977)	*1*, p. *10*; *1.6*, p. *31*; 22.1.2, p. 341
G. S. Agarwal	(b) **79**,	249 (1977)	*13.3*, p. *126*; 13.5.2, p. 149; 20.1, p. 247; 21, p. 276; 22.2.2, p. 354
V. T. Agekyan	(a) **43**,	11 (1977)	6, p. 51; *13.5.1*, p. *143*; 20.1, p. 247; 20.3, p. 267; 22.6, p. 386
M. Aguilar	(b) **84**,	595 (1977)	10.2, p. 92; *20.3*, p. *275*; 22.5.2, p. 381
F. Agulló-López	(b) **84**,	595 (1977)	10.2, p. 92; *20.3*, p. *275*; 22.5.2, p. 381
A. K. Agyei	(b) **80**,	509 (1977)	*1.2*, p. *15*
	(b) **82**,	565 (1977)	*1.2*, p. *17*; 22, p. 327
Ah Mee Hor	see under letter "H"		
M. A. Ahmed	(a) **41**,	295 (1977)	18.1, p. 214; *18.2*, p. *220*; 22.9, p. 416
J.-P. Aicardi	(a) **39**,	125 (1977)	13.4, p. 130; *20.3*, p. *267*; 22.4.1, p. 362
	(a) **42**,	545 (1977)	*13.1*, p. *121*; *20.3*, p. *272*; 22.4.1, p. 363
	(a) **43**,	191 (1977)	6, p. 59; 13.4, p. 137; *20.3*, p. *273*; 22.4.1, p. 363
Nguyen Ai Viet	see under letter "V"		
A. F. Akkerman	(b) **83**,	K145 (1977)	*11*, p. *98*; 21, p. 285; 22.1.2, p. 342
Z. Al Shaieb	see under letter "S"		
J. B. Aladekomo	(b) **80**,	333 (1977)	*13.5.1*, p. *144*; *20.3*, p. *268*; 22.9, p. 414
G. M. Aladzhadzhyan	(a) **43**,	399 (1977)	*1*, p. *10*
M. A. Alario Franco	see under letter "F"		
A. Alazhazhian	(a) **44**,	621 (1977)	13.4, p. 140; *14.3*, p. *172*; 22, p. 330
J. Albers	(a) **39**,	K49 (1977)	*1.2*, p. *14*; 8, p. 64; 14.4, p. 189; 22.9, p. 414
G. Albertini	(a) **44**,	127 (1977)	*1.1*, p. *13*; 22.1.2, p. 343
A. van Alboom	(a) **44**,	215 (1977)	18.1, p. 215; *18.3*, p. *230*; 22.8.2, p. 412
G. Albrecht	(a) **42**,	K143 (1977)	*14.2*, p. *163*; 21, p. 283
	(a) **44**,	K99 (1977)	*14.2*, p. *164*; 21, p. 287
	(a) **44**,	577 (1977)	*14.2*, p. *164*; 22.6, p. 391
R. Albrecht	(a) **39**,	621 (1977)	*10.2*, p. *84*; *12.2*, p. *104*; 22.5.2, p. 376
	(a) **40**,	147 (1977)	*10.2*, p. *85*; *12.2*, p. *104*; 22.5.2, p. 376
E. Alcalá	(a) **44**,	717 (1977)	*10.2*, p. *93*; *20,1*, p. *264*; 22.5.3, p. 384
R. Alcalá	(b) **81**,	333 (1977)	*13.4*, p. *133*; *20.1*, p. *254*; 22.5.3, p. 384
S. Al-Dallal	(a) **44**,	183 (1977)	*1.5*, p. *30*; *3*, p. *43*; 22.4.2, p. 366
A. Aldea	(b) **79**,	365 (1977)	6, p. 51; 13.4, p. 129; *14.3*, p. *165*; 22, p. 320

	(b) **81**, 409 (1977)	6, p. 51; 13.4, p. 129; *14.3*, p. *165*; 22, p. 320
C. Aldrich	(b) **82**, 663 (1977)	*13.5.1*, p. *146*; 13.5.3, p. 152; 22.5, p. 371; 22.6, p. 390
I. V. Aleksandrov	(b) **83**, K173 (1977)	9, p. 70; *13*, p. *111*; *18*, p. *212*
L. N. Aleksandrov	(a) **39**, 327 (1977)	*1.5*, p. *29*; *3*, p. *42*; 22.1.1, p. 333
	(a) **44**, 11 (1977)	*1.5*, p. *29*; *3*, p. *42*; 14.3.1, p. 173; 22.1.1, p. 333; 22.1.2, p. 337; 22.2.1, p. 350
N. E. Alekseevskii	(a) **44**, K79 (1977)	*8*, p. *68*; *14.2*, p. *164*; 18, p. 212; 22.8, p. 400
G. A. Alers	(a) **41**, 241 (1977)	*12.1*, p. *101*; 21.1, p. 293
V. G. Aleshin	(b) **83**, 537 (1977)	*13*, p. *111*; 17, p. 207; 22, p. 328
P. Alexopoulos	(a) **41**, K75 (1977)	*12.1*, p. *101*; 21, p. 280
R. R. Alfano	(b) **83**, 249 (1977)	*13.5.1*, p. *146*; *20.3*, p. *272*; 22.9, p. 418
L. C. R. Alfred	(b) **82**, 467 (1977)	*13.1*, p. *120*; *14*, p. *156*; 21, p. 282; 22, p. 326
M. Ali	(a) **40**, 605 (1977)	10.2, p. 86; 13.4, p. 133; *20.3*, p. *269*; 22.6, p. 387
L. A. Alieva	(b) **81**, K47 (1977)	13.5.1, p. 145; *14.3.4*, p. *185*; *20.1*, p. *254*; 22.8, p. 398
L. N. Alieva	(b) **82**, 705 (1977)	*6*, p. *58*; *20.1*, p. *258*; 22.3, p. 358
G. Allais	(a) **40**, 181 (1977)	*14.4.1*, p. *194*
K. R. Allakhverdiev	(a) **39**, K111 (1977)	1.2, p. 15; *6*, p. *53*; *20.1*, p. *250*; 22.3.1, p. 360
G. Allan	(b) **81**, 681 (1977)	*13.4*, p. *134*; 14.4, p. 191; 21.1, p. 293
P. J. Alonso	(b) **81**, 333 (1977)	*13.4*, p. *133*; *20.1*, p. *254*; 22.5.3, p. 384
U. von Alpen	(b) **79**, 763 (1977)	*13.4*, p. *131*; *14.3*, p. *166*; 20.1, p. 249; 22.1.3, p. 345
I. Alstrup	(b) **80**, 301 (1977)	*13.4*, p. *132*; 22.1.2, p. 338
V. Ya. Altshul	(b) **82**, K99 (1977)	*13.1*, p. *120*; 22.5, p. 371
T. S. Altshuler	(b) **80**, K109 (1977)	*19*, p. *237*; 21.1, p. 291
I. K. Amanis	(a) **41**, 385 (1977)	11, p. 95; *19*, p. *239*; 22.6, p. 388
Amar Nath Nigam	see under letter "N"	
M. A. Amato	(b) **81**, 395 (1977)	*13.5.4*, p. *153*; 14.3, p. 167; 18.2, p. 219; 22, p. 324
C.-G. d'Ambly	(a) **40**, K135 (1977)	*2*, p. *36*; *18.3*, p. *229*; 21, p. 280; 21.1, p. 292
G. Ambrazevičius	(b) **82**, K45 (1977)	*13.1*, p. *119*; *20.1*, p. *257*; 22.8, p. 399
S. Amelinckx	(a) **39**, 383 (1977)	*1.2*, p. *15*; 1.4, p. 24; 22, p. 322
	(a) **40**, 343 (1977)	*1.2*, p. *15*; 1.4, p. 25; 22, p. 323
	(a) **40**, 531 (1977)	*1.3*, p. *21*; *1.4*, p. *25*; 21.1, p. 292; 21.6, p. 315
	(a) **40**, 669 (1977)	*1.4*, p. *25*; 10.2, p. 87; 22, p. 324
	(a) **41**, 51 (1977)	*1.4*, p. *25*; 22.7, p. 394
	(a) **41**, 129 (1977)	*1.2*, p. *16*; *1.4*, p. *25*; 22.6, p. 388
	(a) **41**, 555 (1977)	*1.4*, p. *26*; *10.2*, p. *88*; 22.6, p. 388
	(a) **42**, 319 (1977)	1.3, p. 22; *1.4*, p. *26*; 22, p. 326
	(a) **42**, 565 (1977)	*1.1*, p. *12*; *1.2*, p. *18*; 22, p. 327
	(a) **43**, 133 (1977)	*1.3*, p. *22*; 10.1, p. 80; 21, p. 284; 21.1, p. 297
	(a) **43**, 541 (1977)	*1.4*, p. *27*; 10.2, p. 91; 22, p. 329
	(a) **43**, 553 (1977)	*1.3*, p. *23*; *1.4*, p. *28*; 21.1, p. 298; 21.6, p. 317
	(a) **43**, K137 (1977)	*1.1*, p. *13*; 1.4, p. 28; 22, p. 329
	(a) **44**, 411 (1977)	1.3, p. 22; *1.4*, p. *26*; 22, p. 326
Cao-Xuân An	see under letter "C"	
E. Anachkova	(b) **82**, 677 (1977)	*6*, p. *57*; 14.4.1, p. 196; *20.1*, p. *258*; 22.8, p. 399
	(b) **84**, 401 (1977)	1.2, p. 20; 6, p. 60; *14.4.1*, p. *197*; *20.1*, p. *262*; 22.8, p. 400
T. Anagnostopoulos	(a) **40**, K109 (1977)	*18.1*, p. *214*; *18.4*, p. *232*; 21.4, p. 312; 21.6, p. 315
H. R. Anand	(b) **84**, 227 (1977)	*5*, p. *49*; 22.5.2, p. 381
K. V. Anand	(a) **44**, 313 (1977)	*14.3.1*, p. *175*; *20.3*, p. *274*; 22.4.1, p. 364
E. E. Anderson	(b) **82**, 343 (1977)	*13.4*, p. *13.6*; *14.1*, p. *158*; 18, p. 211; 21.1, p. 295
L. O. Andersson	(b) **82**, K9 (1977)	*19*, p. *240*; 22.8.1, p. 405

W. Andrä	(a) **42**, 227 (1977)	*18.2.1*, p. *226*; 21.1, p. 295
G. A. Andreev	(a) **41**, 679 (1977)	*10.2*, p. *89*; 22.5.2, p. 378
A. Anedda	(b) **84**, K87 (1977)	*13.5.1*, p. *148*; 20.1, p. 263; 22.5, p. 372
J. F. Angress	(b) **79**, K107 (1977)	*6*, p. *52*; *20.1*, p. *249*; 22.5.4, p. 385
Vo Hong Anh	(b) **83**, 395 (1977)	*6*, p. *59*; *13.5.2*, p. *150*; 20, p. 246; 22, p. 328
	(b) **84**, 115 (1977)	*6*, p. *60*; *13.5*, p. *142*; 20, p. 246; 22.4.1, p. 364
M. K. Anis	(a) **44**, K5 (1977)	*3*, p. *43*; 22.3, p. 359
H.-J. Anklam	(a) **44**, 161 (1977)	*14.3.3*, p. *181*; *14.3.4*, p. *187*; 22, p. 330
A. I. Anselm	(b) **80**, 691 (1977)	*10.2*, p. *86*; *13.5.1*, p. *144*; 22.5.2, p. 377
E. Antic	(b) **81**, 463 (1977)	*13.4*, p. *134*; 20.1, p. 255; 22.8.2, p. 411
A. A. Antipin	(b) **81**, 287 (1977)	*13.4*, p. *133*; 19, p. 238; 20.3, p. 270; 22.5.4, p. 385
J. Antoine	(a) **41**, 423 (1977)	*19*, p. *239*; 22.6.1, p. 393
F. Antonangeli	(a) **42**, K41 (1977)	*14.1*, p. *159*; 21, p. 282
E. Antončik	(b) **79**, 605 (1977)	*5*, p. *47*; 13.4, p. 130; 21.7, p. 319
G. Antonioli	(b) **81**, 665 (1977)	13.5.1, p. 146; *20.1*, p. *255*; 22.3, p. 358
V. A. Antonov	(b) **81**, 719 (1977)	*1.2*, p. *16*; *1.3*, p. *21*; 21.1, p. 293; 21.6, p. 316
V. A. Antonov	(a) **41**, 45 (1977)	10.2, p. 87; 13.4, p. 133; *20.1*, p. *254*; *20.3*, p. *270*; 22.8.1, p. 404
	(a) **41**, K127 (1977)	*20.1*, p. *256*; 20.2, p. 265; *20.3*, p. *271*; 22.8.1, p. 405
S. Anzai	(a) **40**, K199 (1977)	*14.1*, p. *158*; *18.3*, p. *229*; 21.7, p. 319
V. B. Anzin	(a) **42**, 385 (1977)	12.2, p. 105; 13.1, p. 120; *16*, p. *203*; *20.1*, p. *257*; 22.1.3, p. 346
Y. Aoki	(a) **42**, 469 (1977)	*1.3*, p. *22*; *1.4*, p. *27*; 21.1, p. 296
F. Appel	(a) **42**, 61 (1977)	*1.4*, p. *26*; 10.2, p. 89; 22.6, p. 389
T. Arai	(b) **81**, 185 (1977)	6, p. 55; *20.1*, p. *253*; 23, p. 420
S. Arajs	(b) **82**, 343 (1977)	*13.4*, p. *136*; *14.1*, p. *158*; 18, p. 211; 21.1, p. 295
E. T. Arakawa	(b) **84**, 367 (1977)	*13.5.2*, p. *150*
H. Arend	(b) **83**, K45 (1977)	*10.2*, p. *90*; *19*, p. *241*; 22.8, p. 399
S. K. Arif	(a) **41**, 585 (1977)	*5*, p. *48*; 18.2, p. 220; 21.1, p. 294; 21.4, p. 312
T. Arizumi	(a) **41**, K181 (1977)	*2*, p. *37*; *14.3.1*, p. *174*; 22.1.1, p. 334
	(a) **42**, 769 (1977)	*11*, p. *97*; *14.3.2*, p. *177*; 16, p. 203; 22.1.2, p. 341
S. Arnold	(b) **80**, 333 (1977)	*13.5.1*, p. *144*; *20.3*, p. *268*; 22.9, p. 414
B. A. Aronzon	(b) **79**, 753 (1977)	*14.3*, p. *166*; 18, p. 209; *20*, p. *244*; 22.2.3, p. 356
P. A. Arsenev	(a) **41**, 45 (1977)	10.2, p. 87; 13.4, p. 133; *20.1*, p. *254*; *20.3*, p. *270*; 22.8.1, p. 404
	(a) **41**, K127 (1977)	*20.1*, p. *256*; 20.2, p. 265; *20.3*, p. *271*; 22.8.1, p. 405
	(a) **42**, K19 (1977)	*20.1*, p. *257*; 20.2, p. 266; *20.3*, p. *271*; 22.8.1, p. 405
	(a) **42**, K183 (1977)	20.1, p. 259; *20.2*, p. *266*; 20.3, p. 272; 22.8.1, p. 406
	(a) **43**, K15 (1977)	13.4, p. 137; *20.2*, p. *266*; 22.8, p. 399
	(a) **44**, 51 (1977)	*20.1*, p. *262*; 20.2, p. 266; *20.3*, p. *274*; 22.8.1, p. 407
E. K. Arushanov	(a) **43**, K159 (1977)	13.1, p. 122; *15*, p. *199*; 22.3.1, p. 360
K. Asai	(a) **42**, K7 (1977)	*14.3.3*, p. *180*; 22.2.1, p. 352
J. C. Ashley	(b) **83**, K159 (1977)	*13.5*, p. *142*; 13.5.2, p. 150; 21, p. 285
M. Kh. Ashurov	(a) **42**, 101 (1977)	*1.3*, p. *22*; 13.4, p. 136; 20.1, p. 257; 20.3, p. 271; 22.8.2, p. 411
G. J. Ashwell	(b) **79**, 629 (1977)	*14.1*, p. *157*; *14.3*, p. *165*; 15, p. 198; 18.1, p. 213, 22.9, p. 414
B. Astie	(a) **44**, K59 (1977)	*10.1*, p. *82*; *18.2*, p. *224*; 21.1.1, p. 306
G. Athanassiadis	(a) **40**, 425 (1977)	*1.3*, p. *21*; 5, p. 48; 21, p. 279; 21.1.1, p. 303

H. F. Aubauer	(a) **43**, 261 (1977)	*1.2*, p. *18*; 4, p. 45
	(a) **43**, 601 (1977)	*1.2*, p. *19*; 4, p. 46; 21.1, p. 298
G. R. Augst	(b) **83**, 55 (1977)	*10.2*, p. *90*; 22.5.2, p. 380
A. Authier	(a) **41**, K9 (1977)	*1.4*, p. *26*; *10*, p. *72*
G. K. Averkieva	(a) **39**, 453 (1977)	13.4, p. 131; *20.3*, p. *268*; 22.8, p. 396
A. D. Avezov	(a) **44**, K87 (1977)	*20.3*, p. *274*; 22.8.1, p. 407
R. Ayroles	(a) **44**, 485 (1977)	*1.1*, p. *13*
R. Ayyar	(a) **40**, 691 (1977)	*1*, p. *9*; 1.4, p. 25; 22.8, p. 397
E. Aziz	(a) **44**, K53 (1977)	*10.1*, p. *82*; 21.6, p. 317
J. Azoulay	(b) **81**, 295 (1977)	6, p. 55; *14.4.1*, p. *195*; 22.8, p. 398
M. M. Babaev	(b) **84**, 473 (1977)	6, p. 60; 13.4, p. 139; *14.3.4*, p. *188*; *15*, p. *199*; 22, p. 330
Yu. A. Babanov	(b) **79**, 423 (1977)	*13*, p. *107*; 20, p. 244; 21, p. 277
	(b) **80**, 73 (1977)	*13*, p. *108*; 20, p. 244; 21, p. 278
B. Babić	(a) **41**, K133 (1977)	*1.2*, p. *17*; 18.4, p. 233; 22.6, p. 388
E. Babić	(a) **44**, 339 (1977)	*1.2*, p. *20*; *2*, p. *40*; 21.1, p. 299; 21.1.1, p. 306
G. Babonas	(b) **82**, K45 (1977)	*13.1*, p. *119*; *20.1*, p. *257*; 22.8, p. 399
M. Bacci	(b) **82**, 169 (1977)	*13.4*, p. *135*; *20.1*, p. *257*; 22.6, p. 389
J. P. Bachheimer	(a) **41**, 673 (1977)	*1.2*, p. *17*; 22.6, p. 388
G. Bacquet	(b) **82**, K9 (1977)	*19*, p. *240*; 22.8.1, p. 405
J. P. Badiali	(b) **80**, 533 (1977)	*2*, p. *35*; 8, p. 65; 21, p. 279
S. Badrinarayanan	(a) **43**, 653 (1977)	*9*, p. *71*; 21.6, p. 317
Kh. S. Bagdasarov	(a) **39**, 541 (1977)	*13.4*, p. *131*; 20.2, p. 265; 20.3, p. 268; 22.8, p. 396
	(b) **80**, K17 (1977)	*13.1*, p. *116*; 20.1, p. 251; 22.5.3, p. 383
	(a) **40**, 643 (1977)	6, p. 55; *13.4*, p. *133*; 20.3, p. 269; 22.8.1, p. 403
	(a) **44**, 51 (1977)	*20.1*, p. *262*; 20.2, p. 266; *20.3*, p. *274*; 22.8.1, p. 407
R. Bair	(a) **41**, K1 (1977)	*10.1*, p. *77*; 11, p. 95; 21.1, p. 293
K. K. Bajaj	(b) **82**, 663 (1977)	*13.5.1*, p. *146*; 13.5.3, p. 152; 22.5, p. 371; 22.6, p. 390
H. Bakker	(b) **79**, K1 (1977)	*9*, p. *69*
	(a) **39**, K143 (1977)	*10.1*, p. *76*; *11*, p. *94*; 21.1, p. 290
	(b) **82**, 581 (1977)	*9*, p. *70*; 10.1, p. 79; 21, p. 283
	(a) **42**, 537 (1977)	*9*, p. *70*; 21, p. 283; 21.1, p. 296
S. Balasubramanian	(b) **84**, 171 (1977)	*13.4*, p. *138*; *14.3*, p. *171*; 22.1.1, p. 336; 22.1.2, p. 343
J. A. Baldwin, Jr.	(a) **39**, K17 (1977)	*10.2*, p. *84*; *18.3*, p. *228*; 22.8.2, p. 409
	(a) **42**, K123 (1977)	*18.2*, p. *222*; 21.1.1, p. 305
Yu. I. Balkarei	(b) **82**, 127 (1977)	*6*, p. *57*; *13.5.4*, p. *153*; 18.2, p. 221; 22, p. 326
A. Balogh	(b) **81**, K81 (1977)	*13*, p. *110*; 21, p. 281; 22.1.3, p. 345
I. Balogh	(a) **39**, K123 (1977)	*14.3.3*, p. *178*; *14.3.4*, p. *184*; 22.2.1, p. 350
R. von Baltz	(b) **79**, 185 (1977)	*13.5.3*, p. *151*; 22.1.2, p. 337
A. Balzarotti	(a) **42**, K41 (1977)	*14.1*, p. *159*; 21, p. 282
L. Bányai	(b) **79**, 365 (1977)	6, p. 51; 13.4, p. 129; *14.3*, p. *165*; 22, p. 320
	(b) **81**, 409 (1977)	6, p. 51; 13.4, p. 129; *14.3*, p. *165*; 22, p. 320
T. Banys	(a) **41**, 177 (1977)	13.4, p. 134; *14.3.4*, p. *185*; 22.1.3, p. 345
J. J. Bara	(a) **43**, K41 (1977)	*5*, p. *49*; *18.1*, p. *215*; 22.8.1, p. 407
	(a) **44**, 325 (1977)	*1.1*, p. *13*; 5, p. 49; *18.3*, p. *230*; 22.8.2, p. 413
	(a) **44**, 737 (1977)	*1*, p. *10*; 22.8.2, p. 413
	(a) **44**, K107 (1977)	*5*, p. *50*; 22.8.1, p. 408
D. Barančok	(a) **44**, K177 (1977)	*2*, p. *41*; *19*, p. *243*; 22, p. 331
P. G. Baranov	(b) **79**, K27 (1977)	*19*, p. *235*; 20.1, p. 248; 20.3, p. 267; 22.5.2, p. 375
B. Baranowski	(a) **40**, K35 (1977)	*18.2*, p. *219*; 21.1, p. 291

J. M. Baranowski	(b) **81**, 629 (1977)	*13.4*, p. *134*; 16, p. 202; 20.1, p. 255; 22.4.2, p. 366
	(b) **84**, 361 (1977)	13.4, p. 138; *20.1*, p. *261*; *20.3*, p. *274*; 22.4, p. 361
P. I. Baranskii	(b) **81**, 709 (1977)	*14.3*, p. *168*; 15, p. 199; 22.1.1, p. 334
V. I. Barbashov	(a) **44**, K39 (1977)	*10.2*, p. *92*; *12.2*, p. *106*; 22.5.2, p. 381
G. E. Barberis	(b) **81**, K27 (1977)	*19*, p. *238*; 22.8, p. 398
M. Bardashova	(b) **80**, K89 (1977)	*13.1*, p. *116*; 22, p. 324
Y. Barjhoux	(b) **80**, 313 (1977)	*18.1*, p. *213*; 22.5.4, p. 385
I. B. Barkan	(a) **44**, K91 (1977)	*14.3*, p. *172*; 22.8.1, p. 408
Á. Barna	(a) **41**, 81 (1977)	*1.5*, p. *29*; 2, p. 36; 22.1.2, p. 339
	(a) **43**, K135 (1977)	1.5, p. 30; *1.6*, p. *32*; 2, p. 39; 22.6, p. 391
P. B. Barna	(a) **41**, 81 (1977)	*1.5*, p. *29*; 2, p. 36; 22.1.2, p. 339
R. D. Barnard	(b) **83**, 555 (1977)	*13.2*, p. *125*; 13.4, p. 137; 14.1, p. 159; 15, p. 199; 21.6, p. 317
K. Bärner	(b) **80**, 451 (1977)	13.1, p. 166; *14.4*, p. *190*; *20.1*, p. *252*; 22.8, p. 397
	(b) **84**, 385 (1977)	*13.1*, p. *122*; 14.1, p. 159; 18.4, p. 234; *20.1*, p. *261*; 21.1, p. 299
D. M. Barnett	(b) **84**, 741 (1977)	*10*, p. *74*
A. Barone	(a) **41**, 393 (1977)	*14.2*, p. *162*
J. Barrau	(b) **80**, 461 (1977)	*13.5.1*, p. *144*; 20.3, p. 269; 22.1.2, p. 338
R. Barrio	(a) **42**, K1 (1977)	*11*, p. *96*; 22.6, p. 389
L. Bartelsen	(b) **81**, 471 (1977)	10.2, p. 88; *19*, p. *239*; 22.1.2, p. 340
W. Barth	(a) **39**, 249 (1977)	*13.4*, p. *130*; 16, p. 201; 20.1, p. 248; 22.1.1, p. 333
S. K. Barthwal	(a) **40**, 247 (1977)	*2*, p. *35*; *14.3.1*, p. *173*; 15, p. 198; 21.1, p. 291; 21.6, p. 315
K. Bartkowski	(a) **42**, 351 (1977)	*8*, p. *66*; 21.7, p. 319
B. Di Bartolo	(a) **39**, 147 (1977)	13.4, p. 130; *20.3*, p. *267*; 22.5.3, p. 383
	(a) **43**, K45 (1977)	*20.2*, p. *266*; 22.8.1, p. 407
J. Barton	(a) **39**, 259 (1977)	5, p. 47; *18.2*, p. *217*; 21.1, p. 288; 21.1.1, p. 302
I. Barvík	(b) **84**, 691 (1977)	*13.4*, p. *139*; *20.1*, p. *263*; 22.4.3, p. 367
V. G. Baryakhtar	(b) **80**, K37 (1977)	*18.4*, p. *232*; 19, p. 236; 22, p. 323
V. G. Baryshevskii	(b) **82**, 403 (1977)	*1.1*, p. *12*; *11*, p. *96*
V. K. Bashenov	(b) **80**, K17 (1977)	*13.1*, p. *116*; 20.1, p. 251; 22.5.3, p. 383
	(b) **80**, K89 (1977)	*13.1*, p. *116*; 22, p. 324
	(b) **81**, K55 (1977)	*13.1*, p. *117*; 22.5, p. 371
	(b) **82**, K95 (1977)	*13.1*, p. *120*; 22.3, p. 358
	(b) **82**, K99 (1977)	*13.1*, p. *120*; 22.5, p. 371
I. O. Bashkin	(b) **83**, 517 (1977)	*1.2*, p. *19*; 14.1, p. 159; 21.4, p. 313
E. M. Baskin	(b) **83**, K97 (1977)	*14.3.4*, p. *186*; 16, p. 204; 22.2, p. 348
L. M. Baskin	(a) **42**, 757 (1977)	13.3, p. 127; *14.3.4*, p. *186*; 22, p. 327
H. Bässler	(b) **81**, 521 (1977)	*13.5.1*, p. *145*; 16, p. 202; 22.9, p. 416
G. Bastard	(b) **79**, 585 (1977)	*13.1*, p. *115*; 18, p. 209; 20.1, p. 249; 22.4.4, p. 368
	(b) **80**, 641 (1977)	*13.4*, p. *132*; 14.3, p. 167; 22.4.4, p. 368
P. Basu	(a) **41**, 317 (1977)	13.3, p. 126; *14.3.3*, p. *179*; *16*, p. *202*; 22.1.2, p. 339; 22.2.1, p. 351
P. K. Basu	(b) **79**, K157 (1977)	*14.3*, p. *166*; 18, p. 210; 22, p. 322
D. N. Batchelder	(a) **40**, 279 (1977)	10.2, p. 85; *20.1*, p. *252*; 20.3, p. 269; 22.9, p. 415
F. L. Battye	(b) **80**, 425 (1977)	*13.1*, p. *116*; 17, p. 206; 22.9, p. 415
H. J. Bauer	(a) **40**, K35 (1977)	*18.2*, p. *219*; 21.1, p. 291
J. Bauer	(a) **39**, 173 (1977)	*1.5*, p. *29*; 13.1, p. 114; 20.1, p. 248; 22.2, p. 348
	(a) **39**, 411 (1977)	*2*, p. *34*; 13.1, p. 115; 20.1, p. 249; 22, p. 322

D. Bäuerle	(a) **42**, K119 (1977)	*20.1*, p. *259*; 22.9, p. 417
	(b) **83**, 99 (1977)	*1.2*, p. *18*; 6, p. 58; *14.4.1*, p. *196*; 20.1, p. 259; 22.8.1, p. 406
I. Baumann	(b) **81**, K55 (1977)	*13.1*, p. *117*; 22.5, p. 371
J. F. Baumard	(a) **39**, 373 (1977)	10.2, p. 84; *15*, p. *198*; 22.6, p. 386
I. J. R. Baumvol	(b) **79**, K65 (1977)	*13.4*, p. *129*; 14.4, p. 189; 22.8.1, p. 402
L. Bauriegel	(a) **41**, K193 (1977)	*1*, p. *9*; 18.2, p. 220; 18.3, p. 229; 21.1, p. 294; 22.8.2, p. 411
J. Beauvillain	(a) **44**, 485 (1977)	*1.1*, p. *13*
H. Becherer	(a) **41**, 75 (1977)	*14.3.3*, p. *179*; *20.3*, p. *270*; 22.8, p. 398
F. Bechstedt	(b) **80**, 225 (1977)	*6*, p. *53*; 13.1, p. 115; 13.5.1, p. 144; *20.1*, p. *251*
	(b) **81**, 211 (1977)	*13.5.1*, p. *145*; *20.3*, p. *270*; 22.4.1, p. 363; 22.5, p. 370
	(b) **83**, 239 (1977)	2, p. 38; *6*, p. *58*; 13.1, p. 121; *20.1*, p. *259*; 22, p. 328
C. R. Becker	(b) **84**, 619 (1977)	6, p. 60; *20.1*, p. *262*; 22.7, p. 395
P. J. Becker	(b) **84**, 551 (1977)	13.4, p. 139; *18.4*, p. *234*; *20.1*, p. *262*; 22.8.1, p. 408
P. Becla	(a) **42**, 721 (1977)	13.1, p. 121; *14.3.1*, p. *174*; 22.4.4, p. 369
A. A. Bednyakov	(b) **84**, K79 (1977)	*11*, p. *98*; 22.1, p. 332
M. N. Bedrisova	(a) **39**, 683 (1977)	*18.4*, p. *231*; 22.8.2, p. 410
T. Bedyńska	(a) **39**, 725 (1977)	*10.2*, p. *85*; 22.1.2, p. 338
U. Behn	(b) **84**, 649 (1977)	*13.1*, p. *122*; 14.1, p. 160
H. Behrendt	(b) **83**, 223 (1977)	*14.3.4*, p. *186*; 22.1.3, p. 346
G. L. Belenkii	(b) **82**, 705 (1977)	*6*, p. *58*; *20.1*, p. *258*; 22.3, p. 358
M. A. Belogolovskii	(b) **79**, K17 (1977)	6, p. 52; *14.1*, p. *157*; *14.2*, p. *161*; 21, p. 276
A. G. Belous	(a) **44**, 247 (1977)	5, p. 49; *14.4.1*, p. *197*; 18.2, p. 223; *18.4*, p. *234*; 22.8, p. 400
M. V. Belousov	(b) **80**, 417 (1977)	*6*, p. *54*; 20.1, p. 252; 22.8.1, p. 403
K. P. Belov	(a) **41**, K123 (1977)	*18.2*, p. *220*; 22.8.2, p. 411
V. S. Belovol	(b) **82**, 213 (1977)	*1.2*, p. *17*; *18.2*, p. *221*; 21.4, p. 312
G. N. Belozerskii	(b) **79**, K125 (1977)	*5*, p. *47*; 18, p. 209; 22.8.2, p. 409
V. Belzner	(a) **39**, 183 (1977)	*10.2*, p. *83*; 22.5.2, p. 375
S. I. Ben-Abraham	(b) **84**, 435 (1977)	9, p. 71; *10*, p. *74*
S. Benci	(b) **81**, 603 (1977)	*13.4*, p. *134*; 20.1, p. 255; 20.3, p. 270; 22.5.2, p. 378
V. Z. Bengus	(a) **44**, 457 (1977)	*10.2*, p. *93*; 22.5.2, p. 381
H. Benning	(a) **43**, 633 (1977)	*10.2*, p. *91*; 19, p. 242; 22.8, p. 400
S. Berar	(b) **81**, 245 (1977)	*13*, p. *109*; 18.2, p. 219
G. Berg	(a) **39**, 265 (1977)	*10.2*, p. *83*; 19, p. 235; 22.5.2, p. 375
	(a) **42**, 73 (1977)	*10.2*, p. *89*; 22.5.2, p. 379
H. Berg	(a) **40**, 559 (1977)	*18.2*, p. *219*; 18.4, p. 232; 22.8, p. 397
F. Bergauer	(b) **81**, K91 (1977)	13.1, p. 118; *18*, p. *211*; *20.1*, p. *255*; 22.2.1, p. 352
E. A. van den Bergen	(a) **44**, 517 (1977)	1, p. 10; *9*, p. *71*; 22.5.2, p. 382
J. Bergmann	(b) **80**, K55 (1977)	*6*, p. *54*; *20*, p. *244*; 22.8.1, p. 403
G. O. Berim	(b) **79**, 489 (1977)	*13.5.1*, p. *143*; *19*, p. *235*
	(a) **40**, K53 (1977)	*19*, p. *237*; 21.2, p. 308
N. F. Berk	(b) **83**, K1 (1977)	*18*, p. *212*; *20.3*, p. *272*; 22.9, p. 418
	(b) **84**, 27 (1977)	*13.5.1*, p. *147*; *19*, p. *242*; 22.9, p. 419
F. Bernhard	(b) **84**, K79 (1977)	*11*, p. *98*; 22.1, p. 332
Hj. Bernhardt	(a) **40**, 257 (1977)	16, p. 201; *20.1*, p. *252*; *20.3*, p. *269*; 22.8.1, p. 403
P. Bernstein	(a) **40**, 159 (1977)	*18.3*, p. *229*; 22.8.2, p. 410
T. Bernstein	(a) **41**, 283 (1977)	*11*, p. *95*; 14.3.2, p. 176; 22, p. 324
E. F. Bertaut	(a) **39**, 57 (1977)	*18.4*, p. *231*; 22.6, p. 386

K.-H. Berthel	(a) **44**, 277 (1977)	1.3, p. 23; *3*, p. *43*; 14.2, p. 164; 21.1, p. 299
	(a) **44**, 499 (1977)	*10.1*, p. *82*; 14.2, p. 164; 21.1, p. 300
C. Berthet-Colominas	(a) **43**, 81 (1977)	*1*, p. *10*; *18.4*, p. *233*; 22.9, p. 418
V. I. Betekhtin	(a) **39**, 621 (1977)	*10.2*, p. *84*; *12.2*, p. *104*; 22.5.2, p. 376
	(a) **40**, 147 (1977)	*10.2*, p. *85*; *12.2*, p. *104*; 22.5.2, p. 376
H. Bethge	(a) **42**, 61 (1977)	*1.4*, p. *26*; 10.2, p. 89; 22.6, p. 389
K. Betzler	(a) **42**, K119 (1977)	*20.1*, p. *259*; 22.9, p. 417
A. van den Beukel	(a) **41**, 513 (1977)	*10.1*, p. *78*; 21, p. 281; 21.1, p. 294
P. A. Bezirganyan	(a) **40**, K77 (1977)	*1*, p. *9*
	(a) **43**, 399 (1977)	*1*, p. *10*
	(a) **43**, K183 (1977)	*1*, p. *10*
B. V. Beznosikov	(a) **44**, 429 (1977)	*18.2*, p. *224*; *18.4*, p. *234*; 22.8, p. 400
C. M. Bhandari	(b) **84**, K125 (1977)	*18*, p. *212*
S. S. Bhandari	(a) **41**, K89 (1977)	*5*, p. *48*; 21, p. 280; 21.7, p. 319
G. C. Bhar	(a) **41**, 331 (1977)	*20*, p. *245*; 22.8, p. 398
S. Bharati	(a) **43**, 653 (1977)	*9*, p. *71*; 21.6, p. 317
	(a) **44**, 391 (1977)	*9*, p. *71*; 21.6, p. 317
D. P. Bhattacharya	(a) **39**, 49 (1977)	*13*, p. *107*; *14.3.4*, p. *183*; 22, p. 321
K. Bhattacharya	(a) **41**, 317 (1977)	13.3, p. 126; *14.3.3*, p. *179*; *16*, p. *202*; 22.1.2, p. 339; 22.2.1, p. 351
H. Bialas	(a) **42**, 125 (1977)	1.5, p. 30; *7*, p. *62*; *19*, p. *240*; 21.1, p. 295
D. Bianchi	(b) **81**, 665 (1977)	13.5.1, p. 146; *20.1*, p. *255*; 22.3, p. 358
A. Bianconi	(a) **42**, K41 (1977)	*14.1*, p. *159*; 21, p. 282
Z. Biegański	(a) **44**, 91 (1977)	*8*, p. *67*; 18, p. 212; 21.4, p. 313
J. Biellmann	(a) **40**, 503 (1977)	*20.1*, p. *253*; 22.6, p. 387
S. W. Biernacki	(b) **79**, 683 (1977)	*13.4*, p. *130*; 19, p. 236; 22.4.2, p. 365
	(b) **81**, K5 (1977)	*13.4*, p. *133*; 22.4.2, p. 365
	(b) **84**, 699 (1977)	*13.4*, p. *139*; 18.1, p. 215; *20.1*, p. *263*
M. Biget	(a) **40**, 101 (1977)	*11*, p. *95*; 21.4, p. 311
Z. Bílek	(a) **44**, 117 (1977)	*10.1*, p. *81*; 12.1, p. 103; 21.1, p. 299
H. R. Bilger	(a) **40**, K85 (1977)	*14.3*, p. *167*; 22, p. 323
H. Bill	(b) **79**, 259 (1977)	*10.2*, p. *83*; 19, p. 235; 22.5.3, p. 383
	(a) **41**, 163 (1977)	13.4, p. 134; *19*, p. *239*; 22, p. 324
L. Billard	(b) **79**, 307 (1977)	*18.2*, p. *217*
	(b) **82**, 269 (1977)	*1.2*, p. *17*; *18.2*, p. *221*; 21.1, p. 295; 21.1.1, p. 304
	(b) **84**, 335 (1977)	*1.3*, p. *23*; *18.2*, p. *223*; 21.1.1, p. 306
R. Bindemann	(a) **43**, 529 (1977)	*13.4*, p. *138*; *20.3*, p. *273*; 22.2.2, p. 354
M. Binder	(a) **40**, 621 (1977)	11, p. 95; *13.5.1*, p. *144*; *20.3*, p. *269*; 22.9, p. 415
U. Birkholz	(a) **39**, 467 (1977)	*13.4*, p. *131*; 14.3, p. 166; 16, p. 201; 20.3, p. 268; 22, p. 322
	(a) **41**, K163 (1977)	*1.6*, p. *31*; 22, p. 325
J. L. Birman	(b) **80**, 509 (1977)	*1.2*, p. *15*
	(b) **82**, 565 (1977)	*1.2*, p. *17*; 22, p. 327
L. Biste	(a) **39**, 173 (1977)	*1.5*, p. *29*; 13.1, p. 114; 20.1, p. 248; 22.2, p. 348
A. B. Biswas	(a) **44**, 45 (1977)	*1*, p. *10*; 22.8.2, p. 412
	(a) **44**, 787 (1977)	*14.4.1*, p. *197*; 20.1, p. 264; 22.8.1, p. 408
M. Bitterlich	(b) **80**, K85 (1977)	*2*, p. *34*; *18.3*, p. *228*; 21.1, p. 290; 21.4, p. 311
H. Bittner	(a) **41**, K101 (1977)	*18.2*, p. *220*; 21.1, p. 294; 21.1.1, p. 304
A. Bivas	(b) **83**, K101 (1977)	*13.5*, p. *141*; *20.3*, p. *273*; 22.5, p. 372
	(b) **84**, 235 (1977)	*13.5.1*, p. *148*; *20.3*, p. *274*; 22.5, p. 372
	(b) **84**, 673 (1977)	*13.5.1*, p. *148*; *20.3*, p. *275*; 22.5, p. 372
	(b) **84**, 731 (1977)	*13.5.1*, p. *148*; *20.1*, p. *263*; 22.5, p. 372
W. Bizzaro	(b) **84**, 27 (1977)	*13.5.1*, p. *147*; *19*, p. *242*; 22.9, p. 419

G. Blasse	(a) **39**, K157 (1977)	*20.3*, p. *268*; 22.8.1, p. 403
	(a) **44**, K163 (1977)	*20.3*, p. *275*; 22.5, p. 372
W. Blau	(b) **81**, 535 (1977)	13.1, p. 118; *18.2*, p. *220*; *18.4*, p. *233*; 21.1.1, p. 303
	(b) **82**, 495 (1977)	13.1, p. 120; *20*, p. *245*; 21.1, p. 296; 21.1.1, p. 304
R. Blinc	(a) **40**, K121 (1977)	14.4.1, p. 195; *19*, p. *238*; 22.8, p. 398
	(a) **42**, 315 (1977)	*14.4.1*, p. *196*; *19*, p. *240*; 22.8.1, p. 405
M. D. Blokh	(b) **83**, K97 (1977)	*14.3.4*, p. *186*; *16*, p. *204*; 22.2, p. 348
D. I. Blokhintsev	(b) **82**, 627 (1977)	*1.6*, p. *31*; 11, p. 96
D. Bloor	(a) **39**, 607 (1977)	1.1, p. 11; *20.1*, p. *250*; 22.9, p. 414
	(a) **40**, 279 (1977)	10.2, p. 85; *20.1*, p. *252*; 20.3, p. 269; 22.9, p. 415
N. A. Blum	(a) **41**, 481 (1977)	2, p. 37; *20.1*, p. *256*; 22.1, p. 332
A. Blumen	(b) **83**, 425 (1977)	*13.1*, p. *121*; 14.4.1, p. 196; 22.5, p. 371
K. Blüthner	(a) **44**, K99 (1977)	*14.2*, p. *164*; 21, p. 287
	(a) **44**, 577 (1977)	*14.2*, p. *164*; 22.6, p. 391
H. J. Blythe	(a) **43**, K87 (1977)	10.1, p. 80; *11*, p. *97*; *18.2*, p. *222*; 21.1.1, p. 306
A. Bobák	(b) **83**, K169 (1977)	*2*, p. *39*; *18.2*, p. *223*; 21, p. 285
E. A. Bobrova	(b) **82**, 237 (1977)	*14.3*, p. *169*; *16*, p. *203*; 22.1.3, p. 345
P. Boch	(a) **43**, 625 (1977)	1.3, p. 23; *12.1*, p. *102*; 21, p. 286; 21.3, p. 310
B. Bochu	(a) **41**, 459 (1977)	*18.4*, p. *233*; 22.8.1, p. 405
R. Bodenberger	(a) **44**, K7 (1977)	*18.2*, p. *224*; 21.1, p. 299; 21.4, p. 313
I. V. Bodnar	(a) **41**, 669 (1977)	*13.1*, p. *118*; *20.1*, p. *256*; 22.2.4, p. 357
	(b) **84**, K65 (1977)	*6*, p. *60*; *20.1*, p. *262*; 22.8, p. 400
K. W. Böer	(a) **40**, 355 (1977)	14.3.3, p. 178; 14.3.4, p. 183; *16*, p. *201*; 20.1, p. 247; 22, p. 320; 22.4.1, p. 362
E. Boesman	(b) **79**, K23 (1977)	*19*, p. *235*; 22.5.2, p. 375
	(b) **82**, K69 (1977)	*19*, p. *241*; 22.6, p. 390
A. Boeuf	(a) **44**, 127 (1977)	*1.1*, p. *13*; 22.1.2, p. 343
B. F. Bogacz	(a) **44**, K107 (1977)	*5*, p. *50*; 22.8.1, p. 408
Yu. V. Bogatyrev	(a) **41**, K151 (1977)	*11*, p. *96*; *14.4*, p. *191*; 22, p. 325; 22.6, p. 389
G. A. Bogomolova	(a) **40**, 643 (1977)	6, p. 55; *13.4*, p. *133*; 20.3, p. 269; 22.8.1, p. 403
J. Bohm	(a) **39**, 517 (1977)	*1.4*, p. *24*; *10.2*, p. *84*; 22.8.2, p. 409
	(a) **43**, 71 (1977)	3, p. 42; 13.4, p. 137; 20.1, p. 260; *20.2*, p. *266*; *20.3*, p. *272*; 22.8, p. 399
M. Böhm	(a) **41**, 535 (1977)	*14.3*, p. *169*; 22.8.1, p. 405
W. Bohmeyer	(a) **41**, K27 (1977)	*14.3.4*, p. *185*; 22.1.3, p. 345
J. N. Boland	(a) **39**, 631 (1977)	*1.4*, p. *24*; *10.2*, p. *84*; 22.8, p. 396
W. Bollmann	(a) **39**, 477 (1977)	*10.2*, p. *84*; 14.3, p. 166; 20.1, p. 250; 22.8.1, p. 402
	(a) **40**, 49 (1977)	*10.2*, p. *85*; 22.5.1, p. 373; 22.5. 2, p. 376; 22.5.3, p. 383
	(a) **40**, 83 (1977)	*10.2*, p. *85*; *16*, p. *201*; 22.8.1, p. 403
	(a) **40**, 409 (1977)	*10.2*, p. *86*; 14.4, p. 190; 22.5.3, p. 383
	(a) **43**, 175 (1977)	*10.2*, p. *90*; 22.5.2, p. 380; 22.5.3, p. 384
D. Bolze	(a) **39**, 173 (1977)	*1.5*, p. *29*; 13.1, p. 114; 20.1, p. 248; 22.2, p. 348
V. L. Bonch-Bruevich	(b) **84**, 99 (1977)	*13*, p. *112*; 14.3, p. 171; 22, p. 329
E. Bonetti	(a) **39**, 661 (1977)	*10.1*, p. *76*; *12.1*, p. *101*; 21, p. 278
	(a) **44**, K31 (1977)	*10.1*, p. *82*; *12.1*, p. *103*; 21, p. 287
J. E. Bonnet	(a) **40**, 101 (1977)	*11*, p. *95*; 21.4, p. 311
R. Bonnet	(a) **40**, 599 (1977)	*1*, p. *9*; *1.4*, p. *25*; 21.1, p. 292
U. Bonse	(a) **43**, 487 (1977)	*1.1*, p. *13*; 22.1.2, p. 342
	(a) **44**, K131 (1977)	*1*, p. *10*
K. Borggreen	(a) **39**, 223 (1977)	*1.4*, p. *24*; 10, p. 72
S. Bornmann	(a) **41**, 505 (1977)	*3*, p. *42*; 22.8.2, p. 411
	(a) **43**, 277 (1977)	3, p. 43; *18.3*, p. *230*; 22.8.2, p. 412
S. M. Borodachev	(b) **84**, 33 (1977)	*14.1*, p. *159*; 18.4, p. 233; 21, p. 286

Yu. G. Borodko	(b) **83**, K141 (1977)	13, p. 111; 17, p. 207; *20*, p. *246*; 22.9, p. 418
	(a) **43**, K73 (1977)	*20.1*, p. *260*; 22.9, p. 418
R. Borrelly	(a) **39**, 525 (1977)	*15*, p. *198*; 21.1, p. 289; 21.1.1, p. 302
V. V. Borshch	(a) **44**, K15 (1977)	*20.1*, p. *262*; 22.3, p. 359
G. Borstel	(b) **80**, 109 (1977)	*13.5*, p. *141*; *20.1*, p. *250*; 22.6, p. 386
	(b) **83**, 11 (1977)	*6*, p. *51*; *13.5*, p. *141*; 20.1, p. 247; 22, p. 320; 22.6, p. 386
J. von Borzeszkowski	(b) **79**, 525 (1977)	6, p. 52; 13.1, p. 115; *14.3.4*, p. *183*; 22.1.2, p. 337
A. van den Bosch	(a) **41**, 225 (1977)	*11*, p. *95*; *14.2*, p. *162*; 18.1, p. 214; 21.1.1, p. 303; 21.5, p. 314
H. N. Bose	(a) **44**, K127 (1977)	*1.5*, p. *30*; 14.3.1, p. 175; 16, p. 205; 22, p. 331
L. Bosi	(b) **82**, K49 (1977)	*10.2*, p. *89*; *13.4*, p. *136*; 20.3, p. 271; 22.5.2, p. 379
O. Bostanjoglo	(a) **40**, K97 (1977)	*18.2.1*, p. *226*; 21.1.1, p. 303
	(a) **42**, K103 (1977)	*18.2.1*, p. *226*; 21.1.1, p. 305
H. Böttger	(b) **80**, 569 (1977)	2, p. 35; 6, p. 54; 13, p. 108; 13.5.3, p. 151; *14.3*, p. *167*; 22, p. 323
	(b) **81**, 97 (1977)	2, p. 36; 6, p. 55; 13, p. 109; 13.5.3, p. 151; *14.3*, p. *167*; 22, p. 324
	(b) **81**, 433 (1977)	2, p. 37; 6, p. 56; 13, p. 109; 13.5.3, p. 151; *14.3*, p. *168*; 22, p. 325
	(b) **84**, 575 (1977)	*2*, p. *40*; *6*, p. *60*
R. Böttger	(a) **43**, K131 (1977)	*10.2*, p. *91*; *20.1*, p. *261*; 22.6, p. 391
B. Boucher	(a) **40**, 197 (1977)	*2*, p. *35*; *18.2*, p. *219*; 21.4, p. 311; 21.6, p. 315
	(a) **42**, K165 (1977)	*2*, p. *38*; *18.2*, p. *222*; 21.4, p. 313; 21.6, p. 316
V. A. M. Brabers	(a) **40**, K169 (1977)	*19*, p. *238*; 22.8.2, p. 410
F. G. Brady Moreira	see under letter "M"	
B. Brandt	(a) **43**, K41 (1977)	*5*, p. *49*; *18.1*, p. *215*; 22.8.1, p. 407
E. H. Brandt	(b) **79**, 479 (1977)	*14.2*, p. *161*
	(b) **84**, 269 (1977)	*14.2*, p. *163*
	(b) **84**, 637 (1977)	*14.2*, p. *164*
W. Brandt	(b) **81**, 609 (1977)	*13*, p. *109*; 22.5.2, p. 378; 22.6, p. 388
J. Bras	(a) **41**, 595 (1977)	*1.2*, p. *17*; *1.3*, p. *22*; 21, p. 281; 21.1.1, p. 304
M. Brauer	(b) **81**, 147 (1977)	*13*, p. *109*; *14.3.4*, p. *185*; 22.2.1, p. 351
W. Braune	(b) **79**, 501 (1977)	*13.2*, p. *124*; 14.1, p. 157; 21.7, p. 319
R. Braunstein	(b) **80**, 451 (1977)	13.1, p. 116; *14.4*, p. *190*; *20.1*, p. *252*; 22.8, p. 397
E. Bredner	(a) **43**, 159 (1977)	13.5.1, p. 147; *20.3*, p. *272*; 22.1.3, p. 346
F.-W. Breitbarth	(a) **41**, 151 (1977)	13.3, p. 126; 14.3.2, p. 176; *14.3.4*, p. *185*; 22.9, p. 415
B. Březina	(b) **79**, K35 (1977)	*20.1*, p. *248*; 22.8.1, p. 402
S. H. van den Brink	(a) **41**, 513 (1977)	*10.1*, p. *78*; 21, p. 281; 21.1, p. 294
D. Brinkmann	(b) **83**, K45 (1977)	*10.2*, p. *90*; *19*, p. *241*; 22.8, p. 399
G. K. Bristow	(b) **82**, K81 (1977)	*13.2*, p. *124*; *14.1*, p. *159*; 21, p. 283
W. Brodkorb	(a) **44**, 161 (1977)	*14.3.3*, p. *181*; *14.3.4*, p. *187*; 22, p. 330
A. M. Brodskii	(b) **83**, 331 (1977)	*17*, p. *207*; 21, p. 284; 22, p. 328
	(b) **83**, 633 (1977)	*13.3*, p. *128*; 20.1, p. 260; 21, p. 285
F. Brosens	(b) **80**, 99 (1977)	*14*, p. *155*
	(b) **81**, 551 (1977)	*14.4*, p. *191*
	(b) **81**, 557 (1977)	*4*, p. *45*; 6, p. 56; 21, p. 280; 21.2, p. 308
	(b) **82**, 117 (1977)	*14*, p. *155*
	(b) **82**, 439 (1977)	*13.5.3*, p. *151*
H. Bross	(b) **80**, 137 (1977)	*6*, p. *53*; 21.2, p. 308
A. D. Brothers	(b) **80**, 201 (1977)	12.2, p. 104; 13.1, p. 115; *13.5.1*, p. *144*; 22.5, p. 370
M. Brousseau	(b) **80**, 461 (1977)	*13.5.1*, p. *144*; 20.3, p. 269; 22.1.2, p. 338

K. L. Brower	(a) **41,** 637 (1977)	*10.2,* p. *88*; 11, p. 95; *19,* p. *240*; 22.1.2, p. 340
D. Broz	(a) **41,** 255 (1977)	*1.2,* p. *16*; 14.3, p. 168; 22.3, p. 358
L. Brú	(a) **40,** 167 (1977)	*10.1,* p. *76*; 21.1, p. 290
	(a) **43,** K187 (1977)	*10.1,* p. *81*; 21.1, p. 298
M. Bruchholz	(a) **44,** K13 (1977)	*10.2,* p. *92*; 22.2.2, p. 355
H.-P. Brückner	(a) **42,** 295 (1977)	*1.2,* p. *17*; *3,* p. *42*; 14.3, p. 170; 18.1, p. 214; 22.6.1, p. 393
W. Brückner	(a) **40,** K131 (1977)	*1.2,* p. *16*; *18.1,* p. *214*; 22.6.1, p. 393
	(a) **42,** 295 (1977)	*1.2,* p. *17*; *3,* p. *42*; 14.3, p. 170; 18.1, p. 214; 22.6.1, p. 393
	(a) **44,** K95 (1977)	*1.2,* p. *20*; 5, p. 49; 14.3, p. 172; 22.6, p. 391
H.-G. Brühl	(a) **39,** 133 (1977)	*1.1,* p. *11*; 1.2, p. 14; 22.2.4, p. 357
H. J. de Bruin	(b) **82,** 581 (1977)	*9,* p. *70*; 10.1, p. 79; 21, p. 283
O. Brümmer	(a) **39,** 401 (1977)	*10.1,* p. *75*; 21.1, p. 289
	(a) **39,** K95 (1977)	*10.1,* p. *76*; 21, p. 278; 21.1, p. 289
	(b) **80,** K95 (1977)	*13.1,* p. *117*; 17, p. 206; 20, p. 245; 22.7, p. 394
	(a) **42,** K15 (1977)	*10.1,* p. *79*; 21, p. 282
	(b) **83,** 257 (1977)	*13.1,* p. *121*; 17, p. 207; 20, p. 245; 21.1, p. 297
U. Brunsmann	(a) **42,** K79 (1977)	*17,* p. *206*; 22.5.2, p. 379
	(a) **43,** 519 (1977)	*17,* p. *207*; 22.5.2, p. 380
V. V. Bryksin	(b) **80,** 569 (1977)	2, p. 35; 6, p. 54; 13, p. 108; 13.5.3, p. 151; *14.3,* p. *167*; 22, p. 323
	(b) 81, 97 (1977)	2, p. 36 ;6, p. 55; 13, p. 109; 13.5.3, p. 151; *14.3,* p. *167*; 22, p. 324
	(b) 81, 433 (1977)	2, p. 37; 6, p. 56; 13, p. 109; 13.5.3, p. 151; *14.3,* p. *168*; 22, p. 325
R. Bubáková	(a) **39,** 725 (1977)	*10.2,* p. *85*; 22.1.2, p. 338
V. T. Bublik	(a) **39,** 345 (1977)	1.2, p. 14; *1.3,* p. *21*; 22.4.4, p. 368
J. Buchar	(a) **44,** 117 (1977)	*10.1,* p. *81*; 12.1, p. 103; 21.1, p. 299
G. Buchheim	(a) **44,** 585 (1977)	2, p. 41; *14.3.1,* p. *175*; *14.3.2,* p. *177*; 22.1.2, p. 344
U. Büchner	(b) **80,** 615 (1977)	13.1, p. 116; *14.4,* p. *190*; 22.6.1, p. 393
	(b) **81,** 227 (1977)	13.1, p. 117; *20.1,* p. *253*; 22.1, p. 332; 22.2, p. 348
	(b) **83,** 493 (1977)	*13.1,* p. *122*; 14.4, p. 192; 22, p. 328
O. Buck	(a) **42,** 285 (1977)	*10.1,* p. *79*; 11, p. 96; 12.1, p. 102; 21.1, p. 295
I. S. Buda	(b) **81,** 709 (1977)	*14.3,* p. *168*; 15, p. 199; 22.1.1, p. 334
G. Buhr	(b) **82,** 297 (1977)	*14,* p. *155*
S. J. H. Bukhari	(a) **40,** 605 (1977)	10.2, p. 86; 13.4, p. 133; *20.3,* p. *269*; 22.6, p. 387
Yu. V. Bulgakov	(b) **84,** K79 (1977)	*11,* p. *98*; 22.1, p. 332
D. St. P. Bunbury	(a) **41,** 585 (1977)	*5,* p. *48*; 18.2, p. 220; 21.1, p. 294; 21.4, p. 312
E. Burattini	(a) **42,** K41 (1977)	*14.1,* p. *159*; 21, p. 282
V. A. Buravikhin	(a) **40,** 11 (1977)	*1.2,* p. *14*; 1.4, p. 24; *1.5,* p. *29*; 18.2, p. 217; 21.1, p. 288; 21.1.1, p. 302
A. F. Burenkov	(b) **79,** K161 (1977)	*11,* p. *94*; 22.5.2, p. 375
J. Burgos	(b) **83,** 249 (1977)	*13.5.1,* p. *146*; *20.3,* p. *272*; 22.9, p. 418
F. Burkhardt	(a) **39,** K63 (1977)	*11,* p. *94*; 22.1.2, p. 337
A. Buroff	(a) **40,** K195 (1977)	*2,* p. *36*; *9,* p. *70*; 22, p. 324
G. Burrafato	(b) **81,** 579 (1977)	*14.2,* p. *162*
K. H. J. Buschow	(a) **40,** 447 (1977)	*1.2,* p. *16*; *18.4,* p. *232*; 22.8.1, p. 403
G. Busse	(a) **43,** K103 (1977)	12.2, p. 106; *20.1,* p. *260*; 22.5.2, p. 380
T. I. Butaeva	(a) **39,** 541 (1977)	*13.4,* p. *131*; 20.2, p. 265; 20.3, p. 268; 22.8, p. 396
	(a) **40,** K69 (1977)	*20.2,* p. *265*; 22.8.2, p. 410
	(a) **41,** K199 (1977)	*20.2,* p. *265*; 22.8.2, p. 410
E. Butter	(a) **43,** K95 (1977)	*14.3,* p. *171*; 22.2.2, p. 354

C. Butti	(a) **42**, 545 (1977)	*13.1*, p. *121*; *20.3*, p. *272*; 22.4.1, p. 363
J. Buxo	(a) **42**, 173 (1977)	14.3.2, p. 176; *14.3.3*, p. *180*
Yu. F. Bydin	(a) **39**, K53 (1977)	*11*, p. *94*; *18.2*, p. *217*; 21.1, p. 289
V. N. Bykov	(a) **40**, K89 (1977)	*9*, p. *69*; *11*, p. *95*; 21, p. 280; 21.1, p. 292
J. G. Byrne	(a) **41**, K75 (1977)	*12.1*, p. *101*; 21, p. 280
	(a) **42**, K175 (1977)	*10.1*, p. *80*; 21, p. 284
P. Byszewski	(b) **84**, K83 (1977)	*13.4*, p. *140*; *14.3*, p. *172*; 22.4.3, p. 367
J. M. Cabrera	(a) **41**, K119 (1977)	*10.2*, p. *89*; 20.1, p. 256; 22.5.2, p. 378
G. Le Caer	(a) **40**, 425 (1977)	*1.3*, p. *21*; 5, p. 48; 21, p. 279; 21.1.1, p. 303
M. Cahoreau	(a) **42**, 449 (1977)	*10*, p. *73*; *12*, p. *100*
V. B. Campos	(b) **81**, 657 (1977)	*13.5.2*, p. *149*; 14.4, p. 191; 21, p. 281
V. Canevari	(b) **81**, 665 (1977)	13.5.1, p. 146; *20.1*, p. *255*; 22.3, p. 358
Cao-Xŭân An	(a) **43**, K69 (1977)	6, p. 59; *12.2*, p. *106*; 22.5.3, p. 384
V. Čápek	(b) **81**, 571 (1977)	*13.4*, p. *134*; 20.1, p. 255; 22.8.2, p. 411
P. Čapková	(a) **42**, 681 (1977)	*6*, p. *58*; *8*, p. *66*; 21, p. 283; 21.3, p. 310
V. Capozzi	(a) **40**, 93 (1977)	*13.4*, p. *132*; *14.3.4*, p. *184*; 22.3, p. 358
M. Cardona	(b) **79**, 269 (1977)	*6*, p. *51*; 12.2, p. 104; *20.1*, p. *247*; 22, p. 320
G. S. Cargill III	(a) **40**, 37 (1977)	*2*, p. *35*; 22.1.1, p. 333
A. Carl	(a) **43**, 443 (1977)	*1.2*, p. *19*; 1.3, p. 23; 8, p. 67; 21.1, p. 297
D. Carles	(a) **44**, 97 (1977)	2, p. 40; 13.4, p. 139; 14.3.1, p. 175; *16*, p. *204*; 22.1.3, p. 347
M. T. Carlos	(b) **82**, K103 (1977)	*10.2*, p. *90*; *20.1*, p. *258*; 22.5.4, p. 385
G. M. Carneiro	(b) **81**, 79 (1977)	*13.5.4*, p. *153*; 18.2, p. 219; 21, p. 280
P. E. Caro	(b) **81**, 463 (1977)	*13.4*, p. *134*; 20.1, p. 255; 22.8.2, p. 411
M. Casalboni	(b) **84**, 23 (1977)	*13.4*, p. *138*; *20.3*, p. *274*; 22.5.2, p. 381
L. M. Caspers	(a) **39**, K143 (1977)	*10.1*, p. *76*; *11*, p. *94*; 21.1, p. 290
	(a) **40**, 235 (1977)	*10.1*, p. *76*; 21, p. 279
A. Castaldini	(a) **43**, K205 (1977)	10.2, p. 91; *13.4*, p. *138*; 14.3, p. 171; 22.1.1, p. 336
B. G. Caswell	(a) **44**, K47 (1977)	*14.3.3*, p. *181*; *16*, p. *205*; 22, p. 330; 22.4.1, p. 364
R. M. Catchings	(a) **39**, K71 (1977)	*2*, p. *33*; *18*, p. *209*; 22.8, p. 396
	(a) **39**, K101 (1977)	*2*, p. *34*; *19*, p. *236*; 22.8, p. 396
A. Cavallini	(a) **43**, K205 (1977)	10.2, p. 91; *13.4*, p. *138*; 14.3, p. 171; 22.1.1, p. 336
B. Čelustka	(a) **41**, 255 (1977)	*1.2*, p. *16*; 14.3, p. 168; 22.3, p. 358
Hilda A. Cerdeira	(b) **83**, 607 (1977)	13.1, p. 122; *14.4*, p. *192*; *20*, p. *246*; 22, p. 329
V. I. Chaikin	(a) **43**, 311 (1977)	*14.3.2*, p. *177*; *16*, p. *204*; 22.2.3, p. 356
I. A. Chaikovskii	(b) **80**, K5 (1977)	*14.1*, p. *158*; 21.7, p. 319
	(b) **80**, 697 (1977)	*6*, p. *55*; 22, p. 323
	(b) **82**, 391 (1977)	*13.5.2*, p. *149*; 22, p. 326
S. K. Chakrabarti	(b) **82**, 557 (1977)	*1.2*, p. *17*; *12.2*, p. *105*; 22.5.2, p. 379
D. K. Chakrabarty	(a) **44**, 787 (1977)	*14.4.1*, p. *197*; 20.1, p. 264; 22.8.1, p. 408
A. N. Chakravarti	(b) **79**, K31 (1977)	13.1, p. 114; *14.3*, p. *165*; 18, p. 209; 22, p. 320
	(a) **44**, K111 (1977)	*14.3.2*, p. *177*; 22.1.2, p. 343
M. A. Chalabyan	(a) **42**, 273 (1977)	*18.2.1*, p. *226*; 21.1, p. 295; 21.1.1, p. 304
D. Chambaere	(a) **43**, 619 (1977)	*5*, p. *49*; 22.8.2, p. 412
A. Chamberod	(b) **82**, 269 (1977)	*1.2*, p. *17*; *18.2*, p. *221*; 21.1, p. 295; 21.1.1, p. 304
	(b) **84**, 335 (1977)	*1.3*, p. *23*; *18.2*, p. *223*; 21.1.1, p. 306
B. P. Chandra	(a) **40**, 403 (1977)	*12.2*, p. *104*; *20.3*, p. *269*; 22.9, p. 415
K. Chandra	(b) **80**, K1 (1977)	*13.4*, p. *132*; 22.5.3, p. 383
	(a) **43**, 357 (1977)	*1.3*, p. *22*; 5, p. 49; 21.1.1, p. 305
	(a) **44**, 443 (1977)	*5*, p. *50*; 22.8.2, p. 413
R. Chandra	(b) **83**, 273 (1977)	*5*, p. *49*; 22.8, p. 399

C. P. Chang	(a) **39,** 549 (1977)	*10.1,* p. *75*; 21, p. 277
K. A. Chao	(b) **84,** 747 (1977)	*13,* p. *113*; 18.2, p. 224; 18.4, p. 234
I. M. Chapnik	(a) **39,** K135 (1977)	*13,* p. *108*; 21.1, p. 289; 21.1.1, p. 302; 22.7, p. 394
	(a) **41,** K71 (1977)	*14.3,* p. *168*; 22, p. 325
G. Chapuis	(a) **43,** 203 (1977)	*1.2,* p. *18*; 22.9, p. 418
G. Chassagne	(a) **40,** 629 (1977)	*10.2,* p. *87*; 20.1, p. 253; 22.5.2, p. 377
	(a) **41,** 183 (1977)	*10.2,* p. *87*; 22.5.2, p. 378
S. Chatterjee	(b) **81,** 493 (1977)	*13.1,* p. *118*; 22.4.1, p. 363
D. Chattopadhyay	(b) **81,** K51 (1977)	13.1, p. 117; *14.3,* p. *168*; 22, p. 324
P. K. Chaturvedi	(a) **40,** K191 (1977)	*14.3.3,* p. *179*
A. K. Chaudhuri	(a) **44,** K127 (1977)	*1.5,* p. *30*; 14.3.1, p. 175; 16, p. 205; 22, p. 331
S. Chaudhuri	(a) **43,** K9 (1977)	14.1, p. 159; *18.1,* p. *215*; *18.2,* p. *222*; 21.1, p. 297
A. S. Chaves	(b) **79,** 731 (1977)	*6,* p. *52*; 14.4, p. 189; 20.1, p. 249; 22.8.1, p. 402
V. N. Chebotar	(b) **80,** K5 (1977)	*14.1,* p. *158*; 21.7, p. 319
V. V. Chebotaru	(b) **83,** K149 (1977)	13.5.1, p. 147; *20.1,* p. *261*; 22.3, p. 358
T. I. Chekmasova	(a) **44,** K155 (1977)	*1.2,* p. *20*; *14.4.1,* p. *197*; 19, p. 243; 22.8.1, p. 408
J. R. Chelikowsky	(b) **82,** 357 (1977)	*13.1,* p. *119*; 14, p. 155; 22, p. 326
A. Chełkowski	(b) **81,** K95 (1977)	*18.1,* p. *214*; *19,* p. *239*; 22.8.2, p. 411
C. H. Chen	(b) **83,** 347 (1977)	*13.1,* p. *121*; 14.4, p. 192; 22.1.2, p. 341
C. W. Chen	(a) **39,** 549 (1977)	*10.1,* p. *75*; 21, p. 277
G. L. Chen	(b) **83,** 115 (1977)	*6,* p. *58*; *10,* p. *73*
I. V. Chepeleva	(b) **82,** 189 (1977)	2, p. 37; *19,* p. *240*; 22.8, p. 399
F. G. Cherkasov	(a) **40,** K53 (1977)	*19,* p. *237*; 21.2, p. 308
N. A. Chernoplekov	(b) **79,** 105 (1977)	*6,* p. *51*; 21.1, p. 288
A. P. Chernov	(b) **82,** 189 (1977)	2, p. 37; *19,* p. *240*; 22.8, p. 399
I. Chernov	(a) **43,** K201 (1977)	*10.2,* p. *91*; *11,* p. *98*; 22.1.2, p. 342
M. A. Chernov	(a) **43,** 363 (1977)	*1,* p. *10*; 1.2, p. 18; 22.3, p. 358
V. L. Chernov	(b) **84,** K79 (1977)	*11,* p. *98*; 22.1, p. 332
T. Chiba	(b) **81,** 609 (1977)	*13,* p. *109*; 22.5.2, p. 378; 22.6, p. 388
H. Chihara	(a) **41,** K17 (1977)	*1.2,* p. *16*; *19,* p. *239*; 22.9, p. 416
M. Chiriac	(a) **44,** K63 (1977)	*18.2,* p. *224*; 18.4, p. 234; 21.1, p. 299
E. V. Chisler	(b) **79,** 347 (1977)	*6,* p. *51*; *20.1,* p. *248*; 22, p. 320
V. Chiu	(b) **83,** 365 (1977)	*6,* p. *59*; *10,* p. *73*
E. Chock	(b) **80,** 451 (1977)	13.1, p. 116; *14.4,* p. *190*; *20.1,* p. *252*; 22.8, p. 397
B. K. Choi	(a) **43,** 111 (1977)	*5,* p. *49*; *18.2,* p. *222*; 21.1.1, p. 305; 21.4, p. 313
K. L. Chopra	(a) **40,** 247 (1977)	*2,* p. *35*; *14.3.1,* p. *173*; 15, p, 198; 21.1, p. 291; 21.6, p. 315
D. R. Choudhury	(a) **44,** K111 (1977)	*14.3.2,* p. *177*; 22.1.2, p. 343
A. Chovet	(a) **43,** 213 (1977)	*14.3,* p. *170*; 22.1.1, p. 335; 22.1.2, p. 341; 22.2.3, p. 356
A. K. Chowdhury	(b) **79,** K31 (1977)	13.1, p. 114; *14.3,* p. *165*; 18, p. 209; 22, p. 320
	(a) **44,** K111 (1977)	*14.3.2,* p. *177*; 22.1.2, p. 343
A. N. Christensen	(a) **43,** 81 (1977)	*1,* p. *10*; *18.4,* p. *233*; 22.9, p. 418
V. Christoph	(b) **80,** K117 (1977)	*14.1,* p. *158*; *14.3,* p. *167*
	(b) **81,** K125 (1977)	*14.1,* p. *158*; 21, p. 281
	(b) **82,** 297 (1977)	*14,* p. *155*
F. A. Chudnovskii	(a) **41,** 125 (1977)	*1.2,* p. *16*; 14.3, p. 168; 22.6, p. 388
G. P. Chugunova	(b) **81,** 69 (1977)	*5,* p. *48*
D. Y. Chung	(a) **43,** 329 (1977)	*7,* p. *63*; *12.2,* p. *106*; 22.6.1, p. 393
M. Chybicki	(a) **39,** 271 (1977)	14.3.1, p. 173; *14.3.3,* p. *178*; 22.9, p. 414
M. Ciechanowicz-Rutkowska	(a) **43,** 53 (1977)	*1.2,* p. *18*; 22.8, p. 399
J. V. Cieminski	(a) **40,** K153 (1977)	*14.4.1,* p. *195*; 22.8.1, p. 404

A. Cingolani	(a) **40**, 93 (1977)	*13.4*, p. *132*; *14.3.4*, p. *184*; 22.3, p. 358
J. Cisowski	(a) **41**, K59 (1977)	12.2, p. 105; *14.3*, p. *168*; 22.3.1, p. 360
A. Čížek	(a) **43**, 589 (1977)	*10.2*, p. *91*; 22, p. 329
J. A. Cocks	(b) **83**, 365 (1977)	*6*, p. *59*; *10*, p. *73*
M. L. Cohen	(b) **82**, 357 (1977)	*13.1*, p. *119*; 14, p. 155; 22, p. 326
L. B. Coleman	(b) **82**, K117 (1977)	*1.1*, p. *12*; 22.9, p. 417
J. Collet	(b) **80**, 461 (1977)	*13.5.1*, p. *144*; 20.3, p. 269; 22.1.2, p. 338
A. Collomb	(a) **41**, 459 (1977)	*18.4*, p. *233*; 22.8.1, p. 405
R. Collongues	(a) **41**, 423 (1977)	*19*, p. *239*; 22.6.1, p. 393
A. Conan	(a) **41**, 629 (1977)	13.1, p. 118; *14.3*, p. *169*; *15*, p. *199*; 22, p. 325
M. Condat	(a) **40**, K183 (1977)	*10*, p. *72*
	(a) **41**, K5 (1977)	*1.4*, p. *26*; *10.1*, p. *78*; 21, p. 280
H. van Cong	(b) **80**, K45 (1977)	*14.3*, p. *166*; 22.2.3, p. 356
J. W. Corbett	(b) **81**, 637 (1977)	6, p. 56; *13.4*, p. *134*; 22.1.2, p. 340
	(a) **41**, 637 (1977)	*10.2*, p. *88*; 11, p. 95; *19*, p. *240*; 22.1.2, p. 340
J. M. Corbett	(a) **44**, 353 (1977)	*1.1*, p. *13*; 21.6, p. 317
A. Coret	(b) **83**, 599 (1977)	*13.5.1*, p. *147*; *16*, p. *204*; 22.9, p. 418
P. G. McCormick	see under letter "M"	
J. Cornelis	(b) **81**, 557 (1977)	*4*, p. *45*; 6, p. 56; 21, p. 280; 21.2, p. 308
W. D. Corner	(a) **43**, 251 (1977)	*14.1*, p. *159*; 21.4, p. 313
P. R. Couchman	(a) **42**, K47 (1977)	*12.1*, p. *102*; *12.2*, p. *105*; 21, p. 282; 22.5.2, p. 379; 23, p. 420
Y. Couder	(b) **81**, 615 (1977)	13.1, p. 118; *18*, p. *210*; *20.1*, p. *255*; 22.4.4, p. 368
	(b) **82**, 149 (1977)	13.1, p. 118; *18*, p. *211*; *20.1*, p. *256*; 22.4.4, p. 368
J. J. Couderc	(a) **41**, 595 (1977)	*1.2*, p. *17*; *1.3*, p. *22*; 21, p. 281; 21.1.1, p. 304
J. E. Cousins	(b) **79**, 645 (1977)	*19*, p. *235*; 21, p. 277
M. Couzi	(a) **41**, 271 (1977)	*1.2*, p. *16*; 6, p. 56; 20.1, p. 254; 22.9, p. 416
J. H. Crawford, Jr.	(b) **79**, 559 (1977)	10.2, p. 84; *20.3*, p. *267*; 22.5.2, p. 375
	(a) **39**, 669 (1977)	*10.2*, p. *84*; 19, p. 236; *20.3*, p. *268*; 22.6, p. 386
	(a) **42**, K137 (1977)	11, p. 97; *20.1*, p. *259*; *20.3*, p. *272*; 22.8.2, p. 412
R. K. Crawford	(a) **43**, 611 (1977)	*1.2*, p. *19*; 23, p. 420
V. H. Crawford	(b) **81**, K75 (1977)	*18.4*, p. *233*; 19, p. 239; 22.9, p. 416
D. C. Creagh	(a) **39**, 705 (1977)	*20*, p. *244*; 21.3, p. 310
M. Crişan	(b) **80**, K121 (1977)	1.2, p. 16; *14.2*, p. *162*; *18.2*, p. *219*
S. Cristoloveanu	(a) **39**, 109 (1977)	14.3, p. 165; *15*, p. *198*; 22.2.3, p. 356
	(a) **43**, 213 (1977)	*14.3*, p. *170*; 22.1.1, p. 335; 22.1.2, p. 341; 22.2.3, p. 356
L. E. Cross	(a) **41**, K83 (1977)	*14.4.1*, p. *195*; 22.8.1, p. 404
G. Cubiotti	(b) **79**, 655 (1977)	*2*, p. *33*; *13*, p. *107*; 14, p. 155; 22, p. 321
A. M. Cucolo	(a) **42**, K111 (1977)	*14.2*, p. *162*
I. Curelaru	(b) **79**, 277 (1977)	*13.1*, p. *114*; 20, p. 244; 21.1, p. 288
C.-P. Czaya	(b) **82**, 595 (1977)	*13.4*, p. *136*; 14.3.2, p. 177; 22.1.3, p. 346
G. Czjzek	(b) **84**, 611 (1977)	*5*, p. *50*; 14.2, p. 164; *18.2*, p. *224*; 21.4, p. 313
L. Dąbrowski	(a) **39**, 643 (1977)	*5*, p. *47*; 22.8.2, p. 409
	(a) **41**, 191 (1977)	*5*, p. *48*; 18.2, p. 219; 21.1.1, p. 303
L. Dagens	(b) **84**, 311 (1977)	*4*, p. *46*; *13*, p. *112*; 21.1, p. 299; 21.6, p. 317
J. Z. Damm	(b) **80**, K59 (1977)	*20.3*, p. *268*; 22.5.2, p. 376
H. Danan	(a) **42**, 227 (1977)	*18.2.1*, p. *226*; 21.1, p. 295
W. Dander	(b) **80**, 173 (1977)	*11*, p. *94*; *14.1*, p. *157*; 21.1, p. 290
Dang Huy Lam	see under letter "L"	
J. N. Daou	(a) **40**, 101 (1977)	*11*, p. *95*; 21.4, p. 311
A. Daoud	(a) **41**, 271 (1977)	*1.2*, p. *16*; 6, p. 56; 20.1, p. 254; 22.9, p. 416

A. Dargys	(a) **41**, 177 (1977)	13.4, p. 134; *14.3.4*, p. *185*; 22.1.3, p. 345
M. P. Das	(b) **83**, K165 (1977)	2, p. 39; *13*, p. *111*; 21, p. 285
S. K. Das	(b) **83**, 615 (1977)	*6*, p. *59*; 21, p. 285; 21.1, p. 297; 21.6, p. 317
N. Dass	(b) **80**, K1 (1977)	*13.4*, p. *132*; 22.5.3, p. 383
Nguyen Nhu Dat	(b) **82**, K39 (1977)	*13*, p. *110*; *20*, p. *245*
	(b) **84**, 115 (1977)	*6*, p. *60*; *13.5*, p. *142*; 20, p. 246; 22.4.1, p. 364
T. Datta	(b) **81**, 281 (1977)	*14*, p. *155*; 15, p. 198; 22.9, p. 415
A. Dauger	(a) **42**, 687 (1977)	*1.3*, p. *22*; 12.1, p. 102; 21, p. 283; 21.1, p. 296
	(a) **43**, 625 (1977)	1.3, p. 23; *12.1*, p. *102*; 21, p. 286; 21.3, p. 310
C. Dauwe	(a) **40**, 153 (1977)	*8*, p. *65*; 21, p. 279
M. David	(b) **79**, 215 (1977)	*1.1*, p. *11*
	(b) **80**, 477 (1977)	*1.1*, p. *11*
	(b) **84**, 133 (1977)	*1.1*, p. *13*
T. Sh. Davitashvili	(a) **40**, K113 (1977)	8, p. 65; *10.2*, p. *87*; *20.1*, p. *253*; 22.5.2, p. 377
V. Yu. Davydov	(b) **79**, 347 (1977)	*6*, p. *51*; *21.0*, p. *248*; 22, p. 320
M. P. Davydova	(b) **81**, 287 (1977)	*13.4*, p. *133*; 19, p. 238; 20.3, p. 270; 22.5.4, p. 385
B. Dayal	(b) **81**, 403 (1977)	*6*. p, *55*; 22.5.2, p. 377
Udayan De	(b) **79**, K87 (1977)	*9*, p. *69*
P. Deák	(b) **79**, K89 (1977)	*10.2*, p. *84*; 22.1.2, p. 338
R. J. Deck	(b) **81**, 281 (1977)	*14*, p. *155*; 15, p. 198; 22.9, p. 415
E. Dedieu	(a) **42**, 449 (1977)	*10*, p. *73*; *12*, p. *100*
A. Defrain	(b) **80**, 533 (1977)	*2*, p. *35*; 8, p. 65; 21, p. 279
J. Degauque	(a) **44**, K59 (1977)	*10.1*, p. *82*; *18.2*, p. *224*; 21.1.1, p. 306
G. Dehe	(a) **39**, K115 (1977)	5, p. 47; *18.3*, p. *228*; 22.6, p. 386
I. Ya. Dekhtyar	(a) **42**, K15 (1977)	*10.1*, p. *79*; 21, p. 282
	(b) **83**, 323 (1977)	*13*, p. *110*; 21, p. 284
	(a) **43**, 589 (1977)	*10.2*, p. *91*; 22, p. 329
A. Delahaigue	(b) **83**, 139 (1977)	*6*, p. *58*; *14.4.1*, p. *196*; 20.1, p. 259; 22.9, p. 417
A. Delapalme	(a) **44**, 713 (1977)	*18*, p. *212*; 22.8, p. 401
D. Delaunay	(a) **41**, 629 (1977)	13.1, p. 118; *14.3*, p. *169*; *15*, p. *199*; 22, p. 325
P. Delavignette	(a) **41**, 51 (1977)	*1.4*, p. *25*; 22.7, p. 394
C. J. Delbecq	(b) **80**, 471 (1977)	*10.2*, p. *86*; 20.1, p. 252; 22.5.2, p. 377
P. Delhaes	(b) **80**, 125 (1977)	*8*, p. *64*; 18.1, p. 213; 22.9, p. 414
R. Delhez	(a) **44**, 517 (1977)	1, p. 10; *9*, p. *71*; 22.5.2, p. 382
K. M. Demchuk	(b) **82**, 59 (1977)	*13.1*, p. *118*; 14.3, p. 169; 15, p. 199; 22.2.3, p. 356
V. B. Demin	(b) **82**, K59 (1977)	*10.1*, p. *78*; 21, p. 282
D. Demus	(a) **41**, K67 (1977)	*2*, p. *36*; *14.4*, p. *191*; 22.9, p. 416
J.-M. Derochette	(a) **39**, 281 (1977)	*1.6*, p. *31*; *17.1*, p. *208*; 21, p. 277
U. V. Desnica	(a) **39**, K33 (1977)	*11*, p. *94*; *14.3*, p. *165*; 22.4.1, p. 362
	(b) **83**, K41 (1977)	*10.2*, p. *90*; 11, p. 97; 22, p. 328
	(a) **44**, K29 (1977)	*3*, p. *43*; 22.3.1, p. 360
F. Dettmann	(a) **43**, 283 (1977)	*14.2*, p. *163*
	(a) **44**, 577 (1977)	*14.2*, p. *164*; 22.6, p. 391
W. J. Devlin	(a) **41**, 117 (1977)	*14.3.3*, p. *179*; 22.1.1, p. 334; 22.1.2, p. 339; 22.2.1, p. 351; 22.2.2, p. 354
J. T. Devreese	(b) **80**, 99 (1977)	*14*, p. *155*
	(b) **81**, 551 (1977)	*14.4*, p. *191*
	(b) **81**, 557 (1977)	*4*, p. *45*; 6, p. 56; 21, p. 280; 21.2, p. 308
	(b) **82**, 117 (1977)	*14*, p. *155*
	(b) **82**, 439 (1977)	*13.5.3*, p. *151*
I. Dézsi	(b) **81**, K81 (1977)	*13*, p. *110*; 21, p. 281; 22.1.3, p. 345
C. V. Dharmadhikari	(a) **42**, K99 (1977)	*17.1*, p. *208*; 22, p. 327
M. Dhwaj Singh	see under letter "S"	

J. Diehl	(a) **43**, K175 (1977)	10.1, p. 81; 11, p. 98; *18.2*, p. *223*; 21.1.1, p. 306
	(a) **43**, K179 (1977)	10.1, p. 81; *11*, p. *98*; 12.1, p. 102; *18.2*, p. *223*; 21.1.1, p. 306
Diep-The-Hunc	(b) **82**, 383 (1977)	1.6, p. 31; *13.5.4*, p. *153*; *18.3*, p. *229*
	(b) **84**, K159 (1977)	1.6, p. 32; *13.5.4*, p. *154*; 18.4, p. 234
T. Dietl	(b) **82**, 717 (1977)	*13.1*, p. *120*; 14.3, p. 170; *15*, p. *199*; 22.4.3, p. 367
B. Dietrich	(a) **43**, K131 (1977)	*10.2*, p. *91*; *20.1*, p. *261*; 22.6, p. 391
U. Dillner	(a) **44**, 197 (1977)	*14.1*, p. *160*; *14.3.1*, p. *175*; 21.7, p. 319
Troung Dinh Thanh	see under letters "Tr"	
J. Dixmier	(a) **43**, K61 (1977)	*2*, p. *39*; 22.1.3, p. 346
G. Dlubek	(a) **39**, 401 (1977)	*10.1*, p. *75*; 21.1, p. 289
	(a) **39**, K95 (1977)	*10.1*, p. *76*; 21, p. 278; 21.1, p. 289
	(a) **42**, K15 (1977)	*10.1*, p. *79*; 21, p. 282
V. P. Dobrego	(a) **44**, 435 (1977)	*11*, p. *99*; *13.4*, p. *140*; 22.1.1, p. 336
N. M. Dobrovolskii	(a) **44**, K79 (1977)	*8*, p. *68*; *14.2*, p. *164*; 18, p. 212; 22.8, p. 400
V. M. Dobryak	(a) **41**, 443 (1977)	*10.2*, p. *88*; 22.5.2, p. 378
T. Singh Dod	(b) **80**, 703 (1977)	13.4, p. 132; *14.3.4*, p. *184*; *20.1*, p. *252*; 22, p. 323
D. G. Dolgopolov	(b) **83**, 341 (1977)	*13*, p. *111*; *18.1*, p. *215*; 22.1.3, p. 346
G. Dolino	(a) **41**, 673 (1977)	*1.2*, p. *17*; 22.6, p. 388
E. Donato	(b) **79**, 655 (1977)	*2*, p. *33*; *13*, p. *107*; 14, p. 155; 22, p. 321
J. Donecker	(b) **81**, 199 (1977)	13.5.1, p. 145; *18*, p. *210*; *20.1*, p. *253*; 22.1.2, p. 339; 22.2.2, p. 354
H. J. Donnert	(a) **44**, 679 (1977)	*10.2*, p. *93*; *20.1*, p. *264*; 22.5.2, p. 382
L. W. Doremus	(a) **43**, 305 (1977)	*1.1*, p. *12*; *1.2*, p. *18*; 8, p. 67; 22.1.3, p. 346
M. Dorikens	(a) **40**, 153 (1977)	*8*, p. *65*; 21, p. 279
L. Dorikens-Vanpraet	(a) **40**, 153 (1977)	*8*, p. *65*; 21, p. 279
B. Dorner	(b) **80**, 167 (1977)	*6*, p. *53*; 14.4.1, p. 194; 22.8, p. 397
H. Dötsch	(a) **39**, 589 (1977)	*18.3*, p. *228*; 22.8.2, p. 409
	(b) **82**, 11 (1977)	13.5.4, p. 153; *18.3*, p. *228*; 19, p. 235; 22.8.2, p. 409
I. N. Douglas	(b) **79**, K99 (1977)	13.4, p. 131; *18*, p. *209*; *20.1*, p. *249*; 22.5, p. 370; 22.5.3, p. 383
J. C. Doukhan	(b) **79**, 763 (1977)	*13.4*, p. *131*; *14.3*, p. *166*; 20.1, p. 249; 22.1.3, p. 345
M. Doyama	(b) **80**, 193 (1977)	*10.1*, p. *76*; 21, p. 278
	(b) **80**, K131 (1977)	*4*, p. *45*; *10.1*, p. *77*; 21, p. 279
	(b) **82**, 369 (1977)	*11*, p. *96*; 21, p. 282
J. Drabecki	(a) **42**, 359 (1977)	*18.2*, p. *221*; 21.1.1, p. 304
U. Draeger	(a) **43**, K113 (1977)	*18.2.1*, p. *226*; 21.1.1, p. 306
G. Dräger	(b) **80**, K95 (1977)	*13.1*, p. *117*; 17, p. 206; 20, p. 245; 22.7, p. 394
	(b) **83**, 257 (1977)	*13.1*, p. *121*; 17, p. 207; 20, p. 245; 21.1, p. 297
N. A. Drozdov	(b) **83**, K137 (1977)	10.2, p. 91; *20.3*, p. *273*; 22.1.2, p. 342
H. Drulis	(a) **40**, K11 (1977)	*19*, p. *237*; 21, p. 279
M. Drulis	(a) **44**, 91 (1977)	*8*, p. *67*; 18, p. 212; 21.4, p. 313
M. Dryś	(a) **42**, K159 (1977)	*14.2*, p. *163*; 21, p. 284
K. S. Dubey	(b) **79**, K119 (1977)	*6*, p. *53*; *8*, p. *64*; 22.1.1, p. 333
	(b) **81**, K83 (1977)	*6*, p. *56*; *8*, p. *65*; 22, p. 325
	(b) **82**, K63 (1977)	6, p. 57; *8*, p. *66*; 22, p. 326
	(b) **84**, 69 (1977)	*6*, p. *59*; 8, p. 67; 22.7, p. 395
	(b) **84**, K57 (1977)	*6*, p. *60*; 8, p. 67; 22.8.1, p. 407
M. Dubiel	(a) **39**, 265 (1977)	*10.2*, p. *83*; 19, p. 235; 22.5.2, p. 375
G. T. Dubovka	(a) **41**, K63 (1977)	*18.2*, p. *220*; 21.1, p. 293; 21.1.1, p. 303
I. Ya. Dubovskaya	(b) **82**, 403 (1977)	*1.1*, p. *12*; *11*, p. *96*
E. Dudziak	(a) **42**, 721 (1977)	13.1, p. 121; *14.3.1*, p. *174*; 22.4.4, p. 369
J. F. Dufresne	(a) **42**, 751 (1977)	*18.2*, p. *222*; 21.1.1, p. 305

G. Duggan	(b) **83,** 543 (1977)	*13.4*, p. *137*; 14.4, p. 192; 22.1.1, p. 335
V. R. Dumke	(a) **44,** K133 (1977)	*10.2*, p. *93*; 20.1, p. 264; 22.5.2, p. 382
L. H. Dunleavy	(a) **43,** K87 (1977)	10.1, p. 80; *11*, p. *97*; *18.2*, p. *222*; 21.1.1, p. 306
M. Dupont	(b) **80,** 533 (1977)	*2*, p. *35*; 8, p. 65; 21, p. 279
D. Durand	(a) **40.** 629 (1977)	*10.2*, p. *87*; 20.1, p. 253; 22.5.2, p. 377
	(b) **81,** 443 (1977)	*6*, p. *56*; *20.1*, p. *254*; 22.5.2, p. 378
	(a) **41,** 183 (1977)	*10.2*, p. *87*; 22.5.2, p. 378
U. A. Durasova	(a) **42,** K171 (1977)	*18.2.1*, p. *226*; 21.1.1, p. 305
R. Durný	(a) **44,** K177 (1977)	*2*, p. *41*; *19*, p. *243*; 22, p. 331
G. B. Dürr	(b) **81,** 609 (1977)	*13*, p. *109*; 22.5.2, p. 378; 22.6, p. 388
H. Düsterhöft	(a) **39,** K147 (1977)	*17*, p. *206*; 21, p. 278; 21.1, p. 290; 22.1.2, p. 338
R. C. Du Varney	(a) **42,** 609 (1977)	*13.4*, p. *137*; 19, p. 241; 22.6, p. 390
Vu Duy Phach	see under letter "P"	
L. N. Dwivedi	(b) **84,** K125 (1977)	*18*, p. *212*
K. S. Dy	(b) **81,** K111 (1977)	*2*, p. *37*; *13*, p. *110*
P. N. Dyachkov	(b) **80,** 681 (1977)	*4*, p. *44*; *13.1*, p. *116*; 22.1.2, p. 339; 22.1.3, p. 345; 22.2, p. 348; 22.4, p. 361
D. van Dyck	(a) **39,** 383 (1977)	*1.2*, p. *15*; 1.4, p. 24; 22, p. 322
	(a) **40,** 343 (1977)	*1.2*, p. *15*; 1.4, p. 25; 22, p. 323
	(a) **40,** 669 (1977)	*1.4*, p. *25*; 10.2, p. 87; 22, p. 324
	(a) **41,** 555 (1977)	*1.4*, p. *26*; *10.2*, p. *88*; 22.6, p. 388
	(a) **42,** 319 (1977)	1.3, p. 22; *1.4*, p. *26*; 22, p. 326
	(a) **43,** 541 (1977)	*1.4*, p. *27*; 10.2, p. 91; 22, p. 329
	(a) **44,** 411 (1977)	1.3, p. 22; *1.4*, p. *26*; 22, p. 326
T. D. Dzhafarov	(a) **42,** 11 (1977)	*9*, p. *69*; 10.2, p. 83; 14.3.1, p. 173; 14.3.3, p. 178; 20.2, p. 265; 22.1, p. 332; 22.2, p. 348; 22.4, p. 361
A. P. Dzhotyan	(b) **79,** K143 (1977)	*13.5.1*, p. *144*; 22, p. 322
W. Ecke	(a) **42,** 157 (1977)	*14.3.1*, p. *174*; *16*, p. *203*; 22.1.2, p. 340
L. Eglinski	(a) **41,** 177 (1977)	13.4, p. 134; *14.3.4*, p. *185*; 22.1.3, p. 345
H. Eichele	(a) **43,** 465 (1977)	13.4, p. 138; *20.3*, p. *273*; 22.9, p. 418
F. Eichhorn	(a) **40,** 205 (1977)	*1*, p. *9*; *11*, p. *95*; 22.6, p. 387
K. Eichler	(a) **44,** 499 (1977)	*10.1*, p. *82*; 14.2, p. 164; 21.1, p. 300
W. Eichler	(a) **42,** K107 (1977)	13.4, p. 137; *15*, p. *199*; 22.1.3, p. 346
F. O. Eiramshyan	(a) **43,** K183 (1977)	*1*, p. *10*
J. Eisner	(a) **43,** K1 (1977)	3, p. 43; *14.4*, p. *192*; 22.9, p. 418
H. M. Eissa	(a) **43,** 343 (1977)	*1.4*, p. *27*; *10.1*, p, *80*; 21.1, p, 297
W. Ekardt	(b) **80,** 51 (1977)	*13.5.1*, p. *144*; 20.1, p. 250; 20.3, p. 268; 22.5, p. 370
	(b) **84,** 293 (1977)	*13.5.1*, p. *148*; 20.1, p. 261; 22.2.1, p. 353; 22.2.2, p. 355
D. D. Eley	(b) **79,** 629 (1977)	*14.1*, p. *157*; *14.3*, p. *165*; 15, p. 198; 18.1, p. 213 22.9, p. 414
K. Elk	(b) **81,** 179 (1977)	2, p. 36; 13, p. 109; *18.2*, p. *219*
	(b) **81,** K125 (1977)	*14.1*, p. *158*; 21, p. 281
W. D. Ellenson	(b) **83,** K115 (1977)	*1.2*, p, *19*; *6*, p. *59*; 12.2, p. 106; 20.1, p. 260; 22.8.1, p. 407
A. F. Ellervee	(b) **82,** 91 (1977)	13.4, p. 135; *20.3*, p. *271*; 22.4.1, p. 363; 22.6, p. 389
K. Ellmer	(b) **83,** K129 (1977)	*11*, p. *98*
S. H. El-Mahalawy	(b) **79,** 713 (1977)	13.1, p. 115; *14.3*, p. *166*; 22, p. 321
E. El-Mamoun	(b) **82,** 617 (1977)	*13*, p. *110*
K. Elsaesser	(a) **39,** 249 (1977)	*13.4*, p. *130*; 16, p. 201; 20.1, p. 248; 22.1.1, p. 333
V. N. Emelyanov	(b) **81,** 719 (1977)	*1.2*, p. *16*; *1.3*, p. *21*; 21.1, p. 293; 21.6, p. 316

U. EMILIANI	(b) **81,** 665 (1977)	13.5.1, p. 146; *20.1*, p. *255*; 22.3, p. 358
V. N. ENAKII	(a) **40,** 643 (1977)	6, p. 55; *13.4*, p. *133*; 20.3, p. 269; 22.8.1, p. 403
R. ENDERLEIN	(b) **80,** 225 (1977)	*6*, p. *53*; 13.1, p. 115; 13.5.1, p. 144; *20.1*, p. *251*
	(a) **42,** 629 (1977)	*14.3.4*, p. *185*; *20.1*, p. *258*; 22.2.4, p. 357
	(b) **83,** 239 (1977)	2, p. 38; *6*, p. *58*; 13.1, p. 121; *20.1*, p. *259*; 22, p. 328
P. ENGELS	(b) **82,** 145 (1977)	*1.2*, p. *17*; 6, p. 57; 20.1, p. 256; 22.8, p. 399
D. ENGEMANN	(b) **79,** 195 (1977)	*2*, p. *33*; *20.3*, p. *267*; 22.1.2, p. 337
TH. ENGLERT	(b) **81,** 119 (1977)	*13.3*, p. *126*; 14.3.2, p. 176; 22.1.3, p. 345
I. A. ENTIN	(a) **39,** 327 (1977)	*1.5*, p. *29*; *3*, p. *42*; 22.1.1, p. 333
M. V. ENTIN	(b) **83,** K97 (1977)	*14.3.4*, p. *186*; *16*, p. *204*; 22.2, p. 348
	(a) **44,** K91 (1977)	*14.3*, p. *172*; 22.8.1, p. 408
Y. EPELBOIN	(a) **41,** K9 (1977)	*1.4*, p. *26*; *10*, p. *72*
	(a) **44,** 651 (1977)	*10.2*, p. *93*; *12.2*, p. *106*; 22.8, p. 400
E. M. EPSHTEIN	(b) **80,** 697 (1977)	*6*, p. *55*; 22, p. 323
	(b) **82,** 391 (1977)	*13.5.2*, p. *149*; 22, p. 326
O. ERB	(a) **41;** 535 (1977)	*14.3*, p. *169*; 22.8.1, p. 405
V. V. EREMENKO	(b) **81,** 351 (1977)	*13.4*, p. *133*; 18.4, p. 232; 19, p. 238; 22.8.1, p. 404
M. I. EREMETS	(a) **42,** 385 (1977)	12.2, p. 105; 13.1, p. 120; *16*, p. *203*; *20.1*, p. *257*; 22.1.3, p. 346
M. V. EREMIN	(b) **79,** 775 (1977)	4, p. 44; *13.4*, p. *131*; 20.1, p. 249; 22.5.3, p. 383
	(b) **80,** 579 (1977)	1, p. 9; *13*, p. *108*
	(b) **82,** 221 (1977)	1, p. 10; *13*, p. *110*; 22.5.4, p. 385; 22.6, p. 389
	(b) **84,** 803 (1977)	13.4, p. 140; *19*, p. *243*; 22.8.1, p. 408
C. ERHARD	(a) **39,** 419 (1977)	*1.5*, p. *29*; *3*, p. *42*; 22.9, p. 414
T. ERICSSON	(a) **42,** K65 (1977)	*5*, p. *48*; 21.1, p. 296; 21.1.1, p. 304
R. ERLANDSSON	(a) **41,** K163 (1977)	*1.6*, p. *31*; 22, p. 325
O. P. ERMOLAEV	(a) **44,** 435 (1977)	*11*, p. *99*; *13.4*, p. *140*; 22.1.1, p. 336
H.-G. ERNST	(a) **43,** K95 (1977)	*14.3*, p. *171*; 22.2.2, p. 354
M. E. ERTL	(a) **44,** K123 (1977)	*14.3.3*, p. *182*; 15, p. 200; 22.7, p. 395
B. ESCAIG	(a) **40,** 393 (1977)	*10.2*, p. *86*; 22.9, p. 415
H. ESCHRIG	(b) **79,** 283 (1977)	*6*, p. *51*; 21.3, p. 310
U. ESSMANN	(a) **43,** 151 (1977)	1.4, p. 27; *14.2*, p. *163*; 21, p. 284
D. ESTEVE	(a) **42,** 173 (1977)	14.3.2, p. 176; *14.3.3*, p. *180*
YU. Z. ESTRIN	(a) **42,** 391 (1977)	7, p. 62; *10*, p. *73*; *12*, p. *100*
B. ETLINGER	(a) **44,** K29 (1977)	*3*, p. *43*; 22.3.1, p. 360
E. EVANGELISTA	(a) **39,** 661 (1977)	*10.1*, p. *76*; *12.1*, p. *101*; 21, p. 278
	(a) **44,** K31 (1977)	*10.1*, p. *82*; *12.1*, p. *103*; 21, p. 287
B. L. EVANS	(b) **79,** 713 (1977)	13.1, p. 115; *14.3*, p. *166*; 22, p. 321
R. A. EVARESTOV	(b) **79,** 743 (1977)	*4*, p. *44*; *13*, p. *107*; 22, p. 321
N. V. FADEEVA	(a) **44,** 247 (1977)	5, p. 49; *14.4.1*, p. *197*; 18.2, p. 223; *18.4*, p. *234*; 22.8, p. 400
A. FAESSLER	(b) **84,** 561 (1977)	*20*, p. *246*; 22.2, p. 349
M. FAGOT	(a) **41,** 595 (1977)	*1.2*, p. *17*; *1.3*, p. *22*; 21, p. 281; 21.1.1, p. 304
M. FÄHNLE	(b) **79,** 139 (1977)	*14.2*, p. *161*; 21, p. 276
	(b) **83,** 433 (1977)	*14.2*, p. *163*
	(b) **84,** 245 (1977)	*11*, p. *98*; *14.2*, p. *163*; 21, p. 286
	(b) **84,** 541 (1977)	*11*, p. *99*; *14.2*, p. *164*; 21, p. 287
H. J. FALGE	(b) **80,** 109 (1977)	*13.5*, p. *141*; *20.1*, p. *250*; 22.6, p. 386
	(b) **83,** 11 (1977)	*6*, p. *51*; *13.5*, p. *141*; 20.1, p. 247; 22, p. 320; 22.6, p. 386
M. L. FALIN	(b) **84,** K29 (1977)	*19*, p. *242*; 22.5, p. 372
W. F. FALLMANN	(a) **42,** 525 (1977)	*14.3.3*, p. *180*; *14.3.4*, p. *185*; 18, p. 212; 22.2.1, p. 352

Gy. Farkas	(a) **39**, K25 (1977)	*17*, p. *206*; 21.6, p. 315
J. L. Farvacque	(b) **79**, 763 (1977)	*13.4*, p. *131*; *14.3*, p. *166*; 20.1, p. 249; 22.1.3, p. 345
	(b) **80**, 361 (1977)	*10*, p. *72*; 13.1, p. 115; *14*, p. *155*
	(b) **80**, 433 (1977)	*10.2*, p. *86*; *14.3*, p. *167*; 22, p. 323; 22.1.3, p. 345
R. Fastenau	(a) **40**, 235 (1977)	*10.1*, p. *76*; 21, p. 279
M. Fayard	(a) **41**, K5 (1977)	*1.4*, p. *26*; *10.1*, p. *78*; 21, p. 280
R. G. Fedchenko	(a) **42**, K15 (1977)	*10.1*, p. *79*; 21, p. 282
A. A. Fedii	(b) **81**, 287 (1977)	*13.4*, p. *133*; 19, p. 238; 20.3, p. 270; 22.5.4, p. 385
A. Fedorov	(a) **44**, 747 (1977)	*16*, p. *205*; *20.3*, p. *275*; 22.3, p. 359
V. A. Fedorov	(a) **39**, 541 (1977)	*13.4*, p. *131*; 20.2, p. 265; 20.3, p. 268; 22.8, p. 396
	(a) **40**, K69 (1977)	*20.2*, p. *265*; 22.8.2, p. 410
	(a) **41**, K199 (1977)	*20.2*, p. *265*; 22.8.2, p. 410
V. P. Fedorov	(b) **84**, 107 (1977)	13.4, p. 138; *13.5.1*, p. *148*; *20.1*, p. *261*; 22.9, p. 419
N. V. Fedoseeva	(a) **44**, 429 (1977)	*18.2*, p. *224*; *18.4*, p. *234*; 22.8, p. 400
A. Feher	(a) **39**, K67 (1977)	*14.1*, p. *157*; 14.2, p. 161; 21, p. 277
	(b) **81**, K19 (1977)	*14.1*, p. *158*; 21.4, p. 312
I. V. Fekeshgazi	(a) **44**, K15 (1977)	*20.1*, p. *262*; 22.3, p. 359
C. Feldman	(a) **41**, 481 (1977)	2, p. 37; *20.1*, p. *256*; 22.1, p. 332
K. Feldmann	(b) **79**, 283 (1977)	*6*, p. *51*; 21.3, p, 310
C. P. Fender	(a) **41**, 85 (1977)	13.1, p. 117; *14.3.4*, p. *185*; 22.2.4, p. 357
V. V. Fenin	(a) **42**, K19 (1977)	*20.1*, p. *257*; 20.2, p. 266; *20.3*, p. *271*; 22.8.1, p. 405
	(a) **42**, K183 (1977)	20.1, p. 259; *20.2*, p. *266*; 20.3, p. 272; 22.8.1, p. 406
	(a) **43**, K15 (1977)	13.4, p. 137; *20.2*, p. *266*; 22.8, p. 399
	(a) **44**, 51 (1977)	*20.1*, p. *262*; 20.2, p. 266; *20.3*, p. *274*; 22.8.1, p. 407
L. G. Ferreira	(b) **83**, 229 (1977)	*13*, p. *110*; *18*, p. *212*
G. Ferron	(a) **44**, 549 (1977)	*10.1*, p. *82*; 12.1, p. 103; 21.1.1, p. 307
A. Feu	(a) **41**, 361 (1977)	*18.3*, p. *229*; 22.8.2, p. 411
T. Figielski	(b) **83**, 93 (1977)	*10.2*, p. *90*; 13.4, p. 137; *16*, p. *204*; 22.1.2, p. 341
L. I. Filina	(a) **39**, 717 (1977)	*13.4*, p. *131*; *14.3*, p. *166*; 22.2.1, p. 350
G. Finger	(b) **83**, K45 (1977)	*10.2*, p. *90*; *19*, p. *241*; 22.8, p. 399
J. Fink	(b) **84**, 611 (1977)	*5*, p. *50*; 14.2, p. 164; *18.2*, p. *224*; 21.4, p. 313
V. A. Finkel	(b) **82**, 213 (1977)	*1.2*, p. *17*;*18.2*, p. *221*; 21.4, p. 312
T. R. Finlayson	(b) **82**, K81 (1977)	*13.2*, p. *124*; *14.1*, p. *159*; 21, p. 283
H. J. Fischbeck	(b) **84**, 603 (1977)	*13*, p. *113*; 14, p. 156
E. Fischer	(b) **80**, 629 (1977)	6, p. 54; *19*, p. *237*; 21.1, p. 291
F. Fischer	(a) **39**, 499 (1977)	*10.2*, p. *84*; *14.4*, p. *189*; 20.1, p. 250; 22.5.2, p. 375
H. Fischer	(a) **43**, 529 (1977)	*13.4*, p. *138*; *20.3*, p. *273*; 22.2.2, p. 354
H. P. Fischer	(a) **44**, 669 (1977)	*13.2*, p. *125*; 21.1, p. 300
R. Fischer	(b) **79**, 195 (1977)	*2*, p. *33*; *20.3*, p. *267*; 22.1.2, p. 337
	(b) **79**, 539 (1977)	*2*, p. *33*; 14.3.1, p. 173; *16*, p. *201*; 22.1.2, p. 337
I. I. Fishchuk	(b) **83**, K29 (1977)	*2*, p. *38*; *14.3*, p. *170*
A. Ya. Fishman	(b) **82**, K59 (1977)	*10.1*, p. *78*; 21, p. 282
H.-J. Fitting	(a) **42**, K75 (1977)	*17*, p. *206*; 22.6, p. 389
	(a) **43**, 185 (1977)	*11*, p. *97*; 21, p. 285; 21.1, p. 297; 21.3, p. 310; 21.6, p. 317
I. P. Fittipaldi	(b) **80**, 385 (1977)	1.2, p. 15; *18.1*, p. *213*; *18.4*, p. *231*

T. A. Fjeldly	(a) **43**, K65 (1977)	*9*, p. *70*; *19*, p. *241*; 22.5, p. 371
K. Flachbart	(b) **81**, K19 (1977)	*14.1*, p. *158*; 21.4, p. 312
R. Flagmeyer	(a) **40**, 113 (1977)	*1.1*, p. *11*; 21, p. 279; 21.1, p. 290
S. Flandrois	(b) **80**, 125 (1977)	*8*, p. *64*; 18.1, p. 213; 22.9, p. 414
A. M. Flank	(a) **41**, 207 (1977)	*1.2*, p. *16*; *1.3*, p. *21*; 21, p. 280; 21.6, p. 316
H. Flietner	(a) **43**, K99 (1977)	*13.3*, p. *127*; 22.1.2, p. 342
R. Florian	(b) **82**, 145 (1977)	*1.2*, p. *17*; 6, p. 57; 20.1, p. 256; 22.8, p. 399
J. Foct	(a) **40**, 425 (1977)	*1.3*, p. *21*; 5, p. 48; 21, p. 279; 21.1.1, p. 303
H. Föll	(a) **39**, 561 (1977)	*10.1*, p. *76*; 11, p. 94; 21.1, p. 289
	(a) **41**, 113 (1977)	*1.1*, p. *11*; 22.8, p. 398
M. P. Fontana	(b) **81**, 603 (1977)	*13.4*, p. *134*; 20.1, p. 255; 20.3, p. 270; 22.5.2, p. 378
V. M. Fomin	(b) **79**, 595 (1977)	*14.3.4*, p. *183*; *20.1*, p. *249*; 22, p. 321
A. Fort	(b) **83**, 599 (1977)	*13.5.1*, p. *147*; *16*, p. *204*; 22.9, p. 418
M. A. Fortes	(b) **82**, 377 (1977)	*1.1*, p. *11*; *1.4*, p. *26*
E. Fortin	(b) **84**, K87 (1977)	*13.5.1*, p. *148*; 20.1, p. 263; 22.5, p. 372
D. R. Franceschetti	(a) **43**, K169 (1977)	*14.3.2*, p. *177*; 14.3.4, p. 187
M. A. Alario Franco	(a) **43**, 81 (1977)	*1*, p. *10*; *18.4*, p. *233*; 22.9, p. 418
W. Frank	(b) **82**, 523 (1977)	*10*, p. *73*; *12*, p. *100*
	(a) **42**, 511 (1977)	*10.1*, p. *79*; 11, p. 97; *12.1*, p. *102*; 21.1.1, p. 304
E. L. Frankevich	(a) **40**, 655 (1977)	13.5.1, p. 144; *16*, p. *202*; 22.9, p. 415
	(b) **84**, 513 (1977)	*13.5.1*, p. *148*; *19*, p. *243*; 22.9, p. 419
J. J. M. Franse	(a) **40**, 479 (1977)	*8*, p. *65*; 10.1, p. 77; 21.1, p. 292
P. Franzosi	(b) **84**, K131 (1977)	*13.2*, p. *125*; 21.1, p. 300
	(b) **84**, K137 (1977)	*13.2*, p. *125*; 21.1, p. 300; 21.6, p. 318
A. Freiberg	(b) **81**, 359 (1977)	*6*, p. *55*; *20.3*, p. *270*; 22.5.2, p. 377
A. Freudenberger	(b) **84**, 195 (1977)	*13.1*, p. *122*; *13.2*, p. *125*; 14.3, p. 171; 22.7, p. 395
G. Freydank	(a) **44**, K13 (1977)	*10.2*, p. *92*; 22.2.2, p. 355
B. Frick	(a) **41**, K185 (1977)	*10.2*, p. *89*; *19*, p. *240*; 22.2.1, p. 352
V. M. Fridkin	(a) **39**, 193 (1977)	*14.4.1*, p. *194*; *16*, p. *201*; 22.8.1, p. 402
M. Friedrich	(b) **80**, K127 (1977)	*19*, p. *237*; 22.8.1, p. 403
L. Fritsche	(b) **79**, 85 (1977)	*13.1*, p. *114*; 13.2, p. 124; 21.4, p. 311
F. Fröhlich	(a) **39**, 265 (1977)	*10.2*, p. *83*; 19, p. 235; 22.5.2, p. 375
	(a) **42**, 73 (1977)	*10.2*, p. *89*; 22.5.2, p. 379
	(a) **42**, 79 (1977)	*12*, p. *100*
B. Fromm	(a) **40**, 63 (1977)	*2*, p. *35*; *14.3.1*, p. *173*; 22, p. 323
	(a) **41**, K47 (1977)	*14.3.4*, p. *185*; 22, p. 324
G. Frommeyer	(a) **41**, 653 (1977)	*18*, p. *211*; 21.1, p. 294; 21.6, p. 316
B. I. Fuchs	(a) **40**, K61 (1977)	*14.3.4*, p. *184*; 22.1.1, p. 334
F. Fujimoto	(a) **43**, 269 (1977)	*11*, p. *97*; 13, p. 111; 21, p. 285
M. Fujimoto	(a) **42** K7 (1977)	*14.3.3*, p. *180*; 22.2.1, p. 352
H. Fujiyasu	(a) **42**, K25 (1977)	*14.3.2*, p. *177*; *14.3.3*, p. *180*; 22.2.1, p. 352
M. Fumeron	(a) **42**, 687 (1977)	*1.3*, p. *22*; 12.1, p. 102; 21, p. 283; 21.1, p. 296
J. E. Furneaux	(b) **83**, 305 (1977)	*13.5.1*, p. *147*; 18.1, p. 215; 22.1.1, p. 335
G. N. Fursey	(a) **42**, 757 (1977)	13.3, p. 127; *14.3.4*, p. *186*; 22, p. 327
W. Füssel	(a) **43**, K99 (1977)	*13.3*, p. *127*; 22.1.2, p. 342
I. Gabaš	(a) **43**, 351 (1977)	*14.3.2*, p. *177*; *16*, p. *204*; 22.4.3, p. 367
R. G. Gabrielyan	(b) **83**, K173 (1977)	9, p. 70; *13*, p. *111*; *18*, p. *212*
R. Z. Gabrielyan	(a) **43**, K183 (1977)	*1*, p. *10*
J. C. Gâcon	(a) **39**, 147 (1977)	13.4, p. 130; *20.3*, p. *267*; 22.5.3, p. 383
V. V. Gagulin	(a) **44**, 247 (1977)	5, p. 49; *14.4.1*, p. *197*; 18.2, p. 223; *18.4*, p. *234*; 22.8, p. 400
U. Gahn	(a) **40**, 463 (1977)	*1.2*, p. *16*; *1.3*, p. *21*; 21, p. 279
J. A. Gaj	(b) **83**, K133 (1977)	13.5.1, p. 147; *20.1*, p. *261*; 22.4.4, p. 369

R. R. Gałązka	(b) **84**, K83 (1977)	*13.4*, p. *140*; *14.3*, p. *172*; 22.4.3, p. 367
A. A. Galkin	(b) **80**, K37 (1977)	*18.4*, p. *232*; 19, p. 236; 22, p. 323
G. N. Galkin	(b) **82**, 237 (1977)	*14.3*, p. *169*; *16*, p. *203*; 22.1.3, p. 345
M. Le Gall	(b) **80**, 279 (1977)	13, p. 108; *18.3*, p. *228*; *20*, p. *244*; 22.8.2, p. 410
F. M. Galperin	(b) **81**, 545 (1977)	*13*, p. *109*; 18.2, p. 220; 21.1, p. 293
M. Galtier	(a) **42**, 141 (1977)	*1.1*, p. *12*; 21.6, p. 316
H. Gamari-Seale	(a) **40**, K105 (1977)	*18.1*, p. *214*; *18.4*, p. *232*; 21.4, p. 311
	(a) **41**, 147 (1977)	*18.3*, p. *229*; 22.8.2, p. 411
T. T. Gan	(a) **43**, 565 (1977)	*14.3.1*, p. *175*; *14.3.4*, p. *187*; 22.9, p. 418
M. Gandais	(a) **44**, 651 (1977)	*10.2*, p. *93*; *12.2*, p. *106*; 22.8, p. 400
B. N. Ganguly	(b) **79**, 617 (1977)	1.2, p. 14; *6*, p. *52*; 12.2, p. 104; 20.1, p. 249; 22.8.1, p. 402
	(b) **83**, K115 (1977)	*1.2*, p. *19*; *6*, p. *59*; 12.2, p. 106; 20.1, p. 260; 22.8.1, p. 407
V. V. Gann	(a) **41**, 225 (1977)	*11*, p. *95*; *14.2*, p. *162*; 18.1, p. 214; 21.1.1, p. 303; 21.5, p. 314
J. Gardavský	(b) **84**, 691 (1977)	*13.4*, p. *139*; *20.1*, p. *263*; 22.4.3, p. 367
R. Garg	(b) **84**, K17 (1977)	*19*, p. *242*; 22.8.1, p. 407
V. K. Garg	(b) **80**, 63 (1977)	*10.2*, p. *85*; *12.2*, p. *104*; 22.5.2, p. 376
	(b) **82**, 139 (1977)	*1.1*, p. *11*; 12.2, p. 105; 22.5.2, p. 379
	(b) **82**, 325 (1977)	*4*, p. *45*; *12*, p. *100*
	(b) **82**, 481 (1977)	*4*, p. *45*; *12.2*, p. *105*; 22.5.2, p. 379
	(b) **84**, K17 (1977)	*19*, p. *242*; 22.8.1, p. 407
I. A. Garifullin	(a) **40**, K11 (1977)	*19*, p. *237*; 21, p. 279
A. K. Garrison	(a) **42**, 609 (1977)	*13.4*, p. *137*; 19, p. 241; 22.6, p. 390
E. Garška	(a) **43**, 699 (1977)	*7*, p. *63*; *14.3.2*, p. *177*; 22.4.1, p. 364
R. K. Gartia	(a) **42**, K155 (1977)	*10.2*, p. *90*; *20.3*, p. *272*; 22.5.2, p. 380
	(a) **44**, K21 (1977)	*10.2*, p. *92*; *20.3*, p. *274*; 22.5.2, p. 381
S. A. Garyainov	(b) **82**, K95 (1977)	*13.1*, p. *120*; 22.3, p. 358
E. M. Gasanov	(a) **44**, K87 (1977)	*20.3*, p. *274*; 22.8.1, p. 407
J. Gastaldi	(a) **43**, 425 (1977)	1.2, p. 19; *1.4*, p. *27*; *10.1*, p. *80*; 21.1, p. 297
T. M. Gasymov	(b) **84**, 473 (1977)	6, p. 60; 13.4, p. 139; *14.3.4*, p. *188*; *15*, p. *199*; 22, p. 330
C. Gault	(a) **43**, 625 (1977)	1.3, p. 23; *12.1*, p. *102*; 21, p. 286; 21.3, p. 310
M. Gawish	(a) **43**, 99 (1977)	2, p. 39; *14.4*, p. *192*; 22.8.1, p. 406
N. E. Geacintov	(a) **40**, 621 (1977)	11, p. 95; *13.5.1*, p. *144*; *20.3*, p. *269*; 22.9, p. 415
W. Gębicki	(b) **80**, 307 (1977)	*6*, p. *53*; *20.1*, p. *251*; 22.8, p. 397
T. O. Gegechkori	(a) **43**, 513 (1977)	6, p. 59; *13*, p. *111*; *14.3.4*, p. *187*
W. Gehlhoff	(b) **80**, 549 (1977)	*19*, p. *237*
	(a) **41**, K21 (1977)	*19*, p. *239*; 22.1.2, p. 339
G. A. Gehring	(b) **81**, 167 (1977)	6, p. 55; 13.5.1, p. 145; *20.1*, p. *253*; 22.8.1, p. 404
H.-D. Geiler	(a) **41**, K171 (1977)	9, p. 70; *11*, p. *96*; 22.1.2, p. 340
V. Geist	(a) **40**, 113 (1977)	*1.1*, p. *11*; 21, p. 279; 21.1, p. 290
P. V. Geld	(b) **84**, 465 (1977)	4, p. 46; *13.1*, p. *122*; 22.3, p. 359
I. Kh. Geller	(a) **44**, K71 (1977)	*2*, p. *40*; *14.3.4*, p. *187*; 22.1.3, p. 347
H. P. Gemünd	(a) **40**, K97 (1977)	*18.2.1*, p. *226*; 21.1.1, p. 303
	(a) **42**, K103 (1977)	*18.2.1*, p. *226*; 21.1.1, p. 305
A. Geoffroy	(a) **41**, 499 (1977)	*10.2*, p. *88*; *20.3*, p. *270*; 22.8.1, p. 405
S. Georgescu	(b) **83**, 71 (1977)	*19*, p. *241*; 22.5.2, p. 380
A. N. Georgobiani	(b) **80**, 589 (1977)	*13.4*, p. *132*; 16, p. 202; 20.1, p. 252; 20.3, p. 269; 22.4.1, p. 362
K. H. Georgy	(a) **44**, K53 (1977)	*10.1*, p. *82*; 21.6, p. 317
A. Gérard	(a) **39**, 425 (1977)	5, p. 47; *18.2*, p. *218*; 22, p. 322
B. Gerlach	(b) **79**, 613 (1977)	*13.5.1*, p. *144*; 22, p. 321

E. GERLACH	(b) **81**, 153 (1977)	*13.4*, p. *133*; *14*, p. *155*
U. GERLACH	(a) **40**, K131 (1977)	*1.2*, p. *16*; *18.1*, p. *214*; 22.6.1, p. 393
	(a) **42**, 295 (1977)	*1.2*, p. *17*; *3*, p. *42*; 14.3, p. 170; 18.1, p. 214; 22.6.1, p. 393
D. GERLICH	(b) **81**, 295 (1977)	6, p. 55; *14.4.1*, p. *195*; 22.8, p. 398
K. H. GERMANN	(b) **80**, 159 (1977)	*18.4*, p. *231*; *20.1*, p. *251*; 22.6, p. 387
H. P. GESERICH	(b) **80**, 119 (1977)	1.4, p. 24; *14.1*, p. *157*; *20.1*, p. *250*; 22, p. 322
R. GEVERS	(b) **79**, 215 (1977)	*1.1*, p. *11*
	(b) **80**, 477 (1977)	*1.1*, p. *11*
	(a) **41**, 51 (1977)	*1.4*, p. *25*; 22.7, p. 394
	(b) **84**, 133 (1977)	*1.1*, p. *13*
M. N. GHATIKAR	(b) **82**, 603 (1977)	*20*, p. *245*; 22.6, p. 390
V. GH. GHEORGIU	(a) **40**, 29 (1977)	*9*, p. *69*; 10.2, p. 85; 14.3, p. 167; 22.2.2, p. 354
S. GHOSH	(a) **41**, 249 (1977)	*14.3.4*, p. *185*; 22.2.3, p. 356
G. GIAQUINTA	(b) **81**, 579 (1977)	*14.2*, p. *162*
J. GIBER	(b) **79**, K71 (1977)	*10.2*, p. *83*; *13.4*, p. *129*; 22.1.2, p. 337
	(b) **79**, K89 (1977)	*10.2*, p. *84*; 22.1.2, p. 338
N. K. GILRA	(a) **40**, 685 (1977)	*10*, p. *72*
M. GIORDANO	(b) **84**, 305 (1977)	18.1, p. 215; *19*, p. *242*; 22.8, p. 400
P. E. GIRARD	(b) **83**, K7 (1977)	*20.1*, p. *259*; 22.7, p. 395
G. GIRO	(a) **42**, K179 (1977)	*16*, p. *203*; *20.1*, p. *259*; 20.3, p. 272; 22.9, p. 417
	(a) **43**, K193 (1977)	*16*, p. *204*; 20.1, p. 261; 20.3, p. 274; 22.9, p. 419
V. N. GITSOVICH	(b) **79**, K125 (1977)	*5*, p. *47*; 18, p. 209; 22.8.2, p. 409
E. S. GIULIANO	(b) **79**, 655 (1977)	*2*, p. *33*; *13*, p. *107*; 14, p. 155; 22, p. 321
H. GLAEFEKE	(a) **42**, K75 (1977)	*17*, p. *206*; 22.6, p. 389
	(a) **43**, 185 (1977)	*11*, p. *97*; 21, p. 285; 21.1, p. 297; 21.3, p. 310; 21.6, p. 317
G. A. GLEDHILL	(b) **79**, K107 (1977)	*6*, p. *52*; *20.1*, p. *249*; 22.5.4, p. 385
K. D. GLINCHUK	(a) **41**, 659 (1977)	*13.4*, p. *135*; *20.3*, p. *271*; 22.2.1, p. 352
	(a) **44**, 777 (1977)	*13.4*, p. *140*; *20.3*, p. *275*; 22.2.1, p. 353
D. GLÖTZEL	(b) **79**, 85 (1977)	*13.1*, p. *114*; 13.2, p. 124; 21.4, p. 311
E. GMELIN	(b) **79**, 763 (1977)	*13.4*, p. *131*; *14.3*, p. *166*; 20.1, p. 249; 22.1.3, p. 345
P. F. GOBIN	(a) **40**, 519 (1977)	*10.1*, p. *77*; 21.1, p. 292
R. GOBLIRSCH	(a) **44**, 593 (1977)	*18.1*, p. *216*; *18.2*, p. *224*; 21.1, p. 300
I. G. GOCHEV	(b) **84**, 785 (1977)	*8*, p. *68*; *18.4*, p. *234*; 22.9, p. 419
E. M. GODZHAEV	(b) **81**, K47 (1977)	13.5.1, p. 145; *14.3.4*, p. *185*; *20.1*, p. *254*; 22.8, p. 398
O. GOEDE	(b) **81**, 235 (1977)	13.1, p. 117; *13.5.1*, p. *145*; *20.1*, p. *253*; 22.4.1, p. 363
C. M. GOEL	(b) **81**, 403 (1977)	*6*, p. *55*; 22.5.2, p. 377
S. C. GOEL	(b) **79**, K103 (1977)	*8*, p. *64*; 21, p. 277; 21.1, p. 289; 21.6, p. 315
	(b) **84**, 423 (1977)	*8*, p. *68*; 21.1, p. 300
L. VAN GOETHEM	(a) **41**, 129 (1977)	*1.2*, p. *16*; *1.4*, p. *25*; 22.6, p. 388
	(a) **42**, 319 (1977)	1.3, p. 22; *1.4*, p. *26*; 22, p. 326
	(a) **44**, 411 (1977)	1.3, p. 22; *1.4*, p. *26*; 22, p. 326
A. GOLDMANN	(b) **80**, 425 (1977)	*13.1*, p. *116*; 17, p. 206; 22.9, p. 415
	(b) **81**, 9 (1977)	*13.1*, p. *114*; 13.5.1, p. 143; *20.1*, p. *247*; 22.5, p. 370; 22.5.1, p. 373
D. GOLDSCHMIDT	(a) **41**, 283 (1977)	*11*, p. *95*; 14.3.2, p. 176; 22, p. 324
I. V. GOLOSOVSKII	(a) **39**, 683 (1977)	*18.4*, p. *231*; 22.8.2, p. 410
S. I. GOLUBOV	(b) **79**, 79 (1977)	13.4, p. 129; *13.5.1*, p. *143*; 20.3, p. 267; 22.9, p. 414
G. I. GONCHARENKO	(a) **41**, 307 (1977)	13.1, p. 117; 13.4, p. 134; *14.3*, p. *168*; 22.3, p. 358

I. N. GONCHARUK	(b) **79**, 347 (1977)	*6*, p. *51*; *20.1*, p. *248*; 22, p. 320
P. GONDI	(a) **39**, 661 (1977)	*10.1*, p. *76*; *12.1*, p. *101*; 21, p. 278
	(a) **43**, K205 (1977)	10.2, p. 91; *13.4*, p. *138*; 14.3, p. 171; 22.1.1, p. 336
	(a) **44**, 333 (1977)	*10.1*, p. *82*; 12.1, p. 103; 21, p. 287
	(a) **44**, K31 (1977)	*10.1*, p. *82*; *12.1*, p. *103*; 21, p. 287
R. GONZÁLEZ	(a) **40**, 167 (1977)	*10.1*, p. *76*; 21.1, p. 290
	(a) **40**, K19 (1977)	*10.1*, p. *77*; 21.1, p. 291
	(a) **43**, K187 (1977)	*10.1*, p. *81*; 21.1, p. 298
M. GOODARZ	(b) **83**, 555 (1977)	*13.2*, p. *125*; 13.4, p. 137; 14.1, p. 159; 15, p. 199; 21.6, p. 317
I. K. GOPALAKRISHNAN	(a) **43**, K31 (1977)	*1.2*, p. *19*; *14.3*, p. *170*; 22.8.1, p. 406
I. GORCZYCA	(b) **82**, 279 (1977)	*13.1*, p. *119*; 22.2.3, p. 356
A. YA. GORDON	(b) **79**, 495 (1977)	*14.4.1*, p. *194*; 22.8.1, p. 402
M. J. GORINGE	(a) **39**, K85 (1977)	*1.4*, p. *24*; *10.2*, p. *85*; 22.5.2, p. 376
A. VAN GORKUM	(a) **40**, 235 (1977)	*10.1*, p. *76*; 21, p. 279
YU. I. GORKUN	(a) **41**, 339 (1977)	11, p. 95; *13.4*, p. *134*; *14.3.3*, p. *179*; 22.1.1, p. 334
P. GÖRNERT	(a) **41**, 263 (1977)	*3*, p. *42*; 22.8.2, p. 411
	(a) **41**, 505 (1977)	*3*, p. *42*; 22.8.2, p. 411
	(a) **43**, 277 (1977)	3, p. 43; *18.3*, p. *230*; 22.8.2, p. 412
E. B. GOROKHOV	(a) **43**, 659 (1977)	*1.5*, p. *30*; 2, p. 39; 22.1.1, p. 336; 22.6, p. 391
W. GORZKOWSKI	(b) **79**, K151 (1977)	*13.5.1*, p. *144*; 22, p. 322; 22.4.1, p. 362; 22.4.2, p. 365
P. GOSAR	(b) **79**, 175 (1977)	*10.2*, p. *83*; 22, p. 320
J. R. GOSSELIN	(a) **40**, K25 (1977)	*5*, p. *47*; 15, p. 198; *18.1*, p. *213*; 22.8.2, p. 410
Y. GOTOH	(a) **42**, 469 (1977)	*1.3*, p. *22*; *1.4*, p. *27*; 21.1, p. 296
V. GOTTSCHALCH	(a) **44**, K13 (1977)	*10.2*, p. *92*; 22.2.2, p. 355
G. GÖTZ	(a) **41**, K171 (1977)	9, p. 70; *11*, p. *96*; 22.1.2, p. 340
R. D. GOULD	(a) **41**, 439 (1977)	14.3.3, p. 180; *17*, p. *206*; 22.6, p. 388
D. GOURIER	(a) **41**, 423 (1977)	*19*, p. *239*; 22.6.1, p. 393
A. GOUSKOV	(a) **39**, 65 (1977)	1.2, p. 14; *13.4*, p. *129*; *14.3*, p. *165*; 22.3.1, p. 360
L. GOUSKOV	(a) **39**, 65 (1977)	1.2, p. 14; *13.4*, p. *129*; *14.3*, p. *165*; 22.3.1, p. 360
B. GOUTIERS	(a) **40**, 583 (1977)	*7*, p. *62*; *14.3.4*, p. *184*; 22.4.1, p. 362
A. GOVAERT	(a) **43**, 619 (1977)	*5*, p. *49*; 22.8.2, p. 412
V. G. GOVORKOV	(a) **43**, K49 (1977)	*10.2*, p. *90*; 22.6, p. 390
T. GOWOREK	(b) **81**, 565 (1977)	*10.2*, p. *88*; 22.9, p. 416
	(b) **83**, K91 (1977)	*10.2*, p. *90*; 22.9, p. 418
	(b) **84**, K49 (1977)	*10.2*, p. *91*; 22.9, p. 419
P. GOY	(b) **80**, 207 (1977)	*13.1*, p. *115*; *19*, p. *236*; 21.7, p. 319
A. K. GOYAL	(b) **81**, K63 (1977)	*6*, p. *56*; 12.2, p. 104; 22, p. 324
D. R. GOYAL	(a) **41**, 323 (1977)	2, p. 36; *14.4*, p. *190*; 22.8, p. 398
R. P. GOYAL	(b) **79**, K115 (1977)	*6*, p. *52*; 22.6, p. 386
	(b) **81**, 409 (1977)	*6*, p. *55*; 22.6, p. 387
S. C. GOYAL	(b) **79**, K115 (1977)	*6*, p. *52*; 22.6, p. 386
	(b) **81**, 409 (1977)	*6*, p. *55*; 22.6, p. 387
	(b) **81**, K63 (1977)	*6*, p. *56*; 12.2, p. 104; 22, p. 324
	(b) **83**, 509 (1977)	*4*, p. *46*; *12.2*, p. *106*; 22, p. 328; 22.1, p. 332; 22.1.1, p. 335; 22.1.2, p. 342; 22.2, p. 348; 22.3, p. 358; 22.4, p. 361
U. GRADMANN	(a) **39**, 41 (1977)	*18.2.1*, p. *226*; 18.4, p. 231; 21.1, p. 288; 21.1.1, p. 302
	(a) **44**, 539 (1977)	*1.5*, p. *30*; *3*, p. *43*; 21.1.1, p. 307
M. H. GRADOWCZYK	(b) **81**, 647 (1977)	*10*, p. *72*

W. GRAEFF	(a) **43**, 487 (1977)	*1.1*, p. *13*; 22.1.2, p. 342
	(a) **44**, K131 (1977)	*1*, p. *10*
J. GRAMMATIKAKIS	(a) **40**, 141 (1977)	*14.3*, p. *167*; *18*, p. *210*; 22.8.2, p. 410
S. GRANDE	(b) **84**, 283 (1977)	*1.2*, p. *19*; 19, p. 242; 22.9, p. 419
F. GRANDJEAN	(a) **39**, 425 (1977)	5, p. 47; *18.2*, p. *218*; 22, p. 322
F. GRANZER	(a) **39**, 183 (1977)	*10.2*, p. *83*; 22.5.2, p. 375
U. M. GRASSANO	(b) **84**, 23 (1977)	*13.4*, p. *138*; *20.3*, p. *274*; 22.5.2, p. 381
E. GRATZ	(b) **82**, 159 (1977)	13.1, p. 118; 14.1, p. 158; *18.2*, p. *221*; *18.4*, p. *233*; 21.1, p. 295; 21.4, p. 312
P. GRAU	(a) **42**, 79 (1977)	*12*, p. *100*
E. DE GRAVE	(a) **43**, 619 (1977)	*5*, p. *49*; 22.8.2, p. 412
E. M. GRAY	(a) **40**, K73 (1977)	14.1, p. 158; *18.2*, p. *219*; 21.1, p. 291
V. A. GRAZHULIS	(a) **43**, 407 (1977)	10.2, p. 91; *13.4*, p. *138*; *14.3*, p. *171*; 22.1.2, p. 342
	(a) **44**, 107 (1977)	10.2, p. 91; *13.4*, p. *139*; *14.3*, p. *171*; 22.1.2, p. 343
V. I. GREBENNIKOV	(b) **79**, 423 (1977)	*13*, p. *107*; 20, p. 244; 21, p. 277
	(b) **80**, 73 (1977)	*13*, p. *108*; 20, p. 244; 21, p. 278
	(b) **83**, 281 (1977)	*13*, p. *110*; 20, p. 246; 21, p. 284
	(b) **83**, 383 (1977)	*13*, p. *111*; 20, p. 246; 21, p. 285
	(b) **84**, 773 (1977)	*13*, p. *113*; 20, p. 246; 21, p. 287
I. GREGORA	(a) **43**, K109 (1977)	*2*, p. *39*; *20.1*, p. *260*; 22.1, p. 332
T. GREGORKIEWICZ	(a) **40**, K127 (1977)	*19*, p. *238*; 22.1.1, p. 334
W. GRELLMANN	(a) **42**, 79 (1977)	*12*, p. *100*
V. S. GRIGOREVA	(a) **39**, 453 (1977)	13.4, p. 131; *20.3*, p. *268*; 22.8, p. 396
E. GRILLI	(a) **40**, 69 (1977)	*20.3*, p. *268*; 22.8, p. 397
E. S. GRINBERG	(b) **82**, 397 (1977)	*18.1*, p. *214*; *19*, p. *240*
V. A. GRITSENKO	(a) **44**, K167 (1977)	*2*, p. *41*; *14.3.4*, p. *188*; 22, p. 331
P. GROH	(a) **42**, 511 (1977)	*10.1*, p. *79*; 11, p. 97; *12.1*, p. *102*; 21.1.1, p. 304
M. GROSBRAS	(a) **42**, 449 (1977)	*10*, p. *73*; *12*, p. *100*
H, E. GROSKREUTZ	(a) **44**, K181 (1977)	*5*, p. *50*; 22.8.1, p. 408
J. G. GROSS	(b) **83**, 161 (1977)	*13.1*, p. *121*; *20.1*, p. *259*; 22.5, p. 371
P. GROSSE	(b) **82**, 457 (1977)	*13.1*, p. *120*; *20.1*, p. *257*; 22.1.3, p. 346
R. GRÖSSINGER	(a) **44**, 593 (1977)	*18.1*, p. *216*; *18.2*, p. *224*; 21.1, p. 300
E. GROUBERT	(a) **40**, 583 (1977)	*7*, p. *62*; *14.3.4*, p. *184*; 22.4.1, p. 362
K. I. GROZDEV	(b) **84**, 785 (1977)	*8*, p. *68*; *18.4*, p. *234*; 22.9, p. 419
J. B. GRUN	(b) **82**, 335 (1977)	*13.5.1*, p. *146*; *20.3*, p. *271*; 22.5, p. 371
	(b) **84**, 235 (1977)	*13.5.1*, p. *148*; *20.3*, p. *274*; 22.5, p. 372
	(b) **84**, 673 (1977)	*13.5.1*, p. *148*; *20.3*, p. *275*; 22.5, p. 372
	(b) **84**, 731 (1977)	*13.5.1*, p. *148*; *20.1*, p. *263*; 22.5, p. 372
P. J. GRUNDY	(a) **41**, 65 (1977)	*18.2.1*, p. *226*; 21.1, p. 293; 21.4, p. 312
R. GRÜNLER	(a) **42**, 149 (1977)	*13.4*, p. *136*; *14.3.3*, p. *180*; 22.6, p. 389
K. U. GUGA	(a) **43**, 311 (1977)	*14.3.2*, p. *177*; *16*, p. *204*; 22.2.3, p. 356
S. GUHA	(a) **41**, 249 (1977)	*14.3.4*, p. *185*; 22.2.3, p. 356
	(a) **43**, 89 (1977)	6, p. 59; 13, p. 111; *14.3.4*, p. *186*; 22.2.1, p. 353
J. P. GUILLOT	(a) **42**, 687 (1977)	*1.3*, p. *22*; 12.1, p. 102; 21, p. 283; 21.1, p. 296
M. GUILLOT	(b) **80**, 279 (1977)	13, p. 108; *18.3*, p. *228*; *20*, p. *244*; 22.8.2, p. 410
Y. GULDNER	(b) **81**, 615 (1977)	13.1, p. 118; *18*, p. *210*; *20.1*, p. *255*; 22.4.4, p. 368
	(b) **82**, 149 (1977)	13.1, p. 118; *18*, p. *211*; *20.1*, p. *256*; 22.4.4, p. 368
A. P. GUPTA	(b) **82**, K13 (1977)	*4*, p. *45*; 12.2, p. 105; 22.5.2, p. 379
I. J. GUPTA	(b) **80**, 353 (1977)	*6*, p. *53*; *8*, p. *64*; 23, p. 420
	(b) **84**, K95 (1977)	*6*, p. *61*; *8*, p. *68*; 23, p. 420
L. R. GUPTA	(b) **83**, 293 (1977)	*5*, p. *49*; 6, p. 58; 21, p. 284

O. P. GUPTA	(b) **81**, 275 (1977)	*6*, p. *55*; *8*, p. *65*; 21.2, p. 308
S. N. GUPTA	(b) **80**, 83 (1977)	*20*, p. *244*; 21.4, p. 311
	(b) **82**, 603 (1977)	*20*, p. *245*; 22.6, p. 390
W. GUSE	(b) **80**, 615 (1977)	13.1, p. 116; *14.4*, p. *190*; 22.6.1, p. 393
G. D. GUSEINOV	(b) **81**, K47 (1977)	13.5.1, p. 145; *14.3.4*, p. *185*; *20.1*, p. *254*; 22.8, p. 398
N. GUSKOS	(b) **80**, K25 (1977)	*19*, p. *236*; 22.5.3, p. 383
	(b) **83**, K79 (1977)	*6*, p. *59*; *19*, p. *241*; 22.5.3, p. 384
W. GÜTH	(a) **39**, 249 (1977)	*13.4*, p. *130*; 16, p. 201; 20.1, p. 248; 22.1.1, p. 333
M. G. GUTIERREZ	(b) **81**, K27 (1977)	*19*, p. *238*; 22.8, p. 398
A. GUTSZE	(a) **43**, K41 (1977)	*5*, p. *49*; *18.1*, p. *215*; 22.8.1, p. 407
C. GUTTEL	(b) **81**, 463 (1977)	*13.4*, p. *134*; 20.1, p. 255; 22.8.2, p. 411
P. GUYOT	(a) **41**, 577 (1977)	*1.4*, p. *26*; 10.1, p. 78; 21, p. 281; 21.1, p. 294
M. GUZZI	(a) **40**, 69 (1977)	*20.3*, p. *268*; 22.8, p. 397
M. M. GVICHIYA	(a) **41**, K43 (1977)	*9*, p. *70*; 21.1, p. 293
I. GYÉMÁNT	(b) **82**, 651 (1977)	*4*, p. *45*; *10.1*, p. *79*; 21.2, p. 309
O. GZOWSKI	(a) **41**, 173 (1977)	2, p. 36; *5*, p. *48*; 22.8.1, p. 404
HA VINH TAN	see under letter "T"	
H. B. HAANSTRA	(a) **39**, K7 (1977)	*1.4*, p. *24*; *14.4.1*, p. *194*; 22.8.1, p. 402
C. HAAS	(b) **80**, 321 (1977)	*6*, p. *53*; *20.1*, p. *251*; 22, p. 322; 22.5, p. 370
S. HAASE	(b) **83**, K83 (1977)	13.5.2, p. 150; *18*, p. *212*; *20.1*, p. *260*; 22.2.1, p. 353
A. A. HAAV	(b) **80**, 255 (1977)	4, p. 44; *6*, p. *53*; 22.5, p. 370
H. U. HABERMEIER	(a) **39**, 203 (1977)	10.1, p. 75; *14.2*, p. *161*; 18, p. 209; 21, p. 276
	(a) **43**, 151 (1977)	1.4, p. 27; *14.2*, p. *163*; 21, p. 284
H. J. HACKELÖER	(b) **80**, 235 (1977)	*10.2*, p. *85*; 19, p. 236; 22.5.2, p. 376
V. HADEK	(b) **81**, 281 (1977)	*14*, p. *155*; 15, p. 198; 22.9, p. 415
A. HADNI	(a) **39**, 419 (1977)	*1.5*, p. *29*; *3*, p. *42*; 22.9, p. 414
D. HADŽIAHMETOVIĆ	(b) **83**, 479 (1977)	*13.5.1*, p. *147*; 20.1, p. 260; 20.3, p. 273
R. HAFERL	(a) **39**, 157 (1977)	*5*, p. *47*; 18.1, p. 213; 18.2, p. 217; 21, p. 276; 21.1, p. 288; 21.1.1, p. 302
	(a) **42**, 739 (1977)	*5*, p. *49*; *18.2*, p. *222*; 21.1.1, p. 305; 21.4, p. 312
	(a) **43**, 111 (1977)	*5*, p. *49*; *18.2*, p. *222*; 21.1.1, p. 305; 21.4, p. 313
	(a) **43**, 437 (1977)	*5*, p. *49*; *18.2*, p. *223*; 21.1.1, p. 306; 21.4, p. 313
W. HAGEN	(a) **42**, K95 (1977)	*1.4*, p. *27*; *10.1*, p. *80*; 21.1.1, p. 305
A. L. HAGSTRÖM	(b) **83**, 77 (1977)	*13.3*, p. *127*; 17, p. 207; 21.1, p. 297
H. HAKEN	(b) **84**, 253 (1977)	*13.5.1*, p. *148*; 19, p. 242; 22.9, p. 419
M. HALBWACHS	(a) **42**, 707 (1977)	*9*, p. *70*; *10.1*, p. *80*; 21.1, p. 296; 21.6, p. 316
N. C. HALDER	(a) **39**, 213 (1977)	*1.5*, p. *29*; *2*, p. *33*; 13.1, p. 115; 14.3.1, p. 173; 22.1.1, p. 333
	(b) **80**, 273 (1977)	*2*, p. *34*; *19*, p. *236*; 21, p. 278
A. F. HALFF	(a) **40**, 511 (1977)	*16*, p. *202*; 22.5, p. 370; 22.5.4, p. 385
K. E. HALLER	(b) **81**, 57 (1977)	*10.2*, p. *87*; 20.1, p. 253; 22.5.2, p. 377
M. HAMERA	(b) **79**, 683 (1977)	*13.4*, p. *130*; 19, p. 236; 22.4.2, p. 365
	(b) **81**, K5 (1977)	*13.4*, p. *130*; 22.4.2, p. 365
S. A. HAMID	(a) **43**, K29 (1977)	*14.3*, p. *170*; 22.8.1, p. 406
K. HAMILTON	(b) **81**, K23 (1977)	*5*, p. *48*; 22.1.3, p. 345
R. N. HAMM	(b) **84**, 367 (1977)	*13.5.2*, p. *150*
S. HANADA	(a) **40**, 589 (1977)	*10.1*, p. *77*; *12.1*, p. *101*; 21.1, p. 292
Y. HANAWA	(a) **43**, K127 (1977)	*1.2*, p. *19*; 21.1, p. 298
T. HÄNSEL	(b) **84**, K43 (1977)	*6*, p. *60*; *13.5.1*, p. *148*; 20.3, p. 274; 22.3, p. 359
O. P. HANSEN	(b) **82**, 133 (1977)	*15*, p. *199*; 21.7, p. 319
H. HAPP	(b) **79**, 473 (1977)	6, p. 52; *14.4.1*, p. *194*; 20.1, p. 248; 22.8.1, p. 402

S. Haridoss	(a) **41**, 649 (1977)	*10.2*, p. *88*; *14.4*, p. *191*; 22.5.2, p. 378
A. Haring	(b) **81**, K11 (1977)	*10.2*, p. *87*; *17*, p. *206*; 22.5.2, p. 378
J. B. Hartmann	(b) **84**, 721 (1977)	*8*, p. *68*; 11, p. 99; 22.5.4, p. 385
U. Hartmann	(a) **42**, 667 (1977)	*14.3.3*, p. *181*; 14.3.4, p. 186; 22, p. 327
	(a) **43**, 103 (1977)	*14.3.1*, p. *175*; *14.3.3*, p. *181*; 22.6, p. 390
M. Hartmanová	(a) **41**, 679 (1977)	*10.2*, p. *89*; 22.5.2, p. 378
H. W. den Hartog	(a) **41**, 95 (1977)	*10.2*, p. *87*; 20.1, p. 254; 22.8, p. 398
	(a) **43**, 633 (1977)	*10.2*, p. *91*; 19, p. 242; 22.8, p. 400
G. Hartung	(a) **44**, K159 (1977)	*16*, p. *205*; *19*, p. *243*; 22.1.2, p. 344
J. Härtwig	(a) **42**, 495 (1977)	*1.1*, p. *12*; 22.1.2, p. 341
	(a) **43**, K131 (1977)	*10.2*, p. *91*; *20.1*, p. *261*; 22.6, p. 391
F. H. Hashmi	(a) **40**, 605 (1977)	10.2, p. 86; 13.4, p. 133; *20.3*, p. *269*; 22.6, p. 387
G. A. Hassan	(a) **40**, 43 (1977)	*10.1*, p. *76*; *12.1*, p. *101*; 21.1, p. 290
	(a) **41**, 541 (1977)	*10.1*, p. *78*; 21, p. 281; 21.3, p. 310
	(a) **42**, K115 (1977)	*12.1*, p. *102*; 21.1, p. 297
T. Hata	(a) **42**, K71 (1977)	*7*, p. *63*; *20.1*, p. *257*; 22.4.1, p. 363
T. Hatano	(a) **42**, K25 (1977)	*14.3.2*, p. *177*; *14.3.3*, p. *180*; 22.2.1, p. 352
W. E. Hatfield	(b) **81**, 379 (1977)	*18*, p. *210*
	(b) **81**, K75 (1977)	*18.4*, p. *233*; 19, p. 239; 22.9, p. 416
	(b) **82**, 69 (1977)	*8*, p. *66*; *18*, p. *211*
H.-D. Hattendorff	(a) **42**, 489 (1977)	*11*, p. *96*; *20*, p. *245*; 21.6, p. 316
K. Hattori	(b) **82**, 489 (1977)	*2*, p. *38*; 18.1, p. 215; 18.2, p. 221; 18.3, p. 229
W. Haubenreisser	(b) **83**, 129 (1977)	*1.6*, p. *32*; *3*, p. *42*
	(a) **44**, 161 (1977)	*14.3.3*, p. *181*; *14.3.4*, p. *187*; 22, p. 330
M. Haubold	(a) **42**, 579 (1977)	*9*, p. *70*; *11*, p. *97*; 22.2.4, p. 357
H. Haug	(b) **82**, 531 (1977)	*13.5.1*, p. *146*; *20.2*, p. *266*; 22.4.1, p. 363
G. Hausch	(a) **40**, 55 (1977)	*12.1*, p. *101*; 18.4, p. 232; 21.1, p. 290; 21.1.1, p. 303
	(a) **41**, K35 (1977)	*8*, p. *65*; *18.4*, p. *232*; 21.1, p. 293
H. Häusler	(a) **40**, K15 (1977)	*2*, p. *35*; *16*, p. *202*; 22.7, p. 394
R. L. Havill	(b) **80**, K141 (1977)	*2*, p. *36*; *19*, p. *237*; 21, p. 279
M. B. Haxton	(b) **83**, 147 (1977)	*6*, p. *58*; 22.5, p. 371
	(b) **83**, 529 (1977)	*6*, p. *59*; 22.5, p. 372
D. R. Hay	(b) **79**, 299 (1977)	*6*, p. *51*; 21.1, p. 288
J. Hayashi	(a) **41**, 139 (1977)	*1.1*, p. *11*; 6, p. 56; 22.2.1, p. 351
A. Heidemann	(a) **39**, K37 (1977)	*18.4*, p. *231*; 22.6, p. 386
R. Heidemann	(a) **42**, 657 (1977)	14.3.3, p. 180; *14.3.4*, p. *186*; 22.1.2, p. 341
	(a) **44**, 191 (1977)	*14.3.3*, p. *181*; *14.3.4*, p. *187*; 22.1.2, p. 343
K. Heidrich	(b) **84**, 629 (1977)	*13.1*, p. *122*; *20.1*, p. *263*; 22.5.4, p. 385
U. Heilig	(a) **39**, K29 (1977)	*13.3*, p. *126*; 22.1.1, p. 333
M. Hein	(a) **41**, K13 (1977)	*10.2*, p. *88*; *19*, p. *239*; 22.2.1, p. 351
M. P. Hemkar	(b) **79**, 289 (1977)	*6*, p. *51*; 12.1, p. 101; 21.2, p. 308
	(b) **81**, 275 (1977)	*6*, p. *55*; *8*, p. *65*; 21.2, p. 308
	(b) **82**, K29 (1977)	*13.2*, p. *124*; 21, p. 282; 21.2, p. 309
F. Henneberger	(b) **79**, K81 (1977)	*13.5.1*, p. *143*; *20.1*, p. *248*; 22.4.1, p. 362
	(b) **81**, 211 (1977)	*13.5.1*, p. *145*; *20.3*, p. *270*; 22.4.1, p. 363; 22.5, p. 370
	(b) **83**, 439 (1977)	*13.5.1*, p. *147*; *20.3*, p. *273*; 22.4.1, p. 364; 22.5, p. 371
K. Henneberger	(b) **79**, K81 (1977)	*13.5.1*, p. *143*; *20.1*, p. *248*; 22.4.1, p. 362
	(b) **83**, 439 (1977)	*13.5.1*, p. *147*; *20.3*, p. *273*; 22.4.1, p. 364; 22.5, p. 371
K. Hennig	(b) **79**, 283 (1977)	*6*, p. *51*; 21.3, p. 310
	(a) **40**, 205 (1977)	*1*, p. *9*; *11*, p. *95*; 22.6, p. 387
P. Hennl	(b) **84**, K151 (1977)	*10.2*, p. *92*; *20.1*, p. *263*; 22.5.2, p. 381
C. Henriet-Iserentant	(a) **44**, 215 (1977)	18.1, p. 215; *18.3*, p. *230*; 22.8.2, p. 412

M. Henry	(a) **44**, 601 (1977)	*5*, p. *50*; 22.8.1, p. 408
S. R. Herd	(a) **44**, 363 (1977)	*2*, p. *40*; *18.2.1*, p. *227*; 21.4, p. 313
R. Hergt	(a) **43**, 277 (1977)	3, p. 43; *18.3*, p. *230*; 22.8.2, p. 412
A. M. Hermann	(b) **81**, 281 (1977)	*14*, p. *155*; 15, p. 198; 22.9, p. 415
	(b) **82**, K117 (1977)	*1.1*, p. *12*; 22.9, p. 417
J. M. Herreros	(b) **83**, K21 (1977)	*10.2*, p. *90*; 22.5.2, p. 380
H. Herrmann	(b) **81**, 513 (1977)	*6*, p. *57*; *20.1*, p. *258*; 22.8, p. 399
K. H. Herrmann	(b) **82**, 237 (1977)	*14.3*, p. *169*; *16*, p. *203*; 22.1.3, p. 345
	(b) **83**, 465 (1977)	*16*, p. *204*; 22.8, p. 400
R. Herrmann	(b) **80**, 207 (1977)	*13.1*, p. *115*; *19*, p. *236*; 21.7, p. 319
	(b) **80**, K101 (1977)	*14.4*, p. *190*; 22.7, p. 394
	(b) **81**, K135 (1977)	*13.2*, p. *124*; 22.7, p. 394
	(b) **83**, 521 (1977)	*14.4*, p. *192*; 22.7, p. 395
B. Hertel	(a) **41**, K1 (1977)	*10.1*, p. *77*; 11, p. 95; 21.1, p. 293
S. Hess	(b) **79**, 501 (1977)	*13.2*, p. *124*; 14.1, p. 157; 21.7, p. 319
H. Hesse	(a) **42**, K119 (1977)	*20.1*, p. *259*; 22.9, p. 417
A. H. Heuer	(a) **39**, 301 (1977)	*10.2*, p. *83*; 22.6, p. 386
W. F. W. M. van Heugten	(b) **82**, 501 (1977)	6, p. 57; *10.1*, p. *79*; 21, p. 282
G. P. M. van den Heuvel	(a) **39**, K157 (1977)	*20.3*, p. *268*; 22.8.1, p. 403
	(a) **44**, K163 (1977)	*20.3*, p. *275*; 22.5, p. 372
J. Heydenreich	(a) **42**, 47 (1977)	*1.4*, p. *26*; *10*, p. 73
M. Hietschold	(b) **83**, 209 (1977)	*13.3*, p. *127*; 21, p. 284
Nguyen Van Hieu	(b) **82**, K39 (1977)	*13*, p. *110*; *20*, p. *245*
	(b) **84**, 417 (1977)	13.1, p. 122; *14.3.4*, p. *188*; *20.1*, p. *262*; 22.2, p. 349; 22.4, p. 361
L. Hildisch	(a) **39**, 133 (1977)	*1.1*, p. *11*; 1.2, p. 14; 22.2.4, p. 357
R. M. Hill	(a) **39**, 615 (1977)	2, p. 34; *13.4*, p. *131*; *14.3.1*, p. *173*
G. Hilscher	(a) **44**, 593 (1977)	*18.1*, p. *216*; *18.2*, p. *224*; 21.1, p. 300
F.-J. Himpsel	(b) **81**, 503 (1977)	*13.1*, p. *118*; 17, p. 206; 23, p. 420
K. Hinode	(b) **82**, 369 (1977)	*11*, p. *96*; 21, p. 282
O. Hipólito	(b) **81**, 657 (1977)	*13.5.2*, p. *149*; 14.4, p. 191; 21, p. 281
A. Hiraki	(b) **80**, K113 (1977)	*2*, p. *35*; *19*, p. *237*; 21.6, p. 315; 22.1.2, p. 339
	(b) **82**, K123 (1977)	*2*, p. *35*; *19*, p. *237*; 21.6, p. 315; 22.1.2, p. 339
Ken-ichi Hirano	(a) **43**, K57 (1977)	*9*, p. *70*; 21.1.1, p. 305
M. Hirose	(a) **42**, 337 (1977)	2, p. 37; 13.1, p. 120; *14.3.1*, p. *174*; 15, p. 199; *16*, p. *203*; 22.1.2, p. 340
	(a) **42**, 483 (1977)	*13.3*, p. *127*; 14.3.3, p. 180; 22.2.1, p. 352
V. Hivert	(a) **42**, 511 (1977)	*10.1*, p. *79*; 11, p. 97; *12.1*, p. *102*; 21.1.1, p. 304
V. Hizhnyakov	(b) **82**, K89 (1977)	*6*, p. *58*; 10.2, p. 89; *20.3*, p. *271*
P. Hlídek	(b) **84**, K119 (1977)	*20.1*, p. *263*; 22.8.2, p. 413
L. T. Ho	(b) **81**, K99 (1977)	*13.4*, p. *134*; *20.1*, p. *255*; 22.1.2, p. 340
T. X. Hoai	(b) **83**, 465 (1977)	*16*, p. *204*; 22.8, p. 400
Le Manh Hoang	(b) **84**, 361 (1977)	13.4, p. 138; *20.1*, p. *261*; *20.3*, p. *274*; 22.4, p. 361
Hoang Mi Shin see under letter "S"		
B. E. Hobbs	(a) **39**, 631 (1977)	*1.4*, p. *24*; *10.2*, p. *84*; 22.8, p. 396
L. W. Hobbs	(a) **39**, K85 (1977)	*1.4*, p. *24*; *10.2*, p. *85*; 22.5.2, p. 376
	(a) **40**, 629 (1977)	*10.2*, p. *87*; 20.1, p. 253; 22.5.2, p. 377
	(a) **41**, 183 (1977)	*10.2*, p. *87*; 22.5.2, p. 378
H.-J. Höbler	(b) **80**, K21 (1977)	13.1, p. 116; *20.1*, p. *251*; 22.8, p. 397
	(b) **81**, K59 (1977)	*13.1*, p. *117*; *20.1*, p. *254*; 22.8, p. 398
S. Hodorowicz	(a) **43**, 53 (1977)	*1.2*, p. *18*; 22.8, p. 399
W. Hoerstel	(a) **41**, K27 (1977)	*14.3.4*, p. *185*; 22.1.3, p. 345
	(b) **83**, 223 (1977)	*14.3.4*, p. *186*; 22.1.3, p. 346

M. A. Hoetjes-Eijkel (a) **42**, 537 (1977) *9*, p. *70*; 21, p. 283; 21.1, p. 296
H. J. Hoffmann (a) **44**, 237 (1977) *13.4*, p. *139*; *14.3*, p. *171*; 22.4.1, p. 364
S. K. Hoffmann (b) **82**, K85 (1977) *1.2*, p. *17*; *14.4.1*, p. *196*; 19, p. 241; 22.9, p. 417
A. Hofstaetter (b) **84**, 93 (1977) 18.4, p. 234; *19*, p. *242*; 22.8.1, p. 407
C. A. Hogarth (a) **41**, 439 (1977) 14.3.3, p. 180; *17*, p. *206*; 22.6, p. 388
C. Hohlfeld (a) **44**, K79 (1977) *8*, p. *68*; *14.2*, p. *164*; 18, p. 212; 22.8, p. 400
D. Hohlwein (a) **39**, 57 (1977) *18.4*, p. *231*; 22.6, p. 386
A. Holas (b) **81**, K107 (1977) *6*, p. *56*; 21.1, p. 294
T. Hollstein (b) **82**, 545 (1977) *20.1*, p. *258*; 21.1, p. 296; 21.6, p. 316
(b) **83**, K49 (1977) *20.1*, p. *260*; 21, p. 284; 21.6, p. 316
G. E. Holmberg (a) **39**, 669 (1977) *10.2*, p. *84*; 19, p. 236; *20.3*, p. *268*; 22.6, p. 386
W. B. Holzapfel (b) **83**, 99 (1977) *1.2*, p. *18*; 6, p. 58; *14.4.1*, p. *196*; 20.1, p. 259; 22.8.1, p. 406
H.-G. Holzhütter (b) **79**, K13 (1977) *13*, p. *107*; 22.3.1, p. 360; 22.4.4, p. 368
(b) **81**, 139 (1977) *13*, p. *109*; *14.3.4*, p. *184*
B. Hönerlage (b) **83**, K101 (1977) *13.5*, p. *141*; *20.3*, p. *273*; 22.5, p. 372
(b) **84**, 235 (1977) *13.5.1*, p. *148*; *20.3*, p. *274*; 22.5, p. 372
(b) **84**, 731 (1977) *13.5.1*, p. *148*; *20.1*, p. *263*; 22.5, p. 372
Vo Hong Anh see under letter "A"
G. Honjo (a) **41**, 523 (1977) *14.4.1*, p. *195*; 22.8.1, p. 405
(a) **42**, 257 (1977) *14.4.1*, p. *196*; 22.8.1, p. 405
(a) **44**, 147 (1977) *14.4.1*, p. *197*; 22.8.1, p. 407
Ah Mee Hor (a) **44**, 725 (1977) *10.2*, p. *93*; 22.5.2, p. 382
H. Hora (b) **80**, 143 (1977) *13.5.2*, p. *149*; *20.2*, p. *265*
(b) **83**, 355 (1977) *13.5.2*, p. *149*; *20.2*, p. *265*
W. Hörig (b) **80**, K21 (1977) 13.1, p. 116; *20.1*, p. *251*; 22.8, p. 397
M. Horobiowski (b) **79**, K147 (1977) *14.2*, p. *161*; 18, p. 209; 21, p. 277
G. Horowitz (a) **39**, 533 (1977) *20.1*, p. *250*; 22.2.1, p. 350
Z. Gy. Horváth (a) **39**, K25 (1977) *17*, p. *206*; 21.6, p. 315
J. L. Horwood (a) **40**, K25 (1977) *5*, p. *47*; 15, p. 198; *18.1*, p. *213*; 22.8.2, p. 410
P. Höschl (a) **43**, 705 (1977) 7, p. *63*; *14.3*, p. *171*; 16, p. 204; 22.4.3, p. 367
K. Hoselitz (a) **44**, K191 (1977) *2*, p. *41*; *18.2*, p. *225*; 21.1.1, p. 307
J. Hoshen (b) **81**, 479 (1977) *13.5.1*, p. *145*; *20.1*, p. *255*; 22.9, p. 416
N. M. Hosseini (a) **40**, K185 (1977) *14.3.3*, p. *179*; 22.8.2, p. 410
J. Th. M. de Hosson (a) **40**, 293 (1977) *10.1*, p. *77*; 21.1, p. 291
M. J. Howes (a) **41**, 117 (1977) *14.3.3*, p. *179*; 22.1.1, p. 334; 22.1.2, p. 339; 22.2.1, p. 351; 22.2.2, p. 354
C. H. Hsu (b) **81**, 281 (1977) *14*, p. *155*; 15, p. 198; 22.9, p. 415
M. Huber (a) **40**, K101 (1977) *1.4*, p. *25*; 22, p. 324
A. Hubert (a) **44**, K7 (1977) *18.2*, p. *224*; 21.1, p. 299; 21.4, p. 313
A. Hübler (b) **79**, K71 (1977) *10.2*, p. *83*; *13.4*, p. *129*; 22.1.2, p. 337
K. Hübner (a) **40**, 133 (1977) 2, p. 35; *4*, p. *44*; *13.1*, p. *116*; 14.4, p. 189; 20.1, p. 251; 22.6, p. 387
(a) **40**, 487 (1977) 2, p. 36; *4*, p. *45*; *13.1*, p. *117*; 14.4, p. 190; 20.1, p. 252; 22.6, p. 387
(a) **42**, 501 (1977) 2, p. 38; *4*, p. *45*; *13.1*, p. *120*; 14.4, p. 191; 20.1, p. 258; 22.6, p. 390
M. E. van Hulle (a) **39**, 253 (1977) *10.2*, p. *83*; *14.4*, p. *189*; 22.5.1, p. 373
(a) **40**, K57 (1977) *10.2*, p. *86*; *14.4*, p. *190*; 22.5.1, p. 373
(a) **40**, K173 (1977) *10.2*, p. *87*; *14.4*, p. *190*; 22.5.1, p. 373
(a) **44**, 229 (1977) *10.2*, p. *92*; *14.4*, p. *193*; 22.5.1, p. 374
J. Hüller (a) **43**, 505 (1977) *11*, p. *98*; *14.3.2*, p. *177*; 22.1.2, p. 342
Diep-The-Hung see under letter "D"
H. Hussinger (a) **44**, 525 (1977) *14.4.1*, p. *197*; 22.8.1, p. 408
G. Hut (b) **80**, 235 (1977) *10.2*, p. *85*; 19, p. 236; 22.5.2, p. 376
Dang Huy Lam see under letter "L"

W. J. Huybrechts	(b) **81**,	585 (1977)	*13.5.3*, p. *151*; *20.1*, p. *255*; 22, p. 325
	(b) **82**,	589 (1977)	*13.5.3*, p. *152*; *20.1*, p. *258*; 22, p. 327
G. Iadonisi	(b) **80**,	245 (1977)	*13.3*, p. *126*; 21, p. 278
E. M. Ibragimova	(a) **41**,	45 (1977)	10.2, p. 87; 13.4, p. 133; *20.1*, p. *254*; *20.3*, p. *270*; 22.8.1, p. 404
S. Ichimura	(a) **43**,	685 (1977)	13.5.1, p. 147; *20.3*, p. *273*; 22.3, p. 359
M. Ida	(a) **42**,	K7 (1977)	*14.3.3*, p. *180*; 22.2.1, p. 352
V. M. Ievlev	(a) **44**,	423 (1977)	*3*, p. *43*; *10.2*, p. *92*; 22.5.2, p. 381
A. A. Ignatov	(b) **81**,	341 (1977)	*13.5.2*, p. *149*
D. Ihle	(b) **80**,	619 (1977)	6, p. 54; *14*, p. *155*
A. Ihlenfeld	(a) **39**,	K147 (1977)	*17*, p. *206*; 21, p. 278; 21.1, p. 290; 22.1.2, p. 338
H. Ihrig	(a) **39**,	K7 (1977)	*1.4*, p. *24*; *14.4.1*, p. *194*; 22.8.1, p. 402
T. Iida	(b) **79**,	391 (1977)	6, p. 52; *13.4*, p. *129*; 20.1, p. 248; 22.5.2, p. 375
M. Ikeya	(b) **79**,	559 (1977)	10.2, p. 84; *20.3*, p. *267*; 22.5.2, p. 375
P. Ikonomou	(a) **40**,	K109 (1977)	*18.1*, p. *214*; *18.4*, p. *232*; 21.4, p. 312; 21.6, p. 315
E. Ildstad	(a) **43**,	K65 (1977)	*9*, p. *70*; *19*, p. *241*; 22.5, p. 371
I. A. Iliev	(b) **84**,	569 (1977)	13.1, p. 122; *13.5.1*, p. *148*; *20.1*, p. *262*; 22.4.2, p. 366
G. Ilonca	(a) **43**,	387 (1977)	*14.2*, p. *163*
A. G. Ilyashenko	(a) **39**,	717 (1977)	*13.4*, p. *131*; *14.3*, p. *166*; 22.2.1, p. 350
E. I. Ilyashenko	(a) **44**,	771 (1977)	*18.2*, p. *225*; 22.8.2, p. 413
M. Inagaki	(a) **41**,	451 (1977)	*5*, p. *48*; 18.2, p. 220; 21.1, p. 294; 21.1.1, p. 303
S. P. Ionov	(b) **79**,	403 (1977)	*13.5*, p. *141*
	(b) **80**,	K9 (1977)	*13*, p. *108*; 22, p. 323
	(b) **80**,	K13 (1977)	*13*, p. *108*; 13.5, p. 141; 22, p. 323
	(b) **81**,	671 (1977)	*13*, p. *109*
	(b) **81**,	691 (1977)	*13*, p. *109*; 13.5, p. 141; 22, p. 325
	(b) **84**,	55 (1977)	*13*, p. *112*; 13.5, p. 142; 22, p. 329
	(b) **84**,	485 (1977)	*13.5*, p. *142*; *18*, p. *212*
	(b) **84**,	493 (1977)	*13*, p. *112*; 13.4, p. 139
	(b) **84**,	K1 (1977)	*13*, p. *112*; 13.5, p. 142; 22, p. 330
G. V. Ionova	(b) **80**,	K9 (1977)	*13*, p. *108*; 22, p. 323
	(b) **81**,	671 (1977)	*13*, p. *109*
	(b) **84**,	493 (1977)	*13*, p. *112*; 13.4, p. 139
F. Iova	(b) **84**,	817 (1977)	*10.2*, p. *92*; *20.1*, p. *263*; 22.5, p. 372
M. Z. Iqbal	(a) **41**,	K155 (1977)	*13.4*, p. *135*; *16*, p. *202*; 22.2.2, p. 354
K. Irie	(a) **39**,	313 (1977)	*13.4*, p. *130*; *14.4.1*, p. *194*; 22.5.4, p. 385
W. Irler	(b) **82**,	243 (1977)	*1.2*, p. *17*; 5, p. 48; 18.1, p. 214; 22.8, p. 399
J. C. Irwin	(a) **43**,	K151 (1977)	*20.1*, p. *261*; 22.3, p. 359
A. Ishida	(a) **40**,	K93 (1977)	*14.3.3*, p. *179*; 22.2.1, p. 351
M. Z. Ismailov	(b) **81**,	K47 (1977)	13.5.1, p. 145; *14.3.4*, p. *185*; *20.1*, p. *254*; 22.8, p. 398
A. Isorē	(a) **42**,	235 (1977)	*10.1*, p. *79*; 11, p. 96; *12.1*, p. *101*; 21.1, p. 295
V. A. Isupov	(a) **40**,	K153 (1977)	*14.4.1*, p. *195*; 22.8.1, p. 404
S. Itoh	(a) **42**,	K25 (1977)	*14.3.2*, p. *177*; *14.3.3*, p. *180*; 22.2.1, p. 352
E. M. Itskovich	(a) **39**,	229 (1977)	*14.3.2*, p. *176*; 14.3.4, p. 183; 22.5.1, p. 373
Yu. M. Ivanchenko	(b) **79**,	K17 (1977)	6, p. 52; *14.1*, p. *157*; *14.2*, p. *161*; 21, p. 276
G. A. Ivanov-Smolenskii	(a) **40**,	189 (1977)	*18.2*, p. *218*; 18.4, p. 232; 21.1, p. 291
E. A. Ivanova	(b) **79**,	347 (1977)	*6*, p. *51*; *20.1*, p. *248*; 22, p. 320
Y. Iwama	(a) **41**,	451 (1977)	*5*, p. *48*; 18.2, p. 220; 21.1, p. 294; 21.1.1, p. 303
K. Iwamura	(a) **39**,	189 (1977)	*8*, p. *64*; 10.1, p. 75; *14.1*, p. *157*; 21.2, p. 308
O. Izumi	(a) **40**,	589 (1977)	*10.1*, p. *77*; *12.1*, p. *101*; 21.1, p. 292

G. Jackle	(a) **43**, K151 (1977)	*20.1*, p. *261*; 22.3, p. 359
G. Jacobs	(b) **79**, K75 (1977)	10.2, p. 83; *20.3*, p. *267*; 22.5.3, p. 383
	(b) **83**, 471 (1977)	*10.2*, p. *90*; 20.3, p. 273; 22.5.2, p. 380
J. Jacobs	(a) **40**, 575 (1977)	13.4, p. 132; *13.5.1*, p. *144*; *19*, p. *237*; 22.4.2, p. 365
	(a) **40**, K117 (1977)	*10.2*, p. *87*; *19*, p. *238*; 22.4.2, p. 365
K. Jacobs	(a) **44**, 675 (1977)	*3*, p. *43*; *10.2*, p. *93*; 22.2.1, p. 353
P. W. M. Jacobs	(a) **44**, 725 (1977)	*10.2*, p. *93*; 22.5.2, p. 382
G. Jacobsen	(a) **44**, K67 (1977)	*16*, p. *205*; 22.4.1, p. 364
K. Jagannadham	(a) **42**, 439 (1977)	*10*, p. *73*
	(a) **44**, 633 (1977)	*10*, p. *74*
D. Jäger	(a) **42**, 657 (1977)	14.3.3, p. 180; *14.3.4*, p. *186*; 22.1.2, p. 341
	(a) **44**, 191 (1977)	*14.3.3*, p. *181*; *14.3.4*, p. *187*; 22.1.2, p. 343
E. Jäger	(b) **80**, K81 (1977)	*2*, p. *34*; *18.3*, p. *228*; 21.1, p. 290; 21.4, p. 311
	(b) **80**, K85 (1977)	*2*, p. *34*; *18.3*, p. *228*; 21.1, p. 290; 21.4, p. 311
L. Jahn	(a) **42**, 191 (1977)	*18.3*, p. *229*; 22.8.2, p. 412
Amitabh Jain	(a) **40**, K31 (1977)	*10.2*, p. *86*; 14.3.3, p. 179; 22.1.2, p. 338
A. K. Jain	(b) **83**, K17 (1977)	*19*, p. *241*; 22.8.1, p. 406
B. K. Jain	(a) **43**, 357 (1977)	*1.3*, p. *22*; 5, p. 49; 21.1.1, p. 305
	(a) **44**, 443 (1977)	*5*, p. *50*; 22.8.2, p. 413
K. Jain	(a) **42**, 403 (1977)	*14.4*, p. *191*; 20.1, p. 257; 22.9, p. 417
R. K. Jain	(a) **42**, 221 (1977)	*13.1*, p. *120*; *14.3*, p. *169*; 22.1.1, p. 335; 22.1.2, p. 340
M. Janai	(a) **42**, 729 (1977)	*2*, p. *38*; 14.3.1, p. 174; *16*, p. *203*; 22, p. 327
M. Janda	(a) **40**, 225 (1977)	*14.3.1*, p. *173*; 22.1.3, p. 345
S. Jandl	(b) **83**, K7 (1977)	*20.1*, p. *259*; 22.7, p. 395
J. A. Janik	(a) **43**, 53 (1977)	*1.2*, p. *18*; 22.8, p. 399
J. M. Janik	(a) **43**, 53 (1977)	*1.2*, p. *18*; 22.8, p. 399
Š. Jánoš	(a) **39**, K67 (1977)	*14.1*, p. *157*; 14.2, p. 161; 21, p. 277
	(b) **81**, K19 (1977)	*14.1*, p. *158*; 21.4, p. 312
B. K. Jap	(a) **43**, K37 (1977)	*1.1*, p. *12*; 21, p. 285
F. Jaque	(b) **83**, K21 (1977)	*10.2*, p. *90*; 22.5.2, p. 380
	(b) **84**, 595 (1977)	10.2, p. 92; *20.3*, p. *275*; 22.5.2, p. 381
K. Jarašiunas	(a) **44**, 793 (1977)	*14.3.4*, p. *188*; 20, p. 246; 22.1.2, p. 344; 22.4.2, p. 366
T. Jarlborg	(b) **79**, 277 (1977)	*13.1*, p. *114*; 20, p. 244; 21.1, p. 288
B. Jarosz	(a) **40**, K149 (1977)	*10.2*, p. *87*; 22.5.2, p. 377
S. R. Jawalekar	(a) **41**, K31 (1977)	*14.3.1*, p. *174*; 22.4.2, p. 365
J. Ječný	(a) **44**, 557 (1977)	*6*, p. *61*; 22.1.3, p. 347
A. Jędrzejczak	(b) **82**, 717 (1977)	*13.1*, p. *120*; 14.3, p. 170; *15*, p. *199*; 22.4.3, p. 367
C. D. Jeffries	(b) **83**, 305 (1977)	*13.5.1*, p. *147*; 18.1, p. 215; 22.1.1, p. 335
J. Jelonek	(a) **43**, K199 (1977)	*18.2*, p. *223*; 22.8.2, p. 412
M. L. Jenkins	(a) **39**, 103 (1977)	*10.1*, p. *75*; 21.1, p. 288
B. Jeżowska-Trzebiatkowska	(a) **41**, 547 (1977)	13.4, p. 135; *20.3*, p. *270*; 22.8.1, p. 405
Y. Jinno	(b) **81**, K71 (1977)	*2*, p. *36*; *19*, p. *238*; 22.1.1, p. 334; 22.1.2, p. 339
D. S. Joag	(a) **42**, K99 (1977)	*17.1*, p. *208*; 22, p. 327
H. Johansen	(b) **83**, 187 (1977)	*13.2*, p. *124*; *14.1*, p. *159*; 21, p. 284
L. I. Johansson	(b) **83**, 77 (1977)	*13.3*, p. *127*; 17, p. 207; 21.1, p. 297
W. John	(b) **82**, K111 (1977)	*2*, p. *38*; *13*, p. *110*; 22, p. 327
M. L. Johnson	(a) **41**, K75 (1977)	*12.1*, p. *101*; 21, p. 280
	(a) **42**, K175 (1977)	*10.1*, p. *80*; 21, p. 284
B. K. Jones	(a) **40**, K185 (1977)	*14.3.3*, p. *179*; 22.8.2, p. 410
L. H. Jones	(a) **42**, K53 (1977)	*14.4*, p. *191*; *20*, p. *245*; 22.8.1, p. 406
A. K. Jonscher	(b) **83**, 585 (1977)	2, p. 39; *14.4*, p. *192*; 14.4.1, p. 196; 22, p. 329; 22.9, p. 418

	(b) **84**, 150 (1977)	2, p. 39; *14.4*, p. *192*; 14.4.1, p. 196; 22, p. 329; 22.9, p. 419
J. A. Jordão	(a) **40**, 109 (1977)	7, p. 62; *10.2*, p. *85*; 22.5.2, p. 376
D. K. Joshi	(a) **43**, K31 (1977)	*1.2*, p. *19*; *14.3*, p. *170*; 22.8.1, p. 406
J. C. Joubert	(a) **41**, 459 (1977)	*18.4*, p. *233*; 22.8.1, p. 405
J. Jouglar	(a) **41**, 561 (1977)	6, p. 56; 8, p. 66; *10.2*, p. *88*; 22.2.1, p. 352
C. Jourdan	(a) **43**, 425 (1977)	1.2, p. 19; *1.4*, p. *27*; *10.1*, p. *80*; 21.1, p. 297
P. Jouve	(b) **83**, 139 (1977)	*6*, p. *58*; *14.4.1*, p. *196*; 20.1, p. 259; 22.9, p. 417
T. Joy	(a) **40**, K145 (1977)	*18.2*, p. *219*; 21.1, p. 292; 21.7, p. 319
A. Jucys	(a) **43**, 699 (1977)	*7*, p. *63*; *14.3.2*, p. *177*; 22.4.1, p. 364
T. Jung	(b) **79**, 501 (1977)	*13.2*, p. *124*; 14.1, p. 157; 21.7, p. 319
B. Junghans	(a) **44**, 585 (1977)	2, p. 41; *14.3.1*, p. *175*; *14.3.2*, p. *177*; 22.1.2 p. 344
G. Jungk	(b) **82**, K113 (1977)	*13.1*, p. *120*; 20.1, p. 258; 22.1.1, p. 335
	(a) **42**, 267 (1977)	*20.1*, p. *257*
M. Jurisch	(b) **80**, K95 (1977)	*13.1*, p. *117*; 17, p. 206; 20, p. 245; 22.7, p. 394
	(a) **44**, 277 (1977)	1.3, p. 23; *3*, p. *43*; 14.2, p. 164; 21.1, p. 299,
	(a) **44**, 499 (1977)	*10.1*, p. *82*; 14.2, p. 164; 21.1, p. 300
I. S. Kabanov	(a) **44**, K155 (1977)	*1.2*, p. *20*; *14.4.1*, p. *197*; 19, p. 243; 22.8.1, p. 408
O. Kabisch	(a) **40**, K163 (1977)	1.2, p. 16; *10.1*, p. *77*; 21, p. 280; 21.1, p. 292
L. A. Kachanova	(a) **39**, K11 (1977)	*1.6*, p. *31*; *10.2*, p. *83*; 12.2, p. 104; 22.5.2, p. 375
S. Kachi	(a) **39**, K37 (1977)	*18.4*, p. *231*; 22.6, p. 386
Z. S. Kachlishvili	(a) **39**, 49 (1977)	*13*, p. *107*; *14.3.4*, p. *183*; 22, p. 321
	(a) **43**, 513 (1977)	6, p. 59; *13*, p. *111*; *14.3.4*, p. *187*
J. Kaczér	(a) **39**, 351 (1977)	1.2, p. 14; *18.2*, p. *217*; 22, p. 321
K. Kaczmarska	(b) **81**, K95 (1977)	*18.1*, p. *214*; *19*, p. *239*; 22.8.2, p. 411
G. Kaden	(a) **44**, K117 (1977)	*13.3*, p. *128*; *14.3.3*, p. *182*; 22, p. 330
M. S. Kagan	(a) **40**, K61 (1977)	*14.3.4*, p. *184*; 22.1.1, p. 334
H. G. Kahle	(b) **84**, 167 (1977)	1.2, p. 19; *8*, p. *67*; *18.4*, p. *234*; 22.8.1, p. 407
O. A. Kaibyshev	(a) **44**, 65 (1977)	10.1, p. 81; 21, p. 286; 21.3, p. 310
	(a) **44**, 477 (1977)	1.4, p. 28; *10.1*, p. *82*; 21, p. 287; 21.3, p. 310
Sh. Kaida	(a) **39**, 331 (1977)	*16*, p. *201*; 22.4.4, p. 368
M. M. Kaila	(a) **44**, K135 (1977)	*10.2*, p. *93*; *14.3*, p. *172*; 22.2.3, p. 356
F. Kajzar	(b) **80**, K75 (1977)	*18.2*, p. *218*; 21.1, p. 290
V. T. Kalinnikov	(b) **81**, 379 (1977)	*18*, p. *210*
	(b) **81**, K75 (1977)	*18.4*, p. *233*; 19, p. 239; 22.9, p. 416
	(b) **82**, 69 (1977)	*8*, p. *66*; *18*, p. *211*
J. Kalinowski	(b) **81**, 521 (1977)	*13.5.1*, p. *145*; 16, p. 202; 22.9, p. 416
Yu. H. Kalnin	(b) **84**, 521 (1977)	*9*, p. *71*; *10*, p. *74*
R. Kamel	(a) **41**, K113 (1977)	*10.1*, p. *78*; *12.1*, p. *101*; 21.1.1, p. 304
	(a) **44**, K53 (1977)	*10.1*, p. *82*; 21.6, p. 317
E. Kamieniecki	(a) **39**, 249 (1977)	*13.4*, p. *130*; 16, p. 201; 20.1, p. 248; 22.1.1, p. 333
A. A. Kaminskii	(a) **33**, 541 (1977)	*13.4*, p. *131*; 20.2, p. 265; 20.3, p. 268; 22.8, p. 396
	(a) **40**, K69 (1977)	*20.2*, p. *265*; 22.8.2, p. 410
	(a) **40**, 643 (1977)	6, p. 55; *13.4*, p. *133*; 20.3, p. 269; 22.8.1, p. 403
	(a) **41**, K199 (1977)	*20.2*, p. *265*; 22.8.2, p. 410
	(a) **43**, 71 (1977)	3, p. 42; 13.4, p. 137; 20.1, p. 260; *20.2*, p. *266*; *20.3*, p. *272*; 22.8, p. 399
V. F. Kaminskii	(a) **44**, 77 (1977)	*1*, p. *10*; *14*, p. *156*; 18.1, p. 215; 22.9, p. 419
F. Kammler	(b) **81**, 179 (1977)	2, p. 36; 13, p. 109; *18.2*, p. *219*

T. KANDA	(a) **42**, 769 (1977)	*11*, p. *97*; *14.3.2*, p. *177*; 16, p. 203; 22.1.2, p. 341
T. KANEKO	(b) **83**, 485 (1977)	*11*, p. *98*; 13.3, p. 127; 21.6, p. 317
O. KANERT	(b) **80**, 235 (1977)	*10.2*, p. *85*; 19, p. 236; 22.5.2, p. 376
P. L. KANITKAR	(a) **42**, K99 (1977)	*17.1*, p. *208*; 22, p. 327
A. KANT	(b) **79**, 359 (1977)	*4*, p. *44*; 12.2, p. 104; 22.5.2, p. 375
	(b) **84**, 427 (1977)	*4*, p. *46*; *8*, p. *68*; 20.1, p. 262; 22.5.2, p. 381
M. G. KAPLUNOV	(b) **83**, K141 (1977)	13, p. 111; 17, p. 207; *20*, p. *246*; 22.9, p. 418
	(a) **43**, K73 (1977)	*20.1*, p. *260*; 22.9, p. 418
H. KAPPERT	(a) **42**, 553 (1977)	*14.3*, p. *170*; 14.3.1, p. 174; 22.1.2, p. 341
	(a) **43**, 119 (1977)	*11*, p. *97*; *14.3*, p. *170*; 22.1.2, p. 341
D. KAPOR	(b) **79**, K111 (1977)	*14.4.1*, p. *194*
D. V. KAPOR	(b) **84**, K25 (1977)	*13.5*, p. *142*
S. KAPPHAN	(a) **42**, K119 (1977)	*20.1*, p. *259*; 22.9, p. 417
R. KARCHER	(b) **79**, 519 (1977)	*19*, p. *235*; 21.1, p. 289
	(b) **84**, 189 (1977)	*18*, p. *212*; *19*, p. *242*; 21, p. 286
M. D. KARKHANAVALA	(a) **43**, K31 (1977)	*1.2*, p. *19*; *14.3*, p. *170*; 22.8.1, p. 406
YU. I. KARKHANIN	(a) **39**, 717 (1977)	*13.4*, p. *131*; *14.3*, p. *166*; 22.2.1, p. 350
S.-E. KARLSSON	(a) **41**, K163 (1977)	*1.6*, p. *31*; 22, p. 325
	(b) **83**, 77 (1977)	*13.3*, p. *127*; 17, p. 207; 21.1, p. 297
H. P. KARNTHALER	(b) **81**, 191 (1977)	*10*, p. *72*
A. G. KAROZA	(b) **84**, K65 (1977)	*6*, p. *60*; *20.1*, p. *262*; 22.8, p. 400
D. V. KARPO	(b) **83**, K125 (1977)	*13.5.1*, p. *147*
W. KARTHE	(b) **80**, K127 (1977)	*19*, p. *237*; 22.8.1, p. 403
	(b) **81**, 323 (1977)	*13.4*, p. *133*; *19*, p. *238*; 22.8.1, p. 404
A. L. KARTUZHANSKII	(a) **43**, 321 (1977)	*13.4*, p. *137*; *20.3*, p. *273*; 22.5.1, p. 373
V. A. KASHCHEI	(b) **81**, K131 (1977)	*14.2*, p. *162*
L. KASPER	(b) **80**, 425 (1977)	*13.1*, p. *116*; 17, p. 206; 22.9, p. 415
J. KASPRZAK	(a) **41**, K25 (1977)	*19*, p. *239*; 22.8.1, p. 404
R. KASSING	(a) **42**, 657 (1977)	14.3.3, p. 180; *14.3.4*, p. *186*; 22.1.2, p. 341
	(a) **44**, 191 (1977)	*14.3.3*, p. *181*; *14.3.4*, p. *187*; 22.1.2, p. 343
A. KASTEN	(b) **84**, 551 (1977)	13.4, p. 139; *18.4*, p. *234*; *20.1*, p. *262*; 22.8.1, p. 408
G. KÄSTNER	(a) **43**, 167 (1977)	*1.4*, p. *27*; *3*, p. *43*; 20.3, p. 273; 22.2.4, p. 357
K.-H. KATERBAU	(a) **39**, 103 (1977)	*10.1*, p. *75*; 21.1, p. 288
	(a) **41**, K1 (1977)	*10.1*, p. *77*; 11, p. 95; 21.1, p. 293
A. K. KATHURIA	(a) **41**, K189 (1977)	*2*, p. *37*; *14.3*, p. *169*; 22.1.3, p. 345
H. KATO	(a) **39**, 331 (1977)	*16*, p. *201*; 22.4.4, p. 368
	(a) **42**, 769 (1977)	*11*, p. *97*; *14.3.2*, p. *177*; 16, p. 203; 22.1.2, p. 341
A. A. KATSNELSON	(b) **82**, 701 (1977)	*1.3*, p. *22*; 21.1, p. 296
R. KATZSCHMANN	(a) **40**, K161 (1977)	*13.1*, p. *117*; *20.1*, p. *253*; 22.4.2, p. 365
A. G. KAVETSKII	(a) **43**, 321 (1977)	*13.4*, p. *137*; *20.3*, p. *273*; 22.5.1, p. 373
K. KAWASAKI	(b) **79**, 413 (1977)	*18.1*, p. *213*; 18.2, p. 217; 18.4, p. 231
E. M. KAZARYAN	(a) **42**, 409 (1977)	*7*, p. *62*; *10*, p. *73*
L. M. KAZIMENKO	(a) **42**, 191 (1977)	*18.3*, p. *229*; 22.8.2, p. 412
K. O. KECHECHYAN	(b) **83**, K105 (1977)	10.2, p. 91; *13.5.1*, p. *147*; 22, p. 329
K. KECK	(b) **82**, 595 (1977)	*13.4*, p. *136*; 14.3.2, p. 177; 22.1.3, p. 346
W. KEILIG	(a) **40**, K135 (1977)	*2*, p. *36*; *18.3*, p. *229*; 21, p. 280; 21.1, p. 292
L. D. KEKHVA	(a) **43**, 321 (1977)	*13.4*, p. *137*; *20.3*, p. *273*; 22.5.1, p. 373
T. E. KEKHVA	(a) **43**, 321 (1977)	*13.4*, p. *137*; *20.3*, p. *273*; 22.5.1, p. 373
S. M. KELSO	(b) **83**, 305 (1977)	*13.5.1*, p. *147*; 18.1, p. 215; 22.1.1, p. 335
L. KEMTER	(a) **40**, 385 (1977)	*10.2*, p. *86*; 22.5.2, p. 377
G. KERYER	(b) **80**, 125 (1977)	*8*, p. *64*; 18.1, p. 213; 22.9, p. 414
A. R. KESSEL	(b) **79**, 489 (1977)	*13.5.1*, p. *143*; *19*, p. *235*

F. R. KESSLER	(b) **82**, K35 (1977)	13.1, p. 119; *18*, p. *211*; *20.1*, p. *257*; 22.1.1, p. 335
	(a) **43**, 159 (1977)	13.5.1, p. 147; *20.3*, p. *272*; 22.1.3, p. 346
G. G. KHALYULLIN	(b) **80**, 709 (1977)	*18.1*, p. *214*; 21, p. 279
H. R. KHAN	(b) **84**, K33 (1977)	*14.2*, p. *163*; *19*, p. *242*; 21, p. 286; 21.6, p. 317
M. A. KHAN	(b) **83**, 161 (1977)	*13.1*, p. *121*; *20.1*, p. *259*; 22.5, p. 371
Y. KHAN	(a) **39**, 79 (1977)	1, p. 9; 13.1, p. 114; *14.2*, p. *161*; 20.1, p. 248; 22.6.1, p. 393
E. G. KHARAKHASHYAN	(b) **80**, K109 (1977)	*19*, p. *237*; 21.1, p. 291
	(a) **40**, K11 (1977)	*19*, p. *237*; 21, p. 279
	(a) **40**, K53 (1977)	*19*, p. *237*; 21.2, p. 308
R. P. KHARE	(a) **44**, 627 (1977)	13.4, p. 140; *14.3.1*, p. *175*; 22.6, p. 392
YU. I. KHARKATS	(b) **83**, 313 (1977)	*1.4*, p. *27*; *10*, p. *73*
V. N. KHARKYANEN	(b) **81**, K103 (1977)	*13*, p. *110*
A. I. KHARLAMOV	(b) **83**, 537 (1977)	*13*, p. *111*; 17, p. 207; 22, p. 328
H. L. KHAROO	(b) **81**, 275 (1977)	*6*, p. *55*; *8*, p. *65*; 21.2, p. 308
B. KHELIFA	(b) **83**, 139 (1977)	*6*, p. *58*; *14.4.1*, p. *196*; 20.1, p. 259; 22.9, p. 417
M. L. KHIDEKEL	(a) **44**, 77 (1977)	*1*, p. *10*; *14*, p. *156*; 18.1, p. 215; 22.9, p. 419
YU. P. KHIMICH	(b) **79**, K125 (1977)	*5*, p. *47*; 18, p. 209; 22.8.2, p. 409
H. E. KHODENKOV	(a) **41**, K39 (1977)	*18.2*, p. *220*
V. A. KHODYREV	(a) **43**, 59 (1977)	*11*, p. *97*; 21, p. 284
NGUYEN THE KHOI	(b) **83**, K133 (1977)	13.5.1, p. 147; *20.1*, p. *261*; 22.4.4, p. 369
V. A. KHOKHLOV	(a) **44**, 429 (1977)	*18.2*, p. *224*; *18.4*, p. *234*; 22.8, p. 400
D. I. KHOMSKII	(b) **79**, 441 (1977)	*18.4*, p. *231*; 22.8, p. 396
M. M. KHRUSHCHOV	(b) **82**, 701 (1977)	*1.3*, p. *22*; 21.1, p. 296
L. A. KHUTORNAYA	(a) **42**, 433 (1977)	*10.2*, p. *90*; *20.3*, p. *272*; 22.9, p. 417
F. A. KHVAYA	(b) **82**, 701 (1977)	*1.3*, p. *22*; 21.1, p. 296
Y. KHWAJA	(b) **83**, 413 (1977)	13.1, p. 121; *18.2*, p. *223*; *18.4*, p. *233*; 21.1.1, p. 306
B.-P. KIETIS	(a) **39**, 601 (1977)	*7*, p. *62*; *14.3.4*, p. *183*; 22.4.1, p. 362
	(a) **43**, 705 (1977)	*7*, p. *63*; *14.3*, p. *171*; 16, p. 204; 22.4.3, p. 367
T. KIICHI	(a) **41**, K17 (1977)	*1.2*, p. *16*; *19*, p. *239*; 22.9, p. 416
A. KIKUCHI	(a) **41**, 289 (1977)	*10.2*, p. *88*; 11, p. 95; *20.1*, p. *254*; 22.5.2, p. 378
Y. KIMURA	(a) **43**, K141 (1977)	*1.2*, p. *19*; 8, p. 67; 13, p. 112; *14.2*, p. *163*; 21.1, p. 298
R. KIND	(a) **44**, 661 (1977)	*1.2*, p. *20*; 6, p. 61; 22.9, p. 419
R. KINK	(b) **84**, K61 (1977)	*20.3*, p. *274*; 23, p. 420
W. I. KINNEY	(a) **40**, 37 (1977)	*2*, p. *35*; 22.1.1, p. 333
A. A. KIRAKOSSYAN	(b) **83**, K105 (1977)	10.2, p. 91; *13.5.1*, p. *147*; 22, p. 329
H.-H. KIRCHNER	(a) **42**, K147 (1977)	*17*, p. *207*; 22.6, p. 390
H. O. K. KIRCHNER	(b) **80**, 441 (1977)	*10*, p. *72*
	(b) **81**, 191 (1977)	*10*, p. *72*
A. KIRFEL	(a) **40**, 447 (1977)	*1.2*, p. *16*; *18.4*, p. *232*; 22.8.1, p. 403
D. L. KIRK	(a) **40**, 265 (1977)	*9*, p. *69*; 22.6, p. 387
	(a) **44**, 643 (1977)	*9*, p. *71*; 10.2, p. 93; *14.4*, p. *193*; 22.6, p. 392
A. P. KISELEV	(a) **43**, 321 (1977)	*13.4*, p. *137*; *20.3*, p. *273*; 22.5.1, p. 373
V. F. KISELEV	(a) **42**, K61 (1977)	*14.3.1*, p. *174*; *14.3.2*, p. *177*; 22.9, p. 417
N. KISHIMOTO	(b) **80**, K113 (1977)	*2*, p. *35*; *19*, p. *237*; 21.6, p. 315; 22.1.2, p. 339
	(b) **82**, K123 (1977)	*2*, p. *35*; *19*, p. *237*; 21.6, p. 315; 22.1.2, p. 339
R. KISHORE	(b) **79**, K9 (1977)	*2*, p. *33*; *13*, p. *107*; 22, p. 320
	(b) **80**, K33 (1977)	*2*, p. *34*; 13.5, p. 141; *18.4*, p. *232*
	(b) **84**, K75 (1977)	*18.2*, p. *223*; 21, p. 286
A. KISIEL	(a) **42**, K41 (1977)	*14.1*, p. *159*; 21, p. 282
	(b) **83**, K35 (1977)	*14.3.4*, p. *186*; *20.1*, p. *259*; 22.1.1, p. 335
T. A. KITAEVA	(a) **43**, K73 (1977)	*20.1*, p. *260*; 22.9, p. 418

M. Kitagawa	(b) **79**,	321 (1977)	*9*, p. *69*; 11, p. 94; 22.1.2, p. 337
	(b) **82**,	643 (1977)	*13.5.2*, p. *149*; 21, p. 283
E. Kittinger	(a) **42**,	K33 (1977)	*2*, p. *37*; *7*, p. *62*; 22.1.3, p. 346
	(a) **44**,	K35 (1977)	*2*, p. *40*; *8*, p. *68*; 22.1.3, p. 347
J. Klamut	(b) **80**,	K71 (1977)	*18.2*, p. *218*
B. Klar	(a) **44**,	127 (1977)	*1.1*, p. *13*; 22.1.2, p. 343
M. Klein	(b) **83**,	451 (1977)	*11*, p. *97*
R. Klein	(a) **44**,	K13 (1977)	*10.2*, p. *92*; 22.2.2, p. 355
K. Kleinstück	(b) **82**,	495 (1977)	13.1, p. 120; *20*, p. *245*; 21.1, p. 296; 21.1.1, p. 304
Z. Klem	(a) **42**,	459 (1977)	*18.2*, p. *221*
W. Klement, Jr.	(a) **39**,	K45 (1977)	*1.2*, p. *14*; 22.6, p. 386
V. G. Kleparskii	(a) **44**,	771 (1977)	*18.2*, p. *225*; 22.8.2, p. 413
J. G. Kliava	(a) **41**,	385 (1977)	11, p. 95; *19*, p. *239*; 22.6, p. 388
K. Kliefoth	(a) **42**,	K133 (1977)	*14.3.3*, p. *181*; 22.2.1, p. 352
V. A. Klimov	(a) **41**,	679 (1977)	*10.2*, p. *89*; 22.5.2, p. 378
K.-D. Klinge	(a) **41**,	K171 (1977)	9, p. 70; *11*, p. *96*; 22.1.2, p. 340
M. I. Klinger	(b) **79**,	9 (1977)	13.1, p. 114; *13.5.3*, p. *151*; *14.3*, p. *165*; 15, p. 198; 18.3, p. 228; 22.8.2, p. 409
	(b) **83**,	353 (1977)	13.1, p. 114; *13.5.3*, p. *151*; *14.3*, p. *165*; 15, p. 198; 18.3, p. 228; 22.8.2, p. 409
K. von Klitzing	(b) **81**,	119 (1977)	*13.3*, p. *126*; 14.3.2, p. 176; 22.1.3, p. 345
A. A. Klochikhin	(b) **80**,	43 (1977)	*6*, p. *53*; 13.1, p. 115; *20.1*, p. *250*
S. I. Klokishner	(a) **40**,	643 (1977)	6, p. 55; *13.4*, p. *133*; 20.3, p. 269; 22.8.1. p. 403
B. Klotz	(a) **43**,	283 (1977)	*14.2*, p. *163*
J. Kluge	(b) **81**,	199 (1977)	13.5.1, p. 145; *18*, p. *210*; *20.1*, p. *253*; 22.1.2, p. 339; 22.2.2, p. 354
E. Klugmann	(a) **39**,	K53 (1977)	*11*, p. *94*; *18.2*, p. *217*; 21.1, p. 289
Th. Klupsch	(b) **83**,	129 (1977)	*1.6*, p. *32*; *3*, p. *42*
R. Kmiéc	(b) **79**,	K137 (1977)	*5*, p. *47*; 22.8.1, p. 402
H. Knauer	(a) **44**,	303 (1977)	13.4, p. 139; *14.3.3*, p. *181*; 22, p. 330
K. Kneipp	(b) **80**,	K55 (1977)	*6*, p. *54*; *20*, p. *244*; 22.8.1, p. 403
D. I. M. Knottnerus	(a) **41**,	95 (1977)	*10.2*, p. *87*; 20.1, p. 254; 22.8, p. 398
Yu. V. Knyazev	(b) **80**,	11 (1977)	13.1, p. 114; *20.1*, p. *247*; 21.4, p. 311
U. Köbler	(b) **84**,	189 (1977)	*18*, p. *212*; *19*, p. *242*; 21, p. 286
P. Kocevar	(b) **84**,	681 (1977)	*6*, p. *60*; 22.1.1, p. 336; 22.1.2, p. 344
E. E. Koch	(b) **81**,	503 (1977)	*13.1*, p. *118*; 17, p. 206; 23, p. 420
S. Koch	(b) **82**,	531 (1977)	*13.5.1*, p. *146*; *20.2*, p. *266*; 22.4.1, p. 363
B. I. Kochelaev	(b) **81**,	747 (1977)	*13.5.1*, p. *146*
E. V. Kochergin	(b) **79**,	K17 (1977)	6, p. 52; *14.1*, p. *157*; *14.2*, p. *161*; 21, p. 276
K. Kocot	(b) **81**,	629 (1977)	*13.4*, p. *134*; 16, p. 202; 20.1, p. 255; 22.4.2, p. 366
F. P. Koffyberg	(b) **80**,	669 (1977)	2, p. 35; 13.1, p. 116; *20.1*, p. *252*; 22.8, p. 397
Sh. M. Kogan	(a) **39**,	11 (1977)	*13.4*, p. *129*; *16*, p. *201*; 22, p. 320; 22.1.1, p. 333; 22.1.2, p. 337; 22.2.1, p. 350
M. Kohen	(a) **40**,	519 (1977)	*10.1*, p. *77*; 21.1, p. 292
W. Kohl	(a) **43**,	K95 (1977)	*14.3*, p. *171*; 22.2.2, p. 354
H. Köhler	(b) **80**,	K29 (1977)	13.1, p. 116; *14.3*, p. *166*; 20.1, p. 251; 22.7, p. 394
	(b) **84**,	195 (1977)	*13.1*, p. *122*; *13.2*, p. *125*; 14.3, p. 171; 22.7, p. 395
	(b) **84**,	619 (1977)	*6*, p. *60*; *20.1*, p. *262*; 22.7, p. 395
H.-J. Köhler	(a) **44**,	K99 (1977)	*14.2*, p. *164*; 21, p. 287
V. G. Kohn	(a) **42**,	415 (1977)	*1*, p. *10*; *1.6*, p. *31*; 22.1.2, p. 341
W. Koj	(a) **44**,	675 (1977)	*3*, p. *43*; *10.2*, p. *93*; 22.2.1, p. 353
K. Kojima	(a) **39**,	491 (1977)	*1.6*, p. *31*; *10.1*, p. *75*; 21.1.1, p. 302

M. Kolář	(b) **80**, 521 (1977)	*13.3*, p. *126*; *21.1*, p. *291*
	(b) **83**, 625 (1977)	*13.3*, p. *128*
E. Kolley	(b) **79**, 325 (1977)	*2*, p. *33*; 6, p. 51; *14*, p. *155*; 21, p. 276
	(b) **79**, 461 (1977)	*2*, p. *33*; 6, p. 52; 13, p. 107; *14.1*, p. *157*; 21, p. 277
	(b) **81**, 735 (1977)	*13*, p. *109*; 21, p. 281
	(b) **84**, K147 (1977)	*13*, p. *113*; 18.1, p. 216; 18.2, p. 224
W. Kolley	(b) **79**, 325 (1977)	*2*, p. *33*; 6, p. 51; *14*, p. *155*; 21, p. 276
	(b) **79**, 461 (1977)	*2*, p. *33*; 6, p. 52; 13, p. 107; *14.1*, p. *157*; 21, p. 277
	(b) **81**, 735 (1977)	*13*, p. *109*; 21, p. 281
	(b) **84**, K147 (1977)	*13*, p. *113*; 18.1, p. 216; 18.2, p. 224
O. S. Kolotov	(a) **42**, K171 (1977)	*18.2.1*, p. *226*; 21.1.1, p. 305
F. F. Komarov	(b) **79**, K161 (1977)	*11*, p. *94*; 22.5.2, p. 375
Č. Koňák	(b) **79**, K35 (1977)	*20.1*, p. *248*; 22.8.1, p. 402
T. Kondow	(b) **81**, 157 (1977)	*8*, p. *65*; 22.9, p. 415
N. König	(b) **79**, 237 (1977)	*9*, p. *69*; 10.1, p. 75; 18.2, p. 217; 21.1, p. 288
S. König	(a) **41**, K67 (1977)	*2*, p. *36*; *14.4*, p. *191*; 22.9, p. 416
Yu. V. Konobeev	(b) **79**, 79 (1977)	13.4, p. 129; *13.5.1*, p. *143*; 20.3, p. 267; 22.9, p. 414
	(a) **40**, K89 (1977)	*9*, p. *69*; *11*, p. *95*; 21, p. 280; 21.1, p. 292
J. Konstantinovič	(a) **41**, K133 (1977)	*1.2*, p. *17*; 18.4, p. 233; 22.6, p. 388
N. I. Konyushkina	(a) **43**, 639 (1977)	*17*, p. *207*; 22.5.2, p. 380
R. Kopelman	(b) **81**, 479 (1977)	*13.5.1*, p. *145*; *20.1*, p. *255*; 22.9, p. 416
U. Kh. Kopvillem	(b) **83**, 109 (1977)	7, p. 63; *14*, p. *156*
S. L. Korableva	(b) **81**, 287 (1977)	*13.4*, p. *133*; 19, p. 238; 20.3, p. 270; 22.5.4, p. 385
I. Ya. Korenblit	(b) **79**, K95 (1977)	1.2, p. 14; *13*, p. *107*; *18.2*, p. *218*; 22, p. 321
A. Ya. Korets	(b) **84**, 107 (1977)	13.4, p. 138; *13.5.1*, p. *148*; *20.1*, p. *261*; 22.9, p. 419
C. Korn	(a) **44**, 755 (1977)	*1.3*, p. *23*; 19, p. 243; 21.1, p. 300
E. V. Kornelsen	(a) **39**, K143 (1977)	*10.1*, p. *76*; *11*, p. *94*; 21.1, p. 290
	(a) **40**, 235 (1977)	*10.1*, p. *76*; 21, p. 279
A. Korner	(b) **80**, 441 (1977)	*10*, p. *72*
	(b) **81**, 191 (1977)	*10*, p. *72*
A. A. Kornienko	(b) **79**, 775 (1977)	4, p. 44; *13.4*, p. *131*; 20.1, p. 249; 22.5.3, p. 383
A. A. Kornyshev	(a) **39**, 229 (1977)	*14.3.2*, p. *176*; 14.3.4, p. 183; 22.5.1, p. 373
	(a) **39**, 573 (1977)	*14.3.2*, p. *176*; 14.3.4, p. 183
	(b) **84**, 125 (1977)	*13*, p. *112*; *14.4*, p. *192*
V. L. Korobanov	(a) **41**, 339 (1977)	11, p. 95; *13.4*, p. *134*; *14.3.3*, p. *179*; 22.1.1, p. 334
V. D. Korolev	(b) **82**, 251 (1977)	13.4, p. 135; *19*, p. *240*; 22.6, p. 389
	(b) **84**, 443 (1977)	13.4, p. 139; *19*, p. *243*; 22.6, p. 391
L. I. Koroleva	(a) **41**, K123 (1977)	*18.2*, p. *220*; 22.8.2, p. 411
V. S. Korolkov	(b) **81**, 387 (1977)	*13.4*, p. *133*; 20.1, p. 254; 22.8.2, p. 411
F. P. Korshunov	(a) **41**, K151 (1977)	*11*, p. *96*; *14.4*, p. *191*; 22, p. 325; 22.6, p. 389
A. V. Korshunov	(b) **84**, 107 (1977)	13.4, p. 138; *13.5.1*, p. *148*; *20.1*, p. *261*; 22.9, p. 419
T. F. M. Kortekaas	(a) **40**, 479 (1977)	*8*, p. *65*; 10.1, p. 77; 21.1, p. 292
H.-J. Kos	(a) **41**, K157 (1977)	*10.2*, p. *89*; 13.4, p. 135; *20.1*, p. *256*; 22.5.2, p. 378
	(a) **44**, 259 (1977)	10.2, p. 92; *16*, p. *205*; 22.1.2, p. 343
	(a) **44**, 505 (1977)	*10.2*, p. *93*; 20.1, p. 263; *20.3*, p. *275*; 22.5.2 p. 381
Yu. V. Kosichkin	(a) **42**, 385 (1977)	12.2, p. 105; 13.1, p. 120; *16*, p. *203*; *20.1*, p. *257*; 22.1.3, p. 346
R. Kosicik	(a) **40**, K135 (1977)	*2*, p. *36*; *18.3*, p. *229*; 21, p. 280; 21.1, p. 292

M. E. Kost	(b) **83**,	517 (1977)	*1.2*, p. *19*; 14.1, p. 159; 21.4, p. 313
Yu. P. Kostikov	(a) **39**,	437 (1977)	*1.2*, p. *15*; 2, p. 34; 5, p. 47; 22, p. 322
D. Kostopoulos	(a) **40**,	141 (1977)	*14.3*, p. *167*; *18*, p. *210*; 22.8.2, p. 410
G. Kostorz	(a) **41**,	577 (1977)	*1.4*, p. *26*; 10.1, p. 78; 21, p. 281; 21.1, p. 294
N. Kosuch	(b) **84**,	561 (1977)	*20*, p. *246*; 22.2, p. 349
K. Kosuge	(a) **39**,	K37 (1977)	*18.4*, p. *231*; 22.6, p. 386
Yu. E. Kotelnikov	(b) **81**,	747 (1977)	*13.5.1*, p. *146*
I. M. Kotina	(a) **42**,	397 (1977)	*13.4*, p. *136*; *14.3.3*, p. *180*; 22, p. 326
A. I. Kotov	(a) **43**,	K73 (1977)	*20.1*, p. *260*; 22.9, p. 418
B. V. Kotov	(a) **40**,	655 (1977)	13.5.1, p. 144; *16*, p. *202*; 22.9, p. 415
V. S. Koval	(a) **44**,	K15 (1977)	*20.1*, p. *262*; 22.3, p. 359
M. V. Kovalchuk	(a) **42**,	415 (1977)	*1*, p. *10*; *1.6*, p. *31*; 22.1.2, p. 341
T. P. Kovalenko	(a) **44**,	457 (1977)	*10.2*, p. *93*; 22.5.2, p. 381
V. A. Kovarskii	(b) **80**,	K5 (1977)	*14.1*, p. *158*; 21.7, p. 319
E. K. Kovev	(a) **42**,	415 (1977)	*1*, p. *10*; *1.6*, p. *31*; 22.1.2, p. 341
K. Koyama	(b) **83**,	485 (1977)	*11*, p. *98*; 13.3, p. 127; 21.6, p. 317
B. Kozarzewski	(b) **82**,	205 (1977)	*18*, p. *211*; 22.7, p. 394
E. V. Kozlov	(b) **81**,	719 (1977)	*1.2*, p. *16*; *1.3*, p. *21*; 21.1, p. 293; 21.6, p. 316
S. N. Kozlov	(a) **42**,	115 (1977)	*13.3*, p. *127*; 14.3.2, p. 176; 22, p. 326
V. A. Kozlov	(a) **42**,	191 (1977)	*18.3*, p. *229*; 22.8.2, p. 412
J. A. Kozubowski	(a) **43**,	535 (1977)	*1.1*, p. *13*
P. Kramer	(a) **43**,	K95 (1977)	*14.3*, p. *171*; 22.2.2, p. 354
U. Krämer	(a) **44**,	499 (1977)	*10.1*, p. *82*; 14.2, p. 164; 21.1, p. 300
R. Kranold	(a) **40**,	K161 (1977)	*13.1*, p. *117*; *20.1*, p. *253*; 22.4.2, p. 365
G. L. Krasko	(b) **80**,	341 (1977)	1.3, p. 21; *4*, p. *44*; 21, p. 278
	(b) **80**,	713 (1977)	*1*, p. *9*; *4*, p. *44*
	(b) **81**,	263 (1977)	*1*, p. *9*; *4*, p. *45*; 21.7, p. 319; 22.1.1, p. 334; 22.1.2, p. 339
P. Kratochvíl	(a) **40**,	K165 (1977)	*10.1*, p. *77*; 12.1, p. 101; 21.1, p. 293; 21.6, p. 315
L. Kraus	(a) **39**,	K161 (1977)	*2*, p. *34*; *18.2*, p. *218*; 19, p. 236; 22.8, p. 397
C. Krause	(a) **44**,	675 (1977)	*3*, p. *43*; *10.2*, p. *93*; 22.2.1, p. 353
H. Krause	(a) **42**,	149 (1977)	*13.4*, p. *136*; *14.3.3*, p. *180*; 22.6, p. 389
A. F. Kravchenko	(a) **40**,	321 (1977)	*14.3.4*, p. *184*; *16*, p. *202*; 22.2.1, p. 351
	(a) **42**,	647 (1977)	*13.4*, p. *137*; *14.3.1*, p. *174*; 16, p. 203; 22.2.1, p. 352
K. Kreher	(a) **43**,	529 (1977)	*13.4*, p. *138*; *20.3*, p. *273*; 22.2.2, p. 354
U. Kreibig	(b) **82**,	545 (1977)	*20.1*, p. *258*; 21.1, p. 296; 21.6, p. 316
	(b) **83**,	K49 (1977)	*20.1*, p. *260*; 21, p. 284; 21.6, p. 316
H. Kresse	(a) **41**,	K67 (1977)	*2*, p. *36*; *14.4*, p. *191*; 22.9, p. 416
G. Kretzschmar	(a) **39**,	K139 (1977)	*14.3.3*, p. *178*; *20.3*, p. *268*; 22.2.1, p. 351
E. W. Kreutz	(a) **40**,	415 (1977)	*13.3*, p. *126*; *14.3.2*, p. *176*; 22.2.3, p. 356
	(a) **42**,	K11 (1977)	*1.1*, p. *12*; *1.6*, p. *31*; 21.6, p. 316
M. G. Krimnus	(b) **82**,	695 (1977)	*12.2*, p. *105*; *20.1*, p. *258*; 22.5, p. 371; 22.8.1, p. 406
I. Krishan	(a) **41**,	323 (1977)	2, p. 36; *14.4*, p. *190*; 22.8, p. 398
B. S. Krishnamurthy	(b) **80**,	603 (1977)	*6*, p. *54*; *14.3.4*, p. *184*; 22.1.2, p. 338
P. Krispin	(a) **40**,	K81 (1977)	*13.4*, p. *132*; *14.3.3*, p. *179*; 22.2.2, p. 354
S. Kritzinger	(a) **39**,	239 (1977)	*10.1*, p. *75*; 21, p. 276
H. Krizek	(a) **43**,	251 (1977)	*14.1*, p. *159*; 21.4, p. 313
R. Kroggel	(a) **40**,	K163 (1977)	1.2, p. 16; *10.1*, p. *77*; 21, p. 280; 21.1, p. 292
H. Kronmüller	(b) **79**,	139 (1977)	*14.2*, p. *161*; 21, p. 276
	(b) **79**,	237 (1977)	*9*, p. *69*; 10.1, p. 75; 18.2, p. 217; 21.1, p. 288
	(b) **82**,	523 (1977)	*10*, p. *73*; *12*, p. *100*
	(a) **43**,	K81 (1977)	10.1, p. 80; *11*, p. *97*; *18.2*, p. *222*; 21.1.1, p. 306
D. Kropman	(a) **44**,	K1 (1977)	10.2, p. 92; *19*, p. *243*; 22.1.2, p. 343

F. Kroupa	(a) **42**, 713 (1977)	*12.1*, p. *102*; 21.1.1, p. 305
	(b) **84**, 725 (1977)	*10.1*, p. *82*; 21.1.1, p. 307
H. Krüger	(b) **80**, K101 (1977)	*14.4*, p. *190*; 22.7, p. 394
B. Krukowska-Fulde	(b) **83**, K155 (1977)	*13.4*, p. *138*; 16, p. 204; 22.5.3, p. 384
	(a) **44**, K43 (1977)	*14.3*, p. *172*; *20.1*, p. *262*; 22.5.3, p. 384
	(a) **44**, K83 (1977)	*20.3*, p. *274*; 22.5.3, p. 384
Yu. V. Krylnikov	(a) **39**, 437 (1977)	*1.2*, p. *15*; 2, p. 34; 5, p. 47; 22, p. 322
V. A. Krylov	(b) **84**, 215 (1977)	*13.4*, p. *138*; *20.1*, p. *261*; 22.5.3, p. 384
I. V. Krylova	(a) **43**, 639 (1977)	*17*, p. *207*; 22.5.2, p. 380
M. Krystek	(a) **40**, K65 (1977)	*13.4*, p. *132*; *20.3*, p. *269*; 22, p. 323
M. Kryszewski	(a) **42**, 305 (1977)	*2*, p. *37*; 14.4, p. 191; 22.9, p. 417
Yu. Kryuchkov	(a) **43**, K201 (1977)	*10.2*, p. *91*; *11*, p. *98*; 22.1.2, p. 342
S. Krzemiński	(b) **80**, 651 (1977)	8, p. 65; *13*, p. *108*; *18.2*, p. *219*; 18.4, p. 232
	(b) **83**, K63 (1977)	*18.2*, p. *222*; *18.4*, p. *233*
S. S. Kubakaddi	(b) **80**, 603 (1977)	*6*, p. *54*; *14.3.4*, p. *184*; 22.1.2, p. 338
A. Kubový	(a) **40**, 225 (1977)	*14.3.1*, p. *173*; 22.1.3, p. 345
F. Kuchar	(b) **79**, 115 (1977)	*14.3*, p. *165*; 22.2.1, p. 350; 22.2.2, p. 354
J. Kudrnovský	(b) **84**, 325 (1977)	13.1, p. 122; *14.1*, p. *159*; 21, p. 286
	(b) **84**, K21 (1977)	*14.1*, p. *159*; 21, p. 286
K. I. Kugel	(b) **79**, 441 (1977)	*18.4*, p. *231*; 22.8, p. 396
G. Kühn	(b) **80**, K21 (1977)	13.1, p. 116; *20.1*, p. *251*; 22.8, p. 397
	(b) **81**, K59 (1977)	*13.1*, p. *117*; *20.1*, p. *254*; 22.8, p. 398
	(a) **41**, K13 (1977)	*10.2*, p. *88*; *19*, p. *239*; 22.2.1, p. 351
W. Kühn	(a) **42**, 629 (1977)	*14.3.4*, p. *185*; *20.1*, p. *258*; 22.2.4, p. 357
A. Kühnel	(b) **84**, 653 (1977)	*14.4.1*, p. *197*; 18.2, p. 224
G. Kühnel	(a) **41**, 75 (1977)	*14.3.3*, p. *179*; *20.3*, p. *270*; 22.8, p. 398
G. Kuka	(b) **79**, 501 (1977)	*13.2*, p. *124*; 14.1, p. 157; 21.7, p. 319
E. F. Kukovitskii	(b) **80**, K109 (1977)	*19*, p. *237*; 21.1, p. 291
M. A. Kumakhov	(b) **84**, 41 (1977)	*11*, p. *98*; 13, p. 112; 22.1.2, p. 342
	(b) **84**, 581 (1977)	*11*, p. *99*; 13, p. 112; 22.1.2, p. 344
A. Kumar	(a) **41**, 323 (1977)	2, p. 36; *14.4*, p. *190*; 22.8, p. 398
M. Kumar	(b) **83**, 155 (1977)	*6*, p. *58*; *20.1*, p. *259*; 22.5.2, p. 380
Mahendra Kumar	(b) **82**, K29 (1977)	*13.2*, p. *124*; 21, p. 282; 21.2, p. 309
V. Kumar	(b) **80**, K1 (1977)	*13.4*, p. *132*; 22.5.3, p. 383
	(b) **82**, 577 (1977)	*18.1*, p. *215*; *18.2*, p. *221*
M. Kumeda	(b) **81**, K71 (1977)	*2*, p. *36*; *19*, p. *238*; 22.1.1, p. 334; 22.1.2, p. 339
A. Y. S. Kung	(b) **79**, 663 (1977)	*10.2*, p. *84*; 13.4, p. 130; 20.3, p. 268; 22.5.2, p. 375
V. Kunigėlis	(a) **43**, 699 (1977)	*7*, p. *63*; *14.3.2*, p. *177*; 22.4.1, p. 364
H. Küppers	(a) **39**, K49 (1977)	*1.2*, p. *14*; 8, p. 64; 14.4, p. 189; 22.9, p. 414
J. Kuriata	(b) **80**, K25 (1977)	*19*, p. *236*; 22.5.3, p. 383
	(b) **83**, K79 (1977)	*6*, p. *59*; *19*, p. *241*; 22.5.3, p. 384
H.-D. Kürsten	(a) **39**, 517 (1977)	*1.4*, p. *24*; *10.2*, p. *84*; 22.8.2, p. 409
K. Kurumada	(a) **40**, K93 (1977)	*14.3.3*, p. *179*; 22.2.1, p. 351
V. V. Kurylev	(a) **42**, K61 (1977)	*14.3.1*, p. *174*; *14.3.2*, p. *177*; 22.9, p. 417
E. F. Kustov	(b) **81**, 421 (1977)	*13*, p. *109*
	(a) **41**, 379 (1977)	*20.1*, p. *256*; *20.3*, p. *270*; 22.8.1, p. 405
R. Kužel	(a) **40**, 309 (1977)	*14.3.2*, p. *176*; 14.3.3, p. 178; 22.4.3, p. 367
H. Kuzmany	(a) **40**, 75 (1977)	*7*, p. *62*; *14.3.2*, p. *176*; 14.3.3, p. 178; 16, p. 201; 22.4.1, p. 362; 22.8.1, p. 403
V. K. Kvachadze	(a) **40**, K113 (1977)	8, p. 65; *10.2*, p. *87*; *20.1*, p. *253*; 22.5.2, p. 377
V. V. Kveder	(a) **43**, 407 (1977)	10.2, p. 91; *13.4*, p. *138*; *14.3*, p. *171*; 22.1.2, p. 342
	(a) **44**, 107 (1977)	10.2, p. 91; *13.4*, p. *139*; *14.3*, p. *171*; 22.1.2, p. 343

I. Łabędzka	(a) **44**, K103 (1977)	2, p. 40; *14.1*, p. *160*; *18.2*, p. *224*; 21.1, p. 299; 21.4, p. 313
R. Labusch	(a) **41**, 431 (1977)	*13.4*, p. *135*; *16*, p. *202*; 20.1, p. 256; 22.1.1, p. 334
	(a) **42**, 165 (1977)	13.1, p. 119; *13.4*, p. *136*; *16*, p. *203*; 22.1.1, p. 335
C. Lacour	(a) **40**, K179 (1977)	*1.4*, p. *25*; *3*, p. *42*; 22.8.2, p. 410
J. Laforest	(a) **40**, K105 (1977)	*18.1*, p. *214*; *18.4*, p. *232*; 21.4, p. 311
S. Lagomarsino	(a) **44**, 127 (1977)	*1.1*, p. *13*; 22.1.2, p. 343
G. Lahajnar	(a) **42**, 315 (1977)	*14.4.1*, p. *196*; *19*, p. *240*; 22.8.1, p. 405
S. Lakeou	(a) **43**, 213 (1977)	*14.3*, p. *170*; 22.1.1, p. 335; 22.1.2, p. 341; 22.2.3, p. 356
J. Lakner	(b) **81**, 593 (1977)	13.2, p. 124; *18*, p. *210*; 21, p. 280
K. N. Lakshminarayan	(a) **41**, 323 (1977)	2, p. 36; *14.4*, p. *190*; 22.8, p. 398
Dang Huy Lam	(a) **42**, 675 (1977)	*14.3*, p. *170*; 16, p. 203; 22.4.2, p. 366
	(a) **44**, 83 (1977)	*14.3.3*, p. *181*; *14.3.4*, p. *187*; 22.4.2, p. 366
J. van Landuyt	(a) **39**, 383 (1977)	*1.2*, p. *15*; 1.4, p. 24; 22, p. 322
	(a) **40**, 343 (1977)	*1.2*, p. *15*; 1.4, p. 25; 22, p. 323
	(a) **41**, 51 (1977)	*1.4*, p. *25*; 22.7, p. 394
	(a) **41**, 129 (1977)	*1.2*, p. *16*; *1.4*, p. *25*; 22.6, p. 388
	(a) **42**, 319 (1977)	1.3, p. 22; *1.4*, p. *26*; 22, p. 326
	(a) **42**, 565 (1977)	*1.1*, p. *12*; *1.2*, p. *18*; 22, p. 327
	(a) **43**, K137 (1977)	*1.1*, p. *13*; 1.4, p. 28; 22, p. 329
	(a) **44**, 411 (1977)	1.3, p. 22; *1.4*, p. *26*; 22, p. 326
G. Landwehr	(b) **81**, 119 (1977)	*13.3*, p. *126*; 14.3.2, p. 176; 22.1.3, p. 345
I. G. Lang	(b) **80**, 691 (1977)	*10.2*, p. *86*; *13.5.1*, p. *144*; 22.5.2, p. 377
H. Lange	(a) **42**, 629 (1977)	*14.3.4*, p. *185*; *20.1*, p. *258*; 22.2.4, p. 357
	(b) **84**, 569 (1977)	13.1, p. 122; *13.5.1*, p. *148*; *20.1*, p. *262*; 22.4.2, p. 366
J. Lange	(a) **42**, K75 (1977)	*17*, p. *206*; 22.6, p. 389
A. Lanjouw	(a) **41**, 95 (1977)	*10.2*, p. *87*; 20.1, p. 254; 22.8, p. 398
M. Lannoo	(b) **81**, 681 (1977)	*13.4*, p. *134*; 14.4, p. 191; 21.1, p. 293
V. Lantto	(b) **79**, 699 (1977)	*9*, p. *69*; *17*, p. *206*; 21, p. 277; 21.1, p. 289; 21.6, p. 315
S. Lara	(b) **82**, K21 (1977)	*13.1*, p. *119*; 22, p. 326
	(b) **84**, K143 (1977)	*12.2*, p. *106*; *13.1*, p. *123*; 22, p. 330; 22.6, p. 391
A. C. McLaren	see under letter "M"	
G. Laslaz	(a) **41**, 577 (1977)	*1.4*, p. *26*; 10.1, p. 78; 21, p. 281; 21.1, p. 294
K. Łątka	(b) **79**, K137 (1977)	*5*, p. *47*; 22.8.1, p. 402
E. E. Laufer	(a) **40**, 599 (1977)	*1*, p. *9*; *1.4*, p. *25*; 21.1, p. 292
A. Laugier	(a) **41**, 561 (1977)	6, p. 56; 8, p. 66; *10.2*, p. *88*; 22.2.1, p. 352
J. Laugsch	(b) **81**, 167 (1977)	6, p. 55; 13.5.1, p. 145; *20.1*, p. *253*; 22.8.1, p. 404
C. Lawther	(a) **44**, 313 (1977)	*14.3.1*, p. *175*; *20.3*, p. *274*; 22.4.1, p. 364
	(a) **44**, 693 (1977)	*20.3*, p. *275*; 22.4.2, p. 366
V. N. Lazukin	(b) **82**, 189 (1977)	2, p. 37; *19*, p. *240*; 22.8, p. 399
Le Manh Hoang	see under letter "H"	
J. Lebech	(b) **79**, 723 (1977)	*13.2*, p. *124*; 14.1, p. 157; 21.2, p. 308; 21.6, p. 315
	(b) **81**, K31 (1977)	13.2, p. 124; *14.1*, p. *158*
	(b) **83**, 187 (1977)	*13.2*, p. 124; *14.1*, p. *159*; 21, p. 284
J. Leciejewicz	(b) **80**, K137 (1977)	*13*, p. *108*; *20.1*, p. *252*; 22.5, p. 370
A. Leclerc	(b) **81**, K15 (1977)	*19*, p. *238*; 22.8.1, p. 404
K. H. Lee	(a) **39**, 669 (1977)	*10.2*, p. *84*; 19, p. 236; *20.3*, p. *268*; 22.6, p. 386
	(a) **42**, K137 (1977)	11, p. 97; *20.1*, p. *259*; *20.3*, p. *272*; 22.8.2, p. 412
K. O. Lee	(a) **43**, 565 (1977)	*14.3.1*, p. *175*; *14.3.4*, p. *187*; 22.9, p. 418

Y.-H. Lee	(a) **41**, 637 (1977)	*10.2*, p. *88*; 11, p. 95; *19*, p. *240*; 22.1.2, p. 340
J. J. Legendre	(a) **40**, K101 (1977)	*1.4*, p. *25*; 22, p. 324
R. Leihkauf	(b) **84**, 575 (1977)	*2*, p. *40*; *6*, p. *60*
E. B. Leiko	(a) **44**, 285 (1977)	*10.1*, p. *82*; 21, p. 287
F. Leis	(b) **82**, 545 (1977)	*20.1*, p. *258*; 21.1, p. 296; 21.6, p. 316
	(b) **83**, K49 (1977)	*20.1*, p. *260*; 21, p. 284; 21.6, p. 316
A. M. Lejus	(b) **84**, K105 (1977)	*6*, p. *61*; *20.1*, p. *263*; 22.6, p. 391
A. V. Lelyakov	(b) **82**, 711 (1977)	6, p. 58; *13.5.1*, p. *146*; *20.3*, p. *271*; 22.5.1, p. 373
L. F. Lemmens	(b) **80**, 99 (1977)	*14*, p. *155*
	(b) **81**, 551 (1977)	*14.4*, p. *191*
	(b) **82**, 117 (1977)	*14*, p. *155*
	(b) **82**, 439 (1977)	*13.5.3*, p. *151*
V. Lemos	(a) **39**, 65 (1977)	1.2, p. 14; *13.4*, p. *129*; *14.3*, p. *165*; 22.3.1, p. 360
P. Lenglart	(b) **80**, 361 (1977)	*10*, p. *72*; 13.1, p. 115; *14*, p. *155*
	(b) **80**, 433 (1977)	*10.2*, p. *86*; *14.3*, p. *167*; 22, p. 323; 22.1.3, p. 345
R. Lenk	(b) **82**, 303 (1977)	*1.6*, p. *31*
M. C. Leroy	(a) **39**, 73 (1977)	*2*, p. *33*; 19, p. 235; 22.6.1, p. 393
V. I. Lesin	(b) **84**, 513 (1977)	*13.5.1*, p. *148*; *19*, p. *243*; 22.9, p. 419
T. N. Letova	(a) **42**, K171 (1977)	*18.2.1*, p. *226*; 21.1.1, p. 305
A. Lettington	(b) **83**, 543 (1977)	*13.4*, p. *137*; 14.4, p. 192; 22.1.1, p. 335
C. Levade	(a) **44**, K59 (1977)	*10.1*, p. *82*; *18.2*, p. *224*; 21.1.1, p. 306
M. Levalois	(a) **40**, 181 (1977)	*14.4.1*, p. *194*
A. A. Levin	(b) **80**, 681 (1977)	*4*, p. *44*; *13.1*, p. *116*; 22.1.2, p. 339; 22.1.3, p. 345; 22.2, p. 348; 22.4, p. 361
N. L. Levshin	(a) **42**, K61 (1977)	*14.3.1*, p. *174*; *14.3.2*, p. *177*; 22.9, p. 417
S. Lewonczuk	(b) **83**, 161 (1977)	*13.1*, p. *121*; *20.1*, p. *259*; 22.5, p. 371
J.-P. Leyris	(a) **39**, 125 (1977)	13.4, p. 130; *20.3*, p. *267*; 22.4.1, p. 362
	(a) **43**, 191 (1977)	6, p. 59; 13.4, p. 137; *20.3*, p. *273*; 22.4.1, p. 363
Li-Shing Li	(a) **42**, 125 (1977)	1.5, p. 30; *7*, p. *62*; *19*, p. *240*; 21.1, p. 295
M. Siu Li	(b) **82**, K103 (1977)	*10.2*, p. *90*; 20.1, p. 258; 22.5.4, p. 385
Y. Li	(a) **43**, 329 (1977)	*7*, p. *63*; *12.2*, p. *106*; 22.6.1, p. 393
T. M. Lifshits	(a) **39**, 11 (1977)	*13.4*, p. *129*; *16*, p. *201*; 22, p. 320; 22.1.1, p. 333; 22.1.2, p. 337; 22.2.1, p. 350
A. S. Lileev	(a) **39**, K105 (1977)	*18.2*, p. *218*; 21.1, p. 289; 21.4, p. 311
	(a) **40**, 125 (1977)	*18.2*, p. *218*; 21.1, p. 290; 21.4, p. 311
P. Lilienkamp	(a) **43**, 479 (1977)	*14.3.1*, p. *175*; *14.3.4*, p. *187*; 22.1.2, p. 342
C. A. S. Lima	(b) **80**, 57 (1977)	7, p. 62; *13.5.2*, p. *149*; 18, p. 210; 22.2.1, p. 351
S. Limmer	(b) **84**, 283 (1977)	*1.2*, p. *19*; 19, p. 242; 22.9, p. 419
P.-A. Lindgård	(b) **82**, 449 (1977)	13, p. 110; *18.1*, p. *214*; *18.4*, p. *233*; 21.1, p. 296; 21.4, p. 312
D. Linzen	(a) **43**, 277 (1977)	3, p. 43; *18.3*, p. *230*; 22.8.2, p. 412
H. Lipko	(a) **40**, K39 (1977)	*18.3*, p. *229*; 22.8.2, p. 410
H. Lippmann	(a) **44**, 585 (1977)	2, p. 41; *14.3.1*, p. *175*; *14.3.2*, p. *177*; 22.1.2, p. 344
B. Lippold	(a) **40**, 205 (1977)	*1*, p. *9*; *11*, p. *95*; 22.6, p. 387
B. S. Lisenker	(a) **40**, 321 (1977)	*14.3.4*, p. *184*; *16*, p. *202*; 22.2.1, p. 351
	(a) **42**, 647 (1977)	*13.4*, p. *137*; *14.3.1*, p. *174*; 16, p. 203; 22.2.1, p. 352
V. A. Lisovenko	(a) **42**, 433 (1977)	*10.2*, p. *90*; *20.3*, p. *272*; 22.9, p. 417
A. A. Lisyanskii	(b) **79**, K17 (1977)	6, p. 52; *14.1*, p. *157*; *14.2*, p. *161*; 21, p. 276
V. I. Litvinov	(b) **79**, K131 (1977)	*1.2*, p. *15*; *13*, p. *108*
V. V. Litvintsev	(a) **40**, 11 (1977)	*1.2*, p. *14*; 1.4, p. 24; *1.5*, p. *29*; 18.2, p. 217; 21.1, p. 288; 21.1.1, p. 302
S. T. Liu	(a) **41**, K83 (1977)	*14.4.1*, p. *195*; 22.8.1, p. 404

J. Livage	(a) **39**,	73 (1977)	*2*, p. *33*; 19, p. 235; 22.6.1, p. 393
	(a) **41**,	423 (1977)	*19*, p. *239*; 22.6.1, p. 393
J. D. Livingston	(a) **44**,	295 (1977)	*1.4*, p. *28*; 14.2, p. 164; 21, p. 287
R. Lobo	(b) **81**,	657 (1977)	*13.5.2*, p. *149*; 14.4, p. 191; 21, p. 281
W. Lochmann	(a) **40**,	285 (1977)	*6*, p. *54*; *13.1*, p. *116*; 20, p. 244; 22.2.1, p. 351
	(a) **42**,	181 (1977)	*6*, p. *57*; *20*, p. *245*; 22.2.1, p. 352
A. Lodder	(b) **82**,	K23 (1977)	*10.1*, p. *78*; 21, p. 282
H. Löffler	(a) **40**,	K163 (1977)	1.2, p. 16; *10.1*, p. *77*; 21, p. 280; 21.1, p. 292
A. S. Logginov	(a) **41**,	371 (1977)	*14.3.3*, p. *179*; *20.2*, p. *265*; 22.2.4, p. 357
A. A. Loginov	(b) **84**,	83 (1977)	*13.5.1*, p. *147*; *13.5.4*, p. *154*
E. Loh	(b) **79**,	795 (1977)	*13.4*, p. *131*; *20.1*, p. *249*; 22.5.3, p. 383
	(b) **81**,	411 (1977)	*13.4*, p. *131*; *20.1*, p. *249*; 22.5.3, p. 383
A. Lôhmus	(b) **84**,	K61 (1977)	*20.3*, p. *274*; 23, p. 420
S. Lokanathan	(b) **83**,	273 (1977)	*5*, p. *49*; 22.8, p. 399
L.-A. Lompre	(a) **39**,	K25 (1977)	*17*, p. *206*; 21.6, p. 315
K. A. Long	(b) **79**,	155 (1977)	*6*, p. *51*; 8, p. 64; *14.1*, p. *157*; 15, p. 198; 21.5, p. 314
J. Loos	(b) **84**,	457 (1977)	*18.3*, p. *230*; 22.8.2, p. 413
P. Loose	(a) **42**,	K119 (1977)	*20.1*, p. *259*; 22.9, p. 417
J. Loriers	(a) **44**,	K147 (1977)	*18.3*, p. *230*; 22.8.2, p. 413
A. Lörinczy	(a) **43**,	K135 (1977)	1.5, p. 30; *1.6*, p. *32*; 2, p. 39; 22.6, p. 391
A. Lösche	(b) **84**,	283 (1977)	*1.2*, p. *19*; 19, p. 242; 22.9, p. 419
W. Löser	(b) **83**,	299 (1977)	2, p. 38; *8*, p. *66*; *18.4*, p. *233*
V. A. Lovchikov	(b) **79**,	743 (1977)	*4*, p. *44*; *13*, p. *107*; 22, p. 321
J. E. Lowther	(b) **82**,	K41 (1977)	*2*, p. *37*; *6*, p. *57*; 22.9, p. 416
M. Lozev	(a) **39**,	K1 (1977)	14.3.1, p. 173; *14.3.4*, p. *183*; 22, p. 321
M. Lubecka	(b) **82**,	107 (1977)	*14.3*, p. *169*; *18.2*, p. *220*; 22.6, p. 389
V. S. Lubimov	(b) **79**,	403 (1977)	*13.5*, p. *141*
	(b) **80**,	K13 (1977)	*13*, p. *108*; 13.5, p. 141; 22, p. 323
	(b) **81**,	691 (1977)	*13*, p. *109*; 13.5, p. 141; 22, p. 325
	(b) **84**,	K1 (1977)	*13*, p. *112*; 13.5, p. 142; 22, p. 330
	(b) **84**,	485 (1977)	*13.5*, p. *142*; *18*, p. *212*
G. Lübke	(a) **43**,	K131 (1977)	*10.2*, p. *91*; *20.1*, p. *261*; 22.6, p. 391
A. Lubomirska-Wittlin	(b) **83**,	K155 (1977)	*13.4*, p. *138*; 16, p. 204; 22.5.3, p. 384
	(a) **44**,	K43 (1977)	*14.3*, p. *172*; *20.1*, p. *262*; 22.5.3, p. 384
A. Lucasson	(a) **40**,	101 (1977)	*11*, p. *95*; 21.4, p. 311
P. Lucasson	(a) **40**,	101 (1977)	*11*, p. *95*; 21.4, p. 311
K. Lüders	(b) **79**,	519 (1977)	*19*, p. *235*; 21.1, p. 289
	(b) **84**,	189 (1977)	*18*, p. *212*; *19*, p. *242*; 21, p. 286
	(b) **84**,	K33 (1977)	*14.2*, p. *163*; *19*, p. *242*; 21, p. 286; 21.6, p. 317
Z. Ludmer	(a) **43**,	695 (1977)	*20.3*, p. *274*; 22.9, p. 419
W. Ludwig	(a) **43**,	K77 (1977)	*13*, p. *111*; *14.3.4*, p. *186*; 22.6, p. 391
A. Luft	(a) **44**,	285 (1977)	*10.1*, p. *82*; 21, p. 287
V. N. Lugovoi	(b) **79**,	131 (1977)	*14.3*, p. *165*; *20*, p. *244*
P. Lukáč	(a) **41**,	K197 (1977)	*10.1*, p. *78*; 21, p. 282; 21.6, p. 316
F. Lukeš	(b) **84**,	K113 (1977)	13.1, p. 123; *14.3.4*, p. *188*; *20.1*, p. *263*; 22.2.1, p. 353
T. Lukes	(b) **83**,	197 (1977)	2, p. 38; *13*, p. *110*; *20.1*, p. *259*; 22, p. 327
B. A. Lukiyanets	(b) **81**,	K1 (1977)	*13.5.1*, p. *145*; *14.4*, p. *190*
E. Łukowiak	(a) **41**,	547 (1977)	13.4, p. 135; *20.3*, p. *270*; 22.8.1, p. 405
N. B. Lukyanchikova	(a) **41**,	299 (1977)	*14.3.3*, p. *179*; *20.3*, p. *270*; 22.4.1, p. 363
L. N. Lukyanova	(a) **43**,	K159 (1977)	13.1, p. 122; *15*, p. *199*; 22.3.1, p. 360
A. Lupei	(b) **83**,	71 (1977)	*19*, p. *241*; 22.5.2, p. 380
V. Lupei	(b) **83**,	71 (1977)	*19*, p. *241*; 22.5.2, p. 380
A. A. Lushnikov	(b) **82**,	685 (1977)	*13.5.2*, p. *149*; 21, p. 283
	(b) **83**,	377 (1977)	*13.3*, p. *127*; 13.5.2, p. 150; *20.1*, p. *260*
F. Luty	(b) **81**,	443 (1977)	*6*, p. *56*; *20.1*, p. *254*; 22.5.2, p. 378

T. Luty	(b) **83**, 565 (1977)	*13.5.3*, p. *152*; 14.3, p. 171; 22, p. 328
M. Lutz	(b) **82**, 457 (1977)	*13.1*, p. *120*; *20.1*, p. *257*; 22.1.3, p. 346
M. Luzar	(a) **40**, K121 (1977)	14.4.1, p. 195; *19*, p. *238*; 22.8, p. 398
R. Luzzi	(b) **80**, 181 (1977)	*13.5*, p. *141*; 18.1, p. 213; 20.1, p. 251; 22.2.1, p. 351
	(b) **83**, 607 (1977)	13.1, p. 122; *14.4*, p. *192*; *20*, p. *246*; 22, p. 329
O. I. Lvov	(a) **42**, 757 (1977)	13.3, p. 127; *14.3.4*, p. *186*; 22, p. 327
V. S. Lysenko	(a) **41**, 339 (1977)	11, p. 95; *13.4*, p. *134*; *14.3.3*, p. *179*; 22.1.1, p. 334
A. Lyubchenko	(a) **44**, 747 (1977)	*16*, p. *205*; *20.3*, p. *275*; 22.3, p. 359
R. B. Lyubovskii	(a) **44**, 77 (1977)	*1*, p. *10*; *14*, p. *156*; 18.1, p. 215; 22.9, p. 419
H. Maaref	(b) **80**, 461 (1977)	*13.5.1*, p. *144*; 20.3, p. 269; 22.1.2, p. 338
J. R. Macdonald	(a) **43**, K169 (1977)	*14.3.2*, p. *177*; 14.3.4, p. 187
R. Mach	(a) **42**, K187 (1977)	*14.3.4*, p. *186*; 16, p. 204; 22.5, p. 371
A. Machová	(a) **42**, 713 (1977)	*12.1*, p. *102*; 21.1.1, p. 305
A. J. H. Macke	(a) **39**, 117 (1977)	*20.3*, p. *267*; 22.8, p. 396
A. T. Macrander	(a) **43**, 611 (1977)	*1.2*, p. *19*; 23, p. 420
E. G. Madatova	(a) **43**, 589 (1977)	*10.2*, p. *91*; 22, p. 329
K. Maeda	(a) **43**, 289 (1977)	*10.1*, p. *80*; *14.2*, p. *163*; 21, p. 285
M. Maeda	(a) **43**, 393 (1977)	*1.3*, p. *22*; *18.2*, p. *223*; 21.1, p. 297; 21.1.1, p. 306
W. Maenhout-Van der Vorst	(a) **39**, 253 (1977)	*10.2*, p. *83*; *14.4*, p. *189*; 22.5.1, p. 373
	(a) **40**, K173 (1977)	*10.2*, p. *87*; *14.4*, p. *190*; 22.5.1, p. 373
	(a) **44**, 229 (1977)	*10.2*, p. *92*; *14.4*, p. *193*; 22.5.1, p. 374
L. I. Magarill	(b) **83**, K97 (1977)	*14.3.4*, p. *186*; *16*, p. *204*; 22.2, p. 348
S. Mager	(a) **40**, 497 (1977)	*18*, p. *210*; 21.1, p. 292; 21.1.1, p. 303
	(b) **81**, 535 (1977)	13.1, p. 118; *18.2*, p. *220*; *18.4*, p. *233*; 21.1.1, p. 303
S. H. El-Mahalawy	see under letter "E"	
P. S. Mahesh	(b) **83**, 155 (1977)	*6*, p. *58*; *20.1*, p. *259*; 22.5.2, p. 380
S. A. Mahmoud	(a) **41**, K113 (1977)	*10.1*, p. *78*; *12.1*, p. *101*; 21.1.1, p. 304
K. Maier	(a) **44**, 567 (1977)	*9*, p. *71*; 10.1, p. 82; 21.1, p. 300
G. L. Mailyan	(b) **80**, 543 (1977)	*6*, p. *54*; *14.4.1*, p. *194*
E. Majchrowska	(b) **81**, K67 (1977)	*13.1*, p. *117*; 14.3, p. 168; 22.4.4, p. 368
E. Majerniková	(b) **79**, 339 (1977)	2, p. 33; *13.5.1*, p. *143*; *20.1*, p. *248*; 22, p. 320
	(b) **84**, K109 (1977)	*2*, p. *40*; *13*, p. *113*; 22, p. 330
C. A. Majid	(a) **40**, 605 (1977)	10.2, p. 86; 13.4, p. 133; *20.3*, p. *269*; 22.6, p. 387
E. F. Makarov	(b) **79**, 403 (1977)	*13.5*, p. *141*
	(b) **80**, K9 (1977)	*13*, p. *108*; 22, p. 323
	(b) **80**, K13 (1977)	*13*, p. *108*; 13.5, p. 141; 22, p. 323
	(b) **81**, 671 (1977)	*13*, p. *109*
	(b) **81**, 691 (1977)	*13*, p. *109*; 13.5, p. 141; 22, p. 325
	(b) **84**, 55 (1977)	*13*, p. *112*; 13.5, p. 142; 22, p. 329
	(b) **84**, K1 (1977)	*13*, p. *112*; 13.5, p. 142; 22, p. 330
	(b) **84**, 485 (1977)	*13.5*, p. *142*; *18*, p. *212*
	(b) **84**, 493 (1977)	*13*, p. *112*; 13.4, p. 139
A. G. Makhanek	(b) **81**, 387 (1977)	*13.4*, p. *133*; 20.1, p. 254; 22.8.2, p. 411
A. B. Makhnovetskii	(b) **80**, 341 (1977)	1.3, p. 21; *4*, p. *44*; 21, p. 278
	(b) **80**, 713 (1977)	*1*, p. *9*; *4*, p. *44*
	(b) **81**, 263 (1977)	*1*, p. *9*; *4*, p. *45*; 21.7, p. 319; 22.1.1, p. 334; 22.1.2, p. 339
S. Makolągwa	(a) **39**, 643 (1977)	*5*, p. *47*; 22.8.2, p. 409
G. I. Makovetskii	(a) **39**, K127 (1977)	*18.4*, p. *231*; 22, p. 322

V. V. MAKSIMENKO	(b) **82**,	685 (1977)	*13.5.2*, p. *149*; 21, p. 283
	(b) **83**,	377 (1977)	*13.3*, p. *127*; 13.5.2, p. 150; *20.1*, p. *260*
S. K. MAKSIMOV	(a) **39**,	459 (1977)	*10.2*, p. *84*; 22.1.2, p. 338
J. C. MALAURENT	(a) **43**,	K61 (1977)	*2*, p. *39*; 22.1.3, p. 346
H. S. MALDONADO	(b) **81**,	K27 (1977)	*19*, p. *238*; 22.8, p. 398
S. K. MALIK	(a) **43**,	317 (1977)	1.1, p. 12; *18.2*, p. *222*; 21.1, p. 297; 21.5, p. 314
J. MALINOWSKI	(a) **43**,	675 (1977)	13.3, p. 128; *14.3.4*, p. *187*; 22.5.1, p. 374
M. MALINOWSKI	(a) **41**,	K55 (1977)	*8*, p. *65*; *14.4*, p. *191*; 22.8.1, p. 404
R. SH. MALKOVICH	(b) **82**,	421 (1977)	*9*, p. *70*
I. A. MALTSEVA	(a) **39**,	453 (1977)	13.4, p. 131; *20.3*, p. *268*; 22.8, p. 396
V. K. MALYUTENKO	(a) **43**,	311 (1977)	*14.3.2*, p. *177*; *16*, p. *204*; 22.2.3, p. 356
S. D. MAMCHICH	(a) **43**,	K155 (1977)	*10.1*, p. *81*; 21.1, p. 298
H. S. MAMEDOV	(a) **40**,	K7 (1977)	*1*, p. *9*; 22.3.1, p. 360
A. MAMONTOV	(a) **43**,	K201 (1977)	*10.2*, p. *91*; *11*, p. *98*; 22.1.2, p. 342
L. A. MANAKOVA	(b) **79**,	403 (1977)	*13.5*, p. *141*
	(b) **80**,	K13 (1977)	*13*, p. *108*; 13.5, p. 141; 22, p. 323
	(b) **81**,	691 (1977)	*13*, p. *109*; 13.5, p. 141; 22, p. 325
	(b) **84**,	55 (1977)	*13*, p. *112*; 13.5, p. 142; 22, p. 329
	(b) **84**,	K1 (1977)	*13*, p. *112*; 13.5, p. 142; 22, p. 330
J. P. MANCEAU	(b) **80**,	125 (1977)	*8*, p. *64*; 18.1, p. 213; 22.9, p. 414
N. A. MANCINI	(b) **81**,	579 (1977)	*14.2*, p. *162*
M. MANFREDI	(b) **81**,	603 (1977)	*13.4*, p. *134*; 20.1, p. 255; 20.3, p. 270; 22.5.2, p. 378
LE MANH HOANG	see under letter "H"		
C. V. MANJUNATH	(a) **44**,	687 (1977)	14.4.1, p. 197; *19*, p. *243*; 22.8.1, p. 408
R. MAŃKA	(b) **83**,	K69 (1977)	*13.5.3*, p. *152*; 14.4.1, p. 196; *15*, p. *199*; 22.8.1, p. 406
A. MANOOGIAN	(b) **81**,	K15 (1977)	*19*, p. *238*; 22.8.1, p. 404
	(b) **84**,	K7 (1977)	*19*, p. *242*; 22.8.1, p. 407
N. B. MANSON	(b) **82**,	473 (1977)	13.4, p. 136; *18*, p. *211*; *20*, p. *245*; 22.6, p. 390
F. M. MANSY	(a) **41**,	K113 (1977)	*10.1*, p. *78*; *12.1*, p. *101*; 21.1.1, p. 304
M. J. MARCINKOWSKI	(a) **41**,	213 (1977)	*1.4*, p. *26*; *10*, p. *72*
	(a) **42**,	209 (1977)	*1.4*, p. *26*; 10, p. 73
	(a) **42**,	439 (1977)	*10*, p. *73*
	(a) **44**,	633 (1977)	*10*, p. *74*
	(a) **44**,	703 (1977)	*10*, p. *74*
P. DI MARCO	(a) **42**,	K179 (1977)	*16*, p. *203*; *20.1*, p. *259*; 20.3, p. 272; 22.9, p. 417
	(a) **43**,	K193 (1977)	*16*, p. *204*; 20.1, p. 261; 20.3, p. 274; 22.9, p. 419
L. MĂRCULESCU	(b) **80**,	265 (1977)	13.4, p. 131; *20.3*, p. *268*; 22.5.2, p. 376
S. F. MARENKIN	(a) **39**,	445 (1977)	*16*, p. *201*; *20.1*, p. *250*; 22.3, p. 358
S. I. MARENNIKOV	(a) **44**,	K91 (1977)	*14.3*, p. *172*; 22.8.1, p. 408
J. MARIEN	(a) **39**,	281 (1977)	*1.6*, p. *31*; *17.1*, p. *208*; 21, p. 277
V. MARIGLIANO RAMAGLIA	see under letter "R"		
M. M. MARINKOVIĆ	(b) **81**,	245 (1977)	*13*, p. *109*; 18.2, p. 219
G. MARIOTTO	(a) **40**,	93 (1977)	*13.4*, p. *132*; *14.3.4*, p. *184*; 22.3, p. 358
R. S. MARKIEWICZ	(b) **83**,	305 (1977)	*13.5.1*, p. *147*; 18.1, p. 215; 22.1.1, p. 335
	(b) **83**,	659 (1977)	*13.5.1*, p. *147*; 14.3, p. 171; 22.1.1, p. 335
S. MARKLUND	(b) **80**,	301 (1977)	*13.4*, p. *132*; 22.1.2, p. 338
J. C. MARMEGGI	(a) **39**,	57 (1977)	*18.4*, p. *231*; 22.6, p. 386
Ž. MAROHNIĆ	(a) **44**,	339 (1977)	*1.2*, p. *20*; *2*, p. *40*; 21.1, p. 299; 21.1.1, p. 306
YU. E. MARONCHUK	(a) **40**,	321 (1977)	*14.3.4*, p. *184*; *16*, p. *202*; 22.2.1, p. 351
	(a) **42**,	647 (1977)	*13.4*, p. *137*; *14.3.1*, p. *174*; 16, p. 203; 22.2.1, p. 352
J. MARSDEN	(b) 80,	K141 (1977)	*2*, p. *36*; *19*, p. *237*; 21, p. 279

A. H. Marshak	(a) **42**, 279 (1977)	13.4, p. 136; *14.3*, p. *169*; 22, p. 326
D. B. Marshall	(a) **41**, 231 (1977)	*1.4*, p. *26*; 10.2, p. 87; 22.8, p. 398
S. A. Marshall	(b) **82**, 311 (1977)	*19*, p. *240*; 22.5.2, p. 379
J. J. Martin	(a) **39**, K131 (1977)	*20.1*, p. *250*; 22.5.2, p. 376
	(b) **84**, 721 (1977)	*8*, p. *68*; 11, p. 99; 22.5.4, p. 385
T. P. Martin	(b) **81**, 185 (1977)	6, p. 55; *20.1*, p. *253*; 23, p. 420
	(b) **81**, K41 (1977)	6, p. 55; *20.1*, p. *254*; 22.5.2, p. 378
W.-A. Martin	(a) **44**, K159 (1977)	*16*, p. *205*; *19*, p. *243*; 22.1.2, p. 344
M. Martinelli	(b) **84**, 305 (1977)	18.1, p. 215; *19*, p. *242*; 22.8, p. 400
J. Martinez	(b) **79**, K65 (1977)	*13.4*, p. *129*; 14.4, p. 189; 22.8.1, p. 402
D. Marton	(b) **79**, K71 (1977)	*10.2*, p. *83*; *13.4*, p. *129*; 22.1.2, p. 337
	(b) **79**, K89 (1977)	*10.2*, p. *84*; 22.1.2, p. 338
D. I. Marvakov	(b) **82**, K95 (1977)	*13.1*, p. *120*; 22.3, p. 358
S. I. Masharov	(b) **84**, 33 (1977)	*14.1*, p. *159*; 18.4, p. 233; 21, p. 286
A. I. Maslennikov	(b) **81**, 351 (1977)	*13.4*, p. *133*; 18.4, p. 232; 19, p. 238; 22.8.1, p. 404
T. B. Massalski	(b) **81**, 157 (1977)	*8*, p. *65*; 22.9, p. 415
G. Massé	(a) **39**, 125 (1977)	13.4, p. 130; *20.3*, p. *267*; 22.4.1, p. 362
	(a) **42**, 545 (1977)	*13.1*, p. *121*; *20.3*, p. *272*; 22.4.1, p. 363
	(a) **43**, 191 (1977)	6, p. 59; 13.4, p. 137; *20.3*, p. *273*; 22.4.1, p. 363
T. Masumoto	(a) **44**, K151 (1977)	*2*, p. *41*; *18.2*, p. *225*; 21.1, p. 301; 21.1.1, p. 307
T. A. McMath	see under letters "Mc"	
P. C. Mathur	(a) **39**, K59 (1977)	*14.3.3*, p. *178*; 22.2.1, p. 350
	(a) **41**, K189 (1977)	*2*, p. *37*; *14.3*, p. *169*; 22.1.3, p. 345
	(a) **42**, 791 (1977)	13.5.3, p. 152; *14.3*, p. *170*; 22.4.2, p. 366
	(a) **44**, 401 (1977)	*14.3.3*, p. *181*; *14.3.4*, p. *187*; 22.2.1, p. 353
E. A. Matson	(a) **41**, K93 (1977)	*1.5*, p. *29*; *3*, p. *42*; 22.1.2, p. 339
T. Matsumura	(a) **43**, 685 (1977)	13.5.1, p. 147; *20.3*, p. *273*; 22.3, p. 359
H. Matsunami	(a) **39**, K79 (1977)	*20.3*, p. *267*; 22.3, p. 358
T. Matsushita	(a) **41**, 139 (1977)	*1.1*, p. *11*; *6*, p. *56*; 22.2.1, p. 351
J. W. Matthews	(a) **41**, 347 (1977)	*10.2*, p. *88*; 22.8.2, p. 411
P. Matthys	(b) **79**, K23 (1977)	*19*, p. *235*; 22.5.2, p. 375
	(b) **82**, K69 (1977)	*19*, p. *241*; 22.6, p. 390
E. E. Matyas	(a) **42**, K129 (1977)	*1.2*, p. *18*; 22.2.4, p. 357
M. Matyáš	(b) **81**, K115 (1977)	*2*, p. *37*; *19*, p. *239*; 22, p. 325
W. Matz	(b) **79**, 283 (1977)	*6*, p. *51*; 21.3, p. 310
	(a) **40**, 205 (1977)	*1*, p. *9*; *11*, p. *95*; 22.6, p. 387
A. Mauger	(b) **84**, 761 (1977)	*13.1*, p. *123*; 18.1, p. 216; 18.2, p. 224; 22, p. 330
W. May	(a) **39**, 65 (1977)	1.2, p. 14; *13.4*, p. *129*; *14.3*, p. *165*; 22.3.1, p. 360
S. Mazkedian	(a) **44**, 127 (1977)	*1.1*, p. *13*; 22.1.2, p. 343
Z. Mazurak	(a) **41**, 547 (1977)	13.4, p. 135; *20.3*, p. *270*; 22.8.1, p. 405
P. G. McCormick	(a) **41**, 513 (1977)	*10.1*, p. *78*; 21, p. 281; 21.1, p. 294
A. C. McLaren	(a) **39**, 631 (1977)	*1.4*, p. *24*; *10.2*, p. *84*; 22.8, p. 396
	(a) **41**, 231 (1977)	*1.4*, p. *26*; 10.2, p. 87; 22.8, p. 398
T. A. McMath	(a) **43**, K151 (1977)	*20.1*, p. *261*; 22.3, p. 359
M. V. Medvedev	(b) **79**, 379 (1977)	*18.2*, p. *217*; 21, p. 276
N. N. Medvedev	(b) **83**, 405 (1977)	*2*, p. *39*
Yu. V. Medvedev	(b) **79**, K17 (1977)	6, p. 52; *14.1*, p. *157*; *14.2*, p. *161*; 21, p. 276
Ah Mee Hor	see under letter "H"	
E. E. Meerson	(a) **44**, K167 (1977)	*2*, p. *41*; *14.3.4*, p. *188*; 22, p. 331
M. F. Meftah	(a) **40**, 503 (1977)	*20.1*, p. *253*; 22.6, p. 387
P. C. Mehendru	(a) **42**, 403 (1977)	*14.4*, p. *191*; 20.1, p. 257; 22.9, p. 417
P. Mehendru	(a) **42**, 403 (1977)	*14.4*, p. *191*; 20.1, p. 257; 22.9, p. 417
H. Mehner	(a) **39**, K115 (1977)	5, p. 47; *18.3*, p. *228*; 22.6, p. 386
R. M. Mehra	(a) **41**, K189 (1977)	*2*, p. *37*; *14.3*, p. *169*; 22.1.3, p. 345

S. K. Mehta	(a) **44**,	679 (1977)	*10.2*, p. *93*; *20.1*, p. *264*; 22.5.2, p. 382
V. P. Meiklyar	(b) **84**,	K29 (1977)	*19*, p. *242*; 22.5, p. 372
E. Z. Meilikhov	(b) **79**,	753 (1977)	*14.3*, p. *166*; 18, p. 209; *20*, p. *244*; 22.2.3, p. 356
A. Sh. Mekhtiev	(b) **81**,	K47 (1977)	13.5.1, p. 145; *14.3.4*, p. *185*; *20.1*, p. *254*; 22.8, p. 398
A. Melander	(a) **43**,	223 (1977)	*10*, p. *73*
S. Melone	(a) **44**,	127 (1977)	*1.1*, p. *13*; 22.1.2, p. 343
K. Melzer	(a) **39**,	K115 (1977)	5, p. 47; *18.3*, p. *228*; 22.6, p. 386
H. H. Mende	(a) **42**,	K95 (1977)	*1.4*, p. *27*; *10.1*, p. *80*; 21.1.1, p. 305
H. Menninger	(a) **43**,	167 (1977)	*1.4*, p. *27*; *3*, p. *43*; 20.3, p. 273; 22.2.4, p. 357
W. Mensch	(a) **43**,	K175 (1977)	10.1, p. 81; 11, p. 98; *18.2*, p. *223*; 21.1.1, p. 306
	(a) **43**,	K179 (1977)	10.1, p. 81; *11*, p. *98*; 12.1, p. 102; *18.2*, p. *223*; 21.1.1, p. 306
T. Merceron	(a) **40**,	159 (1977)	*18.3*, p. *229*; 22.8.2, p. 410
	(a) **44**,	K147 (1977)	*18.3*, p. *230*; 22.8.2, p, 413
D. Mergel	(a) **41**,	431 (1977)	*13.4*, p. *135*; *16*, p. *202*; 20.1, p. 256; 22.1.1, p. 334
	(a) **42**,	165 (1977)	13.1, p. 119; *13.4*, p. *136*; *16*, p. *203*; 22.1.1, p. 335
C. Merkel	(b) **83**,	425 (1977)	*13.1*, p. *121*; 14.4.1, p. 196; 22.5, p. 371
J. F. Merklin	(a) **44**,	679 (1977)	*10.2*, p. *93*; *20.1*, p. *264*; 22.5.2, p. 382
J. Merlin	(a) **40**,	519 (1977)	*10.1*, p. *77*; 21.1, p. 292
J. Mertsching	(b) **82**,	289 (1977)	*13.1*, p. *119*; *18*, p. *211*
M. D. Merz	(a) **39**,	697 (1977)	1.2, p. 15; *2*, p. *34*; *4*, p. *44*; 21, p. 278; 22.1 p. 332
U. Meschter	(b) **83**,	521 (1977)	*14.4*, p. *192*; 22.7, p. 395
M. Meshii	(a) **39**,	491 (1977)	*1.6*, p. *31*; *10.1*, p. *75*; 21.1.1, p. 302
E. U. Messerschmidt	(a) **42**,	K187 (1977)	*14.3.4*, p. *186*; 16, p. 204; 22.5, p. 371
U. Messerschmidt	(a) **42**,	61 (1977)	*1.4*, p. *26*; 10.2, p. 89; 22.6, p. 389
B. Mestnik	(b) **84**,	K37 (1977)	*4*, p. *46*; 22, p. 330
G. de Mey	(a) **44**,	K195 (1977)	*14.4*, p. *193*; 22, p. 331
L. P. van der Mey	(b) **82**,	581 (1977)	*9*, p. *70*; 10.1, p. 79; 21, p. 283
N. Meyendorf	(a) **39**,	K95 (1977)	*10.1*, p. *76*; 21, p. 278; 21.1, p. 289
	(a) **42**,	K15 (1977)	*10.1*, p. *79*; 21, p. 282
L. Meyer	(b) **83**,	451 (1977)	*11*, p. *97*
E. N. Mgbenu	(b) **83**,	K75 (1977)	*13.1*, p. *121*; 13.2, p. 125; 21.2, p. 309
Hoang Mi Shin	see under letter "S"		
M. Michaud	(a) **39**,	73 (1977)	*2*, p. *33*; 19, p. 235; 22.6.1, p. 393
B. Michel	(b) **81**,	K87 (1977)	*10*, p. *72*
D. Michel	(b) **84**,	K105 (1977)	*6*, p. *61*; *20.1*, p. *263*; 22.6, p. 391
Y. R. Midha	(b) **79**,	K103 (1977)	*8*, p. *64*; 21, p. 277; 21.1, p. 289; 21.6, p. 315
	(b) **84**,	423 (1977)	*8*, p. *68*; 21.1, p. 300
M. Mijatović	(b) **82**,	193 (1977)	*6*, p. *57*; *18*, p. *211*
V. S. Mikhalenko	(a) **43**,	589 (1977)	*10.2*, p. *91*; 22, p. 329
E. Ya. Mikhlin	(a) **43**,	K23 (1977)	*10.1*, p. *80*; 21.1.1, p. 305
A. Milchev	(b) **79**,	549 (1977)	*4*, p. *44*; 13, p, 107; 22.1, p. 332; 22.1.1, p. 333; 22.1.2, p. 338; 22.2.1, p. 350
F. Milia	(a) **42**,	315 (1977)	*14.4.1*, p. *196*; *19*, p. *240*; 22.8.1, p. 405
B. V. Mill	(a) **39**,	683 (1977)	*18.4*, p. *231*; 22.8.2, p. 410
N. D. J. Miller	(a) **43**,	593 (1977)	*18.3*, p. *230*; 22.8.2, p. 412
S. Milošević	(b) **82**,	193 (1977)	*6*, p. *57*; *18*, p. *211*
B. Milsch	(a) **41**,	K141 (1977)	*19*, p. *240*; 22.9, p. 416
A. Minafra	(a) **40**,	93 (1977)	*13.4*, p. *132*; *14.3.4*, p. *184*; 22.3, p. 358
L. C. M. Miranda	(b) **79**,	K49 (1977)	*6*, p. *52*; *13.5.4*, p. *153*; 18.2, p. 217; 22.8, p. 396
	(b) **80**,	57 (1977)	7, p. 62; *13.5.2*, p. *149*; *18*, p. *210*; 22.2.1, p. 351

	(b) **81**, 395 (1977)	*13.5.4*, p. *153*; 14.3, p. 167; 18.2, p. 219; 22, p. 324
V. S. Mironov	(b) **79**, 403 (1977)	*13.5*, p. *141*
R. H. Misho	(b) **84**, 69 (1977)	*6*, p. *59*; 8, p. 67; 22.7, p. 395
K. D. Misra	(b) **79**, 359 (1977)	*4*, p. *44*; 12.2, p. 104; 22.5.2, p. 375
T. E. Mitchell	(a) **39**, 301 (1977)	*10.2*, p. *83*; 22.6, p. 386
A. V. Mitin	(b) **81**, 69 (1977)	*5*, p. *48*
N. R. Mitra	(b) **83**, 651 (1977)	4, p. 46; *6*, p. *59*; 14.1, p. 159; 21.2, p. 309
	(b) **84**, 797 (1977)	2, p. 40; *14.1*, p. *160*; *15*, p. *200*; 21, p. 287
K. P. Mitrofanov	(a) **44**, 247 (1977)	5, p. 49; *14.4.1*, p. *197*; 18.2, p. 223; *18.4*, p. *234*; 22.8, p. 400
E. J. Mittemeijer	(a) **44**, 517 (1977)	1, p. 10; *9*, p. *71*; 22.5.2, p. 382
L. T. Miyada	(a) **42**, 235 (1977)	*10.1*, p. *79*; 11, p. 96; *12.1*, p. *101*; 21.1, p. 295
J. Mizia	(b) **80**, K75 (1977)	*18.2*, p. *218*; 21.1, p. 290
	(b) **84**, 449 (1977)	*13*, p. *112*; *18.4*, p. *234*
T. Mizutani	(a) **40**, K93 (1977)	*14.3.3*, p. *179*; 22.2.1, p. 351
U. Mizutani	(b) **81**, 157 (1977)	*8*, p. *65*; 22.9, p. 415
G. M. Mladenov	(a) **39**, 321 (1977)	*1.5*, p. *29*
S. Mochizuki	(a) **40**, 119 (1977)	*14.3*, p. *167*; 22.6, p. 387
	(a) **41**, 411 (1977)	*14.3*, p. *168*; 22.6, p. 388
	(a) **41**, 591 (1977)	*14.3*, p. *169*; 22.6, p. 388
M. Mocker	(b) **81**, 139 (1977)	*13*, p. *109*; *14.3.4*, p. *184*
K. P. Mogilnikov	(a) **44**, K167 (1977)	*2*, p. *41*; *14.3.4*, p. *188*; 22, p. 331
E. Mohler	(b) **79**, 509 (1977)	*14.4*, p. *189*; *20.1*, p. *249*; 22.5, p. 370
I. Mojzes	(a) **39**, K123 (1977)	*14.3.3*, p. *178*; *14.3.4*, p. *184*; 22.2.1, p. 350
L. P. Mokhracheva	(b) **84**, 465 (1977)	4, p. 46; *13.1*, p. *122*; 22.3, p. 359
R. T. Molchanova	(a) **39**, 357 (1977)	*1.2*, p. *14*; 12.2, p. 104; 13.5.1, p. 143; *20.1*, p. *248*; 22.3, p. 358
W. Moldenhauer	(a) **42**, 295 (1977)	*1.2*, p. *17*; *3*, p. *42*; 14.3, p. 170; 18.1, p. 214; 22.6.1, p. 393
	(a) **44**, K95 (1977)	*1.2*, p. *20*; 5, p. 49; 14.3, p. 172; 22.6, p. 391
A. G. Moldovan	(a) **43**, 317 (1977)	1.1, p. 12; *18.2*, p. *222*; 21.1, p. 297; 21.5, p. 314
W. Möller	(b) **80**, 119 (1977)	1.4, p. 24; *14.1*, p. *157*; *20.1*, p. *250*; 22, p. 322
R. Moncorgé	(a) **43**, K45 (1977)	*20.2*, p. *266*; 22.8.1, p. 407
J. Monecke	(b) **84**, 665 (1977)	*13.1*, p. *123*; 14, p. 156; 18.4, p. 234
R. Monnier	(b) **79**, 391 (1977)	6, p. 52; *13.4*, p. *129*; 20.1, p. 248; 22.5.2, p. 375
M. Montagna	(a) **40**, 93 (1977)	*13.4*, p. *132*; *14.3.4*, p. *184*; 22.3, p. 358
J. M. Montante	(a) **39**, 213 (1977)	*1.5*, p. *29*; *2*, p. *33*; 13.1, p. 115; 14.3.1, p. 173; 22.1.1, p. 333
T. M. Moravskaya	(b) **83**, K141 (1977)	13, p. 111; 17, p. 207; *20*, p. *246*; 22.9, p. 418
R. Moreira Xavier	see under letter "X"	
F. G. Brady Moreira	(b) **80**, 385 (1977)	1.2, p. 15; *18.1*, p. *213*; *18.4*, p. *231*
M. Moreno	(b) **82**, 669 (1977)	*13.4*, p. *136*; *19*, p. *240*; 22.5, p. 371
J. M. Moret	(a) **41**, 163 (1977)	13.4, p. 134; *19*, p. *239*; 22, p. 324
D. V. Morgan	(a) **41**, 117 (1977)	*14.3.3*, p. *179*; 22.1.1, p. 334; 22.1.2, p. 339; 22.2.1, p. 351; 22.2.2, p. 354
G. J. Morgan	(b) **83**, 543 (1977)	*13.4*, p. *137*; 14.4, p. 192; 22.1.1, p. 335
K. Mori	(a) **42**, 375 (1977)	*10.2*, p. *89*; 20.1, p. 257; 20.2, p. 265; 20.3, p. 271; 22.8.2, p. 412
K. Morigaki	(b) **80**, K113 (1977)	*2*, p. *35*; *19*, p. *237*; 21.6, p. 315; 22.1.2, p. 339
	(b) **82**, K123 (1977)	*2*, p. *35*; *19*, p. *237*; 21.6, p. 315; 22.1.2, p. 339
H. Morishima	(a) **43**, 417 (1977)	11, p. 98; *19*, p. *241*; 22.1.2, p. 342
A. J. Morton	(a) **43**, K123 (1977)	1.3, p. 23; *1.4*, p. *28*; 21.1, p. 298
	(a) **44**, 205 (1977)	*1.3*, p. *23*; *1.4*, p. *28*; 21.1, p. 299
H. Morwinski	(a) **39**, 133 (1977)	*1.1*, p. *11*; 1.2, p. 14; 22.2.4, p. 357
L. L. Moseley	(b) **83**, 197 (1977)	2, p. 38; *13*, p. *110*; *20.1*, p. *259*; 22, p. 327

P. Moser	(a) 42,	511 (1977)	*10.1*, p. *79*; 11, p. 97; *12.1*, p. *102*; 21.1.1, p. 304
	(a) 42,	751 (1977)	*18.2*, p. *222*; 21.1.1, p. 305
A. Mosser	(a) 43,	295 (1977)	*1.6*, p. *32*; 22.6, p. 390
M. F. Mostafa	(a) 41,	295 (1977)	18.1, p. 214; *18.2*, p. *220*; 22.9, p. 416
B. Movaghar	(b) 80,	491 (1977)	2, p. 35; *14.3*, p. *167*; *19*, p. *237*; 22.1.1, p. 334; 22.1.2, p. 338
E. Mrosan	(b) 80,	K49 (1977)	*14.1*, p. *158*; 21.1, p. 290
E. Mugeński	(b) 80,	K59 (1977)	*20.3*, p. *268*; 22.5.2, p. 376
K. Mukasa	(a) 43,	393 (1977)	*1.3*, p. *22*; *18.2*, p. *223*; 21.1, p. 297; 21.1.1, p. 306
V. Yu. Mukhina	(a) 43,	407 (1977)	10.2, p. 91; *13.4*, p. *138*; *14.3*, p. *171*; 22.1.2, p. 342
	(a) 44,	107 (1977)	10.2, p. 91; *13.4*, p. *139*; *14.3*, p. *171*; 22.1.2, p. 343
G. Müller	(a) 42,	579 (1977)	*9*, p. *70*; *11*, p. *97*; 22.2.4, p. 357
H.-R. Müller	(b) 80,	K85 (1977)	*2*, p. *34*; *18.3*, p. *228*; 21.1, p. 290; 21.4, p. 311
H.-U. Müller	(b) 79,	501 (1977)	*13.2*, p. *124*; 14.1, p. 157; 21.7, p. 319
M. Müller	(a) 39,	259 (1977)	5, p. 47; *18.2*, p. *217*; 21.1, p. 288; 21.1.1, p. 302
M. W. Muller	(b) 83,	177 (1977)	*1*, p. *10*; *18.3*, p. *230*; 22.8.2, p. 412
P. Müller	(a) 42,	477 (1977)	*1.2*, p. *18*; *14.2*, p. *162*; 21, p. 283
	(a) 43,	K19 (1977)	*1.2*, p. *18*; *14.2*, p. *163*; 22.8, p. 399
Th. Müller	(a) 40,	471 (1977)	*10.1*, p. *77*; 21.1, p. 291; 21.3, p. 310
V. Müller	(b) 80,	629 (1977)	6, p. 54; *19*, p. *237*; 21.1, p. 291
	(b) 84,	K53 (1977)	2, p. 40; *19*, p. *243*; 21, p. 286
U. Müller-Jahreis	(b) 84,	K79 (1977)	*11*, p. *98*; 22.1, p. 332
W. Müller-Lierheim	(b) 80,	31 (1977)	*6*, p. *53*; 13.5.3, p. 151; 14.4.1, p. 194; 20.1, p. 250; 22.8.1, p. 403
I. Munteanu	(a) 39,	291 (1977)	*13.3*, p. *126*; *14.3.3*, p. *178*; 22, p. 321
K. Murakami	(b) 80,	K113 (1977)	*2*, p. *35*; *19*, p. *237*; 21.6, p. 315; 22.1.2, p. 339
	(b) 82,	K123 (1977)	*2*, p. *35*; *19*, p. *237*; 21.6, p. 315; 22.1.2, p. 339
Y. Muraoka	(a) 42,	369 (1977)	1, p. 10; *5*, p. *48*; *18*, p. *211*; 21.1.1, p. 304
A. Murasik	(b) 80,	K137 (1977)	*13*, p. *108*; *20.1*, p. *252*; 22.5, p. 370
V. I. Muravev	(b) 84,	791 (1977)	*13.4*, p. *140*; 18.1, p. 216; 19, p. 243; 21.1, p. 300
L. E. Murr	(a) 40,	211 (1977)	*1.4*, p. *24*; *20.2*, p. *265*; 21.1. p. 291
C. R. K. Murty	(a) 39,	675 (1977)	11, p. 94; *19*, p. *236*; 22.8.1, p. 402
P. N. Murty	(a) 39,	675 (1977)	11, p. 94; *19*, p. *236*; 22.8.1, p. 402
V. P. Mushinskii	(b) 83,	K149 (1977)	13.5.1, p. 147; *20.1*, p. *261*; 22.3, p. 358
A. M. Mutal	(b) 80,	K89 (1977)	*13.1*, p. *116*; 22, p. 324
P. Muthukumarasamy	(a) 42,	K65 (1977)	*5*, p. *48*; 21.1, p. 296; 21.1.1, p. 304
H. Mutka	(a) 42,	241 (1977)	*11*, p. *96*; 14.3, p. 169; 20.1, p. 257; 22.1.3, p. 346
S. Myhra	(a) 43,	579 (1977)	*10.1*, p. *81*; *11*, p. *98*; 21.1, p. 298
A. Mycielski	(b) 79,	585 (1977)	*13.1*, p. *115*; 18, p. 209; 20.1, p. 249; 22.4.4, p. 368
	(b) 81,	615 (1977)	13.1, p. 118; *18*, p. *210*; *20.1*, p. *255*; 22.4.4, p. 368
	(b) 82,	149 (1977)	13.1, p. 118; *18*, p. *211*; *20.1*, p. *256*; 22.4.4, p. 368
Yu. I. Naberukhin	(b) 83,	405 (1977)	*2*, p. *39*
A. I. Nadezhdinskii	(a) 42,	385 (1977)	12.2, p. 105; 13.1, p. 120; *16*, p. *203*; *20.1*, p. *257*; 22.1.3, p. 346
E. M. Nadgornii	(a) 44,	285 (1977)	*10.1*, p. *82*; 21, p. 287
A. J. Nadolny	(a) 44,	K83 (1977)	*20.3*, p. *274*; 22.5.3, p. 384
F. K. Naehring	(a) 44,	K141 (1977)	1.5, p. 30; *11*, p. *99*; 22.1.2, p. 344
S. Nagakura	(a) 43,	K127 (1977)	*1.2*, p. *19*; 21.1, p. 298

R. Nagarajan	(b) **81**, 107 (1977)	*5*, p. *48*; 22.6.1, p. 393
T. Nagarajan	(b) **82**, 75 (1977)	*10.2*, p. *89*; 22.5.2, p. 379
	(a) **42**, K65 (1977)	*5*, p. *48*; 21.1, p. 296; 21.1.1, p. 304
K. Nahm	(a) **39**, 79 (1977)	1, p. 9; 13.1, p. 114; *14.2*, p. *161*; 20.1, p. 248; 22.6.1, p. 393
E. P. Naiden	(a) **44**, 429 (1977)	*18.2*, p. *224*; *18.4*, p. *234*; 22.8, p. 400
Hideo Nakajima	(a) **43**, K57 (1977)	*9*, p. *70*; 21.1.1, p. 305
N. Nakamura	(a) **41**, K17 (1977)	*1.2*, p. *16*; *19*, p. *239*; 22.9, p. 416
Y. Nakamura	(a) **41**, 451 (1977)	*5*, p. *48*; 18.2, p. 220; 21.1, p. 294; 21.1.1, p. 303
	(a) **42**, 369 (1977)	1, p. 10; *5*, p. *48*; *18*, p. *211*; 21.1.1, p. 304
H. Nakata	(a) **43**, 417 (1977)	11, p. 98; *19*, p. *241*; 22.1.2, p. 342
Nguyen Van Nam	(b) **81**, K59 (1977)	*13.1*, p. *117*; *20.1*, p. *254*; 22.8, p. 398
	(a) **44**, 675 (1977)	*3*, p. *43*; *10.2*, p. *93*; 22.2.1, p. 353
V. S. Nanda	(b) **79**, K103 (1977)	*8*, p. *64*; 21, p. 277; 21.1, p. 289; 21.6, p. 315
	(b) **84**, 423 (1977)	*8*, p. *68*; 21.1, p. 300
S. S. Nandra	(a) **41**, 65 (1977)	*18.2.1*, p. *226*; 21.1, p. 293; 21.4, p. 312
R. Kh. Nani	(b) **82**, 705 (1977)	*6*, p. *58*; *20.1*, p. *258*; 22.3, p. 358
A. Narayanasamy	(a) **42**, K65 (1977)	*5*, p. *48*; 21.1, p. 296; 21.1.1, p. 304
F. S. Nasredinov	(a) **39**, 437 (1977)	*1.2*, p. *15*; 2, p. 34; 5, p. 47; 22, p. 322
T. Nasu	(b) **80**, 675 (1977)	13.5.4, p. 153; *20.1*, p. *252*; 22.6, p. 387
A. N. Nateprov	(a) **43**, K159 (1977)	13.1, p. 122; *15*, p. *199*; 22.3.1, p. 360
Amar Nath Nigam	see under letters "Ni"	
P. Nath	(a) **40**, 247 (1977)	*2*, p. *35*; *14.3.1*, p. *173*; 15, p. 198; 21.1, p. 291; 21.6, p. 315
R. Nath	(a) **44**, 627 (1977)	13.4, p. 140; *14.3.1*, p. *175*; 22.6, p. 392
M. Nauciel-Bloch	(b) **83**, 413 (1977)	13.1, p. 121; *18.2*, p. *223*; *18.4*, p. *233*; 21.1.1, p. 306
S. Naudon	(a) **41**, 207 (1977)	*1.2*, p. *16*; *1.3*, p. *21*; 21, p. 280; 21.6, p. 316
V. M. Naumenko	(b) **81**, 351 (1977)	*13.4*, p. *133*; 18.4, p. 232; 19, p. 238; 22.8.1, p. 404
V. A. Naumov	(b) **84**, 107 (1977)	13.4, p. 138; *13.5.1*, p. *148*; *20.1*, p. *261*; 22.9, p. 419
W. Nazarewicz	(b) **80**, 307 (1977)	*6*, p. *53*; *20.1*, p. *251*; 22.8, p. 397
E. Nebauer	(a) **40**, K195 (1977)	*2*, p. *36*; *9*, p. *70*; 22, p. 324
	(a) **43**, 451 (1977)	*2*, p. *39*; *9*, p. *71*; 14.3.4, p. 187; 22, p. 329
H. Neddermeyer	(b) **80**, 611 (1977)	*13*, p. *108*; 20, p. 244; 21, p. 279; 21.3, p. 310
D. D. Nedeoglo	(b) **80**, 369 (1977)	*13.4*, p. *132*; *14.3*, p. *166*; 22.4.2, p. 365
	(a) **42**, 675 (1977)	*14.3*, p. *170*; 16, p. 203; 22.4.2, p. 366
	(a) **44**, 83 (1977)	*14.3.3*, p. *181*; *14.3.4*, p. *187*; 22.4.2, p. 366
G. N. Neilo	(b) **83**, K59 (1977)	12.2, p. 105; *19*, p. *241*; 22.8.1, p. 406
V. V. Nelaev	(a) **43**, K23 (1977)	*10.1*, p. *80*; 21.1.1, p. 305
E. Nembach	(a) **44**, 669 (1977)	*13.2*, p. *125*; 21.1, p. 300
M. Németh-Sallay	(a) **43**, K135 (1977)	1.5, p. 30; *1.6*, p. *32*; 2, p. 39; 22.6, p. 391
J. P. Nérou	(b) **83**, K7 (1977)	*20.1*, p. *259*; 22.7, p. 395
N. B. Nešković	(a) **41**, K133 (1977)	*1.2*, p. *17*; 18.4, p. 233; 22.6, p. 388
S. Nešpurek	(a) **41**, 619 (1977)	13.4, p. 135; *14.3.4*, p. *185*; 22, p. 325
D. Neubert	(a) **44**, 259 (1977)	10.2, p. 92; *16*, p. *205*; 22.1.2, p. 343
H. Neuhäuser	(a) **43**, 493 (1977)	*10.1*, p. *81*; 21.1, p. 298
H. Neumann	(b) **80**, K21 (1977)	13.1, p. 116; *20.1*, p. *251*; 22.8, p. 397
	(b) **81**, K59 (1977)	*13.1*, p. *117*; *20.1*, p. *254*; 22.8, p. 398
	(a) **43**, K95 (1977)	*14.3*, p. *171*; 22.2.2, p. 354
	(a) **44**, 675 (1977)	*3*, p. *43*; *10.2*, p. *93*; 22.2.1, p. 353
R. D. Newton	(a) **41**, 569 (1977)	*10.2*, p. *88*; 20.1, p. 256; 20.3, p. 270; 22.6, p. 388
Nguyen Ai Viet	see under letter "V"	
Nguyen Nhu Dat	see under letter "D"	
Nguyen The Khoi	see under letter "K"	

Nguyen Van Hieu	see under letter "H"	
Nguyen Van Nam	see under letter "N"	
Nguyen Van Trong	see under letter "T"	
M. Nicol	(b) **79**, 6171 (977)	1.2, p. 14; *6*, p. *52*; 12.2, p. 104; 20.1, p. 249; 22.8.1, p. 402
	(b) **83**, K115 (1977)	*1.2*, p. *19*; *6*, p. *59*; 12.2, p. 106; 20.1, p. 260; 22.8.1, p. 407
W. Niederlag	(b) **82**, 495 (1977)	13.1, p. 120; *20*, p. *245*; 21.1, p. 296; 21.1.1, p. 304
Amar Nath Nigam	(a) **41**, 493 (1977)	*13.3*, p. *127*; *14.1*, p. *158*; 21.1, p. 294
R. R. Nigmatullin	(b) **82**, 397 (1977)	*18.1*, p. *214*; *19*, p. *240*
V. A. Nikitov	(b) **82**, 127 (1977)	*6*, p. *57*; *13.5.4*, p. *153*; 18.2, p. 221; 22, p. 326
P. O. Nilsson	(b) **79**, 277 (1977)	*13.1*, p. *114*; 20, p. 244; 21.1, p. 288
M. Nimis	(b) **82**, K49 (1977)	*10.2*, p. *89*; *13.4*, p. *136*; 20.3, p. 271; 22.5.2, p. 379
R. Nink	(a) **41**, K157 (1977)	*10.2*, p. *89*; 13.4, p. 135; *20.1*, p. *256*; 22.5.2, p. 378
	(a) **44**, 505 (1977)	*10.2*, p. *93*; 20.1, p. 263; *20.3*, p. *275*; 22.5.2, p. 381
J. S. Nkoma	(b) **81**, 307 (1977)	*13.5*, p. *141*; *20.1*, p. *254*; 22.2.2, p. 354
C. Noguet	(a) **42**, 91 (1977)	13.4, p. 136; *16*, p. *203*; 22.6, p. 389
W. Nolting	(b) **79**, 573 (1977)	*13.5.4*, p. *153*; 18.2, p. 218; 22, p. 321
M. Nomoto	(a) **41**, K17 (1977)	*1.2*, p. *16*; *19*, p. *239*; 22.9, p. 416
Th. F. Nonnenmacher	(b) **79**, K153 (1977)	*6*, p. *53*; *8*, p. *64*
	(b) **82**, 257 (1977)	*6*, p. *57*; *8*, p. *66*
L. Nosenzo	(b) **79**, 795 (1977)	*13.4*, p. *131*; *20.1*, p. *249*; 22.5.3, p. 383
	(b) **81**, 411 (1977)	*13.4*, p. *131*; *20.1*, p. *249*; 22.5.3, p. 383
M. M. Noskov	(b) **80**, 11 (1977)	13.1, p. 114; *20.1*, p. *247*; 21.4, p. 311
J. P. Nougier	(a) **40**, K85 (1977)	*14.3*, p. *167*; 22, p. 323
B. V. Novikov	(a) **40**, 575 (1977)	13.4, p. 132; *13.5.1*, p. *144*; *19*, p. *237*; 22.4.2, p. 365
S. R. Novikov	(a) **42**, 397 (1977)	*13.4*, p. *136*; *14.3.3*, p. *180*; 22, p. 326
V. A. Novominskii	(a) **41**, 339 (1977)	11, p. 95; *13.4*, p. *134*; *14.3.3*, p. *179*; 22.1.1, p. 334
P. Novotný	(a) **39**, 351 (1977)	1.2, p. 14; *18.2*, p. *217*; 22, p. 321
	(b) **83**, K87 (1977)	*18.2*, p. *222*; 22, p. 328
A. I. Novozhilov	(a) **44**, K87 (1977)	*20.3*, p. *274*; 22.8.1, p. 407
W. Nowak	(a) **44**, 265 (1977)	*2*, p. *40*; 10.2, p. 92; 22.6, p. 391
R. Nürnberger	(b) **81**, 503 (1977)	*13.1*, p. *118*; 17, p. 206; 23, p. 420
M. Nygren	(a) **43**, 645 (1977)	*1.2*, p. *19*; 8, p. 67; 14.3, p. 171; 18.1, p. 215; 22.6.1, p. 393
T. Ocaña Z.	(a) **42**, 571 (1977)	*1.5*, p. *30*; 21.6, p. 316
H. L. Oczkowski	(a) **40**, K49 (1977)	*14.3.4*, p. *184*; *20.3*, p. *269*; 22.4.2, p. 365
G. Oelgart	(b) **80**, K101 (1977)	*14.4*, p. *190*; 22.7, p. 394
	(b) **81**, K135 (1977)	*13.2*, p. *124*; 22.7, p. 394
	(b) **83**, 521 (1977)	*14.4*, p. *192*; 22.7, p. 395
G. Oertel	(a) **43**, 141 (1977)	1.5, p. 30; *17*, p. *207*; *20.1*, p. *260*; 22.6, p. 390
H. Oesterreicher	(a) **39**, K41 (1977)	*18*, p. *209*; 21, p. 277; 21.1, p. 289
	(a) **39**, K75 (1977)	*18*, p. *209*; *18.4*, p. *231*; 22.7, p. 394
	(a) **39**, K91 (1977)	*18*, p. *210*; 22.7, p. 394
	(a) **40**, K139 (1977)	*18*, p. *210*; 21, p. 280; 21.1.1, p. 303; 21.4, p. 312
	(a) **41**, K101 (1977)	*18.2*, p. *220*; 21.1, p. 294; 21.1.1, p. 304
Z. Ogorelec	(b) **84**, K37 (1977)	*4*, p. *46*; 22, p. 330

S. Ogose	(a) **42**, 337 (1977)	2, p. 37; 13.1, p. 120; *14.3.1*, p. *174*; 15, p. 199; *16*, p. *203*; 22.1.2, p. 340
T. Ohgaku	(a) **42**, K83 (1977)	10.2, p. 89; *14.4*, p. *191*; 22.5.4, p. 385
S. Ohnuma	(a) **44**, K151 (1977)	*2*, p. *41*; *18.2*, p. *225*; 21.1, p. 301; 21.1.1, p. 307
O. Ohtsuki	(a) **42**, K25 (1977)	*14.3.2*, p. *177*; *14.3.3*, p. *180*; 22.2.1, p. 352
Y. H. Ohtsuki	(b) **79**, 321 (1977)	*9*, p. *69*; 11, p. 94; 22.1.2, p. 337
	(b) **82**, 643 (1977)	*13.5.2*, p. *149*; 21, p. 283
H. Okazaki	(a) **42**, 133 (1977)	1.1, p. 12; 22.5.1, p. 373
B. Olés	(b) **83**, K35 (1977)	*14.3.4*, p. *186*; *20.1*, p. *259*; 22.1.1, p. 335
A. M. Oleś	(b) **79**, K61 (1977)	*13.5.4*, p. *153*
	(b) **84**, 747 (1977)	*13*, p. *113*; 18.2, p. 224; 18.4, p. 234
A. C. Oliveira	(b) **84**, K17 (1977)	*19*, p. *242*; 22.8.1, p. 407
M. J. Oliveira	(b) **83**, 299 (1977)	*13*, p. *110*; *18*, p. *212*
V. I. Olkhovskii	(b) **82**, 189 (1977)	2, p. 37; *19*, p. *240*; 22.8, p. 399
T. Olsen	(a) **39**, 167 (1977)	*1.5*, p. *29*; *1.6*, p. *31*; 22.2.1, p. 350
T. Ōmura	(b) **79**, 321 (1977)	*9*, p. *69*; 11, p. 94; 22.1.2, p. 337
S. Onari	(b) **79**, 269 (1977)	*6*, p. *51*; 12.2, p. 104; *20.1*, p. *247*; 22, p. 320
A. B. van Oosterhout	(a) **41**, 607 (1977)	13.4, p. 135; *20.3*, p. *271*; 22.8.1, p. 405
V. N. Opekunov	(a) **43**, 639 (1977)	*17*, p. *207*; 22.5.2, p. 380
H. Oppermann	(a) **41**, 125 (1977)	*1.2*, p. *16*; 14.3, p. 168; 22.6, p. 388
	(a) **42**, 295 (1977)	*1.2*, p. *17*; *3*, p. *42*; 14.3, p. 170; 18.1, p. 214; 22.6.1, p. 393
	(a) **44**, K187 (1977)	*1.2*, p. *20*; 6, p. 61; *20.1*, p. *264*; 22.6, p. 392
H. Oppolzer	(a) **42**, K91 (1977)	*10.2*, p. *90*; 22.1.2, p. 341
S. E. Ormrod	(a) **40**, 265 (1977)	*9*, p. *69*; 22.6, p. 387
	(a) **44**, 643 (1977)	*9*, p. *71*; 10.2, p. 93; *14.4*, p. *193*; 22.6, p. 392
V. M. Orera	(a) **44**, 717 (1977)	*10.2*, p. *93*; *20.1*, p. *264*; 22.5.3, p. 384
H. Ortbauer	(a) **39**, 157 (1977)	*5*, p. *47*; 18.1, p. 213; 18.2, p. 217; 21, p. 276; 21.1, p. 288; 21.1.1, p. 302
M. v. Ortenberg	(b) **82**, K1 (1977)	*13.1*, p. *119*; *19*, p. *240*; 22.1.2, p. 340
I. S. Osadko	(b) **82**, K107 (1977)	*13.4*, p. *136*; 20.1, p. 258; 20.3, p. 272
Y. Osaka	(a) **42**, 337 (1977)	2, p. 37; 13.1, p. 120; *14.3.1*, p. *174*; 15, p. 199; *16*, p. *203*; 22.1.2, p. 340
	(a) **42**, 483 (1977)	*13.3*, p. *127*; 14.3.3, p. 180; 22.2.1, p. 352
V. V. Osiko	(a) **40**, K69 (1977)	*20.2*, p. *265*; 22.8.2, p. 410
	(a) **41**, K199 (1977)	*20.2*, p. *265*; 22.8.2, p. 410
	(a) **42**, 101 (1977)	*1.3*, p. *22*; 13.4, p. 136; 20.1, p. 257; 20.3, p. 271; 22.8.2, p. 411
E. Ostertag	(b) **82**, 335 (1977)	*13.5.1*, p. *146*; *20.3*, p. *271*; 22.5, p. 371
	(b) **83**, K101 (1977)	*13.5*, p. *141*; *20.3*, p. *273*; 22.5, p. 372
	(b) **84**, 673 (1977)	*13.5.1*, p. *148*; *20.3*, p. *275*; 22.5, p. 372
V. S. Ostrovskii	(b) **81**, K103 (1977)	*13*, p. *110*
N. Ōtsuka	(a) **43**, K127 (1977)	*1.2*, p. *19*; 21.1, p. 298
A. Otto	(b) **80**, 289 (1977)	*13.5.2*, p. *149*; 20.1, p. 251; 21, p. 278
G. Otto	(a) **41**, K193 (1977)	*1*, p. *9*; 18.2, p. 220; 18.3, p. 229; 21.1, p. 294; 22.8.2, p. 411
H. H. Otto	(b) **80**, 31 (1977)	*6*, p. *53*; 13.5.3, p. 151; 14.4.1, p. 194; 20.1, p. 250; 22.8.1, p. 403
P. J. Ouseph	(a) **44**, K181 (1977)	*5*, p. *50*; 22.8.1, p. 408
H. Overeijnder	(b) **81**, K11 (1977)	*10.2*, p. *87*; *17*, p. *206*; 22.5.2, p. 378
N. Owada	(b) **80**, K131 (1977)	*4*, p. *45*; *10.1*, p. *77*; 21, p. 279
F. J. Owens	(b) **79**, 623 (1977)	*14.4*, p. *189*; 19, p. 235; 22.8, p. 396
K. Ozawa	(a) **40**, K199 (1977)	*14.1*, p. *158*; *18.3*, p. *229*; 21.7, p. 319
	(a) **41**, 289 (1977)	*10.2*, p. *88*; 11, p. 95; *20.1*, p. *254*; 22.5.2, p. 378
Yu. V. Ozerov	(b) **80**, 589 (1977)	*13.4*, p. *132*; 16, p. 202; 20.1, p. 252; 20.3, p. 269; 22.4.1, p. 362

G. Paasch	(b) **83**, 209 (1977)	*13.3*, p. *127*; 21, p. 284
	(b) **83**, 501 (1977)	*13*, p. *111*; *14.4*, p. *192*
S. Pace	(a) **42**, K111 (1977)	*14.2*, p. *162*
D. Pacheco	(a) **43**, K45 (1977)	*20.2*, p. *266*; 22.8.1, p. 407
O. M. Pachev	(b) **80**, K9 (1977)	*13*, p. *108*; 22, p. 323
	(b) **84**, 493 (1977)	*13*, p. *112*; 13.4, p. 139
B. D. Padalia	(b) **80**, 83 (1977)	*20*, p. *244*; 21.4, p. 311
	(b) **82**, 603 (1977)	*20*, p. *245*; 22.6, p. 390
H. Paerschke	(b) **82**, 633 (1977)	*6*, p. *57*; 20.1, p. 258
J. L. Page	(b) **79**, 203 (1977)	13.4, p. 129; *20.1*, p. *247*; 22.8.2, p. 409
V. Paidar	(a) **42**, 713 (1977)	*12.1*, p. *102*; 21.1.1, p. 305
	(a) **43**, K165 (1977)	*9*, p. *71*; 21.1.1, p. 306
N. L. Pakhomova	(a) **42**, 191 (1977)	*18.3*, p. *229*; 22.8.2, p. 412
A. K. Pal	(a) **43**, K9 (1977)	14.1, p. 159; *18.1*, p. *215*; *18.2*, p. *222*; 21.1, p. 297
L. I. Palaki	(b) **83**, K149 (1977)	13.5.1, p. 147; *20.1*, p. *261*; 22.3, p. 358
E. Palaniyandi	(b) **84**, 171 (1977)	*13.4*, p. *138*; *14.3*, p. *171*; 22.1.1, p. 336; 22.1.2, p. 343
S. K. Palazov	(a) **40**, 29 (1977)	*9*, p. *69*; 10.2, p. 85; 14.3, p. 167; 22.2.2, p. 354
S. Panchanadeeswaran	(a) **42**, K175 (1977)	*10.1*, p. *80*; 21, p. 284
K. P. Pande	(a) **42**, 615 (1977)	*14.3.3*, p. *180*; *14.3.4*, p. *185*; 22.4.1, p. 363
V. V. Panfilov	(a) **39**, 357 (1977)	*1.2*, p. *14*; 12.2, p. 104; 13.5.1, p. 143; *20.1*, p. *248*; 22.3, p. 358
O. S. Panwar	(a) **41**, 323 (1977)	2, p. 36; *14.4*, p. *190*; 22.8, p. 398
L. Pardi	(b) **84**, 305 (1977)	18.1, p. 215; *19*, p. *242*; 22.8, p. 400
B. Pardo	(a) **42**, 141 (1977)	*1.1*, p. *12*; 21.6, p. 316
N. N. Parfenova	(a) **39**, 683 (1977)	*18.4*, p. *231*; 22.8.2, p. 410
P. Parikh	(b) **79**, 299 (1977)	*6*, p. *51*; 21.1, p. 288
J. Parisot	(a) **44**, 549 (1977)	*10.1*, p. *82*; 12.1, p. 103; 21.1.1, p. 307
R. D. Parmentier	(a) **42**, K111 (1977)	*14.2*, p. *162*
F. Paschke	(a) **42**, 525 (1977)	*14.3.3*, p. *180*; *14.3.4*, p. *185*; 18, p. 212; 22.2.1, p. 352
L. Pasemann	(b) **84**, 175 (1977)	*13.1*, p. *122*; 22.1.2, p. 343
R. Pässler	(b) **83**, K55 (1977)	*13.4*, p. *137*; 22, p. 328
	(b) **83**, K111 (1977)	*13.4*, p. *137*; 22, p. 329
	(b) **84**, K163 (1977)	*13.4*, p. *137*; 22, p. 329
G. Paternò	(a) **41**, 393 (1977)	*14.2*, p. *162*
N. L. Pathak	(a) **42**, 403 (1977)	*14.4*, p. *191*; 20.1, p. 257; 22.9, p. 417
A. A. Patrin	(b) **83**, K137 (1977)	10.2, p. 91; *20.3*, p. *273*; 22.1.2, p. 342
J. D. Patterson	(b) **79**, K165 (1977)	*18.2*, p. *218*
C. Patuelli	(a) **44**, 333 (1977)	*10.1*, p. *82*; 12.1, p. 103; 21, p. 287
P. Paufler	(b) **79**, 283 (1977)	*6*, p. *51*; 21.3, p. 310
	(a) **40**, 471 (1977)	*10.1*, p. *77*; 21.1, p. 291; 21.3, p. 310
	(a) **44**, 499 (1977)	*10.1*, p. *82*; 14.2, p. 164; 21.1, p. 300
S. Paul	(b) **83**, 645 (1977)	*10.2*, p. *91*; 22.5.2, p. 380
M. Paulus	(a) **40**, K179 (1977)	*1.4*, p. *25*; *3*, p. *42*; 22.8.2, p. 410
H. J. Pauwels	(a) **44**, K195 (1977)	*14.4*, p. *193*; 22, p. 331
V. A. Pavlov	(a) **43**, K5 (1977)	*10.1*, p. *80*; 21.1.1, p. 305
V. V. Pavlovich	(b) **80**, 697 (1977)	*6*, p. *55*; 22, p. 323
	(b) **82**, 391 (1977)	*13.5.2*, p. *149*; 22, p. 326
D. Pavuna	(a) **44**, 339 (1977)	*1.2*, p. *20*; *2*, p. *40*; 21.1, p. 299; 21.1.1, p. 306
J. M. Pawlikowski	(a) **40**, 613 (1977)	13.3, p. 126; *14.3.4*, p. *184*; 22.4.4, p. 368
	(b) **81**, K67 (1977)	*13.1*, p. *117*; 14.3, p. 168; 22.4.4, p. 368
	(a) **42**, 721 (1977)	13.1, p. 121; *14.3.1*, p. *174*; 22.4.4, p. 369
O. P. Pchelyakov	(a) **43**, 659 (1977)	*1.5*, p. *30*; 2, p. 39; 22.1.1, p. 336; 22.6, p. 391
N. L. Pecherkina	(a) **43**, K5 (1977)	*10.1*, p. *80*; 21.1.1, p. 305

A. T. Pędziwiatr	(a) **43**, K41 (1977)	*5*, p. *49*; *18.1*, p. *215*; 22.8.1, p. 407
	(a) **44**, 325 (1977)	*1.1*, p. *13*; 5, p. 49; *18.3*, p. *230*; 22.8.2, p. 413
G. S. Pekar	(a) **41**, 299 (1977)	*14.3.3*, p. *179*; *20.3*, p. *270*; 22.4.1, p. 363
S. I. Pekar	(b) **82**, 83 (1977)	*13.5.1*, p. *146*; *20.1*, p. *256*; 22, p. 325
I. Pelah	(b) **81**, 295 (1977)	6, p. 55; *14.4.1*, p. *195*; 22.8, p. 398
I. Pelant	(a) **42**, K57 (1977)	7, p. 63; *12.2*, p. *105*; *20.1*, p. *257*; 22.5.1, p. 373
E. Peljo	(b) **80**, 255 (1977)	4, p. 44; *6*, p. *53*; 22.5, p. 370
J. Pelleg	(b) **84**, 435 (1977)	9, p. 71; *10*, p. *74*
J. M. Pelletier	(a) **39**, 525 (1977)	*15*, p. *198*; 21.1, p. 289; 21.1.1, p. 302
J. Pelzl	(b) **82**, 145 (1977)	*1.2*, p. *17*; 6, p. 57; 20.1, p. 256; 22.8, p. 399
A. Peña	(a) **41**, K119 (1977)	*10.2*, p. *89*; 20.1, p. 256; 22.5.2, p. 378
J. Pennisi	(b) **81**, 579 (1977)	*14.2*, p. *162*
J. M. Peransin	(a) **40**, 583 (1977)	*7*, p. *62*; *14.3.4*, p. *184*; 22.4.1, p. 362
P. R. R. Pereira	(b) **79**, K49 (1977)	*6*, p. *52*; *13.5.4*, p. *153*; 18.2, p. 217; 22.8, p. 396
	(b) **81**, 79 (1977)	*13.5.4*, p. *153*; 18.2, p. 219; 21, p. 280
	(b) **81**, 395 (1977)	*13.5.4*, p. *153*; 14.3, p. 167; 18.2, p. 219; 22, p. 324
N. F. Perelman	(b) **80**, K5 (1977)	*14.1*, p. *158*; 21.7, p. 319
J. Perez	(a) **39**, 651 (1977)	7, p. 62; *10.1*, p. *76*; 21, p. 278
P. Perfetti	(a) **42**, K41 (1977)	*14.1*, p. *159*; 21, p. 282
Y. E. Perlin	(b) **80**, 703 (1977)	13.4, p. 132; *14.3.4*, p. *184*; *20.1*, p. *252*; 22, p. 323
	(a) **40**, 643 (1977)	6, p. 55; *13.4*, p. *133*; 20.3, p. 269; 22.8.1, p. 403
M. Pernet	(a) **43**, 81 (1977)	*1*, p. *10*; *18.4*, p. *233*; 22.9, p. 418
E. Pernoux	(a) **39**, 525 (1977)	*15*, p. *198*; 21.1, p. 289; 21.1.1, p. 302
R. Perret	(a) **41**, 271 (1977)	*1.2*, p. *16*; 6, p. 56; 20.1, p. 254; 22.9, p. 416
F. Perrot	(b) **81**, 205 (1977)	*10.1*, p. *77*; 21.1, p. 308
J. di Persio	(a) **40**, 393 (1977)	*10.2*, p. *86*; 22.9, p. 415
R. Perthel	(a) **40**, K135 (1977)	*2*, p. *36*; *18.3*, p. *229*; 21, p. 280; 21.1, p. 292
P. Pertsch	(a) **44**, 577 (1977)	*14.2*, p. *164*; 22.6, p. 391
P. Peshev	(a) **44**, K25 (1977)	*1.2*, p. *20*; 8, p. 67; 22.8.1, p. 407
S. A. Peskovatskii	(b) **82**, 251 (1977)	13.4, p. 135; *19*, p. *240*; 22.6, p. 389
	(b) **84**, 443 (1977)	13.4, p. 139; *19*, p. *243*; 22.6, p. 391
K. Pesz	(b) **83**, 565 (1977)	*13.5.3*, p. *152*; 14.3, p. 171; 22, p. 328
B. Petelenz	(b) **81**, 49 (1977)	*13.5.1*, p. *145*; 22.9, p. 415
P. Petelenz	(b) **79**, 61 (1977)	*13.5.1*, p. *143*; 20.1, p. 247; 22, p. 320
	(b) **81**, 49 (1977)	*13.5.1*, p. *145*; 22.9, p. 415
	(b) **83**, 169 (1977)	*13.5.1*, p. *146*; 22.9, p. 417
L. A. Pétermann	(a) **41**, K147 (1977)	*14.4.1*, p. *195*; 22.8, p. 399
G. Petite	(a) **39**, K25 (1977)	*17*, p. *206*; 21.6, p. 315
E. Petri	(b) **80**, 289 (1977)	*13.5.2*, p. *149*; 20.1, p. 251; 21, p. 278
J. F. Petroff	(a) **43**, 473 (1977)	*1.1*, p. *13*; 22.1.2, p. 342
A. G. Petrosyan	(a) **39**, 541 (1977)	*13.4*, p. *131*; 20.2, p. 265; 20.3, p. 268; 22.8, p. 396
	(a) **40**, K69 (1977)	*20.2*, p. *265*; 22.8.2, p. 410
	(a) **41**, K199 (1977)	*20.2*, p. *265*; 22.8.2, p. 410
K. Petrov	(a) **44**, K25 (1977)	*1.2*, p. *20*; 8, p. 67; 22.8.1, p. 407
V. P. Petrov	(a) **41**, 379 (1977)	*20.1*, p. *256*; *20.3*, p. *270*; 22.8.1, p. 405
D. S. Petrova	(a) **41**, 45 (1977)	10.2, p. 87; 13.4, p. 133; *20.1*, p. *254*; *20.3* p. *270*; 22.8.1, p. 404
	(a) **41**, 379 (1977)	*20.1*, p. *256*; *20.3*, p. *270*; 22.8.1, p. 405
	(a) **41**, K127 (1977)	*20.1*, p. *256*; 20.2, p. 265; *20.3*, p. *271*; 22.8.1, p. 405
P. Petrovič	(a) **39**, K67 (1977)	*14.1*, p. *157*; 14.2, p. 161; 21, p. 277
J. Petrů	(b) **84**, 325 (1977)	13.1, p. 122; *14.1*, p. *159*; 21, p. 286
	(b) **84**, K21 (1977)	*14.1*, p. *159*; 21, p. 286
B. V. Petukhov	(a) **42**, 391 (1977)	7, p. 62; *10*, p. *73*; *12*, p. *100*

B. Petzel	(a) **41**, K133 (1977)	*14.3.3*, p. *181*; 22.2.1, p. 352
A. Pevala	(a) **41**, K175 (1977)	10.1, p. 78; *14.2*, p. *162*; 21, p. 281
A. Pfaller	(a) **43**, K103 (1977)	12.2, p. 106; *20.1*, p. *260*; 22.5.2, p. 380
H. Pfeiffer	(b) **83**, 129 (1977)	*1.6*, p. *32*; *3*, p. *42*
	(a) **43**, 277 (1977)	3, p. 43; *18.3*, p. *230*; 22.8.2, p. 412
	(a) **44**, 221 (1977)	*3*, p. *43*; 22.8.2, p. 413
H. Pfleiderer	(a) **43**, 479 (1977)	*14.3.1*, p. *175*; *14.3.4*, p. *187*; 22.1.2, p. 342
Vu Duy Phach	(b) **83**, K101 (1977)	*13.5*, p. *141*; *20.3*, p. *273*; 22.5, p. 372
	(b) **84**, 235 (1977)	*13.5.1*, p. *148*; *20.3*, p. *274*; 22.5, p. 372
	(b) **84**, 731 (1977)	*13.5.1*, p. *148*; *20.1*, p. *263*; 22.5, p. 372
M. T. Pham	(a) **43**, 505 (1977)	*11*, p. *98*; *14.3.2*, p. *177*; 22.1.2, p. 342
A. Philipp	(b) **79**, 115 (1977)	*14.3*, p. *165*; 22.2.1, p. 350; 22.2.2, p. 354
J. C. Phillips	(b) **79**, 111 (1977)	*18.1*, p. *213*; 22.9, p. 414
	(b) **80**, 395 (1977)	*1*, p. *9*; *14*, p. *155*; 22.9, p. 415
	(b) **83**, 123 (1977)	*1.2*, p. *18*; 1.4, p. 27; 12.2, p. 105; 22.9, p. 417
R. Pickenhain	(b) **79**, 549 (1977)	*4*, p. *44*; 13, p. 107; 22.1, p. 332; 22.1.1, p. 333; 22.1.2, p. 338; 22.2.1, p. 350
J. Piekoszewski	(a) **39**, 643 (1977)	*5*, p. *47*; 22.8.2, p. 409
	(a) **41**, 191 (1977)	*5*, p. *48*; 18.2, p. 219; 21.1.1, p. 303
A. R. Piercy	(a) **44**, K5 (1977)	*3*, p. *43*; 22.3, p. 359
A. Pierrefeu	(b) **80**, 167 (1977)	*6*, p. *53*; 14.4.1, p. 194; 22.8, p. 397
A. Pietraszko	(a) **41**, K55 (1977)	*8*, p. *65*; *14.4*, p. *191*; 22.8.1, p. 404
W. Pietrenko	(a) **41**, 197 (1977)	*14.3*, p. *168*; 22.1.2, p. 339
J. Pietrzak	(a) **41**, K25 (1977)	*19*, p. *239*; 22.8.1, p. 404
A. Pinczuk	(b) **83**, 99 (1977)	*1.2*, p. *18*; 6, p. 58; *14.4.1*, p. *196*; 20.1, p. 259; 22.8.1, p. 406
P. Pineau	(a) **39**, 73 (1977)	*2*, p. *33*; 19, p. 235; 22.6.1, p. 393
L. Pintschovius	(b) **80**, 119 (1977)	1.4, p. 24; *14.1*, p. *157*; *20.1*, p. *250*; 22, p. 322
J. Piqueras	(a) **40**, 167 (1977)	*10.1*, p. *76*; 21.1, p. 290
	(a) **40**, K19 (1977)	*10.1*, p. *77*; 21.1, p. 291
	(a) **43**, K187 (1977)	*10.1*, p. *81*; 21.1, p. 298
M. Pirić	(b) **83**, 479 (1977)	*13.5.1*, p. *147*; 20.1, p. 260; 20.3, p. 273
F. V. Pirogov	(b) **84**, 521 (1977)	*9*, p. *71*; *10*, p. *74*
T. I. Pirozhkova	(a) **42**, 397 (1977)	*13.4*, p. *136*; *14.3.3*, p. *180*; 22, p. 326
P. Pîrvu	(b) **80**, K121 (1977)	1.2, p. 16; *14.2*, p. *162*; *18.2*, p. *219*
D. I. Piskunov	(a) **39**, 459 (1977)	*10.2*, p. *84*; 22.1.2, p. 338
B. T. Plachenov	(a) **43**, 321 (1977)	*13.4*, p. *137*; *20.3*, p. *273*; 22.5.1, p. 373
E. Płaczek-Popko	(b) **81**, K67 (1977)	*13.1*, p. *117*; 14.3, p. 168; 22.4.4, p. 368
	(a) **42**, 721 (1977)	13.1, p. 121; *14.3.1*, p. *174*; 22.4.4, p. 369
V. P. Plakhtii	(a) **39**, 683 (1977)	*18.4*, p. *231*; 22.8.2, p. 410
N. M. Plakida	(b) **80**, 543 (1977)	*6*, p. *54*; *14.4.1*, p. *194*
	(b) **82**, 627 (1977)	*1.6*, p. *31*; 11, p. 96
T. S. Plaskett	(a) **41**, 347 (1977)	*10.2*, p. *88*; 22.8.2, p. 411
A. Platau	(b) **83**, 77 (1977)	*13.3*, p. *127*; 17, p. 207; 21.1, p. 297
W. Plesiewicz	(a) **41**, 403 (1977)	*13.4*, p. *135*; 14.3, p. 168; *16*, p. *202*; 22.1.1, p. 334
B. J. Pletka	(a) **39**, 301 (1977)	*10.2*, p. *83*; 22.6, p. 386
M. V. Plotnikova	(a) **44**, 247 (1977)	5, p. 49; *14.4.1*, p. *197*; 18.2, p. 223; *18.4*, p. *234*; 22.8, p. 400
P. Podini	(b) **81**, 665 (1977)	13.5.1, p. 146; *20.1*, p. *255*; 22.3, p. 358
B. Pödör	(a) **39**, K123 (1977)	*14.3.3*, p. *178*; *14.3.4*, p. *184*; 22.2.1, p. 350
D. E. Pogarev	(b) **80**, 417 (1977)	*6*, p. *54*; 20.1, p. 252; 22.8.1, p. 403
Ya. M. Pogosyan	(a) **42**, 273 (1977)	*18.2.1*, p. *226*; 21.1, p. 295; 21.1.1, p. 304
E. P. Pokatilov	(b) **79**, 595 (1977)	*14.3.4*, p. *183*; *20.1*, p. *249*; 22, p. 321
K. I. Pokhodnya	(b) **83**, K141 (1977)	13, p. 111; 17, p. 207; *20*, p. *246*; 22.9, p. 418
	(a) **43**, K73 (1977)	*20.1*, p. *260*; 22.9, p. 418
V. A. Pokoeva	(b) **82**, 421 (1977)	*9*, p. *70*

S. V. Pokrovskaya	(a) **43,**	659 (1977)	*1.5*, p. *30*; 2, p. 39; 22.1.1, p. 336; 22.6, p. 391
C. A. Poldy	(b) **82,**	159 (1977)	13.1, p. 118; 14.1, p. 158; *18.2*, p. *221*; *18.4*, p. *233*; 21.1, p. 295; 21.4, p. 312
I. Pollini	(b) **79,**	231 (1977)	*13.1*, p. *114*; *20.1*, p. *247*; 22.5, p. 370
M. Połomska	(a) **41,**	K55 (1977)	*8*, p. *65*; *14.4*, p. *191*; 22.8.1, p. 404
V. I. Poltoratskii	(b) **83,**	341 (1977)	*13*, p. *111*; *18.1*, p. *215*; 22.1.3, p. 346
S. A. Polyakov	(a) **41,**	K93 (1977)	*1.5*, p. *29*; *3*, p. *42*; 22.1.2, p. 339
H. E. Ponath	(b) **80,**	K55 (1977)	*6*, p. *54*; *20*, p. *244*; 22.8.1, p. 403
B. K. Ponomarev	(a) **40,**	189 (1977)	*18.2*, p. *218*; 18.4, p. 232; 21.1, p. 291
E. G. Ponyatovskii	(b) **83,**	517 (1977)	*1.2*, p. *19*; 14.1, p. 159; 21.4, p. 313
M. Pope	(b) **80,**	333 (1977)	*13.5.1*, p. *144*; *20.3*, p. *268*; 22.9, p. 414
	(a) **40,**	621 (1977)	11, p. 95; *13.5.1*, p. *144*; *20.3*, p. *269*; 22.9, p. 415
	(b) **83,**	249 (1977)	*13.5.1*, p. *146*; *20.3*, p. *272*; 22.9, p. 418
A. I. Popov	(a) **44,**	K71 (1977)	*2*, p. *40*; *14.3.4*, p. *187*; 22.1.3, p. 347
A. S. Popov	(a) **41,**	687 (1977)	*14.3*, p. *169*; 22.2.4, p. 357
B. N. Popov	(a) **39,**	193 (1977)	*14.4.1*, p. *194*; *16*, p. *201*; 22.8.1, p. 402
V. A. Popov	(b) **84,**	83 (1977)	*13.5.1*, p. *147*; *13.5.4*, p. *154*
S. Popović	(a) **41,**	255 (1977)	*1.2*, p. *16*; 14.3, p. 168; 22.3, p. 358
S. P. S. Porto	(b) **79,**	731 (1977)	*6*, p. *52*; 14.4, p. 189; 20.1, p. 249; 22.8.1, p. 402
V. S. Postnikov	(a) **39,**	K21 (1977)	*14.2*, p. *161*; 21, p. 277
	(a) **44,**	423 (1977)	*3*, p. *43*; *10.2*, p. *92*; 22.5.2, p. 381
V. V. Postnikov	(a) **39,**	K21 (1977)	*14.2*, p. *161*; 21, p. 277
A. V. Potemkin	(a) **42,**	K19 (1977)	*20.1*, p. *257*; 20.2, p. 266; *20.3*, p. *271*; 22.8.1, p. 405
	(a) **42,**	K183 (1977)	20.1, p. 259; *20.2*, p. *266*; 20.3, p. 272; 22.8.1, p. 406
	(a) **43,**	K15 (1977)	13.4, p. 137; *20.2*, p. *266*; 22.8, p. 399
	(a) **44,**	51 (1977)	*20.1*, p. *262*; 20.2, p. 266; *20.3*, p. *274*; 22.8.1, p. 407
I. V. Potykevich	(a) **44,**	K15 (1977)	*20.1*, p. *262*; 22.3, p. 359
I. Potykievich	(a) **44,**	747 (1977)	*16*, p. *205*; *20.3*, p. *275*; 22.3, p. 359
D. Powers	(a) **41,**	K79 (1977)	*9*, p. *70*
J. Prakash	(a) **39,**	499 (1977)	*10.2*, p. *84*; *14.4*, p. *189*; 20.1, p. 250; 22.5.2, p. 375
R. Prakash	(b) **83,**	509 (1977)	*4*, p. *46*; *12.2*, p. *106*; 22, p. 328; 22.1, p. 332; 22.1.1, p. 335; 22.1.2, p. 342; 22.2, p. 348; 22.3, p. 358; 22.4, p. 361
S. Prakash	(b) **79,**	787 (1977)	13.4, p. 131; *14.1*, p. *157*; 21.6, p. 315
S. V. Prants	(b) **83,**	109 (1977)	7, p. 63; *14*, p. *156*
B. Prasad	(b) **80,**	379 (1977)	*6*, p. *54*; *8*, p. *64*; 21.2, p. 308
A. A. Predvoditelev	(a) **43,**	639 (1977)	*17*, p. *207*; 22.5.2, p. 380
F. H. Preston	(a) **39,**	607 (1977)	1.1, p. 11; *20.1*, p. *250*; 22.9, p. 414
	(a) **40,**	279 (1977)	10.2, p. 85; *20.1*, p. *252*; 20.3, p. 269; 22.9, p. 415
B. Prévot	(a) **40,**	503 (1977)	*20.1*, p. *253*; 22.6, p. 387
A. I. Pristupa	(b) **84,**	513 (1977)	*13.5.1*, p. *148*; *19*, p. *243*; 22.9, p. 419
V. D. Prochukhan	(a) **39,**	453 (1977)	13.4, p. 131; *20.3*, p. *268*; 22.8, p. 396
A. D. Prokhorov	(b) **83,**	K59 (1977)	12.2, p. 105; *19*, p. *241*; 22.8.1, p. 406
A. M. Prokhorov	(a) **40,**	K69 (1977)	*20.2*, p. *265*; 22.8.2, p. 410
	(a) **41,**	K199 (1977)	*20.2*, p. *265*; 22.8.2, p. 410
A. V. Prokhorovich	(a) **41,**	659 (1977)	*13.4*, p. *135*; *20.3*, p. *271*; 22.2.1, p. 352
	(a) **44,**	777 (1977)	*13.4*, p. *140*; *20.3*, p. *275*; 22.2.1, p. 353
V. Prosser	(b) **84,**	K119 (1977)	*20.1*, p. *263*; 22.8.2, p. 413
P. Puliti	(a) **44,**	127 (1977)	*1.1*, p. *13*; 22.1.2, p. 343

D. S. Puri	(b) **80**,	63 (1977)	*10.2*, p. *85*; *12.2*, p. *104*; 22.5.2, p. 376
	(b) **82**,	325 (1977)	*4*, p. *45*; *12*, p. *100*
	(b) **82**,	481 (1977)	*4*, p. *45*; *12.2*, p. *105*; 22.5.2, p. 379
R. K. Puri	(b) **83**,	293 (1977)	*5*, p. *49*; 6, p. 58; 21, p. 284
T. Pusch	(a) **40**,	K15 (1977)	*2*, p. *35*; *16*, p. *202*; 22.7, p. 394
D. I. Pushkarov	(b) **80**,	295 (1977)	*8*, p. *64*; 10, p. 72; 13.5, p. 141
	(b) **81**,	703 (1977)	*13.5.4*, p. *153*; *18.2*, p. *220*
Kh. I. Pushkarov	(b) **81**,	703 (1977)	*13.5.4*, p. *153*; *18.2*, p. *220*
H. Puszkarski	(b) **83**,	K25 (1977)	*13.3*, p. *127*; *18.2.1*, p. *226*; 22.8.2, p. 412
A. Rabinovitch	(b) **84**,	435 (1977)	9, p. 71; *10*, p. *74*
I. A. Rachkov	(a) **40**,	K1 (1977)	*10.2*, p. *85*; 22.5.2, p. 377
	(a) **41**,	443 (1977)	*10.2*, p. *88*; 22.5.2, p. 378
S. I. Radautsan	(a) **43**,	K159 (1977)	13.1, p. 122; *15*, p. *199*; 22.3.1, p. 360
S. Radhakrishna	(a) **41**,	649 (1977)	*10.2*, p. *88*; *14.4*, p. *191*; 22.5.2, p. 378
	(a) **44**,	345 (1977)	*10.2*, p. *92*; 20.1, p. 262; 22.5.2, p. 381
	(a) **44**,	741 (1977)	*10.2*, p. *93*; 13.4, p. 140; 22.5.2, p. 382
N. K. Radhakrishnan	(a) **44**,	45 (1977)	*1*, p. *10*; 22.8.2, p. 412
A. P. Radlinski	(b) **84**,	503 (1977)	*13.4*, p. *139*; 16, p. 205; 20.1, p. 262; 22.4.2, p. 366
G. Radnóczi	(a) **41**,	81 (1977)	*1.5*, p. *29*; 2, p. 36; 22.1.2, p. 339
A. M. Radwan	(b) **79**,	167 (1977)	*1.3*, p. *21*; 10.1, p. 75; 13, p. 107; 21.1, p. 288; 21.3, p. 310
	(b) **82**,	507 (1977)	*13.1*, p. *120*; *20*, p. *245*; 21, p. 283; 22, p. 326
V. K. Raev	(a) **41**,	K39 (1977)	*18.2*, p. *220*
J. Rafałowicz	(a) **42**,	351 (1977)	*8*, p. *66*; 21.7, p. 319
R. C. Rai	(b) **79**,	289 (1977)	*6*, p. *51*; 12.1, p. 101; 21.2, p. 308
H. Raidt	(a) **43**,	167 (1977)	*1.4*, p. *27*; *3*, p. *43*; 20.3, p. 273; 22.2.4, p. 357
T. Rajasekharan	(a) **39**,	485 (1977)	*18.2*, p. *218*; *19*, p. *236*; 22.8, p. 396
Yu. V. Rakitin	(b) **80**,	579 (1977)	1, p. 9; *13*, p. *108*
	(b) **81**,	379 (1977)	*18*, p. *210*
	(b) **81**,	K75 (1977)	*18.4*, p. *233*; 19, p. 239; 22.9, p. 416
	(b) **82**,	69 (1977)	*8*, p. *66*; *18*, p. *211*
	(b) **82**,	221 (1977)	1, p. 10; *13*, p. *110*; 22.5.4, p. 385; 22.6, p. 389
K. Raksányi	(b) **80**,	K105 (1977)	*10.2*, p. *86*; 19, p. 237; *20.1*, p. *252*; 22.5.2, p. 377
K. V. S. Rama Rao	see under letters "Rao"		
V. Marigliano Ramaglia	(b) **80**,	245 (1977)	*13.3*, p. *126*; 21, p. 278
V. Ramamurthy	(a) **40**,	K191 (1977)	*14.3.3*, p. *179*
A. Ramanand	(a) **42**,	247 (1977)	6, p. 57; *8*, p. *66*; *12.1*, p. *102*; 21.4, p. 312
P. Ramasamy	(b) **82**,	75 (1977)	*10.2*, p. *89*; 22.5.2, p. 379
S. Ramasamy	(b) **82**,	75 (1977)	*10.2*, p. *89*; 22.5.2, p. 379
O. G. Ramer	(b) **79**,	313 (1977)	*18.2.1*, p. *226*; 19, p. 235; 22.8.2, p. 409
V. V. Randoshkin	(a) **44**,	771 (1977)	*18.2*, p. *225*; 22.8.2, p. 413
A. Ranfagni	(b) **84**,	393 (1977)	*13.4*, p. *138*; 20.1, p. 261; 22.5.4, p. 385; 22.6, p. 391
Ratna Rani	(a) **41**,	493 (1977)	*13.3*, p. *127*; *14.1*, p. *158*; 21.1, p. 294
G. N. Rao	(b) **84**,	9 (1977)	*13.4*, p. *129*; *18.2*, p. *217*; 21.1, p. 288; 21.1.1, p. 302; 21.4, p. 311
K. V. S. Rama Rao	(a) **39**,	485 (1977)	*18.2*, p. *218*; *19*, p. *236*; 22.8, p. 396
K. Suryanarayana Rao	(a) **39**,	K119 (1977)	*14.4*, p. *189*; 22.5.2, p. 376
K. V. Rao	(a) **39**,	K119 (1977)	*14.4*, p. *189*; 22.5.2, p. 376
Meera Rao	(a) **39**,	K81 (1977)	*1.1*, p. *11*; *1.2*, p. *15*; 14.3, p. 166; 22, p. 322
M. K. Rao	(a) **41**,	K31 (1977)	*14.3.1*, p. *174*; 22.4.2, p. 365
R. Ramji Rao	(a) **42**,	247 (1977)	6, p. 57; *8*, p. *66*; *12.1*, p. *102*; 21.4, p. 312
P. Rapps	(a) **44**,	173 (1977)	*1.4*, p. *28*; *10*, p. *73*

W. Rasp	(a) **42**, K147 (1977)	*17*, p. *207*; 22.6, p. 390
	(a) **43**, K117 (1977)	1.6, p. 32; 13.3, p. 128; *17*, p. *207*; 22.6, p. 391
H. Ratajczak	(a) **44**, K103 (1977)	2, p. 40; *14.1*, p. *160*; *18.2*, p. *224*; 21.1, p. 299; 21.4, p. 313
Ch. J. Raub	(b) **84**, K33 (1977)	*14.2*, p. *163*; *19*, p. *242*; 21, p. 286; 21.6, p. 317
H. Rauch	(a) **43**, 487 (1977)	*1.1*, p. *13*; 22.1.2, p. 342
R. Rauch	(a) **41**, K97 (1977)	*10.2*, p. *89*; *20.1*, p. *256*; 20.3, p. 271; 22.5.3, p. 384
M. Rautenberg	(b) **81**, 153 (1977)	*13.4*, p. *133*; *14*, p. *155*
L. A. Rebane	(b) **81**, 57 (1977)	*10.2*, p. *87*; 20.1, p. 253; 22.5.2, p. 377
	(b) **81**, 359 (1977)	*6*, p. *55*; *20.3*, p. *270*; 22.5.2, p. 377
H. S. Reehal	(b) **80**, K63 (1977)	*13.4*, p. *132*; *14.3.4*, p. *184*
	(b) **84**, K13 (1977)	2, p. 40; *13.4*, p. *138*; *14.3.4*, p. *187*; 22.8, p. 400
C. Regnaut	(b) **80**, 533 (1977)	*2*, p. *35*; 8, p. 65; 21, p. 279
E. Reguzzoni	(b) **79**, 795 (1977)	*13.4*, p. *131*; *20.1*, p. *249*; 22.5.3, p. 383
	(b) **81**, 411 (1977)	*13.4*, p. *131*; *20.1*, p. *249*; 22.5.3, p. 383
A. Rehfeld	(a) **40**, K161 (1977)	*13.1*, p. *117*; *20.1*, p. *253*; 22.4.2, p. 365
W. Rehm	(b) **79**, 539 (1977)	*2*, p. *33*; 14.3.1, p. 173; *16*, p. *201*; 22.1.2, p. 337
W. Rehwald	(a) **42**, 699 (1977)	*12.2*, p. *105*; 22, p. 327
A. Reich	(a) **42**, 345 (1977)	*18.2*, p. *221*
P. Reiche	(a) **39**, 517 (1977)	*1.4*, p. *24*; *10.2*, p. *84*; 22.8.2, p. 409
	(a) **43**, 71 (1977)	3, p. 42; 13.4, p. 137; 20.1, p. 260; *20.2*, p. *266*; *20.3*, p. *272*; 22.8, p. 399
W. Reichelt	(a) **41**, 125 (1977)	*1.2*, p. *16*; 14.3, p. 168; 22.6, p. 388
	(a) **44**, K95 (1977)	*1.2*, p. *20*; 5, p. 49; 14.3, p. 172; 22.6, p. 391
	(a) **44**, K187 (1977)	*1.2*, p. *20*; 6, p. 61; *20.1*, p. *264*; 22.6, p. 392
M. Reiffers	(a) **39**, K67 (1977)	*14.1*, p. *157*; 14.2, p. 161; 21, p. 277
S. P. Reifman	(b) **79**, K57 (1977)	*13.5.3*, p. *151*; 22, p. 321
S. M. Seyed Reihani	(a) **39**, 651 (1977)	7, p. 62; *10.1*, p. *76*; 21, p. 278
H. Reimer	(a) **44**, K117 (1977)	*13.3*, p. *128*; *14.3.3*, p. *182*; 22, p. 330
U. Reinhold	(a) **44**, 531 (1977)	*11*, p. *99*; 18.3, p. 230; 22.8.2, p. 413
G. U. Reinsperger	(a) **43**, K77 (1977)	*13*, p. *111*; *14.3.4*, p. *186*; 22.6, p. 391
F. G. dos Reis	(b) **83**, 607 (1977)	13.1, p. 122; *14.4*, p. *192*; *20*, p. *246*; 22, p. 329
A. Reklaitis	(b) **83**, K121 (1977)	*14.3.4*, p. *186*; 22.2.3, p. 356
R. Reksnys	(a) **39**, 601 (1977)	*7*, p. *62*; *14.3.4*, p. *183*; 22.4.1, p. 362
	(a) **43**, 705 (1977)	*7*, p. *63*; *14.3*, p. *171*; 16, p. 204; 22.4.3, p. 367
A. Rembaum	(b) **81**, 281 (1977)	*14*, p. *155*; 15, p. 198; 22.9, p. 415
D. Renard	(a) **42**, 141 (1977)	*1.1*, p. *12*; 21.6, p. 316
P. Rennert	(b) **79**, 167 (1977)	*1.3*, p. *21*; 10.1, p. 75; 13, p. 107; 21.1, p. 288; 21.3, p. 310
	(b) **83**, 501 (1977)	*13*, p. *111*; *14.4*, p. *192*
R. Rentzsch	(a) **43**, 231 (1977)	*20.3*, p. *273*; 22.1.1, p. 335
L. Resca	(b) **81**, 129 (1977)	*13.1*, p. *117*; 22.6, p. 387
R. Resta	(b) **81**, 129 (1977)	*13.1*, p. *117*; 22.6, p. 387
T. Rewaj	(b) **80**, K25 (1977)	*19*, p. *236*; 22.5.3, p. 383
	(b) **83**, K79 (1977)	*6*, p. *59*; *19*, p. *241*; 22.5.3, p. 384
F. Reynaud	(a) **43**, 239 (1977)	1.2, p. 18; *1.3*, p. *22*
C. L. Reynolds, Jr.	(a) **42**, K47 (1977)	*12.1*, p. *102*; *12.2*, p. *105*; 21, p. 282; 22.5.2, p. 379; 23, p. 420
	(a) **42**, K151 (1977)	*8*, p. *66*; 22.5.2, p. 379
S. M. Rezende	(b) **80**, 385 (1977)	1.2, p. 15; *18.1*, p. *213*; *18.4*, p. *231*
B. Rheinländer	(a) **39**, K139 (1977)	*14.3.3*, p. *178*; *20.3*, p. *268*; 22.2.1, p. 351
	(b) **83**, K83 (1977)	13.5.2, p. 150; *18*, p. *212*; *20.1*, p. *260*; 22.2.1, p. 353
H. D. Riccius	(a) **42**, K53 (1977)	*14.4*, p. *191*; *20*, p. *245*; 22.8.1, p. 406
C.-E. Richter	(a) **42**, 579 (1977)	*9*, p. *70*; *11*, p. *97*; 22.2.4, p. 357

J. Richter	(b) **79**, K139 (1977)	*2*, p. *34*; 13, p. 108; *18.2*, p. *218*
	(a) **44**, 303 (1977)	13.4, p. 139; *14.3.3*, p. *181*; 22, p. 330
K. Richter	(a) **44**, 499 (1977)	*10.1*, p. *82*; 14.2, p. 164; 21.1, p. 300
W. Richter	(b) **84**, 619 (1977)	*6*, p. *60*; *20.1*, p. *262*; 22.7, p. 395
	(a) **44**, K99 (1977)	*14.2*, p. *164*; 21, p. 287
R. de Ridder	(a) **39**, 383 (1977)	*1.2*, p. *15*; 1.4, p. 24; 22, p. 322
	(a) **40**, 343 (1977)	*1.2*, p. *15*; 1.4, p. 25; 22, p. 323
	(a) **40**, 669 (1977)	*1.4*, p. *25*; 10.2, p. 87; 22, p. 324
	(a) **41**, 555 (1977)	*1.4*, p. *26*; *10.2*, p. *88*; 22.6, p. 388
	(a) **42**, 319 (1977)	1.3, p. 22; *1.4*, p. *26*; 22, p. 326
	(a) **43**, 133 (1977)	*1.3*, p. *22*; 10.1, p. 80; 21, p. 284; 21.1, p. 297
	(a) **43**, 541 (1977)	*1.4*, p. *27*; 10.2, p. 91; 22, p. 329
	(a) **44**, 411 (1977)	1.3, p. 22; *1.4*, p. *26*; 22, p. 326
C. Rigaux	(b) **79**, 585 (1977)	*13.1*, p. *115*; 18, p. 209; 20.1, p. 249; 22.4.4, p. 368
	(b) **81**, 615 (1977)	13.1, p. 118; *18*, p. *210*; *20.1*, p. *255*; 22.4.4, p. 368
	(b) **82**, 149 (1977)	13.1, p. 118; *18*, p. *211*; *20.1*, p. *256*; 22.4.4, p. 368
L. Rimlinger	(a) **40**, 425 (1977)	*1.3*, p. *21*; 5, p. 48; 21, p. 279; 21.1.1, p. 303
J. Ringeissen	(b) **83**, 161 (1977)	*13.1*, p. *121*; *20.1*, p. *259*; 22.5, p. 259
L. G. Ripley	(a) **40**, K25 (1977)	*5*, p. *47*; 15, p. 198; *18.1*, p. *213*; 22.8.2, p. 410
D. Ríos	(a) **42**, K1 (1977)	*11*, p. *96*; 22.6, p. 389
I. G. Ritchie	(a) **42**, 751 (1977)	*18.2*, p. *222*; 21.1.1, p. 305
R. H. Ritchie	(b) **83**, K159 (1977)	*13.5*, p. *142*; 13.5.2, p. 150; 21, p. 285
	(b) **84**, 367 (1977)	*13.5.2*, p. *150*
G. Robbrecht	(a) **39**, 425 (1977)	5, p. 47; *18.2*, p. *218*; 22, p. 322
	(a) **43**, 619 (1977)	*5*, p. *49*; 22.8.2, p. 412
V. E. Rodionov	(a) **41**, 659 (1977)	*13.4*, p. *135*; *20.3*, p. *271*; 22.2.1, p. 352
M. Rohr	(a) **42**, 477 (1977)	*1.2*, p. *18*; *14.2*, p. *162*; 21, p. 283
	(a) **43**, K19 (1977)	*1.2*, p. *18*; *14.2*, p. *163*; 22.8, p. 399
A. L. Roitburd	(a) **40**, 333 (1977)	*1.2*, p. *15*; 1.4, p. 25; 21.1.1, p. 303
F. A. Rojas	(b) **81**, 647 (1977)	*10*, p. *72*
N. G. Romanov	(b) **79**, K27 (1977)	*19*, p. *235*; 20.1, p. 248; 20.3, p. 267; 22.5.2, p. 375
N. Romeo	(a) **40**, K49 (1977)	*14.3.4*, p. *184*; *20.3*, p. *269*; 22.4.2, p. 365
E. Ronander	(a) **39**, 239 (1977)	*10.1*, p. *75*; 21, p. 276
G. Röpke	(b) **80**, K117 (1977)	*14.1*, p. *158*; *14.3*, p. *167*
	(b) **82**, 297 (1977)	*14*, p. *155*
	(b) **82**, 617 (1977)	*13*, p. *110*
U. Röpke	(a) **40**, K135 (1977)	*2*, p. *36*; *18.3*, p. *229*; 21, p. 280; 21.1, p. 292
H. Röppischer	(a) **40**, 575 (1977)	13.4, p. 132; *13.5.1*, p. *144*; *19*, p. *237*; 22.4.2, p. 365
	(a) **40**, K117 (1977)	*10.2*, p. *87*; *19*, p. *238*; 22.4.2, p. 365
P. Röschmann	(b) **82**, 11 (1977)	13.5.4, p. 153; *18.3*, p. *228*; 19, p. 235; 22.8.2, p. 409
M. Rosenberg	(a) **39**, 79 (1977)	1, p. 9; 13.1, p. 114; *14.2*, p. *161*; 20.1, p. 248; 22.6.1, p. 393
J. Rosenthal	(b) **83**, K1 (1977)	*18*, p. *212*; *20.3*, p. *272*; 22.9, p. 418
	(b) **84**, 27 (1977)	*13.5.1*, p. *147*; *19*, p. *242*; 22.9, p. 419
H. Rosin	(a) **44**, K13 (1977)	*10.2*, p. *92*; 22.2.2, p. 355
V. Roskovec	(b) **83**, K87 (1977)	*18.2*, p. *222*; 22, p. 328
M. Rösler	(b) **83**, 85 (1977)	*13.5*, p. *141*; 13.5.1, p. 146; 13.5.2, p. 149; 13.5.3, p. 152; 22.2.1, p. 353; 22.4.1, p. 363; 22.4.2, p. 366; 22.4.3, p. 367
V. I. Roslyakov	(a) **43**, 59 (1977)	*11*, p. *97*; 21, p. 284
M. Rossinelli	(a) **42**, 699 (1977)	*12.2*, p. *105*; 22, p. 327

G. Roth	(b) **79**, 519 (1977)	*19*, p. *235*; 21.1, p. 289
L. M. Roth	(b) **81**, 637 (1977)	6, p. 56; *13.4*, p. *134*; 22.1.2, p. 340
M. Roth	(a) **41**, 577 (1977)	*1.4*, p. *26*; 10.1, p. 78; 21, p. 281; 21.1, p. 294
H. Roth-Seefrid	(b) **80**, 137 (1977)	*6*, p. *53*; 21.1, p. 308
G. Rother	(b) **79**, 473 (1977)	6, p. 52; *14.4.1*, p. *194*; 20.1, p. 248; 22.8.1, p. 402
M. Rotter	(a) **40**, K169 (1977)	*19*, p. *238*; 22.8.2, p. 410
E. Roudaut	(a) **39**, K151 (1977)	*18.3*, p. *228*; 22.7, p. 394
D. Roy	(b) **83**, 615 (1977)	*6*, p. *59*; 21, p. 285; 21.1, p. 297; 21.6, p. 317
U. Roy	(a) **44**, K25 (1977)	*1.2*, p. *20*; 8, p. 67; 22.8.1, p. 407
M. A. Rozenblat	(a) **44**, K75 (1977)	*18.2.1*, p. *227*
M. Rožmarin	(a) **42**, 315 (1977)	*14.4.1*, p. *196*; *19*, p. *240*; 22.8.1, p. 405
L. G. Rubin	(b) **82**, K1 (1977)	*13.1*, p. *119*; *19*, p. *240*; 22.1.2, p. 340
A. I. Rubinshtein	(b) **84**, 125 (1977)	*13*, p. *112*; *14.4*, p. *192*
Yu. V. Rud	(a) **39**, 453 (1977)	13.4, p. 131; *20.3*, p. *268*; 22.8, p. 396
P. S. Rudman	(a) **41**, 283 (1977)	*11*, p. *95*; 14.3.2, p. 176; 22, p. 324
	(a) **42**, 729 (1977)	*2*, p. *38*; 14.3.1, p. 174; *16*, p. *203*; 22, p. 327
A. S. Rudnev	(a) **43**, 59 (1977)	*11*, p. *97*; 21, p. 284
A. F. Rudolph	(a) **41**, 177 (1977)	13.4, p. 134; *14.3.4*, p. *185*; 22.1.3, p. 345
K. Ruebenbauer	(b) **79**, K137 (1977)	*5*, p. *47*; 22.8.1, p. 402
	(b) **84**, 611 (1977)	*5*, p. *50*; 14.2, p. 164; *18.2*, p. *224*; 21.4, p. 313
M. Rueff	(b) **80**, 215 (1977)	*13.4*, p. *131*; *20.1*, p. *251*
	(b) **81**, 511 (1977)	*6*, p. *56*; *13.4*, p. *134*
H. Rüfer	(a) **42**, 519 (1977)	13.5.1, p. 146; *20.2*, p. *266*; *20.3*, p. *272*; 22.4.4, p. 368
R. Ruggeri	(b) **79**, 655 (1977)	*2*, p. *33*; *13*, p. *107*; 14, p. 155; 22, p. 321
J. C. C. de Ruiter	(b) **82**, K23 (1977)	*10.1*, p. *78*; 21, p. 282
H. Runge	(a) **39**, 141 (1977)	*11*, p. *94*; *14.3.3*, p. *178*; 22.1.2, p. 337
	(a) **39**, 595 (1977)	*11*, p. *94*; 22.1.2, p. 338
W. Ruppel	(a) **42**, 519 (1977)	13.5.1, p. 146; *20.2*, p. *266*; *20.3*, p. *272*; 22.4.4, p. 368
L. Rusiniak	(a) **40**, K127 (1977)	*19*, p. *238*; 22.1.1, p. 334
N. Rusović	(a) **44**, 609 (1977)	*12.1*, p. *103*; 21.1, p. 300
M. Russo	(a) **41**, 393 (1977)	*14.2*, p. *162*
F. Rustichelli	(a) **44**, 127 (1977)	*1.1*, p. *13*; 22.1.2, p. 343
H. Rüthlein	(b) **82**, 595 (1977)	*13.4*, p. *136*; 14.3.2, p. 177; 22.1.3, p. 346
Ž. Ružić-Toroš	(a) **41**, 255 (1977)	*1.2*, p. *16*; 14.3, p. 168; 22.3, p. 358
D. J. Ruzicka	(a) **39**, 167 (1977)	*1.5*, p. *29*; *1.6*, p. *31*; 22.2.1, p. 350
C. Rybka	(b) **81**, 565 (1977)	*10.2*, p. *88*; 22.9, p. 416
	(b) **83**, K91 (1977)	*10.2*, p. *90*; 22.9, p. 418
	(b) **84**, K49 (1977)	*10.2*, p. *91*; 22.9, p. 419
V. M. Ryzhkovskii	(a) **39**, K127 (1977)	*18.4*, p. *231*; 22, p. 322
A. V. Sachenko	(a) **43**, 311 (1977)	*14.3.2*, p. *177*; *16*, p. *204*; 22.2.3, p. 356
K. Sadananda	(a) **44**, 633 (1977)	*10*, p. *74*
K. Saermark	(b) **79**, 723 (1977)	*13.2*, p. *124*; 14.1, p. 157; 21.2, p. 308; 21.6, p. 315
	(b) **81**, K31 (1977)	13.2, p. 124; *14.1*, p. *158*
	(b) **83**, 187 (1977)	*13.2*, p. *124*; *14.1*, p. *159*; 21, p. 284
H. Saha	(a) **41**, 317 (1977)	13.3, p. 126; *14.3.3*, p. *179*; *16*, p. *202*; 22.1.2, p. 339; 22.2.1, p. 351
A. Sakalas	(a) **39**, 601 (1977)	*7*, p. *62*; *14.3.4*, p. *183*; 22.4.1, p. 362
	(a) **43**, 705 (1977)	*7*, p. *63*; *14.3*, p. *171*; 16, p. 204; 22.4.3, p. 367
A. Sakamoto	(a) **40**, 109 (1977)	7, p. 62; *10.2*, p. *85*; 22.5.2, p. 376
M. Sakoh	(a) **44**, 511 (1977)	*1.2*, p. *20*; *8*, p. *68*; 21.1.1, p. 307
V. P. Sakun	(b) **84**, 513 (1977)	*13.5.1*, p. *148*; *19*, p. *243*; 22.9, p. 419

T. Sakurai	(a) **41**,	411 (1977)	*14.3*, p. *168*; 22.6, p. 388
E. Yu. Salaev	(b) **82**,	705 (1977)	*6*, p. *58*; *20.1*, p. *258*; 22.3, p. 358
K. Salama	(a) **41**,	241 (1977)	*12.1*, p. *101*; 21.1, p. 293
N. Saleh	(a) **43**,	99 (1977)	2, p. 39; *14.4*, p. *192*; 22.8.1, p. 406
M. A. Salehi-Manshadi	(a) **40**,	K43 (1977)	*16*, p. *202*; 22.4.1, p. 362
K. Salewski	(a) **39**,	41 (1977)	*18.2.1*, p. *226*; 18.4, p. 231; 21.1, p. 288; 21.1.1, p. 302
S. R. Salinas	(b) **83**,	229 (1977)	*13*, p. *110*; *18*, p. *212*
L. K. Samanta	(a) **41**,	331 (1977)	*20*, p. *245*; 22.8, p. 398
D. Samaras	(a) **41**,	459 (1977)	*18.4*, p. *233*; 22.8.1, p. 405
J. R. Sambles	(b) **79**,	645 (1977)	*19*, p. *235*; 21, p. 277
A. Samoć	(a) **39**,	337 (1977)	*13.4*, p. *130*; 14.3.4, p. 183; 14.4, p. 189; 22.9, p. 414
M. Samoć	(a) **39**,	337 (1977)	*13.4*, p. *130*; 14.3.4, p. 183; 14.4, p. 189; 22.9, p. 414
G. Samoggia	(b) **79**,	795 (1977)	*13.4*, p. *131*; *20.1*, p. *249*; 22.5.3, p. 383
	(b) **81**,	411 (1977)	*13.4*, p. *131*; *20.1*, p. *249*; 22.5.3, p. 383
M. I. Samoilovich	(a) **44**,	K87 (1977)	*20.3*, p. *274*; 22.8.1, p. 407
A. A. Samokhvalov	(b) **79**,	9 (1977)	13.1, p. 114; *13.5.3*, p. *151*; *14.3*, p. *165*; 15, p. 198; 18.3, p. 228; 22.8.2, p. 409
	(b) **83**,	353 (1977)	13.1, p. 114; *13.5.3*, p. *151*; *14.3*, p. *165*; 15, p. 198; 18.3, p. 228; 22.8.2, p. 409
H. Sampaio	(a) **39**,	65 (1977)	1.2, p. 14; *13.4*, p. *129*; *14.3*, p. *165*; 22.3.1, p. 360
J. R. Sandercock	(a) **42**,	699 (1977)	*12.2*, p. *105*; 22, p. 327
S. Santucci	(b) **84**,	305 (1977)	18.1, p. 215; *19*, p. *242*; 22.8, p. 400
A. E. Sapega	(a) **42**,	585 (1977)	*13.3*, p. *127*; *14.3.3*, p. *180*; 22.1.3, p. 346
J. Saraie	(a) **39**,	331 (1977)	*16*, p. *201*; 22.4.4, p. 368
V. S. Sardaryan	(a) **42**,	409 (1977)	*7*, p. *62*; *10*, p. *73*
C. K. Sarkar	(b) **79**,	K157 (1977)	*14.3*, p. *166*; 18, p. 210; 22, p. 322
K. K. Sarkar	(b) **81**,	409 (1977)	*6*, p. *55*; 22.6, p. 387
S. K. Sarkar	(b) **83**,	263 (1977)	4, p. 45; *6*, p. *58*
	(b) **83**,	615 (1977)	*6*, p. *59*; 21, p. 285; 21.1, p. 297; 21.6, p. 317
A. Q. Sarker	(b) **79**,	567 (1977)	*13.4*, p. *130*; 14.3, p. 165; 22.1.2, p. 338
S. E. Sarkisov	(a) **40**,	K69 (1977)	*20.2*, p. *265*; 22.8.2, p. 410
	(a) **41**,	K199 (1977)	*20.2*, p. *265*; 22.8.2, p. 410
	(a) **43**,	71 (1977)	3, p. 42; 13.4, p. 137; 20.1, p. 260; *20.2*, p. *266*; *20.3*, p. *272*; 22.8, p. 399
G. Sarrabayrouse	(a) **42**,	173 (1977)	14.3.2, p. 176; *14.3.3*, p. *180*
T. Sarrazin	(a) **43**,	459 (1977)	*1.4*, p. *27*; 10.1, p. 80; 21.1, p. 298
S. Saterlie	(a) **41**,	K75 (1977)	*12.1*, p. *101*; 21, p. 280
	(a) **42**,	K175 (1977)	*10.1*, p. *80*; 21, p. 284
F. G. Satkiewicz	(a) **41**,	481 (1977)	2, p. 37; *20.1*, p. *256*; 22.1, p. 332
Y. Sato	(b) **82**,	611 (1977)	*5*, p. *48*; 11, p. 96; 22.5.3, p. 384
T. Satow	(a) **40**,	303 (1977)	*2*, p. *35*; 18.1, p. 213; 22.8, p. 397
	(a) **41**,	417 (1977)	*2*, p. *37*; 21.1.1, p. 303; 22.1.1, p. 334
	(b) **84**,	353 (1977)	*2*, p. *39*; *18*, p. *212*; 22, p. 330
	(a) **44**,	731 (1977)	*2*, p. *41*; 18.1, p. 216; 22.8, p. 401
C. B. Satterthwaite	(a) **43**,	K147 (1977)	8, p. 67; 13, p. 112; *14.2*, p. *163*; 21, p. 286
M. Sauvage	(a) **40**,	K101 (1977)	*1.4*, p. *25*; 22, p. 324
	(a) **43**,	473 (1977)	*1.1*, p. *13*; 22.1.2, p. 342
I. Savatinova	(b) **82**,	677 (1977)	*6*, p. *57*; 14.4.1, p. 196; *20.1*, p. *258*; 22.8, p. 399
	(b) **84**,	401 (1977)	1.2, p. 20; 6, p. 60; *14.4.1*, p. *197*; *20.1*, p. *262*; 22.8, p. 400
Ö. Sävborg	(a) **43**,	645 (1977)	*1.2*, p. *19*; 8, p. 67; 14.3, p. 171; 18.1, p. 215; 22.6.1, p. 393
V. I. Savenko	(a) **39**,	K11 (1977)	*1.6*, p. *31*; *10.2*, p. *83*; 12.2, p. 104; 22.5.2, p. 375

B. D. Sawicka	(b) **80,**	K41 (1977)	2, p. 34; *5*, p. *47*; 21.1.1, p. 302
	(a) **41,**	173 (1977)	2, p. 36; *5*, p. *48*; 22.8.1, p. 404
J. A. Sawicki	(b) **80,**	K41 (1977)	2, p. 34; *5*, p. *47*; 21.1.1, p. 302
	(a) **41,**	173 (1977)	2, p. 36; *5*, p. *48*; 22.8.1, p. 404
A. Scacco	(b) **84,**	23 (1977)	*13.4*, p. *138*; *20.3*, p. *274*; 22.5.2, p. 381
A. Scalabrin	(b) **79,**	731 (1977)	*6*, p. *52*; 14.4, p. 189; 20.1, p. 249; 22.8.1, p. 402
G. Schaack	(b) **80,**	559 (1977)	1.2, p. 15; *6*, p. *54*; *8*, p. *65*; 14.4.1, p. 194; 22.9, p. 415
H. Schaber	(b) **81,**	K41 (1977)	6, p. 55; *20.1*, p. *254*; 22.5.2, p. 378
H. E. Schaefer	(b) **80,**	173 (1977)	*11*, p. *94*; *14.1*, p. *157*; 21.1, p. 290
W. Schäfer	(a) **40,**	447 (1977)	*1.2*, p. *16*; *18.4*, p. *232*; 22.8.1, p. 403
	(b) **84,**	629 (1977)	*13.1*, p. *122*; *20.1*, p. *263*; 22.5.4, p. 385
R. Schalge	(a) **39,**	517 (1977)	*1.4*, p. *24*; *10.2*, p. *84*; 22.8.2, p. 409
G. Schanz	(b) **80,**	629 (1977)	6, p. 54; *19*, p. *237*; 21.1, p. 291
	(b) **84,**	K53 (1977)	2, p. 40; *19*, p. *243*; 21, p. 286
A. Scharmann	(a) **41,**	535 (1977)	*14.3*, p. *169*; 22.8.1, p. 405
	(a) **42,**	K79 (1977)	*17*, p. *206*; 22.5.2, p. 379
	(a) **43,**	519 (1977)	*17*, p. *207*; 22.5.2, p. 380
	(b) **84,**	93 (1977)	18.4, p. 234; *19*, p. *242*; 22.8.1, p. 407
K. Scheerschmidt	(a) **42,**	47 (1977)	*1.4*, p. *26*; 10, p. 73
G. Scheiber	(b) **80,**	119 (1977)	1.4, p. 24; *14.1*, p. *157*; *20.1*, p. *250*; 22, p. 322
C. Scherer	(b) **80,**	313 (1977)	*18.1*, p. *213*; 22.5.4, p. 385
W. Scheu	(b) **82,**	523 (1977)	*10*, p. *73*; *12*, p. *100*
W. Schiller	(b) **82,**	297 (1977)	*14*, p. *155*
R. Schimko	(a) **42,**	579 (1977)	*9*, p. *70*; *11*, p. *97*; 22.2.4, p. 357
A. Schlachetzki	(a) **41,**	103 (1977)	*14.3.3*, p. *179*; 22.2.1, p. 351
P. Schley	(a) **42,**	667 (1977)	*14.3.3*, p. *181*; 14.3.4, p. 186; 22, p. 327
P. Schlotter	(b) **81,**	521 (1977)	*13.5.1*, p. *145*; 16, p. 202; 22.9, p. 416
M. Schlüter	(b) **82,**	357 (1977)	*13.1*, p. *119*; 14, p. 155; 22, p. 326
H. Schmid	(a) **41,**	K147 (1977)	*14.4.1*, p. *195*; 22.8, p. 399
W. Schmid	(b) **84,**	529 (1977)	*13.5.1*, p. *148*; *20.3*, p. *275*; 22.1.2, p. 344
A. Schmidt	(a) **44,**	K141 (1977)	1.5, p. 30; *11*, p. *99*; 22.1.2, p. 344
G. Schmidt	(a) **40,**	K153 (1977)	*14.4.1*, p. *195*; 22.8.1, p. 404
H. Schmidt	(b) **84,**	611 (1977)	*5*, p. *50*; 14.2, p. 164; *18.2*, p. *224*; 21.4, p. 313
M. Schmidt	(b) **79,**	533 (1977)	*13.4*, p. *130*; 22.2.1, p. 350
N. Schmidt	(a) **41,**	157 (1977)	*14.4.1*, p. *195*; 22.8, p. 398
V. Schmidt	(a) **39,**	621 (1977)	*10.2*, p. *84*; *12.2*, p. *104*; 22.5.2, p. 376
	(a) **40,**	147 (1977)	*10.2*, p. *85*; *12.2*, p. *104*; 22.5.2, p. 376
W. Schmidt	(a) **39,**	133 (1977)	*1.1*, p. *11*; 1.2, p. 14; 22.2.4, p. 357
W. D. Schmidt	(a) **42,**	K143 (1977)	*14.2*, p. *163*; 21, p. 283
R. Schmucker	(b) **79,**	479 (1977)	*14.2*, p. *161*
	(b) **80,**	89 (1977)	10.1, p. 76; *14.2*, p. *161*
D. Schneider	(a) **42,**	73 (1977)	*10.2*, p. *89*; 22.5.2, p. 379
F. Schneider	(a) **39,**	K115 (1977)	5, p. 47; *18.3*, p. *228*; 22.6, p. 386
J. Schneider	(a) **39,**	K161 (1977)	*2*, p. *34*; *18.2*, p. *218*; 19, p. 236; 22.8, p. 397
K. F. Schneider	(a) **41,**	653 (1977)	*18*, p. *211*; 21.1, p. 294; 21.6, p. 316
W. Schnelle	(a) **44,**	197 (1977)	*14.1*, p. *160*; *14.3.1*, p. *175*; 21.7, p. 319
W. Schnitker	(b) **82,**	K35 (1977)	13.1, p. 119; *18*, p. *211*; *20.1*, p. *257*; 22.1.1, p. 335
T. Schober	(a) **43,**	443 (1977)	*1.2*, p. *19*; 1.3, p. 23; 8, p. 67; 21.1, p. 297
G. Schoeck	(a) **40,**	271 (1977)	*10.1*, p. *76*; 21, p. 279; 21.1, p. 291
J. Scholz	(b) **79,**	525 (1977)	6, p. 52; 13.1, p. 115; *14.3.4*, p. *183*; 22.1.2, p. 337
J. Schöneich	(a) **44,**	K141 (1977)	1.5, p. 30; *11*, p. *99*; 22.1.2, p. 344
E. Schönherr	(b) **79,**	269 (1977)	*6*, p. *51*; 12.2, p. 104; *20.1*, p. *247*; 22, p. 320
J. Schoonman	(a) **40,**	511 (1977)	*16*, p. *202*; 22.5, p. 370; 22.5.4, p. 385
J. Schreiber	(b) **81,**	371 (1977)	*14.4*, p. *190*; *14.4.1*, p. *195*; 22.8.1, p. 404

M. Schreiber	(b) **84**,	629 (1977)	*13.1*, p. *122*; *20.1*, p. *263*; 22.5.4, p. 385
H. Schreyer	(a) **43**,	K81 (1977)	10.1, p. 80; *11*, p. *97*; *18.2*, p. *222*; 21.1.1, p. 306
K. Schröder	(a) **40**,	K145 (1977)	*18.2*, p. *219*; 21.1, p. 292; 21.7, p. 319
E. Schubert	(a) **39**,	133 (1977)	*1.1*, p. *11*; 1.2, p. 14; 22.2.4, p. 357
E. Schuberth	(b) **84**,	K91 (1977)	*1.2*, p. *20*; 6, p. 60; 20.1, p. 263; 23, p. 420
W. Schülke	(b) **80**,	K67 (1977)	*13.2*, p. *124*; 21.2, p. 308
	(b) **82**,	229 (1977)	*13.1*, p. *119*; *13.2*, p. *124*; 21.2, p. 308; 22.1.2, p. 340
E. Schuller	(b) **80**,	109 (1977)	*13.5*, p. *141*; *20.1*, p. *250*; 22.6, p. 386
F. K. Schulte	(b) **79**,	149 (1977)	*13.3*, p. *126*; 21, p. 276
	(b) **79**,	K5 (1977)	8, p. 64; *13.3*, p. *126*; 18, p. 209; 21, p. 276
D. Schultze	(a) **43**,	71 (1977)	3, p. 42; 13.4, p. 137; 20.1, p. 260; *20.2*, p. *266*; *20.3*, p. *272*; 22.8, p. 399
G. Schulz	(a) **43**,	493 (1977)	*10.1*, p. *81*; 21.1, p. 298
W. Schüppel	(a) **43**,	277 (1977)	3, p. 43; *18.3*, p. *230*; 22.8.2, p. 412
M. Schwab	(b) **84**,	167 (1977)	1.2, p. 19; *8*, p. *67*; *18.4*, p. *234*; 22.8.1, p. 407
R. Schwabe	(b) **84**,	K43 (1977)	*6*, p. *60*; *13.5.1*, p. *148*; 20.3, p. 274; 22.3, p. 359
W. Schwander	(a) **41**,	K167 (1977)	*20*, p. *245*; 21.1, p. 294
M. Schwartz	(b) **83**,	115 (1977)	*6*, p. *58*; *10*, p. *73*
G. Schwarz	(a) **42**,	579 (1977)	*9*, p. *70*; *11*, p. *97*; 22.2.4, p. 357
E. Schwarzer	(b) **84**,	253 (1977)	*13.5.1*, p. *148*; 19, p. 242; 22.9, p. 419
L. Schweitzer	(b) **80**,	491 (1977)	*2*, p. 35; *14.3*, p. *167*; *19*, p. *237*; 22.1.1, p. 334; 22.1.2, p. 338
N. Schwentner	(b) **81**,	503 (1977)	*13.1*, p. *118*; 17, p. 206; 23, p. 420
M. Schwoerer	(a) **43**,	465 (1977)	13.4, p. 138; *20.3*, p. *273*; 22.9, p. 418
G. H. Schwuttke	(a) **42**,	553 (1977)	*14.3*, p. *170*; 14.3.1, p. 174; 22.1.2, p. 341
	(a) **43**,	43 (1977)	3, p. 42; 14.3.3, p. 181; *16*, p. *204*; 20.1, p. 260; 22.1.2, p. 341
	(a) **43**,	119 (1977)	*11*, p. *97*; *14.3*, p. *170*; 22.1.2, p. 341
G. B. Scott	(b) **79**,	203 (1977)	13.4, p. 129; *20.1*, p. *247*; 22.8.2, p. 409
J. Šedivý	(a) **42**,	681 (1977)	*6*, p. *58*; *8*, p. *66*; 21, p. 283; 21.3, p. 310
B. Sedlák	(a) **40**,	K169 (1977)	*19*, p. *238*; 22.8.2, p. 410
A. Seeger	(a) **42**,	285 (1977)	*10.1*, p. *79*; 11, p. 96; 12.1, p. 102; 21.1, p. 295
K. Seeger	(b) **79**,	115 (1977)	*14.3*, p. *165*; 22.2.1, p. 350; 22.2.2, p. 354
D. Segers	(a) **40**,	153 (1977)	*8*, p. *65*; 21, p. 279
K. H. Segsa	(a) **41**,	K21 (1977)	*19*, p. *239*; 22.1.2, p. 339
P. Seidel	(a) **44**,	303 (1977)	13.4, p. 139; *14.3.3*, p. *181*; 22, p. 330
M. Sekinobu	(a) **42**,	K25 (1977)	*14.3.2*, p. *177*; *14.3.3*, p. *180*; 22.2.1, p. 352
H. Selbach	(b) **80**,	235 (1977)	*10.2*, p. *85*; 19, p. 236; 22.5.2, p. 376
M. Selg	(b) **84**,	K61 (1977)	*20.3*, p. *274*; 23, p. 420
A. Selkin	(b) **83**,	47 (1977)	*13.5.1*, p. *146*; 20.1, p. 259; 22.4.1, p. 363
M. Semary	(a) **41**,	295 (1977)	18.1, p. 214; *18.2*, p. *220*; 22.9, p. 416
O. S. Semerdzhyan	(a) **43**,	399 (1977)	*1*, p. *10*
P. K. Sen	(a) **43**,	89 (1977)	6, p. 59; 13, p. 111; *14.3.4*, p. *186*; 22.2.1, p. 353
R. Send	(a) **41**,	157 (1977)	*14.4.1*, p. *195*; 22.8, p. 398
I. Senff	(a) **42**,	K183 (1977)	20.1, p. 259; *20.2*, p. *266*; 20.3, p. 272; 22.8.1, p. 406
	(a) **43**,	K15 (1977)	13.4, p. 137; *20.2*, p. *266*; 22.8, p. 399
S. Sengupta	(b) **83**,	263 (1977)	4, p. 45; *6*, p. *58*
	(b) **83**,	615 (1977)	*6*, p. *59*; 21, p. 285; 21.1, p. 297; 21.6, p. 317
	(b) **83**,	645 (1977)	*10.2*, p. *91*; 22.5.2, p. 380
F. Seppi	(a) **41**,	K75 (1977)	*12.1*, p. *101*; 21, p. 280
T. I. Serebryakova	(b) **83**,	537 (1977)	*13*, p. *111*; 17, p. 207; 22, p. 328
P. P. Seregin	(a) **39**,	437 (1977)	*1.2*, p. *15*; 2, p. 34; 5, p. 47; 22, p. 322
J. Sériot	(a) **33**,	147 (1977)	13.4, p. 130; *20.3*, p. *267*; 22.5.3, p. 383
R. Serneels	(b) **79**,	215 (1977)	*1.1*, p. *11*
	(a) **41**,	51 (1977)	*1.4*, p. *25*; 22.7, p. 394

J. Serughetti	(a) **40**,	629 (1977)	*10.2*, p. *87*; 20.1, p. 253; 22.5.2, p. 377
	(a) **41**,	183 (1977)	*10.2*, p. *87*; 22.5.2, p. 378
B. R. Sethi	(a) **42**,	791 (1977)	13.5.3, p. 152; *14.3*, p. *170*; 22.4.2, p. 366
J. P. Šetrajčić	(b) **79**,	433 (1977)	*6*, p. *52*; *13.5.1*, p. *143*
	(b) **84**,	K25 (1977)	*13.5*, p. *142*
L. A. Sevastianova	(a) **44**,	247 (1977)	5, p. 49; *14.4.1*, p. *197*; 18.2, p. 223; *18.4*, p. *234*; 22.8, p. 400
S. M. Seyed Reihani	see under letter "R"		
K. Shahi	(a) **41**,	11 (1977)	9, p. 69; 10.2, p. 83; *14.4*, p. *189*; 15, p. 198; 22.5.1, p. 373; 22.8, p. 396; 22.9, p. 414
N. Shahtahmasebi	(b) **83**,	543 (1977)	*13.4*, p. *137*; 14.4, p. 192; 22.1.1, p. 335
Z. Al Shaieb	(a) **39**,	K131 (1977)	*20.1*, p. *250*; 22.5.2, p. 376
G. M. Shakhlevich	(a) **39**,	K127 (1977)	18.4, p. 231; 22, p. 322
A. M. Shalaev	(a) **42**,	K29 (1977)	*10.1*, p. *79*; 21.1, p. 295
Jai Shanker	(b) **82**,	K13 (1977)	*4*, p. *45*; 12.2, p. 105; 22.5.2, p. 379
B. B. Sharma	(a) **40**,	691 (1977)	*1*, p. *9*; 1.4, p. 25; 22.8, p. 397
K. C. Sharma	(b) **82**,	577 (1977)	*18.1*, p. *215*; *18.2*, p. *221*
M. N. Sharma	(b) **79**,	359 (1977)	*4*, p. *44*; 12.2, p. 104; 22.5.2, p. 375
	(b) **82**,	K53 (1977)	*8*, p. *66*; 22.5.2, p. 379
	(b) **84**,	427 (1977)	*4*, p. *46*; *8*, p. *68*; 20.1, p. 262; 22.5.2, p. 381
O. P. Sharma	(b) **82**,	K13 (1977)	*4*, p. *45*; 12.2, p. 105; 22.5.2, p. 379
	(a) **42**,	791 (1977)	13.5.3, p. 152; *14.3*, p. *170*; 22.4.2, p. 366
T. P. Sharma	(b) **83**,	155 (1977)	*6*, p. *58*; *20.1*, p. *259*; 22.5.2, p. 380
Vijender Sharma	(a) **39**,	K59 (1977)	*14.3.3*, p. *178*; 22.2.1, p. 350
	(a) **44**,	401 (1977)	*14.3.3*, p. *181*; *14.3.4*, p. *187*; 22.2.1, p. 353
G. Sharp-Dent	(b) **79**,	645 (1977)	*19*, p. *235*; 21, p. 277
A. L. Shashimohan	(a) **44**,	787 (1977)	*14.4.1*, p. *197*; 20.1, p. 264; 22.8.1, p. 408
E. Shatkovskii	(a) **42**,	111 (1977)	15, p. 199; *16*, p. *203*; 22.2.1, p. 352
L. A. Shcherbina	(b) **81**,	709 (1977)	*14.3*, p. *168*; 15, p. 199; 22.1.1, p. 334
E. D. Shchukin	(a) **39**,	K11 (1977)	*1.6*, p. *31*; *10.2*, p. *83*; 12.2, p. 104; 22.5.2, p. 375
I. F. Shchegolev	(a) **44**,	77 (1977)	*1*, p. *10*; *14*, p. *156*; 18.1, p. 215; 22.9, p. 419
M. I. Sheboul	(b) **80**,	51 (1977)	*13.5.1*, p. *144*; 20.1, p. 250; 20.3, p. 268; 22.5, p. 370
S. S. Sheinin	(a) **44**,	353 (1977)	*1.1*, p. *13*; 21.6, p. 317
M. Sheinkman	(a) **44**,	747 (1977)	*16*, p. *205*; *20.3*, p. *275*; 22.3, p. 359
M. K. Sheinkman	(a) **39**,	717 (1977)	*13.4*, p. *131*; *14.3*, p. *166*; 22.2.1, p. 350
	(a) **41**,	299 (1977)	*14.3.3*, p. *179*; *20.3*, p. *270*; 22.4.1, p. 363
A. V. Shekoyan	(a) **42**,	409 (1977)	*7*, p. *62*; *10*, p. *73*
V. K. Shemetova	(a) **44**,	K71 (1977)	*2*, p. *40*; *14.3.4*, p. *187*; 22.1.3, p. 347
A. P. Sherstyakov	(a) **42**,	647 (1977)	*13.4*, p. *137*; *14.3.1*, p. *174*; 16, p. 203; 22.2.1, p. 352
V. N. Sherstyakova	(a) **42**,	647 (1977)	*13.4*, p. *137*; *14.3.1*, p. *174*; 16, p. 203; 22.2.1, p. 352
V. I. Shevchenko	(b) **83**,	323 (1977)	*13*, p. *110*; 21, p. 284
V. Ya. Shevchenko	(a) **39**,	445 (1977)	*16*, p. *201*; *20.1*, p. *250*; 22.3, p. 358
	(a) **41**,	307 (1977)	13.1, p. 117; 13.4, p. 134; *14.3*, p. *168*; 22.3, p. 358
R. P. Shibaeva	(a) **44**,	77 (1977)	*1*, p. *10*; *14*, p. *156*; 18.1, p. 215; 22.9, p. 419
M. Shiga	(a) **42**,	369 (1977)	1, p. 10; *5*, p. *48*; *18*, p. *211*; 21.1.1, p. 304
D. S. Shim	(b) **79**,	731 (1977)	*6*, p. *52*; 14.4, p. 189; 20.1, p. 249; 22.8.1, p. 402
A. Shimizu	(b) **80**,	K113 (1977)	*2*, p. *35*; *19*, p. *237*; 21.6, p. 315; 22.1.2, p. 339
	(b) **82**,	K123 (1977)	*2*, p. *35*; *19*, p. *237*; 21.6, p. 315; 22.1.2, p. 339
M. Shimizu	(a) **44**,	511 (1977)	*1.2*, p. *20*; *8*, p. *68*; 21.1.1, p. 307
T. Shimizu	(b) **81**,	K71 (1977)	*2*, p. *36*; *19*, p. *238*; 22.1.1, p. 334; 22.1.2, p. 339
Hoang Mi Shin	(a) **41**,	299 (1977)	*14.3.3*, p. *179*; *20.3*, p. *270*; 22.4.1, p. 363
S. Shindo	(a) **43**,	295 (1977)	*1.6*, p. *32*; 22.6, p. 390

A. M. Shirokov	(a) **42**, 385 (1977)	12.2, p. 105; 13.1, p. 120; *16*, p. *203*; *20.1*, p. *257*; 22.1.3, p. 346
F. S. Shishiyanu	(a) **40**, 29 (1977)	*9*, p. *69*; 10.2, p. 85; 14.3, p. 167; 22.2.2, p. 354
Shi-yu-Wu	see under letter "W"	
G. I. Shklarevskaya	(a) **41**, K197 (1977)	*10.1*, p. *78*; 21, p. 282; 21.6, p. 316
B. I. Shklovskii	(b) **83**, K11 (1977)	*2*, p. *38*; *14.3.1*, p. *175*; 22.1, p. 332
I. S. Shlimak	(a) **43**, 231 (1977)	*20.3*, p. *273*; 22.1.1, p. 335
G. M. Shmelev	(b) **80**, 697 (1977)	*6*, p. *55*; 22, p. 323
	(b) **82**, 391 (1977)	*13.5.2*, p. *149*; 22, p. 326
M. T. Shpak	(a) **42**, 433 (1977)	*10.2*, p. *90*; *20.3*, p. *272*; 22.9, p. 417
K. A. Shoaib	(a) **40**, 605 (1977)	10.2, p. 86; 13.4, p. 133; *20.3*, p. *269*; 22.6, p. 387
V. G. Shteinberg	(a) **42**, K87 (1977)	*10.1*, p. *79*; *12.1*, p. *102*; 21.1, p. 296
V. M. Shulga	(b) **82**, 251 (1977)	13.4, p. 135; *19*, p. *240*; 22.6, p. 389
	(b) **84**, 443 (1977)	13.4, p. 139; *19*, p. *243*; 22.6, p. 391
A. A. Shultin	(b) **80**, 417 (1977)	*6*, p. *54*; 20.1, p. 252; 22.8.1, p. 403
E. V. Shvedov	(a) **44**, 423 (1977)	*3*, p. *43*; *10.2*, p. *92*; 22.5.2, p. 381
Radhey Shyam	(a) **41**, K189 (1977)	*2*, p. *37*; *14.3*, p. *169*; 22.1.3, p. 345
W. A. Sibley	(a) **41**, 569 (1977)	*10.2*, p. *88*; 20.1, p. 256; 20.3, p. 270; 22.6, p. 388
P. Sickert	(a) **39**, 401 (1977)	*10.1*, p. *75*; 21.1, p. 289
J. M. de Siebenthal	(b) **79**, 259 (1977)	*10.2*, p. *83*; 19, p. 235; 22.5.3, p. 383
D. Siebert	(a) **41**, K185 (1977)	*10.2*, p. *89*; *19*, p. *240*; 22.2.1, p. 352
V. Siegel	(a) **42**, K147 (1977)	*17*, p. *207*; 22.6, p. 390
	(a) **43**, K117 (1977)	1.6, p. 32; 13.3, p. 128; *17*, p. *207*; 22.6, p. 391
W. Siegel	(a) **41**, 75 (1977)	*14.3.3*, p. *179*; *20.3*, p. *270*; 22.8, p. 398
K. Sierański	(b) **81**, K67 (1977)	*13.1*, p. *117*; 14.3, p. 168; 22.4.4, p. 368
	(a) **42**, 721 (1977)	13.1, p. 121; *14.3.1*, p. *174*; 22.4.4, p. 369
M. Sieskind	(a) **40**, 503 (1977)	*20.1*, p. *253*; 22.6, p. 387
M. A. Sigal	(a) **42**, 775 (1977)	*18.2*, p. *222*; *19*, p. *241*; 22.8.2, p. 412
I. Sigalas	(a) **41**, 585 (1977)	*5*, p. *48*; 18.2, p. 220; 21.1, p. 294; 21.4, p. 312
E. Sigmund	(b) **80**, 215 (1977)	*13.4*, p. *131*; *20.1*, p. *251*
	(b) **81**, 511 (1977)	*6*, p. *56*; *13.4*, p. *134*
R. Silbermann	(b) **81**, 119 (1977)	*13.3*, p. *126*; 14.3.2, p. 176; 22.1.3, p. 345
A. Šileika	(b) **82**, K45 (1977)	*13.1*, p. *119*; *20.1*, p. *257*; 22.8, p. 399
N. I. Silkin	(b) **84**, 803 (1977)	13.4, p. 140; *19*, p. *243*; 22.8.1, p. 408
E. L. Sill	(a) **39**, K131 (1977)	*20.1*, p. *250*; 22.5.2, p. 376
V. M. Silonov	(b) **82**, 701 (1977)	*1.3*, p. *22*; 21.1, p. 296
A. V. Simashkevich	(a) **42**, 675 (1977)	*14.3*, p. *170*; 16, p. 203; 22.4.2, p. 366
	(a) **44**, 83 (1977)	*14.3.3*, p. *181*; *14.3.4*, p. *187*; 22.4.2, p. 366
J. P. Simon	(a) **41**, K107 (1977)	*10.1*, p. *78*; 21, p. 281; 21.1, p. 294
F. Simone	(b) **81**, 579 (1977)	*14.2*, p. *162*
A. J. Simonov	(b) **82**, 685 (1977)	*13.5.2*, p. *149*; 21, p. 283
	(b) **83**, 377 (1977)	*13.3*, p. *127*; 13.5.2, p. 150; *20.1*, p. *260*
Z. Šimša	(a) **40**, K169 (1977)	*19*, p. *238*; 22.8.2, p. 410
A. K. Singh	(a) **43**, 357 (1977)	*1.3*, p. *22*; 5, p. 49; 21.1.1, p. 305
	(a) **44**, 443 (1977)	*5*, p. *50*; 22.8.2, p. 413
D. P. Singh	(a) **40**, 315 (1977)	*18*, p. *210*; *20.2*, p. *265*; 22.2.3, p. 356
H. Singh	(a) **40**, 691 (1977)	*1*, p. *9*; 1.4, p. 25; 22.8, p. 397
J. Singh	(b) **79**, 787 (1977)	13.4, p. 131; *14.1*, p. *157*; 21.6, p. 315
Jai Singh	(b) **82**, 263 (1977)	*13.1*, p. *119*; 14.3, p. 169; 22.9, p. 416
M. Dhwaj Singh	(b) **84**, 375 (1977)	*6*, p. *60*; 8, p. 67; 22.1.1, p. 336; 22.1.2, p. 343
N. Singh	(b) **79**, 787 (1977)	13.4, p. 131; *14.1*, p. *157*; 21.6, p. 315
O. Singh	(b) **82**, 179 (1977)	*7*, p. *62*; 14.2, p. 162; 21, p. 282
R. N. Singh	(b) **83**, 651 (1977)	4, p. 46; *6*, p. *59*; 14.1, p. 159; 21.2, p. 309
T. Singh Dod	see under letter "D"	
Vijay A. Singh	(b) **81**, 637 (1977)	6, p. 56; *13.4*, p. *134*; 22.1.2, p. 340

A. K. Singhvi	(b) **84**,	9 (1977)	*13.4*, p. *129*; *18.2*, p. *217*; 21.1, p. 288; 21.1.1, p. 302; 21.4, p. 311
N. D. Sinh	(a) **43**,	K99 (1977)	*13.3*, p. *127*; 22.1.2, p. 342
A. P. B. Sinha	(a) **43**,	653 (1977)	*9*, p. *71*; 21.6, p. 317
	(a) **44**,	391 (1977)	*9*, p. *71*; 21.6, p. 317
N. L. P. Sinha	(a) **44**,	K127 (1977)	*1.5*, p. *30*; 14.3.1, p. 175; 16, p. 205; 22, p. 331
P. Sinha	(b) **81**,	493 (1977)	*13.1*, p. *118*; 22.4.1, p. 363
S. P. Sinitsa	(a) **44**,	K167 (1977)	*2*, p. *41*; *14.3.4*, p. *188*; 22, p. 331
N. N. Sirota	(a) **41**,	669 (1977)	*13.1*, p. *118*; *20.1*, p. *256*; 22.2.4, p. 357
E. I. Sirotinin	(a) **43**,	59 (1977)	*11*, p. *97*; 21, p. 284
J. De Sitter	(a) **43**,	619 (1977)	*5*, p. *49*; 22.8.2, p. 412
M. Siu Li	see under letter "L"		
V. S. Sivasankar	(a) **44**,	345 (1977)	*10.2*, p. *92*; 20.1, p. 262; 22.5.2, p. 381
	(a) **44**,	741 (1977)	*10.2*, p. *93*; 13.4, p. 140; 22.5.2, p. 382
V. P. Sivkov	(a) **39**,	437 (1977)	*1.2*, p. *15*; 2, p. 34; 5, p. 47; 22, p. 322
G. Sixt	(a) **43**,	119 (1977)	*11*, p. *97*; *14.3*, p. *170*; 22.1.2, p. 341
N. L. Sizova	(a) **40**,	K1 (1977)	*10.2*, p. *85*; 22.5.2, p. 377
	(a) **41**,	443 (1977)	*10.2*, p. *88*; 22.5.2, p. 378
P. Skalicky	(a) **43**,	473 (1977)	*1.1*, p. *13*; 22.1.2, p. 342
T. Skóskiewicz	(b) **79**,	K147 (1977)	*14.2*, p. *161*; 18, p. 209; 21, p. 277
Ž. M. Škrbić	(b) **83**,	K125 (1977)	*13.5.1*, p. *147*
V. A. Skrebnev	(b) **81**,	727 (1977)	*19*, p. *239*
M. J. Škrinjar	(b) **79**,	433 (1977)	*6*, p. *52*; *13.5.1*, p. *143*
	(b) **83**,	K125 (1977)	*13.5.1*, p. *147*
	(b) **84**,	K25 (1977)	*13.5*, p. *142*
	(b) **84**,	K101 (1977)	*13.5.1*, p. *148*
A. I. Skvortsov	(a) **41**,	225 (1977)	*11*, p. *95*; *14.2*, p. *162*; 18.1, p. 214; 21.1.1, p. 303; 21.5, p. 314
P. Sladký	(a) **42**,	K57 (1977)	7, p. 63; *12.2*, p. *105*; *20.1*, p. *257*; 22.5.1, p. 373
J. Slak	(a) **40**,	K121 (1977)	14.4.1, p. 195; *19*, p. *238*; 22.8, p. 398
	(a) **42**,	315 (1977)	*14.4.1*, p. *196*; *19*, p. *240*; 22.8.1, p. 405
N. Slavin	(a) **43**,	K201 (1977)	*10.2*, p. *91*; *11*, p. *98*; 22.1.2, p. 342
A. Ślawska-Waniewska	(a) **40**,	K39 (1977)	*18.3*, p. *229*; 22.8.2, p. 410
A. W. Sleeswyk	(b) **80**,	235 (1977)	*10.2*, p. *85*; 19, p. 236; 22.5.2, p. 376
P. S. Smertenko	(a) **43**,	333 (1977)	*14.3*, p. *170*; 14.3.4, p. 186; 22, p. 328
O. P. Smirnov	(a) **39**,	683 (1977)	*18.4*, p. *231*; 22.8.2, p. 410
G. F. Smirnova	(a) **41**,	669 (1977)	*13.1*, p. *118*; *20.1*, p. *256*; 22.2.4, p. 357
	(b) **84**,	K65 (1977)	*6*, p. *60*; *20.1*, p. *262*; 22.8, p. 400
B. J. Smith	(a) **40**,	K31 (1977)	*10.2*, p. *86*; 14.3.3, p. 179; 22.1.2, p. 338
D. L. Smith	(a) **44**,	381 (1977)	*14.3.2*, p. *177*; *14.3.3*, p. *181*; 22.2.1, p. 353
D. S. Smith	(a) **42**,	K53 (1977)	*14.4*, p. *191*; *20*, p. *245*; 22.8.1, p. 406
I. O. Smith	(a) **39**,	583 (1977)	*10*, p. *72*
J. H. Smith	(a) **40**,	K73 (1977)	14.1, p. 158; *18.2*, p. *219*; 21.1, p. 291
T. F. Smith	(b) **82**,	K81 (1977)	*13.2*, p. *124*; *14.1*, p. *159*; 21, p. 283
M. Snykers	(a) **41**,	51 (1977)	*1.4*, p. *25*; 22.7, p. 394
A. A. Sobol	(a) **42**,	101 (1977)	*1.3*, p. *22*; 13.4, p. 136; 20.1, p. 257; 20.3, p. 271; 22.8.2, p. 411
E. Sobula	(b) **80**,	K71 (1977)	*18.2*, p. *218*
M. S. Sodha	(a) **40**,	315 (1977)	*18*, p. *210*; *20.2*, p. *265*; 22.2.3, p. 356
F. Söffge	(a) **42**,	621 (1977)	*8*, p. *66*; 12.1, p. 102; 18.2, p. 221; 21.1, p. 296
W. Sohn	(a) **44**,	237 (1977)	*13.4*, p. *139*; *14.3*, p. *171*; 22.4.1, p. 364
Ya. M. Soifer	(a) **42**,	K87 (1977)	*10.1*, p. *79*; *12.1*, p. *102*; 21.1, p. 296
I. A. Sokolik	(a) **40**,	655 (1977)	13.5.1, p. 144; *16*, p. *202*; 22.9, p. 415
L. V. Sokolov	(a) **43**,	659 (1977)	*1.5*, p. *30*; 2, p. 39; 22.1.1, p. 336; 22.6, p. 391
O. B. Sokolov	(b) **79**,	423 (1977)	*13*, p. *107*; 20, p. 244; 21, p. 277
	(b) **80**,	73 (1977)	*13*, p. *108*; 20, p. 244; 21, p. 278
	(b) **83**,	281 (1977)	*13*, p. *110*; 20, p. 246; 21, p. 284

	(b) **83**,	383 (1977)	*13*, p. *111*; 20, p. 246; 21, p. 285
	(b) **84**,	773 (1977)	*13*, p. *113*; 20, p. 246; 21, p. 287
V. I. Sokolov	(a) **39**,	683 (1977)	*18.4*, p. *231*; 22.8.2, p. 410
H. Solbrig	(b) **82**,	303 (1977)	*1.6*, p. *31*
	(b) **82**,	429 (1977)	*1.6*, p. *31*
G. I. Solovev	(b) **79**,	105 (1977)	*6*, p. *51*; 21.1, p. 288
S. P. Solovev	(a) **44**,	247 (1977)	5, p. 49; *14.4.1*, p. *197*; 18.2, p. 223; *18.4*, p. *234*; 22.8, p. 400
V. E. Solovev	(a) **41**,	371 (1977)	*14.3.3*, p. *179*; *20.2*, p. *265*; 22.2.4, p. 357
G. Solt	(b) **82**,	651 (1977)	*4*, p. *45*; *10.1*, p. *79*; 21.2, p. 309
J. Sołtys	(a) **41**,	K51 (1977)	*1.3*, p. *21*; 21, p. 280; 21.1, p. 293
T. Soma	(b) **79**,	K53 (1977)	*4*, p. *44*; *13*, p. *107*; 21, p. 276
	(b) **82**,	319 (1977)	*6*, p. *57*; 8, p. 66; 22.1.1, p. 335; 22.1.2, p. 340
K.-H. Sommer	(b) **82**,	303 (1977)	*1.6*, p. *31*
R. B. Somoano	(b) **81**,	281 (1977)	*14*, p. *155*; 15, p. 198; 22.9, p. 415
	(b) **82**,	K117 (1977)	*1.1*, p. *12*; 22.9, p. 417
M. Sondergeld	(b) **81**,	253 (1977)	*13.5.1*, p. *145*; *20.1*, p. *254*; 22.4.2, p. 365
	(b) **81**,	451 (1977)	*13.5.1*, p. *145*; *20.1*, p. *255*; 22.4.2, p. 366
T. Soovik	(b) **84**,	K61 (1977)	*20.3*, p. *274*; 23, p. 420
J. Sosnowski	(a) **42**,	K159 (1977)	*14.2*, p. *163*; 21, p. 284
N. Sotnik	(a) **42**,	K11 (1977)	*1.1*, p. *12*; *1.6*, p. *31*; 21.6, p. 316
J. C. Souillat	(a) **39**,	147 (1977)	13.4, p. 130; *20.3*, p. *267*; 22.5.3, p. 383
Z. Šourek	(a) **39**,	725 (1977)	*10.2*, p. *85*; 22.1.2, p. 338
M. de Souza	(b) **80**,	597 (1977)	*10.2*, p. *86*; 14.4, p. 190; 22.5.2, p. 377
	(a) **40**,	109 (1977)	7, p. 62; *10.2*, p. *85*; 22.5.2, p. 376
	(b) **82**,	K103 (1977)	*10.2*, p. *90*; *20.1*, p. *258*; 22.5.4, p. 385
J. Spałek	(b) **79**,	K61 (1977)	*13.5.4*, p. *153*
	(b) **82**,	107 (1977)	*14.3*, p. *169*; *18.2*, p. *220*; 22.6, p. 389
	(b) **84**,	747 (1977)	*13*, p. *113*; 18.2, p. 224; 18.4, p. 234
I. P. Spevakova	(a) **44**,	429 (1977)	*18.2*, p. *224*; *18.4*, p. *234*; 22.8, p. 400
H. Spiering	(b) **82**,	243 (1977)	*1.2*, p. *17*; 5, p. 48; 18.1, p. 214; 22.8, p. 399
T. Spruijt	(a) **42**,	537 (1977)	*9*, p. *70*; 21, p. 283; 21.1, p. 296
R. Srinivasan	(a) **44**,	687 (1977)	14.4.1, p. 197; *19*, p. *243*; 22.8.1, p. 408
G. P. Srivastava	(b) **80**,	657 (1977)	*6*, p. *54*; *8*, p. *65*
J. K. Srivastava	(b) **81**,	107 (1977)	*5*, p. *48*; 22.6.1, p. 393
K. K. Srivastava	(a) **41**,	323 (1977)	2, p. 36; *14.4*, p. *190*; 22.8, p. 398
O. N. Srivastava	(a) **39**,	K81 (1977)	*1.1*, p. *11*; *1.2*, p. *15*; 14.3, p. 166; 22, p. 322
P. L. Srivastava	(b) **83**,	651 (1977)	4, p. 46; *6*, p. *59*; 14.1, p. 159; 21.2, p. 309
	(b) **84**,	797 (1977)	2, p. 40; *14.1*, p. *160*; *15*, p. *200*; 21, p. 287
R. S. Srivastava	(b) **80**,	379 (1977)	*6*, p. *54*; *8*, p. *64*; 21.2, p. 308
Z. M. Stadnik	(a) **43**,	K41 (1977)	*5*, p. *49*; *18.1*, p. *215*; 22.8.1, p. 407
	(a) **44**,	325 (1977)	*1.1*, p. *13*; 5, p. 49; *18.3*, p. *230*; 22.8.2, p. 413
B. Staliński	(a) **40**,	K11 (1977)	*19*, p. *237*; 21, p. 279
S. Stamenković	(b) **80**,	661 (1977)	*6*, p. *55*; *14.4.1*, p. *195*
A. A. Stanchev	(a) **39**,	K1 (1977)	14.3.1, p. 173; *14.3.4*, p. *183*; 22, p. 321
F. Stangler	(a) **41**,	K137 (1977)	*10.1*, p. *78*; *12.1*, p. *101*; 21, p. 281
J. Stankowski	(b) **82**,	K85 (1977)	*1.2*, p. *17*; *14.4.1*, p. *196*; 19, p. 241; 22.9, p. 417
E. Steichele	(a) **42**,	621 (1977)	*8*, p. *66*; 12.1, p. 102; 18.2, p. 221; 21.1, p. 296
E. F. Steigmeier	(b) **80**,	167 (1977)	*6*, p. *53*; 14.4.1, p. 194; 22.8, p. 397
W. Steiner	(a) **39**,	157 (1977)	*5*, p. *47*; 18.1, p. 213; 18.2, p. 217; 21, p. 276; 21.1, p. 288; 21.1.1, p. 302
	(a) **39**,	K105 (1977)	*18.2*, p. *218*; 21.1, p. 289; 21.4, p. 311
	(a) **40**,	125 (1977)	*18.2*, p. *218*; 21.1, p. 290; 21.4, p. 311
	(a) **42**,	739 (1977)	*5*, p. *49*; *18.2*, p. *222*; 21.1.1, p. 305; 21.4, p. 312
	(a) **43**,	111 (1977)	*5*, p. *49*; *18.2*, p. *222*; 21.1.1, p. 305; 21.4, p. 313
	(a) **43**,	437 (1977)	*5*, p. *49*; *18.2*, p. *223*; 21.1.1, p. 306; 21.4, p. 313
	(a) **44**,	593 (1977)	*18.1*, p. *216*; *18.2*, p. *224*; 21.1, p. 300

K. H. Steinmetz	(a) **44**, 669 (1977)	*13.2*, p. *125*; 21.1, p. 300
S. I. Stenin	(a) **43**, 659 (1977)	*1.5*, p. *30*; 2, p. 39; 22.1.1, p. 336; 22.6, p. 391
D. Stephan	(a) **40**, 113 (1977)	*1.1*, p. *11*; 21, p. 279; 21.1, p. 290
J. D. Stephenson	(a) **39**, 89 (1977)	*1.4*, p. *24*; 10.1, p. 75; 21.1.1, p. 302
J. Šternberk	(b) **83**, K87 (1977)	*18.2*, p. *222*; 22, p. 328
A. Stesmans	(b) **79**, 645 (1977)	*19*, p. *235*; 21, p. 277
W. Stetter	(b) **79**, 269 (1977)	*6*, p. *51*; 12.2, p. 104; *20.1*, p. *247*; 22, p. 320
R. J. Stewart	(a) **41**, 577 (1977)	*1.4*, p. *26*; 10.1, p. 78; 21, p. 281; 21.1, p. 294
K. Stierstadt	(a) **42**, 621 (1977)	*8*, p. *66*; 12.1, p. 102; 18.2, p. 221; 21.1, p. 296
H.-J. Stöhr	(a) **49**, 477 (1977)	*10.2*, p. *84*; 14.3, p. 166; 20.1, p. 250; 22.8.1, p. 402
N. Stojadinović	(a) **43**, K91 (1977)	*14.3.3*, p. *181*
S. D. Stojanović	(b) **79**, 433 (1977)	*6*, p. *52*; *13.5.1*, p. *143*
	(b) **80**, 661 (1977)	*6*, p. *55*; *14.4.1*, p. *195*
	(b) **84**, K101 (1977)	*13.5.1*, p. *148*
J. P. Stokes	(b) **80**, 471 (1977)	*10.2*, p. *86*; 20.1, p. 252; 22.5.2, p. 377
A. L. Stolov	(b) **81**, 287 (1977)	*13.4*, p. *133*; 19, p. 238; 20.3, p. 270; 22.5.4, p. 385
N. A. Stolwijk	(b) **79**, K1 (1977)	*9*, p. *69*
	(a) **42**, 537 (1977)	*9*, p. *70*; 21, p. 283; 21.1, p. 296
H. J. Stolz	(b) **80**, 289 (1977)	*13.5.2*, p. *149*; 20.1, p. 251; 21, p. 278
H. Stolze	(b) **82**, 457 (1977)	*13.1*, p. *120*; *20.1*, p. *257*; 22.1.3, p. 346
R. Straubel	(a) **42**, 273 (1977)	*18.2.1*, p. *226*; 21.1, p. 295; 21.1.1, p. 304
	(a) **43**, K113 (1977)	*18.2.1*, p. *226*; 21.1.1, p. 306
W. Stręk	(a) **41**, 547 (1977)	13.4, p. 135; *20.3*, p. *270*; 22.8.1, p. 405
J. P. Strømsheim	(a) **39**, 167 (1977)	*1.5*, p. *29*; *1.6*, p. *31*; 22.2.1, p. 350
K. De Strooper	(a) **39**, 425 (1977)	5, p. 47; *18.2*, p. *218*; 22, p. 322
	(a) **39**, 431 (1977)	*18.1*, p. *213*; 18.3, p. 228; 18.4, p. 231; 22.8.2, p. 409
	(a) **44**, 215 (1977)	18.1, p. 215; *18.3*, p. *230*; 22.8.2, p. 412
E. E. Strumban	(b) **80**, 589 (1977)	*13.4*, p. *132*; 16, p. 202; 20.1, p. 252; 20.3, p. 269; 22.4.1, p. 362
H. Strunk	(a) **40**, 385 (1977)	*10.2*, p. *86*; 22.5.2, p. 377
M. Stubičar	(a) **44**, 339 (1977)	*1.2*, p. *20*; *2*, p. *40*; 21.1, p. 299; 21.1.1, p. 306
V. A. Stukan	(a) **39**, 445 (1977)	*16*, p. *201*; *20.1*, p. *250*; 22.3, p. 358
	(a) **41**, 307 (1977)	13.1, p. 117; 13.4, p. 134; *14.3*, p. *168*; 22.3, p. 358
J. Stuke	(b) **79**, 539 (1977)	*2*, p. *33*; 14.3.1, p. 173; *16*, p. *201*; 22.1.2, p. 337
I. Stulíková	(a) **41**, K197 (1977)	*10.1*, p. *78*; 21, p. 282; 21.6, p. 316
D. Subašić	(a) **44**, 339 (1977)	*1.2*, p. *20*; *2*, p. *40*; 21.1, p. 299; 21.1.1, p. 306
A. V. Subbotin	(a) **40**, K89 (1977)	*9*, p. *69*; *11*, p. *95*; 21, p. 280; 21.1, p. 292
S. I. Subbotin	(a) **39**, 357 (1977)	*1.2*, p. *14*; 12.2, p. 104; 13.5.1, p. 143; *20.1*, p. *248*; 22.3, p. 358
S. Subramanian	(a) **39**, 675 (1977)	11, p. 94; *19*, p. *236*; 22.8.1, p. 402
M. Sudo	(a) **43**, 685 (1977)	13.5.1, p. 147; *20.3*, p. *273*; 22.3, p. 359
M. Suezawa	(a) **40**, 437 (1977)	*10.2*, p. *86*; *14.4.1*, p. *195*; 22.8.1, p. 403
K. Sugawara	(b) **81**, 313 (1977)	*18.1*, p. *214*; 19, p. 238; 21.4, p. 312
	(b) **84**, 709 (1977)	*19*, p. *243*
T. Sugeta	(a) **42**, K7 (1977)	*14.3.3*, p. *180*; 22.2.1, p. 352
A. Sügis	(a) **44**, K1 (1977)	10.2, p. 92; *19*, p. *243*; 22.1.2, p. 343
N. M. Suleymanov	(a) **40**, K11 (1977)	*19*, p. *237*; 21, p. 279
C. Sułkowski	(b) **82**, 657 (1977)	*14.2*, p. *162*; 21, p. 283
P. Süptitz	(a) **40**, K195 (1977)	*2*, p. *36*; *9*, p. *70*; 22, p. 324
	(a) **43**, 451 (1977)	*2*, p. *39*; *9*, p. *71*; 14.3.4, p. 187; 22, p. 329
P. Suortti	(b) **80**, 255 (1977)	4, p. 44; *6*, p. *53*; 22.5, p. 370
R. A. Suris	(a) **40**, K61 (1977)	*14.3.4*, p. *184*; 22.1.1, p. 334

G. M. SURMAVA	(a) **41**, K43 (1977)	*9*, p. *70*; 21.1, p. 293
K. SURYANARAYANA RAO	see under letter "R"	
T. SUSKI	(b) **80**, 31 (1977)	*6*, p. *53*; 13.5.3, p. 151; 14.4.1, p. 194; 20.1, p. 250; 22.8.1, p. 403
K.-E. SÜSSE	(b) **82**, 633 (1977)	*6*, p. *57*; 20.1, p. 258
J. SUWALSKI	(a) **39**, 643 (1977)	*5*, p. *47*; 22.8.2, p. 409
	(a) **41**, 191 (1977)	*5*, p. *48*; 18.2, p. 219; 21.1.1, p. 303
A. SUZUKI	(a) **39**, K79 (1977)	*20.3*, p. *267*; 22.3, p. 358
K. SUZUKI	(a) **42**, 643 (1977)	*13.5.2*, p. *149*; 21, p. 283
M. SUZUKI	(a) **42**, 133 (1977)	*1.1*, p. *12*; 22.5.1, p. 373
	(a) **42**, K71 (1977)	*7*, p. *63*; *20.1*, p. *257*; 22.4.1, p. 363
T. SUZUKI	(a) **42**, 337 (1977)	2, p. 37; 13.1, p. 120; *14.3.1*, p. *174*; 15, p. 199; *16*, p. *203*; 22.1.2, p. 340
	(a) **43**, 289 (1977)	*10.1*, p. *80*; *14.2*, p. *163*; 21, p. 285
Y. SUZUKI	(a) **41**, 417 (1977)	*2*, p. *37*; 21.1.1, p. 303; 22.1.1, p. 334
I. SVARE	(a) **43**, K65 (1977)	*9*, p. *70*; *19*, p. *241*; 22.5, p. 371
I. V. SVECHKAREV	(b) **83**, 341 (1977)	*13*, p. *111*; *18.1*, p. *215*; 22.1.3, p. 346
S. V. SVECHNIKOV	(a) **43**, 333 (1977)	*14.3*, p. *170*; 14.3.4, p. 186; 22, p. 328
P. SVOBODA	(b) **80**, 441 (1977)	*10*, p. *72*
W. E. SWARTZ, JR.	(a) **39**, 213 (1977)	*1.5*, p. *29*; *2*, p. *33*; 13.1, p. 115; 14.3.1, p. 173; 22.1.1, p. 333
CH. E. SWENBERG	(a) **40**, 621 (1977)	11, p. 95; *13.5.1*, p. *144*; *20.3*, p. *269*; 22.9, p. 415
	(b) **83**, 249 (1977)	*13.5.1*, p. *146*; *20.3*, p. *272*; 22.9, p. 418
	(b) **83**, K1 (1977)	*18*, p. *212*; *20.3*, p. *272*; 22.9, p. 418
J. SWORAKOWSKI	(a) **39**, 337 (1977)	*13.4*, p. *130*; 14.3.4, p. 183; 14.4, p. 189; 22.9, p. 414
	(a) **41**, 619 (1977)	13.4, p. 135; *14.3.4*, p. *185*; 22, p. 325
G. F. SYRYKH	(b) **79**, 105 (1977)	*6*, p. *51*; 21.1, p. 288
A. SZADKOWSKI	(b) **83**, K155 (1977)	*13.4*, p. *138*; 16, p. 204; 22.5.3, p. 384
	(a) **44**, K43 (1977)	*14.3*, p. *172*; *20.1*, p. *262*; 22.5.3, p. 384
C. SZAFRAŃSKI	(a) **41**, 547 (1977)	13.4, p. 135; *20.3*, p. *270*; 22.8.1, p. 405
Z. SZALMASSY	(a) **40**, 75 (1977)	*7*, p. *62*; *14.3.2*, p. *176*; 14.3.3, p. 178; 16, p. 201; 22.4.1, p. 362; 22.8.1, p. 403
J. SZATKOWSKI	(b) **81**, K67 (1977)	*13.1*, p. *117*; 14.3, p. 168; 22.4.4, p. 368
	(a) **42**, 721 (1977)	13.1, p. 121; *14.3.1*, p. *174*; 22.4.4, p. 369
L. S. SZCZEPANIAK	(b) **82**, K85 (1977)	*1.2*, p. *17*; *14.4.1*, p. *196*; 19, p. 241; 22.9, p. 417
M. SZCZYGIELSKI	(a) **43**, K199 (1977)	*18.2*, p. *223*; 22.8.2, p. 412
J. SZCZYRBOWSKI	(b) **79**, K43 (1977)	*2*, p. *33*; *20.1*, p. *248*; 22.1, p. 332; 22.2, p. 348
I. C. SZÉP	(a) **43**, K135 (1977)	1.5, p. 30; *1.6*, p. *32*; 2, p. 39; 22.6, p. 391
W. A. SZILVA	(a) **40**, 211 (1977)	*1.4*, p. *24*; *20.2*, p. *265*; 21.1, p. 291
B. SZPUNAR	(b) **82**, 205 (1977)	*18*, p. *211*; 22.7, p. 394
	(b) **82**, 449 (1977)	13, p. 110; *18.1*, p. *214*; *18.4*, p. *233*; 21.1, p. 296; 21.4, p. 312
Z. SZÜCS	(b) **79**, 519 (1977)	*19*, p. *235*; 21.1, p. 289
	(b) **84**, 189 (1977)	*18*, p. *212*; *19*, p. *242*; 21, p. 286
	(b) **84**, K33 (1977)	*14.2*, p. *163*; *19*, p. *242*; 21, p. 286; 21.6, p. 317
W. SZUSKIEWICZ	(b) **79**, 691 (1977)	13.1, p. 115; *20.1*, p. *249*; 22.4.3, p. 367
	(b) **81**, K119 (1977)	*13.1*, p. *118*; 20.1, p. 255; 22.4.3, p. 367
J. SZYMASZEK	(a) **42**, K159 (1977)	*14.2*, p. *163*; 21, p. 284
M. SZYMOŃSKI	(b) **81**, K11 (1977)	*10.2*, p. *87*; *17*, p. *206*; 22.5.2, p. 378
A. SZYTUŁA	(a) **44**, 325 (1977)	*1.1*, p. *13*; 5, p. 49; *18.3*, p. *230*; 22.8.2, p. 413

C. A. TAFT	(b) **82**, K21 (1977)	*13.1*, p. *119*; 22, p. 326
	(b) **84**, K143 (1977)	*12.2*, p. *106*; *13.1*, p. *123*; 22, p. 330; 22.6, p. 391
M. M. TAGYEV	(a) **39**, K111 (1977)	1.2, p. 15; *6*, p. *53*; *20.1*, p. *250*; 22.3.1, p. 360
R. TAHARA	(a) **41**, 451 (1977)	*5*, p. *48*; 18.2, p. 220; 21.1, p. 294; 21.1.1, p. 303

R. A. Tahir-Kheli	(b) **79**, 93 (1977)	1.2, p. 14; *1.3*, p. *21*
	(b) **79**, 413 (1977)	*18.1*, p. *213*; 18.2, p. 217; 18.4, p. 231
	(b) **80**, 385 (1977)	1.2, p. 15; *18.1*, p. *213*; *18.4*, p. *231*
Yu. M. Tairov	(a) **43**, 363 (1977)	*1*, p. *10*; 1.2, p. 18; 22.3, p. 358
S. Takács	(a) **41**, K175 (1977)	10.1, p. 78; *14.2*, p. *162*; 21, p. 281
S. Takahashi	(a) **42**, 201 (1977)	*10.1*, p. *79*; *18.2*, p. *221*; 21.1, p. 295; 21.1.1, p. 304
	(a) **42**, 529 (1977)	*10.1*, p. *79*; *18.2*, p. *221*; 21.1, p. 296; 21.1.1, p. 304
O. Takai	(b) **80**, 193 (1977)	*10.1*, p. *76*; 21, p. 278
	(b) **80**, K131 (1977)	*4*, p. *45*; *10.1*, p. *77*; 21, p. 279
K. Takanaka	(b) **84**, 813 (1977)	*4*, p. *46*; 21, p. 287
K. Takasaki	(a) **40**, 173 (1977)	6, p. 54; 13.1, p. 116; *14.3.2*, p. *176*; 22.8, p. 397
H. Takei	(a) **40**, 437 (1977)	*10.2*, p. *86*; *14.4.1*, p. *195*; 22.8.1, p. 403
N. Takeuchi	(a) **42**, K83 (1977)	10.2, p. 89; *14.4*, p. *191*; 22.5.4, p. 385
S. Takeuchi	(a) **43**, 289 (1977)	*10.1*, p. *80*; *14.2*, p. *163*; 21, p. 285
K. Takita	(b) **80**, 149 (1977)	*14.3*, p. *166*; 22.8, p. 397
Y. I. Talanov	(a) **40**, K53 (1977)	*19*, p. *237*; 21.2, p. 308
P. L. Talwar	(a) **42**, 791 (1977)	13.5.3, p. 152; *14.3*, p. *170*; 22.4.2, p. 366
A. Tamaševičius	(a) **42**, 111 (1977)	15, p. 199; *16*, p. *203*; 22.2.1, p. 352
Ha Vinh Tan	(b) **84**, 345 (1977)	*6*, p. *60*; *13*, p. *112*
K. Tanaka	(a) **42**, 235 (1977)	*10.1*, p. *79*; 11, p. 96; *12.1*, p. *101*; 21.1, p. 295
S. Tanaka	(b) **80**, 149 (1977)	*14.3*, p. *166*; 22.8, p. 397
	(a) **40**, 173 (1977)	6, p. 54; 13.1, p. 116; *14.3.2*, p. *176*; 22.8, p. 397
T. Tanaka	(a) **39**, 331 (1977)	*16*, p. *201*; 22.4.4, p. 368
	(a) **39**, K79 (1977)	*20.3*, p. *267*; 22.3, p. 358
A. Tanga	(b) **84**, 23 (1977)	*13.4*, p. *138*; *20.3*, p. *274*; 22.5.2, p. 381
E. Tani	(a) **39**, 373 (1977)	10.2, p. 84; *15*, p. *198*; 22.6, p. 386
S. Tanigawa	(b) **82**, 369 (1977)	*11*, p. *96*; 21, p. 282
T. L. Tansley	(a) **44**, 449 (1977)	13.4, p. 140; *14.3.3*, p. *182*; 22.2.1, p. 353; 22.2.4, p. 357
M. Tapiero	(a) **42**, 91 (1977)	13.4, p. 136; *16*, p. *203*; 22.6, p. 389
V. A. Taranets	(a) **43**, 363 (1977)	*1*, p. *10*; 1.2, p. 18; 22.3, p. 358
R. Taranko	(b) **82**, K5 (1977)	*13.1*, p. *119*; *17*, p. *206*; 21, p. 282
L. Tarricone	(a) **40**, K49 (1977)	*14.3.4*, p. *184*; *20.3*, p. *269*; 22.4.2, p. 365
C. Tatsuyama	(a) **43**, 685 (1977)	13.5.1, p. 147; *20.3*, p. *273*; 22.3, p. 359
M. Taut	(b) **82**, 507 (1977)	*13.1*, p. *120*; *20*, p. *245*; 21, p. 283; 22, p. 326
F. N. Tavadze	(a) **41**, K43 (1977)	*9*, p. *70*; 21.1, p. 293
K. N. R. Taylor	(a) **43**, 251 (1977)	*14.1*, p. *159*; 21.4, p. 313
E. Tegeler	(b) **84**, 561 (1977)	*20*, p. *246*; 22.2, p. 349
I. Tehver	(b) **82**, K89 (1977)	*6*, p. *58*; 10.2, p. 89; *20.3*, p. *271*
D. Teitel	(a) **44**, 755 (1977)	*1.3*, p. *23*; 19, p. 243; 21.1, p. 300
V. T. Telepa	(b) **80**, K37 (1977)	*18.4*, p. *232*; 19, p. 236; 22, p. 323
R. V. Telesnin	(a) **42**, K171 (1977)	*18.2.1*, p. *226*; 21.1.1, p. 305
	(a) **44**, 771 (1977)	*18.2*, p. *225*; 22.8.2, p. 413
V. A. Telezhkin	(b) **81**, K139 (1977)	*13.4*, p. *135*; 22.1.2, p. 340
G. van Tendeloo	(a) **39**, 383 (1977)	*1.2*, p. *15*; 1.4, p. 24; 22, p. 322
	(a) **40**, 343 (1977)	*1.2*, p. *15*; 1.4, p. 25; 22, p. 323
	(a) **40**, 531 (1977)	*1.3*, p. *21*; *1.4*, p. *25*; 21.1, p. 292; 21.6, p. 315
	(a) **40**, 669 (1977)	*1.4*, p. *25*; 10.2, p. 87; 22, p. 324
	(a) **41**, 555 (1977)	*1.4*, p. *26*; *10.2*, p. *88*; 22.6, p. 388
	(a) **42**, 319 (1977)	1.3, p. 22; *1.4*, p. *26*; 22, p. 326
	(a) **42**, 565 (1977)	*1.1*, p. *12*; *1.2*, p. *18*; 22, p. 327
	(a) **43**, 133 (1977)	*1.3*, p. *22*; 10.1, p. 80; 21, p. 284; 21.1, p. 297
	(a) **43**, 541 (1977)	*1.4*, p. *27*; 10.2, p. 91; 22, p. 329

	(a) **43**, 553 (1977)	*1.3*, p. *23*; *1.4*, p. *28*; 21.1, p. 298; 21.6, p. 317
	(a) **43**, K137 (1977)	*1.1*, p. *13*; 1.4, p. 28; 22, p. 329
	(a) **44**, 411 (1977)	1.3, p. 22; *1.4*, p. *26*; 22, p. 326
E. I. Terukov	(a) **40**, K157 (1977)	*14.3.1*, p. *173*; 22.6, p. 387
	(a) **41**, 125 (1977)	*1.2*, p. *16*; 14.3, p. 168; 22.6, p. 388
	(a) **44**, K187 (1977)	*1.2*, p. *20*; 6, p. 61; *20.1*, p. *264*; 22.6, p. 392
W. Teubner	(a) **40**, 63 (1977)	*2*, p. *35*; *14.3.1*, p. *173*; 22, p. 323
	(a) **41**, K47 (1977)	*14.3.4*, p. *185*; 22, p. 324
R. Teworte	(a) **43**, 487 (1977)	*1.1*, p. *13*; 22.1.2, p. 342
A. P. Thakoor	(a) **40**, 247 (1977)	*2*, p. *35*; *14.3.1*, p. *173*; 15, p. 198; 21.1, p. 291; 21.6, p. 315
J. Thakur	(b) **84**, 797 (1977)	2, p. 40; *14.1*, p. *160*; *15*, p. *200*; 21, p. 287
Troung Dinh Than	(b) **80**, 149 (1977)	*14.3*, p. *166*; 22.8, p. 397
Nguyen The Khoi	see under letter "K"	
D. Theis	(b) **79**, 125 (1977)	*13.1*, p. *114*; 13.5.1, p. 143; *20.1*, p. *247*; 22.4.1, p. 362; 22.4.2, p. 365
J. Thery	(a) **41**, 423 (1977)	*19*, p. *239*; 22.6.1, p. 393
W. Thielemann	(b) **81**, K91 (1977)	13.1, p. 118; *18*, p. *211*; *20.1*, p. *255*; 22.2.1, p. 352
V. G. Thiessen	(a) **40**, 189 (1977)	*18.2*, p. *218*; 18.4, p. 232; 21.1, p. 291
A. R. Thölén	(a) **39**, 223 (1977)	*1.4*, p. *24*; 10, p. 72
B. Thomas	(b) **79**, 509 (1977)	*14.4*, p. *189*; *20.1*, p. *249*; 22.5, p. 370
C. B. Thomas	(b) **84**, K13 (1977)	2, p. 40; *13.4*, p. *138*; *14.3.4*, p. *187*; 22.8, p. 400
P. Thomas	(a) **41**, 81 (1977)	*1.5*, p. *29*; 2, p. 36; 22.1.2, p. 339
R. Thomas	(a) **39**, 419 (1977)	*1.5*, p. *29*; *3*, p. *42*; 22.9, p. 414
W. R. Thorpe	(a) **39**, 583 (1977)	*10*, p. *72*
F. Thuselt	(b) **79**, K39 (1977)	*13.4*, p. *129*; 13.5.1, p. 143; 22.2, p. 348
	(b) **80**, 499 (1977)	*13.4*, p. *132*; *20.3*, p. *269*; 22, p. 323
	(b) **81**, 89 (1977)	*13.4*, p. *133*; *20.3*, p. *269*; 22, p. 324
	(b) **84**, K43 (1977)	*6*, p. *60*; *13.5.1*, p. *148*; 20.3, p. 274; 22.3, p. 359
B. Thuss	(a) **40**, K131 (1977)	*1.2*, p. *16*; *18.1*, p. *214*; 22.6.1, p. 393
O. J. A. Tiainen	(a) **42**, 241 (1977)	*11*, p. *96*; 14.3, p. 169; 20.1, p. 257; 22.1.3, p. 346
W. Tikvic	(a) **40**, 271 (1977)	*10.1*, p. *76*; 21, p. 279; 21.1, p. 291
H.-J. Tiller	(a) **41**, 151 (1977)	13.3, p. 126; 14.3.2, p. 176; *14.3.4*, p. *185*; 22.9, p. 415
P. Tillmanns	(a) **44**, 539 (1977)	*1.5*, p. *30*; *3*, p. *43*; 21.1.1, p. 307
V. V. Timofeenko	(b) **80**, K17 (1977)	*13.1*, p. *116*; 20.1, p. 251; 22.5.3, p. 383
	(b) **81**, K55 (1977)	*13.1*, p. *117*; 22.5, p. 371
	(b) **82**, K99 (1977)	*13.1*, p. *120*; 22.5, p. 371
M. I. Timoshechkin	(a) **40**, K69 (1977)	*20.2*, p. *265*; 22.8.2, p. 410
	(a) **41**, K199 (1977)	*20.2*, p. *265*; 22.8.2, p. 410
	(a) **42**, 101 (1977)	*1.3*, p. *22*; 13.4, p. 136; 20.1, p. 257; 20.3, p. 271; 22.8.2, p. 411
Yu. Timoshnikov	(a) **43**, K201 (1977)	*10.2*, p. *91*; *11*, p. *98*; 22.1.2, p. 342
P. Tischer	(a) **40**, 551 (1977)	*14.3.4*, p. *184*; 22.8.2, p. 410
J. M. Titman	(b) **80**, K141 (1977)	*2*, p. *36*; *19*, p. *237*; 21, p. 279
A. V. Titov	(a) **44**, 247 (1977)	5, p. 49; *14.4.1*, p. *197*; 18.2, p. 223; *18.4*, p. *234*; 22.8, p. 400
N. N. Tkachenko	(a) **41**, 299 (1977)	*14.3.3*, p. *179*; *20.3*, p. *270*; 22.4.1, p. 363
Yu. B. Tkachenko	(a) **44**, K39 (1977)	*10.2*, p. *92*; *12.2*, p. *106*; 22.5.2, p. 381
V. D. Tkachev	(b) **83**, K137 (1977)	10.2, p. 91; *20.3*, p. *273*; 22.1.2, p. 342
	(a) **44**, 435 (1977)	*11*, p. *99*; *13.4*, p. *140*; 22.1.1, p. 336
J. Todorovič	(a) **44**, 325 (1977)	*1.1*, p. *13*; 5, p. 49; *18.3*, p. *230*; 22.8.2, p. 413
R. Tognato	(a) **39**, 661 (1977)	*10.1*, p. *76*; *12.1*, p. *101*; 21, p. 278
	(a) **44**, 333 (1977)	*10.1*, p. *82*; 12.1, p. 103; 21, p. 287
V. V. Tokii	(a) **44**, 767 (1977)	*10*, p. *74*

K. B. Tolpygo	(b) **81**, K139 (1977)	*13.4*, p. *135*; 22.1.2, p. 340
K. Tomala	(b) **84**, 611 (1977)	*5*, p. *50*; 14.2, p. 164; *18.2*, p. *224*; 21.4, p. 313
I. Tomáš	(a) **42**, 459 (1977)	*18.2*, p. *221*
	(a) **43**, 381 (1977)	*18.2*, p. *223*
Z. Tomkowicz	(a) **44**, 325 (1977)	*1.1*, p. *13*; 5, p. 49; *18.3*, p. *230*; 22.8.2, p. 413
A. Toneva	(a) **44**, 621 (1977)	13.4, p. 140; *14.3*, p. *172*; 22, p. 330
E. Török	(a) **40**, 55 (1977)	*12.1*, p. *101*; 18.4, p. 232; 21.1, p. 290; 21.1.1, p. 303
G. Torres	(a) **42**, K1 (1977)	*11*, p. *96*; 22.6, p. 389
B. S. Tošić	(b) **79**, 433 (1977)	*6*, p. *52*; *13.5.1*, p. *143*
	(b) **81**, 245 (1977)	*13*, p. *109*; 18.2, p. 219
	(b) **83**, 479 (1977)	*13.5.1*, p. *147*; 20.1, p. 260; 20.3, p. 273
J. Toušek	(a) **43**, 351 (1977)	*14.3.2*, p. *177*; *16*, p. *204*; 22.4.3, p. 367
J. Toušková	(a) **40**, 309 (1977)	*14.3.2*, p. *176*; 14.3.3, p. 178; 22.4.3, p. 367
L. Tóth	(a) **41**, 81 (1977)	*1.5*, p. *29*; 2, p. 36; 22.1.2, p. 339
K. D. Tovstyuk	(b) **79**, K131 (1977)	*1.2*, p. *15*; *13*, p. *108*
	(b) **81**, K1 (1977)	*13.5.1*, p. *145*; *14.4*, p. *190*
M. G. Townsend	(a) **40**, K25 (1977)	*5*, p. *47*; 15, p. 198; *18.1*, p. *213*; 22.8.2, p. 410
U. Traxlmayr	(a) **42**, 525 (1977)	*14.3.3*, p. *180*; *14.3.4*, p. *185*; 18, p. 212; 22.2.1, p. 352
H.-R. Trebin	(b) **81**, 527 (1977)	13.1, p. 118; *13.5.3*, p. *151*; 22.2, p. 348; 22.4, p. 361
A. Treindl	(b) **80**, 159 (1977)	*18.4*, p. *231*; *20.1*, p. *251*; 22.6, p. 387
L. Treitinger	(a) **40**, 551 (1977)	*14.3.4*, p. *184*; 22.8.2, p. 410
R. J. Tremblay	(a) **40**, K25 (1977)	*5*, p. *47*; 15, p. 198; *18.1*, p. *213*; 22.8.2, p. 410
G. Trendel	(b) **84**, 629 (1977)	*13.1*, p. *122*; *20.1*, p. *263*; 22.5.4, p. 385
J. Treusch	(b) **84**, 629 (1977)	*13.1*, p. *122*; *20.1*, p. *263*; 22.5.4, p. 385
M. M. Tribel	(a) **40**, 655 (1977)	13.5.1, p. 144; *16*, p. *202*; 22.9, p. 415
N. Triem	(a) **41**, K171 (1977)	9, p. 70; *11*, p. *96*; 22.1.2, p. 340
V. I. Trifonov	(a) **39**, 445 (1977)	*16*, p. *201*; *20.1*, p. *250*; 22.3, p. 358
	(a) **41**, 307 (1977)	13.1, p. 117; 13.4, p. 134; *14.3*, p. *168*; 22.3, p. 358
S. K. Trikha	(b) **80**, 353 (1977)	*6*, p. *53*; *8*, p. *64*; 23, p. 420
	(b) **84**, K95 (1977)	*6*, p. *61*; *8*, p. *68*; 23, p. 420
S. Trimper	(b) **82**, K75 (1977)	*1.2*, p. *17*; 2, p. 38; 14.4.1, p. 196
M. F. Trinkler	(b) **79**, 49 (1977)	*13.4*, p. *129*; *20.3*, p. *267*; 22.5.2, p. 375
B. C. Tripathi	(b) **82**, 603 (1977)	*20*, p. *245*; 22.6, p. 390
R. S. Tripathi	(b) **83**, 197 (1977)	2, p. 38; *13*, p. *110*; *20.1*, p. *259*; 22, p. 327
V. K. Tripathi	(a) **40**, 315 (1977)	*18*, p. *210*; *20.2*, p. *265*; 22.2.3, p. 356
S. O. Troia	(b) **81**, 579 (1977)	*14.2*, p. *162*
R. Trojko	(a) **44**, K29 (1977)	*3*, p. *43*; 22.3.1, p. 360
E. Trojnar	(b) **79**, K147 (1977)	*14.2*, p. *161*; 18, p. 209; 21, p. 277
Nguyen Van Trong	(b) **83**, 395 (1977)	*6*, p. *59*; *13.5.2*, p. *150*; 20, p. 246; 22, p. 328
	(b) **84**, 345 (1977)	*6*, p. *60*; *13*, p. *112*
Troung Dinh Thanh	see under letters "Th"	
A. Truţia	(b) **84**, 817 (1977)	*10.2*, p. *92*; *20.1*, p. *263*; 22.5, p. 372
C. Tsallis	(b) **79**, 451 (1977)	*13.4*, p. *130*; *14.2*, p. *161*
V. I. Tscherbak	(a) **40**, K89 (1977)	*9*, p. *69*; *11*, p. *95*; 21, p. 280; 21.1, p. 292
I. M. Tsidilkovskii	(b) **82**, 59 (1977)	*13.1*, p. *118*; 14.3, p. 169; 15, p. 199; 22.2.3, p. 356
G. A. Tsintsadze	(b) **83**, K59 (1977)	12.2, p. 105; *19*, p. *241*; 22.8.1, p. 406
V. A. Tskhai	(b) **84**, 465 (1977)	4, p. 46; *13.1*, p. *122*; 22.3, p. 359
I. Tsolovski	(a) **44**, K25 (1977)	*1.2*, p. *20*; 8, p. 67; 22.8.1, p. 407
B. S. Tsukerblat	(b) **80**, 703 (1977)	13.4, p. 132; *14.3.4*, p. *184*; *20.1*, p. *252*; 22, p. 323
S. Tsunekawa	(a) **40**, 437 (1977)	*10.2*, p. *86*; *14.4.1*, p. *195*; 22.8.1, p. 403
V. F. Tsvetkov	(a) **43**, 363 (1977)	*1*, p. *10*; 1.2, p. 18; 22.3, p. 358

Z. Tucholski	(a) **41**,	191 (1977)	*5*, p. *48*; 18.2, p. 219; 21.1.1, p. 303
B. Tuck	(a) **39**,	509 (1977)	*9*, p. *69*; 22.2, p. 348
A. F. Tulinov	(a) **43**,	59 (1977)	*11*, p. *97*; 21, p. 284
I. I. Tupizin	(b) **82**,	99 (1977)	*1.1*, p. *11*; 22, p. 325
V. I. Turchaninkov	(a) **41**,	339 (1977)	11, p. 95; *13.4*, p. *134*; *14.3.3*, p. *179*; 22.1.1, p. 334
E. A. Turov	(b) **83**,	281 (1977)	*13*, p. *110*; 20, p. 246; 21, p. 284
	(b) **83**,	383 (1977)	*13*, p. *111*; 20, p. 246; 21, p. 285
	(b) **84**,	773 (1977)	*13*, p. *113*; 20, p. 246; 21, p. 287
V. Tutovan	(a) **44**,	K63 (1977)	*18.2*, p. *224*; 18.4, p. 234; 21.1, p. 299
Y. Uchida	(a) **43**,	269 (1977)	*11*, p. *97*; 13, p. 111; 21, p. 285
J. P. Udalov	(a) **41**,	379 (1977)	*20.1*, p. *256*; *20.3*, p. *270*; 22.8.1, p. 405
R. Uecker	(a) **43**,	71 (1977)	3, p. 42; 13.4, p. 137; 20.1, p. 260; *20.2*, p. *266*; *20.3*, p. *272*; 22.8, p. 399
Y. Ueda	(a) **39**,	K37 (1977)	*18.4*, p. *231*; 22.6, p. 386
O. Uemura	(a) **40**,	303 (1977)	*2*, p. *35*; 18.1, p. 213; 22.8, p. 397
	(a) **41**,	417 (1977)	*2*, p. *37*; 21.1.1, p. 303; 22.1.1, p. 334
	(b) **84**,	353 (1977)	*2*, p. *39*; *18*, p. *212*; 22, p. 330
	(a) **44**,	731 (1977)	*2*, p. *41*; 18.1, p. 216; 22.8, p. 401
K.-D. Ufert	(a) **40**,	K157 (1977)	*14.3.1*, p. *173*; 22.6, p. 387
	(a) **42**,	187 (1977)	*11*, p. *96*; *14.3.1*, p. *174*; 22.6, p. 389
V. A. Ulanov	(b) **84**,	K29 (1977)	*19*, p. *242*; 22.5, p. 372
H.-J. Ullrich	(a) **40**,	113 (1977)	*1.1*, p. *11*; 21, p. 279; 21.1, p. 290
	(a) **44**,	277 (1977)	1.3, p. 23; *3*, p. *43*; 14.2, p. 164; 21.1, p. 299
	(a) **44**,	499 (1977)	*10.1*, p. *82*; 14.2, p. 164; 21.1, p. 300
K. Ulmer	(a) **41**,	113 (1977)	*1.1*, p. *11*; 22.8, p. 398
	(a) **41**,	K167 (1977)	*20*, p. *245*; 21.1, p. 294
W. Ulrici	(b) **84**,	215 (1977)	*13.4*, p. *138*; *20.1*, p. *261*; 22.5.3, p. 384
	(b) **84**,	155 (1977)	*13.4*, p. *140*; 20.1, p. 263; 22.5.3, p. 384
Y. Umakoshi	(a) **43**,	667 (1977)	*10.1*, p. *81*; 12.1, p. 102; 21.1, p. 298
G. R. Umarov	(b) **83**,	K93 (1977)	*4*, p. *45*; *13.1*, p. *121*; 22, p. 328
K. Unger	(b) **79**,	K39 (1977)	*13.4*, p. *129*; 13.5.1, p. 143; 22.2, p. 348
	(b) **81**,	89 (1977)	*13.4*, p. *133*; *20.3*, p. *269*; 22, p. 324
H.-G. Unruh	(a) **44**,	525 (1977)	*14.4.1*, p. *197*; 22.8.1, p. 408
E. J. Unterhorst	(b) **80**,	629 (1977)	6, p. 54; *19*, p. *237*; 21.1, p. 291
	(b) **84**,	K53 (1977)	2, p. 40; *19*, p. *243*; 21, p. 286
G. C. Upreti	(b) **83**,	K17 (1977)	*19*, p. *241*; 22.8.1, p. 406
R. Urao	(a) **40**,	567 (1977)	*1.4*, p. *25*; *1.5*, p. *29*; 21.1, p. 292
M. I. Urbakh	(b) **83**,	633 (1977)	*13.3*, p. *128*; 20.1, p. 260; 21, p. 285
B. Urban	(b) **79**,	185 (1977)	*13.5.3*, p. *151*; 22.1.2, p. 337
R. Ure	(a) **41**,	K75 (1977)	*12.1*, p. *101*; 21, p. 280
N. B. Urli	(b) **83**,	K41 (1977)	*10.2*, p. *90*; 11, p. 97; 22, p. 328
A. A. Urusovskaya	(a) **40**,	K1 (1977)	*10.2*, p. *85*; 22.5.2, p. 377
	(a) **41**,	443 (1977)	*10.2*, p. *88*; 22.5.2, p. 378
A. I. Ushakov	(a) **40**,	11 (1977)	*1.2*, p. *14*; 1.4, p. 24; *1.5*, p. *29*; 18.2, p. 217; 21.1, p. 288; 21.1.1, p. 302
R. Vaglio	(a) **41**,	393 (1977)	*14.2*, p. *162*
S. N. Vaidya	(a) **43**,	K31 (1977)	*1.2*, p. *19*; *14.3*, p. *170*; 22.8.1, p. 406
J. M. Vail	(b) **79**,	663 (1977)	*10.2*, p. *84*; 13.4, p. 130; 20.3, p. 268; 22.5.2, p. 375
J. Vaitkus	(a) **44**,	793 (1977)	*14.3.4*, p. *188*; 20, p. 246; 22.1.2, p. 344; 22.4.2, p. 366
D. Vajda	(a) **42**,	K37 (1977)	*14.4.1*, p. *196*; 22.8.1, p. 406

P. Vajda	(a) **40**, 101 (1977)	*11*, p. *95*; 21.4, p. 311
S. A. Vakhidov	(a) **44**, K87 (1977)	*20.3*, p. *274*; 22.8.1, p. 407
Sh. A. Vakhidov	(a) **41**, 45 (1977)	10.2, p. 87; 13.4, p. 133; *20.1*, p. *254*; *20.3*, p. *270*; 22.8.1, p. 404
R. Z. Valiev	(a) **44**, 65 (1977)	*10.1*, p. *81*; 21, p. 286; 21.3, p. 310
	(a) **44**, 477 (1977)	1.4, p. 28; *10.1*, p. *82*; 21, p. 287; 21.3, p. 310
H. J. L. van der Valk	(b) **80**, 321 (1977)	*6*, p. *53*; *20.1*, p. *251*; 22, p. 322; 22.5, p. 370
V. Valvoda	(a) **44**, 557 (1977)	*6*, p. *61*; 22.1.3, p. 347
Nguyen Van Hieu	see under letter "H"	
Nguyen Van Nam	see under letter "N"	
Nguyen Van Trong	see under letter "T"	
E. R. Vance	(b) **82**, 473 (1977)	13.4, p. 136; *18*, p. *211*; *20*, p. *245*; 22.6, p. 390
R. Vandenberghe	(a) **44**, 215 (1977)	18.1, p. 215; *18.3*, p. *230*; 22.8.2, p. 412
A. Vanderhaeghe	(b) **79**, K75 (1977)	10.2, p. 83; *20.3*, p. *267*; 22.5.3, p. 383
	(b) **83**, 471 (1977)	*10.2*, p. *90*; 20.3, p. 273; 22.5.2, p. 380
G. Vanderschaeve	(a) **43**, 459 (1977)	*1.4*, p. *27*; 10.1, p. 80; 21.1, p. 298
M. Vanhaelst	(b) **79**, K23 (1977)	*19*, p. *235*; 22.5.2, p. 375
	(b) **82**, K69 (1977)	*19*, p. *241*; 22.6, p. 390
R. Vanleerberghe	(a) **44**, 215 (1977)	18.1, p. 215; *18.3*, p. *230*; 22.8.2, p. 412
A. M. Varaprasad	(a) **44**, 787 (1977)	*14.4.1*, p. *197*; 20.1, p. 264; 22.8.1, p. 408
D. M. Vardanyan	(a) **43**, 399 (1977)	*1*, p. *10*
J. Varma	(a) **41**, K89 (1977)	*5*, p. *48*; 21, p. 280; 21.7, p. 319
R. C. DuVarney	see under letter "D"	
F. Varret	(a) **44**, 601 (1977)	*5*, p. *50*; 22.8.1, p. 408
A. R. Vasconcellos	(b) **80**, 181 (1977)	*13.5*, p. *141*; 18.1, p. 213; 20.1, p. 251; 22.2.1, p. 351
L. N. Vasilev	(a) **39**, 437 (1977)	*1.2*, p. *15*; 2, p. 34; 5, p. 47; 22, p. 322
A. Vasquez	(b) **79**, K65 (1977)	*13.4*, p. *129*; 14.4, p. 189; 22.8.1, p. 402
C. Vautier	(a) **44**, 97 (1977)	2, p. 40; 13.4, p, 139; 14.3.1, p. 175; *16*, p. *204*; 22.1.3, p. 347
A. van Veen	(a) **39**, K143 (1977)	*10.1*, p. *76*; *11*, p. *94*; 21.1, p. 290
	(a) **40**, 235 (1977)	*10.1*, p. *76*; 21, p. 279
E. N. Velikaya	(a) **42**, 433 (1977)	*10.2*, p. *90*; *20.3*, p. *272*; 22.9, p. 417
Yu. N. Venevtsev	(a) **44**, 247 (1977)	5, p. 49; *14.4.1*, p. *197*; 18.2, p. 223; *18.4*, p. *234*; 22.8, p. 400
H. Venghaus	(b) **81**, 221 (1977)	13.1, p. 117; 13.5.2, p. 149; *20.1*, p. *253*; 22.1, p. 332
K. A. Verkhovskaya	(a) **39**, 193 (1977)	*14.4.1*, p. *194*; *16*, p. *201*; 22.8.1, p. 402
M. P. Verma	(b) **80**, 63 (1977)	*10.2*, p. *85*; *12.2*, p. *104*; 22.5.2, p. 376
	(b) **82**, 139 (1977)	*1.1*, p. *11*; 12.2, p. 105; 22.5.2, p. 379
	(b) **82**, 325 (1977)	*4*, p. *45*; *12*, p. *100*
	(b) **82**, 481 (1977)	*4*, p. *45*; *12.2*, p. *105*; 22.5.2, p. 379
Yu. P. Veshchunov	(b) **79**, K27 (1977)	*19*, p. *235*; 20.1, p. 248; 20.3, p. 267; 22.5.2, p. 375
G. C. Vezzoli	(a) **43**, 305 (1977)	*1.1*, p. *12*; *1.2*, p. *18*; 8, p. 67; 22.1.3, p. 346
J. L. Vicent	(b) **84**, K71 (1977)	13, p. 112; *14.1*, p. *160*; *18.2.1*, p. *226*; 21.1, p. 299
J. Victoria	(a) **40**, 167 (1977)	*10.1*, p. *76*; 21.1, p. 290
	(a) **40**, K19 (1977)	*10.1*, p. *77*; 21.1, p. 291
	(a) **43**, K187 (1977)	*10.1*, p. *81*; 21.1, p. 298
Nguyen Ai Viet	(b) **84**, 417 (1977)	13.1, p. 122; *14.3.4*, p. *188*; *20.1*, p. *262*; 22.2, p. 349; 22.4, p. 361
C. Viger	(a) **44**, 97 (1977)	2, p. 40; 13.4, p. 139; 14.3.1, p. 175; *16*, p. *204*; 22.1.3, p. 347
V. P. Vijayavargiya	(b) **80**, 83 (1977)	*20*, p. *244*; 21.4, p. 311
	(b) **82**, 603 (1977)	*20*, p. *245*; 22.6, p. 390
V. S. Vikhnin	(b) **79**, 495 (1977)	*14.4.1*, p. *194*; 22.8.1, p. 402

P. Viktorovitch	(a) **39**, 109 (1977)	14.3, p. 165; *15*, p. *198*; 22.2.3, p. 356
G. Viliani	(b) **84**, 393 (1977)	*13.4*, p. *138*; 20.1, p. 261; 22.5.4, p. 385; 22.6, p. 391
P. Villemain	(b) **82**, 269 (1977)	*1.2*, p. *17*; *18.2*, p. *221*; 21.1, p. 295; 21.1.1, p. 304
	(b) **84**, 335 (1977)	*1.3*, p. *23*; *18.2*, p. *223*; 21.1.1, p. 306
A. Vincent	(a) **39**, 651 (1977)	7, p. 62; *10.1*, p. *76*; 21, p. 278
Ha Vinh Tan	see under letter "T"	
M. Vinnal	(a) **44**, K1 (1977)	10.2, p. 92; *19*, p. *243*; 22.1.2, p. 343
M. A. Vinnik	(a) **42**, 191 (1977)	*18.3*, p. *229*; 22.8.2, p. 412
V. S. Vinogradov	(b) **80**, 589 (1977)	*13.4*, p. *132*; 16, p. 202; 20.1, p. 252; 20.3, p. 269; 22.4.1, p. 362
A. V. Vinokurov	(b) **81**, 287 (1977)	*13.4*, p. *133*; 19, p. 238; 20.3, p. 270; 22.5.4, p. 385
A. S. Vishnevskii	(a) **43**, 589 (1977)	*10.2*, p. *91*; 22, p. 329
R. Visocekas	(a) **41**, 499 (1977)	*10.2*, p. *88*; *20.3*, p. *270*; 22.8.1, p. 405
V. Vitkauskas	(a) **43**, 699 (1977)	*7*, p. *63*; *14.3.2*, p. *177*; 22.4.1, p. 364
B. Vitt	(b) **84**, 93 (1977)	18.4, p. 234; *19*, p. *242*; 22.8.1, p. 407
D. Vivien	(a) **41**, 423 (1977)	*19*, p. *239*; 22.6.1, p. 393
K. M. van Vliet	(a) **42**, 279 (1977)	13.4, p. 136; *14.3*, p. *169*; 22, p. 326
Vo Hong Anh	see under letter "A"	
M. Vodă	(b) **84**, 817 (1977)	*10.2*, p. *92*; *20.1*, p. *263*; 22.5, p. 372
C. M. Vodenicharov	(a) **39**, K1 (1977)	14.3.1, p. 173; *14.3.4*, p. *183*; 22, p. 321
	(a) **41**, 487 (1977)	*14.3.1*, p. *174*; *14.3.3*, p. *180*; 22.9, p. 416
	(a) **42**, 785 (1977)	*13.1*, p. *121*; *14.3.1*, p. *174*; 14.3.4, p. 186; 22, p. 327; 22.6, p. 390; 22.9, p. 417
M. Vodenicharova	(a) **41**, 487 (1977)	*14.3.1*, p. *174*; *14.3.3*, p. *180*; 22.9, p. 416
H. Vogel	(b) **82**, 243 (1977)	*1.2*, p. *17*; 5, p. 48; 18.1, p. 214; 22.8, p. 399
F. Voigt	(a) **41**, 505 (1977)	*3*, p. *42*; 22.8.2, p. 411
	(a) **43**, 277 (1977)	3, p. 43; *18.3*, p. *230*; 22.8.2, p. 412
G. Voigt	(a) **43**, 167 (1977)	*1.4*, p. *27*; *3*, p. *43*; 20.3, p. 273; 22.2.4, p. 357
J. Voigt	(b) **79**, K81 (1977)	*13.5.1*, p. *143*; *20.1*, p. *248*; 22.4.1, p. 362
	(b) **83**, 439 (1977)	*13.5.1*, p. *147*; *20.3*, p. *273*; 22.4.1, p. 364; 22.5, p. 371
N. N. Voinova	(a) **43**, K49 (1977)	*10.2*, p. *90*; 22.6, p. 390
V. A. Volkov	(b) **84**, 33 (1977)	*14.1*, p. *159*; 18.4, p. 233; 21, p. 286
H. D. Vollmer	(b) **79**, 249 (1977)	*13.3*, p. *126*; 13.5.2, p. 149; 20.1, p. 247; 21, p. 276; 22.2.2, p. 354
V. A. Voloshin	(b) **82**, 695 (1977)	*12.2*, p. *105*; *20.1*, p. *258*; 22.5, p. 371; 22.8.1, p. 406
M. P. Volotzkoi	(a) **41**, 125 (1977)	*1.2*, p. *16*; 14.3, p. 168; 22.6, p. 388
V. A. Volpyas	(a) **39**, K53 (1977)	*11*, p. *94*; *18.2*, p. *217*; 21.1, p. 289
V. V. Vorobev	(b) **82**, 213 (1977)	*1.2*, p. *17*; *18.2*, p. *221*; 21.4, p. 312
Yu. V. Vorobev	(a) **39**, 717 (1977)	*13.4*, p. *131*; *14.3*, p. *166*; 22.2.1, p. 350
Yu. K. Voronko	(a) **42**, 101 (1977)	*1.3*, p. *22*; 13.4, p. 136; 20.1, p. 257; 20.3, p. 271; 22.8.2, p. 411
S. A. Vorontsov	(b) **83**, 341 (1977)	*13*, p. *111*; *18.1*, p. *215*; 22.1.3, p. 346
M. A. Vorotyntsev	(a) **39**, 229 (1977)	*14.3.2*, p. *176*; 14.3.4, p. 183; 22.5.1, p. 373
	(a) **39**, 573 (1977)	*14.3.2*, p. *176*; 14.3.4, p. 183
	(b) **84**, 125 (1977)	*13*, p. *112*; *14.4*, p. *192*
W. Maenhout-van der Vorst	see under letter "M"	
V. I. Vovnenko	(a) **41**, 659 (1977)	*13.4*, p. *135*; *20.3*, p. *271*; 22.2.1, p. 352
A. E. de Vries	(b) **81**, K11 (1977)	*10.2*, p. *87*; *17*, p. *206*; 22.5.2, p. 378
Vu Duy Phach	see under letter "P"	

P. L. Vuillermoz	(a) **41**, 561 (1977)	6, p. 56; 8, p. 66; *10.2*, p. *88*; 22.2.1, p. 352
V. B. Vykhodets	(b) **82**, K59 (1977)	*10.1*, p. *78*; 21, p. 282
D. N. Vylegzhanin	(a) **40**, 643 (1977)	6, p. 55; *13.4*, p. *133*; 20.3, p. 269; 22.8.1, p. 403
C. Wagner	(a) **39**, K63 (1977)	*11*, p. *94*; 22.1.2, p. 337
D. Wagner	(a) **41**, K137 (1977)	*10.1*, p. *78*; *12.1*, p. *101*; 21, p. 281
H. Wagner	(b) **79**, 539 (1977)	*2*, p. *33*; 14.3.1, p. 173; *16*, p. *201*; 22.1.2, p. 337
	(a) **44**, K187 (1977)	*1.2*, p. *20*; 6, p. 61; *20.1*, p. *264*; 22.6, p. 392
M. Wagner	(b) **81**, 511 (1977)	*6*, p. *56*; *13.4*, p. *134*
D. C. Wallace	(b) **81**, 557 (1977)	*4*, p. *45*; 6, p. 56; 21, p. 280; 21.2, p. 308
W. E. Wallace	(a) **43**, 317 (1977)	1.1, p. 12; *18.2*, p. *222*; 21.1, p. 297; 21.5, p. 314
P. J. Walsh	(a) **43**, 305 (1977)	*1.1*, p. *12*; *1.2*, p. *18*; 8, p. 67; 22.1.3, p. 346
F. Walz	(b) **79**, 237 (1977)	*9*, p. *69*; 10.1, p. 75; 18.2, p. 217; 21.1, p. 288
	(a) **40**, 455 (1977)	*14.4*, p. *190*; 14.4.1, p. 195; 22.8.1, p. 403
	(a) **43**, K81 (1977)	10.1, p. 80; *11*, p. *97*; *18.2*, p. *222*; 21.1.1, p. 306
R. Wang	(a) **39**, 697 (1977)	1.2, p. 15; *2*, p. *34*; *4*, p. *44*; 21, p. 278; 22.1, p. 332
B. M. Wanklyn	(a) **40**, 663 (1977)	*20.1*, p. *253*; 22.8.1, p. 404
M. Wardzyńska	(a) **40**, 663 (1977)	*20.1*, p. *253*; 22.8.1, p. 404
H. Warlimont	(a) **44**, 609 (1977)	*12.1*, p. *103*; 21.1, p. 300
A. Warnaar	(a) **40**, 235 (1977)	*10.1*, p. *76*; 21, p. 279
S. Wartewig	(b) **81**, K5 (1977)	*13.4*, p. *133*; 22.4.2, p. 365
	(a) **41**, K13 (1977)	*10.2*, p. *88*; *19*, p. *239*; 22.2.1, p. 351
Y. Waseda	(b) **81**, K37 (1977)	*2*, p. *36*; *14.1*, p. *158*; 21.1, p. 293
W. Wasserbäch	(b) **84**, 205 (1977)	*8*, p. *67*; 10.1, p. 81; 21, p. 286
K. Watanabe	(a) **44**, K151 (1977)	*2*, p. *41*; *18.2*, p. *225*; 21.1, p. 301; 21.1.1, p. 307
S. Watanabe	(a) **42**, 235 (1977)	*10.1*, p. *79*; 11, p. 96; *12.1*, p. *101*; 21.1, p. 295
	(a) **44**, 731 (1977)	*2*, p. *41*; 18.1, p. 216; 22.8, p. 401
A. Watterich	(b) **80**, K105 (1977)	*10.2*, p. *86*; 19, p. 237; *20.1*, p. *252*; 22.5.2, p. 377
M. Wautelet	(a) **39**, 425 (1977)	5, p. 47; *18.2*, p. *218*; 22, p. 322
J. Wawryszczuk	(b) **84**, K49 (1977)	*10.2*, p. *91*; 22.9, p. 419
E.-H. Weber	(a) **43**, 141 (1977)	1.5, p. 30; *17*, p. *207*; *20.1*, p. *260*; 22.6, p. 390
H. W. Weber	(a) **44**, 593 (1977)	*18.1*, p. *216*; *18.2*, p. *224*; 21.1, p. 300
P. Weber	(a) **44**, 577 (1977)	*14.2*, p. *164*; 22.6, p. 391
R. Wedell	(b) **83**, 451 (1977)	*11*, p. *97*
	(b) **83**, K129 (1977)	*11*, p. *98*
	(b) **84**, 581 (1977)	*11*, p. *99*; 13, p. 112; 22.1.2, p. 344
A. Węgrzyn	(b) **82**, 107 (1977)	*14.3*, p. *169*; *18.2*, p. *220*; 22.6, p. 389
C. Weigel	(b) **81**, 637 (1977)	6, p. 56; *13.4*, p. *134*; 22.1.2, p. 340
K. H. Weilacher	(b) **80**, 137 (1977)	*6*, p. *53*; 21.2, p. 308
H. Weinert	(b) **81**, K59 (1977)	*13.1*, p. *117*; *20.1*, p. *254*; 22.8, p. 398
	(b) **83**, K83 (1977)	13.5.2, p. 150; *18*, p. *212*; *20.1*, p. *260*; 22.2.1, p. 353
M. Weller	(a) **43**, K179 (1977)	10.1, p. 81; *11*, p. *98*; 12.1, p. 102; *18.2*, p. *223*; 21.1.1, p. 306
D.-G. Welsch	(b) **82**, 633 (1977)	*6*, p. *57*; 20.1, p. 258
H. Welser	(a) **43**, 437 (1977)	*5*, p. *49*; *18.2*, p. *223*; 21.1.1, p. 306; 21.4, p. 313
G. Wende	(a) **41**, 263 (1977)	*3*, p. *42*; 22.8.2, p. 411
M. Wendt	(a) **41**, 505 (1977)	*3*, p. *42*; 22.8.2, p. 411
	(a) **43**, 277 (1977)	3, p. 43; *18.3*, p. *230*; 22.8.2, p. 412
S. Wendt	(b) **84**, 653 (1977)	*14.4.1*, p. *197*; 18.2, p. 224
F. Werfel	(b) **80**, K95 (1977)	*13.1*, p. *117*; 17, p. 206; 20, p. 245; 22.7, p. 394
	(b) **83**, 257 (1977)	*13.1*, p. *121*; 17, p. 207; 20, p. 245; 21.1, p. 297
J. Wesselinowa	(b) **84**, 653 (1977)	*14.4.1*, p. *197*; 18.2, p. 224

M. A. B. Whitaker	(b) **79**, 673 (1977)	*19*, p. *236*
	(b) **83**, 573 (1977)	*19*, p. *241*; 22.5.2, p. 380
	(b) **84**, 149 (1977)	*19*, p. *242*; 22.5.2, p. 381
G. S. White	(a) **42**, K137 (1977)	11, p. 97; *20.1*, p. *259*; *20.3*, p. *272*; 22.8.2, p. 412
B. G. Whitford	(a) **42**, K53 (1977)	*14.4*, p. *191*; *20*, p. *245*; 22.8.1, p. 406
G. Wiech	(b) **84**, 561 (1977)	*20*, p. *246*; 22.2, p. 349
D. M. Wieliczka	(b) **80**, 201 (1977)	12.2, p. 104; 13.1, p. 115; *13.5.1*, p. *144*; 22.5, p. 370
B. Wielke	(a) **40**, 271 (1977)	*10.1*, p. *76*; 21, p. 279; 21.1, p. 291
E. Wiener-Avnear	(b) **81**, 295 (1977)	6, p. 55; *14.4.1*, p. *195*; 22.8, p. 398
E. Wieser	(a) **39**, 259 (1977)	5, p. 47; *18.2*, p. *217*; 21.1, p. 288; 21.1.1, p. 302
	(a) **40**, 497 (1977)	*18*, p. *210*; 21.1, p. 292; 21.1.1, p. 303
	(b) **81**, 535 (1977)	13.1, p. 118; *18.2*, p. *220*; *18.4*, p. *233*; 21.1.1, p. 303
	(a) **44**, K95 (1977)	*1.2*, p. *20*; 5, p. 49; 14.3, p. 172; 22.6, p. 391
H. D. Wiesinger	(a) **41**, 465 (1977)	*1.3*, p. *21*; *14.2*, p. *162*; 21, p. 281
W. Wiethaup	(a) **43**, 151 (1977)	1.4, p. 27; *14.2*, p. *163*; 21, p. 284
W. Wild	(a) **42**, K75 (1977)	*17*, p. *206*; 22.6, p. 389
	(a) **43**, 185 (1977)	*11*, p. *97*; 21, p. 285; 21.1, p. 297; 21.3, p. 310; 21.6, p. 317
B. Wilhelmi	(b) **82**, 633 (1977)	*6*, p. *57*; 20.1, p. 258
M. Wilkens	(a) **39**, 103 (1977)	*10.1*, p. *75*; 21.1, p. 288
	(a) **39**, 561 (1977)	*10.1*, p. *76*; 11, p. 94; 21.1, p. 289
	(a) **44**, 173 (1977)	*1.4*, p. *28*; *10*, p. *73*
G. Will	(a) **40**, 447 (1977)	*1.2*, p. *16*; *18.4*, p. *232*; 22.8.1, p. 403
C. Willaime	(a) **44**, 651 (1977)	*10.2*, p. *93*; *12.2*, p. *106*; 22.8, p. 400
I. Willert	(a) **40**, K195 (1977)	*2*, p. *36*; *9*, p. *70*; 22, p. 324
	(a) **43**, 451 (1977)	*2*, p. *39*; *9*, p. *71*; 14.3.4, p. 187; 22, p. 329
M. W. Williams	(b) **84**, 367 (1977)	*13.5.2*, p. *150*
R. Williams	(b) **82**, K117 (1977)	*1.1*, p. *12*; 22.9, p. 417
M. R. Willis	(b) **79**, 629 (1977)	*14.1*, p. *157*; *14.3*, p. *165*; 15, p. 198; 18.1, p. 213; 22.9, p. 414
H. Willner	(a) **39**, 79 (1977)	1, p. 9; 13.1, p. 114; *14.2*, p. *161*; 20.1, p. 248; 22.6.1, p. 393
C. H. Wilts	(b) **79**, 313 (1977)	*18.2.1*, p. *226*; 19, p. 235; 22.8.2, p. 409
W. Windsch	(a) **41**, K141 (1977)	*19*, p. *240*; 22.9, p. 416
H. R. Winteler	(a) **41**, 561 (1977)	6, p. 56; 8, p. 66; *10.2*, p. *88*; 22.2.1, p. 352
M. Wintenberger	(a) **44**, 713 (1977)	*18*, p. *212*; 22.8, p. 401
S. Winter	(b) **79**, 637 (1977)	*13.4*, p. *130*; 22.1.1, p. 333
V. Winterfeldt	(b) **80**, 559 (1977)	1.2, p. 15; *6*, p. *54*; *8*, p. *65*; 14.4.1, p. 194; 22.9, p. 415
M. J. Witcomb	(a) **42**, 595 (1977)	*1.4*, p. *27*; *14.2*, p. *162*; 21, p. 283
J. Witters	(b) **79**, 645 (1977)	*19*, p. *235*; 21, p. 277
D. Włosewicz	(a) **42**, 351 (1977)	*8*, p. *66*; 21.7, p. 319
G. Wolf	(a) **44**, K79 (1977)	*8*, p. *68*; *14.2*, p. *164*; 18, p. 212; 22.8, p. 400
H. Wolf	(b) **82**, 303 (1977)	*1.6*, p. *31*
R. Wolf	(a) **40**, 531 (1977)	*1.3*, p. *21*; *1.4*, p. *25*; 21.1, p. 292; 21.6, p. 315
	(a) **42**, 629 (1977)	*14.3.4*, p. *185*; *20.1*, p. *258*; 22.2.4, p. 357
J. P. Wolfe	(b) **83**, 305 (1977)	*13.5.1*, p. *147*; 18.1, p. 215; 22.1.1, p. 335
A. Wolfenden	(a) **44**, K171 (1977)	*12.1*, p. *103*; 21.1, p. 301
K. Woodbridge	(a) **44**, K123 (1977)	*14.3.3*, p. *182*; 15, p. 200; 22.7, p. 395
J. Woods	(a) **40**, K43 (1977)	*16*, p. *202*; 22.4.1, p. 362
	(a) **44**, K47 (1977)	*14.3.3*, p. *181*; *16*, p. *205*; 22, p. 330; 22.4.1, p. 364
	(a) **44**, 693 (1977)	*20.3*, p. *275*; 22.4.2, p. 366
J. Woodward	(b) **79**, 629 (1977)	*14.1*, p. *157*; *14.3*, p. *165*; 15, p. 198; 18.1, p. 213; 22.9, p. 414

T. Wosiński	(b) **83**, 93 (1977)	*10.2*, p. *90*; 13.4, p. 137; *16*, p. *204*; 22.1.2, p. 341
J. G. Wright	(b) **81**, K37 (1977)	*2*, p. *36*; *14.1*, p. *158*; 21.1, p. 293
A. Wronkowski	(a) **43**, K41 (1977)	*5*, p. *49*; *18.1*, p. *215*; 22.8.1, p. 407
Shi-Yu Wu	(b) **83**, 115 (1977)	*6*, p. *58*; *10*, p. *73*
	(b) **83**, 365 (1977)	*6*, p. *59*; *10*, p. *73*
R. Wunderle	(b) **79**, K153 (1977)	*6*, p. *53*; *8*, p. *64*
	(b) **82**, 257 (1977)	*6*, p. *57*; *8*, p. *66*
M. Wurlitzer	(a) **44**, 531 (1977)	*11*, p. *99*; 18.3, p. 230; 22 8.2, p. 413
M. Wuttig	(a) **41**, K79 (1977)	*9*, p. *70*
B. Wysłocki	(a) **42**, 359 (1977)	*18.2*, p. *221*; 21.1.1, p. 304
R. Moreira Xavier	(b) **82**, K21 (1977)	*13.1*, p. *119*; 22, p. 326
	(b) **84**, K143 (1977)	*12.2*, p. *106*; *13.1*, p. *123*; 22, p. 330; 22.6, p. 391
M. Yacamán	(a) **39**, K85 (1977)	*1.4*, p. *24*; *10.2*, p. *85*; 22.5.2, p. 376
M. J. Yacamán	(a) **42**, 571 (1977)	*1.5*, p. *30*; 21.6, p. 316
Y. Yacobi	(b) **83**, 99 (1977)	*1.2*, p. *18*; 6, p. 58; *14.4.1*, p. *196*; 20.1, p. 259; 22.8.1, p. 406
K. Yagi	(a) **41**, 523 (1977)	*14.4.1*, p. *195*; 22.8.1, p. 405
	(a) **42**, 257 (1977)	*14.4.1*, p. *196*; 22.8.1, p. 405
	(a) **44**, 147 (1977)	*14.4.1*, p. *197*; 22.8.1, p. 407
E. B. Yagubskii	(a) **43**, K73 (1977)	*20.1*, p. *260*; 22.9, p. 418
	(a) **44**, 77 (1977)	*1*, p. *10*; *14*, p. *156*; 18.1, p. 215; 22.9, p. 419
M. Yahagi	(a) **39**, 189 (1977)	*8*, p. *64*; 10.1, p. 75; *14.1*, p. *157*; 21.2, p. 308
T. N. Yakimov	(a) **39**, 321 (1977)	*1.5*, p. *29*
J. K. Yakinthos	(a) **39**, K151 (1977)	*18.3*, p. *228*; 22.7, p. 394
	(a) **40**, K105 (1977)	*18.1*, p. *214*; *18.4*, p. *232*; 21.4, p. 311
	(a) **40**, K109 (1977)	*18.1*, p. *214*; *18.4*, p. *232*; 21.4, p. 312; 21.6, p. 315
	(a) **41**, 147 (1977)	*18.3*, p. *229*; 22.8.2, p. 411
	(b) **82**, 349 (1977)	*18.3*, p. *229*; 21.1, p. 295; 21.4, p. 312
N. Yamada	(a) **39**, 331 (1977)	*16*, p. *201*; 22.4.4, p. 368
M. Yamaguchi	(a) **43**, 667 (1977)	*10.1*, p. *81*; 12.1, p. 102; 21.1, p. 298
N. Yamamoto	(a) **41**, 523 (1977)	*14.4.1*, p. *195*; 22.8.1, p. 405
	(a) **42**, 257 (1977)	*14.4.1*, p. *196*; 22.8.1, p. 405
	(a) **44**, 147 (1977)	*14.4.1*, p. *197*; 22.8.1, p. 407
R. Yamamoto	(b) **80**, 193 (1977)	*10.1*, p. *76*; 21, p. 278
	(b) **80**, K131 (1977)	*4*, p. *45*; *10.1*, p. *77*; 21, p. 279
	(b) **84**, 813 (1977)	*4*, p. *46*; 21, p. 287
T. Yamamoto	(a) **43**, K53 (1977)	*1.1*, p. *12*; *1.4*, p. *27*
	(a) **44**, 137 (1977)	*1.4*, p. *28*; *10.1*, p. *81*; 21, p. 287; 21.1, p. 299; 21.6, p. 317; 22.1.2, p. 343
	(a) **44**, 467 (1977)	*1.4*, p. *28*; *10.1*, p. *82*; 21, p. 287; 21.1, p. 300; 21.6, p. 318; 22.1.2, p. 344
I. Y. Yanchev	(a) **41**, 687 (1977)	*14.3*, p. *169*; 22.2.4, p. 357
K. Yang	(a) **42**, 553 (1977)	*14.3*, p. *170*; 14.3.1, p. 174; 22.1.2, p. 341
Y. D. Yao	(b) **82**, 343 (1977)	*13.4*, p. *136*; *14.1*, p. *158*; 18, p. 211; 21.1, p. 295
L. Yarmus	(b) **83**, K1 (1977)	*18*, p. *212*; *20.3*, p. *272*; 22.9, p. 418
	(b) **84**, 27 (1977)	*13.5.1*, p. *147*; *19*, p. *242*; 22.9, p. 419
K. Yasuda	(a) **41**, K181 (1977)	*2*, p. *37*; *14.3.1*, p. *174*; 22.1.1, p. 334
	(a) **42**, 769 (1977)	*11*, p. *97*; *14.3.2*, p. *177*; 16, p. 203; 22.1.2, p. 341
S. P. S. Yen	(b) **81**, 281 (1977)	*14*, p. *155*; 15, p. 198; 22.9, p. 415

S. Yokoyama	(a) **42**, 483 (1977)	*13.3*, p. *127*; 14.3.3, p. 180; 22.2.1, p. 352
A. Yoshida	(a) **41**, K181 (1977)	*2*, p. *37*; *14.3.1*, p. *174*; 22.1.1, p. 334
	(a) **42**, 769 (1977)	*11*, p. *97*; *14.3.2*, p. *177*; 16, p. 203; 22.1.2, p. 341
L. Youdri	(b) **82**, K9 (1977)	*19*, p. *240*; 22.8.1, p. 405
M. L. Young	(a) **41**, 85 (1977)	13.1, p. 117; *14.3.4*, p. *185*; 22.2.4, p. 357
N. B. Yunusov	(b) **84**, 791 (1977)	*13.4*, p. *140*; 18.1, p. 216; 19, p. 243; 21.1, p. 300
S. E. Yurchenko	(a) **44**, K75 (1977)	*18.2.1*, p. *227*
Yu. P. Yurovskii	(b) **79**, 71 (1977)	*14.2*, p. *161*
P. H. Yuster	(b) **80**, 471 (1977)	*10.2*, p. *86*; 20.1, p. 252; 22.5.2, p. 377
V. I. Yuzvak	(a) **44**, K155 (1977)	*1.2*, p. *20*; *14.4.1*, p. *197*; 19, p. 243; 22.8.1, p. 408
A. V. Zaborov	(b) **79**, 379 (1977)	*18.2*, p. *217*; 21, p. 276
M. D. Zahari	(a) **39**, 509 (1977)	*9*, p. *69*; 22.2, p. 348
A. A. Zaitsev	(a) **39**, 345 (1977)	1.2, p. 14; *1.3*, p. *21*; 22.4.4, p. 368
V. I. Zaitsev	(a) **44**, K39 (1977)	*10.2*, p. *92*; *12.2*, p. *106*; 22.5.2, p. 381
Ya. A. Zakharin	(a) **41**, 443 (1977)	*10.2*, p. *88*; 22.5.2, p. 378
R. B. Žakula	(b) **80**, 661 (1977)	*6*, p. *55*; *14.4.1*, p. *195*
C. Zárate	(b) **81**, K27 (1977)	*19*, p. *238*; 22.8, p. 398
A. Zaręba	(b) **83**, K155 (1977)	*13.4*, p. *138*; 16, p. 204; 22.5.3, p. 384
	(a) **44**, K43 (1977)	*14.3*, p. *172*; *20.1*, p. *262*; 22.5.3, p. 384
	(a) **44**, K83 (1977)	*20.3*, p. *274*; 22.5.3, p. 384
W. Zarek	(a) **44**, 325 (1977)	*1.1*, p. *13*; 5, p. 49; *18.3*, p. *230*; 22.8.2, p. 413
M. M. Zaripov	(b) **80**, K109 (1977)	*19*, p. *237*; 21.1, p. 291
	(a) **40**, K11 (1977)	*19*, p. *237*; 21, p. 279
M. I. Závětová	(a) **43**, K109 (1977)	*2*, p. *39*; *20.1*, p. *260*; 22.1, p. 332
G. S. Zavt	(b) **80**, 399 (1977)	*10.2*, p. *85*; 20.1, p. 251; 22.5.2, p. 376
	(b) **81**, 57 (1977)	*10.2*, p. *87*; 20.1, p. 253; 22.5.2, p. 377
F. C. Zawislak	(b) **79**, K65 (1977)	*13.4*, p. *129*; 14.4, p. 189; 22.8.1, p. 402
W. Żdanowicz	(a) **41**, K59 (1977)	12.2, p. 105; *14.3*, p. *168*; 22.3.1, p. 360
E. Zedler	(a) **44**, 499 (1977)	*10.1*, p. *82*; 14.2, p. 164; 21.1, p. 300
A. Zehe	(a) **39**, K139 (1977)	*14.3.3*, p. *178*; *20.3*, p. *268*; 22.2.1, p. 351
B. Žekš	(b) **80**, 385 (1977)	1.2, p. 15; *18.1*, p. *213*; *18.4*, p. *231*
M. G. Zemlyanov	(b) **79**, 105 (1977)	*6*, p. *51*; 21.1, p. 288
E. V. Zhakirov	(a) **40**, K69 (1977)	*20.2*, p. *265*; 22.8.2, p. 410
	(a) **41**, K199 (1977)	*20.2*, p. *265*; 22.8.2, p. 410
N. G. Zhdanova	(a) **40**, K61 (1977)	*14.3.4*, p. *184*; 22.1.1, p. 334
V. Zhelev	(a) **43**, 675 (1977)	13.3, p. 128; *14.3.4*, p. *187*; 22.5.1, p. 374
V. S. Zheleznyi	(a) **39**, K21 (1977)	*14.2*, p. *161*; 21, p. 277
A. P. Zhernov	(b) **79**, 105 (1977)	*6*, p. *51*; 21.1, p. 288
E. A. Zhilinskaya	(b) **82**, 189 (1977)	2, p. 37; *19*, p. *240*; 22.8, p. 399
R. A. Zhitnikov	(b) **79**, K27 (1977)	*19*, p. *235*; 20.1, p. 248; 20.3, p. 267; 22.5.2, p. 375
M. F. Zhvaniya	(a) **40**, K113 (1977)	8, p. 65; *10.2*, p. *87*; *20.1*, p. *253*; 22.5.2, p. 377
E. Ziegler	(a) **41**, 75 (1977)	*14.3.3*, p. *179*; *20.3*, p. *270*; 22.8, p. 398
K. Ziegler	(a) **39**, 467 (1977)	*13.4*, p. *131*; 14.3, p. 166; 16, p. 201; 20.3, p. 268; 22, p. 322
J. P. Zielinger	(a) **42**, 91 (1977)	13.4, p. 136; *16*, p. *203*; 22.6, p. 389
E. Zielińska-Rohozińska	(a) **44**, 59 (1977)	*1.4*, p. *28*; 11, p. 99; 22.1.2, p. 343
J. Zieliński	(b) **82**, K17 (1977)	*13.1*, p. *119*; *18.2*, p. *221*; 21.1, p. 295
M. Zieliński	(a) **42**, 305 (1977)	*2*, p. *37*; 14.4, p. 191; 22.9, p. 417
W. J. Ziętek	(a) **43**, K113 (1977)	*18.2.1*, p. *226*; 21.1.1, p. 306
S. C. Zílio	(b) **80**, 597 (1977)	*10.2*, p. *86*; 14.4, p. 190; 22.5.2, p. 377

K. Zimmermann	(b) **79**, K35 (1977)	*20.1*, p. *248*; 22.8.1, p. 402
R. Zimmermann	(b) **83**, 85 (1977)	*13.5*, p. *141*; 13.5.1, p. 146; 13.5.2, p. 149; 13.5.3, p. 152; 22.2.1, p. 353; 22.4.1, p. 363; 22.4.2, p. 366; 22.4.3, p. 367
W. Zimmermann	(a) **40**, K153 (1977)	*14.4.1*, p. *195*; 22.8.1, p. 404
H. Zinke	(b) **84**, 649 (1977)	*13.1*, p. *122*; 14.1, p. 160
M. Zoaeter	(a) **41**, 629 (1977)	13.1, p. 118; *14.3*, p. *169*; *15*, p. *199*; 22, p. 325
Z. Żołnierek	(b) **80**, K137 (1977)	*13*, p. *108*; *20.1*, p. *252*; 22.5, p. 370
V. V. Zolotarev	(b) **83**, K179 (1977)	*13.5.4*, p. *153*; 18.2, p. 223
I. S. Zolovkina	(b) **79**, 49 (1977)	*13.4*, p. *129*; *20.3*, p. *267*; 22.5.2, p. 375
M. V. Zubova	(a) **44**, 247 (1977)	5, p. 49; *14.4.1*, p. *197*; 18.2, p. 223; *18.4*, p. *234*; 22.8, p. 400
E. P. Zucchelli	(b) **80**, 245 (1977)	*13.3*, p. *126*; 21, p. 278
G. Zvára	(b) **84**, 691 (1977)	*13.4*, p. *139*; *20.1*, p. *263*; 22.4.3, p. 367
	(b) **84**, K119 (1977)	*20.1*, p. *263*; 22.8.2, p. 413
A. V. Zvarykina	(a) **44**, 77 (1977)	*1*, p. *10*; *14*, p. *156*; 18.1, p. 215; 22.9, p. 419
G. L. Zvereva	(a) **44**, 77 (1977)	*1*, p. *10*; *14*, p. *156*; 18.1, p. 215; 22.9, p. 419
I. P. Zvyagin	(b) **83**, 63 (1977)	2, p. 38; 13, p. 110; *14.3*, p. *170*; 15, p. 199; 22, p. 327
A. Zygmunt	(a) **43**, 573 (1977)	18.1, p. 215; *18.4*, p. *233*; 22.8, p. 400
A. N. Zyuganov	(a) **43**, 333 (1977)	*14.3*, p. *170*; 14.3.4, p. 186; 22, p. 328

19. Magnetic Resonance

20. Optical Properties
    20.1 Spectra. Optical Constants (X-Ray Spectra See 20)
    20.2 Lasers
    20.3 Luminescence (X-Ray Spectra See 20)

21. Metals. Alloys
    21.1 Transition Metals (Group Sc to Zn) and Their Alloys
        21.1.1 Fe and Fe Alloys
    21.2 Alkali Metals
    21.3 Alkali-Earth Metals
    21.4 Rare-Earth Metals
    21.5 Actinides
    21.6 Noble Metals
    21.7 Semi-Metals (Semiconducting Alloys See 22.7)

22. Semiconductors and Ionic Crystals
    22.1 Elements
        22.1.1 Germanium
        22.1.2 Silicon
        22.1.3 Group VI Elements
    22.2 III–V Compounds
        22.2.1 Arsenides
        22.2.2 Phosphides
        22.2.3 Antimonides
        22.2.4 Mixed Crystals
    22.3 IV–IV, II–V, and III–VI Compounds
        22.3.1 Mixed Crystals
    22.4 II–VI Compounds
        22.4.1 Sulphides
        22.4.2 Selenides
        22.4.3 Tellurides
        22.4.4 Mixed Crystals
    22.5 Halides
        22.5.1 Silver Halides
        22.5.2 Alkali Halides
        22.5.3 Alkali-Earth Halides
        22.5.4 Mixed Crystals
    22.6 Simple Oxides (CdO, ZnO See 22.4)
        22.6.1 Mixed Crystals
    22.7 Semiconducting Intermetallic Compounds
    22.8 Tertiary and Higher Compounds
        22.8.1 Salts of Oxy-Acids (Carbonates, Phosphates, Silicates Including Mica etc.)
        22.8.2 Spinels, Garnets, Ferrites
    22.9 Organic Semiconductors

23. Solidified Gases

For comments on the use of the classification scheme see p. 6.

Manuscripts and letters for physica status solidi (b) — basic research and physica status solidi (a) — applied research

should be sent to

DDR-102 Berlin, Neue Schönhauser Str. 20

or

Internationaler Buchversand GmbH, 1 Berlin 19, Eschenallee 21.

The object of "physica status solidi (b)" and "physica status solidi (a)" is to publish rapidly important and topical results concerning fundamental research in the field of solid state physics and aspects of application and experimental techniques of fundamental solid state research, respectively. The issues of "physica status solidi (b)" and "physica status solidi (a)" will be published on the 1st and the 16th day of each month, respectively. Either series will contain the following categories:

1. Review articles written by arrangement with the editors. They will be remunerated.
2. Original papers selected from those which have not been submitted for publication elsewhere, after being reviewed. (The latest date for acceptance of contributions is the 8th and the 23rd day of each month for "physica status solidi (b)" and "physica status solidi (a)", respectively, 50 days before the delivery of the journal.)
3. Short notes published in the off-set part of the journal. They are, if necessary after being reviewed, selected by the editorial staff. (The latest date for acceptance of manuscripts for this part is the 12th and the 27th day of each month for "physica status solidi (b)" and "physica status solidi (a)", respectively, 16 days before the delivery of the journal.)
4. Pre-printed titles of papers to be published in the next issue of both series.

Editorial Information

Manuscripts should be submitted preferentially in English. Besides, papers in German, French, and Russian are admitted. The final decision on the incorporation of the papers in the series (b) or (a) is taken by the editorial staff after consultation of the referee. Authors are asked to observe the following rules:

1. Manuscripts have to be written as concisely as possible. As most, 20 and 4 pages (28 lines per page; abstracts, tables, references, figures, and captions of figures included, where 3 figures are counted as about one page) are admitted for original papers and short notes, respectively. All manuscripts should be type-written on one side only. Two copies are required. Badly arranged manuscripts are returned.
2. The author's name, initials, and institution should be given.
3. One abstract of about ten lines in English and one in German, French, or Russian are required for original papers.
4. Because of lack of space conclusions (if any) must not have the character of a summary (i. e. substantially a repetition of the contents of the abstract).
5. Formulae should not be written too small and with pencil. Separate lines for formulae are desirable. Indistinct writing will lead to a return of the manuscript.
6. Drawings should be enclosed on separate sheets of paper, not exceeding the maximum and minimum sizes of 20×30 and $10 \times 15\ cm^2$, respectively. Photographs should be submitted, on white paper, only for review articles and original papers. Duplicates are required.
7. Captions for the figures should be submitted on separate sheets of paper, and special instructions with regard to the size of illustrations should be given on these.
8. References should be numbered in order of appearance and should be given, on a separate sheet, in the form used in physica status solidi.
9. Units, symbols, etc. should be used in accordance with the recommendations given by the IUPAP.
10. Each paper should contain a proposal for the subject classification according to the classification scheme appearing on the second and third cover of each issue.

The authors will be sent galley proofs of review articles and original papers. Proof reading of short notes and pre-printed titles will be done by the editorial staff.

For each original paper, and short note, a total of 75 reprints is available free of charge. The number of reprints free of charge per review article is 25. Requests for additional reprints (against payment) should be given together with the submitted manuscript. **The deadline for orders of additional reprints is the 15th and the 30th day of each month** for "physica status solidi (b)" and "physica status solidi (a)", respectively, two weeks before the delivery of the journal.

Bei Fragen zur Produktsicherheit wenden Sie sich bitte an:
If you have any questions regarding product safety,
please contact:

Walter de Gruyter GmbH
Genthiner Straße 13
10785 Berlin
productsafety@degruyterbrill.com